EXPERIMENTAL MECHANICS
VOLUME 2

Books are to be returned on or before
the last date below.

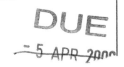

DUE

-5 APR 2000

PROCEEDINGS OF THE 11TH INTERNATIONAL CONFERENCE ON EXPERIMENTAL MECHANICS/OXFORD/UK/24-28 AUGUST 1998

Experimental Mechanics

Advances in Design, Testing and Analysis

Edited by
I.M.Allison
University of Warwick, UK

VOLUME 2

A.A.BALKEMA/ROTTERDAM/BROOKFIELD/1998

This Conference was organised by the British Society for Strain Measurement on behalf of the European Permanent Committee for Experimental Mechanics.

Cover illustration by Chris Donaghue Photography, The Oxford Photo Library.

The texts of the various papers in this volume were set individually by typists under the supervision of each of the authors concerned.

Published by
A.A.Balkema, P.O.Box 1675, 3000 BR Rotterdam, Netherlands
Fax: +31.10.413.5947; E-mail: balkema@balkema.nl; Internet site: http://www.balkema.nl
A.A.Balkema Publishers, Old Post Road, Brookfield, VT 05036-9704, USA
Fax: 802.276.3837; E-mail: info@ashgate.com

For the complete set of two volumes, ISBN 90 5809 014 0
For Volume 1, ISBN 90 5809 015 9
For Volume 2, ISBN 90 5809 016 7

Experimental Mechanics, Allison (ed.) © 1998 Balkema, Rotterdam, ISBN 90 5809 014 0

Table of contents

MATERIALS BEHAVIOUR AND TESTING

Plenary lecture

Materials testing

Elastic plastic fracture

Residual stress

Biomedical engineering aspects of experimental mechanics

Plenary lecture

Experimental Mechanics, Allison (ed.)© 1998 Balkema, Rotterdam, ISBN 90 5809 014 0

The joints of the skeleton – Kinematic pairs, mechanisms or flexible structures?

J.J.O'Connor, T.-W. Lu, J. Feikes, D.R. Wilson & A. Leardini
University of Oxford, Oxford Orthopaedic Engineering Centre, Nuffield Orthopaedic Centre, UK

ABSTRACT: The paper provides extensive evidence that the knee behaves as a flexible mechanism. The joint allows a range of unresisted motion in the unloaded state because ligaments can rotate about their origins and insertions on the bones, without stretching, and because the articular surfaces of the bones can roll and slide upon each other, without indentation. When the joint is loaded, the ligaments stretch under tension and the surfaces indent under compression. Preliminary work suggests that the ankle behaves similarly. On the other hand, the sub-talar joint complex under the ankle requires the application of load to produce motion; motion requires tissue deformation which is recovered when load is removed. The sub-talar joint behaves as a flexible structure. The ligaments at the hip are tight only at the limits of motion and slack elsewhere. They play no part in controlling motion within those limits. The spherical surfaces of the joint are the sole constraints to motion. In the language of mechanism theory, the hip is a higher kinematic pair. This classification of joints should guide the development of appropriate mechanical models.

1. INTRODUCTION

It has been known for centuries that it is the levels of force transmitted by the muscles which determine the loading of bones and joints. Borelli (1679) performed extensive experiments with post-mortem specimens and demonstrated that the forces transmitted by muscle tendons as they span the joints can be much larger than the loads applied by hanging weights on the ends of the limbs. He explained these results using the principle of the lever:

$$M.r = W.\ell \qquad (1)$$

or

$$M = W\ell/r. \qquad (2)$$

The muscle force M balances a load W. As equation 2 suggests, the muscle force may be many times larger than the applied load because its lever arm r (typically the radius of a limb) may be only a fraction of the length of the lever arm ℓ of the load (typically the length of a limb). Contact forces at joints can therefore be large because the muscle forces are large.

This very simple analysis explains in general terms the large levels of hip contact forces, several times body weight, calculated by Paul (1965) and measured with instrumented prostheses by Rydell (1966), English & Kilvington (1979), Davy *et al* (1988) and Bergmann *et al* (1993). Similar measurements (Taylor & Donaldson, 1990, Taylor

et al, 1997) of the axial force in the shaft of a massive proximal femoral prosthesis show that equally large forces are transmitted along the shafts of the bones. Analysis of the results of these experiments (Lu *et al*, 1998a, Lu *et al*, 1998b) has vindicated Pauwels's arguments that bi-articular muscles spanning two joints act as tension bands and play an important role in the transmission of bending moment along the limbs (Pauwels, 1950). Pauwels used simple bending theory to demonstrate that the action of bi-articular muscles can have profound effects on the distribution of stress within the bones. Determination of stress distributions in bone, either experimentally or by the finite element method, requires appropriate simulation of muscle forces.

The simple lever equations (1) and (2) have been displayed to emphasise the point that the geometric factors which control the lengths of the lever arms of the muscle tendons at the joints are prime determinants of the levels of force transmitted by the muscles and therefore, according to Pauwels, of the distributions of stress in the bones. The values of W, ℓ and r in equations (1) and (2) all vary during activity. The values of W and ℓ can be measured in the gait laboratory. The lengths of the muscle lever-arms are determined by the positions of their tendons relative to the articular surfaces and ligaments of the joints, so that understanding the kinematics of joints is key to understanding load transmission through the musculoskeletal system.

2. JOINTS

The joints which give the skeleton its mobility have a number of features in common. The bones make contact through layers of articular cartilage. The bones are held together by the capsules which enclose the joints and, in particular, by the ligaments, some of which appear as condensations of fibres in the capsule and some of which are intra-articular, connecting the bones inside the capsule. The articular surfaces are compressed and the ligaments are stretched under load.

Despite the similarities, the joints differ greatly both in the ranges of movement which they allow and in the details of the movement possible within that range. Considering the lower limb, the hip allows a wide range of unresisted multi-axial motion. The tissues of the joint do not resist passive displacement of the bones from one position to another within that range. The knee and the ankle allow a more restricted range of motion while movement at the sub-talar and some of the more distal joints of the foot is more limited still. These differences in mobility arise not because of differences in the mechanical properties of the tissues but because of differences in the geometry of the articular surfaces and differences in the geometric arrangements of the ligaments and the capsules.

As well as allowing mobility, the joints also have to transmit load. In the presence of external loads due mainly to gravity, muscle forces are required to stabilise the skeleton by suppressing movement at the joints. During activity, muscle forces also initiate and maintain movement and can, when required, achieve precise control of limb position and velocity. The passive structures of each of the joints contribute to load transfer and joint stabilisation through the transmission of compressive stress between the articular surfaces and the transmission of tensile stress along the fibres of the ligaments and capsules.

It is instructive to distinguish between those features of joint behaviour which control mobility and those which contribute to stability and load transfer. It will be shown that the features which control mobility can be deduced mainly from geometric analysis while the study of stability and load transfer obviously also requires mechanics. In the following, the knee joint will be used as a paradigm and its behaviour described through the use of a mathematical model.

3. THE KNEE

Many computer-based mathematical models of the knee joint have been proposed. Hefzy & Grood (1988) described over 100 mathematical or physical models. The mathematical models fall into two main categories: those in which geometric and mechanical analyses are carried out simultaneously and those in which these analyses are carried out sequentially.

The model of Blankevoort et al (1991, 1996, Mommersteeg et al 1996, 1997), a development of that described by Wismans (1980), is an example of the first type. The articular surfaces were considered to be deformable and, in the unloaded state, were represented as polynomial surfaces fitted to experimental data. The ligaments were each represented by a number of distinct fibres, with non-linear force extension characteristics. Equations representing geometric constraints, enforcing coincidence of the contact areas on the articular surfaces of the two bones, and equations of equilibrium of the ligament and contact forces with the external loads were solved simultaneously. Andriacchi et al (1983) and Essinger et al (1989) proposed similar models but with the addition of simulations of the extensor muscles.

An alternative approach is to study geometry first and then mechanics. With this approach, two categories of models can also be found. In the first, a priori assumptions are made about the relative movements of the bones and/or the muscle tendons. Morrison (1968) performed gait analysis on three subjects and calculated muscle, ligament, and contact force values on the assumption that the knee behaves as a simple hinge with a fixed axis of flexion/extension. The orientations of the ligaments and flexor muscle tendons were then deduced from the calculated relative positions of their origins and insertions. The changing inclination of the patellar tendon (which connects the knee cap to the tibia) relative to the tibia was measured from x-ray pictures. Later studies (Ellis et al 1979, Mikosz et al 1988) took account of the rolling movements of the femur on the tibia during flexion/extension by deducing the point of closest approach of the femoral condyle to the tibial plateau. Crowninshield et al (1976) assumed a relationship between femoral rolling, tibial rotation and flexion in the unloaded state as a simulation of the "screw-home" motion first described by Meyer (1853).

The geometric analysis can also be performed by modelling the joint as a mechanical linkage, deducing the relative motions of the bones and ligaments upon each other from first principles. The only experimental input then necessary is that required to define model parameters. The mechanism approach makes it possible to distinguish clearly between those characteristics of the joint which can be deduced from geometric analysis and those which also require mechanics. There is an immediate application to joint replacement and ligament reconstruction since the effects on joint motion of articular surface shapes and their positions on the bones as well as the effects of changes in the origins and insertions and fibre arrangements of the ligaments can be studied. Although the bulk of the work in this area has been two-dimensional, concentrating on the sagittal plane, it will be described in some detail since it demonstrates the potential of the mechanism model approach. Early steps in the development of an analogous three-dimensional model will also be described.

4. MOBILITY OF THE KNEE

4.1 *The Tibiofemoral Joint*

The human knee joint is not a simple hinge, with an axis of rotation fixed relative to both the femur and the tibia. Weber and Weber (1836) first described how the femur rolls backwards and slides forwards on the tibia during flexion, *vice versa* during extension. Strasser (1917) explained these movements by describing the knee as a single degree of freedom mechanism, the femur, the tibia and the two cruciate ligaments forming a four-bar linkage. Kapandji (1970), Menschik (1974) and Huson (1974) developed the model further and Goodfellow and O'Connor (1978) used it to explain the movements of the meniscal bearings in their knee prosthesis.

This simple model shows why passive movement of flexion/extension of the bones upon each other is not resisted. The ligament fibres rotate but do not stretch and the articular surfaces slide upon each other but do not indent so that movement can be accomplished without tissue deformation. The simple four-bar linkage is therefore a model of unresisted passive flexion-extension.

An important feature of the linkage is that its instant centre lies at the point at which the two cruciate fibres cross (O'Connor *et al*, 1989, O'Connor *et al*, 1990a). The model femur flexes and extends about an axis passing through the instant centre. Because the geometry of the linkage changes, the instant centre moves backwards during flexion and forwards during extension. This explains why the knee is not a fixed-axis hinge.

A second feature of the linkage is that the normal to the articular surfaces at their point of contact must pass through the instant centre (O'Connor *et al*, 1989, O'Connor *et al*, 1990a). This is necessary to avoid separation or interpenetration of the surfaces and stretching or slackening of the ligament fibres. To satisfy this condition, the contact point moves from the front of the tibial plateau in extension to the back in flexion so that the model femur rolls backwards and slides forwards on the tibia during flexion. The model both reproduces and explains the observations of Weber & Weber (1836). The common normal theorem allows the calculation of surface shapes and a movement path compatible with an absence of tissue deformation. The coupling of rolling to flexion is one characteristic of a single degree of freedom system.

4.2 *The Patellofemoral Joint*

It has been observed (Elias *et al* 1990, O'Connor *et al* 1989) that the surfaces of the femur which make contact with the patella and those which contact the tibia are both circular in the sagittal plane (ie when seen from the side) but have different centres. The model knee has been developed to include the extensor mechanism, with the patella (knee-cap) (Gill & O'Connor 1996).

The model shows that the muscle tendons rotate about their points of origin and insertion on the bones during movement. It shows that the patellofemoral contact point moves from the distal pole of the patella in extension towards the proximal pole in flexion. The patella therefore rolls as well as slides upon the femur. The rolling movement is necessary because, with no friction, the patellofemoral contact force must lie on the perpendicular to the two articular surfaces at their point of contact and must pass through the point of intersection of the two tendons (Gill & O'Connor 1996).[*] As the joint flexes, the point of intersection of the two tendons moves proximally and the point of contact between the patella and the femur has to follow.

Dye exclusion studies showed that the contact area on the human patella moves proximally during flexion (Goodfellow *et al* 1976). Miller *et al* (1997) confirmed these patterns of contact using a different experimental technique.

4.3 *Strain Patterns in the Ligaments*

The cruciate ligaments of the human knee do not consist each of a single fibre, as in the simplest

---

[*] The concurrence of the tendons and patellofemoral normal, necessary for equilibrium of the patella, is the only contribution of mechanics to the formulation of the geometric model.

models (O'Connor *et al*, 1989). The geometric model has been further developed but with the cruciate ligaments represented in a more lifelike way as continuous arrays of fibres (Lu & O'Connor, 1996a, Zavatsky & O'Connor, 1992a, Zavatsky & O'Connor, 1992b). The anterior fibre of the anterior cruciate ligament (ACL) and the central fibre of the posterior cruciate ligament (PCL) remain isometric, staying just tight over the whole range of flexion of the unloaded joint. The reference lengths of the fibres in each ligament have to be defined in some position of the joint. All the fibres of the ACL and the posterior fibres of the PCL are just tight in extension where they are more or less parallel. The anterior fibres of the PCL are slack in extension but just tight at 120°.

As their attachment areas rotate relative to each other during flexion and extension, the apparent shapes of the model ligaments change as fibres in both ligaments cross and uncross. The changes in apparent shapes of the ligaments are consistent with observations reported by Girgis *et al* (1975) and Friederich *et al* (1992). In the human knee, ligaments twist as well as bend because their attachment areas are not parallel to the axis of rotation of the joint, the axis of rotation having components parallel as well as perpendicular to the ligaments (Zavatsky & O'Connor, 1994).

As the joint flexes, all fibres slacken and buckle while they lie behind the instant centre and tighten again and straighten while they come to lie in front of the instant centre; *vice versa* during extension. Similar patterns of slackening and tightening of individual fibres have also been demonstrated in models of the collateral ligaments (Lu & O'Connor 1996a). The patterns of fibre length change predicted by the model agree well (Zavatsky & O'Connor, 1992a) with those observed by Fuss (1989) and Sapega *et al* (1990). The changes in direction of the model ligament fibres agree well (Lu & O'Connor, 1996b) with the measurements of Herzog and Read (1992). The changes in strain distributions within the ligaments are compatible with the movements of the bones on each other.

5. STABILITY OF THE KNEE

The *stability* of a joint is assessed by the resistance which it offers to movement. *Passive stability* is provided by mechanical interactions between the articular surfaces and the ligaments and is measured by such clinical tests as the Lachman or Drawer tests. *Active stability* is provided by the added effects of myogenic forces when muscles are used to suppress or control movement at the joint in the presence of external loads.

6. PASSIVE STABILITY

6.1 *Passive Resistance to Antero-posterior Translation*

The model joint can be used to simulate the Lachman Test (Lu & O'Connor, 1996a, Zavatsky & O'Connor, 1992b). In the unloaded state, the anterior fibre of the ACL and the central fibre of the PCL are just tight and remain straight; all other fibres are slack and buckled. When the tibia is pulled forwards, the fibres of the ACL straighten and tighten progressively as they are recruited to bear load. There is a similar progressive tightening and recruitment of the fibres of the PCL when the tibia is moved backwards from its position in the unloaded state. As one ligament tightens progressively, the other slackens.

The collateral ligaments behave similarly. The ACL and medial collateral ligament (MCL) tighten together to resist anterior translation of the tibia while the PCL and lateral collateral ligament (LCL) share resistance to posterior translation. The developing patterns of strain in the pairs of ligaments are compatible with shared translations of their attachment areas relative to each other. The contributions of the various ligaments to the passive stability of the knee has been the subject of a considerable body of often contradictory experimental work, recently summarised by Shoemaker and Daniel (1990).

7. FROM GEOMETRY TO MECHANICS

Having determined from geometry the distributions of strains within the ligaments for a specified translation of the tibia, mechanics is needed to determine the values of the external forces required to produce that translation. With the patterns of strain within the ligaments defined, constitutive equations can be used to calculate distributions of stress within the ligaments and then ligament forces (Zavatsky & O'Connor, 1992b). The external forces which have to be applied to balance the ligament and contact forces resulting from any assumed tibial translation can then be calculated. The plotted relationships between the horizontal force applied to the tibia and the resulting horizontal displacement of the tibia (Zavatsky & O'Connor, 1992b) calculated from the model are similar in shape to those obtained experimentally by Markolf *et al* (1976). There is little resistance to the first 2 mm of antero-posterior movements of the tibia in either direction but resistance increases rapidly as more fibres within the ligaments tighten and are recruited to bear load.

The relationships between applied force and tibial displacement are non-linear because (Zavatsky & O'Connor, 1992b) the relationship between tibial displacement and fibre length is non-linear (the cosine rule), and because the relationship between fibre strain and fibre stress is non-linear but mainly because the effective cross-sectional area of a ligament increases as more fibres are recruited to bear load. The relationships vary over the range of flexion because the directions of the ligaments and their patterns of slackness and tightness alter.

When tension forces are developed in the ligaments, they force the articular surfaces against each other, loading the cartilage in compression. If the consequent indentation of the articular surfaces is also taken into account, the joint is more compliant and a preliminary analysis suggests that the force required to achieve any given antero-posterior displacement is reduced by about 20% (Huss et al, 1998). Blankevoort's model also suggests that surface deformation makes only a small contribution to joint compliance in passive or moderate loading conditions (Blankevoort et al 1991, 1996).

8. ACTIVE STABILITY

The passive structures of the knee (the ligaments, capsule, menisci and articular surfaces) suppress most of the possible six degrees of freedom of the tibia relative to the femur, confining the movement to a relatively narrow range of flexion from 0° to 140° with a small range of axial rotation and an even smaller range of abduction/adduction. To be able to stand vertically in the presence of gravity requires muscle forces to suppress the remaining freedom of movement at the joint. Additionally, muscles are used to initiate and maintain movement.

8.1 Lever Arms Lengths

As the knee moves, the lever arms of its muscles vary in length. The muscle tendons spanning the joint rotate about their origins and insertions during flexion and extension. At the same time, the flexion axis of the joint, the instant centre of the linkage model, moves backwards and forwards relative to the tibia. The lever arms of the muscles, the perpendicular distances from the instant centre to the tendons, therefore vary during flexion and

can be calculated from the geometry of the model (Lu & O'Connor, 1996b, O'Connor et al. 1990a, O'Connor et al, 1990b, O'Connor, 1993).

8.2 Muscle Forces

The authors have performed experiments on the human knee, similar to those performed by Borelli (1679). The femur was held at 45° to the horizontal and a weight hung on the end of the tibia was balanced by tension in a wire sewn either to the quadriceps or the hamstrings tendons (O'Connor et al, 1990c). The length of the wire was adjusted to hold the tibia stationary at different flexion angles and the force in the wire necessary to suppress movement was measured in each position.

The measured muscle tendon forces varied over the range of motion of the joint because of changes in the lever-arm of both external load and tendon. The measurements agree well with calculations from the computer model, using model parameters derived from the same specimen. Quadriceps force up to 16 times the applied load and hamstrings force up to 12 times the applied load demonstrate that the levels of muscle force needed to suppress movement at the joints can be much larger than the external loads. The similarity of the measurements and calculations validate the mathematical model and the calculated muscle lever-arm lengths. Large muscle forces give rise to even larger tibio-femoral contact forces.

8.3 Ligament Forces

Forces in muscle tendons can be measured more easily than forces in the ligaments in vitro. Although buckle transducers have been used to measure ligament forces (Lewis et al, 1982), they must necessarily disturb the patterns of fibre crossing and twisting and the patterns of fibre recruitment described above. To evaluate ligament forces, it is generally necessary to resort to calculation. Morrison (1968) studied the gait of three subjects in the laboratory and used a mathematical model to calculate the forces transmitted by the ligaments. He found that the ACL and PCL are each loaded at different phases of the gait cycle, as are the two collateral ligaments.

The tendons do not pass vertically across the knee joint cleft but are usually inclined to the tibial plateau. The forces they exert have components parallel to the tibial plateau. The primary function of a muscle is to balance the flexing or extending effect of an applied load and it is this requirement which mainly determines the magnitude of the muscle force. Any resultant of the muscle force and the load parallel to the tibial plateau is balanced by the ligaments. When the resultant force tends to pull the tibia forwards, the ACL is loaded. When the resultant force tends to pull the tibia backwards, the PCL is loaded. The collateral ligaments share these loads.

Some features of the complementary actions of the two cruciate ligaments was demonstrated during a simulation of isometric quadriceps exercises, when the quadriceps act against a resisting force applied perpendicular to the tibia (Zavatsky & O'Connor, 1993). The ACL is loaded when the joint is flexed less than about 80° with the resistance placed distally on the tibia. The PCL is loaded when the joint is flexed more than 80° or when the resistance is placed proximally on the tibia. There is a combination of resistance placement and flexion angle at which neither the ACL or the PCL is loaded. In these circumstances, the forward pull of the patellar tendon force is exactly balanced by the backwards thrust of the resisting force. If the resistance is placed more distally, the quadriceps force is larger than the resisting load. Near extension, the patella tendon pulls the tibia forwards, loading the ACL. In the highly flexed joint, the patellar tendon pulls the tibia backwards, loading the PCL. For more proximal placement of the resistance, the quadriceps force is relatively small and the PCL is loaded by the backwards thrust of the resisting force.

Ligament loading therefore depends on the line of action of the external forces and on the values of the muscle forces, which in turn depend on the lengths of the lever-arms of the muscle tendons.

8.4 *Comment*

Evaluation of the forces transmitted by ligaments in activity requires prior knowledge of the directions of the muscle tendons and ligaments, obtained from a geometric model. The forces can be calculated directly, assuming the ligaments to be inextensible. When account is taken of tissue deformation, incremental loading must be considered, requiring iterative numerical solution of the non-linear equations at each increment (Zavatsky & O'Connor, 1993), with sequential geometrical and mechanical analysis during each iteration. The muscle, ligament and contact forces are then no longer proportional to the external load (Zavatsky & O'Connor, 1993).

9. CORONAL AND TRANSVERSE PLANES

9.1 *Experimental*

Although analysis of the behaviour of the knee joint in the sagittal plane has been instructive, it cannot account for events in the coronal plane and the effects of axial rotations.

The authors have described experiments on knee specimens which were flexed and extended repeatedly under very small loads (the weight of about 15 cm of distal femur, about 5N) (Wilson, 1996, Wilson *et al*, 1996). When unloaded, the bones followed a single unique track, the neutral track of passive flexion. Displacement of the bones along the track was not resisted. There was no elastic spring-back.

In the unloaded state, axial rotation and abduction/adduction and the translation of a chosen reference point are uniquely coupled to flexion. It was concluded that the unloaded knee has a single degree of unresisted freedom. Meyer (1853) first described the coupling of axial rotation to flexion, commonly called the "screw home mechanism".

In a similar experiment (Wilson *et al*, 1996), the flexion movement of the specimen was stopped at several positions, the tibia was displaced from its position in the unloaded state and was then released. The tibia could be rotated or abducted by the application of external load but, when that load was removed, it returned to its original track (elastic spring-back).

The experiment is a demonstration of the fact that the femur and the tibia follow a unique track relative to each other over the range of movement only in the unloaded state. Displacement of the bones from one position to another along that track is not resisted. Deviation from that track is resisted, requiring the application of external force and deformation of the tissues. The deviation is almost completely recovered when the applied force is quickly removed. Each combination of loads results in its own track.

9.2 *Theoretical*

The authors have begun development of a 3-dimensional analogue of the four-bar linkage, a parallel spatial mechanism like an aircraft cockpit simulator (Wilson, 1996, Wilson *et al*, 1996). In this model, three lines represent isometric fibres in the ACL, PCL and MCL, holding together two pairs of articular surfaces to represent the medial and lateral compartments. The five constraints represented by three ligaments and two pairs of surfaces in contact reduce the number of degrees of freedom of the mechanism to unity, coupling axial rotation and abduction/adduction to flexion. The kinematics of the mechanism were analysed (Wilson, 1996, Wilson *et al*, 1996) using the method devised by Uicker *et al* (1964). The analysis finds the path of unresisted motion which satisfies the geometric constraints that the ligament fibres do not stretch or slacken and that the articular surfaces do not separate or interpenetrate. When it is flexed and extended, the model exhibits the obligatory tibial rotation which is observed in the human knee (the "screw-home mechanism"). During flexion, the tibia rotates internally on the femur and during extension it rotates externally, as described first by Meyer (1853).

Parameters for the model were determined from specimens which had been used in the experiments. The similarity between the calculated and measured tracks of the bones upon each other confirms the hypotheses underlying the model, that the geometrical constraints offered by the three ligaments and the two pairs of articular surfaces can be satisfied by a precise coupling of rotation to flexion and that passive movement can occur without tissue deformation or surface separation.

The movement path and the range of movement of the model joint can be altered by changing the shapes of the articular surfaces or their positions on the bones or by altering the origins or insertions of the ligament fibres. The model can therefore provide a theoretical basis for prosthesis design or ligament reconstruction.

10. THE REAR-FOOT COMPLEX

Preliminary experiments and analysis of the joints between the tibia, the fibula, the talus and the calcaneus in the rear foot suggest that the approach just described for the study of the knee may prove fruitful in the study of other joints (Leardini *et al*, 1996, Leardini *et al*, 1998). Like the knee, each specimen exhibited a unique track characteristic of a single-degree of freedom system, with rotation and pronation coupled to flexion. Most if not all passive motion of the rearfoot complex occurs at the ankle, with virtually no motion at the subtalar joints. By applying external force to the system, it was possible to produce deviations from the neutral track of passive motion, as with the knee. The deviations were fully recovered when the external force was removed.

The experiments suggest that the ankle joint offers one degree of unresisted freedom in the unloaded state whereas the subtalar joints offer none. Motion could be induced at the subtalar joints only by the application of load, the bones returning to a single unique position when load was removed. During passive motion of the rearfoot complex, the fibular-calcaneal ligament and the central fibres of the deltoid ligament remained approximately isometric over the range of motion whereas all other ligament fibres joining the tibia to the calcaneus were slack except at the limits of dorsi- or plantarflexion. Fibres behind the isometric fibres were tight only in dorsiflexion and those in front of the isometric fibres were tight only in plantarflexion. The ligaments joining the talus to the calcaneus and the fibula to the tibia remained tight over the range of passive motion.

These experimental results imply that passive motion of the ankle-subtalar complex is controlled by a single degree of freedom mechanism, the isometric ligament fibres described above and the three sets of articular surfaces at the ankle serving as five constraints to motion. As a first step, we have modelled the complex as a four-bar linkage in the sagittal plane, with the tibia/fibula, the calcaneus/talus, the calcaneofibular ligament and the central fibres of the deltoid ligament acting as the four links in the mechanism (Leardini *et al* 1998). As with the knee model, the instant centre of the linkage lies at the point at which the ligaments cross and the articular surfaces touch at the point where the common normal to the surfaces passes through the instant centre. As a consequence, the surfaces roll as well as slide upon each other, contact moving from the back of the tibial trochlea in plantarflexion, to the front in dorsiflexion. The calculated tracks of the ligament insertions on the calcaneus agree well with experiment.

11. A FUNCTIONAL CLASSIFICATION OF JOINTS AND LIGAMENTS

Joints may be classified according to the number of degrees of unresisted passive motion they allow.

The hip allows three degrees of unresisted freedom, the articular surfaces providing the principal constraints to motion, the ligaments and capsule serving only to resist distraction and to define the limits to unresisted motion, being slack elsewhere. Ligaments are not involved in guiding the surfaces over each other. In the language of kinematics, the hip is a spherical pair.

The knee and the ankle allow only one degree of unresisted freedom, the movements of abduction/adduction, axial rotation and all three components of translation being uniquely coupled to flexion/extension. These joints move as mechanisms, movement being controlled by interactions between certain isometric ligament fibres and sliding movements of the articular surfaces. Appropriate models for the study of the kinematic geometry (Hunt, 1978) of these joints can take the articular surfaces to be rigid and the isometric ligament fibres to be inextensible. The mobility of these joints does not, to a first approximation, depend on the mechanical properties of ligaments or cartilage. The kinematic analysis provides the configuration of the joint structures at any position in the unloaded state, serving as the starting point for subsequent mechanical analysis.

Joints such as the subtalar complex and tibio-fibular joint do not allow unresisted motion, force being required to deform the tissues and produce relative motion of the bones from a single unique configuration in the unloaded state. Such joints behave like flexible structures, their mobility depending critically on the mechanical properties of the tissues.

By the same token, there appear to be three classes of ligament. The ligaments of the hip tighten only at the limits of motion and play no part in the guidance of the articular surfaces on each other or in load transmission within the range of motion. Some ligaments (ACL, PCL, MCL) contain fibres which guide the movements of the bones on each other and other fibres which are slack in most positions but are recruited progressively to bear load. The configuration of these ligaments changes systematically over the range of passive movement. Some ligaments (the talar-calcaneal ligaments) hold the bones together and stretch or slacken to allow movement. These ligaments have a single configuration and remain uniformly tight in the unloaded state.

12. LOCOMOTOR SYSTEM MODELS

Force analysis of the musculoskeletal system requires, first, a geometrical model of the system in the unloaded state, preferably customised through choice of parameters to individual subjects, animated with data from a gait analysis system. Understanding the complex movements of the system and the relative movements of its elements can be helped by the use of sophisticated computer graphics.

A number of such models have been proposed (Chao et al 1996, Delp et al 1990) and show how the bones and the muscles move relative to each other during activity. The parameters of such a model which define the shapes of the bones and the areas of origin and insertion of muscles are precisely the quantities changed by surgery. It is proposed that, by varying parameters, the models could be used to study the effects of degenerative

changes, injury or deformity and to simulate surgery.

However, gait analysis systems may never be accurate enough to reproduce the subtle constraints to motion imposed by the different types of joint. Models of the anatomical structures of the joints, articular surfaces, ligaments and capsule, should be included in the geometric model of the locomotor system. Their inclusion would allow the study of joint stability in activity and the simulation of the effects of injury or degeneration of joint tissue.

Such a model is under development by the authors (Collins & O'Connor 1991, Collins 1995, Lu et al 1998a, Lu et al 1998b). Calculations of the values of axial force in the femur using this model compare well with measurements telemetered from instrumented prostheses (Lu et al 1998a, Lu et al 1998b) in simple experiments both for a two-dimensional locomotor system model (Lu et al 1998b) including the 4-bar linkage knee and for a 3-dimensional model (Lu et al 1998a) including the parallel spatial mechanism knee. These studies are producing experimentally validated values of the forces applied to the bones during certain activities and should help improve the formulation of FEM models of stress distributions in bones.

13. CONCLUSION

Computer-based mathematical modelling of joints leads to a variety of insights into the inter-relationships between the articular surfaces, the ligaments and the muscles. It is important that such mathematical models should be validated against independent *in vivo* and *in vitro* experimental data. Accurate force analysis of the musculo-skeletal system, taking full account of muscle activity, is a necessary preliminary to the study of the response of living tissues to their mechanical environment.

14. ACKNOWLEDGEMENTS
The work of the authors described here was supported by grants from the Arthritis and Rheumatism Council and scholarships from the Government of the Republic of China, the University of Oxford and the Rizzoli Institute of Orthopaedics, Bologna.

REFERENCES
1. Andriacchi, T., Mikosz, R., Hampton, S. and Galante, J. (1983) Model studies of the

stiffness characteristics of the knee joint. *J Biomech* 16, 23-9.

2. Bergmann, G., Graichen, F. and Rohlmann A. (1993) Hip joint loading during walking and running, measured in two patients. *J Biomech* 26, 969-90.

3. Blankevoort, L., Huiskes, R. and de Lange, A. (1988) The envelope of passive knee joint motion. *J Biomech* 21, 705-20.

4. Blankevoort, L., Kuiper, J., Huiskes, R. and Grootenboer, H. (1991) Articular contact in a three-dimensional model of the knee. *J Biomech* 24, 1019-31.

5. Blankevoort, L. and Huiskes, R. (1996) Validation of a three-dimensional model of the knee. *J Biomech* 29, 955-61.

6. Borelli, G.A. (1679) *De motu animalium*. English Translation: Maquet, P. (1989) *On the Movement of Animals*. Springer-Verlag, Berlin.

7. Chao, E.Y.S., Barrance, P., MacWilliams, B., Genda,E. and Li, G. (1996) Application of virtual reality model in orthopaedic research. *Trans Orthop Res Soc*, 708.

8. Collins, J.J. and O'Connor, J.J. (1991) Muscle-ligament interactions at the knee during walking. *Proc Instn Mech Engrs, Part H, J Eng Med* 205, 11-8.

9. Collins, J.J. (1995) The redundant nature of locomotor optimisation laws. *J Biomech* 28, 251-67.

10. Crowninshield, R., Pope, M.H. and Johnson, R.J. (1976). An analytical model of the knee. *J Biomech* 9, 397-405.

11. Davy, D.T., Kotzar, G.M., Brown, R.H., Heiple, K.G. Sr, Goldberg, V.M., Heiple, K.G. Jr., Berilla, J. and Burstein, A.H. (1988) Telemetric force measurements across the hip after total arthroplasty. *J Bone Jt Surg* 70-A, 45-50.

12. Delp, S.L., Loan, J.P., Hoy, M.G. Zajac, F.E., Topp, E.L. and Rosen, J.M. (1990) An interactive graphics-based model of the lower extremity to study orthopaedic surgical procedures. *IEEE Trans Bio Eng* 37(8), 757-67.

13. Elias, G.G., Freeman, M.A.R., Gokçan, E.I. (1990) A correlative study of the geometry and anatomy of the distal femur. *Clin Orthop* 260: 98-103.

14. Ellis, M., Seedhom, B., Amis, A., Dowson, D. and Wright, V. (1979) Forces in the knee joint whilst rising from normal and motorised chairs. *Eng Med* 8, 33-40.

15. English, T.A. and Kilvington, M. (1979) *In vivo* records of hip loads using a femoral implant with telemetric output. *J Biomed Eng* 1, 111-15.

16. Essinger, J., Leyvraz, P., Heegard, J. and Robertson, D. (1989) A mathematical model for the evaluation of the behaviour during flexion of condylar-type knee prostheses. *J Biomech* 22, 1229-41.

17. Friederich, N.F., Muller, W. and O'Brien, W.R. (1992) Klinische Anwendung biomechanischer und funktionell anatomischer Daten am Kniegelenk. *Orthopäde* 21, 41-50.

18. Fuss, F.K. (1989) Anatomy of the cruciate ligaments and their function in extension and flexion of the human knee joint. *Am J Anat* 184, 165-76.

19. Gill, H.S. and O'Connor, J.J. (1996) A bi-articulating two-dimensional computer model of the human patellofemoral joint. *Clin Biomech* 2, 81-9.

20. Girgis, F.G., Marshall, J.L. and Monajem, A.R.S. (1975) The cruciate ligaments of the knee joint. *Clin Orthop* 106, 216-31.

21. Goodfellow, J.W., Hungerford, D. and Zindel, M. (1976) Patello-femoral joint mechanics and pathology: I. Functional anatomy of the patello-femoral joint. *J Bone Jt Surg [Br]*, 58-B, 287-90.

22. Goodfellow, J. and O'Connor, J. (1978) The mechanics of the knee and prosthesis design. *J Bone Jt Surg [Br]* 60-B, 358-69.

23. Hatze, H. (1976) The complete optimisation of a human motion. *Math Biosci* 28, 99-135.

24. Hefzy, M. and Grood, E. (1988) Review of knee models. *Appl Mech Rev* 41, 1-13.

25. Herzog, W. and Read, L.J. (1993) Lines of action and moment arms of the major force-carrying structures crossing the human knee joint. *J Anat* 182, 213-30.

26. Hunt, K.H. (1978) *Kinematic Geometry of Mechanisms*. Oxford University Press, Oxford

27. Huss, R.A., Holstein, H. and O'Connor, J.J. (1998) The effect of cartilage deformation on laxity of the knee joint. *J Eng Med*, under review.

28. Huson, A. (1974) Biomechanische Probleme des Kniegelenks. *Orthopäde* 3, 119-26.

29. Leardini, A., Lu, T.-W., Catani, F. and O'Connor, J.J. (1996) A four-bar linkage model of the ankle. *British Orthopaedic Research Society, September*, Abstracts p.16.

30. Leardini, A., Lu, T.-W., Catani, F. and O'Connor, J.J. (1998) The one degree of freedom nature of the human ankle complex. *J Biomech*, in press.

31. Lewis, J.L., Lew, W.D. and Schmidt, J. (1982) A note on the application and evaluation of the buckle transducer for knee ligament force measurement. *J Biomech Engng* 104, 125-8.

32. Lu, T-W. and O'Connor, J.J. (1996a) Fibre recruitment and shape changes of knee

ligaments during motion: as revealed by a computer graphics-based model. *Proc Instn Mech Engrs, Part H, J Eng Med* 210, 71-9.

33. Lu, T-W. and O'Connor, J.J. (1996b) Lines of action and moment arms of the major force-bearing structures crossing the human knee joint: comparison between theory and experiment. *J Anat* 189, 575-85.

34. Lu, T.-W., O'Connor, J.J., Taylor, S.J.G., Walker, P.S. (1998a) Comparison of telemetered femoral forces with model calculations. *Trans Orthop Res Soc,* **1**:409.

35. Lu, T.-W., O'Connor, J.J., Taylor, S.J.G., Walker, P.S. (1998b) Validation of a lower limb model with *in vivo* femoral forces telemetered from two subjects. *J Biomech,* **31(1)**:63-9, 1998b.

36. Markolf, K.L., Mensch, J.S. and Amstutz, H.C. (1976) Stiffness and laxity of the knee - the contributions of the supporting structures. A quantitative *in-vitro* study. *J Bone Jt Surg [Am]* 58-A, 583-93.

37. Menschik, A. (1974) Mechanik des Kniegelenks - teil 1. *Z Orthop* 112, 481-95

38. Meyer, H. (1853) Die Mechanik des Kniegelenks. *Archiv für Anatomie und Physiologie,* 497-547.

39. Mikosz, R., Andriacchi, T. and Andersson, G. (1988) Model analysis of factors influencing the prediction of muscle forces at the knee. *J Orthop Res* 6, 205-14.

40. Miller, R.K., Murray, D.W., O'Connor, J.J., Goodfellow, J.W. (1997) *In vitro* patellofemoral joint force determined by a non-invasive technique. *Clin Biomech* 12 (1),1-7..

41. Mommersteeg, T.J.A., Huiskes, R., Blankevoort, L., Kooloos, J.G.M. and Kauer, J.M.G. (1996) A global verification study of a quasi-static knee model with multi-bundle ligaments. *J Biomech* 29, 1659-64.

42. Mommersteeg, T.J.A., Huiskes, R., Blankevoort, L., Kooloos, J.G.M. and Kauer, J.M.G. (1997) An inverse dynamics modelling approach to determine the restraining function of human knee ligament bundles. *J Biomech* 30, 139-46.

43. Morrison, J.B. (1968) Bioengineering analysis of force actions transmitted by the knee joint. *BioMed Engng* 90, 164-70.

44. O'Connor, J., Shercliff, T., Biden, E. and Goodfellow, J. (1989) The geometry of the knee in the sagittal plane. *Proc Inst Mech Eng Part H, J Eng Med* 203, 223-33.

45. O'Connor, J.J., Shercliff, T., FitzPatrick, D., Bradley, J.,Daniel, D. Biden, E. and Goodfellow, J. (1990a) Geometry of the knee. Chapter 10. In: *Knee Ligaments: Structure, Function, Injury and Repair.* (Eds Daniel, D.M., Akeson, W.H. and O'Connor, J.J). Raven Press, New York, 163-200.

46. O'Connor, J.J., Shercliff, T., FitzPatrick, D., Biden, E. and Goodfellow, J. (1990b) Mechanics of the Knee. Chapter 11. In: *Knee Ligaments: Structure, Function, Injury, and Repair.* (Eds: Daniel, D.M., Akeson, W.H. and O'Connor, J.J). Raven Press, New York, 201-38.

47. O'Connor, J.J., Biden, E. Bradley, J., FitzPatrick, D., Young, S. and Kershaw, C., Daniel, D. and Goodfellow, J. (1990c) The muscle-stabilized knee. Chapter 12. In: *Knee Ligaments: Structure, Function, Injury, and Repair.* (Eds: Daniel, D.M., Akeson, W.H. and O'Connor, J.J). Raven Press, New York, 239-78.

48. O'Connor, J.J. (1993) Can muscle co-contraction protect knee ligaments after injury or repair? *J Bone Jt Surg [Br]* 75-B, 41-8.

49. Pandy, M. G., Zajac, F. E., Sim, E. and Levine, W.S. (1990) An optimal control model for maximum-height human jumping. *J Biomech* 23, 1185-98.

50. Paul, J.P. (1965) Bio-engineering studies of the forces transmitted by joints: II. Engineering analysis. In: *Biomechanics and Related Bio-Engineering Topics* (Edited by Kenedi, R.M.). Pergamon Press, Oxford, 369-80.

51. Pauwels, F. (1950) Principles of construction of the lower extremity. Their significance for the stressing of the skeleton of the leg. Chapter 6. In: *Biomechanics of the Locomotor Apparatus,* Springer-Verlag, Berlin, 1980. (Original published in *Z. Anat. Entwickl. Gesch.* 1950:114, 525-38).

52. Rydell, N.W. (1966) Forces acting on the femoral head prosthesis. *Acta Orthop Scand* (Suppl.) 88.

53. Sapega, A.A., Moyer, R.A., Schneck, C. and Komalahiranya, N. (1990) Testing for isometry during reconstruction of the anterior cruciate ligament. *J Bone Jt Surg [Am]* 72-A, 259-67.

54. Shoemaker, S.C. and Daniel, D.M. (1990) The limits of knee motion: *in vitro* studies. Chapter 9. In: *Knee Ligaments: Structure, Function, Injury and Repair* (Eds: Daniel, D.M., Akeson, W.H. and O'Connor, J.J). Raven Press, New York, 153-62.

55. Strasser, H. *Lehrbuch der Muskel- und Gelenkmechanik.* Springer, Berlin, 1917.

56. Taylor, S.J. and Donaldson, N. (1990) Instrumenting Stanmore prostheses for long-term strain measurement *in vivo.* In: *Proc of a Workshop on Implantable Telemetry in Orthopaedics* (Eds Bergmann, G., Graichen, F.

and Rohlmann, A.) Freie Universitat, Berlin, 93-102.

57. Taylor, S.J.G., Perry, J.S., Meswania, J.M., Donaldson, N., Walker, P.S. and Cannon, S.R. (1997) Telemetry of forces from proximal femoral replacements and relevance to fixation. *J Biomech* 30, 225-34.

58. Uicker, J., Denavit, J. and Hartenberg, R. (1964) An iterative method for the displacement analysis of spatial mechanisms. *Trans ASME, J Appl Mech* 31, 309-14.

59. Weber, W.E. and Weber, E.F.W. Mechanik der menschlichen Gehwerkzeuge. *in der Dietrichschen Buchhandlung*, Gottingen, 1836.

60. Wilson, D. R., Zavatsky, A. B. and O'Connor, J. J. (1993) Cruciate ligament forces at the knee in gait: parameter sensitivity and effects of ligament elasticity. *International Society of Biomechanics*, Abstracts pp 1466-7.

61. Wilson, D. (1996) *Three-dimensional kinematics of the knee.* D. Phil. thesis, University of Oxford, Oxford, U.K.

62. Wilson, D., Feikes, J., Zavatsky, A., Bayona, F. and O'Connor, J. (1996) The one degree-of-freedom nature of the human knee joint - basis for a kinematic model. *Canadian Society of Biomechanics,* Abstracts pp 194-5.

63. Winters, J. M. and Stark, L. (1987) Muscle models: what is gained and what is lost by varying model complexity. *Biol Cybern* 55, 403-20.

64. Wismans, J., Veldpaus, F. and Janssen, J. (1980) A three-dimensional mathematical model of the knee-joint. *J Biomech* 13, 677-85.

65. Zavatsky, A.B. and O'Connor, J.J. (1992a) A model of human knee ligaments in the sagittal plane. Part I: response to passive flexion *Proc Instn Mech Engrs, Part H, J Engng Med* 206, 125-34.

66. Zavatsky, A.B. and O'Connor, J.J. (1992b) A model of human knee ligaments in the sagittal plane. Part II: fibre recruitment under load. *Proc Instn Mech Engrs, Part H, J Eng Med* 206, 135-45.

67. Zavatsky, A.B. and O'Connor, J.J., (1993). Ligament forces at the knee during isometric quadriceps contractions. *Proc Instn Mech Engrs, Part H, J Engng Med* 207, 7-18.

68. Zavatsky, A.B. and O'Connor, J.J. (1994) Three-dimensional geometrical models of human knee ligaments. *Proc Instn Mech Engrs, Part H, J Engng Med* 208, 229-40.

Biomedical engineering

Experimental Mechanics, Allison (ed.) © 1998 Balkema, Rotterdam, ISBN 90 5809 014 0

Development of simple system for neonatal respiration monitoring

P.Cappa
Department of Mechanics and Aeronautics, University of Rome 'La Sapienza', Italy

S.A.Sciuto
Department of Mechanical and Industrial Engineering, III University of Rome, Italy

ABSTRACT: In this work the feasibility of measuring chest displacement by means of a simple system based on a strain gauge transducer, generally addressed in Biomechanics as "buckle transducer", directly connected to an electroencephalograph is presented. The proposed system, in-house developed, is characterised by realisation easiness, intrinsic reliability and modest cost. Three specimens were realised and tested to determine their linearity (linear correlation coefficient equal to 0.994), repeatability (lesser than ±5%FSO) and mechanical impedance (mean value equal to $143 \cdot 10^{-3}$mm/N and standard deviation of ±4%), that resulted adequate for the foreseen utilisation in the Neurology Department of the Children's Hospital "Bambino Gesù" of Rome. Time dependent tests were also carried out over a 2 hour time-length and results showed a global error lesser than ±1%FSO and a maximum temperature coefficient lesser to ±0.1%FSO/°C. Finally, field tests were conducted at the Neurology Department for several months and the gathered data demonstrated the effectiveness of the proposed system. Currently, the breathing strain gauge measuring system has completely replaced other previously utilised commercially available transducers with total neurologist satisfaction.

1. INTRODUCTION

Premature infants frequently have periods of apnoea (Rolfe 1986, Little 1986) which may cause severe damages to central nervous system or even sudden infant death syndrome even after few minutes, if apnoea remains undetected. Moreover, there are also other pathological respiratory diseases that need documentation of spontaneous respiratory changes as in case of chronic obstructive pulmonary disease, apnoeustic and ataxic respiration, etc. Therefore, in neurology is usual practice (Daly & Pedley 1990) the integration of cerebral activity monitoring, carried out by means of electroencephalograph (EEG), with respiratory activity recording (McIntosh 1983, Niedermeyer & Da Silva 1993) in order to reduce time and strain of nursing staff in infant observation for apnoeic attacks. Nevertheless, most common application of respiratory monitoring only implies the identification of apnoea periods and the setting of an alarm whereas, in other pathologies, the respiration pattern is necessary to combine

frequency values with changes in inhalation depth in order to evaluate also significant decrease in airflow (hypopnea). A variety of techniques can be utilised to indirectly monitor upper airway breathing, thoracoabdominal movement and arterial oxygen such as transthoracic electrical impedance measure, whole-body plethysmography and contacting motion sensors based on strain gauges, air-filled capsule or vest, etc., as described by Neuman (1988). Several other methodologies were recently proposed and applied to monitor apnea in non-intubated subjects, such as (1) utilizing a rapid-response hygrometer positioned in front of the nostril and capable of identifying respiratory phases even in case of tachypnea as high as 60 breaths per minute in infants (Tatara & Tsuzaki 1997, Ma et al. 1995) or (2) the evaluation of transmitted light variation through a bent optical fiber wrapped around the chest for qualitative and quantitative measurement of breathing effort and respiratory rate (Babchenko et al. 1995). However, respiration monitoring instrument reliability often falls down in case of

Neonatal Intensive Care Unit patients because of (1) artifacts due to motion and respiratory movements, in case of motion sensors (2) or oral breathing or crying, in case of sensors placed in front of nostril. At the Neurology Department of the Children Hospital "Bambino Gesù" of Rome it is usual to integrate neonatal patient EEG record with respiratory trend, gathered by means of a piezoelectric force sensor. The Children's Hospital "Bambino Gesù" ($\cong$730 bed-medical facility) is a private and non-profit-making hospital located in the Vatican City, i.e. the independent Papal state within the city of Rome (Italy), and is officially recognised by the Italian Government as a "Research and Care Institute of a Scientific Nature". However, the hereby utilised commercially available sensors generally exhibit intrinsic noticeable mechanical impedance, which often causes patient discomfort, and unacceptable unreliability and cost (upper than 300US$ for each transducer, which can never be repaired in any case and, as a consequence, needs to be replaced with a new one at any damage): for these reasons, at the Clinical Engineering Service (established in 1980 and manages about 3800 electro-medical devices for a global value of about 35 million US$) it was decided to in-house develop a low cost reliable respiratory monitoring system that accomplishes (1) the constraints on reduced mechanical impedance, due to neonatal application, and (2) the specific exigency of being interfaced with the already operating EEGs.

Among the proposed methods, despite its intrinsic proneness to patient movements, a contacting sensor was chosen, such as an electrical resistance strain gauge displacement sensor, for its realisation simplicity, that allowed in-house making and fixing, its intrinsic reliability and modest cost.

Then, the purpose of this paper is twofold: firstly, the Authors wish to determine the metrological performances of the proposed sensor and, in particular, to evaluate mechanical impedance; secondly, they want to verify the actual diagnostic performances during *in vivo* utilisation at the Neurology Department.

2. EXPERIMENTAL SET-UP

The chosen sensor is based on the scheme of a load cell, addressed as "buckle transducer", capable of providing magnitude, as well as patterns, of forces exerted by ventricular walls (Cotten & Bay

1956) or collageneous tissue structures (Piziali et al. 1980). The sensor design is similar to those proposed by Xu et al. (1992) to monitor muscle activity and mainly based on those utilised by Branca et al. (1995) during electrocardiographic analysis.

On the base of the previously indicated sensor, three specimens were realised in harmonic steel short elastic strip (10mm W × 70mm L × 0.2mm T, with 50 mm radius of curvature), see Fig.1. These transducers were equipped with a 90° stacked resistance strain gauge rosette bonded to the centre of the convex surface. Then, the realised transducer was wired in a half bridge arrangement with teflon insulated cables. The insulation resistance of the backing film and the cyanoacrylate cement was measured by means of a strain gauge tester and it always remained higher than 1000MΩ. Finally, the displacement sensor was connected to an elastic band surrounding the patient's chest. In order to operate the EEG downstream amplifier at lower gains and increase their zero stability, the transducer excitation level was set to the maximum value suggested by the manufacturer, with the obvious aim of maximising transducer intrinsic signal to noise ratio.

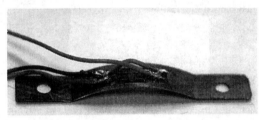

Figure 1. The respiratory transducer with the elastic strip and the amplifier unit.

In consideration of the specific need of interfacing the breathing sensor to the EEG models operating at the previously mentioned Neurology Department, one of the main design constraint is the suitability of transducer output to EEG conditioning system. Furthermore, it is to be kept into account that the breathing transducer will be utilised in a children's hospital, which implies working with neonatal and premature patients, i.e. with particularly reduced chest displacement. For these reasons it was decided to utilise the external input channel, common to all EEG models, after adequately pre-conditioning the signal generated by the breathing transducer by means of an integrated strain gauge conditioner with adjustable gain (linearity lesser than ±0.005%, thermal drift of ±0.25µV/°C, CMRR higher than 140dB). In this case, it was possible to employ the EEG input amplifier both for electrical impedance adaptation and for the intrinsic elevated CMRR (≥92dB).

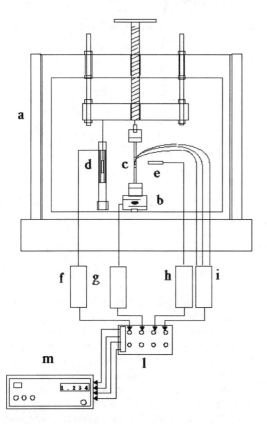

Figure 2. The experimental test set-up for transducer static characterisation: (a) screw testing machine, (b) load cell; (c) specimen test; (d) linear variable differential transformer; (e) Pt100 thermometer; (f) LVDT supply and conditioning unit; (g) DC power supply and strain gauge conditioning unit; (h) Pt100 conditioning unit; (i) breathing transducer conditioning unit; (l) switch unit; (m) digital voltmeter.

Hence, the balance of the trade-off between transducer operating easiness and the previously mentioned constraints, induced to carry out preliminary tests to determine transducer metrological performances. More specifically, due to the nature of the application which is quasi-static, only static investigations were performed. Thus, these tests were conducted to evaluate sensor performances, in terms of mechanical impedance and output range, by setting the strain gauge excitation voltage to 4V DC and the conditioner gain unitary, and its output signal was directly measured with a digital voltmeter. In particular, ten tensile tests were carried out on three different harmonic steel specimens, here named HS1, HS2 and HS3, respectively, by means of a screw testing machine, currently utilised in photoelastic tests. This machine (Fig. 2) was equipped with: (1) a load cell with a full scale of 1.962N conditioned by means of a strain gauge DC conditioning unit (with a linearity of ±0.05%); (2) a linear variable differential transformer with the relative power supply and conditioning unit (with ±1.27mm full scale and a linearity of ±0.25%); (3) a Pt100 thermometer with an accuracy lesser than ±0.1°C to monitor test area temperature variations; (4) a digital voltmeter with a switch unit to gather the transducer outputs. Because of the FS of the available load cell was about two times of the expected load applied during normal breathing, according to Branca (1995), a verification of the entire force measuring system accuracy in the range of 0.5FS was preliminary conducted by means of dead weights specifically prepared by means of an analytical balance (accuracy lesser than $±1·10^{-6}$N) already present at the Hospital. The maximum measured load cell inaccuracy evaluated in five calibration cycles was equal to $4.905·10^{-3}$N.

The breathing transducer were tested with loading cycles, from unloaded specimen to 0.981N with variations of 0.049N. The mean displacement values as a function of the applied load are summarised in Fig.3 for all of the three examined specimens. From an analysis of the obtained results, it emerges a linear correlation coefficient always greater than 0.999 and a global average sensitivity, i.e. the transducer mechanical impedance, is equal to $143·10^{-3}$mm/N±4% (mean value ± standard deviation). As it regards to the measuring system output, i.e. the breathing transducer and the pre-conditioning unit, the average FSO voltage was equal to 0.649mV, which appeared adequate for the EEG input constraints.

Successively, zero-shift and creep, with an applied constant load of 0.490 and 0.981N, were tested over a 2 hour time period for each of the specimens, either at a constant room temperature of about 24°C or imposing 10°C temperature increase, applied by means a thermoregulated oven characterised by a constancy in temperature within ±0.01°C. A global error lesser than 1%FSO for all of the applied load was appraised during ambient temperature tests, while 10°C temperature increase test allowed determining a maximum temperature coefficient lesser than 0.1%FSO/°C, in correspondence of the maximum applied load.

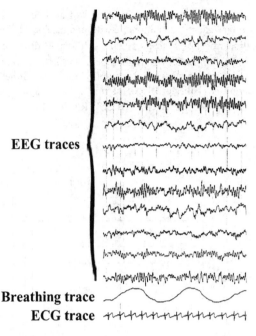

EEG traces

Breathing trace
ECG trace

Figure 4. Polygraphic strip-chart of a neurological study.

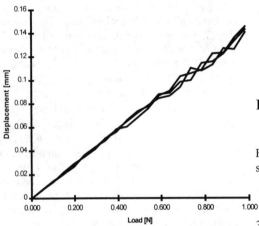

Figure 3. Mean displacement values as a function of the applied load for the three specimen.

Thus, the realised breathing system was commonly adopted at the Neurology Department in routine EEG studies. During six moths test period of continuous utilisation, an extremely wide range of (1) possible pathologies, (2) patient's chest volume and (3) time pattern of respiratory signal were covered. With regard to system reliability and robustness, only one case of failure occurred during test period because of cable damage. As an example, in Fig.4 the strip-chart recorded in a clinical study is reported. In this figure is possible to recognise the EEG traces, the breathing pattern monitored by the proposed measuring system and the electrocardiographic waveform.

Finally, according to Neurologists, the respiratory signals obtained were ever more than satisfactory.

3.   CONCLUSIONS

The here proposed breathing transducer and the pre-conditioning unit were at first metrologically tested, then were assessed when applied with newborn and infant patients during usual neurological tests. The obtained results in static analysis showed good transducer interchangeability and its sufficient unproneness to test area temperature variations. Six month continuous clinical application confirmed the reliability of the system which, given to its low cost and simplicity, could be a useful tool in routing breathing monitoring studies.

REFERENCES

Babchenko A, Turinvenko S, Khanokh B, Nitzan M, Fiber optic sensor for the assessment of breathing effort, (1995) Proceedings of SPIE - The International Society for Optical Engineering Society of Photo-Optical Instrumentation Engineers, Bellingham, WA, USA 2631:64-71.

Branca FP, Calcagnini G, Cerutti G, Mastrantonio F, A simple low-cost transducer for breath detection over 24h in clinical cardiology, (1995) 1995 IEEE Engineering and Biology Society 17th International Conference, Montreal, Sept. 95.

Cotten MdeV, Bay E, Direct measurement of changes in cardiac contractile force: relationship such measurements of stroke work, isometric pressure gradient and other parameters of cardiac functions, (1956) Am J Physiol, 187, 122.

Daly DD, Pedley TA, Current Practice of Clinical Electroencephalography, (1990) Raven Press, New York.

Little GA, Infantile Apnea and Home Monitoring, (1986) NIH Consens Statement, Sep 29-Oct 1, 6(6):1-10.

Ma Y, Ma S, Wang T, Fang W, Air-flow sensor and humidity sensor application to neonatal infant respiration monitoring, (1995) Sensors and Actuators, A: Physical 49(1-2):47-50.

McIntosh N, The monitoring of critically ill neonates, (1983) J Med Engng & Tecnol, 7(3):121-129.

Neuman M, Neonatal Monitoring, Encyclopedia of Medical Devices and Instrumentation, (1988) J Webster, John Wiley, New York.

Niedermeyer E, Da Silva FL, Electroencephalography, (1993) Williams & Wilkins, Baltimore.

Piziali RL, Rastegar J, Nagel DA, Schurman DJ, The contribution of the cruciate ligaments to the load - displacement characteristics of the human knee joint, (1980) ASME J Biomech Eng, 4:467-474.

Rolfe P, Neonatal critical care monitoring, (1986) J Med Engng & Tecnol, 10(3):115-120.

Tatara T, Tsuzaki K, An apnea monitor using a rapid-response hygrometer, (1997) J Clin Monit, 13(1): 5-9.

Xu WS, Butler DL, Stouffer DC, Grood ES, Glos DL, Theoretical analysis of an impiantable force transducer for tendon and ligament structures, (1992) ASME Trans, 114 : 170-177.

Experimental Mechanics, Allison (ed.) © 1998 Balkema, Rotterdam, ISBN 90 5809 014 0

Relaxation of living maple trees

H. Kato, K. Kageyama & M. Tomita
Department of Mechanical Engineering, Saitama University, Japan

ABSTRACT: Bending tests of living maple trees were carried out under different conditions to clarify the effect of the biological activity on the relaxation process of branches. The bending test was done in summer and winter. Effects of the repeated loading were also examined. The branch was cut off from the stem and bent to examine the change of biological activity with time. The bending test was also conducted with dead trees for comparison. When the branch was bent in summer, the decrease in force (resistance against bending) was smaller than that measured in winter. The decrease in force of the living branch was smaller than that of the dead branch. After the branch was cut off from the stem, the force decreased with time and kept almost constant at a time from 10 h to 100 h and then decreased to converge to a constant value. These are explained by the effect of the biological activity of the branch on the relaxation.

1 INTRODUCTION

Recently much attention has been focused on intelligent materials or smart structures. In previous works (Vincenzini 1995, Sarikaya & Aksay 1995), intelligent materials have been developed with models based on the animal: the brain was interpreted as CPU, muscles as actuator and nerves as sensor and transmission lines. The organic structure of the animal is quite complicated and all functions are stopped when the brain is damaged. On the other hand, since plants do not have a brain, whole functions are not destroyed even if considerable part of the body is damaged or removed. Therefore it is desirable to develop intelligent materials based on the plant, which shows intelligence such as self–healing without control of the brain.

The present work is performed to obtain basic knowledge for developing intelligent materials based on the plant. Branches of maple trees were bent and the relaxation process of the branch was examined to clarify self healing properties. Since plants are mainly composed of cellulose, a kind of polymer, when the tree is loaded, it responses rheologically (Boding & Jayne 1993). At that time, the tree also shows biological response. The biological response is different from the rheological one and it keeps the body stable although the rheological response causes larger deflection. Therefore, biological activity can be evaluated if the different response against loading is obtained for living and dead trees.

In a study of plants, there are some difficulties in measuring mechanical behaviors: it is difficult to mount gages on a stem or branches of trees to measure stresses, strains and deformations, and also the measurement is largely influenced by atmospheres such as temperature, humidify and so on. In the present work, a bending stress was not measured by a strain gage adhered on the branch, but a force of the branch to return to an original location was measured by a sensor fixed between branches and a stage. Although the measurement was conducted in a room, the temperature in the room changed since the measurement was continued for weeks or months. Therefore, a relation between outputs of sensors and temperature was measured and data were calibrated at each measurement.

2 EXPERIMENTAL PROCEDURE

2.1 *Fabrication of sensors*

Force and displacement sensors were prepared for measuring the force and the displacement of the tree, respectively. The sensors were made of high carbon steel plates of 0.6 ~ 0.7 mm in thickness as shown in Fig. 1. A root of the displacement sensor was necked to improve sensitivity. The force sensor was made similar to a clip gage.

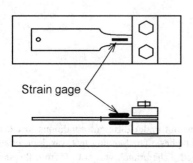

(a) Displacement sensor

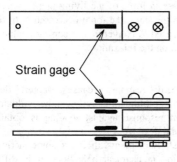

(b) Force sensor

Figure 1   Schematic representation of sensors

Linearity and temperature dependence of outputs of the sensors were examined as follows. To clarify connected to the micrometer through a tungsten wire of 0.1 mm diameter and outputs were measured by giving a displacement of 0 ~ 0.4 mm to an end of the sensor at an interval of 0.01 mm. The maximum load was 0.049 ~ 0.069 N (5 ~ 7 gf) at a displacement of 0.4 mm. The force sensor was hung with a tungsten wire from a rigid pipe and a load up to 4.9 N (500 gf) was applied to the end of the sensor at an interval of 0.245 N (25 gf) though the tungsten wire. Outputs linearly changed with the force or the displacement, and linear regression equations were obtained. It was confirmed when the force was removed from sensors, the output became null.

The temperature dependence of the sensor was measured in a room where the relaxation experiment was performed. The sensors were connected to rigid pipes at a height of 600 mm from the ground, the same height of the branch used in the relaxation experiment, with the tungsten wire. Figure 2 shows changes of outputs of the displacement sensors and temperature in the room with elapsed time. The output of the displacement linearly changed with temperature as shown in Fig. 3. A linear regression equation was obtained for each sensor to calibrate measured values in the relaxation experiment. The output of force sensors also linearly changed with temperature, and the regression equation was obtained.

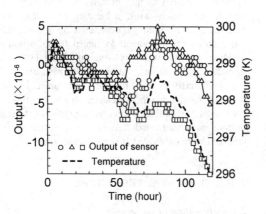

Figure 2   Changes of outputs of displacement sensors and temperature with elapsed time

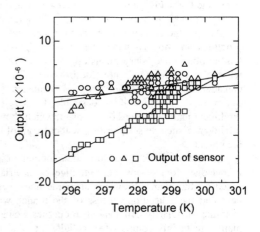

Figure 3   Relation between outputs of displacement sensors and temperature

2.2 Relaxation test

Maple trees of about 1500 mm height with stems of about 30 mm diameter were prepared. Figure 4 shows a setup for the relaxation test. Branches of 4.3 ~ 10.1 mm diameter growing at a height of about 600 mm from the ground were bent at an initial force of 1.4 ~ 2.4 N. The displacement sensors were connected to roots of branches, and the force sensors were connected to branches at a position 100 mm apart from the root of the branch. The displacement and the force were measured every 1 h. Temperature in the room was also measured near the tree for calibration of outputs of the sensors. Since the diameter of the branches used in the measurement was not the same, outputs were normalized by using a relation between the diameter of the branch and the bending force.

The first relaxation test was conducted to examine dependence of the relaxation process on the season: in summer (August ~ September) of higher temperature and higher biological activity, and in winter (November ~ December) of lower temperature and lower activity. The measurement was also done with dead branches for comparison.

The second test was done to examine influences of repeated loading on the relaxation. Since branches were bent several times in the test and the relaxation test was continued for weeks or months, it is probable that the organic structure can be damaged and the deformation character of the branch may be changed after each loading to cause scatter of data. The measurement was performed twice with an interval of 48 h in summer.

Finally, the relaxation test was carried out with branches cut off from the stem to examine duration of the biological activity. Branches were cut off at a root from the stem and fixed to a steel rod horizontally. The force sensor was connected to the branch at a position 100 mm apart from the fixed end with the tungsten wire. The measurement was started within 15 min after the branch was cut off. The initial load was about 2.0 N, and the force was measured every 15 min. To estimate change in water content of the branch with elapsed time, weight of the branch was also measured with other branches of 0.4 ~ 0.2 g in initial mass, which mass was in the range of that of branches used in the relaxation test. The relaxation test was also conducted with a dead branch cut off from the stem and left three months for comparison.

3 RESULTS AND DISCUSSION

3.1 Relaxation in summer and winter

The relaxation test was carried out in summer and winter to examine the effect of the biological activity on the relaxation. Changes in force are shown in Fig. 5. When living branches were bent in

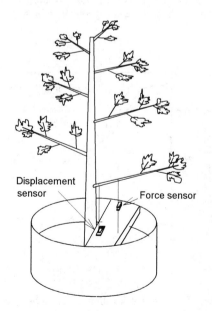

Figure 4   Setup of force and displacement sensors

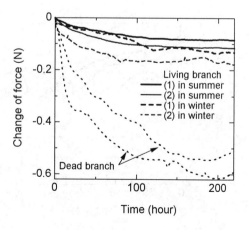

Figure 5   Stress relaxation of living branches in summer and in winter

summer, an amount of the force change was less than that in winter. The decreasing rate of the force in the initial stage was larger in water although the decreasing rate in later stages was same in summer and winter. The change in force of the living branch was also compared with that of the dead branch; the force of the living branch decreased slowly and became constant at a shorter time, but that of the dead branch decreased largely for longer times. These show that the resistance of the branch against the relaxation is dependent on the biological activity.

When the branch is forced to deform, the relaxation occurs to decrease the stress in the branch because of viscoelasticity of the tree, which organic substances typically show. Since the branch is alive, it also shows the biological response. There are two possibilities of the biological response such as,
(1) to reduce a force applied to the branch, the deformation of the branch proceeds.
(2) to keep the deformation and the shape, the branch keeps a force constant.
In the present work, the latter case occurred, and higher the biological activity was, the greater the resistance was, as shown in Fig. 5. In the present work, however, it was not clarified a microscopic mechanism of the response against the outer force, such as at a level of cell.

3.2 Effect of repeated loading

To clarify the effect of the repeated loading on the relaxation, living branches were bent twice. As shown in Fig. 6, the change in force is smaller in the

second loading than that in the first loading up to 50 h, but after 100 h, forces in the first and the second measurements became almost the same. This result shows that the repeated bending did not affect the relaxation process so largely that can be detected under present conditions.

3.3 Relaxation of branch cut off from stem

Changes in force with time after cutting off is shown in Fig. 7. The force of the branch decreased with time and kept almost constant at a time of 10 h to 100 h. Then it decreased again with time. As was expected, the force of the dead tree monotonically decreased and did not show a stay at an intermediate stage. Weight change of the branch was also expected following decreasing water content in the branch after cutting off from the stem. The weight of the branch monotonically decreased with time as shown in Fig. 8, but at a time of 10 h to 100 h, the changing rate was decreased as was observed in the force change.

In Fig. 8, the ratio of the change in weight to the initial weight of the branch was taken as the ordinate. Comparing Figs. 7 and 8, the decrease in force is related to the weight change of the branch, which was caused by decrease in water content in the branch. It can be said when the branch is cut off from the stem, the biological activity remains for a while (for almost 100 h) and it decreases with decreasing the water content contained in the branch.

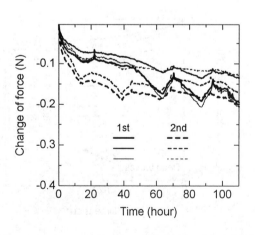

Figure 6   Stress relaxation of living branches at first and second loading

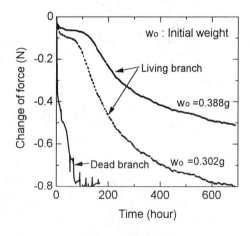

Figure 7   Stress relaxation of branches cut off from stem

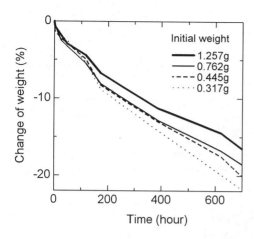

Figure 8   Weight change of branches after cutting off from stem

REFERENCES

Bodig J. & B.A. Jayne: "Mechanics of Wood and Wood Composites", Krieger Pub. Co., Lalabar, Florida, (1993).

Sarikaya M. & I.A. Aksay (eds): "Biominetics, Design and Process of Materials", AIP Press, Woodburg, New York, (1995).

Vincenzini P. (ed): "Intelligent Materials and Systems, Advances in Science and Technology, 10", TECHNA Srl, Faenza, Italy, (1995).

Therefore, if the water content in the branch is kept constant, the biological activity may remain for longer times.

4 CONCLUSIONS

Maple trees were bent and the change in force of the branch was examined under different conditions to clarify the self healing behavior of the living body. The decrease in force of the living branch was smaller than that of the dead branch.   When the branch was bent in summer, the force decreased smaller than that measured in winter.   After the branch was cut off from the stem, the force decreased with time and kept constant at a time of 10 h to 100 h and then decreased to converge to a constant value.   The weight of the branch showed the same tendency as the force.   These results give conclusions when the biological activity is higher, the resistance against the force is larger, and biological activity of the branch remains for few days after cutting off from the stem.

5 ACKNOWLEDGEMENT

The authors appreciate Mr. W. Miyagi and Mr. I. Yanagi for their help in the experiment.   Thanks are also expressed to Mr. H. Koide, Daido Seimitsu Kogyo Ltd. for supplying steel plates for sensor.

Experimental Mechanics, Allison (ed.)© 1998 Balkema, Rotterdam, ISBN 90 5809 014 0

On flight of gadfly

S. Sudo & M. Watanabe
Iwaki Meisei University, Japan

T. Ikohagi, F. Ohta & K. Katagiri
Tohoku University, Sendai, Japan

ABSTRACT: This paper is concerned with the wing beating behavior and the aerodynamic characteristics of a gadfly. A three dimensional motion analysis system revealed the wing motion of the gadfly, *Tabanus chrysurus*. The experimental system composed of two high speed video cameras, a motion grabber system and a personal computer, analyzed the wing motion in three dimensions. The free flight of the gadfly during take off was analyzed. The flapping motion of a living gadfly, attached to a wooden needle with adhesive, was also analyzed by the high speed video camera system. Both motions were compared. The locomotory center of winged insects, such as the gadfly, is the thorax. Corresponding to the contraction and the relaxation of the muscles, the extrinsic skeleton vibrates up and down because the muscles and the wings are functionally connected by the extrinsic skeleton of the pterothorax and the wing articulation. The measurements of displacement of extrinsic skeleton vibration were made by an optical displacement detector. The results of a series of measurements revealed the complexity of the flapping motion in the gadfly flight.

1 INTRODUCTION

In the development of bioengineering, the biomechanical study of the flying function of insects is of fundamental interest and importance with respect to a variety of applications. There are several hundred thousand different kinds of insects. Insects have been developing from the remote past, and their tissue and structure elements have various superior mechanical functions. Flying insects have the ability to fly by wing flapping. Flight is an effective form of transport because it is fast and direct. Therefore, extensive investigations on insect flight have been conducted (Azuma 1992). For example, the hovering motions of the chalcid wasp, *Encarsia formosa*, has been observed, and a new mechanism of lift generation has been proposed (Weis–Fogh 1973). An alternative lift–generating mechanism was proposed for the hovering flight of dragonflies, and the forces were analysed using flow visualization tests and unsteady inviscid flow theory (Savage et al. 1979). New morphological and kinematic data for a variety of insects were presented, and the aerodynamics of hovering insect flight was re–examined (Ellington 1984). The mechanical characteristics of the beating wings of the dragonfly were analysed by the simple method (Azuma et al. 1985), and the theoretical analysis of flight performance at various speeds was carried out (Azuma & Watanabe 1988).

In spite of many investigations, however, there still remains a wide unexplored domain in the flight of the gadfly. Research data on the wing motions of the gadfly are scant, and there are many points which must be clarified.

In the present paper, the results of a three dimensional analysis of the wing flapping motion for the gadfly, *Tabanus chrysurus*, are described. In the analysis, the insect wings are idealized as rigid lines. Flying experiments were performed in free flight and tethered flight. The measurements of extrinsic skeleton vibration were made by the optical displacement detector system, because the muscles and the wings are functionally connected by the extrinsic skeleton of the pterothorax and the wing articulation. The results of a series of measurements showed different types of vibration in wing flapping.

2 EXPERIMENTAL METHODS

The experiments on the gadfly flight were conducted by use of a three dimensional motion analysis system. A block diagram of the experimental apparatus is shown in Figure 1. The experimen-

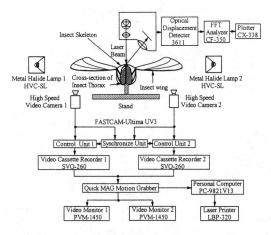

Figure 1. Block diagram of experimental apparatus and measuring devices.

tal system of the three dimensional motion analysis is composed of two high speed video cameras, synchronize unit, two video cassette recorders, a motion grabber, two video monitors, and a personal computer. The three dimensional representation of the gadfly flight can be gained by using two high speed video cameras. A series of frames of free flight of the gadfly were analyzed by the motion grabber and personal computer. A series of frames of wing flapping of the tethered gadfly stuck on the wooden needle with the adhesive, were also analyzed in the same manner. The wooden needle can easily rotate clockwise and counterclockwise, according to the flapping of the wings.

In the experiment on the measurements of displacement of extrinsic skeleton vibration, the live gadfly was also attached to the fixed wooden needle with the adhesive as shown Figure 1. The optical displacement detector was used to measure the displacement of skeleton vibration produced by the contraction and the relaxation of the gadfly muscles. The displacement detector was equipped with a laser diode which, via a lens system, emits a beam of visible light (wavelength:670nm, output power:0.95mW). The output signal from the photodetector gives the position of the measured skeleton surface, relative to the gauge probe. The displacement signal was analyzed by the dual channel FFT analyzer.

3 THE STRUCTURE OF GADFLY

The body of the gadfly is segmented. The segments are grouped in 3 body regions, the head, thorax, and abdomen. The head is the anterior capsulelike body region that bears the eyes, antennae, and mouth parts. The thorax, the middle section of the body (between head and abdomen), is divided into 3 segments: (1) prothorax, (2) mesothorax, and (3) metathorax. Each segment typically bears a pair of legs lateroventrally, and the mesothorax bears a pair of wings dorsolaterally. The wings are located dorsolaterally on the mesothorax. The wing movements are produced largely by changes in the shape of the thorax (Snodgrass 1935). Microscopic observations on the gadfly wing were conducted with a scanning electron microscope. It was observed that wings are clothed in minute hairs. Most of hairs are tilted. Directions of hair tilts suggest flow condition around the wing. Minute hairs reduce the aerodynamic drag, helping to maintain laminar flow conditions.

4 EXPERIMENTAL RESULTS AND DISCUSSION

4.1 *Free flight analysis*

Both kinematics and dynamics of fixed flying *Drosophila melanogaster* wing beats were investigated to understand the basic mechanisms of *Drosophila* flight (Zanker 1990). In his experiments, a steel pin was rigidly attached to the head and thorax with dental cement under cold anaesthesia. The wing movements were reconstructed from artificial slow–motion pictures of tethered flying flies.

In this paper, the properties of a gadfly (*Tobanus chrysurus*) wing motion are described quantitatively. By means of a computerized three–dimensional reconstruction, the variables of the wing–beat cycle, such as wing path and velocities of wing tip, are analysed quantitatively.

In a first step, a measured frame box was digitized by the computer with the motion grabber system for the spatial reconstruction of a three–dimensional coordinate system. In the following step, the gadfly was released into the digitized space. The free flight behavior of the gadfly was recorded on the motion grabber system at 4500 frames per second. The nine characteristic points on the gadfly were digitized. Figure 2 shows nine such characteristics points, numbered from 1 to 9. In Figure 2, 5 and 8 are wingtips, and 6 and 7 show wing roots. Figure 3 shows the three-dimensional coordinate system and the simplified gadfly. The unit of the coordinate scale is meter(m). The numbers show the simplified gadfly, and they correspond to Figure 2. The lines 6 5 and 7 8 correspond to the left and right wings respectively. The square 1 3 4 2 correspond to the head, and the line 4 9 corresponds to a kind of

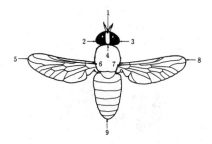

Figure 2. Nine characteristics of the gadfly.

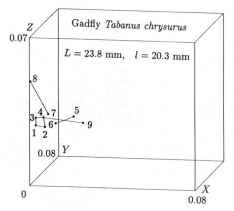

Figure 3. Coordinate system and simplified gadfly.

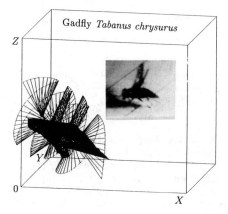

Figure 4. A 3–D view of the wing flapping motion for the gadfly.

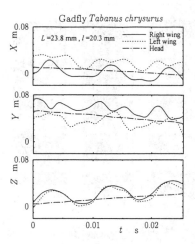

Figure 5. Three components of the wingtip path in Cartesian coordinates.

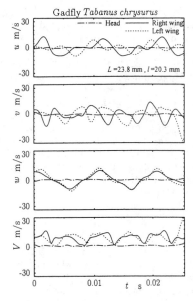

Figure 6. Velocity variations of the wingtip motion for the gadfly.

body line. In Figure 3, L is the body length of gadfly and l is wing length. Figure 4 shows the free flight behavior of the gadfly. The photograph in Figure 4 is a frame showing the gadfly's condition during the analysis. The final stage of gadfly flight in Figure 4 corresponds to Figure 3. The time interval between each line is $1/4500$s. Wing path and flapping of the gadfly during take off from the plain surface can be seen in Figure 4. Figure 5 shows three components of the wingtip path for the gadfly flight. The wingtip paths in the Z coordinate axis show the flapping motion during take off. It can be seen that the flapping motion of the right and left wings is almost the same. The total flapping amplitude is about 33.3mm. On the other hand, the right wingtip path is different from the left wingtip path in the

809

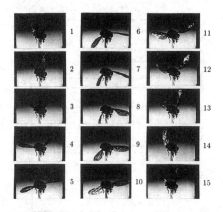

Figure 7. The film sequence of the flapping motion for the gadfly.

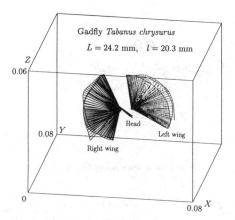

Figure 8. A 3–D view of the wing motion for the tethered gadfly.

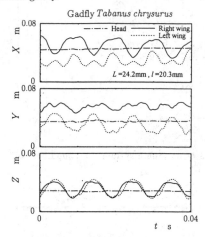

Figure 9. Three components of the wingtip displacement in Cartesian coordinates.

X and Y coordinate axes. In this stage, the gadfly is preparing to turn left. Figure 6 shows the variations of velocity components of the wingtip motion for the gadfly flight. In Figure 6, u, v, w are velocities in the X–, Y– and Z–directions, and V is the three dimensional velocity described as follows

$$V = \sqrt{u^2 + v^2 + w^2} \qquad (1)$$

It can be seen from Figure 6 that the maximum velocity of the flapping motion reaches about 21m/s. It corresponds to Reynolds number $R_e \approx 10^4$. Due to the flapping motion, the frequency of the u and w variations is one–half that of the v and V variations.

4.2 *Fixed flight analysis*

The flapping motion of the tethered gadfly was also analyzed in the same manner as before. The live gadfly was stuck on a wooden needle with adhesive. Figure 7 shows the high speed photographs of the fixed flight of the gadfly. The time interval of each frame is 0.8ms. A period of the gadfly flapping is shown in Figure 7. It can be seen from the number 5 in Figure 7 that the wings are cambered on the downstroke. The corrugated form of the elastic wings is helpful to camber in this stage of flapping. The cambered wing is further improved in aerodynamic performances as compared with the flat one. The wing tends to rotate as a flat plate in number 10. Figure 8 shows wing path and flapping of the tethered gadfly analyzed at 9000 frames per second. Figure 8 shows not a pair of wings but also the body line of the gadfly. Only a one and half period of flapping is shown, because of the avoidance of line congestion. The right wing is different from the

left wing in the flapping motion. In this stage, the gadfly is turning left. Three components of the wingtip path during the fixed gadfly flight are shown in Figure 9. It can be seen that the up and down motion of the wingtip describes nearly a sine curve in the Z–direction displacement. In addition, the left and right wings describe different paths for the wing–beat cycle. The different paths show the gadfly flight is in a state of rotation. Figure 10 shows the velocity variations of the wingtip motion in the X–, Y–, and Z–components and the three dimensional velocity. The velocity fluctuations generated by the beating motion of the gadfly are nearly periodic in the X–, Y–, and Z–components. The basic frequency of the velocity fluctuation in the three component velocity corresponds to the flapping frequency of the gadfly, f_i=100Hz. The flapping frequency of insects is described as follows (Sudo et al. 1996)

$$f_i = Km^{-\frac{1}{6}} \qquad (2)$$

where K is the proportional constant, and m is the insect mass.

There is the phase difference between the right and left wings in Figure 10. The phase and displacement differences between the right and left wings could lead to the rotating motion of the gadfly. The superior frequency of the velocity fluctuation in the three dimensional velocity is twice the flapping frequency as described in the free flight.

If we assume simple flapping for gadflies, three velocity components of right wingtip are described as follows:

$$\left.\begin{array}{l} u_r = a_r \cos \omega t \\ v_r = b_r \cos \omega t \\ w_r = c_r \cos \omega t \end{array}\right\} \qquad (3)$$

where a_r, b_r and c_r are the amplitudes of the velocity fluctuation, ω is the angular frequency of flapping. In the same way, we can describe for left wingtip as follows:

$$\left.\begin{array}{l} u_\ell = a_\ell \cos(\omega t + \delta_1) \\ v_\ell = b_\ell \cos(\omega t + \delta_2) \\ w_\ell = c_\ell \cos(\omega t + \delta_3) \end{array}\right\} \qquad (4)$$

where δ is the phase difference between right and left wingtips. The gadfly may turn by the changing of a, b, c, and δ.

4.3 Extrinsic skeleton vibration

The insect flaps its wings and generates aerodynamic forces. Insect wings, unlike those of flying vertebrates, lack intrinsic musculature and are activated by contraction of the thoracic muscles. The flight muscles of insects were widely examined in the electron microscope (Smith 1965). The muscles and the wings are functionally connected by the extrinsic skeleton of the pterothorax and the wing articulation (Brodsky 1994). Contraction of the dorsoventral muscles causes the tergum to be depressed and the wings to move upward. Contraction of the dorsal longitudinal muscles causes the tergum to arch upward and the wings to move downwards (Snodgrass 1927). The extrinsic skeleton of the pterothorax vibrates up and down. In this paper, the extrinsic skeleton vibration is examined with the optical displacement detector system as shown in Figure 1.

Figure 11 shows an example of the output signal from the optical displacement detector. The

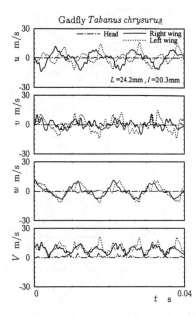

Figure 10. Velocity variations of the wingtip motion for the fixed gadfly.

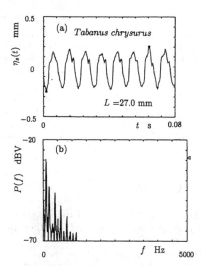

Figure 11. Output signal and power spectrum.

signal in Figure 11(a) corresponds to the extrinsic skeleton vibration during the flapping motion of the gadfly, *Tabanus chrysurus*. In Figure 11, L is the body length of the gadfly. It can be seen that the amplitude of the steady skeleton vibration is about 0.2mm. Figure 11(b) shows the power spectrum of the skeleton vibration. The frequency spectrum has a single dominant peak at f=100Hz, that is, the flapping frequency of

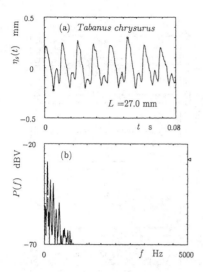

Figure 12. Output signal and power spectrum.

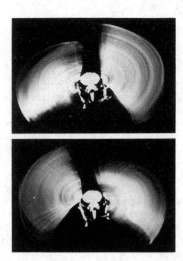

Figure 13. Photographs of the wing beat for a fixed gadfly.

this gadfly is f_i=100Hz.

On the other hand, Figure 12(a) shows the extrinsic skeleton vibration generated during the flapping motion of the same gadfly. It can be seen from Figure 12 that the amplitude and the wave form are different from those in Figure 11. However, the flapping frequency is the same as shown in Figure 12(b). Thus, the extrinsic skeleton vibration of the gadfly showed a variety of oscillation. A variety of oscillations of the gadfly corresponds to various flapping spectra. Figure 13 shows the photographs of the wing beat. It can be seen that the gadfly shows the different flapping motions. That fact is equivalent to Eqs.(3) and (4).

5 CONCLUSIONS

1. The maximum velocity of the wingtip during the flapping motion of gadfly reaches about 21m/s.

2. The wings of gadfly are cambered on the downstroke, and the wing tends to rotate as a flat plate.

3. The gadfly may turn by changing the amplitude during the flapping motion, and of the phase difference between right and left wing.

4. The extrinsic skeleton of gadfly vibrates in various wave forms during the wing beat cycle. A variety of oscillations of a gadfly relate to flight control.

REFERENCES

Azuma,A. 1992. *The biokinetics of flying and swimming.* Tokyo:Springer–Verlag.

Azuma,A., Azuma,S., Watanabe,I. & Furuta,T. 1985. Flight mechanics of a dragonfly. *J.exp.Biol.* 116: 79–107.

Azuma,A. & Watanabe,T. 1988. Flight performance of a dragonfly. *J.exp.Biol.* 137:221–252.

Brodsky,A.K. 1994. *The evolution of insect flight.* Oxford: Oxford University Press.

Elligton,C.P. 1984. The aerodynamics of hovering insect flight, I the quasi–steady analysis. *Phil. Trans.R.Soc.Lond.* B305:1–15.

Savage, S. B., Newman, B. G. & Wong, D. T. 1979. The role of vortices and unsteady effects during the hovering flight of dragonflies. *J.exp.Biol.* 83: 59–77.

Smith,D.S. 1965. The flight muscles of insects. *Scientific American.* 212:76–85.

Snodgrass,R.E. 1927. Morhology and mechanism of the insect thorax. *Smithsonian Misc.Coll.* 80:1–105.

Sudo,S., Tsuyuki,K., Hashimoto,H. Ohta,F. & Katagiri,K. 1996. *Post Conf.Proc. 8th Int.Cong.Exp. Mech., Nashville,10–13 June 1996*:69–74.

Weis–Fogh,T. 1973. Quick estimates of flight fitness in hovering animals, including novel mechanisms for lift production. *J.exp.Biol.* 59:169–230.

Zanker,J.M. 1990. The wing beat of *Drosophila melanogaster. Phil.Trans.R.Soc.Lond.* B327:1–64.

Experimental Mechanics, Allison (ed.) © 1998 Balkema, Rotterdam, ISBN 90 5809 014 0

Behaviour of facial muscular mass due to gravity

D. Pamplona & P. Calux
Department of Civil Engineering, Catholic University PUC-Rio, Brazil

H. I. Weber
Department of Mechanical Engineering, Catholic University PUC-Rio, Brazil

ABSTRACT: Facial ageing is a biological phenomenon but the facial movement of the muscular mass during ageing is also due to mechanics. Ageing appears through decay of palpebral pouches, lateral pouches of face and eyelids, narrowing of lips, formation of the Nasogenian fold, and enlargement of upper lip and ear. Research toward a model of the ageing process is fundamental for improving face-lifting techniques and possibly for finding a way to control the ageing process mechanically. This paper presents a qualitative comparison of the effects of gravity on the face and the pattern of change of the ageing parameters, discussed in previous research. The effect of gravity on the muscular mass of the face was observed, photographed, and measured in a sculpture of a face made of visco-elastic material.

1 INTRODUCTION

The human face changes with age, and some factors that are decisive to this process have been identified in the literature (Pitanguy, I. *et al*, 1977). One main factor is that there is atrophy in osteocartilages, and another is that the skin loses elasticity and thins, causing the wrinkling process and decay of the skin.

The decay of the palpebral pouches, the lateral pouches of the face and of the eyelids, the narrowing of the lips, the formation of the Nasogenian fold and the enlargement of the ear are some factors that appear on the ageing face. The literature on ageing, however, lacks an identification of the mechanical forces related to the ageing process. The research to obtain an adequate model to study the process of ageing is fundamental not only for improving the techniques used in face lifting, but also possibly for finding a way to control the ageing process mechanically.

The skin can be considered an anisotropic and inelastic material (Fung, Y.C., 1980), i.e., there is no single-valued relationship between stress and strain, where relaxation, creep and hysteresis can be observed. The material presents different rates for loading and unloading. To test the skin, preconditioning is used; several cycles of loading are done so that afterwards the stress-strain relationship is repeatable. The pseudo potential energy function of deformation W^*, which is defined for these preconditioned tissues, has the

same thermodynamic meaning as the strain energy function, W, (Fung, Y. C., 1980 and Daly, H.C., 1982). There are other models that can represent the skin and the superficial mass as an elastic-viscoplastic material smas, (Rubin et al., in prep.), where the constitutive equations for visco-plastic materials are modified to model the highly non-linear elastic and rate-dependent inelastic response, exhibited in experiments performed with the skin and smas.

The present study is concerned with a qualitative comparison between the effects of gravity acting on the face and the pattern of change of the ageing parameters that were defined, measured and discussed in our previous work (Pamplona et al, 1996, 1998).

This earlier study was conducted on a group of 40 women, photographed at two different ages, at least five years apart. These women were selected among two hundred middle class Caucasian women, considering several aspects for the selection: general facial morphology; absence of trauma, both physical and psychological, in the period between photos; absence of extreme photo-ageing. All photographs were scanned, digitally processed, and normalised in size. Linear measurements of each patient were divided by the distance between the pupils. Curves starting at the age of 25 and ending at 65 describe the behaviour of the ageing parameters through time. The choice of ageing parameters was based on facial changes that occur over a period of time, so that the preceding photograph of each patient was used as a pattern for the subsequent one. In this way the looks of an aged person could be predicted. Results of this research are used in the present one for comparison.

2 EXPERIMENTAL METHODOLOGY

To understand the behaviour of facial muscular mass due to gravity, a skull was covered with clay by a sculptor, using the thickness of muscular mass that covers a young female human skull. The measurements were obtained in the literature. See Table 1 and Figure 1.

Table 1 Thickness of muscular mass in white women

Supraglabela	03.5 mm
Upper lip margin	09.0 mm
Lower lip margin	10.0 mm
Mental eminence	10.0 mm
Supraorbiatal	07.0 mm
Suborbital	06.0 mm
Frontal eminence	03.5 mm

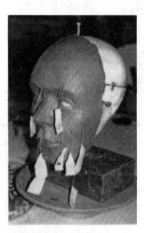

Figure 1. Clay sculpture covering a female skull with pins in the marked points to assure the thickness.

A mold of rubber-like material was made, Figure 2, in such a way that the clay sculpture could be reproduced with 270 grams of a viscous material. The viscous material used is silicon. This material has the elastic and viscous diagrams presented in Figure 3.

The following eight points were marked and measured on the viscous face:

* eyelids (2 points)
* tip of the nose
* lower edge of the nose
* vertical edges of the lips (2 points)
* chin
* contour of the face in the same vertical as the centre of the pupil.

Figure 2. Mold and skull.

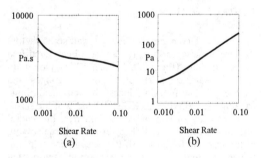

Figure 3. Properties of the viscous material (a) Viscosity (b) Shear Stress.

The viscous sculpture was photographed, every 100 seconds, so that the movement of the marked points could be measured employing the digitalized photographs. A sequence of these photographs can be seen in Figure 4.

The five vertical distances that were measured are:

* from the pupil to the eyelid
* from the lower edge of the nose to the upper extreme of the mouth
* between the vertical edges of the lips
* from the pupil to the chin
* from the pupil to a point in the contour of the face in the same vertical as the centre of the pupil.

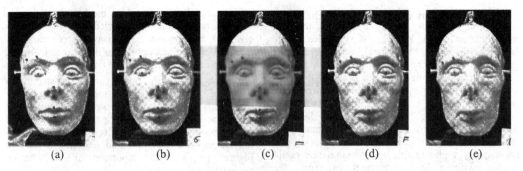

Figure 4. Sequence of photographs showing the movement of the viscous face over a period of time, every 200 seconds.

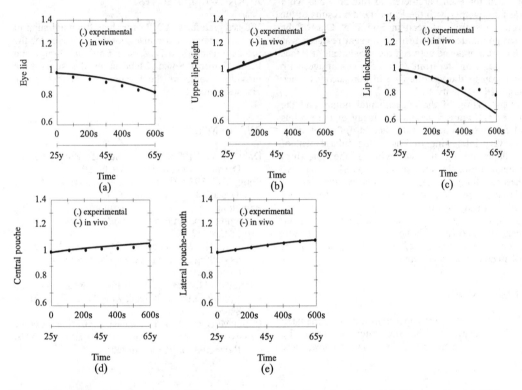

Figure 5. Normalized displacements of the facial parameters in time, for: (a) eyelid; (b) upper lip, (c) thickness of the lip, (d) central pouch and (e) lateral pouch of the face.

3 RESULTS:

The experimental procedure was performed several times, giving ten sequences of data. The measurements were normalised twice. First all the distances of each sequence were divided by the distance between pupils marked on the skull in such a way that there would be no error introduced by the fact that the photographs of the sequences could have been taken from different distances. Secondly each of the five vertical distances, in each sequence, was divided by the corresponding distance in the first photograph, in such a way that all the sizes would be one at the first moment.

Our previous work considered the possibility of predicting the ageing process by offering an initial quantification of this process, as it applies to the face.. The sample for the research was restricted to a group of 40 white female patients with a history of limited exposure to the sun, with ages ranging from 25 to 65. The parameters for each patient were measured at two different ages. The aim of the study was to define a general ageing curve that describes the increase (or decrease) in the ageing parameters at a proportional rate of change. After measuring and normalising a photograph of a person, one could predict, with a known amount of error, the appearance of that person at a different age.

Using Malta 4.0, the general ageing curve obtained for each parameter, in our previous work was put together with the experimental results of the present research, as seen in Figure 5. One of the main points determined in the present research was the correspondence between the movement of a face in 40 years of life, from 25 to 65 years of age, with the decay of the viscous face in 600 seconds when subjected to gravity.

Observation of the experimental points and the curves of Figure 5 shows the decay of the eyelids (a), and of the contour of the face (d-e) through time. One very interesting finding is the enlargement of the upper lip and the narrowing of the lips, (b-c) through time. In our previous work when we observed the narrowing of the lips, we thought that this could be explained by the loss of bones and teeth, with ageing. But even though our skull does not lose bone mass, we observe the same phenomenon (even with the points not fitting so well in this case); we can therefore conclude that the great villain continues to be gravity.

4 DISCUSSION

The main point of this paper is to introduce the need for identification of the mechanical forces in the mechanics of facial ageing. As far as we know, there lacks a complete study of the mechanics of ageing. We understand that the knowledge of this process can lead to improvement in ways of reverting the effects of ageing or even controlling it mechanically. This simulation is the first step for future work.

Results confirm the importance of gravity in the looks of an aged person. Our previous study (Pamplona et al, 1996) presented photographic records of women through their lives. In the present study the qualitative comparison of the observed movement with the measured parameters seen in Figure 5 points out the importance of gravity in the ageing process.

ACKNOWLEDGMENTS

We are grateful to CNPq for their financial support for this study. We are also grateful to Bayer for the viscous material, to Professors Eduardo Daruge and Paulo Roberto Souza Mendes for the support on providing some of the references and testing the viscous material.

REFERENCES

Daly, H.C. 1982. Biomechanics Properties of Dermis, *J. Invest. Derm.* 79:17-19

Fung, Y.C. 1980. On the Pseudo-elasticity of Living Tissues, *Mech. T.,* 5:49-66.

Pamplona, D. et all 1996. Defining and measuring aging parameters, *Appl. Math. Comp.* 78/2,3:217-227.

Pitanguy, I. & Pamplona D. 1998. Numerical modelling of facial ageing, *Plast. and Rec. Surg.* (accepted).

Pitanguy, I. et all 1977. Anatomia do envelhecimento da face, *Rev. Bras. Cir.* 67:79-84.

Rubin M. B. at all., An elastic-viscoplastic model for excited facial skin and smas, *ASME Bioengineering ,* (submitted).

Experimental Mechanics, Allison (ed.)© 1998 Balkema, Rotterdam, ISBN 90 5809 014 0

A load acquisition device for the paddling action on Olympic kayak

N. Petrone
Department of Engineering, University of Ferrara, Italy

M. Quaresimin
Institute of Management and Engineering, University of Padova, Italy

S. Spina
Department of Mechanical Engineering, University of Padova, Italy

ABSTRACT: Aims of the work was the experimental evaluation of forces applied by the athletes to the shell of an Olympic kayak during a typical paddling action. The correct knowledge of the loads applied to the water and transmitted to the kayak through the saddle and the anterior footpad enables the evaluation of the structural behaviour of the boat and the optimisation of the composite structure of the kayak. The present work reports the design of a special set of instrumented saddle and footpad, suitable for Olympic kayaks and stationary training devices, in terms of force decomposition, design of the device and force calibration. The data recorded during paddling actions on a indoor training device were analysed to extract the relevant loads useful for comparison with those applied to the kayak in flatwater trials.

1 INTRODUCTION

Success in Olympic flatwater kayaking is usually related not only to sportive factors such as the skill of the athlete, the training program or the paddling technique, but also to technological factors which can have an influence on the performance levels, such as the paddle profile or the hydrodynamic and structural behaviour of the kayak.

The different levels of performance are related to different techniques or paddle features and correspond to different force systems applied by the athlete to the paddle and transmitted to the kayak.

Furthermore, training sessions are generally performed either in flatwater or by means of indoor training devices that enable to reproduce the paddling action of the athlete and to measure the performance levels. Such devices are usually based on simple rotating wheels moved by ropes and, although quite popular in the sport teams, they are generally far from reproducing the actual kind of motion and resistance encountered in the water by the athletes.

Finally, the optimum design of an Olympic kayak in terms of stiffness and overall behaviour at high performance level could be achieved only by the knowledge of the complete system of forces applied to the kayak by the athlete. These forces can be fundamental not only for a K1 boat, but also for K2 and K4 boats where the synchronism of loads determines the flexural and torsional behaviour of the boat and its hydrodynamic performance.

For these reasons, the present work focuses on the design, the construction and the application of a set of instrumented saddle and footpad, suitable to be applied both to an Olympic kayak and to a commercial paddle-ergometer, to measure the system of forces developed in a typical flatwater trial or in a indoor training session.

1.1 Aims of the work

Object of the work was a state-of-the-art K1 Olympic kayak. The main parts of the kayak structure that receive the athlete are represented in figure 1, together with the internal frame of reference XYZ fixed to the kayak that was adopted following a previous work found in literature (Kendal & Sanders 1992).

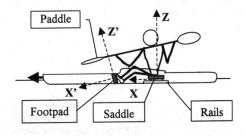

Figure 1. Nomenclature adopted for the kayak parts and definition of the frames of reference.

The access for the athlete to the boat is gained through the cockpit: the athlete sits on an anatomic saddle with the legs slightly bent and his left and right feet acting on a footpad that allow for pushing and pulling, as in a stirrup. The saddle is inserted into two lateral rails enabling the adjustment needed by different athletes. The propelling action is actually developed at the paddle and transmitted through the body and the three contact points to the kayak to win the drag resistance of the water.

Several works have been published about the kayak since it became an Olympic discipline in the 1936: the present shape of the boat was introduced in 1965 and marked the beginning of a scientific approach to the discipline, focusing mainly on training aspects (Celentano et al. 1974, Perri 1976, Kearney et al. 1979) or on biomedical implications of the kayak sport action (Cerretelli 1993). A recent paper (Kendal & Sanders 1992) investigated the technical aspects related to the use of different paddles and their effect on performance, expressed as velocity of the boat in water during a trail. This research, based on high speed film techniques, analysed the athlete's motion at different body joints and the paths of the blade tip, with any evaluation of force levels.

Some development were introduced by the use of an instrumented paddle-shaft with one force channel (Colli et al. 1993), adopted by elite Italian athletes to define a performance index based on the developed forces. The complete screening of forces applied to the kayak has never been applied, at least to the authors' knowledge.

A recent structural optimisation analysis (Zanovello 1997) on a K1 kayak made of composite material, even if eventually successful in terms of specific stiffness achievements, required gross assumptions in the definition of loads applied to the kayak and of its overall behaviour, due to the lack of knowledge in this field.

The present paper reports the design of a six component dynamometric saddle and a four component dynamometric footpad that allow the synchronous acquisition of the force system applied by the athlete's body to the kayak at the three points of contact: the right foot, the left foot and the saddle. Forces applied during training sessions on ergometer were recorded to assess the maximum load levels to be compared with forces in flatwater trials.

1.2 Physical analysis

In the fundamentals of paddling technique, the cyclic action of the athlete is usually split in different phases, to enable detailed analysis and performance optimisation. Each passage of the paddle in the water is a stroke and the stroke frequency in terms of

strokes per minute (str/min) is the reference pace; a complete cycle is the sequence of right and left stroke. In each stroke, the pull phase is the time from blade entry to blade exit on the same side and the glide phase is the time from blade exit to blade entry on the opposite side.

Due to the relative complexity of the athlete's action, a physical simplified analysis of the paddling action was executed to identify the minimum number of meaningful forces to account for.

The modern techniques consist in a pushing action of the foot simultaneous with the pulling on the paddle, at the same side of the stroke: this footpush has to be applied all along the pull time, together with a pulling action of the opposite foot at the footpad. These foot actions counterbalance in some measure the pulling action of the arms in the water and the Z axis moment due to the lateral displacement of the blade tip with respect to the kayak X axis.

This implies that the leg at the stroke side acts as an opening triangle and that a force directed backward is applied to the saddle. Forces to the saddle are applied as resultant of pressure and friction force distribution from the athlete's body, with a resultant vector whose direction and intensity are continuously changing.

The independent force components that are needed to completely describe the athlete-kayak interaction are shown in figure 2 with the positive directions that will be adopted.

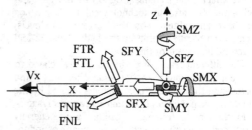

Figure 2. Independent force components considered in the design and their positive directions.

Forces applied by the feet at the footpad were defined as force "normal" (FN) and "tangential" (FT) to the footpad (suffix "R" or "L" means right or left foot respectively) usually tilted of 20° with respect to a vertical plane. The three forces and the three moments at the saddle geometric centre O were defined as indicated in figure 2 and were measured by modifying a standard saddle with the introduction of a lower frame with four load cells interposed.

The total number of force components to evaluate all the six resultants on the kayak resulted equal to ten. This was due to the neglecting of the Y force components at the footpad and to the hypothesis of non influence of the actual foot position on the footpad, due to the restricted degree of motion allowed to the feet within the shell.

2 EXPERIMENTAL METHOD

The experimental method adopted for the definition of the footpad and saddle dynamometer was the introduction of a set of dedicated strain gauge load cells, calibrated as force transducers.

The load cells were solid shafts loaded in bending with rectangular cross section in the gauge length and circular cross section at the clamping end. This allowed to define two decoupled force transducers in two perpendicular planes of the rectangular sections of each cell and to perform the correct orientation of the cells at the circular clamping section.

All the load cells were made of steel and milled from circular to rectangular cross sections in the gauge length. All the strain gauges used were 3/120 LY11 HBM strain gauges with 120 Ω resistance and 3mm grid, applied with cold curing rapid adhesive.

The design was oriented to a popular Italian kayak K1 ADLER made of composite laminates and sandwich structures.

2.1 Footpad dynamometer design

The standard footpad in the kayak is a wooden plate, 180x100 mm and 12 mm thick, fixed at the two ends to the rails inbuilt in the kayak shell. The feet are inserted between the plate and a transverse beam which is again connected to the rails at the ends and allows the application of pulling forces, whereas the pushing action is applied to the plate. The feet are quite close to each other and maintain a small degree of motion in the lateral direction, in such a way that once the foot is posed with the heel to the inner part of the shell, the centre of pressure of the foot is restricted to a very small area on the footpad. The complexity of the structure is increased by the presence of the rudder rod, which has to be moved laterally by the feet and runs into an elongated hole in the footpad.

The need of measuring both the pushing and pulling actions of the left and right foot, imposed a completely new design for the footpad, within the dimensional restraints due to the application to standard rails and the functional restraints related to the relative stiffness needed, the possibility of applying a pulling action and the housing of the rudder rod.

A solution was found by the introduction of a rear stiff aluminium mainframe to be applied from one rail to the other and two independent aluminium pedals shaped with an upper stirrup for the pulling action of the feet. The connection between the two pedals and the rear mainframe was obtained with two independent steel load cells, each instrumented with a couple of perpendicular strain gauge bridges. The pushing action of one foot, acting perpendicular to the pedal surface, was recorded as bending stress along the maximum inertial axis of the rectangular cross section of the cell with a positive sign, whereas the pulling action applied to the upper stirrup was considered negative. The tangential force developed at the pedal, even if lower in magnitude, was recorded as bending stress along the minimum inertial axis of the cross section. This solution enabled to decouple completely the actions of the two feet, however maintaining the possibility of guidance of the rudder and a free degree of orientation of the pedals to comply with different athletes attitudes.

Particular care was taken in the design of the strain gauge bridges. Due to the small degree of lateral movement of the foot during a flatwater trail, only the magnitude of the push-pulling force on each foot was considered to be relevant, thus neglecting the effect of the actual position of the force in the pedal surface. To avoid the influence of this factor, a full bridge connection for the Wheatstone bridge was adopted with four active arms at two different gauge length locations in the top-bottom sides of the rectangular cross section. Due to this arrangement, the bridge output resulted to be proportional only to the magnitude of the push-pull force on the pedal and insensitive to the position of the force on the pedal surface.

2.2 Saddle dynamometer design

The standard saddle was a compact closed composite structure with a stiff anatomic seat in the upper part and two inbuilt aluminium laminas in the lower part that matched with the longitudinal cut present in the rails. This "upside-down hat" structure did not allow for any direct decoupled load measurement: a completely redesigned structure was then introduced, saving the strictly necessary upper part of the composite seat and connecting it by means of interposed load cells to a lower stiff aluminium mainframe reproducing the laminas needed for the connection to the rails.

The system is described in figure 3. The lower frame was a square bolted frame which allowed the housing of three cantilever beam load cells. These horizontal load cells presented at the free end a fork and pin system suitable to be connected with a ball joint. The upper seat was applied to a stiff aluminium plate and presented in the lower side two vertical beams (no.4) and a fork and rod device (no.5). The overall assembly was obtained by connecting the forks of cells no.2 and no.3 to the respective beams no.4 through ball joints, and by applying the connecting rod no.5 to cell no.1.

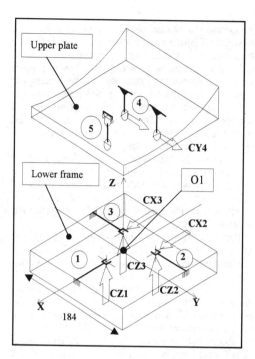

Figure 3. Schematic representation of the saddle dynamometer in a disassembled configuration: load cells numbers and positive directions of forces at the cells are indicated.

This system resulted to be isostatic due to the presence of 3 in-plane joints. Actions applied to the seat resulted in a combined system of vertical Z, longitudinal X and lateral Y forces that can be resolved by the measurement of load cell signals. The lateral load force in Y direction was revealed by a full bridge applied on the two beams (no.4) connected to the upper plate. The geometric centre O1 of the lower frame's diagonal crosspoint was chosen as origin of the saddle frame of reference for evaluation of the moment effect of forces. The force components introduced in figure 2, referred to a seat frame of reference with origin in O, can be evaluated from those in the plane of the cells referred to origin O1 by means of the geometrical dimensions of the final assembly.

The force and moment components referred to point O1 can be evaluated from the six load cell outputs by means of equations (1).

On load cell no.1 only a half bridge sensitive to Z axis force was applied. On load cells no.2 and no.3, two half bridges were defined for each cell, one acting in the vertical bending plane sensitive to Z axis forces and the other acting in the horizontal bending plane sensitive to X axis forces. Load cell no.4 was obtained by means of a full bridge

connection at the two vertical beams in the upper plate, sensitive to Y loads.

$$
\begin{cases}
SFX' = CX2 + CX3 \\
SFY' = CY4 \\
SFZ' = CZ1 + CZ2 + CZ3 \\
SMX' = (CZ2 - CZ3) \cdot a \\
SMY' = (CZ2 + CZ3 - CZ1) \cdot a \\
SMZ' = (CX3 - CX2) \cdot a
\end{cases}
\tag{1}
$$

(where a=19 mm and the definition of each load component is given in figure 2)

2.3 Design assumptions

Due to the lack of data about the load levels applied to the kayak in a sport trail, the design of the load cells was carried out after the introduction of some initial hypotheses about the force magnitudes and directions. It was assumed that during the pull time the forces were reaching the maximum values and that the main action of the athlete was to push forwards on the footpad and to load backwards the saddle in the meantime.

The maximum force applied with one leg in push was assumed to be FN = 1400 N at the footpad, with a maximum pulling action of 400 N. The consequent load on the saddle was assumed to be a force acting in a vertical plane passing through a point P on the seat located at X, Y, Z = -120, -60, 70 mm with respect to the O1 origin; the two components were assumed to be SFX = -500 N and SFZ = -1000 N.

By the inverse solution of equations (1), the forces expected at each load cell were evaluated and the load cell cross sections were defined.

Particular care was taken in the design of the load cells: their arrangement and dimensions were defined in order to achieve reasonable strains at the expected loads, and sufficient static safety margins at the critical sections. A 38 NiCrMo16 steel with 1100 MPa ultimate strength was used for cells no.1,2,3, whereas a C40 medium carbon steel was used for load cell no.4 and for the footpad load cells.

2.4 Device calibration

Each load cell was applied to a restraint system reproducing the actual housing in the frames and loaded by an increasing set of calibrated dead weights. Strain gauge bridges were conditioned with a HBM UPM 100 digital unit and bridge scale factors expressed in N/με were evaluated with several tests to check for output linearity and accuracy. Load cells applied to the footpad were checked for insensitivity to load positions, by applying the same known weight at different positions in the pedal.

After the calibration of each load cell, the saddle dynamometer was assembled and a dummy seat was applied to the upper plate; the overall response of the dynamometer was checked by means of a load device enabling to introduce all the 6 calibration load components at the origin O of the seat reference frame.

3 TESTS AND RESULTS

3.1 *Test procedure*

Test were performed on an indoor training device equipped with the saddle and footpad dynamometers previously described. A laboratory HBM DMCplus digital unit with 6 channels dynamic acquisition capability was used for bridge conditioning and data acquisition.

The sampling rate was 150 Hz for each channel, with a standard 16 bit resolution.

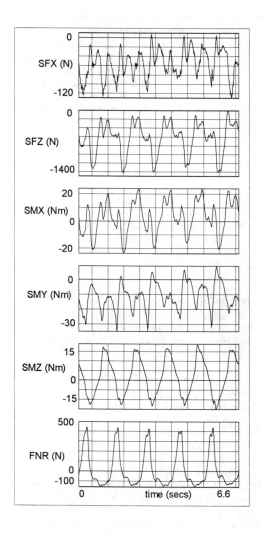

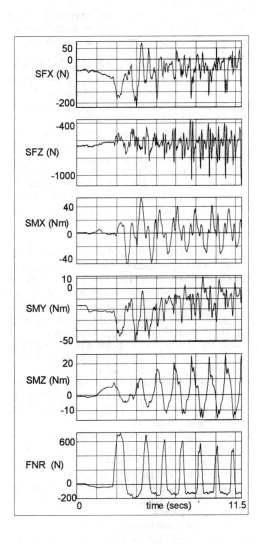

Figure 4. Time based plots of the six synchronous channels recorded on a stationary ergometer during a 90 str/min action, with an expert tester.

Figure 5. Time plot of the six synchronous channels recorded on a stationary ergometer during a simulated starting event, with an expert tester

All the load cell outputs were set to zero when the tester was off the paddle ergometer, so that a mean load due to the athlete's weight was present during the paddling.

The total number of simultaneous channels available in the acquisition unit was six: several combinations of channels were analysed with a consistent right footpad normal force FNR, to cover all the force components.

The recorded signals from all load cells were converted into force units and the resultant forces at the saddle frame of reference O1 were evaluated by means of equations (1).

The tester was a 26 years old athlete of 75 kg of mass and 1,75 m height: he was a well trained mature paddler, with about 4 hours training in flatwater per week.

The tester was asked to paddle synchronous with an audio timer at 30, 60 and 90 str/min: these paces were considered representative of normal indoor training sessions. Higher paces, typical of flatwater trials (100-120 str/min) were not considered due to some functional limits of the paddle-ergometer used. Nevertheless, an intensive starting event was simulated by the tester to reproduce a crucial action of flatwater trials: together with the final rush, usually of lower intensity, the start influences the overall performance and corresponds to the highest expected load levels.

3.2 Results

The preliminary analysis of the test results on all channels allowed to identify the 90 str/min paddling and the starting action as the more significant events in terms of load synchronism and level. Moreover, tangential forces FTL,R on footpad and lateral Y forces on saddle resulted to be of one order of magnitude lower with respect to the others.

The six meaningful load components evaluated for a 90 str/min test are plotted in figure 4 for a sequence of 10 strokes (5 complete cycles).

The simulated starting event is plotted in figure 5. Maximum and minimum load as revealed by the measurements on the paddle ergometer are summarised in Table 1.

The qualitative examination of figure 4 evidences a good uniformity of the load cycles and allows to identify some typical features of each force components that can be traced in figure 5.

The analysis of data listed in Table 1 confirms that loads in a starting event are about twice as much as in the 90 str/min tests.

Finally, the maximum load resulted to be in fair agreement with the load levels which were analytically predicted and used for the design of the footpad and saddle load cells.

Table 1. Maximum and minimum load values for different paddling actions.

Event		SFX [N]	SFZ [N]	SMX [Nm]	SMY [Nm]	SMZ [Nm]	FNR [N]
90 str/min	Max	6	382	23.3	8.7	18.6	444
	Min	-128	-1040	-23.3	-34.7	-15	-152
Start	Max	73	-370	52.5	11.3	25.5	721
	Min	-219	-1110	-45.5	-50.2	-15.3	-196

4 CONCLUSIONS

A new procedure for the measurement of loads applied to an Olympic kayak has been proposed. The design and construction of a four component dynamometric footpad and a six components dynamometric saddle have been described. The system was calibrated and used in paddling tests by an expert athlete and loads at different paces were recorded. Test results obtained on an indoor training device were in agreement with assumptions made in the design of the device.

The measured load values and their combination in typical paddling actions suggest a comparison with data recorded in flatwater kayaking for the tuning of training devices. The combination of the proposed system with an instrumented paddle will bring to a complete monitoring of the paddling action enabling to define some quantitative performance index or style markers for different athletes.

REFERENCES

Celentano, F., G. Cortili, P.E. Di Prampero & P. Cerretelli 1974. Mechanical aspects of rowing. *Journal of Applied Physiology* 36: 642-647.

Cerretelli, P. 1993. *Handbook of sport physiology and muscle work*. Rome, ITALY: Società Editrice Universo. (In Italian).

Colli, R., E. Introini & C. Schermi 1993. L'ergokayak. *Canoa News*. 1. (In Italian).

Kearney, J.T., L. Klein & R. Mann 1979. The elements of style: an analysis of the flatwater canoe and kayak stroke. *Canoe*, 7(3): 18-20.

Kendal, S.J. & R.H. Sanders 1992. The technique of elite flatwater kayak paddlers using the wing paddle. *International Journal of Sport Biomechanics*. 8:233-250.

Perri, O. 1976. L'allenamento in canoa, *M.S. Dissertation*. ISEF. Milan, ITALY. (In Italian).

Zanovello G. 1997. Performance analysis and optimisation of an Olympic kayak. *M.S. Dissertation*, Department of Mechanical Engineering. University of Padova. Padova, ITALY. (In Italian).

Experimental Mechanics, Allison (ed.)© 1998 Balkema, Rotterdam, ISBN 90 5809 014 0

Evaluation of dynamic response of a femur sensor using impact

K.Ueda & A.Umeda

National Research Laboratory of Metrology, Tsukuba, Japan

ABSTRACT: Dynamic response of a femur sensor, which is used for dummies in automobile crash tests, is evaluated by applying the split Hopkinson pressure bar technique or the Kolsky's apparatus. Elastic wave pulses propagating in thin circular bars impart an impulsive input of 10 kN in peak amplitude and 0.7 ms in duration time to the femur sensor placed between the two bars. Dynamic force acting on both sides of the sensor in the axial direction is determined by the measurement of the incident, reflected and transmitted pulses in the bars using strain gauges. The strain gauges on the bars are calibrated in advance using a laser interferometer. It was found that the sensitivity of the sensor is flat and the phase lag is quite small up to 3.9 kHz, that the first resonance takes place at 9.8 kHz, and that the cross-talk output is ±3~4% of the axial input.

1 INTRODUCTION

The precise measurement of dynamic force may be more often required than that of static force in the field of advanced industries. A good example is the measurement of force and moment acting on both passengers (dummies) and chassis in automobile crash tests. Even though various types of force transducers have been developed and used for the dynamic force measurement, current calibration techniques for force transducers are realized only under static loading conditions. It has not been proved yet that the statically determined sensitivity can be extended to the dynamic measurement. It is the first step toward the precise measurement of dynamic force and is of significance to develop a technique for evaluating dynamic response of force transducers.

A promising dynamic calibration technique is under study, where a shaker is used to generate pure sinusoidal motion and accelerometers are used to evaluate the input (Kumme 1992). However, it seems that this vibration method is not substantially useful in a high frequency range above 1 kHz, mainly because the stiffness of load mass and clamping fixture is practically limited (Kumme 1994). In the meantime, the authors proposed another technique, in which elastic deformation of the apparatus is positively utilized in the form of elastic waves to generate impulsive motion and interferometrically calibrated strain gauges are adopted to determine the input (Ueda & Umeda 1994). This method is applicable to a high frequency range up to 10^4 Hz, and can complement the vibration method in terms of frequency.

In the previous study (Ueda & Umeda 1994), however, only thin transducers were examined using the impact method. The thickness of the transducers was 1/50 ~ 1/100 of the wavelength of the incident elastic wave pulses. This report shows that the method can be applicable to thicker transducers than the previous ones, where the ratio of the thickness to the wavelength is approximately 1/11. More specifically, a femur sensor, which is to be mounted on a dummy in the automobile crash tests, has its dynamic response to the axial input evaluated using the impact method.

2 THE EVALUATION METHOD

The basic concept of the impact method is illustrated in Figure 1. The apparatus shown in Figure 1(a) is known as the split Hopkinson pressure bar or the Kolsky's apparatus (Kolsky 1949), and it has been widely used for the investigation on dynamic behavior of materials. We realized that this sort of apparatus can be utilized to apply impulsive input to force transducers.

A test transducer is sandwiched between two thin circular bars. When a projectile impinges on one end of a thin circular bar, namely the incident bar, the edge is locally compressed. The state of compression is propagated down the bar in the form of an elastic wave pulse. Some portion of the pulse is reflected at the other end of the incident bar and bounces back in the opposite direction. The other portion of the pulse is transmitted to the test transducer imposing an im-

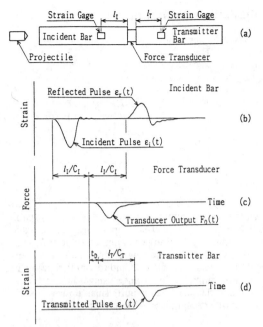

Figure 1. The basic concept of the method.

quired for the pulse to be transmitted through the transducer can not be neglected. This issue will be discussed later.

According the theory of one-dimensional elastic waves, the force $F_I(t)$ acting on the interface between the incident bar and the transducer can be evaluated from the measurement of strain of the incident and the reflected pulses with the following equation.

$$F_I(t) = A_I \rho_I C_I^2 \{\varepsilon_i(t - l_I / C_I) + \varepsilon_r(t + l_I / C_I)\} \quad (1)$$

The force $F_T(t)$ on the transmitter side is given as follows:

$$F_T(t) = A_T \rho_T C_T^2 \varepsilon_t(t + l_T / C_T) \qquad (2)$$

where A, ρ, C are the cross-sectional area of the bars, the density of bar material and the velocity of the longitudinal elastic waves in the bars, respectively, and the subscripts I and T stand for the properties of the incident bar and the transmitter bar, respectively.

Although the conventional SHPB technique adopts the one-dimensional theory described above, actual bars are not ideal one-dimensional media for elastic waves. The influences of wave dispersion and attenuation should be compensated for, so that the force at the ends of the bars can be evaluated more precisely. In addition, the frequency characteristics of the strain measurement system should be taken into account, even though the system usually has a wide frequency band up to 10^5 Hz (Ueda & Umeda 1996). If we compare output of the strain gauges on each bar with the velocity at the end of the bar measured by a laser interferometer prior to the main test, we will have the compensation data for these influences. The setup to obtain the compensation data is shown in Figure 2. The compensation func-

pulsive input on it, and further to another bar (the transmitter bar). If we attach strain gauges on the incident bar at a distance l_I from the end, axial strain caused by the incident pulse $\varepsilon_i(t)$ and the reflected pulse $\varepsilon_r(t)$ can be observed (Figure 1(b)). In the same way, the transmitted pulse $\varepsilon_t(t)$ can be detected by strain gauges placed on the transmitter bar a distance l_T apart from the end (Figure 1(d)). The output of the force transducer will come out at the timing shown in Figure 1(c). If the transducer is thick, time t_D re-

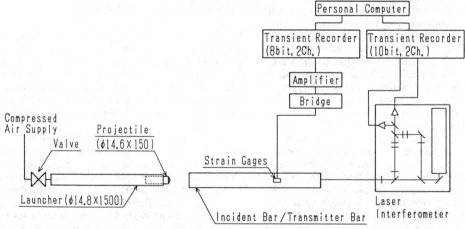

Figure 2. The setup to obtain the compensation data using an interferometer.

tions $G_{Ci}(j\omega)$ to the incident pulse and $G_{Cr}(j\omega)$ to the reflected pulse are determined by comparing output $\varepsilon_i(t)$ and $\varepsilon_r(t)$ of the strain gauges on the incident bar with the velocity of the bar end $v_{LI}(t)$ in the frequency domain.

$$G_{Ci}(j\omega) = L[2\, C_I\, \varepsilon_i(t - l_I\, /\, C_I)]\, /\, L[v_{LI}(t)] \qquad (3)$$

$$G_{Cr}(j\omega) = L[2\, C_I\, \varepsilon_r(t + l_I\, /\, C_I)]\, /\, L[v_{LI}(t)] \qquad (4)$$

where j is the imaginary unit, ω is the angular frequency and $L[\ ]$ is the Laplace transform operator, respectively. The compensation function $G_{Ct}(j\omega)$ to the transmitted pulse is obtained in the same manner by comparing output of the strain gauges on the transmitter bar $\varepsilon_t(t)$ during the passage of the reflected pulse with the interferometrically measured velocity.

$$G_{Ct}(j\omega) = L[2\, C_T\, \varepsilon_t(t + l_T\, /\, C_T)]\, /\, L[v_{LI}(t)] \qquad (5)$$

With these compensation functions, the force on the incident bar side $F_{IC}(t)$ and on the transmitter bar side $F_{TC}(t)$ can be evaluated more precisely.

$$F_{IC}(t) = A_I \rho_I C_I^2 L^{-1}\{L[\varepsilon_i(t - l_I\, /\, C_I)]\, /\, G_{Ci}(j\omega)$$
$$+ L[\varepsilon_r(t + l_I\, /\, C_I)]\, /\, G_{Cr}(j\omega)\} \qquad (6)$$

$$F_{TC}(t) = A_T \rho_T C_T^2 L^{-1}\{L[\varepsilon_t(t + l_T\, /\, C_T)]\, /\, G_{Ct}(j\omega)\} \qquad (7)$$

where $L^{-1}[\ ]$ is the inverse Laplace transform operator.

In the previous report (Ueda & Umeda 1994), force transducers tested were thin enough compared with the wavelength of the incident pulse (1/50 or less). Hence, the time lag due to the transmission or the retroreflection of the pulse within the transducer could be neglected, and indeed timewise change of the force on the incident bar side was almost identi-

cal to that on the transmitter bar side. In the case of thick transducers, however, the force on the incident bar side is not equilibrated to that on the transmitter bar side. The utilization of pulses of long duration, consequently the use of long bars measuring tens of meters, is one of the solutions, but it is not practical. In addition, the frequency band will be limited. Instead of using very long bars, here we define the input to the transducer as an average of two values on both sides.

$$F_{AV}(t) = \{F_{IC}(t) + F_{TC}(t)\}\, /\, 2 \qquad (8)$$

The transfer function $G(j\omega)$, the gain characteristics $g(\omega)$, and the phase characteristics $\phi(\omega)$ of the force transducer are calculated by the comparison of the output $F_O(t)$ with the defined input $F_{AV}(t)$ in the following manner.

$$G(j\omega) = L[F_O(t)]\, /\, L[F_{AV}(t)] \qquad (9)$$

$$g(\omega) = |\, G(j\omega)\, | \qquad (10)$$

$$\phi(\omega) = \arg\{G(j\omega)\} \qquad (11)$$

3 EXPERIMENT

The schematic drawing of the apparatus is shown in Figure 3. A projectile made of aluminum is accelerated through the launcher by compressed air and collides with one end of the incident bar perpendicularly just after it comes out of the tube. Each projectile has a conical tip of 60 degrees, so that the tip is deformed plastically during the collision to generate a long duration pulse with a mild rising edge in the incident bar. Note that the duration time is mainly governed by the plastic deformation of the tip, not by the length of the projectile.

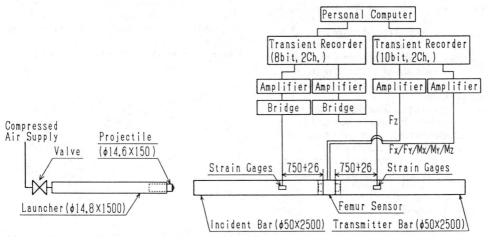

Figure 3. A schematic drawing of the apparatus.

The incident bar and the transmitter bar are made of stainless steel, and each bar is supported by four bearing balls on V-shaped grooves. This configuration allows us to minimize the attenuation of elastic waves. Prior to the testing, the launcher and the bars are aligned coaxially by using a laser beam as a baseline. The tilt of the launcher and the bars with respect to each other is estimated to be less than 1/5000.

With this combination of the projectile and bars, this apparatus can generate an incident pulse having a peak amplitude up to 10^4 N and a duration time of 280 μs, *i.e.* a wavelength of 1400 mm. However, it depends on the structure of the test transducer how much kinetic energy of the incident pulse is transmitted to the transmitter bar through the test transducer and to what extent the effective frequency bandwidth of the transmitted pulse reaches. In the case of a femur sensor tested here, the amplitude, duration time and the effective bandwidth of the transmitted pulse are typically 10 kN, 0.7 ms and up to 11 kHz, respectively.

Two sheets of strain gauges are adhesively bonded to each bar on diametrically opposite sides at a distance 750 mm apart from the end. The gauges are of semiconductor type (Kyowa KSP-1-350-E4), 1 mm in gauge length and 151 in gauge factor. The effective band of amplifiers (Kyowa CDV-230C) is from DC to 200 kHz.

The force transducer tested is called a femur sensor (Robert A. Denton, Inc. Model 1538), which is to be mounted at the location of the thighbone of a dummy in the automobile crash tests in order to detect three orthogonal components of force and three of moment. In this experiment, only the response to the axial force input is evaluated. The amplifier for the femur sensor is of the same type as for the strain gauges. A bias voltage of 2.000 V is provided to the axial force F_z sensing circuit. The femur sensor is mounted between the bars using two pieces of annular shaped adapters which are 26.0 mm in length, 50 mm in outer diameter and 25.45 mm in inner diameter. The adapters are bonded to the ends of the bars using adhesive. Interfaces between the sensor and the adapters are lightly lubricated by applying grease. The two bars are tied with four stripes of vinyl tape to secure the initial contact of the sensor and the adapters.

The output signals of the axial force F_z and one of the other components of force or moment are recorded in a transient recorder with a sampling rate of 1 MHz and a record length of 2048 points. The output of the strain gauges is also recorded in another transient recorder with the same conditions. The two recorders are triggered simultaneously by the signal of strain gauges on the incident bar. After each shot is performed, the stored data is sent to a personal computer and is processed.

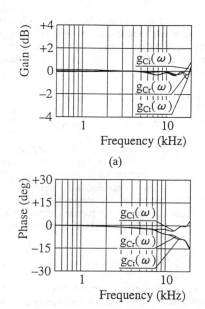

(a)

(b)

Figure 4. The amplitude (a) and the phase (b) of the compensation functions.

4 RESULTS AND DISCUSSION

Figure 4 shows the amplitude and the phase of the compensation functions for the wave dispersion and other effects. As for the amplitude, deviation from the static gauge factor given by the manufacturer is only a few percent up to 20 kHz, and it is clear that the attenuation of elastic waves can be neglected. However, influences which may come from the wave dispersion and the frequency characteristics of the amplifiers are remarkable. Therefore, the compensation using these data is indispensable in order to obtain accurate phase characteristics of the femur sensor, especially in the high frequency range above approximately 8 kHz.

Figure 5 shows the spectra of an incident pulse

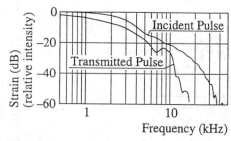

Figure 5. Spectra of an incident pulse and corrsponding transmitted pulse.

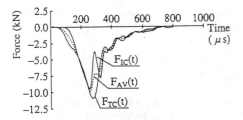

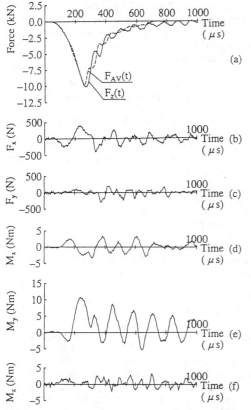

Figure 6. A comparison of force acting on the incident bar side and on the transmitter bar side.

Figure 7. Waveforms of the axial input F_{AV} and the axial output F_z (a), the output of the latheral force components F_x (b), F_y (c) and the output of three components of moment M_x (d), M_y (e), M_z (f).

and corresponding transmitted pulse. While the incident pulse has an effective bandwidth up to around 30 kHz, the spectrum of the transmitted pulse sharply drops at 11 kHz. In other words, the structure of the femur sensor can be considered as a kind of mechanical low-pass filter which does not effectively transmit force of high frequency above about 11 kHz.

Figure 6 compares the force on the incident bar side F_{IC} and on the transmitter bar side F_{TC}. If the transducer is thin enough, say 1/50 or less, compared with the wavelength of the incident pulse, force on both sides of the transducer is equilibrated and coincides well with each other (Ueda & Umeda 1994). In this case, however, the ratio of the thickness to the wavelength is just about 1/11, mainly due to the practical limitation of lengths of bars. We recognize a time lag of about 45 µs between the rising edges of the force on the incident bar side and the transmitter bar side. The behavior of these waveforms of force around the peak is also different, and we understand that the equilibrium is not established even after the incident pulse vanishes. The best way is to use pulses of larger wavelength. However, it requires long bars measuring tens of meters in order to distinguish the incident pulse from the reflected pulse. It is not practical. The second choice would be to take an average of force on both sides as the input to the sensor. The waveform of the average is also shown in Figure 6 with a broken line.

In Figure 7(a), the output F_z of the axial force sensing axis of the femur sensor is compared with the average input F_{AV}. Time lag at the rising edge is hardly detected, but the post-shock oscillation or the ringing in the output still remains after the input almost ceases. In principle, it is necessary to record the entire ringing signal in order to evaluate the frequency response of the sensor accurately. Long bars are preferable in this sense. There should be a compromise between the practical size of the bars and the room available.

Figures 7(b)~(f) show the output signals of two lateral force F_x, F_y and three moment M_x, M_y, M_z components sensing axes. These signals are recorded in different shots, but the condition is almost the same, *i.e.* under the pure axial input of 10 kN in peak amplitude. The cross-talk output in F_x component is ±400 N in amplitude, which is ±4% of the axial input, and its dominant frequency is 4.9 kHz. In the case of F_y, the amplitude is +200,-300 N, which is +2,-3% of the axial input, and the dominant frequency is 8.8 kHz. As for three components of moment M_x, M_y, M_z, the amplitudes are ±3.5 Nm, +11,-6 Nm, +3.5,-2 Nm and the dominant frequencies are 4.9 & 8.8 kHz, 7.8 kHz, 13.7 kHz, respectively. Asymmetry in the output of F_x and F_y is estimated to be 0.2% and 0.9 % of the axial input. From this result, it is clear that the state of pure axial input is achieved.

The frequency response of the femur sensor to the axial input is shown in Figure 8. Since it was not possible to record the output signal entirely until the ringing fades, a single shot did not yield smooth curves of the frequency response function. Thus, we took an average of five shots in the frequency domain to obtain the curves shown in Figure 8. The sensitivity given by the manufacturer is taken as the reference sensitivity (0 dB) in the gain diagram. The

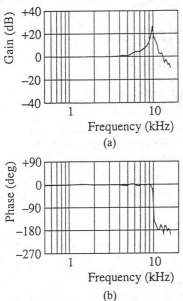

Figure 8. The frequency response of the femur sensor to the axial input (the gain characteristics (a) and the phase characteristics (b)).

sensitivity of the axial force sensing axis is flat within ±5% up to 3.9 kHz and it is consistent with the manufacturer's data. The phase lag is quite small (less than 0.05 rad) in this frequency range. From Figure 8, it can also be seen that the first resonance takes place at a frequency of 9.8 kHz.

This sensor bears the axial compressive force by flanges, while the axial tensile force or the torsional moment is transmitted through pins of about 9.5 mm diameter. It should be kept in mind that the effective frequency band of this sensor to the tension or the torsion may be narrower than that to the compression.

In the practical situations of dynamic force measurement, the force transducer itself constitutes an integral part of a structure to be measured. Therefore, the dynamics of the whole system including transducers should be considered. However, we think it of considerable significance to have the knowledge about the frequency response of sole force transducer under the simple input condition. This method will contribute to the progress of the dynamic force measurement in this sense. Certainly this method has room for improvement, and some of our future tasks are as follows.

(a) To devise a way to record the output signal entirely so that smooth frequency response curves can be obtained in one shot.

(b) To extend the method so that various types of input such as tension, torsion or bending can be applied to the test transducer.

5 CONCLUSIONS

A novel method is proposed for the evaluation of the dynamic response of force transducers, where elastic waves in thin circular bars are used for the generation of dynamic input and interferometrically calibrated strain gauges are adopted for the determination of the input force. The method is applied to a femur sensor and its dynamic response is evaluated under the pure axial input of 10 kN in peak amplitude and 0.7 ms in duration time. The following conclusions can be drawn.

(1) The sensitivity of the axial force sensing axis is constant (within ±5%) and the phase lag is small (less than 0.05 rad) in the frequency range up to 3.9 kHz.

(2) The resonant frequency of the axial force sensing axis is located at 9.8 kHz.

(3) The cross-talk output of two lateral force sensing axes is ±3~4% of the axial input, and that of three moment sensing axes is 2~10 Nm.

(4) The dominant frequencies of the cross-talk output are 4.9 kHz for F_x, 8.8 kHz for F_y, 4.9 and 8.8 kHz for M_x, 7.8 kHz for M_y and 13.7 kHz for M_z, respectively.

The authors are indebted to Mr.A.Fukuda of Japan Automobile Research Institute for the test femur sensor.

REFERENCES

Kolsky, H. 1949. An investigation of the mechanical properties of materials at very high rates of loading. *Proc. Phys. Soc.* B62:676-700.

Kumme, R. 1992. Dynamic investigation of force transducers. *Proc. VII Int. Cong. on Exp. Mech., Las Vegas, 8-11 June 1992*:1282-1288. Bethel: SEM.

Kumme, R. 1994. Error sources in dynamic force calibration. *Proc. XIII IMEKO World Cong., Torino, 5-9 Sept. 1994*:259-264. Padova: CLEUP.

Ueda, K. & A. Umeda 1994. Evaluation of force transducers' dynamic characteristics by impact. *Proc. XIII IMEKO World Cong., Torino, 5-9 Sept. 1994*:265-270. Padova: CLEUP.

Ueda, K. & A. Umeda 1996. Dynamic response of strain gages up to 300 kHz. *Post-Conf. Proc. VIII Int. Cong. on Exp. Mech., Nashville, 10-13 June 1996*:219-224. Bethel: SEM.

Experimental Mechanics, Allison (ed.)© 1998 Balkema, Rotterdam, ISBN 90 5809 014 0

Integrating a photoelastic device into an MRI system for soft tissue mechanics studies

A. Gefen – *Biomedical Engineering Department, Faculty of Engineering, Tel Aviv University, Israel*

M. Megido-Ravid & Y. Itzchak – *Diagnostic Imaging Department, Sheba Medical Center, Israel*

M. Azariah – *Orthopedic Rehabilitation Department, Sheba Medical Center, Israel*

M. Arcan – *Applied Mechanics and Biomechanics, Faculty of Engineering, Tel Aviv University, Israel*

ABSTRACT: Knowledge of local biomechanical properties of living soft tissues is of great importance, as it can indicate the presence of a disease. Changes in tissue stiffness are known to characterize Diabetes Mellitus, some types of breast cancer, etc'. Diagnosis and treatment of these pathologies often involves a Magnetic Resonance Imaging (MRI) examination. A method of measuring mechanical properties of soft tissues *in vivo,* by integrating a photoelastic technique called Contact Pressure Display (CPD) into an MRI system, is presented. A Tissue Stiffness Measurement method (TSM) designed in this work, uses the indentation of a sphere (ϕ=8 mm) into the subject's soft tissue, while the concerned region is scanned by the accurate newly developed MRI system, the Open MRI. The indentation force is obtained using the CPD technique, while the tissue deflection is measured on the MR image. Through this method, a characteristic load-deflection curve is obtained and basic biomechanical properties are evaluated. The TSM method was applied to investigate changes of plantar soft tissue mechanical properties in Diabetes Mellitus. Diabetic feet were scanned to obtain tissue stiffness and indentation patterns, in order to characterize pathological conditions.

1 INTRODUCTION

Identification of soft tissue biomechanical properties is of great interest, since it can provide an indication of a disease presence, as well as basic information about the material. For example, knowing the biomechanical properties of breast tissue is helpful for the diagnosis of breast cancer (Garra et al. 1992).

Measuring the local mechanical properties of a soft tissue *in vivo* is not an easy task. Preliminary estimation of the soft tissue stiffness may be achieved using suction cup devices or pipette aspiration techniques (Aoki et al. 1997). In these methods, a cylindrical pipe is placed on the surface of the subject's tissue, with its axis perpendicular to the surface. Then, the pressure in the pipe is gradually reduced so that the surface portion of the tissue is aspirated into it. The Young modulus is approximated from the aspiration pressure and tissue displacement.

Indentation tests were also used to obtain mechanical properties of various soft tissues (Gefen 1997). These studies make use of the basic concept suggested by Lee and Radok (1960) for measuring stiffness of soft tissues using an indentation of a sphere into the tissue, whereas the Young modulus is calculated from the applied load and measured indentation.

More complex developed methods, generally called Tissue Elastography techniques, use Doppler Ultrasound or Magnetic Resonance Imaging (MRI) systems to visualize soft tissue viscoelastic properties. Internal mechanical excitation (motion of cardiac structures or arterial pulsation) or external vibrational sources of motion produce displacements of the tissue under investigation.

The displacement of different tissues are then analyzed by Doppler velocity measurement, cross-correlation techniques, or image inspection. The elastographic techniques provide not only the structural distribution of stiffness close to the body surface but also in internal soft tissues, where it is impossible to perform aspiration or indentation tests. However, the elastographic methods are not free from problems. The main difficulties encountered when using these methods, are related to the uncertainty of the magnitude and direction of both internal (heart/aorta) and external forces (Garra et al. 1992). The inability to define the direction and magnitude of the applied force limits the ability of these methods to provide quantitative information about the elastic properties of the tissue under investigation.

In the present work, a new method was developed for direct mechanical measurements of local soft tissue properties *in vivo,* near the body surface. This method exploits the indentation concept simplicity and combines it with the state-of-the-art MRI technique, the Open MRI, and with an accurate contact force/pressure measurement technique, the Contact Pressure Display (CPD) method (Arcan & Brull 1980, Arcan 1990, Brosh & Arcan 1994, Gefen 1997). Through this method, the tissue's local load-deflection curve is obtained and its Young modulus calculated. This tissue stiffness measurement method was applied to the investigation of plantar soft tissue biomechanical properties in diabetic feet.

In general, most foot skin injuries in diabetic patients occur on plantar soft tissue, at the highest pressure sites, due to tissue stiffening (Cavanagh & Ulbrecht 1992). MRI scans were found to be very effective in the diagnosis and prognosis process of diabetic patients, to evaluate foot skeletal deformation and damage of soft tissue due to acute infection and ulceration (Horowitz et al. 1993, Cook et al. 1996). The presently suggested MRI/CPD tissue stiffness examination can be integrated into these type of procedures, to provide the clinician with additional parameters describing the disease status. Diabetic patients maybe therefore scanned, using the integrated MRI/CPD techniques, to obtain their tissue biomechanical properties in order to complete and characterize pathological conditions.

2 METHODS

2.1 *The Spherical Indentation Method*

The Tissue Stiffness Measurement method (TSM) developed in this work, is used for indentation of a pin with spherical tips (ϕ=8 mm) into the subject's soft tissue. In this method, two parameters have to be determined in order to find the tissue's shear modulus G. These parameters are δ, the indentation value, and P, the loading force. The indentation δ [mm] of a smooth rigid sphere with radius R into an incompressible material with a shear modulus G [N/mm^2] and Poisson ratio ν=0.5, when a total load P [N] is applied, may be represented by the following relationship (Lee & Radok 1960):

$$(R\delta)^{1.5} = (3RP)/(16G) \qquad (1)$$

In order to quantitatively measure the soft tissue elastic modulus, a simplifying assumption is made so that the soft tissue is considered as an incompressible material. Furthermore, when using the TSM method to measure biomechanical properties of a tissue which is normally subjected to an external load (such as plantar or buttock soft tissues), the tissue may be considered elastic and not viscoelastic because, during normal standing or sitting, one continually changes body balance (Bhatnager et al. 1985). Therefore, the effect of time duration on these tissues may be neglected, resulting in a mechanical definition which is time-independent. Hence, the Young modulus (E) is obtained from the relationship:

$$G=0.5E/(1+\nu) \qquad (2)$$

A TSM device which is based on the above spherical indentation concept, was designed and constructed in order to quantitatively measure plantar soft tissue biomechanical characteristics by integrating the above mentioned photoelastic technique called Contact Pressure Display (CPD) into an Open MRI system.

2.2 *The CPD Technique*

The CPD photoelastic technique uses a birefringent integrated optical sandwich for the simultaneous and instantaneous analysis of contact pressure (Arcan & Brull 1980, Arcan 1990). Through the CPD technique, the local load is transferred to the optical sandwich by a pin with a spherical tip. Concentric isochromatic circles appear on the optical sandwich under the contact point and the maximal diameter is a function of the local contact load; the calibration curve (contact load [N] vs. fringe diameter [mm]) maintains an exponential behavior according to the equation:

$$F=a \cdot e^{b\phi} + c \qquad (3)$$

for a fringe diameter $\phi \neq 0$ and constant characteristics a, b, c of the optical sandwich.

2.3 *The Open MRI Technique*

The Open MRI, also called Interventional MRI (General Electric), is an MRI scanner designed to provide images during surgical operations, in real time. It allows the surgeon to be active while the patient is scanned. The Open MRI scanner uses magnets which generate a magnetic field of 0.5 Tesla. It offers a wide range of scanning possibilities, which allow acquisition of complex images with very high quality, accurate geometry and exact location (Lufkin 1997). Since the TSM instrument is designed to be placed inside the Open MRI scanner, its components are all non-metallic due to the strong magnetic field activated by the scanner. The experiments with the TSM method were performed using the Open MRI at Sheba Medical Center, Tel Aviv, Israel.

2.4 The Tissue Stiffness Measurement Method

The TSM method in this work is specifically designed for plantar soft tissue stiffness measurement by integrating the two previously described techniques: The CPD and the Open MRI. During the tissue stiffness measurement, the subject is scanned using the Open MRI, while he is lying straight. An optical sandwich (111×111mm) is placed in front of the plantar aspect of the foot, using a special construction which allows the optical sandwich position to be adjusted. A rigid pin (8 mm diameter, 20 mm length) with hemispherical ends is free to slide axially, perpendicularly to the optical sandwich. The subject's foot is pressed towards the pin using an elastic band (see Fig. 1).

One tip of the pin indents the plantar soft tissue in the area between the 1st and 2nd metatarsal heads, while the other tip transfers its contact load to the optical sandwich. The area between the medial metatarsal heads was selected for indentation because it contains only soft tissue, with no bone stiffness effects. The subject is scanned to obtain the MR indentation image. The indentation is measured on the MR image using the Open MRI scanner software. As the pin indents the soft tissue, concentric isochromatic circles appear under its contact point with the plate and their maximal diameter is a function of the load transferred to the tissue. The fringe diameter was recorded for each loading case and measured using a calibrated system of circles.

2.5 Subjects

Two normal subjects (subjects #1, #2) and two diabetic patients (subjects #3, #4) volunteered to take part at the indentation tests. The subjects body characteristics are given in Table 1.

Table 1. Body characteristics of normal and diabetic subjects used for indentation tests.

Subject #	Normal Subjects		Diabetic	
	1	2	3	4
Sex	Female	Male	Male	Male
Weight [Kg]	56	84	70	62
Age [years]	24	26	72	84

The indentations of the soft tissues were measured on both coronal and sagittal MR images, in order to compare the qualities of MR indentation images acquired in different planes. Representative MR sagittal indentation images, obtained from normal and diabetic subjects are shown in Fig. 2a, 2b respectively. The indentation areas are clearly seen.

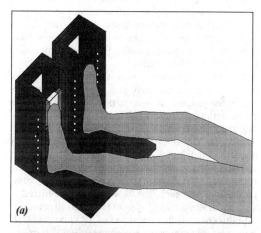

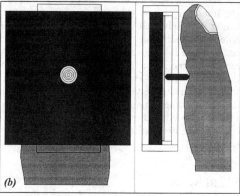

Fig. 1: The Tissue Stiffness Measurement (TSM) method: (a) The subject is positioned inside the Open MRI with his plantar aspect of the foot facing the optical sandwich; (b) One spherical end of the pin indents the soft tissue, while the other end transfers the load to the optical sandwich. As the pin indents the tissue, concentric isochromatic circles appear under its contact point with the optical sandwich.

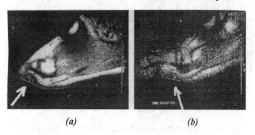

Fig. 2: Representative sagittal MR indentation images: (a) An indentation image obtained from normal subject #2, for a loading force of 1.75 N; (b) An indentation image obtained from diabetic patient #3, for a loading force of 1.75 N.

3 RESULTS AND CONCLUSIONS

Edge detection was performed using an image processing software to examine the indentation area boundary shape in normal and diabetic soft tissues. Fig. 3 presents the normal (subject #2) and diabetic (subject #3) indentation patterns, obtained when a load of 0.85 N was applied. The indentation patterns are both displayed at the same scale.

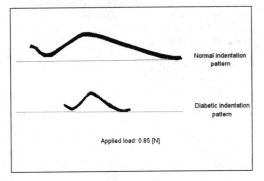

Fig. 3: Normal compared to diabetic indentation patterns.

We note that the diabetic indentation pattern is significantly thinner and sharper than the normal, indicating that the diabetic tissue is stiffer than the normal. The results obtained from the indentation experiments are presented in Fig. 4 as load vs. displacement curve.

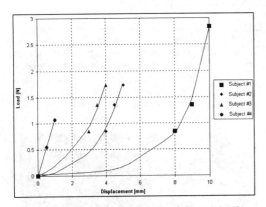

Fig. 4: Results of the indentation experiments: Load vs. displacement curve obtained from all subjects (#1, #2 normal subjects; #3, #4 diabetic patients).

The displacement-load relationship for all subjects may be represented by the equation:

$$\ln P = a + b\delta^{c} \qquad (4)$$

with a correlation coefficient of $r^2 = 0.99$, for a load of P [N], an indentation δ [mm], and constant values a, b, c characterizing the subject's tissue stiffness.

Only the smallest slope zones for each curve in Fig. 4 are considered, since the suggested equation (Eq. 1) is more accurate for relatively small displacements and loads. The resulted plantar soft tissue shear modulus G and Young modulus E, for each of the four subjects, are detailed in Table 2.

Table 2. Plantar soft tissue shear modulus and Young modulus obtained from normal subjects and diabetic patients in the indentation tests.

Subject #	Normal Subjects		Diabetic Patients	
	1	2	3	4
G [N/mm^2]	0.0035	0.0100	0.0153	0.0494
E	0.0106	0.0299	0.0460	0.1481

It is clearly shown that the load-deflection curves obtained from the diabetic patients are characterized by a significantly higher initial slope, and thus a higher elastic modulus, in comparison to the normal subjects. However, this initial work was aimed only to test the proposed method. A comprehensive large scale clinical research, on a sufficient number of subjects with similar body and age characteristics is still in the process, in order to use the load-deflection curve obtained in the TSM method for diagnosis of the exact stage of Diabetes Mellitus development. This work will involve the analysis of a statistically significant database for each stage of the disease.

The accuracy of the resulted data is dependent mostly on the MRI image resolution, or voxel dimensions. As such, the data may be considered as very accurate; the MRI system provided slices of 0.5 mm thickness, which consisted of 2-D pixels of 0.01 mm^2 area, yielding a voxel volume of 0.005 mm^3.

In summary, a new tissue stiffness measurement method of spherical indentation, integrating a photoelastic technique into an Open MRI system, was developed for investigating the biomechanical properties of normal and diabetic plantar soft tissue. The new TSM method described in this work was shown to characterize the biomechanical behavior of the plantar tissue. This method may also be applied to investigate other diseases which involve changes in soft tissue mechanical properties, such as breast cancer.

REFERENCES

Aoki, T., Ohashi, T., Matsumoto, T., & Sato, M. 1997. The pipette aspiration applied to the local stiffness measurement of soft tissues, *Annals of Biomedical Engineering*, 23: 581-587.

Arcan, M. 1990. Non invasive and sensor techniques in contact mechanics: A revolution in progress. *Proc. 9th Int. Conf. on Exp. Mech., Copenhagen, Denmark.* 1: 122-131.

Arcan, M. & Brull, M.A. 1980. An experimental approach to the contact problem between flexible and rigid bodies. *Mech. Res. Comm.* 7: 151-157.

Bhatnager, V., Drury, C.G. & Schiro, S.G. 1985 Posture, postural discomfort and performance. *Human Factors* 27: 189-199

Brosh, T. & Arcan, M. 1994. Toward early detection of the tendency to stress fractures. *Clin. Biomech.* 9: 111-116.

Cavanagh, P.R. & Ulbrecht, J.S. 1992. Biomechanics of the foot in Diabetes Mellitus. In: Levin, M.E., O'Neal, L.W. & Bowker, J.H. (eds), *The Diabetic Foot*: 199-232. Mosby.

Cook, T.A., Rahim, N., Simpson, H.C. & Galland, R.B. 1996 Magnetic resonance imaging in the management of the diabetic foot infection, *Br. J. Surg.* 83: 245-8.

Garra, B.S., Cespedes, E.I., Ophir, J., Spratt, S.T., Zurbier, R.A., Magnant, C.M. & Pennanen, M.F. 1992. Elastography of breast lesions: initial clinical results, *Radiology* 202: 79-86.

Gefen, A. 1997. A Structural Model of the Human Foot as related to the Plantar Pressure Pattern. Applications to Foot Disorders, *M.Sc. Thesis,* Tel Aviv Univ., Israel.

Horowitz, J.D., Durham, J.R., Nease, D.B., Lukens, M.L., Wright, J.G. & Smead, W.L. 1993. Prospective evaluation of magnetic resonance imaging in the management of acute diabetic foot infections. *Ann. Vasc. Surg.*, 7: 44-50.

Lee, E.H. & Radok, J.R.M. 1960. The contact problem for viscoelastic bodies. *J. of Appl. Mech.*, 27: 438-444.

Lufkin, R.B. 1997. Applications in interventional MRI: Current status and new advances. *Medical Imaging International.*, 7: 3-4.

Experimental Mechanics, Allison (ed.)© 1998 Balkema, Rotterdam, ISBN 90 5809 014 0

Evaluation of treatment of flexion contractures of interphalangeal joint

J. Jírová
Institute of Theoretical and Applied Mechanics, Prague, Czech Republic

J. Jíra
Faculty of Transportation Sciences, Prague, Czech Republic

F. Kolář
Institute of Rock Structures and Mechanics, Prague, Czech Republic

ABSTRACT: The study compares two ways (Fig. 1) of treating flexion contractures of the proximal interphalangeal joint (PIP):
- the conservative treatment by gradual splintage, using compressive bandage
- traction using a reposition device.

We have determinated using both numerical and experimental modelling the detailed stress field in the PIP for the conservative method of treatment only (Fig. 1a) which was determinated as the considerably unfavourable one. In Figures 4-5 we can see the compression concentration in the joint which is substantial for analysing of this problem. The whole loading of the articulate cartilage is in fact concentrated on the relatively small area, i.e. it is distributed non-uniformly. In this way the articular cartilage is considerably even unphysiologically, overstressed during the conservative treatment.

1 INTRODUCTION

The purpose of the study was to evaluate the stress state inside the joint structure, in other words the articular cartilage, which suffers the most during treatment flexion contractures of the proximal interphalangeal joint. Flexion contractures are a major therapeutical problem in orthopaedics, plastic and reconstructive surgery of the hand. In most cases the contractures result from profound burns, scald, electric shock, and from secondary healed wounds.

We know two ways of treatment: (i) the conservative treatment by gradual splintage, using compressive bandage, (ii) traction using a reposition device. The difference in loading between the two above mentioned methods of joint reconstruction is outlined by the Figure 1.

2 NUMERICAL MODELLING

2.1 *Frame Structure*

The difference in loadig between the two methods of joint reconstruction is outlined in the Figures 1a-1b. By calculation of the frame structure (Fig. 2) we calculated for the same deflection of the joint in the direction of axis y, i.e. for the same value of stretch on the tendon (F=14.4 N), approximately a tripling

of the force on the joint in the case of the first "conservative" method (Fig. 2a) in comparison with using a reposition device (Fig. 2b).

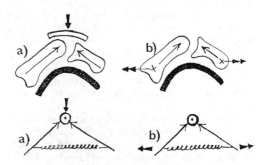

Figure 1. Scheme of two loading types

We represented the phalanxes as hollow tubes with a modulus of elasticity comparable with cortical bone E=17 GPa and $\mu=0.25$. Beam elements (hollow tubes) were mutually hinged connected with no regard to cartilage. The tendon was again substituted by a bar hinged connected to both beams with modulus of elasticity E=20 MPa which was experimentally measured (Jírová & Kafka, 1993). We considered elastic behaviour of all structures, we did

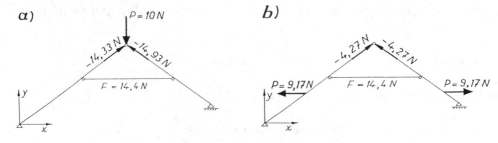

Figure 2. Frame structures

not calculate with viscoelastic behaviour of tendon which in nature is significant. It means that the results are correct only for start of loading. The frame structure was calculated as statically determinate structure with fixed bearing on one side and mobile bearing on the other side (Fig. 2). The results have confirmed the experiences of orthopaedic surgeons, who often report pain and swelling of the joint in cases of conservative method of treatment (Richtr & Ryšavý, 1992). When treatment is accelerated it can result in cartilage necrosis and lamellar chondrolisis (Boyes, 1964).

2.2 *Plane FM Model*

In order to find out the actual stress state of the articulate cartilage it was necessary to create the numerical model of behaviour of the joint structure in the course of loading during treatment. The model was constructed on the basis of the radiograph of a finger (Chao et al. 1989), when the co-ordinates of nodes were transferred into the computer by using ScanJet IIp. For the analysis we used two-dimensional plane elements from the programme ANSYS which is based on the Finite Element Method.

The element was used for plane strain analysis of cortical (E = 17 GPa) and spongy bone tissues (E = 75 MPa). The soft tissue was substituted by a link. As in a pin-jointed structure, no bending of the element was considered. The articular joint was constructed using contact elements which were used to represent contact and sliding between two surfaces in two dimensions. The coefficient of friction between the articular cartilage of a joint is very low (can be as low as 0.0026, about the lowest of any known solid material); hence no friction was considered between the cartilage surfaces.

Because calculation of a problem was very complicated we have used for the solution of this problem a superelement which was a group of previously assembled ANSYS elements of different types and material properties. The reason for substructuring was to reduce computer time and to allow solution of very large task with limited computer resources. For simplifying solution of a problem we considered spongy bone tissue for the both bone structures in the first approximation of our analysis. The task was reduced into one superelement and 1624 contact elements (Fig. 3).

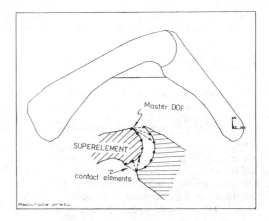

Figure 3. Finite Element Model

3 PHOTOELASTIC MEASURING

It holds generally, and in biomechanics particularly, that it is necessary to verify the theoretical model experimentally. If the experimental results corrolate with the theoretical model, it is relatively easy and fast to analyze a number of different loading variants than to devise a new experimental model for each of them.

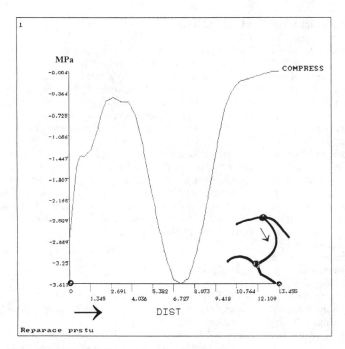

Figure 4. FEM analysis - stress state distribution

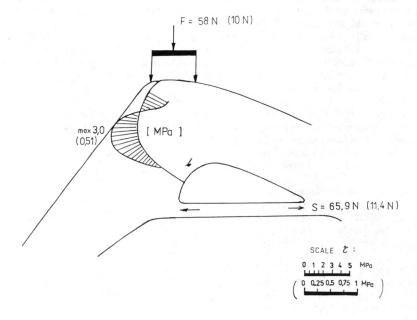

Figure 5. Photoelastic measuring - stress state distribution

We have decided to solve the problem of which reflection photoelastic measurement appears to be the most suitable for to ascertain of stress concentrations in the joint cartilage. We have applied two different approaches to the manufacture of the model: (i) with reference to similarity in the material characteristics, (ii) with reference to similarity in the rigidity characteristics.

We prefer the photoelastic measuring because strain gauges provide data only at the points where they have been mounted, and little is learned about other regions on the surface of the component. Attention was paid to the qualitative assessment of places with stress concentrations in the joint structure. The goal was to find stress peaks and describe the distribution of stresses during experimental simulation of treatment.

We have modelled a structural system consisting of two phalanges of one material, connected by a cartilage of a different material. The modulus of elasticity of the spongy bone was considered in the Finite Element model as 75 MPa and the modulus of elasticity of the cartilage as 25 MPa. Both materials for experimental research models were made in the same ratio of the moduli of elasticity. Two models were made:
1. The whole model was made of epoxy resins of two different moduli of elasticity.
2. The phalanges of the other were made of aluminium and the cartilage of epoxy resin. The design of this model was based on the fact that we were interested primarily in the stress state of the cartilage. While preserving the 1:3 ratio of the moduli of elasticity the epoxy resin modelling the cartilage could have higher modulus of elasticity than used in the first model, as a result of which the material did not have such pronounced rheological properties which is more favourable for the photoelastic measuring. The material has better optical characteristics enabling more accurate measurements.

The third model represents similarity in the rigidity characteristics. The whole model was made of the same epoxy resin with different thickness for the bone and the cartilage.

All models was shaped according to the X-ray photo. The boundary conditions were arranged in accordance with the computational model: fixed hinge at one end, sliding hinge at the other end.

The first two models yielded qualitatively the similar results as the theoretical model used for numerical analysis. In the second aluminium and epoxy model the stress concentration was distributed about a larger area. The third model has enabled the evaluation of the stresses not only in the cartilage, but also in the bone tissue of the phalanges.

4 CONCLUSION

The performed research has proved that the application of the reflection photoelastic measurements to the investigations of local stress fields in the proximal interphalangeal joint in the course of treatment is sufficiently accurate both qualitatively and quantitatively and not connected with excessive requirements of either time or costs. The whole loading of the articulate cartilage, which has been calculated for the frame structure is in fact concentrated on the relatively small area, i.e. it is distributed non-uniformly. Experimental research and computational modelling have showed that the articulate cartilage is considerably even unphysiologically overstressed (Figs 4-5) during the conservative treatment.

REFERENCES

ANSYS 1993. *User's Manual*, Volume I-IV.
Boyes, J.H. 1964. *Bunnel's Surgery of the hand*. Philadelphia. J.B.Lippincot.
Chao, E.Y.S. & An, K.-N. & Cooney III, W.P. & Linscheid, R.L. 1989. *Biomechanics of the Hand*. World Scientific, Singapore•New Jersey•London •Hong Kong.
Jírová, J. & Kafka, V. 1993. Experimental modelling of rheological properties of tendons. Proceedings of 31. Conference EAN, Měřín, ČR, str.151-154 (in Czech).
Richtr, M. & Ryšavý, M. 1992. Use of distraction apparatuses for the therapy of flectural contractions of interphalangeal joints. *Rozhledy v chirurgii*, 71, č. 3-4. (in Czech).

The research has been supported by the project no. 103 / 96 / 0268 of the Grant Agency of the Czech Republic.

Experimental Mechanics, Allison (ed.) © 1998 Balkema, Rotterdam, ISBN 90 5809 014 0

Dynamic mechanical testing of prosthetic acetabular components

A. M. R. New – *The Robert Jones and Agnes Hunt Orthopaedic Hospital, Oswestry, & IRC in Biomedical Materials, Queen Mary and Westfield College, London, UK*

M. Kloss – *The Robert Jones and Agnes Hunt Orthopaedic Hospital, Oswestry, UK & Fachhochschul Studiengaenge Vorarlberg, Dornbirn, Austria*

M. D. Northmore-Ball & J. H. Kuiper – *The Robert Jones and Agnes Hunt Orthopaedic Hospital, Oswestry, UK*

K. E. Tanner – *IRC in Biomedical Materials, Queen Mary and Westfield College, London, UK*

E. Veldkamp – *The Robert Jones and Agnes Hunt Orthopaedic Hospital, Oswestry, UK & Hogeschool Enschede, Netherlands*

ABSTRACT: The mechanics of cemented acetabular reconstruction remain little studied, particularly with respect to hip joint loading during gait, where both the magnitude and the direction of the hip joint force with respect to the pelvic axes vary significantly throughout the gait cycle. To address this problem, relationships between cementing technique and stability of acetabular component fixation in bovine calf acetabula have been investigated using a novel joint simulator. Micromotion between prosthesis and bone at the anterior-superior rim of the acetabulum, even in radiographically "well fixed" cases, was greater than that which could be attributed to elastic deformation of a specimen with a perfectly bonded interface. Thus it appeared that "perfect" bonding was never achieved even under laboratory conditions and that relatively large micromotion was possible in specimens that appeared radiologically well fixed with no apparent radiolucent lines, and thus that finite element models which assume complete bonding of the cement bone interface may be unrealistic.

1. INTRODUCTION

Total hip replacement is a highly successful treatment for disabling disorders of the hip. In most cases, a polyethylene acetabular component (socket) and a metal femoral component (ball) are fixed to the skeleton with a self-polymerising acrylic bone cement. Long term multi-centre studies show the survival rate for well established implant designs fixed in this way to be around 90% at 10 years (Malchau et al. 1993). However, current estimates put the annual number of hip replacements at 38,000 in the UK, 18% of which are revision operations (Biomaterials and Implants Research Advisory Group 1996). Similar trends are seen throughout the world and the number of revisions is forecast to rise for the foreseeable future. Revisions are complicated, costly and in general less successful than the primary operation (Kershaw et al. 1991). Further improvements to the long term performance of the primary hip are therefore highly desirable.

Aseptic loosening, that is loosening of one or both components without accompanying microbial infection, is the most common cause of late failure, with most studies reporting higher rates of loosening for the acetabular component than the femoral component (Wroblewski 1986, Schulte et al. 1993). Mechanical factors, such as the method of fixation of the components to the skeleton, which determines the bonding conditions at the implant-bone interface, appear to strongly affect the success of the reconstruction (Stocks et al. 1995).

The use of bone cement is generally considered to provide greater initial stability than alternative fixation methods (Perona et al. 1992) by filling completely the space between implant and bone and establishing a "microinterlock" between cement and bone. However, the presence of radiolucencies in post-operative radiographs and in specimens prepared in the laboratory (Ohlin & Balkfors 1992, New 1997) indicate the formation of gaps between the cement mantle and bone, possibly as a result of shrinkage or inadequate pressurisation of cement during the implantation procedure and the period while the cement cures. These gaps may allow movement, termed "micromotion", between cement and bone, which research into uncemented implants has shown to encourage the development of a fibrous tissue interlayer rather than a direct prosthesis to bone bond (Pilliar et al. 1986, Burke et al. 1991).

Despite these observations, the mechanics of cemented acetabular reconstruction remain little studied, particularly with respect to the dynamic aspects of hip joint loading during activity where both the magnitude and the direction of the resultant hip joint force vary significantly with respect to the anatomical axes of the pelvis.

Thus, the aim of the present study was to evaluate the stability of acetabular cups implanted in bovine calf acetabula using two cementing techniques which differed in the steps taken to pressurise the bone cement. Micromotion between cup and bone was measured while the reconstructed acetabulum was loaded in a novel joint simulator, designed to approximate the forces acting on the acetabulum during walking.

2. THE JOINT SIMULATOR

The joint simulator, shown in Figure 1, consists of a rigid steel frame (components A) bolted to the base plate of a servo-hydraulic universal testing machine. An aluminium slide block (B) is supported on two shafts (C) with end bearings that run on semi circular rails (D) incorporated in the frame. A stepper motor (E) mounted on the slide block drives two shaft mounted pinion gears (F) that engage with semi-circular racks (G) mounted below the rails. Torque is transmitted from motor to shaft by sprockets and chain (H). A second stepper motor (I) drives a rotating table (J), also via sprockets and chain (K) about an axis perpendicular to the axis of rotation of the slide block.

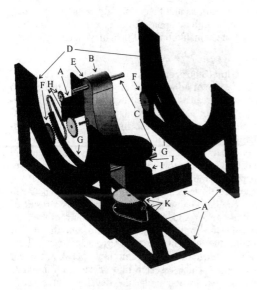

The acetabular specimen under test is potted in a fixture attached to the rotating table such that the centre of rotation of the acetabulum (the geometric centre of the cup bearing surface) is coincident with the intersection of the axes of rotation of the two driven axes of the joint simulator. This intersection is arranged to lie along the line of action of the testing machine actuator such that rotation of the driven axes allows actuator load to be applied along any line passing through the intersection of the two axes

and any point on the hemispherical bearing surface of the acetabular cup.

A loading fixture accepting a modular prosthetic femoral head with a standard (5°40') tapered bore attaches to the actuator of the testing machine. This fixture incorporates a biaxial linear slide that transmits only axial forces perpendicular to the sliding surface. The biaxial slide allows for the inevitable misalignments in the load train introduced when potting the specimens and by specimen deformation under load and avoids non-axial loading of the specimen and the testing machine actuator.

The functional relationships between the components of the simulator are shown in Figure 2. Overall control of the system is provided by the simulator control unit. This unit provides power and command signals to the motors from a programmable control board, which also generates an 8 bit digital output that, after digital-to-analogue conversion and filtering, is used as a command signal for the load servo-controller of the universal testing machine. Thus the motions of the two driven axes and the load applied by the materials testing machine can be synchronised. Arbitrary load and motion profiles may be generated using the computer and downloaded to the control board via an RS232 link. The computer is also fitted with an analogue input/output card (AD1200, Brainboxes, Liverpool, UK) with custom written software to log data from the outputs of the testing machine (load and stroke).

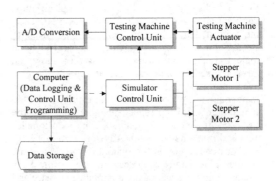

Figure 2: Functional diagram of the simulator.

3. METHODS AND MATERIALS

Six segments of frozen bovine calf pelvis were thawed and cleaned of soft tissues and prepared for cup implantation by reaming the acetabula with a hemispherical "cheese grater" type reamer of 50 mm diameter followed by vigorous brushing with a stiff polyethylene fibre brush. CMW 1 bone cement (De-Puy CMW, Blackpool, UK) was mixed in an open bowl in accordance with the manufacturer's instruc-

Figure 1: The principal mechanical components of the UJR Joint Simulator.

tions. At three minutes from mixing the cement was delivered to the acetabulum. In group 1 (3 acetabula), an UHMWPE cup with 48 mm outside diameter (Ultima, Johnson & Johnson Orthopaedics, New Milton, UK) and 28 mm bearing surface diameter was immediately inserted and held in place until the cement had cured. In group 2 (3 acetabula), an instrumented acetabular cement pressuriser with a 63 mm diameter elastomer seal was used to pressurise the bone cement for approximately one minute and then a cup identical to those in group 1 inserted. Mean and peak pressures for the pressurised group were 41.5 kPa (standard deviation ±18.8 kPa) and 64.7 kPa (±18.8 kPa) respectively, similar to per-operative measurements (New 1997). Post-implantation, a radiograph was taken.

To measure motion of the implanted cup relative to the acetabular rim, a linear variable differential transformer (LVDT, Schlumberger DFg5, RS Components, UK) was mounted on the anterior-superior rim of the acetabulum near the ilium using a Steinmann pin and jubilee clip. The armature of the LVDT was attached to the acetabular cup with a small aluminium bracket. A combined supply/amplifier unit (Entran PS30A, Entran Ltd, Watford, UK) was used to power the LVDT and to amplify its output. The output was then connected to an analogue input of the PC's I/O card via a passive low pass filter with a cut-off frequency of 20 Hz. The LVDT was calibrated after every test using a digital Vernier calliper (Digimatic, Mitutoyo, Andover, UK) which in turn was verified against workshop gauge blocks.

Prepared acetabula were then potted using Wood's metal (Goodfellow Cambridge Ltd, Cambridge, UK) in the fixture attached to the rotating table of the joint simulator as shown in Figure 3.

Figure 3: Potted acetabulum ready for testing. The biaxial slide (A) and the position of the LVDT (B) can be seen.

A combined motion and load cycle simulating normal gait, shown in Figure 4, was applied to the acetabula in groups of ten cycles, steadily increasing the load amplitude in six successive groups until the maximum load amplitude of 2.4 kN was reached. For each specimen five experimental runs were carried out to assess the repeatability of the tests and to accommodate any time dependent conditioning effects. Following testing, a second radiograph of each acetabulum was taken. Between test runs the acetabula were covered with wet paper towels so that desiccation did not occur. Laboratory temperature was 19±3°C at 51±20% relative humidity.

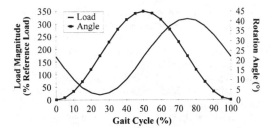

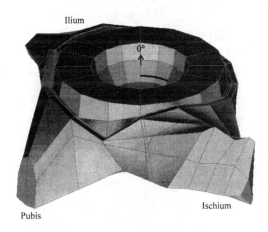

Figure 4: Simulated gait cycle: magnitude and orientation of the hip load and a view of a mounted specimen along the axis of the testing machine actuator showing the locus of the hip joint reaction force vector (thick line on inner surface of cup).

4. RESULTS

Figure 5 shows the average micromotion amplitude per cycle for the final test run at maximum load for the two implantation techniques. Given the limited data points and the scatter, there was no statistically significant difference between the two groups (p = 0.54, two tailed t-test assuming unequal variance). However the scatter was greater for group 1, where no additional cement pressurisation beyond that achieved by cup insertion alone was attempted.

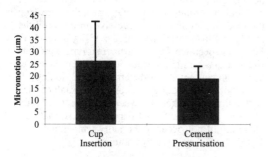

Figure 5: Micromotion at the ilium for two cementing techniques. Error bars represent one standard deviation.

Two cases, by examination of the pre-testing radiographs apparently the worst and best fixed specimens (case 1 and case 2 respectively), are used for illustration. Figure 6 and Figure 7 show the micromotion measured on the first and the last of the five runs respectively for each of these specimens.

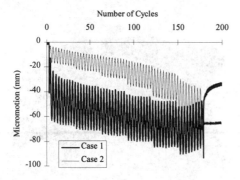

Figure 6: Initial run micromotions for the two cases.

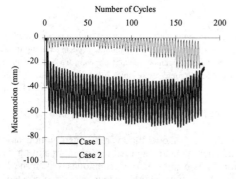

Figure 7: Final run micromotions for the two cases.

Typically the first run with each specimen demonstrated some "settling" or "conditioning" behav-

iour. This was observed as a background drift superimposed on the progressive increase in the amplitude of micromotion with the amplitude of the applied load cycles. Some small background drift was to be expected, since the minimum load was non-zero and thus scaled with the rest of the load cycle. Only in case 2, apparently the best fixed specimen, did this behaviour disappear in the latter tests.

5. DISCUSSION

Modern surgical techniques have lead to a reduction in the rate of clinical aseptic loosening (Malchau & Herberts 1996). In studies of isolated specimens of bone, great improvement in the strength of the cement-bone interface can be achieved by careful cleaning of the bone and pressurisation of the cement such that the cement penetrates the bone pore spaces. A similar effect *in vivo* is considered to be responsible for the improvement in the clinical results.

This experimental study represents a first attempt to compare the stability of complete acetabular specimens reconstructed using different cementing techniques under loading that approximates the *in vivo* situation.

In most tests, relatively large micromotion was detected between cup and bone, up to 40 μm in each cycle for the worst fixed specimen, which exhibited a radiolucent line 14 mm long along the superior margin of the acetabulum. In all cases, micromotion was greater by a factor of 5 than that which could be attributed to the elastic deformation predicted by a finite element model of the bovine calf acetabulum tested in the present configuration and assuming a fully bonded interface (New 1997). Conventional radiographic follow-up of patients with total hip replacements shows that radiolucencies between cement and bone first appear at both the superior and inferior edges of the acetabulum as viewed on an anterior-posterior radiograph and that these radiolucencies progress in time towards the depth of the acetabulum, eventually "cutting out" the cement and cup. The study of autopsy specimens by Schmalzried et al. (1992) indicates that these radiolucencies represent a fibrous tissue layer which is circumferential, although this cannot be detected on an anterior-posterior radiograph. If radiolucencies appear early or are present immediately post operatively the prognosis for the acetabular cup is worse (Strömberg et al. 1996). The present study confirms that the gaps that these radiolucencies represent permit micromotion between cement and bone. Micromotion of more than 150 μm is known to preclude apposition of bone to the surface of uncemented implants, whereas movements of less than 30-40 μm do not prevent apposition (Pilliar et al. 1986, Burke et al.

1991). Since these values are almost universally from animal studies, where healing potential is generally greater than in elderly humans, it may be expected that 30 μm is an optimistic estimate of the maximum tolerable micromotion. In the worst case presented here, this tentative threshold for bone apposition was exceeded.

These experiments also showed progressive settling or migration of the cup and cement into the acetabulum to occur, superimposed on the cyclic micromotion. Patient follow-up shows that, similar to those with early radiolucencies, acetabular cups with greater early migration exhibit earlier clinical signs of aseptic loosening (Stocks et al. 1995). Since the calf acetabula contained un-ossified triradial cartilaginous growth plates, the apparent settling behaviour in the present study may have been due to consolidation of these relatively compliant layers, although the potting of the specimen would be expected to counteract this effect. It is also possible that the settling represents elastic and plastic deformation of the cut bone trabeculae, produced by reaming and not constrained by sufficient interdigitation of cement. In comparison, the initial toe in the stress strain curve of isolated cancellous bone specimens is now thought to be an artefact associated with the cut surface in contact with the loading platens (Keaveny et al. 1994) which can be eliminated by careful embedding of the specimen ends.

6. SUMMARY & CONCLUSIONS

A simulator has been designed to reproduce the forces acting on reconstructed joints when used in conjunction with a servo-hydraulic materials testing machine. This device has been used to investigate the relationship between cementing technique and acetabular component fixation in bovine calf acetabula. Micromotion between prosthesis and bone, measured by a displacement transducer at the superior margin of the acetabulum near the ilium, was found to be less when the bone cement was pressurised in a separate step than when a cup was simply inserted, although this difference was not statistically significant.

Differences in the stability of cups defined qualitatively by radiological examination as "well fixed" or "poorly fixed" were also measurable. Micromotion was found, even in "well fixed" cases, to be far greater than that which could be reasonably attributed to elastic deformation of the intervening material. Thus it seems unlikely that "perfect" bonding was ever achieved even under laboratory conditions and that relatively large micromotion was possible in specimens that appeared radiologically well fixed with no apparent radiolucent lines. These results suggest that secure fixation at the acetabular rim against the edge of the pelvic cortex may not be achievable even with the best current cementing techniques. Finite element models which assume complete bonding of the cement bone interface may also require careful evaluation.

ACKNOWLEDGEMENTS

This work, which forms part of the PhD project of AMRN, was funded by the Edward Smith Estate Charity. The IRC in Biomedical Materials is funded by the United Kingdom Engineering and Physical Sciences Research Council.

REFERENCES

Biomaterials and Implants Research Advisory Group 1996, Report to the United Kingdom Department of Health: 7.

Burke, D.W., Bragdon, C.R., O'Connor, D.O., Jasty M., Haire, T. & Harris, W.H. 1991. Dynamic measurements of interface mechanics in vivo and the effect of micromotion on bone ingrowth into a porous surface device under controlled loads in vivo, 37th Annual Meeting of the Orthopaedic Research Society: 10.

Keaveny, T.M., Guo, X.M., Wachtel, E.F., McMahon, T.A. & Hayes, W.C. 1994. Trabecular bone exhibits fully linear elastic behaviour and yields at low strains. *J. Biomech.* 27: 1127-1136.

Kershaw, C.J., Atkins, R.M., Dodd, C.A.F. & Bulstrode, C.J.K 1991. Revision total hip arthroplasty for aseptic failure: a review of 276 cases. *J. Bone Jt. Surg.* 73B: 564-568.

Malchau, H. & Herberts, P. 1996. Prognosis of total hip replacement surgical and cementing technique in THR: a revision risk study of 134,056 primary operations. 63rd Annual Meeting of the AAOS.

Malchau, H., Herberts, P., Ahnfelt, L. & Johnell, O 1993. Prognosis of total hip replacement. 61st Annual Meeting of the AAOS.

New, A.M.R. 1997. Experimental and finite element studies of acetabular cement pressurisation and socket fixation in total hip replacement. PhD thesis, University of London.

Ohlin, A. & Balkfors, B. 1992. Stability of cemented sockets after 3-14 years. *J. Arthrop.* 7: 87-92.

Perona, P.G., Lawrence, J., Paprosky, W.G., Patwardhan, A.G. & Sartori, M. 1992. Acetabular micromotion as a measure of initial implant stability in primary hip arthroplasty - an in vitro comparison of different methods of initial acetabular com-ponent fixation. *J. Arthrop.* 7: 537-547.

Pilliar, R.M., Lee, J.M. & Maniatopoulos, C. 1986. Observation on the effect of movement on bone ingrowth into porous surfaced implants. *Clin. Orthop. Rel. Res.* 208: 108-113.

Schmalzried, T.P., Kwong, L.M., Jasty, M., Sedlacek, R.C., Haire, T.C., O'Connor, D.A., Bragdon, C.R., Kabo, J.M., Malcolm, A.J. & Harris, W.H. 1992. The mechanism of loosening of cemented acetabular components in total hip arthroplasty - analysis of specimens retrieved at autopsy. *Clin. Orthop. Rel. Res.* 274: 60-78.

Schulte, K.R., Callaghan, J.J., Kelley, S.S. & Johnston, R.C. 1993. The outcome of Charnley total hip arthroplasty with cement after a minimum twenty-year follow-up. *J. Bone Jt. Surg.* 75A: 961-975.

Stocks, G.W., Freeman, M.A.R. & Evans, S.J.W. 1995. Acetabular cup migration: prediction of aseptic loosening. *J. Bone Jt. Surg.* 77B: 853-861.

Strömberg, C.N., Herberts, P., Palmertz, B. & Garellick, G. 1996. Radiographic risk signs for loosening after cemented THA 61 loose stems and 23 loose sockets compared with 42 controls. *Acta Orthop. Scand.* 67: 43-48.

Wroblewski, B.M. 1986. 15-21 year results of the Charnley low-friction arthroplasty. *Clin. Orthop. Rel. Res.* 211: 30-35.

Experimental Mechanics, Allison (ed.)© 1998 Balkema, Rotterdam, ISBN 90 5809 014 0

Numerical and experimental cross-investigation of total hip replacements

Marco Viceconti
Laboratorio di Tecnologia dei Materiali, Istituti Ortopedici Rizzoli, Bologna, Italy

Luca Cristofolini
D.I.E.M., University of Bologna, Italy

Aldo Toni
Orthopaedic Clinic at Istituti Ortopedici Rizzoli, University of Bologna, Italy

ABSTRACT: The new EC regulations for the certification of medical devices call for an extensive pre-clinical validation of new prosthetic designs before clinical trials are started. While experimental *in vitro* protocols are used to validate the final design, Finite Element Analysis (FEA) are commonly used only during the early design phases to verify tentative solutions. In the present study the synergetic use of numerical and experimental methods was used to augment the pre-clinical validation of a new hybrid hip prosthesis. *In vitro* experiments and FEA models were simultaneously designed to be easily cross-correlated. Structural displacements, strain gauge measurements and implant-bone relative micromotions were acquired experimentally and used to validate the FEA models. The FEA models thus validated were used to predict quantities which are difficult to measure experimentally, such as: cement stresses, implant-cement relative micromotion or implant stresses in a variety of load cases. Using the numerical-experimental integrated approach it was possible to detect a potential problem of stress concentration in the cement mantle which would have been hard to identify using traditional experimental-only pre-clinical validation protocols.

1 INTRODUCTION

Computational methods in general (and specifically Finite Element Method, FEM) are some of the most important tools used by engineers in today's practice. Although sometimes FEM is perceived as a particular kind of experimental method, it is basically a numerical method based on the solution of differential equations. In this sense it should be considered more as a theoretical method than as an experimental one (Cook et al. 1989).

Apart from this consideration, the creation of a FEM model always involves a whole set of modelling assumptions and simplifications, which should verified before one can rely on the results deriving from the model. Even if the modelling assumptions are valid, the FEM is still an approximate method (Baguley and Hose, 1997). Therefore, its convergence to the theoretical solution should always be verified.

Several different approaches and strategies may be used to validate a FEM model, the best being a relevant previous experience on similar problems. When there is a lack of such a previous knowledge, the problem is a complex one and the system non-linearities cannot be neglected. The best way to validate the FEM model is then by comparison with experimental measurements taken in experiments simulating conditions identical or relevant to the ones being modelled (Baguley and Hose, 1997).

The largest majority of the simulations encountered in orthopaedic biomechanics falls under these conditions of scarce knowledge of the problem. As this is a relatively young discipline, the experience of the community about most biomechanical problems ranges from moderate to absent, especially when compared to traditional problems (e.g.: pressure vessels). Most of the problems involving a biological system are complex by definition. They are also frequently not completely defined; in other words, the confidence on a large part of the input data is very low. Last but not least biomechanical problems general include all possible sorts of non-linearity: structural, contact, material, etc.

Nevertheless, only in a small fraction of the works published in the literature, the FEM model were validated by an extensive experimental campaign. In most of the cases, strain gauge measurements were used to supply experimental data. Strain gauges are often selected as they are relatively simple to use and yield quantitative information in an easy way. They can provide accurate strain measurement, if some care is taken to reduce the experimental errors (Cristofolini et al. 1997b).

In the present work a research project is described that aimed at the pre-clinical validation of a new hip prosthesis design. To collect all the information needed to evaluate such a complex device both computational and experimental methods were used. Numerical and experimental results were combined to achieve an effective cross-validation between FEM models and *in vitro* experimental measurements.

2   MATERIALS AND METHODS

2.1 The implants

A large variety of pathologies affecting the human hip joint can be treated with the replacement of the affected articulation with an artificial prosthetic joint. Such surgical operation is called total hip replacement. The femoral component of the hip prosthesis is often referred to as "stem". It can be fixed to the hosting bone directly or with the interposition of a layer of bone cement.

The main cause of failure of cemented stems is the long term failure of the cement mantle under cyclic loading. In fact, the loads occurring in daily activities often lead to a progressive destruction of the cement associated with creep-fatigue phenomena. The advantage of using a non cemented stem is that it does not suffer from the risks related to cement mantle failures. The main problem of uncemented stems is primary stability, which is necessary to guarantee a proper long-term integration of the implant.

A new fixation concept was developed for a partially cemented stem (Cement Locked Uncemented, CLU), derived from an anatomic uncemented press-fit stem (AnCAFit, Cremascoli, Italy). To improve its primary stability, the stem was modified by machining two pockets for cement injection in the lateral-proximal area.

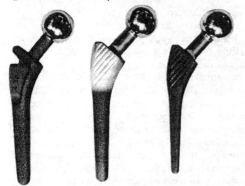

Fig. 1 - Image of the three stem types under investigation. From left to right: (i) the anatomic Ti AnCAFit 9.5 mm stem, which was cemented (CEM); (ii) a larger AnCAFit 14 mm that was press-fitted (AFIT); (iii) the Cement Locked Uncemented (CLU) prototype.

Three stem types were chosen for the present comparative study (fig. 1): an anatomic Ti AnCAFit 9.5 mm stem cemented with PMMA (CEM); an AnCAFit 14 mm stem press-fitted (AFIT); a Cement Locked Uncemented (CLU) prototype, having an identical shape to the AnCAFit 14 mm, except for the cement pockets.

2.2 The composite femur

Synthetic composite models of the human femur (Mod. 3103, Pacific Research Labs) were selected, as they allow reproducible conditions for comparative studies (Viceconti et al., 1996). A previous study (Cristofolini et al., 1996) has shown that they are an excellent substitute to cadaveric specimens.

The femurs were prepared following standard surgical technique to host the prosthetic stems. The CLU stems were inserted in position and bone cement was subsequently injected in the two pockets. The cement used has a Young modulus of 2721±76 MPa (measurements following the ISO 5833 standard). The CEM stems were inserted after filling the medullary canal with bone cement injected and pressurised with a "cement gun" (Cerim, Cremascoli, Italy).

2.3 Heel strike load case

In order to simulate the most relevant loading conditions, two different load cases were considered. The first one was intended as a simulation of the peak loads induced by normal walking (immediately after heel strike, corresponding to the highest peak force). This load case is relevant because is one of the most frequent during the daily activity, and thus affects bone adaptation. The system of forces applied to the femur was chosen based on the literature and represented a compromise between reproducibility of the set-up, and physiological simulation of the selected phase of gait. The design of the setup (fig. 2) was based on an extensive methodological investigation (Cristofolini et al., 1995 and 1996a).

For this part of the investigation, nine femurs were prepared with uniaxial strain gauges (Cristofolini et al., 1997a), at ten optimal locations (Cristofolini et al., 1997b). Repeated measurement were taken disassembling and re-aligning the entire set-up for ten times. The testing protocol showed a repeatability on the same femur of better than 0.8% at all gauge locations, and an overall reproducibility between femurs of better than 11%.

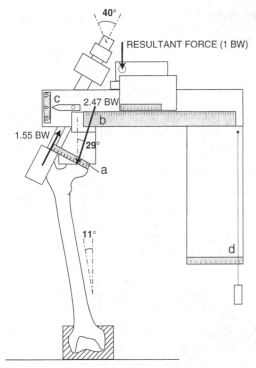

Fig. 2 - Schematic of the loading set-up, including the direction and intensity of the forces is indicated. The rulers to adjust the head centre-abducting force (a) and head centre-resultant force (b) lever arms are reported as well as the goniometers to check the cantilever horizontality (c) and the abducting force direction (d) are indicated. Also indicated are the strain gauge locations on five levels, on the lateral and medial aspects.

2.4 Stair climbing load case

A second loading case was chosen to investigate its effect on the primary stability of hip stems. It was designed to simulate the loads (mainly torsional) acting on the femur during stair climbing. Stability was assessed applying a torque of 18.9 Nm, corresponding to peak during stair climbing (Harman, 1995). The torque was applied for 1000 cycles, corresponding to about 1 month of patient activity (McLeod et al., 1975). The vertical force of the machine was converted into a torque which was transmitted to the femur by means of a custom jig. The femur was constrained distally. Relative micromotion was measured at four locations (proximal-medial, proximal-anterior, mid-stem medial, stem-tip medial). Four precision LVDT's (D5/40G8, RDP Electronics, UK; accuracy better than 1 micron) were used. The transducers were bonded to the cortical bone and reached the stem by means of a 3 mm transcortical hole.

For each of the three stem types, six femurs were prepared. The stems were evaluated in terms of permanent relative (peak to peak) and elastic (peak to valley) displacement.

2.5 FEM model definition

The geometry of the femur-stem system was acquired from a composite femur prepared by an orthopaedic surgeon. A CT scanner was used to sample the geometry. The CT data set was segmented and converted in a solid model. The solid models of the bone, of the cement mantle and that of the prostheses were imported into PATRAN (MacNeal-Schwendler Corp.; Los Angeles, USA) via Initial Graphics Exchange Specifications (IGES) neutral format and used to create a mapped mesh (fig. 3) using a procedure similar to the one described in (McNamara et al., 1996).

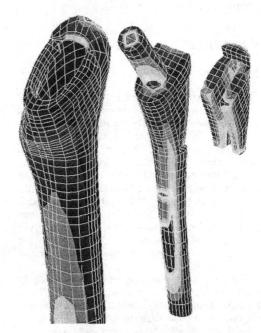

Fig. 3 - Finite Element meshes: from left to right: operated femur, CLU stem, CLU cement mantle meshes. The fringes report the Von Mises stress induced by the HS loading.

Similarly to the experimental work, two load cases were simulated: a heel strike (HS) load case, defined after the stress shielding experimental protocol, and a pure torsion (PT) loading, defined after the stair climbing load case used in the primary stability protocol. The first load case (HS) was used (1) to validate the FE model, (2) to investigate the structural behaviour of the model, and (3) to predict the stresses in the cement. The torsional load case was used first of all (1) to assess the accuracy of the interface modelling, and then (2) to predict cement stresses in the most critical load case for primary stability.

The material properties were indicated by the manufacturers of the bone analogue and of the implants. In both load cases, the boundary conditions were imposed such that all nodes on the most distal slice (cortical bone only) were fixed with zero displacement.

All the models described were solved using the PATRAN/FEA solver. The AFIT model was designed with a gap interface between the bone tissue and the metal. After some preliminary analyses, the CLU model was designed with most of the nodes at the cement-metal interface merged, while the stem-bone interface was modelled with gap elements. Only the most lateral nodes of the cement inlet were kept separated from the stem surface, as in this area a clear detachment of the stem-cement interface was observed.

3   RESULTS

3.1 FEM models validation

The structural displacements experimentally measured at some relevant points, was compared with the values computed by the FE models, in order to evaluate the correctness of assigned boundary conditions. For the HS loading, the medio-lateral displacement of the medial-most points of the calcar and of the most distal section (just above the clamping level) were used as reference. Under the HS load case, both the AFIT and the CLU models predicted a calcar displacement of 7.4 mm (experimental measurements gave 7.5 ± 0.1 mm); distal displacements were always smaller than 0.1 mm both in the experimental measurement, and in the F.E. prediction.

The torsional loading case was then validated by measuring the periosteal rotation of the medial line at the distal-most location, where the transcortical pins are inserted for the micro-motion test. During the experimental measurements a rotation of 1.3 ± 0.2° for both implants was measured. The CLU model predicted a rotation of 1.2°. The AFIT FEM model predicted a rotation of 1.4°.

The predicted surface strains were compared with the experimental strain gauge measurements in order to validate the stress results obtained from the FE models. The AFIT model showed an overall agreement between the strain gauge measurements and the FE predicted strains. The Root Mean Squared Error (RMSE) between the 10 strain gauges measurements and the surface strains predicted at the same nominal locations by the FE model, was 150

με, approximately 5% of the peak measured strain. A similar agreement was found for the CLU results (RMSE = 81 με or 2% of peak strain).

In order to verify the ability of the FE models to accurately represent the non-linear bone-stem contact, the relative micro-motion experimentally recorded by the LVDT transducers was compared with the analogous relative motions FE computed.

Experimental relative micro-motion was recorded at the medio-proximal (MP), antero-proximal (AP) medio-mid stem (M1/2), medio-full stem (M1/1) and axial (A) locations. However, at the M1/2 level of the FEM model, the stem was not contacting the endosteon. Thus, the relative micro-motion would could only be obtained by means of extrapolation. Since this would produce sensible inaccuracies, the data from M1/2 were not used here. The comparison between predicted and bone-stem relative micro-motion is reported in table 1.

	AFIT		CLU	
	FEM	Exp.	FEM	Exp.
MP	83	60 ± 25	13	15 ± 5
AP	31	10 ± 10	0	0 ± 5
$M_{1/2}$	-	50 ± 30		10 ± 5
$M_{1/1}$	44	40 ± 30	25	0 ± 5
A	13	10 ± 10	0	0 ± 5

Tab. 1 - Comparison of LVDT-measured and FE-predicted stem-femur relative micromotion, reported in micrometers, for the AFIT and CLU stems.

3.2 Cement stress

Most of the cement presented a 4 MPa tensile stress or lower, under the PT load case action. Zones of higher stress were found in correspondence of some stress raisers; their intensity was always below 10 MPa. In the HS load case the cement is more stressed than under the PT load case. The inlet connection to the cement pockets incorporates some sharp angles which act as stress risers. In these points the tensile stress reached up to 25 MPa. However, stresses in most of the cement were well below 8.5 MPa.

3.3 Cement-stem micromotion and stem stresses

Another possible concern for the CLU stem design was the relative micro-motion between cement and stem. For both load cases, the model predicted relative motions smaller than 60 microns. These displacement magnitude should easily be accommodated by the elastic deformation of the ridges on the cement surface.

The stresses predicted by the CLU FE model on all simulations were significantly lower than the fatigue limit for the Ti6Al4V alloy (safety factor n = 5.3 for the HS load case; n = 5.5 for the PT load case). Another important remark is that the slots cut in the CLU stem did not produce any relevant stress concentration in comparison with the AFIT cementless stem.

4 CONCLUSIONS

Although contemporary results for total hip arthroplasty have significantly improved in the last 20 years, at least 5-10% implants fail within 10 years from the operation. Thus, the actual implant designs may still be improved significantly. However, with respect to the pioneer years, today it is mandatory to thoroughly investigate any new solution before clinical use.

The pre-clinical validation protocol described combined an *in vitro* biomechanical simulation with finite element analyses. This integration proved itself an effective and reliable approach to this complex matter.

The CLU concept was developed with the objective of improving the primary stability of a cementless stem, while still maintaining the advantages of cementless implants. The prototype design achieved most of the project goals; However, the cement inlets produce an unacceptable stress concentration in the cement. An engineering optimisation of the geometry of the cement inlets and pockets is required before any further step in the validation.

5 REFERENCES

Baguley and Hose, 1997. *How to interpret finite element results*. Glasgow: NAFEMS.

Cook, R.D., Malkus, D.S. and Plesha, M.E. 1989. *Concepts and applications of finit element analysis*. New York: Wiley & Sons.

Cristofolini, L., Viceconti, M., Toni, A. and Giunti, A. 1995. Influence of thigh muscles on the axial strains in a proximal femur during early stance in gait, *J. Biomech.*, 28:617-624.

Cristofolini, L. Viceconti, M. Cappello A. and Toni A. 1996. Mechanical validation of whole bone composite femur models. *J. Biomech.*, 29: 525-535.

Cristofolini, L. and Viceconti, M. 1997a. Comparison of uniaxial and triaxial rosette gauges for strain measurement in the femur. *Experimental Mechanics*, 37(3): 350-354.

Cristofolini, L., McNamara, B.P., Freddi, A. and Viceconti, M. 1997b. In-vitro measured strains in the loaded femur: quantification of experimental error. *J. Strain Anal. Eng. Des.*, 32(3): 193-200.

Harman, M.K., Toni, A., Cristofolini, L. and Viceconti, M. 1995. Initial stability of uncemented hip stems: an in-vitro protocol to measure torsional interface motion. *Med. Eng. & Physics*, 17:163-171.

McLeod, P.C., Kettelkamp, D.B., Srinivasan, V. and Henderson, O.L. 1975. Measurement of repetitive activities of the knee, *J. Biomech.*, 8: 369-373.

McNamara, B.P., Viceconti, M., Cristofolini, L., Toni, A. and Taylor, D. 1996. Experimental and numerical pre-clinical evaluation relating to total hiparthroplasty. In: Middleton, J., Jones, M.J. and Pande, G.N. (ed) *Computer Methods in Biomechanics and Biomedical Engineering*: 1-10. Amsterdam: gordon and Breach.

Viceconti, M., Casali M., Massari B., Cristofolini, L., Bassini S. and Toni, A. 1996. The 'Standardized femur program'. Proposal for a reference geometry to be used for the creation of finite element models of the femur. *J. Biomech.*, 29:1241.

Experimental Mechanics, Allison (ed.) © 1998 Balkema, Rotterdam, ISBN 90 5809 014 0

Study with speckle interferometry of bone-implant interface conditions

J. Simões – *Department of Mechanical Engineering of the University of Aveiro, Portugal*

J. Monteiro & M. Vaz – *Laboratory of Optics and Experimental Mechanics, INEGI, University of Porto, Portugal*

M. Taylor – *Department of Orthopaedics, University of Lund, Sweden & IRC in Biomedical Materials, Queen Mary and Westfield College, UK*

S. Blatcher – *IRC in Biomedical Materials, Queen Mary and Westfield College, UK*

ABSTRACT: The measurement of deformation within cancellous bone is extremely difficult using conventional contact measurement techniques due to the material's highly porous structure and poor mechanical properties. Simões *et al.* (1997) have developed and proven the adaptability of an experimental technique using speckle interferometry to measure displacement patterns on a cancellous bone substitute (commercial polyurethane HEREX® C 70). The aim of this paper is to study the influence of interface conditions (bonded versus press-fit implants) using speckle interferometry. Bovine cancellous bone was used in the experiments with a metal tapered rod simulating the implant, which was inserted centrally in the cancellous bone. The model was sectioned longitudinally to expose the bone-implant interface and the displacement distributions of bonded and press-fit implants were assessed.

1 INTRODUCTION

Micromotion, migration and subsequent aseptic loosening of proximal femoral prostheses are the most common forms of failure of hip replacements. The resulting revision operation is distressing for the patient, technically difficult for the surgeon and costly to the health service. The mechanism by which aseptic loosening occurs is still not clearly established. Many new designs of prostheses are introduced into the orthopaedic market, with little or no pre-clinical test results.

Conventional, clinically based, techniques of implant evaluation can take up to 10 years to identify a poorly performing prosthetic design. In an increasingly competitive market, such information is usually redundant by the time it is published. Recently, migration studies of the progressive motion of the implant within the bone, have shown that subsidence, or the vertical component of migration, can be used to predict the incidence of late aseptic loosening as early as two years post operatively (Freeman and Plante-Bordeneuve, 1994; Kärrholm *et al.*, 1994). This suggests that the mechanism of migration, and hence late aseptic loosening, is a mechanical, rather than a biological process, or is at least triggered by the initial mechanical environment. Therefore, providing a suitable predictor can be found it

should be possible to predict the early migration using *in vitro* experimental or analytical techniques.

The failure of the soft spongy bone, cancellous bone, supporting an implant may be one of the contributing factors to migration and loosening. The cancellous bone stress distribution of the intact and implanted femur has been studied using finite element analysis (FEA) and a number of implant configurations have been analysed (Taylor *et al.*, 1995a; 1995b). In comparison to clinical migration and survivorship data, the FEA results showed a good correlation between the predicted peak cancellous bone stresses and the observed mean subsidence, as measured two years post-operatively (Taylor *et al.*, 1995a). The implant designs which generated the lowest cancellous bone stresses, were those which migrated the least, whereas the implants which generated high cancellous bone stresses, migrated the most. Based on these results, the initial cancellous bone stresses may be used to predict the likelihood of early migration and hence the incidence of late aseptic loosening (Taylor *et al.*, 1995a). To have more confidence in the predictions, the finite element models need to be validated by comparison with *in vitro* experimental results.

Simões *et al.* (1997) developed an experimental technique to measure the displacement field surrounding a bone-implant interface replicate. In this study, commercial polyurethane foam (commercial HEREX® C 70 from manufacturer AIREX AG Speciality Foams) was used as bone substitute due to its similar ultimate compressive strength/Young's modulus ratio of cancellous bone. However, differences still exist at the macroscopic and microscopic level between the two structures.

The results published by these authors show that the displacement pattern for the bonded and press-fit situation are different, suggesting that the load applied to the implant is transferred to the surrounding bone substitute in different ways. A close qualitative correlation was observed between the experimental and the numerical results.

Since it was a preliminary study, no attempt was made by the authors to replicate the real structure of cancellous bone. The only objective within this study was to assess if the speckle interferometry technique was suitable to measure displacements, and therefore deformations, within a structure with similar cancellous bone characteristics.

In this study, the influence of the interface conditions on the characterisation of the displacement field in the immediate vacinity of a bone-implant interface was assessed.

2 ESPI METHOD

Conventional methods of measuring surface strain and displacement utilise strain gauges, dial gauges and other mechanical and electrical sensing devices. These measuring methods can be highly sensitive and accurate, but give information in limited regions of the structure. To obtain the strain or displacement distribution of bigger areas it is necessary to use full field techniques, like Moiré and Speckle. Moiré techniques show in-plane displacements distributions in a continuous manner over the entire surface but its sensitivity depends on the grid pitch. Some problems can arise on the production of high frequency grids on the object surface with the Moire method.

The speckle technique can also be used to measure in-plane displacement distributions. It is a non-contact field technique that makes use of the speckle patterns resulting from a coherent illumination of a rough surface. This technique has been widely used for accurate measurements of the deformation patterns on structures when subjected to mechanical or other stresses. One of its advantages relies on its high sensitivity; displacements of the order of the wavelength can be assessed. The results of this method are obtained in the form of fringe patterns, each fringe

corresponding to points having the same displacement in the direction of the sensitivity vector. The use of electronic recording media to record the speckle patterns gives the technique known as Electronic Speckle Pattern Interferometry (ESPI). The image processing techniques allow one to build a very flexible set-up to measure displacement fields.

The study of porous structures, like cancellous bone, using optical methods has not been done, although, Bay (1995) used texture correlation to measure the displacement and strain patterns within samples of trabecular bone. This technique is a modification of digital image correlation, a method for analysis of deformation in objects marked with random surface speckle. Due to the cancellous bone structure characteristics, and instead of surface speckle, the trabecular pattern itself is used as a basis for correlation.

3 MATERIALS AND METHODS

Two fresh bovine femurs were used. The femurs were sectioned in the frontal plane and the cancellous bone of the proximal femur was fixed with an epoxy resin to the test rig. Figure 1 shows the bone specimen and the implant used. The test rig used by Simões et al. (1997) was optimised to correctly measure the applied forces and to avoid boundary effects in the displacement measurements. The rig was designed to enclose the bone, and to have sufficient rigidity to avoid rigid body displacements due to applied load to the implant. For the bonded situation, the implant was glued with a very thin layer to the bone. It was observed that the glue did not penetrate the cancellous bone and therefore did not reinforce it at the contact area. The force was applied by imposing a constant displacement to the implant.

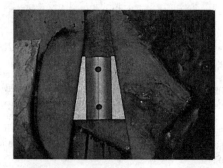

Figure 1. Bone and implant used to study the influence of different interface conditions.

A set-up with four directions of illumination was used to measure the in-plane displacement field at the surrounding bone due to the axial movement of the implant (figure 2). In this set-up a liquid crystal retarder (LCR) was used to shift the illumination from one direction to the other. This set-up is a duplication of the one proposed by Ennos (1968) to assess in-plane displacements. In this case, using a special routine of image processing, it is possible to record two different references and obtain displacement distributions in two different directions.

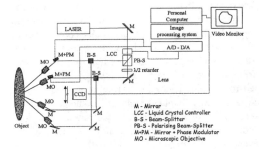

Figure 2. The experimental set-up used to measure the in-plane displacement.

Within the holograms fringe patterns each fringe represents points of an equal displacement in the sensitivity vector direction. An image processing system was used to calculate the phase. Figure 3 shows the region of the cancellous bone, adjacent to the implant that was used to study the influence of the interface conditions on the displacement field induced by the implant.

4 RESULTS

The displacement fields in the region of the bone surrounding the bonded and the press-fit implants were assessed in the axial and transverse directions of the implants. For both, the press-fit and bonded implant, a constant displacement was applied to the implant and it was observed that qualitatively the fringe pattern did not depend on the displacement value. The fringe patterns for the bonded implant, in the axial and transverse directions, are shown in figure 4 and 5 respectively.

The results for the press-fit implant are shown in figure 6 and 7 (axial and transverse direction).

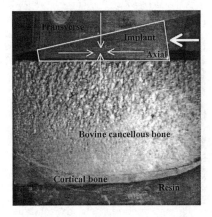

Figure 3. Bone adjacent to the implant where the fringes were assessed.

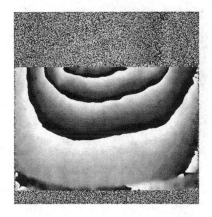

Figure 4. Fringe pattern of the axial displacements.

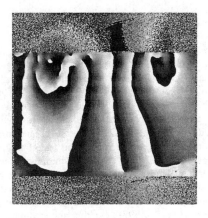

Figure 5. Fringe pattern of the transverse displacements.

Figure 6. Fringe pattern of the axial displacements.

Figure 7 – Fringe pattern of the transverse displacements.

5 DISCUSSION

It has been speculated that the failure of the soft spongy bone or cancellous bone supporting an implant is one of the contributing factors to migration and loosening (Taylor *et al.*, 1995a; 1995b). However, the majority of *in vitro* experimental and analytical studies have concentrated on the cortical bone stress and strain distributions, despite the cancellous bone being the weakest material of any bone-prosthesis construct. Very few studies have been done to assess the stress-strain distribution due to implants inserted in cancellous bone. The preliminary study done by Simões *et al.* (1997) clearly shows that the experimental technique to obtain the stress-strain distribution within porous material is suitable, and

therefore applicable to the examination of the mechanical behaviour of a structure similar to cancellous bone. The study presented in this communication confirms the adaptability of the ESPI technique to study deformation patterns within cancellous bone.

The experimental set-up was sensitive to in-plane displacements. The sensitivity vector is obtained by the intersection of the plane defined by the two beams with the surface of the bone, and the fringe pattern shows the displacement distribution in each of these directions. By using the set-up shown in figure 2, it was possible to analyse the fringe patterns in both directions without rotating the bone-implant ·interface. Equal spaced parallel fringes perpendicular to the sensitivity vector corresponds to a constant displacement field. Fringe's spatial frequency changes in the presence of deformation. Fringes, which are not perpendicular to the sensitivity vector, are due to rotations, which induce corresponding displacements in the direction of the sensitivity vector. So, the deformation field is codified in the spatial frequency variation of the fringe pattern and can be calculated by numerical differentiation.

The fringe patterns for the bonded implant are similar, in both directions, to those obtained in the previous work done with foam, but for the press-fit implant, some differences were observed in the transverse direction. To the study here presented, it can be seen (figure 5 and 7) that the fringe patterns for the bonded and press-fit implants are similar, which may be due to the different implant design used in this study. The implant used in our previous study (Simões *et al.*, 1997) was conical, whereas the one used in this study is prismatic (figure 1). We also observed that the press-fit and the bonded implant performed in a similar way, although a higher force was necessary to be applied to the bonded implant to obtain an equal number of fringes.

For the bonded implant, the points near the implant move together with the surrounding cancellous bone, since it is glued to it. Stress concentration zones were observed due to the implant edges. For the axial direction, it can be seen that the fringe pattern for the bonded interface is uniformly distributed, where for the press-fit interface, a localised stress gradient is observed at the implant edge.

Relative to the bonded implant, the press-fit implant generated much higher displacements. The bonded implant provokes a high displacement gradient at its tip, whereas for the press-fit implant this is not so pronounced. In all the tests the cancellous-cortical bone boundary effect is clearly evident. At the moment no numerical differentiation of the ESPI results is available. However, a FEM analysis was performed to compare the displacement field of the bonded implant with the one obtained with ESPI. The simulation was done only for the bonded implant, since the coefficient of friction of the press-fit interface is unknown. Figure 8 shows the FEM results for the axial (a) and transverse (b) direction.

(a)   (b)

Figure 8 - Iso-displacement field for the bonded implant.

It can be observed, comparing figure 8 with figure 4 and 5, that a good agreement was achieved between the ESPI and FEM data.

Using image processing techniques, with phase shift, the phase distribution was calculated for all of the study area. As the phase is the relevant information of the fringe pattern, it is possible to calculate the displacements from the phase maps and therefore the deformations.

This comparison shows that the ESPI experimental method has a great potential to study in more detail the influence of the interface bonding conditions of bonded and unbonded implants. The method also allows the numerical value of the displacement distribution to be obtained, which will be the aim of a next study, and hence validate in the near future, the finite element model.

6 CONCLUSIONS

This study has shown that ESPI can be used to examine the deformation of cancellous bone. Further work is now in progress to perform a quantitative comparison of the experimental results and the finite element predictions.

Specimens of bovine cancellous bone will be tested to obtain its mechanical characteristics to accurately simulate the unbonded interface conditions with a finite element model, using contact elements with friction.

Concerning the main aim of this study, it was observed that the interface bonding conditions play an important role on the cancellous bone displacement distribution and hence on the resulting stress and strain patterns. A detailed knowledge of the influence of the interface conditions, between the implant and the surrounding bone, on the displacement of the prosthesis within the surrounding media would allow us to speculate, with more certainty, on the mechanism of load transfer.

ACKNOWLEDGEMENTS

We would like to thank Prof. Joaquim Silva Gomes, director of the Optical and Experimental Mechanics Laboratory (LOME) for the facilities given to develop the experimental measurements.

REFERENCES

Bay, K. B., " Texture correlation: a method for the measurement of detailed strain distributions within trabecular bone", *J. of Orthopaedic Res.*, 13, 258, 1995.

Ennos, A. E., "Measurement of in-plane surface strain by hologram interferometry", *Journal of Scientific Instruments (Journal of Physics E)*, Series 2, 1, 731-734, 1968.

Freeman, M. A. R. & Plante-Bordeneuve, P., "Early migration and late aseptic failure of proximal femoral prostheses", *J. Bone Joint Surg.*, 76B, 432-438, 1994.

Kärrholm, J., Borssën, B., Löwenhielm, G. & Snorrason, F., "Does early migration of femoral prostheses matter?", *J. Bone Joint Surg.*, 76B, 912-917, 1994.

Simões, J. A. O., Vaz, M. A., Chousal, J. A., Taylor, M. & Blatcher, S., "Speckle interferometry to measure the strain distribution within porous materials", *International Conference on Advanced Technology in Experimental Mechanics*, The Japan Society of Mechanical Engineers, Japan, 423-428, 1997.

Taylor, M., Tanner, K. E., Freeman, M. A. R. & Yettram, A. L., "Finite element modelling - A predictor of implant survival?", *J. Mat. Sci.: Mat. Med.*, 6, 808-812, 1995a.

Taylor, M., Tanner, K. E., Freeman, M. A. R. & Yettram, A. L., "Cancellous bone stresses surrounding the femoral component of a hip prosthesis: An elastic-plastic finite element analysis", *Med. Eng. Phys.*, 17, 544-550, 1995b.

Experimental Mechanics, Allison (ed.)© 1998 Balkema, Rotterdam, ISBN 90 5809 014 0

Uncemented hip prostheses: An in-vitro analysis of torsional stability

L.Cristofolini
DIEM, Engineering Faculty, University of Bologna & Laboratory for Biomaterials Technology, Rizzoli Orthopaedic Institutes, Italy

L.Monti
DIEM, Engineering Faculty, University of Bologna, Italy

M.Viceconti
Laboratory for Biomaterials Technology, Rizzoli Orthopaedic Institutes, Bologna, Italy

ABSTRACT: Torsional load occurring in daily activities is the most severe threat for primary stability and success of uncemented hip stems. Prosthetic behaviour under torsional loads must hence be analysed prior to perform in-vivo tests. Even though much effort has been done, it is hard to find in the literature an in-vitro protocol that is accurate, physiological and reproducible at the same time.

An in-vitro procedure was developed to study primary stability by measuring the relative micromotion at the bone-stem interface in simulated stair ascent. Measurements were taken at four locations by means of a transcortical pin attached to a LVDT.

To achieve a relevant protocol, physiological load cycles were simulated (including proximal-to-distal axial force, torsional and bending moments). The methodologies of placing both the specimen under the testing machine and the sensors on the specimen were standardized and tested for reproducibility.

The effects of the loading frequency and of the strain distribution across the cortical bone were investigated. Measurement errors were of 1.8 μm between load cycles and of 4.9 μm for repeated set-ups, which compare favourably with the literature.

INTRODUCTION

Scientific investigations (1-7) on the biomechanics of the hip before and after THA (Total Hip Arthroplasty) have demonstrated that the most common mechanical problem of uncemented implants is the failure of the stem fixation. In the case of successful implantations primary stability is achieved by surgical preparation of the bone cavity to optimally fit the prosthesis and by mechanical press-fit and interlock. Secondary stability is the result of a healing process (osteointegration), where bone formation leads to intimate contact between the bone and certain parts of the implant surface. The osteointegration process is adversed by the presence of excessively high relative micromovements at the bone-stem interface (9). The acceptable micromotion values have not been precisely established yet. Previous studies (7,10-11) suggest however that the relative displacement at the interface between implant and bone should be between 28 and 150 μm, i.e. lower than the pore size. The amount of relative motion immediately after the operation, depends on (9): the bone quality, the accuracy of the bone bed preparation, the loading conditions, the implant geometry and the implant surface texture.

The efforts made to improve primary fixation tend basically towards 4 directions: 1) the improvement of the surgical techniques; 2) the identification of all the physical activities which expose the operated articulation to risky movements/loading conditions which inhibit the osteointegration process; 3) the optimisation of the surface texture and characteristics and 4) the mechanical and geometrical properties of the stems.

In-vivo (5,12,13,16), clinical (18-20) and in-vitro (21) studies have demonstrated that the activities which involve large hip flexion moments (such as stair climbing) generate large hip loads (23). Torque promotes loosening (2,22,24) and fracture (19) of the stems and cannot be neglected when studying the mechanics of the femur after THA (21). Thus, rotational motion can be regarded as a significant parameter to evaluate the effect of the design and mechanical features on the success of the implant itself. These results have been confirmed by in-vivo studies using instrumented implants (12-17).

The goal of the present investigation was to develop an experimental in-vitro biomechanical method to measure the interface motion between an uncemented stem and the femur. The target is to design a method that:

1) simulates the physiological loading conditions that are most severe in terms of torsional stability;
2) is reproducible and repeatable;
3) is precise and sensitive to the stem properties.

REVIEW OF THE LITERATURE

A critical analysis of the literature was made before designing the protocol (6,9,10,22,25-40). The aspects we focused on are the following:
1) the reference system definition;
2) the loading system, including: a) the constraints applied; b) the loading cycle;
3) the measurement system, including: a) the transducer type; b) the transducer anchorage device; c) the transducers position on the specimen; d) the type of displacement observed.

The weak point is most commonly the reference system. This is defined in the majority of the protocols, except in four cases (6,32,37,38) where it is not specified whether the vertical axis is collinear to the femur or to the stem long axis. Only in half of the works it is possible to determine to which direction the applied forces, the sensor positioning and the recorded displacements are referred. These reference systems are femur-aligned or stem-aligned in equal proportions. A stem-aligned reference system makes the test dependent on the stem geometry and anteversion angle after the insertion: thus, the direction of the reference vertical axis with respect to the fixed femoral shaft changes with the specimen and with the magnitude of the applied load. In short, only 1/3 of the works are based on completely and uniquely identifiable (hence reproducible) reference system and specimen positioning.

The large variety of loading conditions suggested by different authors can be resumed in: 1) load cycles of fixed amplitude to measure the bone-stem motion (employed by the majority); 2) load cycles of increasing amplitudes until a given failure or loosening criterion is reached; 3) monotonic ramp to individuate the weakest section (25) or the stress transfer mechanism (31). The load magnitudes and composition (purely torsional, purely axial or variably directed load) are heterogeneous, making the results incomparable. The variety on the number of applied cycles per single test session reflects the diversity of the load composition and of the experimental goals. The presence or the absence (in most of the cases) of a preconditioning set of cycles strongly influences the recordings.

The specimen constraints represent the second major problem. Physiologically, the femur is over-constrained by the joint reactions, the ligaments and the muscle forces. For testing purposes it is necessary to design an isostatic set-up that allows a direct control on the forces applied. The importance of this fact is clear from the work of Pawluk et al. (41). They found enormous differences in strains (up to 70%) between the isostatically-constrained and the over-constrained femur. As the femur bends under the applied loads, the constraints will react with force components that are unpredictable and uncontrolled.

Therefore, as an unknown system of forces is applied, the results will be completely unpredictable. Cristofolini (45) reported that in a large number of papers he reviewed, it was not clear what constraints were applied to one or to both extremities of the femur.

Before developing a sensing set-up, we looked at how other authors recorded a relative interface rotation. Therefore we observed that: 1) when the map of the sensors on the specimen is not uniquely identified, the repeatability of the protocol is lost. 2) The same happens if the two points between which relative displacement is recorded are not exactly defined. 3) If these two points are not close to one another, the recording is the sum of the relative bone-stem motion and of the deformation of the portion of bone in between. 4) Due to the geometric complexity of the prosthesis, a single measuring point is not enough to give an appropriate information on the relative displacement of the stem with respect to the femur (40). 5) Because of their structure, most displacement transducers sense displacements along an assigned direction. If the mobile element and the external frame are forced to a non-coaxial relative motion (because fixed to two points that move on a non-linear path), the measurements are invalidated by misalignment errors.

Among those protocols that measure relative interface motions between the stem and the femur (6, 9,22,28-40), six (29,31-33,35) record the rigid body motion of the stem with respect to the femur, making the gross hypothesis that both the femur and the prosthesis behave like rigid bodies when loaded.

On the other hand, two kinds of relative displacement can be recorded locally at the interface at each cycle. One (permanent displacement) is the amount of motion caused by a loading cycle that is not recovered before the application of the subsequent cycle. The final migration is the sum of the contribution of each cycle; these contributions tend towards zero as the stem stabilises into the femoral canal. The other motion recorded (cyclic motion) is the one recovered at each load application and which remains constant over the entire test. In most of the cases the cyclic displacement is measured, since it can be regarded as an indicator of the state of stimulation of the interface, beyond problems of gross stem migration.

The data recorded and their statistical analysis give a significative information on the efficacy and the precision of the implemented protocols. In almost half of the cases, no statistical error estimate is given. In five cases (25,26,28,35,40) the maximum coefficient of variation lies between 20 and 50%.

In conclusion, the in-vitro protocols analysed succeed at the most in confirming the order of magnitude of the in-vivo results. Moreover, the research groups have never used the same loading conditions in an attempt to give comparable results, hence a testing protocol standardisation is far from being achieved.

MATERIALS AND METHODS

Three synthetic composite femurs (Pacific Research Labs. Inc., Vashon Island, WA) were used for this study. Previous investigations (10,45) have demonstrated that the rotational stability of the prosthesis and the sensitivity to the stem design features are of the same order of magnitude in synthetic and human femurs. Synthetic models keep the same stiffness properties (42,45) while allowing the reduction of the specimen variability compared to human bones, thus obtaining a better sensitivity for comparative studies.

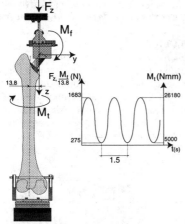

Fig 1: The femur with the reference system, the applied forces and the waveform of the dynamic load applied during the tests.

A femur-aligned reference system was marked on each specimen in a reproducible way (8,40) - Fig 1.

A collarless anatomic uncemented hip stem (An.C.A.Fit 14, Cremascoli, Italy) was implanted on each femur using standard surgical techniques (40).

Four LVDT (Linear Variable Displacement Transducers, D5/40G8, RDP Electronics, Wolverhampton, UK) were used to measure the relative rotational motion between the stem and the femur. These transducers measure linear displacements in the range of ± 1 mm with an accuracy better than 1 μm. They were connected to the data acquisition system of the loading machine through four amplifiers (S7M, RDP Electronics).

Based on the reference system, each of the 4 measuring points was determined in a reproducible way (Fig 2). A hole (∅ 1 mm) was drilled in the prosthesis and a second one (∅ 4 mm), coaxial with the former, was drilled in the cortical bone.

Each sensor was then mounted transcortically (Fig 3). Procedures and holding fixtures were studied to accurately position all the components mentioned

above. This way, the two points between which the micromotion was sensed, were placed one on the prosthesis and the other one on the femur, at a negligible but controllable distance.

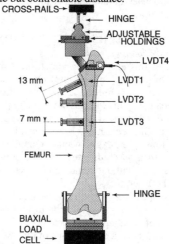

Fig 2: The specimen mounted under the servoidraulic testing machine with the 4 LVDT attached.

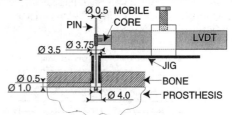

Fig 3: The section of the sensor set-up measuring the bone-stem interface relative motion.

The femurs were then placed under a biaxial servoidraulic testing machine (858 Minibionix, MTS Corporation). The constraints were designed so as to transmit all and only the load components previously described (Fig 1). Great care was taken to avoid any over constraining: the femur was held between two hinges, having the rotational axis perpendicular to one another (thus, no undesirable moment was applied). A system of two cross rails was mounted proximally to ensure that no additional horizontal force component was applied.

Prior to perform the tests described in this paper, each specimen was subjected to 1000 cycle preconditioning in order to stabilise the stems in their positions. This way, the variability between specimens was further reduced by excluding the possibility of loosening and failure during the tests.

Each loading cycle included an axial force, a bending and a torsional moment in order to reproduce the physiological stair climbing activity (13). The three components were sinusoidally varying between the values shown in Fig 1. At each cycle the LVDT outputs were recorded at maximal and minimal torsion. This way for each loading cycle the cyclic motion under dynamic load was calculated as the difference between the outputs at the maximum and minimum torques.

TEST PARAMETERS OPTIMISATION
• *Effect of the loading frequency*

The aim of this test session was to determine the influence of the loading frequency on the magnitude of the micromovements at the interface. We investigated the optimal length of the interval between cycles, defined as the trade-off value that preserves the worst loading conditions (higher displacements), while minimising the experimental execution time.

One specimen was subjected to three consecutive 100 cycle tests without removing the sensors. The intervals between the loading cycles were of 0, 7.5 and 15s respectively. The relative displacements, their mean values, standard deviations and coefficients of variation were calculated.

Results: The mean displacements of the 15s interval test were taken as a reference (Fig 4). The mean displacements at the different locations of the 0s-interval test were only 43 - 59% of those recorded in the 15s-interval test; for the 7.5s-interval test no significant difference was found in the recordings with respect to the 15s-interval.

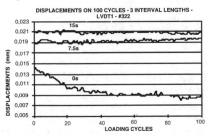

Fig 4: Micromovements recorded at 15s- 7.5s- and 0s- interval between cycles. #322 specimen.

• *Effect of the deformation across the cortical bone*

In response of an applied cyclic load, the bone undergoes an elastic deformation. If the relative motion is measured between a point on the prosthesis and a point on the outer surface of the bone, the deformation is superimposed to the relative displacement, thus affecting the recordings.

To determine the deformation characteristics across the cortical bone and to quantify its effects on the measurements, three consecutive 100 cycle tests were conducted on each specimen without removing the sensors. A FE Model (42) was used for cross-comparison and to support this analysis.

The anchorage of the LVDT was fixed with cyanoacrilate to the external surface of the bone. The pin attached to the LVDT sensed the displacements corresponding to the elastic deformations at three controlled depths across the cortex in three separate

test sessions: 0.5 mm under the external surface, 0.5 mm over the bone-stem interface and midway from

Results: The results are reported in Tab I for the four LVDT and for the specimen with the worst behaviour in term of coefficient of variation. The values given by the FE Model (42) are larger than the data recorded (reaching 9 μm at the interface). However the FE and the in-vitro analysis agree on the order of magnitude and on the distribution of the deformation across the cortical bone, which is not negligible when recording small interface motions.

	MV range (μm)	SD (μm)
External surface	0.9 - 1.4	0.2 - 0.6
Midway	1.4 - 2.4	0.4 - 1.9
Close to stem surface	2.6 - 4.1	0.2 - 2.3

Tab I. The mean displacements under dynamic load across the cortex at 3 different depths in one femur.

VALIDATION
• *Estimation of the measurement chain error*

In order to estimate the variability among repetitive measurements, the same controlled displacement (43 μm), externally imposed at the interface of a 10 cm segment of femur and a coaxial Aluminium cylinder simulating the prosthesis, was sensed with the measurement chain (anchorage + LVDT + amplifier).

Fig 5: Mechanism used to estimate the error of the chain anchorage + LVDT + amplifier.

This was achieved as shown in Fig 5. Before each measurement, the components of the LVDT anchorage system were removed, then placed again in their identical position. The entire procedure was repeated ten times.

Results: When the ten repetitions were performed without removing any component of the anchorage system, the coefficient of variation was 3.5%. When removing the transcortical pin and the screw attached to the tip of the LVDT (Fig 3), the coefficient of variation raised to 4%. Finally, when removing each component of the set-up (pin, screw, LVDT and jig) the coefficient of variation was 4.5%.

• *Measurement error in dynamic loading conditions*

The purpose of this test was to assess the robustness of the LVDT anchorage system when the specimen is loaded. One implanted femur was subjected to 1000 cycles. For each LVDT the relative displacement was calculated every 100 cycles.

Results: The coefficient of variation was between 2.4 and 6.7% among the ten groups of 100 cycles. The displacements tended to decrease from the first to the last loading cycle, due to the progressive stabilisation of the prosthesis in its position. For this reason the data distribution around the mean values does not give a good statistical representation of the measurement error. A linear regression analysis allows to individuate the straight line that best approximate this trend. Thus, the regression residuals represent more significantly the measurement error. The mean values of the residuals were typically of 0.2 - 0.8 μm, in all cases lower than 2.3 μm (Tab II).

Range among the four LVDT	
MV - mean value (μm)	14.8 - 65.4
SD - standard deviation (μm)	0.9 - 5.3
CV - coefficient of variation (%)	2.4 - 6.7
Regression residuals MV (μm)	0.2 - 0.9
Regression residuals maximum (μm)	0.4 - 2.3

Tab II. Estimation of the measurement error in dynamic loading conditions. The last two rows refer to the residuals of the original ten measurements with respect to the estimated linear regression values.

• *Repeatability of the protocol*

To verify the repeatability of the measuring system, a series of 10 tests was performed on each of the 3 specimens, each consisting in 100 loading cycles. Before each test the transducers were removed, then placed again on the specimen. This way for each specimen and for each sensor 10 mean values were obtained.

Range among the four LVDT	
MV - mean value (μm)	16.5 - 43.4
SD - standard deviation (μm)	1.1 - 3.5
CV - coefficient of variation (%)	2.6 - 21.0
Regression residuals MV (μm)	0.7 - 1.8
Regression residuals maximum (μm)	1.8 - 4.9

Tab III. Assessment of the protocol repeatability. Data for the worst femur in term of coefficient of variation. The mean values are calculated among the 10 mean values, obtained from 10 tests of 100 cycles each. The last two rows refer to the residuals of the linear regression based on the 10 mean values.

Results: The final result of these experiments is the error of the measurement chain plus the LVDT anchorage system to the specimen. Again, a regression analysis was performed (see Tab III).

The mean value of the residuals was between 0.2 and 1.8 μm in all cases except for LVDT4, for which the mean residual value in one specimen reached 5 μm (Fig 2).

DISCUSSION

In this paper we presented the development and the validation of an in-vitro protocol for the analysis of the torsional stability of uncemented femoral stems.

The goal was to increase the reproducibility and the repeatability. Hence particular care was taken in the standardisation of the procedure steps and in the reduction of the sources of experimental variability.

This has lead to the following choices: use of synthetic composite femurs; standardised surgical implantation technique; standardised reference system definition; well-posed specimen holding system.

Besides, the protocol had to simulate the physiological loading conditions that are most risky for the rotational stability of the implant. The final set-up of the test was fixed to 1000 loading cycles to reproduce the amount of torsional load that the prosthesis undergoes in the first three months postoperative (44).

Precision and sensitivity have been obtained through: 1) use of high-precision transducers; 2) use of standardised and uniquely defined sensor anchorage system to the specimen; 3) evaluation of the measurement error and analysis of its causes; 4) reduction of the sources of error.

Obviously a certain amount of error is still present, due to the approximation of the arc to the chord, as the LVDT measures a linear displacement instead of a rotation.

The first test session on the protocol optimisation has demonstrated that the presence of a resting interval between successive loading cycles allows the specimen to recover all (or part of) the elastic deformation. This generates higher shear motion at the femur-implant interface, hence worse loading conditions. Moreover the resting interval interposition makes the test more physiological: a patient would not climb 2000 stairs (1000 per leg) one after the other at a pace of one step every 1.5s. The best trade-off between experimental time and close reproduction of the worst loading condition has proved to be obtained with 7.5s interval between cycles.

The results on the deformation across the cortical bone have proved that at the interface level the motion due to the deformation is of 2.6 - 4.1 μm, thus not negligible. When the relative motion is not measured close to the interface, the recordings under dynamic load are affected by an error of some micrometers, which may be unacceptable for precise measurements.

Finally, the three validation test sessions have proved the excellent behaviour of the measurement set-up under dynamic load. The precision found disassembling the whole set-up is not significantly worse than the one of the LVDT alone, as the measurement error is always less than 4.9 μm (as it results from the test session on the repeatability of the protocol). This error is acceptable if we consider that the relative rotational displacement typically ranges from 50 to 500 μm.

BIBLIOGRAPHY

1. Gruen T.A., McNeice G.M., Amstutz H.C.: "'Modes of failure' of cemented stem-type femoral components", Clin Orthop Rel Res 141: 17-27, 1979
2. Crowninshield R.D., Johnston R.C., Andrews J.G., Brand R.A.: "A biomechanical investigation of the human hip", J Biomech, 11: 75-85, 1978
3. Berme N., Paul J.P.: "Load actions transmitted by implants", J Biomech 12: 268-272, 1979
4. Charnley J.W., Kettlewell J.: "The elimination of slip between prosthesis and femur", J Bone Jt Surg 47-B: 56, 1965
5. Andriacchi T.P., Galante J.O.: "A study of lower limb mechanics during stair climbing", J Bone Jt Surg 62A: 749-757, 1980
6. Burke D.W., O'Connor D.O., Zalensky E.B., Jasty M., Harris W.H.: "Micromotion of cemented and uncemented femoral components", J Bone Jt Surg, Vol 73-B: 33-37, 1991
7. Pilliar R.M., Lee J.M., Maniatopoulos C.: "Observations on the effect of movement on bone ingrowth into porous-surfaced implants", Clin Orthop Res Soc, 208: 108-113, 1986
8. Ruff C.B., Hayes W.C.: "Cross-sectional geometry at pecos pueblo femora and tibiae - a biomechanical investigation: I. method and general patterns of variation", American J Phys Anthrop 60: 359-381, 1983
9. Schneider E., Kinast C., Eulenberger J., Wyder D., Eskilsson G., Perren S.M.: "A comparative study of the initial stability of cementless hip prostheses", Clin Orthop Rel Res 248: 200-210, 1989
10. Ebramzadeh E., McKellop H., Wilson M., Sarmiento A.: "Design factors affecting micromotion of porous-coated and low modulus total hip prostheses", J Bone Jt Surg 12(2): 464-465, 1988
11. Rydell N.W.: "Forces acting on the femoral head prostheses", Acta Orthop Scand: 96-125, Munksgaard, Copenhagen, 1966
12. Bergmann G., Graichen F., Rohlmann A.: "Hip joint loading during walking and running, measured in two patients", J Biomech 26: 969-990, 1993
13. Bergmann G., Graichen F., Rohlmann A.: "Is stair-case walking a risk for the fixation of hip implants?", J Biomech 28(5): 535-553, 1995
14. Hodge W.A., Carlsson K.L., Fijan R.S., Riley P.O., HArris W.H., Mann RW.: "Human in-vivo acetabular pressure measurement:18 month follow-up", Orthop Res Soc: 320
15. Davy D.T., Kotzar G.M., Brown R.H., Heiple K.G., Goldberg V.M., Heiple K.G. JR., Berilla J., Burstein A.: "Telemetric force measurements across the hip after total hip after total arthroplasty", J Bone Jt Surg, 70-A: 45-50, 1988
16. Hodge W.A., Carlsson K.L., Fijan R.S., Riley P.O., Harris W.H., Mann R.W.: "Contact pressures from an instrumented hip endoprosthesis", J Bone Jt Surg 71-A(9): 1378-1385, 1989
17. English T.A.: "In vivo measurement of hip forces using a telemetric method", J Bone Jt Surg 61-B(3), 1979
18. Mjöberg B., Hansson L.I., Selvik G.: "Instability of total hip prosthesis at rotational stress: a roengten stereophotogrammetric study", Acta Orthop Scand 55: 504-506, 1984
19. Wroblewski B.M.: "The mechanism of fracture of the femoral prosthesis in THR", Internat Orthop 3: 137-139, 1979
20. Paul J.P., MC Grouther D.A.: "Forces transmitted at the hip and knee joint of normal and disabled persons during a range of activities", Acta Orthop Belg, 41: 78-88, 1975
21. Cheal E.J., Spector M., Hayes W.C.: "Role of loads and prosthesis material properties on the mechanics of the proximal femur after total hip..", J Orthop Res 10: 405-422, 1992
22. Sugiyama H., Whteside L.A., Kaiser A.D.: "Examination of rotational fixation of the femoral component in total hip arthroplasty", Clin Orthop Rel Res, 249: 122-128, 1989
23. Shelley F.J., Anderson D.D., Kolar M.J., Miller M.C., Rubash H.E.: "Physical modelling of hip joint forces in stair", Proc Instn Mech Engrs [H]: J Eng Med 210: 65-68, 1996
24. McKellop H., Ebramzadeh E., Niederer P.G., Sarmiento A.: "Comparison of the stability of press-fit hip prosthesis femoral stems using a synthetic model femur", J Orthop Res 9: 297-305, 1991
25. Martens M., Van Audekercke R., Delport P., De Meester P., Mulier J.C.: "The mechanical characteristics of the long bones of the lower extremity in torsional loading", J Biomech 13: 667-676, 1980
26. Markolf K.L., Amstutz H.C., Hirschowitz D.L.: "The effect of calcar contact on femoral component...", J Bone Jt Surg 62-A: 1315-1323, 1980
27. Allen J., Phillips T.W., McDonald P.: "Mechanical analysis of the rotational fixation of cementless femoral stems", J Bone Jt Surg 10(1): 89-90, 1986
28. Walker P.S., Schneeweist D., Murphy S., Nelson P.: "Strains and micromotions of press-fit femoral stem prostheses", J Biomech 20: 693-702, 1987
29. Schneider E., Eulenberger J., Steiner W., Wyder D., Friedman R.J., Perren S.M.: "Experimental method for the in-vitro testing of the initial stability of cementless hip prosthesis", J Biomech 22: 735-744, 1989
30. Nunn D., Freeman M.A.R., Tanner K.E., Bonfield W.: "Torsional stability of the femoral component of hip arthroplasty", J Bone Jt Surg 71-B: 452-455, 1989
31. Maloney W.J, Jasty M., Burke D.W., O'Connor D.O., Zalenskie B., Brangdon C., Harris W.H.: "Biomechanical and histological investigation of cemented total hip arthroplasty", Clin Orthop Rel Res 249: 129-140, 1989
32. Gustilo R.B., Bechtold J.E., Giacchetto J., Kyle R.F.: "Rational, experience and results of long-stem femoral prostehsis", Clin Orthop Rel Res 249: 159-168, 1989
33. Kenneth J.: "In-vitro study of initial stability of a conical collared femoral component", J of Arthropl 7, 1992
34. Gilbert J.L., Bloomfeld R.S., Lautenschlager E.P., Wixson R.L.: "A computer-based biomechanical analysis of the three-dimensional motion of cementless hip prostheses", J Biomech 25: 329-340, 1992
35. Phillips T.W., Nguyen L.T., Munro S.D.: "Loosening of cementless femoral stems: a biomoechanical analysis of immediate fixation with loading vertical, femur horizontal", J Biomech 24(1): 37-48, 1991
36. Sugiyama H., Whiteside L.A., Engh C.A.: "Torsional fixation of the femoral component in total hip arthroplasty", Clin Orthop and Rel Res 275: 187-193, 1992
37. Callaghan J.J., Fulghum C.S., Glisson R.R., Strenn S.K.: "The effect of femoral stem geometry on interface motion in uncemented porous-coated prostheses", J Bone Jt Surg 74-A: 839-848, 1992
38. Davey J.R., O'Connor D.O., Burke D.W., Harris W.H.: "Femoral component offset: its effect on micromotion in stance and stairclimbing loading", Trans 35th Ann Meet Orthop Res Soc, Las Vegas: 409, 1989
39. Ohl M.D., Whiteside L.A., McCarthy D.S., White S.E.: "Torsional fixation of a modular femoral hip component", Clin Orthop Rel Res 287: 135-141, 1993
40. Harman M.K., Toni A., Cristofolini L., Viceconti M.: "Initial stability of uncemented hip stems: an in-vitro protocol to measure torsional interface motion", Med Eng Phys 17(3): 163-171, 1995
41. Pawluk R.J., Greer J., Tzitzikalakis G.I., Michelsen C.B.: "The effect of experimental models on femoral strain distribution", 30th Ann Meet Orthop Soc Book of Abstracts, Atlanta: 54, 1984
42. Toni A., Viceconti M., Cristofolini L., Acquisti G., Schneirer R.: "Torsional stability of total hip arthroplasty: in-vitro and fem analysis with new trends for the future" Comp Meth Biomech Biomed Eng 1998 (in press)
43. Keaveny T.M., Bartel D.L.: "Fundamental load transfer patterns for press-fit, surface-treated intramedullary fixation stems", J Biomech 27: 1147-1157, 1994
44. McLeod P.C., Kettelkamp D.B., Srinivasam V., Henderson O.L.: "Measurement of repetitive activities of the knee", J Biomech 8: 369-373, 1975
45. Cristofolini L. "A critical analysis of stress shielding evaluation of hip prostheses" Crit Rev in Biomed Eng, 1998 (in press)

862

Materials behaviour and testing

Plenary lecture

Experimental Mechanics, Allison (ed.)© 1998 Balkema, Rotterdam, ISBN 90 5809 014 0

Aeroengine materials: Past, present and future

N.E.Glover & M.A.Hicks
Rolls-Royce plc, Derby, UK

ABSTRACT: Advanced materials have been a key technology throughout the history of the aeroengine. The drive for performance has demanded materials with ever greater high temperature strength whilst at the same time having the lowest possible density. Current materials are however now nearing the limits of their potential. New materials are emerging which offer a step change in capability, however these technologies must demonstrate their value within the current cost dominated market place.

This paper reviews the development of aeroengine materials drawing on examples of components for which materials, or associated manufacturing technologies have led to significant improvements in performance. Materials for the future will be discussed, including both the evolutionary development of existing systems and the so called revolutionary material classes. Not neglected are the manufacturing and modelling technologies which will underpin any increase in materials performance or reductions in development lead times.

1 INTRODUCTION

Since the development of the aircraft gas turbine in the 1930s dramatic improvements have been made in both thrust-to-weight ratio and specific fuel consumption. These have only been possible as a result of continual advances in both materials and manufacturing technologies. This situation will not change in the immediate future. The continuing drive for performance makes ever greater demands upon engine materials providing an ongoing challenge to materials scientists and engineers.

The service requirements placed upon aeroengine materials are extremely severe and often call for compromise. High strength at temperature is critical but must be coupled with toughness and the lowest possible density. In addition materials technology must increasingly be employed in an efficient and cost-effective manner in an ever more cost-conscious market place.

2 1930s TO THE PRESENT DAY

The composition of today's aeroengine is very different from that of fifty years ago. Steels, which formerly dominated, have been displaced due to their limited high temperature strength, whilst aluminium alloys have virtually disappeared from the engine. In their place have emerged two major alloy classes; the nickel based superalloys, with superb high temperature capability, and the titanium alloys offering a combination of mechanical strength and low density, figure 1.

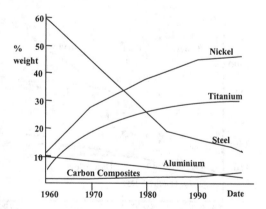

Figure 1. Trends in Jet Engine Materials Usage

The introduction of titanium and nickel blading into the compressor stages, for example, in place of aluminium and steel has allowed increases in compressor exit temperature of over 300°C. Steels now remain only in applications such as bearings and shafts where their high strength and surface hardenability are advantageous. The last fifty years have also seen the introduction of composite materials. Although not yet fulfilling early predictions these materials now make up a large proportion of the engine structure and hold considerable promise for the future.

The following discussion outlines the major developments leading to the current state of the art in aeroengine materials and assesses potential materials for the future. Illustrations are drawn from three key engine modules; the high pressure turbine, the fan and the high pressure compressor. These applications cover a wide range of temperatures and control three parameters crucial to the performance of an aeroengine: the turbine entry temperature, the fan diameter and the compressor delivery temperature. They also highlight the integrated approach necessary to materials development, component design and manufacturing method.

2.1 *The Nickel Based Superalloys*

Nickel based superalloys have come to dominate the high temperature stages of aeroengines, from the high pressure compressor through the combustor and turbine stages to the exhaust outlet. This success arises from their unique ability to operate at high stresses at temperatures in excess of 85% of their melting temperature. To achieve this the nickel based superalloys combine a remarkable retention of mechanical properties, arising from the Ni_3Al, γ' strengthening precipitate with superb oxidation resistance conferred by a mixed Cr_2O_3 and Al_2O_3 scale.

Two components which have driven the development of the nickel based superalloys are the high pressure turbine blade and disc.

2.2 *The High Pressure Turbine Blade*

Turbine blade design represents a conflict between aerodynamic efficiency, weight, vibration control, integrity and cost. Blades are required to withstand centrifugal loads of up to ten tonnes while operating in a gas stream at temperatures in excess of the melting point of the alloy. Operation in this environment makes severe demands on both the mechanical properties and environmental stability of the blade system and is only possible through the close integration of design, materials and manufacturing.

Turbine blade evolution has illustrated the hand in hand advance of materials and manufacturing techniques. This development and the increase in temperature capability achieved is shown in figure 2.

As higher strength alloys were developed the manufacture of blades via a forged route became increasingly difficult. The higher levels of alloying elements necessary to increase strength both raise the solvus temperature of the strengthening precipitate and lower the melting point of the alloy. This narrows the forging range to the point where it becomes impossible to forge without the danger of incipient melting.

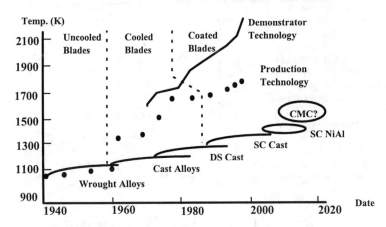

Figure 2. Progress in Turbine Materials and Technologies and Associated Increases in Turbine Entry Temperatures (K)

The introduction of investment casting provided an alternative economic route for the production of near net shape blades. Techniques were developed to allow such castings to be manufactured incorporating a complex system of air passages to achieve both internal and surface film cooling, enabling metal temperatures to be kept significantly lower than the surrounding gas stream.

A step change in temperature capability was realised through the introduction of directional solidification, eliminating transverse grain boundaries, a source of weakness in a creep dominated application. It was then a natural progression to the total elimination of grain boundaries via single crystal casting. This incidentally gives additional benefits through the elimination of the need for grain boundary strengthening elements and the resultant increase in the alloy's melting temperature.

In a single crystal casting the crystal orientation is critical. The <100> orientation which arises naturally from the casting process gives a significant advantage under thermal fatigue conditions due to its low elastic modulus. Seeded casting also makes possible the control of transverse orientation. This enables the selection of elastic moduli to restrict high frequency resonance which may give rise to high cycle fatigue.

With increasing blade operating temperatures the intrinsic resistance of the metal to environmental attack is no longer sufficient. Protective surface coatings are then required to give the necessary oxidation and corrosion resistance.

The first class of coatings, employed extensively on current turbine blades, are diffusion coatings such as the aluminides. Enhanced oxidation resistance comes from the introduction of an aluminium rich surface layer by a pack cementation process. The addition of platinum and precious metals can reduce degradation by diffusion and further enhance oxidation resistance.

More recently the MCrAlY overlay coatings have been developed, where M represents Ni or Co. Here chromium and aluminium provide oxidation resistance while the presence of yttrium improves scale adhesion. These coatings are applied by plasma spraying and offer improvements in both hot corrosion resistance and mechanical properties over diffusion coatings.

Overlay coatings are also employed as bond coats to prevent spalling of the latest category of coatings, the thermal barrier coatings (TBCs). TBCs are low thermal conductivity ceramics which prevent the flow of heat from the gas stream to the underlying blade. This maximises the benefit obtained from blade cooling and hence offers a potential increase in operating temperature of over 100°C.

Originally magnesium zirconate but more recently of yttria stabilised zirconia, TBCs have been used in the combustion chambers of the RB211 since 1975. However, advances in bond coat technology and ceramic deposition have allowed their use to be extended to rotating parts.

Coatings are applied by electron beam physical vapour deposition, EBPVD, to develop the columnar grain microstructure necessary to resist thermal and mechanical strains, particularly around the blades leading and trailing edges. The coating system must be durable since any breach would expose the blade to gas stream temperatures which could not be endured by the metal alone having serious consequences on the component life. Bond coats are therefore a key technology and the subject of much current research. The bond coat provides an oxidation resistant layer between metal and TBC thus preventing the formation of alumina which ultimately leads to spalling. The bond coat also has a role in controlling thermal expansion mismatches between alloy and coating.

Full exploitation of coating technology to obtain the maximum benefit requires the integral consideration of the coating in the component design process. The coating system must be compatible with the requirements of aerodynamics, mechanical integrity and blade cooling and must be achieved in a cost-effective manner suitable for mass production. All of these considerations have lead to the unique, cost-effective TBC system designed for use in Rolls-Royce Trent engines.

The total effect of all the advances in blade technology outlined above, coupled with alloy development from the first to the third generation alloys, has been to increase metal temperatures by around 300°C over the last fifty years: a figure which can be doubled when the temperature of the gas stream itself is being considered.

2.3 The High Pressure Turbine Disc.

For disc applications as for blades, traditional steels have been superseded by nickel based superalloys because of their exceptional high temperature mechanical properties. Turbine discs operate at lower temperatures than blades, as they are not in the direct gas path leaving the combustor, however they must attain the most stringent levels of mechanical integrity. The failure of a turbine disc may not be contained by the engine casing and would seriously hazard the aircraft.

The development of disc alloys has traditionally been driven by the High Pressure Turbine Disc. The objective always being for a hotter disc with an equivalent cyclic life, requiring higher strength materials. The development of nickel alloys within Rolls-Royce has allowed the current Trent 700 to run at a 40% higher stress and 100°C hotter than the original RB211-22B. Further increases are being realised from the introduction of high strength U720Li in the EJ200 and the BR700 series.

Recently the High Pressure Compressor, HPC, disc has begun to require the same capability. As a result the new Rolls-Royce powder alloy RR1000 will make its entry into service as an HPC material in future Trent engines.

The design of a turbine disc material such as RR1000 highlights the trade-off necessary between the various critical mechanical properties, figure 3. The disc is required to contain high tensile hoop stresses and to be resistant to low cycle fatigue; both considerations which would lead to a high strength, fine grain microstructure. The material is also however required to have a slow crack growth rate and a high creep strength, particularly at the rim. These factors would be adversely affected by the selection of a fine grain microstructure. The development of an alloy composition and its processing route is therefore always to some extent a compromise.

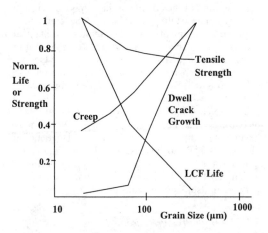

Figure 3. Schematic of Mechanical Property Trades in Nickel Disc Alloys - (after J. C. Williams & J. M. Theret, 3rd ASM Int. Paris Conf. 1997)

To assure mechanical integrity requires a thorough understanding of materials behaviour at all engine operating conditions and a strictly controlled manufacturing route. Traditionally the manufacture of turbine discs has been via a cast and wrought route, however as for blades the development of more advanced materials has led to difficulties. Again with higher alloy contents segregation at the ingot stage becomes problematic. The solution has been to move to a powder processing route involving the atomisation of a molten stream of metal in an inert atmosphere. Here rapid solidification and the fine powder size restricts segregation. Consolidation is achieved by HIPping followed by extrusion to provide a fully dense billet for subsequent forging. This method has now been adopted across the industry.

2.4 Titanium Alloys

Titanium alloys are very attractive materials for aerospace offering high strength to intermediate temperatures at a density almost half that of steel or nickel based alloys. As a result they have been adopted widely in the fan and compressor stages of aeroengines for both disc and blade applications.

The most common alloy in current use is Ti-6Al-4V which has been applied to industrial as well as aerospace applications for over forty years. This single alloy accounts for about two thirds of all titanium usage worldwide.

This is not to say that the development of advanced titanium alloys has remained static. As with nickel alloys there is an ever present drive to apply titanium at higher and higher temperatures. This has led to the development of high temperature alloys such as IMI834, figure 4, a field in which the UK currently has a world lead.

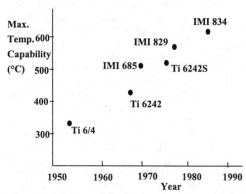

Figure 4. Titanium Alloy Development

2.5 Titanium for Disc Applications

In 1960/61 titanium alloys began replacing steel for compressor discs in the JT3D and Olympus 320 engines. Since then their dominance has increased to the stage where the Trent 700 has an entirely titanium compressor drum.

Both fan and compressor discs are made from Ti 6-4 up to its temperature limit of approximately 350°C. As with nickel, the application of higher strength alloys is problematic because of the concomitant reduction in defect tolerance.

Nonetheless recent processing advances have allowed the established alloy Ti 6-2-4-6 to be produced with a defect tolerant microstructure and with a useable strength significantly higher than 6-4. This enables weight savings in excess of 15% in a typical intermediate pressure compressor.

The development of high temperature alloys such as IMI834 has extended the range of Titanium alloys significantly up to 630°C. This has been achieved through the careful optimisation of the balance of primary alpha and transformed beta to give the elusive combination of high tensile strength, fatigue resistance and enhanced creep behaviour. As a result a wholly titanium compressor drum, such as that in the Trent 700, can be designed, giving further weight gains over drums using nickel based alloys in the later stages.

Into the future the drive for reduced weight and increased temperature will continue. This will lead to the introduction of integrally bladed discs (blisks) realising significant cost and weight savings. The elimination of blade fixing features could lead to weight savings as large as 50%. For a small engine blisks may be produced from a single forging but larger engines will require the use of linear friction bonding technology to attach the aerofoils to a separate disc forging. The use of integral blading will also raise issues of reparability when aerofoils sustain foreign object damage, FOD. Here again linear friction bonding will be a key technology.

Further advances in the application of titanium alloys may be achieved through the introduction of titanium matrix composite technology. This allows the extension of the blisk concept one stage further and the introduction of the bling which will be discussed later.

2.6 Titanium Blades

The Rolls-Royce wide chord fan blade is perhaps the best example of the application of titanium, coupled with advanced processing techniques, to give a significant service advantage.

Fan blades are subject to all of the same design pressures as a turbine blade with the added requirement of resistance to foreign object impact. For many years civil turbofans used narrow chord solid titanium blades with clappers at mid span which linked to form a stiffening ring. These clappers however impede the air flow and so reduce efficiency. If the clappers are to be removed a wide chord design is required to retain mechanical stability. This gives the additional benefit of reduced numbers of blades per set. At the same time there is a requirement for increased fan diameter. A critical contributor to engine thrust, this has increased over the years from 86" in the RB211 to 110" in the Trent 800 making ever more stringent demands upon the mechanical strength of the blade.

The Rolls-Royce solution to this problem is the hollow wide chord titanium fan blade, which delivers an enhanced mechanical performance and a net weight decrease.

The first hollow fan blades were manufactured by a brazing technique and consisted of two titanium skins reinforced with a honeycomb core. These have been superseded by the current, advanced diffusion bonded and super-plastically formed blades (SPF/DB), with integral stiffening. This process is an excellent example of the vital integration of materials and manufacturing technology, exploiting the SPF and DB capability of fine grain Ti 6-4.

The component is manufactured from three sheets of material representing the two outer skins and the internal corrugated structure. An inhibitor is applied, to define the internal structure, and then the three pieces bonded in a high temperature pressure vessel. The blade is twisted and the cavity inflated at SPF temperatures, using an inert gas in a shaped die to yield its final aerofoil shape. The total process results in bonds with properties equivalent to the parent material and an internal stiffening structure which bears its share of the centrifugal load. Compared to a clapper design this gives a fan module which is 24% lighter, an overall engine weight benefit of 7%, with a significant increase in FOD resistance over competitor designs.

The process is also eminently suitable to the manufacture of state of the art blades with complex swept geometries and may in the future be combined with titanium metal matrix composite technology to bring about further advances in fan blade materials.

An alternative approach to the design of wide chord fan blades is that taken by G.E., whose G.E.90 has a carbon composite fan blade. However this requires a significant amount of titanium sheathing in order to satisfy the demands of airworthiness and can be seen as convergent with the hollow titanium fan blade utilising the composite as an internal stiffening structure.

3 CURRENT TECHNOLOGY DRIVERS

Progress in materials technology has, in the past, largely been driven by the desire for enhanced performance. Today however the business climate has begun to change.

The global military market has seen a steady decline in turnover partially as a result of the end of the cold war. The demand for high technology levels and increased thrust to weight ratio for military engines still exists but with increasing emphasis on both lifetime and first costs.

At the same time the civil market has become increasingly competitive with three major manufacturers competing for market share. Historically the struggle for competitive advantage has centred around reduced fuel consumption with rapid gains being made in the early years largely through the introduction of bypass engines. Today this improvement is slowing down, for example the three engines offered for the Boeing 777 all give a fuel burn within about 3%, figure 5.

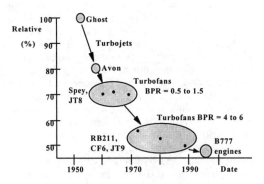

Figure 5. Specific Fuel Consumption Trend with Time.

The world civil airline business is in a period of rapid recovery and growth following the recession of the late 1980s and early 1990s but airlines still remain under severe financial constraints. This translates to an unrelenting pressure on aeroengine prices and costs. Improved performance is still desirable but future competitive advantage is increasingly dependent upon unit cost and product development time.

Consequently current technology drivers are increasingly cost based. Performance dominated military requirements no longer drive the development of new technology with civil engines benefiting from the spin off. Today's drivers are dominated by the civil engine market and the financial pressure exerted by the airlines. The major pressure is therefore for cost-effective technologies to enable further thrust growth, fuel consumption reduction and unit cost reduction. In such a climate it is increasingly important to identify technologies which wherever possible are dual use.

The requirement for reduced product development times will necessitate a parallel reduction in timescales for materials development. Together with the desire to reduce development costs this will see the increasing importance of modelling techniques both to guide research, cutting down on

empirical testing, and to better exploit materials in service.

The improved materials required for future aeroengines may come from either of two routes; the further development of conventional alloys or the application of new materials. Both of these approaches; the so called evolutionary development of existing alloys and the revolutionary development of new materials are discussed in greater detail below.

4 EVOLUTIONARY MATERIALS

Current nickel and titanium alloys have been developed to the point where further advances are becoming increasingly difficult and expensive to achieve. At the same time research budgets are ever more limited. New materials are now becoming available which offer a step change in capability, however before these materials can displace conventional alloys they must demonstrate the ability to deliver cost-effective performance benefits to the whole engine system. Increased unit costs and operating costs can negate any economic benefit derived from improved cycles.

4.1 The Development of Conventional Alloys

Efforts to further exploit conventional materials continue although increases in temperature capability and strength beyond today's levels will be difficult to achieve. The nickel based superalloys are ultimately limited by the melting point of nickel while conventional titanium alloys are reaching limits imposed by both creep considerations and long term stability.

It seems likely that future developments will concentrate on the adaptation of existing materials to niche applications and the production of cheaper alloys with performance parity. Such cost reductions are important in a situation where raw materials account for approximately 30% of the cost of a turbine disc and 20% of a blade. The standardisation of materials can also contribute significantly to reduced costs throughout the supply chain. Engine suppliers benefit through the need to provide support for fewer materials and the reduction of raw material price which results from the increase in production volumes.

Further advances in performance will require a detailed understanding of material behaviour and processing and will benefit greatly from the application of computer modelling techniques. Such an understanding may also contribute to the more complete exploitation of existing properties and the safe employment of standard, cheaper, materials to higher stress levels.

4.2 *Modelling Techniques.*

Computer modelling and simulation will have a central role in any future alloy development programme.

A holistic modelling strategy is being developed whereby thermodynamic models can be used to select promising alloy compositions, kinetic models can then tailor the processing route to produce designed microstructures and the resulting properties can be predicted using mechanical models, figure 6. This enables the behaviour of a new material to be predicted from its chemical composition and processing route in a 'virtual design process' which reduces empiricism and hence yields dramatic benefits in programme costs and timescales.

Process models also play a vital role in the optimisation of manufacturing routes.

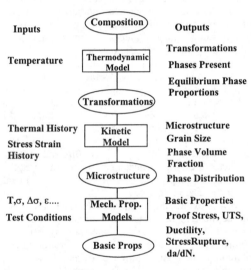

Figure 6. Modelling Strategy

Of the models required to support this strategy thermodynamic models and process models are already established while kinetic and mechanical property models are under development.

4.3 *Thermodynamic Models.*

Thermodynamic models, allowing the prediction of equilibrium alloy structures from chemistry, have been developed by Rolls-Royce since the early 1990s. The volume fraction of each of the phases present can be determined, together with data such as phase compositions and solvus temperatures. The capability exists to model nickel alloys, nickel aluminides, titanium alloys and titanium aluminides. Extensive thermodynamic databases have been generated to provide the input to these models. The nickel database for example incorporates fourteen elements.

Rolls-Royce has made extensive use of this modelling capability during the development of the disc alloy RR1000 and in support of several other blade and disc alloys.

4.4 *Process Modelling*

A considerable capability for process modelling exists, mostly utilising finite element techniques. This encompasses a large range of processes throughout the manufacturing sequence, key examples of which are outlined below.

Melting processes such as Vacuum Induction Melting (VIM) and Vacuum Arc Remelting (VAR) can be modelled with the goal of improving process control and hence reducing the occurrence of defects.

The modelling of thermal evolution and grain growth during casting is possible and is applied particularly to single crystal casting and directional solidification.

Modelling of compressor and fan blade manufacture, as well as nickel and titanium disc forging, enables both temperature and strain distribution to be studied and the process optimised to produce the desired microstructure. These models can be used to predict residual stresses and distortion, and are incorporated in the selection of heat treatment.

Modelling of welding processes including electron beam, inertia and linear friction welding is under development. These models will again predict both residual stresses and the subsequent component distortion.

Consolidation processes used during the manufacture of MMC blings, discussed below, can also be modelled, again with the goal of predicting residual stresses and optimising the manufacturing process.

4.5 *Neural Network Models*

Recent developments in the use of neural networks suggest an alternative approach to materials modelling. Neural nets offer the potential to predict the properties of new alloys from basic variables such as composition by exploiting the vast database of existing alloy properties.

Some success has been achieved with this method for the prediction of tensile strength in nickel based superalloys although modelling complex phenomena such as creep and fatigue may well prove more difficult.

5 REVOLUTIONARY MATERIALS.

There are two major classes of material with the potential to replace conventional metallics; the intermetallics and composite materials.

If the potential of these materials is to be realised a cost-effective performance benefit will need to be demonstrated. Central to this will be the existence of an established manufacturing base with an economic processing route together with adequate design and lifing methodologies.

5.1 Composite Materials

The aerospace industry is drawn towards composite materials because of their high specific strength and stiffness. Of the three groups the polymer matrix composites (PMCs) currently account for almost all the composite materials used in aeroengines, despite being limited to relatively low temperatures, generally less than 150°C. Metal matrix composites (MMCs) are attractive materials for application at intermediate temperatures whereas ceramic matrix composites (CMCs) offer exciting possibilities for very high temperature applications where loads are modest.

If all these classes of composite materials achieve their full potential then the gas turbine engine of the 2020s could be a largely composite engine. Polymer matrix systems could constitute much of the nacelle, fan system, shaft support structures, casings and some stators. The intermediate and high pressure compressor rotors will be MMC while the combustor can, nozzles and some of the rear structure will exploit the advantages of CMCs. There are however real constraints to the application of composites to more sensitive components. Analysis methodologies lack the maturity of those for metal structures while the necessary understanding of fatigue and lifing is not yet established. Issues such as long term service exposure, fire resistance and impact tolerance will also all require further study.

With all composite materials there is a need to establish a large scale manufacturing base if mature manufacturing processes are to be developed with the concomitant reduction in costs. The business case for investment in such a manufacturing base must however be justified by volume projections for the market and currently this is not the case. Indeed it is uncertain whether the aerospace industry itself would ever require the volumes necessary to constitute a sufficient market. In addition the generation of other applications and markets will be difficult before the manufacturing base is in place and the viability of the technology has been demonstrated.

Polymer matrix composites have been used in aeroengines since the early 1960s. In fact glass reinforced epoxy constituted 40% by volume of the RB162, including the compressor stators and rotor blades. This became the only engine to enter service with a substantially composite compressor as an auxiliary power unit for improved take-off in the Trident IIIB airliner. Meanwhile interest in carbon fibre composites was stimulated by their successful application for fan blades on some Rolls-Royce Conway engines for VC10 airliners.

The usage of composites accelerated in the 1970s with the arrival of the high bypass ratio engine where the nacelle became a substantial structure in its own right. Composites, in particular carbon composites, offer a low weight solution to the design of the nacelle, where they may constitute 25% of the weight and 20% of the cost of the whole engine.

Today on smaller engines such as the BRR700 the bypass duct is a sophisticated composite structure used as a structural member supporting the engine in the airframe. Such applications are beginning to edge composites into the field of structural parts.

Further expansion of these materials into the core of the engine and components critical to its successful operation will require a greater maturity from both the materials and the associated design methodologies. The higher temperatures will demand the development of a high temperature matrix system capable of being economically moulded into complex geometries. Fire and damage tolerance are also barriers to be overcome, yet the rewards of replacing heavier expensive metal fabrications are large.

MMCs extend the potential temperature range for the application of composite materials. The most promising metal matrix composite system for aeroengines is silicon carbide fibre-reinforced titanium. The high performance of the fibres means that this system can yield a 50% increase in strength over current high temperature titanium alloys and a twofold improvement in stiffness together with a reduction in density. This offers an intermediate temperature system for the compressor section of the engine to replace conventional titanium alloys. Here blades, vanes, casings and discs are all potential applications.

Titanium fan blades have long been considered as possible candidates for fibre reinforcement. Increased strength would allow larger blades with higher rotational speeds giving greater thrust. A reduced number of lighter blades could also be envisaged with knock-on weight savings for discs and casings. The use of fibre stiffening reinforcement also enables the blade to be tuned for optimum performance and to eliminate unwanted vibration modes. Unfortunately although the required manufacturing technology is compatible

Schematic sections showing the effects on disc design and weight savings through the use of TiMMC

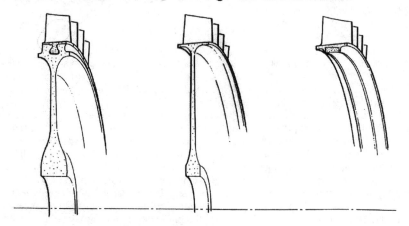

Figure 7. The Evolution of the MMC Bling

with the hollow SPF/DB fan blade process the costs of such a route are at present prohibitive.

The application which is seen to yield the greatest benefit from the application of TMMCs is the compressor bling, an integrally bladed ring. Here the bore of the disc can be eliminated, as the hoop stress can be born by the fibre reinforced rim alone, giving very significant weight savings of the order of 40% over a conventional titanium blisk design. Figure 7 illustrates a bling, together with a blisk and a conventional disc design. The present state of development of bling technology is that demonstration components have been manufactured and the Allison Engine Company (part of the Rolls-Royce Group) has successfully run the first demonstration of TMMC blings in an engine.

This application is ideal for unidirectionally reinforced TMMCs since the radial loads in the weaker transverse direction can be kept low whilst exploiting the good longitudinal strength. Hence the need to form the fibres into complex shapes, a technology that is at present a challenge, is avoided. The most promising manufacturing route for critical components at present is metal coated fibre processing. The metal coating is applied directly to the silicon carbide fibre using an electron beam physical vapour deposition process. Consolidation is then achieved by HIPping. This yields a near perfect fibre distribution and eliminates "touching fibre" defects.

One key technology is that of fibre coatings. These are required in order to prevent chemical interaction between the fibre and the matrix resulting in a low strength brittle material. All commercially available fibres currently have a protective surface coating, generally based on layers of Carbon.

Ceramic matrix composites (CMCs) offer potentially significant temperature advantages over metals, together with a density typically one third that of nickel. CMCs exhibit very high levels of specific stiffness compared to the competitor superalloys, but this must be weighed against a lack of ductility and the consequently low defect tolerance. Whilst a CMC does retain a degree of specific strength to higher temperatures than metals, in absolute terms the useable strength is very low, figure 8. If CMCs are ever to make a substantial entry into aeroengine applications there is a need to evolve designs to exploit this combination of high stiffness and low strength.

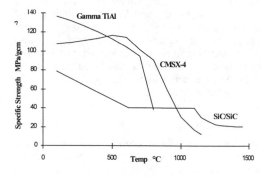

Figure 8. Comparison of Useable Strength of CMC and Metal

The main attraction of CMCs is their ability to withstand high temperatures for long times. However in reality today's systems e.g. SiC/SiC or SiC/Al$_2$O$_3$ have a temperature capability of only 1200°C for extended times with occasional excursions to 1400°C. This is barely beyond the capability of many current single crystal coated turbine blade systems and will need to be significantly increased if these materials are to make a serious challenge for such applications.

As a result CMCs have, to-date, only been considered for niche applications as high temperature components at low structural loads, where their temperature capability can be exploited in order to avoid the need for cooling of metal components. This can yield a cost benefit for the overall engine even where the use of the CMC is not justified on a component for component basis.

CMCs may be employed for this kind of application within the next two to five years, enabling in service experience to be gained, facilitating ultimate wider application. One example of such an application is the turbine tip seal segments which have been recently tested in a Rolls-Royce Trent engine. Here the hardness and wear resistance of the ceramic material can be exploited, the need for cooling removed and a lower weight component realised. CMCs also show promise for applications as turbine stators however their use for rotating components is still a long way off, significantly further in fact than was commonly projected ten years ago. The rewards of such a technology nonetheless remain very high.

In addition to mechanical limitations there are significant cost barriers to be overcome before ceramic matrix composite components can become economically viable. A CMC component costs substantially more than the equivalent metal part, despite the raw materials themselves being cheap. This cost penalty arises largely out of processing difficulties. The current manufacturing route for ceramic matrix composites is via chemical vapour deposition and requires advanced textile handling techniques. Fibre manufacture, moulding and processing are all expensive requiring a large equipment investment. Machining is also problematic requiring a near net shape route and constraining production to simple shapes. These problems are compounded by the fact that the material has no scrap value.

5.2 Intermetallics.

Two main categories of intermetallics are available as superalloy successor materials: titanium aluminides for use at lower temperatures and nickel aluminides for higher temperature components. Both offer potential for cost competitive improvements in performance through significant density reductions. However the reluctance to accept the risk of designing with a material which is intrinsically brittle must be overcome.

5.3 Titanium Aluminides.

The gamma titanium aluminides are at present the most advanced of the successor materials. Their potential as a high temperature structural material has long been recognised despite problems with low temperature brittleness. They have a density approximately half that of current alloys and hence excellent specific properties, particularly stiffness. Recently advances have been made through the development of a refined two phase gamma / alpha 2 structure and alloying modifications which allows the realisation of ductility values of up to 3%.

A current temperature capability of approximately 750°C, limited by high temperature degradation, which in the future development may be extended to 850°C, makes them suitable for back end compressor and turbine applications. Here their inherently non-burning nature makes them particularly attractive for stator vanes, where the use of titanium alloys is limited by the occurrence of titanium fires. The application of aluminides to such components would lead to significant weight savings. Indeed a recent design study predicted a total engine weight reduction of up to 4%. At this level of mass saving continued development is justified on the basis of the anticipated reduction in life cycle costs and the improved functionality of the engine/airframe combination.

Although the first application of these materials will be in static and rotating compressor aerofoils, their application to more critical components may become possible as manufacturing and engine running experience is accrued. Use will initially be in military applications but will inevitably cascade into the civil arena. The largest advantage may well be obtained from the application of aluminides to low pressure turbine blades. Here a saving of over 250lbs can be achieved through reduced blade weight and the subsequent reductions in disc and shaft weights.

Economically titanium aluminides benefit from cheap raw materials and good castability to complex shapes. An investment casting route can therefore be used, exploiting existing technology and expertise to provide an intrinsically inexpensive processing route. One drawback is that machining is currently difficult and hence scrap rates high. It is anticipated however that as the process matures these scrap rates and hence the cost per component will be reduced to the point where TiAl components will demonstrate a cost advantage over conventional wrought components such as compressor blades, possibly reaching around half the cost per component. It is likely by contrast that the business case for replacing cast components such as turbine blades will remain more marginal.

Looking further to the future it is possible to see potential for the application of titanium aluminides as a matrix material in metal matrix composites. In this context the intermetallic would be a replacement for conventional titanium alloys. This would present some difficult technical challenges for both processing and coating systems to minimise fibre-matrix mismatch.

5.4 Nickel Aluminides.

The development of nickel aluminides is currently much less advanced than titanium aluminides. Indeed, consensus is yet to be reached on the final alloy system although attention is concentrated on NiAl based systems.

Nickel aluminides again offer a significant reduction in density compared with metallic materials, coupled with a high melting point. They may therefore be considered for use in turbine applications in the temperature range 700-1200°C. The full exploitation of the high intrinsic melting point is unfortunately limited by the fact that the addition of alloying elements to increase high temperature strength typically degrades the room temperature ductility.

With regard to processing and cost, nickel aluminides have much in common with the titanium aluminides. Again raw materials are cheap and a cast route can be utilised with the resultant cost benefit. The inherent brittleness of the material leads to problems with machining but this is to some extent alleviated by the a high thermal conductivity allowing reduced cooling and simpler component geometries. As before the development of a robust, mature processing route will be required before costs are reduced to the levels required if the material is to be a serious competitor to the superalloys.

6 CONCLUSIONS

Historically the development of aeroengine materials has been driven largely by the military requirement for ever increasing performance and hence higher temperatures and specific strengths. This has lead to the current dominance of the nickel-based superalloys and titanium alloys. Now however the situation is changing. The desire for increased performance is still there but aeroengine technology is increasingly driven by the cost based requirements of the civil market. Competitive advantage from further reductions in specific fuel consumption is

becoming increasingly difficult and costly to achieve and hence, increasingly, it is cost and timescale which win contracts.

New materials are being developed which offer a step change in capability over today's alloy systems but considerable barriers still remain before they will find widespread application. Consequently the incremental development of existing alloys remains vital. In this context modelling techniques will become increasingly important to optimise materials and manufacturing processes and reduce development costs and cycle times

In the light of current economic constraints new technology must be able to demonstrate the potential for a clear competitive advantage before its adoption. Past predictions of usage for new materials are now looking optimistic. Titanium and nickel alloys will undoubtedly be with us for some time to come.

Materials testing

Experimental Mechanics, Allison (ed.) © 1998 Balkema, Rotterdam, ISBN 90 5809 014 0

Behaviour of corrugated cardboard in the linear range for biaxial tension tests

S. Jazouli, P. Duval & F. Brémand
Laboratoire de Mécanique des Solides, Université de Poitiers, CNRS UMR 6610, France

ABSTRACT: The most common use of corrugated board is packaging. Some laboratories have developed new tests in order to model the behavior of the board sheet in the following situations : buckling of vertical partitions and resistance to the vertical compression of the folds. Interpreting these tests is complicated by the need for a numeric treatment of the data and does not allow us to come out with a pattern of behavior. When analyzing their constituents and their process of manufacture, board seems to belong to the group of orthotropic materials. If we assume a linear elastic behavior, the problem we will attempt to solve will consist of finding five mechanical constants and confirming the orthotropy conditions. We suggest using a classic technique which has never been used in the study of board : tension. In order to limit the number of tests, we have devised a two dimension method.

1 INTRODUCTION

Corrugated board is one of the most used packing materials. It is mainly to manufacture boxes for the transportation and the protection of goods. Its mechanical properties are given by its special sandwich structure composed of two face sheets of paper called liners and a corrugated core called fluting. The liners provide stiffness and strength. Its features (widely recognized) fully justify the interest of engineers and the present scientific research works. A deep knowledge of its physical and mechanical properties could present a best choice of the type of board panel for a given use. The board is usually considered as an orthotropic material. The orthotropic directions are called in this paper L for machine direction and T for cross direction in reference to its paper constituents.

The current knowledge is based on experimental normalized tests. For the most part, this material is analyzed from vertical compressive loadings made on empty boxes. In these conditions not only the buckling of the material is viewed but also the influence of the corners can be obtained (Hahn et al., 1992 ; Renham, 1996). The aim of this kind of tests is the evaluation of a critical compressive load Fc assuming that the box breaks if the loading is greater than Fc. Manufacturers use the expression given by McKee et al. (1963) which gives the vertical compressive strength according to the mechanical features and the collapse load of each panel when simply supported.

The main goal of this study is the experimental measurement of the orthotropic parameters for this type of material. Our first idea was the development of the shadow moiré technique applied to the deformation measurement of a corrugated panel submitted to an anticlastic bending (Mauvoisin et al., 1994). For this work, we have introduced a phase shifting procedure in order to improve the sensitivity of such a technique associated with an automatic phase unwrapping algorithm (Brémand, 1994). The accuracy reached one hundredth of a millimeter which is sufficient for the corrugated board. Then the values of the out-of plane displacement were used in an identification process by finite-element method. Unfortunately, the numerical algorithm does not correctly converge to results in good agreement with the orthotropic condition. It means that several sets of numerical solution lead to the same experimental deflection fields. In fact this is due because of the anticlastic bending does not correctly solicit the two Poisson's ratios. Nevertheless, the shear modulus and the two Young's moduli are correctly estimated. Two solutions can be brought to this problem. First of all, instead of using the values of the out-of plane displacement, it is .possible to write a new algorithm with the local curvature of the panel. This can be made with the use of a bi-harmonic interpolation by using the diffuse finite-element method. The second possibility is to change the experimental loading. To reach this goal, we propose to use tension tests.

To the best of our knowledge, no experimental study of tension tests made on corrugated board has yet been carried out. In order to get the four parameters of the orthotropic laws, we need, at least, three uniaxial tension tests with several directions of orthotropy. The problem is the non-contact and non-disturbing strain measurement at the surface of the panel. In the first part of this article we introduce an automatic technique using a spectral interpolation of a grid marked at the surface. The experimental data issued from the three tensile tests leads to the determination of the two Young's moduli, the two Poisson's ratios and the shear modulus. But we have seen that three different specimens are necessary. So we have considered a biaxial tension test which can only use one specimen. To reach this goal, we have developed a new experimental site and the geometry of the specimen. The new shape has been obtained by finite-element method according to the previous experimental data. Finally, we present the whole biaxial tension procedure we have developed and the mechanical parameters which are calculated by linear interpolation on the stress-strain curves.

2 EXPERIMENTAL PROCEDURE

2.1 *Orthotropic constitutive laws*

We suppose that the behavior of corrugated board can be modeled for a first approximation by orthotropic's law. It means that five mechanical parameters must be measured. In the referential associated with the orthotropic directions L and T oriented by the angle θ from the loading directions 1 and 2, the constitutive equation can be expressed as :

$$
\begin{Bmatrix} \varepsilon_{11} \\ \varepsilon_{22} \\ \varepsilon_{12} \end{Bmatrix} = \begin{pmatrix} \dfrac{1}{E_L} & -\dfrac{v_{TL}}{E_T} & 0 \\ -\dfrac{v_{LT}}{E_L} & \dfrac{1}{E_T} & 0 \\ 0 & 0 & \dfrac{1}{2G_{LT}} \end{pmatrix} \begin{Bmatrix} \sigma_{11} \\ \sigma_{22} \\ \sigma_{12} \end{Bmatrix} \quad (1)
$$

with $\dfrac{v_{LT}}{E_L} = \dfrac{v_{TL}}{E_T}$ (orthotropic condition) (2)

In the referential (O, 1, 2) associated to the loading axis, the relationship (1) becomes :

$$
\begin{Bmatrix} \varepsilon_{11} \\ \varepsilon_{22} \\ \varepsilon_{12} \end{Bmatrix} = \begin{pmatrix} \dfrac{1}{E_1} & -\dfrac{v_{21}}{E_2} \\ -\dfrac{v_{12}}{E_1} & \dfrac{1}{E_2} \\ \dfrac{\eta_1}{E_1} & \dfrac{\mu_2}{E_2} \end{pmatrix} \begin{Bmatrix} \sigma_{11} \\ \sigma_{22} \end{Bmatrix} \quad (3)
$$

With $c = \cos\theta$ and $s = \sin\theta$, one can get :

$$
\frac{1}{E_1} = \frac{c^4}{E_L} + \frac{s^4}{E_T} + c^2 s^2 \left(\frac{1}{G_{LT}} - 2\frac{v_{LT}}{E_L} \right)
$$

$$
\frac{1}{E_2} = \frac{s^4}{E_L} + \frac{c^4}{E_T} + c^2 s^2 \left(\frac{1}{G_{LT}} - 2\frac{v_{TL}}{E_T} \right)
$$

$$
\frac{v_{21}}{E_2} = \frac{v_{LT}}{E_L}(c^4 + s^4) - c^2 s^2 \left(\frac{1}{E_L} + \frac{1}{E_T} - \frac{1}{G_{LT}} \right) \quad (4)
$$

$$
\frac{\eta_1}{E_1} = -2c\,s\left\{ \frac{c^2}{E_L} - \frac{s^2}{E_T} + (c^2 - s^2)\left(\frac{v_{TL}}{E_T} - \frac{1}{2G_{LT}} \right) \right\}
$$

$$
\frac{\mu_2}{E_2} = -2c\,s\left\{ \frac{s^2}{E_L} - \frac{c^2}{E_T} - (c^2 - s^2)\left(\frac{v_{TL}}{E_T} - \frac{1}{2G_{LT}} \right) \right\}
$$

From equation (4) and for $\theta = 45°$ we have :

$$
\frac{1}{G_{LT}} = \frac{1}{E_{45}} - \frac{1}{E_L} - \frac{1}{E_T} + 2\frac{v_{LT}}{E_L} \quad (5)
$$

2.2 *Method of strain measurement*

Most of the classical methods of strain measurement are inappropriate for board because they require the preparation of the surface. It is important to make sure that the method does not actually influence the results, so it must be carried out with a non-contact and non-disturbing evaluation. Furthermore, it must perform a two dimension analysis. For that respect, we have used a grid method using a numerical spectral interpolation (Brémand et al., 1992).

Deformations are obtained by comparing the initial and the deformed geometry of a grid marked on the specimen. An inside point $M(X_1, X_2)$ of the initial state moves to $m(x_1, x_2)$ in the final state. The linear transformation between both states is only function on the geometrical parameters (the two line spacings p_1 and p_2 and the two orientations α_1 and α_2, the upper indexes i and d indicate the initial and deformed state). One can note that the two directions of the initial grid are assumed to be perpendicular as shown on figure 1.

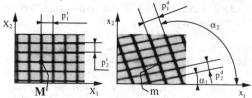

Figure 1. Deformation of a square grid

So one can write the following relationships :

$$
\begin{cases} x_1 = \dfrac{p_1^d}{p_1^i}\dfrac{\cos\alpha_1}{\sin(\alpha_2 - \alpha_1)}X_1 + \dfrac{p_2^d}{p_2^i}\dfrac{\cos\alpha_2}{\sin(\alpha_2 - \alpha_1)}X_2 \\[3mm] x_2 = \dfrac{p_1^d}{p_1^i}\dfrac{\sin\alpha_1}{\sin(\alpha_2 - \alpha_1)}X_1 + \dfrac{p_2^d}{p_2^i}\dfrac{\sin\alpha_2}{\sin(\alpha_2 - \alpha_1)}X_2 \end{cases} \quad (6)
$$

Let us recall the kinematics of large deformations. The components of the matrix which represents the gradient transformation tensor F is then given by :

$$F = \begin{pmatrix} \dfrac{p_1^d}{p_1^i} \dfrac{\cos\alpha_1}{\sin(\alpha_2-\alpha_1)} & \dfrac{p_2^d}{p_2^i} \dfrac{\cos\alpha_2}{\sin(\alpha_2-\alpha_1)} \\[2ex] \dfrac{p_1^d}{p_1^i} \dfrac{\sin\alpha_1}{\sin(\alpha_2-\alpha_1)} & \dfrac{p_2^d}{p_2^i} \dfrac{\sin\alpha_2}{\sin(\alpha_2-\alpha_1)} \end{pmatrix} \qquad (7)$$

And the matrix of the de Green-Lagrange tensor E is obtained from F and leads to the knowledge of the longitudinal strain ε_{11}, the transverse strain ε_{22} and shear strain ε_{12} by :

$$E = \frac{1}{2}(^tF\,F - I) \qquad E = \begin{pmatrix} \varepsilon_{11} & \varepsilon_{12} \\ \varepsilon_{12} & \varepsilon_{22} \end{pmatrix} \qquad (8)$$

The images of the grating are recorded with the help of an imaging system in function on time. The deformations are then obtained from the calculation of the spectrum of each image using a FFT procedure and a spectral interpolation (Brémand & Lagarde, 1988) on the first order peak leads to an accuracy sufficient for the board. In order to be precise enough (0.01% of deformation), a minimal number of lines (around 10) must be taken into account. Of course the size of the measuring base depends on the line density of the grating used for the board. In this case, the grid is obtained by an ink stamp applied on the surface of the specimen shown for example on figure 4a. The line density is about 1 line per millimeter.

2.3 Stress evaluation

Now we have to define exactly what is a stress for corrugated board. Of course, everybody knows that stress is load divided by area. In the case of this complex material, the surface depends on the orthotropy direction θ. For the machine direction, the surface is given by the thickness of each liner multiplied by the width of the specimen. While the cross direction is strained, the corrugated core participates to the strength. In fact we can get the relationship giving the surface S by :

$$S = t_1\ell + t_2\ell + t_{cor}\ell_{cor}\sin\theta \qquad (9)$$

where t_1, t_2 and t_{cor} are respectively the thicknesses of the two liners and the fluting, ℓ is the width of the specimen and ℓ_{cor} is the total length of the fluting obtained from the following expression :

$$\ell_{cor} = \int_0^1 \sqrt{1 + \left(\frac{df(x)}{dx}\right)^2}\,dx \; ; \; f(x) = \frac{t}{2}\cos\left(\frac{2\pi x}{p(\theta)}\right) \qquad (10)$$

where t is the distance between the two liners and p(θ) is the pitch of the fluting.

2.4 First evaluation by uniaxial tension tests

For each test, the displacement is imposed with a speed on each axis equal to 1 mm/min in order to remain in the elastic part of the material during the loading phase. The atmospheric conditions have been controlled : the temperature was 22 °C and the hygrometry was 60% humidity. From relations 4 and 5, we can observe that three one dimension tension tests should give enough information to solve the problem, i.e. to measure the five components of the constitutive law. A first experiment is carried out in the longitudinal direction (θ=0°) giving E_L and v_{LT} by using a linear interpolation of the strain stress curves. The second test corresponding to θ=90° gives E_T and v_{TL}. The modulus of rigidity E_{45} is obtained from a third experiment when θ=45°. All these experimental results are shown on figure 2.

Now by using equation (5), one can obtain the shear modulus equal to : 1300 MPa. The orthotropic condition is well verified and the ratio $E_L/E_T = 3.6$ shows the orthotropic character of the corrugated board.

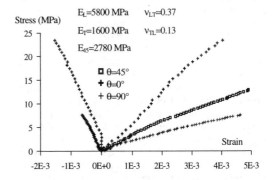

Figure 2. Experimental curves for three one dimension tension tests (θ=0°, θ=45°, θ=90°)

2.5 Experimental site for biaxial tension

As one can see the three uniaxial tests constitute a good approach for the determination of the mechanical parameters of an orthotopy material. Therefore, several specimen are required which is a source of errors for the corrugated board principally due to its inhomogeneity and its non linear mechanical response. To avoid this disadvantage, it is possible to analyze a crossed-shape specimen in a two dimension testing machine. But at first, we have to define the geometry of our specimen. In the literature related to biaxial specimen used in tension, generally we have noted that some slots were produced in the arms of a cruciform model. Among the first investigation we can note the geometry used by Monch with araldite (Monch & Galster, 1963). In this case, eight slots are produced on each of the four arms of the specimen. A photoelastic determination has proved that the stress field is homogeneous in the central zone and some stress concentrations appear at the end of the slot. Similar shapes have been used for steel (Hayhurst, 1973 ; Lin & Hayhurst, 1993) but the thickness in the central zone is then reduced in order to compensate the lack of matter in the slots. In the opposite way, we have increased the thickness in the arm by introducing lamellae during a molding operation with polymers (Kelly, 1976). An optimal shape (Demmerle & Boehler, 1993) has been defined by using a finite-element analysis for orthotropic materials as composite. The cruciform specimen has also been adapted for leather (Brémand & Lagarde, 1987) and paper (Urruty, 1995). Since the corrugated board is similar to these materials, we have investigated the shape by finite-element method and by using the mechanical properties obtained from the three uniaxial tension tests presented on figure 2. Two shapes are compared on figure 3, without slots and with three slots. With symmetry condition, the calculation has been only made on a quarter of the specimen and the orthotropy direction ($\theta=0$) is parallel to the x-axis. Furthermore the same load ($\sigma=200$MPa) has been applied on each arm.

On figure 3, we can observe that the stress fields σ_{xx} and σ_{yy} are more homogenous with the slotted arms. The stress values in the central zone and in the arms are also equivalent and can be identified to the applied stress. Furthermore we can note a good stress uncoupling between the two loading axes. The shear stress is very small and indicates the presence of a biaxial stress state in the central part of the specimen. Practically we have used a similar shape with three slots made with a cutter. This geometry is shown on figure 4. The specimen is embedded at the ends of each axis with the help of four clips in Plexiglas.

In order to study the behavior of board when subjected to a two dimension loading state, we have developed our own testing machine (Fig. 5). The maximal loading has been limited to 5000 N and we have imposed that the center of the specimen should stay in the same position during the test, this is imposed by the strain measurement method. This requires a central symmetry in the geometry of the testing machine. Furthermore, we have also chosen to separate the control for each direction in order to apply independent loadings. Practically, we have chosen four stepper motors controlled by a micro-computer. Each of them introduces a maximal displacement of 50 mm. On each axis, some mechanical and opto-electronic sensors have been introduced to give the two extreme positions and the

middle position. The command allows the control in displacement and effort in order to apply to any kind of loading.

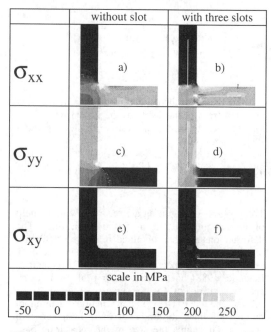

	without slot	with three slots
σ_{xx}	a)	b)
σ_{yy}	c)	d)
σ_{xy}	e)	f)

scale in MPa

-50 0 50 100 150 200 250

Figure 3. Comparison on stress field between two crossed shapes by finite-element method

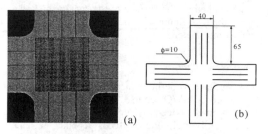

Figure 4. Geometry of the specimen and the grating

Figure 5. Experimental site

3 EXPERIMENTAL RESULTS, DISCUSSION

Although the uniaxial tension test procedure seems to be promising, it requires three different specimens which is a source of errors due to certain inhomogeneities of the materials even if they come from the same plate. The experimental conditions can also change during the whole experiment. Thus we have used the crossed shape and made two-dimension tension tests. The first idea was then to impose at the same time a stress σ_1 and a stress σ_2 evolving linearly in function of time but not at the same rate. In this case, the two stresses are reliable together by a proportional coefficient. The resolution of the mechanical problem (equation 4) is impossible since we can not separate the Young's modulus and the Poisson's ratio. To avoid this lack of information, we have to use two independent loadings. For the very first time, only the stress σ_1 is applied linearly with time. Then this level of load is maintained constant and the stress σ_2 increases (Figs 6, 8). Two types of specimen have been analyzed with two directions of orthotropy ($\theta=0°$ and $\theta=-21°$). The stress-strain curves are shown on Figures 7, 9. One can note a linear response between the stress and the strain allowing a linear interpolation which is similar to equation 3 :

$$A\,\sigma_{11} + B\sigma_{22} = \varepsilon_{11}$$
$$C\,\sigma_{11} + D\sigma_{22} = \varepsilon_{22} \qquad (13)$$
$$E\,\sigma_{11} + F\sigma_{22} = \varepsilon_{12}$$

For example, the determination of the constants A and B is made by considering the function ε_{11} of two variables σ_1 and σ_2. The linear interpolation is made by a least square method. The calculated slopes of each curves are given equations 14 and 16.

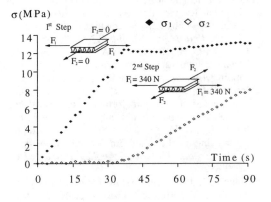

Figure 6. Loadings σ_1 and σ_2 for $\theta=0°$

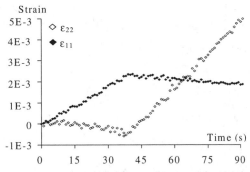

Figure 7. Longitudinal and transverse strain ($\theta=0°$)

$$\begin{Bmatrix} \varepsilon_{11} \\ \varepsilon_{22} \end{Bmatrix} = \begin{bmatrix} 1.9388\,E\text{-}4 & -8.44\,E\text{-}5 \\ -7.242\,E\text{-}5 & 7.2105\,E\text{-}4 \end{bmatrix} \begin{Bmatrix} \sigma_{11} \\ \sigma_{22} \end{Bmatrix} \qquad (14)$$

$$\begin{cases} E_L = 5156\,\text{MPa} & E_T = 1390\,\text{MPa} \\ v_{LT} = 0.37 & v_{TL} = 0.12 \end{cases} \qquad (15)$$

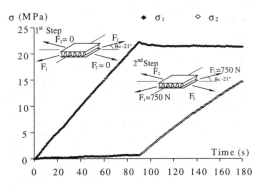

Figure 8. Loadings σ_1 and σ_2 for $\theta=-21°$

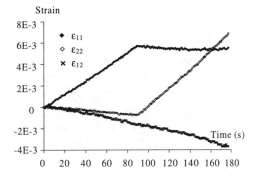

Figure 9. Longitudinal, transverse and shear strains for $\theta=-21°$

885

$$\begin{Bmatrix} \varepsilon_{11} \\ \varepsilon_{22} \\ \varepsilon_{12} \end{Bmatrix} = \begin{bmatrix} 2.707\,E-4 & -5.399\,E-5 \\ -6.094\,E-5 & 4.718\,E-4 \\ -8.106\,E-5 & -9.952\,E-5 \end{bmatrix} \begin{Bmatrix} \sigma_{11} \\ \sigma_{22} \end{Bmatrix} \qquad (16)$$

$$\begin{cases} E_L = 4148\ \text{MPa} & E_T = 1958\ \text{MPa} \\ \nu_{LT} = 0.248 & \nu_{TL} = 0.10 \\ G_{LT} = 1220\ \text{MPa} \end{cases} \qquad (17)$$

As we can see, tests in the direction of orthotropy directly give the two Young's moduli and the two Poisson's ratios but they do not lead to any information on the shear modulus (equation 15). Therefore, a two dimension tension test made out-of axis ($\theta \neq 0°$) is absolutely required for the determination of the five parameters (equation 17). We can note that the orthotropy condition is verified for both cases around 16% which is natural for this case of material. Nevertheless the coefficients between both approach are slightly different. This is due to the experimental conditions because the specimens, of course, were issued from two different board panels.

CONCLUSION

A biaxial apparatus has been developed for testing corrugated board under biaxial tension tests. The geometry of the specimen has been studied by finite-element method with the mechanical characteristics obtained from three uniaxial tension tests. We have seen that three slots made on each arm of the crossed-shape allow a good stress uncoupling between the two loading axes. In this paper, a non-contact and non-disturbing strain measurement technique is presented. It uses a numerical spectral analysis of a grid marked with an ink stamp at the surface of the specimen. This is well adapted to corrugated board.

A method for identifying the five parameters of the orthotropic behavior law board panel is presented. This method employs one biaxial tension test with independent loadings in each direction. When the tension test is made in the orthotropic directions, only the two Young's moduli and the two Poisson's ratio can be determined. We have shown that an out-of axis tension test can lead to the determination not only of the four previous parameters, but also of the shear modulus.

ACKNOWLEDGMENT

The authors are grateful for the financial support provided by the Région Poitou-Charentes and the société APE for the experimental site development. We would also thank the société OTOR GODARD for providing the corrugated board panels.

REFERENCES

Brémand, F. & A. Lagarde 1987. Sur une forme simple d'éprouvettes de traction biaxiale pour élastomère. Application à la détermination de lois de comportement mécanique et optico-mécanique en grandes déformations. C. R. Acad. Sci. Paris II. 305 : 929-934.

Brémand, F. & A. Lagarde 1988. Analyse spectrale bidimensionnelle d'un réseau de traits croisés. Application à la mesure de grandes et petites déformations. C. R. Acad. Sci. Paris II. 307 : 683-688.

Brémand, F., J.C. Dupré & A. Lagarde 1992. Non-contact and non-disturbing local strain measurement methods. European Journal of Mechanics A/Solids. 11(3) : 349-366.

Brémand F. 1994. A phase unwraping technique for object relief determination. Optics and Lasers in Eng. 21 : 49-60.

Demmerle, S. & J.P. Boehler 1993. Optimal design of biaxial tensile cruciform specimens. Jour. Mech. Phys. Sol. 41(1) : 143-181.

Hahn E.K., A. de Ruvo, B.S. Westerling & L.A. Carlson 1992. Compressive strength of edge-loaded corrugated board panels. Exp. Mech. 32(3): 259-264

Hayhurst D.R. 1973. A biaxial tension creep-rupture testing machine. Journal of strain analysis. 8 : 119-123.

Kelly D.A. 1976. Problems in creep testing under biaxial stress systems. Journal of Strain Analysis. 11(1) : 1-6

Lin J. & D.R. Hayhurst 1993. The development of a bi-axial tension test facility and its uses to establish constitutive equations for leather. Europ. Jour. Mech. A/Solids. 12(4) : 493-507.

Mauvoisin G., F. Brémand & A. Lagarde 1994. 3D shape reconstruction by phase shifting shadow moiré. Applied Optics. 33(11) : 2163-2169.

McKee R.C., J.W. Gander & J.R. Whachuta 1963. Compression strength formula for corrugated boxes. Paperboard Packaging. 48(8) : 149-159

Monch E. & D. Galster 1963. A method for producing a defined uniform biaxial tensile stress field. Britain Journal of Applied Physics. 14 : 810-812.

Renham M. 1996. Test fixture for eccentricity and stiffness of corrugated board. Exp. Mech. 36(3): 262-268.

Urruty J.P. 1995. Développement d'un système d'analyse d'images en surface des papiers et films. Applications en métrologie mécanique. Thèse de Doctotat de l'Université de BORDEAUX I.

Experimental Mechanics, Allison (ed.) © 1998 Balkema, Rotterdam, ISBN 90 5809 014 0

A simple model for uniaxial testing of fiber reinforced concrete

H. Stang & S. Bendixen
Department of Structural Engineering and Materials, Technical University of Denmark, Denmark

ABSTRACT: Fibers are added to concrete primarily in order to increase the fracture toughness of the material. In the design of fiber reinforced concrete structures information from three or four point bending tests is often used in terms of toughness indices. This is typically the case in traditional design methods for slabs on grade. However, much more information is contained in the so-called stress-crack opening relationship and many structural models (including non-linear FEM) use this relationship directly.

The stress-crack opening relationship can in principle be determined by direct uniaxial tension tests when it is assumed that only one crack is formed and the crack surfaces remain parallel or almost parallel during crack opening so that measurement of stress and crack opening can be performed through averaging. This, however, requires that both ends of the test specimen are clamped, that the specimen is sufficiently short and that the testing machine is stiff with regard to rotation. These requirements are not all fulfilled in a practical test.

It is the purpose of the present paper to investigate under which conditions the uniaxial test can be used to directly derive information about the stress-crack opening relationship.

The paper contains a theoretical modeling of the uniaxial test using a simple analytical/numerical approach. With this model it is investigated whether the material input can be retrieved from average measurements. Further, the analysis is compared with detailed experimental measurements including measurement of the rotation of crack surfaces as function of crack opening and measurements of the rotational stiffness of the clamping fixture of the testing machine. Finally, the validity of the uniaxial test under standard conditions is assessed.

1 INTRODUCTION

Fracture mechanics of concrete has undergone a very important development during the last 20 to 30 years. One of the major achievements in this development has been the formulation of the so-called fictitious crack model (FCM) originally suggested by Hiller-borg (Hillerborg, Modéer & Petersson 1976).

The model includes the concept of a tension soften-ing fracture process zone existing ahead of the stress free crack. This fracture process zone is envisioned as a fictitious crack with certain crack closing forces acting on its surfaces. These crack closing forces are assumed to act in such a way that the stress inten-sity factor vanishes at the tip of the extended fictitious crack. Thus a criterion for the stability and growth of such cracks is related entirely to the crack clos-ing forces. The FCM is obviously strongly related to both the Barenblatt and the Dugdale cohesive crack models, however in two important aspects the FCM differs from these models. First of all the crack clos-ing forces are not constant. They decrease from the

tensile strength of the material f_t^u at the tip of the fic-titious crack to zero at the tip of the stress free crack. Thus, the magnitude of the crack closing forces de-pends on the magnitude of the crack opening, and the distribution of stress along the fictitious crack is governed by the crack opening profile. Secondly, the length of the process zone is significant compared to a characteristic length of most conventional concrete structures. Both circumstances emphasize the need to be able to obtain knowledge of the relationship between stresses along the fictitious crack and crack opening.

In the FCM the relation between crack closing stresses σ_c and the opening of the fictitious crack w – the so-called stress-crack opening relationship – is considered a material characteristic. Outside the crack the material is assumed to be linear elastic. Thus the following material parameters are funda-mental in fracture mechanical description of concrete: the Young's modulus E, the tensile strength, f_t^u and the stress-crack opening ($\sigma - w$) relationship denoted $\sigma_c(w)$.

In plain concrete The $\sigma - w$ relationship is often characterized in terms of the area under the curve, known as the fracture energy, G_f:

$$G_f = \int_0^\infty \sigma_c(w)\,dw \qquad (1)$$

together with an analytical (e.g. a linear or bi-linear) approximation of the true stress-crack opening curve. Note furthermore that the shape of stress-crack opening curve (i.e. the stress-crack opening curve normalized with respect to the tensile strength)is fairly independent of the concrete type in question (Hordijk 1991), (Stang & Aarre 1992).

Fiber Reinforced Concrete (FRC) has been developed with the specific aim of increasing the fracture energy of concrete rather than to increase stiffness and strength which is the typical situation in other fiber reinforced composite materials. This means that the same fracture mechanical framework known from conventional concrete can be used in FRC-materials as well. On the other hand, the shape of the stress-crack opening curve is complex and greatly influenced by the type and amount of fiber used (the pull-out of fibers being the primary mechanism behind the stress transfer). Furthermore the shape is essential for the understanding of the mechanical behaviour of the material. Finally, the fibers are still carrying load across the crack even for very large crack openings - crack openings which are usually not relevant in structural design situations. In the light of these observations it follows that the fracture energy G_f as defined above is not a very useful tool in the characterization of material toughness and that the stress-crack opening curve itself has to be referred to and used in design situations.

Some discussion is currently taking place regarding whether the deformation-controlled uniaxial tensile test is the most suitable way to determine experimentally the fracture parameters of concrete and FRC, furthermore, if a uniaxial tensile test is performed how should it be designed? Hordijk (1991) carried out a detailed investigation of the influence of specimen size, geometry and the boundary conditions on the results of uniaxial fracture tests on plain concrete. He concluded that the uniaxial fracture test is the only test which yields directly all relevant fracture parameters even though care must be taken to minimize effects of structural behaviour of the test specimen on the measured results. These structural effects can never be completely eliminated, however some guidelines were set up by Hordijk (1991) for design of deformation controlled uniaxial tensile tests:

- use smallest possible cross-sectional area, i.e. 4-5 times the maximum aggregate size

- for the specimen length use two to three times the dimension of the cross-section

- the rotational stiffness of the loading platens must be high compared to the specimen rotational stiffness

Van Mier argues (Van Mier, Schlangen & Vervuurt 1996) that the use of fixed boundary conditions (high rotational stiffness of the loading platens) promotes the formation of several crack planes resulting in an artificial increase in crack density. Thus, the measured fracture toughness will be artificially increased. It is argued that hinges should be used as boundary conditions for uniaxial tensile test specimens, allowing for the determination of a lower bound of the fracture energy. On the other hand, determination of fracture parameters other than the fracture energy from a test using hinges as boundary conditions is not straightforward.

Considering the fact that the *shape* of the stress-crack opening relationship is of primary importance in FRC and that any other type of boundary condition than fixed boundaries makes determination of curve shape very difficult it is believed that direct tensile tests should be performed taking the test design considerations mentioned above into consideration. In the following a method for evaluating a given test setup is given. Furthermore it will be shown that a test setup can be established in a standard universal testing machine which can produce reliable results for the complete stress-crack opening relationship.

2 TEST SETUP

In order to determine the stress-crack opening relationship experimentally deformation controlled tensile tests are conducted on notched specimens. To eliminate the pre-stressing inevitably introduced in the specimen when using conventional grips, for improved alignment and for maximum stiffness - especially with respect to rotation - special fixtures have to be developed which allow the test specimen to be glued directly into the test machine.

At different research laboratories around the world experimental setups have been developed to allow for this kind of testing. At LCPC in France the test specimens are glued onto aluminum rods of the same diameter. The rods are fixed to the frame and actuator of a hydraulic universal testing machine. The crack opening is measured by means of an extensometer made of three displacement transducers (LVDTs) arranged at 120° and supported by aluminum rings in contact with the specimen through elastic heads (Casanova & Rossi 1997).

Figure 1. The test setup used for uniaxial fracture mechanical testing of plain and fiber reinforced concrete at DTU.

At The Technical University of Denmark, Department of Structural Engineering and Materials a similar test setup has been developed. The notched cylindrical test specimen is glued to interchangeable cylindrical steel blocks of the same diameter as the test specimen and these steel blocks can be fixed rigidly to the testing machine. The specimen is instrumented with three standard Instron extensometers arranged at 120° with a gauge length of 25.0 mm, measuring across the notch. The signals from the three extensometers are averaged and used as the feed back signal in the closed loop testing. Previously, rectangular specimens with notches at two opposite sides where used, instrumented using two extensometers (Li, Stang & Krenchel 1993).

Cylindrical test specimens are preferred since this geometry allows testing of both cast test specimens and cores from structures reflecting the in-situ fiber orientation and matrix quality. Typically Ø130 mm cylinders with a Ø110 mm cross section at the notch are used for concretes containing up to 25 mm aggregate. The cast test specimens are cut to a suitable length before testing. The full test setup is shown in Fig. 1 where the test specimen is instrumented not only with extensometers but also with three displacement transducers (LVDTs) in order to obtain independent non-averaged measurements which can be used for checking the rotation of the crack surfaces.

The stress-crack opening relationship is extracted from load and average extensometer deformation data. Stress is referred to the notched cross section, while the crack opening is calculated from the average extensometer deformation by assuming that the material outside the crack unloads linearly with a stiffness corresponding to the initial stiffness observed in uniaxial tension, see eg. (Petersson 1981) or (Hordijk, Van Mier & Reinhardt 1989).

3 SIMPLE MODEL FOR UNIAXIAL TEST

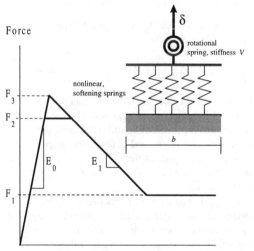

Figure 2. The non-linear, softening spring characteristics simulating the tensile behavior of a plain or fiber reinforced concrete specimen in tension along with an outline of the model.

In the following a simple model for the uniaxial tension test will be described. The model consists of two infinitely stiff plates of width b connected by a series of non-linear, softening springs. One plate is connected to a linear hinge with rotational stiffness

V. Finally, a separation δ of the two plates is prescribed in steps and the total force and rotation of the plates is determined in a numerical calculation. The springs model the linear elastic material volumes between the crack surface and the controlling LVDTs as well as the fictitious crack. The spring characteristics are outlined in Fig. 2 together with an outline of the model. A few springs with a little less strength (max. load F_2 instead of F_3) are introduced in one side in order to introduce softening from one side.

It is possible to identify the following dimensionless properties governing the behavior of the model:

$$\frac{\delta}{b}, \frac{F_1}{E_0 b}, \frac{F_2}{F_1}, \frac{F_3}{F_1}, \frac{E_1}{E_0}, \frac{V}{E_0 b^2} \quad (2)$$

In the present investigation two types of springs are being considered: one corresponding to a "brittle" behavior and one corresponding to a "tough" behavior. Expressed in terms of the dimension-less parameters a uniaxial test setup with the brittle material (corresponding to plain concrete) could be characterized by:

$$\frac{F_1}{E_0 b} = 1.0\,10^{-4}, \frac{F_2}{F_1} = 2.95, \frac{F_3}{F_1} = 3, \frac{E_1}{E_0} = -1 \quad (3)$$

while the tensile test of the tough FRC material could be characterized by:

$$\frac{F_1}{E_0 b} = 1.0\,10^{-4}, \frac{F_2}{F_1} = 2.95, \frac{F_3}{F_1} = 3, \frac{E_1}{E_0} = -0.2 \quad (4)$$

For each of the concretes the simulated dimensionless load-deformation relation is determined in order to compare the load-deformation response of the test setup with the input load-deformation response expressed as spring characteristics. A parameter study was carried out to investigate the influence of the rotational stiffness.

In Fig. 3 the dimension-less load-deformation relation for the brittle material is shown along with the rotation-deformation relation for the following values of the dimension-less stiffness: $\frac{V}{E_0 b^2} = 0.1, 0.05, 0.01, 0.005$. The curve corresponding to the highest rotational stiffness is a direct replica of the spring characteristics (see Fig. 2), however the shape of the load-deformation curve is becoming increasingly distorted as the stiffness is decreased. Also the corresponding curves showing the relative rotation of the two fixtures show increasing rotation as the stiffness decreases. The distortion causes the initial part of the descending force-deformation curve to become more steep followed by the formation of a plateau which suddenly drops to the input level as the rotation suddenly drops to zero. The curves become

more steep and the plateau is extended as the rotational stiffness of the test setup is decreased.

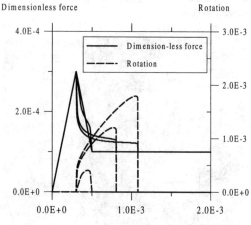

Figure 3. The simulated force-deformation relationships and the corresponding rotation-deformation relationships for four different values of the dimensionless rotational stiffness: 0.1, 0.05, 0.01, 0.005. The brittle material characteristics are considered here.

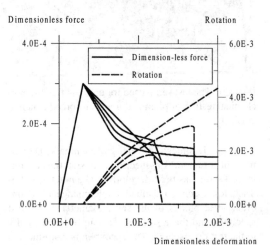

Figure 4. The simulated force-deformation relationships and the corresponding rotation-deformation relationships for four different values of the dimensionless rotational stiffness: 0.05, 0.01, 0.005, 0.001. The tough material characteristics are considered here.

In Fig. 4 the load-deformation relation for the tough material is shown along with the rotation-deformation relation for the following values of the dimensionless stiffness: $\frac{V}{E_0 b^2} = 0.05, 0.01, 0.005, 0.001$. Again the curve corresponding to the highest rotational stiffness is a direct replica of the spring characteristics, and again the curves are distorted as the stiffness decreases. It is noteworthy that the toughness of the material stabilizes the test in such a way that the correct determination of spring characteristics can be done with less stiff test setup (0.05 versus 0.1) and an acceptable determination in the case of the tough material can be obtained with a stiffness of 0.01 compared to 0.05 in the case of the brittle material.

4 EXPERIMENTAL VERIFICATION

A series of tests have been carried out at Department of Structural Engineering and Materials at DTU, including a determination of the rotational stiffness of the testing machine.

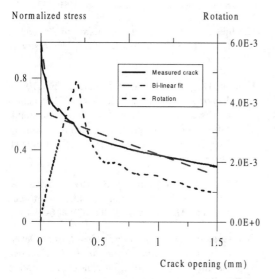

Figure 5. The stress-crack opening relationship and rotation observed in uniaxial testing of fiber reinforced concrete. Shown is also a bi-linear fit of the experimentally observed stress-crack opening relationship.

One test was performed on concrete with high concentration of steel fiber (2 % by volume) and a relatively low tensile strength of 2 MPa. The $\sigma - w$ relationship and the rotation of the crack are showed in Fig. 5. The $\sigma - w$ relationship is obtained as described above, i.e. as the average stress related to the cross section between notches versus the average crack opening. The rotation is obtained from measurements with the LVDTs shown in Fig. 1. As it can be seen the tested FRC is very ductile with a stress of

40 % of the tensile strength when the crack opening is 1 mm. It is clear that even though we have a low tensile strength and a very ductile concrete a significant rotation is observed. Furthermore, it is interesting to see that the shape of the measured $\sigma - w$ relationship corresponds very well to the slightly distorted curves of the model: steep part, plateau and sudden drop.

To investigate the influence of the rotation on the $\sigma - w$ relationship the measured relationship is considered the true material response and a bi-linear approximation to this true response is made as shown in Fig. 5. By assuming that the crack surfaces are plane it is possible to calculate the crack opening at any point on the crack surface $\hat{w}$ as function of the measured average crack opening w and rotation ϕ. Then using the $\sigma - w$ relationship assumed to be the true material response it is simple to calculate the expected response from the test σ_c^{test} by integration:

$$\sigma_c^{test}(w) = \frac{1}{A} \int_A \sigma_c(\hat{w}(w, \phi)) \, dA \qquad (5)$$

In this example the integration is done using numerical integration. This can be done at all data points throughout the test and a predicted curve $\sigma_c^{test}(w)$ will appear.

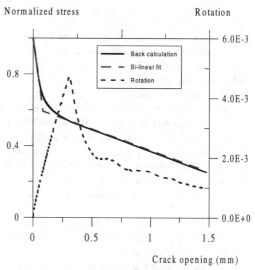

Figure 6. The bi-linear input together with the expected stress-crack opening relation calculated taking the shown rotation into account.

The bi-linear input curve and the back calculated $\sigma_c^{test}(w)$ curve are shown together in Fig. 6 It is observed that the calculated curve is close to the bi-linear input $\sigma - w$ relationship which in turn is fitted to the original average measurements. Thus it is concluded that the true material $\sigma - w$ relationship is close to the one measured under the test.

The rotational stiffness of the testing machine was found by loading the testing machine eccentrically and measuring the corresponding angel of the loading grips. This procedure has given the following estimate of the dimension-less parameters of the analytical model:

$$\frac{F_1}{E_0 b} = 2.0 \, 10^{-5}, \frac{F_2}{F_1} = 1.95, \frac{F_3}{F_1} = 2, \frac{E_1}{E_0} = -0.05 \quad (6)$$

together with a rotational stiffness of

$$\frac{V}{E_0 b^2} = 0.012 \quad (7)$$

A model simulation with these parameters clearly predicts that the obtained measurements should be very close to the original input. Thus, the analytical model confirms the experimental findings.

5 CONCLUSIONS AND DISCUSSION

A simple analytical model has been set up to predict the reliability of uniaxial testing to determine the stress-crack opening relationship of concrete and fiber reinforced concrete.

The model can be used to investigate an experimental setup if the rotational stiffness is known.

The model predicts reliable conditions for finite rotational stiffness of the testing machine.

The model predicts less strict requirements on the stiffness of the testing machine as toughness of the tested material increases.

The experimental setup used at Department of Structural Engineering and Materials at the Technical University of Denmark has been investigated in details involving a back calculation of the measured response taking measured rotations into account. The investigations show that reliable results for fiber reinforced concrete can be obtained with the present setup.

Predictions by the analytical model confirm the experimental findings, and even the shape of the distortion seems to be reproduced.

The model does not take into account the effect of specimen eccentricity and inhomogeneity of material. Thus, it can be expected that the model gives a lower bound on the machine stiffness requirement.

6 ACKNOWLEDGEMENTS

The support from the research programme *Design Methods for Fiber Reinforced Cement Based Composite Materials* funded by the *Materials Technology Development Program, MUP2* is gratefully acknowledged by both authors.

REFERENCES

Casanova, P. & Rossi, P. (1997). Analysis and design of steel fiber reinforced concrete beams, *ACI Structural. J.* **94**(5): 595–602.

Hillerborg, A., Modéer, M. & Petersson, P. (1976). Analysis of crack formation and crack growth in concrete by means of fracture mechanics and finite elements, *Cem. Concr. Res.* **6**(6): 773–782.

Hordijk, D. (1991). *Local Approach to Fatigue of Concrete*, PhD thesis, Technical University of Delft.

Hordijk, D., Van Mier, J. & Reinhardt, H. (1989). *Material Properties*, Vol. Fracture Mechanics of Concrete Structures, from theory to applications. Ed. L. Elfgren, Chapman and Hall, 11 New Fetter Lane, London EC4P 4EE, England, pp. 67–127.

Li, V., Stang, H. & Krenchel, H. (1993). Micromechanics of crack bridging in fiber reinforced concrete, *Mat. and Struc.* **26**(162): 486–494.

Petersson, P.-E. (1981). *Crack Growth and Development of Fracture Zones in Plain Concrete and Similar Materials.*, PhD thesis, Division of Building Materials, Lund Institute of Technology.Report TVBM - 1006.

Stang, H. & Aarre, T. (1992). Evaluation of crack width in frc with convential reinforcement, *Cement & Concrete Comp.* **14**(2): 143–154.

Van Mier, J., Schlangen, E. & Vervuurt, A. (1996). Tensile cracking in concrete and sandstone: Part 2 - effect of boundary rotations., *Mat. Struc.* **29**(186): 87–96.

Experimental Mechanics, Allison (ed.)© 1998 Balkema, Rotterdam, ISBN 90 5809 014 0

A sensor for measuring eigenstresses in concrete

B. F. Dela & H. Stang
Department of Structural Engineering and Materials, Technical University of Denmark, Denmark

ABSTRACT: A sensor developed for measuring eigenstresses due to shrinkage during hardening of cement paste is presented. The stresses are measured utilizing the spherical shape of the sensor and an embedded pressure sensitive resistance alloy called manganin®. The sensor can be applied to various kinds of measurements where eigenstress development is of interest. Results from tests carried out with the sensor embedded in a high shrinkage cement paste are presented. A simple linear elastic model is applied for estimating the stresses in the sensor.

1 INTRODUCTION

Concrete can be considered a composite material consisting of inhomogeneities (aggregates) embedded in an aging matrix (cement paste). During the early hardening process the cement paste undergoes significant phase changes due to chemical reactions. These changes are associated with the development of stiffness and strength as well as a significant shrinkage. Shrinkage of the matrix taking place after the development of stiffness produces eigenstresses due to the restraining effect of the aggregates on the deformation of the matrix.

Eigenstress is a generic name given to self-equilibrated internal stresses caused by nonelastic (stress free) strains in bodies which are free from any other external force and surface constraint (Mura 1982). Hence, residual stress can be considered a special kind of eigenstress.

For most concrete structures durability is closely related to concrete porosity. Especially in large construction works high demands are placed upon the durability of the concrete. In modern concrete microsilica is often added to the cement paste in order to increase strength and reduce porosity. However, the presence of microsilica increases the autogenous shrinkage of cement paste leading to an increase of eigenstresses in the concrete. Also the brittleness of the cement paste is increased resulting in a situation where eigenstress induced microcracks are likely to form in the paste surrounding the aggregates.

Although the influence of microcracks on durability of concrete structures is not established it is evident that formation of microcracks will influence the mechanical behavior of the concrete material. It is therefore important to be able to control the risk of microcracking due to shrinkage of cement paste.

A number of analytical models have been set up for determining the state of eigenstress in case of a spherical inhomogeneity embedded in a homogeneous matrix undergoing a volumetric deformation. Examples are (Shu 1963), (Golterman 1994), and (Garboczi 1997). However, the material behavior of cement paste is complex in the hardening phase, and thus, the stress state around the inhomogeneity is difficult to determine theoretically. Therefore, it is necessary to make direct measurements of the stress state around the inhomogeneity in order to get information about the state of eigenstresses in concrete. Due to the phase change of cement paste it is not possible to measure directly the stress state in the matrix. However, with knowledge of the stresses inside the inhomogeneity the stress state in the matrix can easily be determined mathematically. Nevertheless, only a few measurements of aggregate stresses have been carried out previously.

An electrical strain sensor was developed by Nielsen (1971). The main interest of this study was the aggregate stresses under (long time) loading. Only a small part of the study was based on eigenstresses in the aggregates.

Another sensor based on a mercury thermometer was developed by Stang (1996). The basic idea in this work was to use the mercury container of the thermometer as a volumetric strain sensor. By removing the outer protective tube on the thermometer the mercury container can be considered as an ellipsoidal inhomogeneity connected to a capillary tube when cast inside cement paste. The pressure applied on the mercury container when shrinkage in the paste develops will lead to a deformation and thus a volume change of the container. This volume change can be measured as an apparent temperature change in the mercury level on the capillary tube. By calibrating the apparent temperature change as a function of surrounding pressure it is possible to determine the actual contact stresses on the mercury container in cement paste.

The mercury container has several disadvantages. First of all the measurements have to be carried out manually. Since the measurements vary rapidly with time it is preferable to be able to carry out measurements continuously. This requires automatic measurements. Furthermore, the thermometer sensor is not suited for field application.

In order to overcome the disadvantages of the thermometer sensor a stress sensor is currently under development. The sensor is based on a spherical geometry representing an inhomogeneity (the aggregate). The stress state inside spherical inhomogeneity embedded in a matrix undergoing a uniform change of volume (like shrinkage) will be hydrostatic (Eshelby 1957).

The measuring technique is based on changes in resistance of a manganin® wire. Manganin is a resistance alloy with a low strain sensitivity whereas the sensitivity to changes in surrounding hydrostatic pressure is fairly high. By combining the knowledge of the stress state inside a spherical inhomogeneity (according to linear elastic solutions) with the properties of the manganin wire it is possible to produce a sensor capable of measuring the stresses in a spherical aggregate.

In the type of sensor presented here the manganin wire is glued to a diametral plane inside the spherical sensor.

A simple calibration in a pressure chamber is needed in order to determine the changes in resistance of the wire as a function of applied hydrostatic pressure. Furthermore, a compensation for the difference in overall stiffness between the aggregate and the sensor is necessary for the correct interpretation of eigenstresses.

2 DESCRIPTION OF STRESS SENSOR

The sensor is based on a special resistance alloy called manganin. The resistance of a manganin wire changes with the change in surrounding hydrostatic pressure. The stress state inside a spherical inhomogeneity embedded in an infinite matrix that undergoes an overall shrinkage will be hydrostatic (Eshelby 1957). By placing a manganin wire inside a spherical sensor it is possible to measure the changes in resistance as a function of the hydrostatic pressure inside the sensor. Since the stress field is hydrostatic, placement and direction of the manganin wire has no influence on the measurements.

In the sensors presented here, a manganin wire (diameter: 0.05 mm, length: 500 mm) is stretched back and forth on the diametral plane between two hemispheres of glass. A photograph of two sensors, one prior to gluing the two hemispheres together and the other completed, is shown in Figure 1.

Figure 1. Photograph showing two sensors - one complete sensor and one prior to gluing the two hemispheres together. The diameter of the sphere is 18 mm

The sensor is made of glass because of the stiffness of glass being in the same range as that of granite - a typical aggregate used in concrete. The coefficient of thermal expansion is low for glass which is also beneficial to the measurements. However, glass is known to create alkali-silica reactions in concrete, thus, if long term measurements are required, they should be carried out with sensors made of materials with a high stability in alkali environment which is typical for cement based materials.

By gluing the wire in between two hemispheres it is perfectly insulated from the aggressive alkali environment. In an environment with a pH of 12-14 and high humidity most electrical equipment will be unstable if not protected in a proper way. The sensors are expected to operate over a long period of time and it is therefore necessary to protect the wire.

A simple calibration of the sensor can be carried out by placing the sensor inside an oil pressure chamber. In Figure 2 the results of a calibration of two sensors and one free, loosely wound manganin wire are shown. In the figure the applied hydrostatic pressure is shown as a function of changes in electrical signal from the sensors.

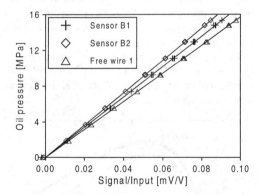

Figure 2. Result of calibration of two sensors and a free manganin wire

The calibration of the sensors has been carried out within the range of expected stresses since eigenstresses in concrete of more than 16 MPa are not expected. Note that the change in resistance of the manganin wire is linearly dependent on the hydrostatic pressure in the pressure chamber. The coefficient of correlation for each linear fit is given in Table 1.

The slope of the curves in Figure 2 are used as calibration factors for the sensors. The calibration factors are given in Table 1 together with the initial resistance of each wire.

Table 1. Calibration of sensors

	initial resistance [Ω]	calibration factor [MPa]	coefficient of correlation [-]
Sensor B1	118.3	$169.6 \cdot 10^3$	0.999899
Sensor B2	109.6	$181.1 \cdot 10^3$	0.999984
Free wire	161.6	$157.1 \cdot 10^3$	0.999966

The changes in resistance are almost the same for a given change in hydrostatic pressure no matter if the wire is glued to a linear elastic material or it is put directly into the pressure chamber.

An investigation of the temperature dependency has been carried out. The measurements were made in a closed box at temperatures varying from 22°C to 40°C. Inside the box an electrical heater with a fan was placed. The electrical heater was controlled by a thermostat. Three sensors where exposed to two temperature cycles. The signals were reset at the beginning of each cycle. A constant temperature distribution throughout each sensor was ensured by keeping a constant temperature for 30 minutes before measuring. The results from the temperature calibration are shown in Figure 3. The temperature dependency according to the specifications of the manganin alloy is $1 \cdot 10^{-5} \text{ K}^{-1}$.

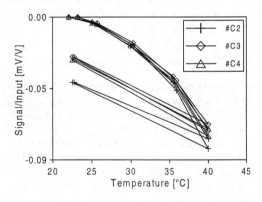

Figure 3. Temperature dependency of 3 sensors

The temperature calibration of the 3 sensors show the same trends. The temperature dependency during heating is parabolic. No measurements were made in order to determine the dependency in the cooling phase. However, readings were made after cooling the sensor down to 22°C. These readings show an irreversible dependency of the temperature of approximately 1/3 of the full signal at 40°C. The irreversible temperature dependency of sensor C2 is approximately -0.04 mV/V and for the two sensors C3 and C4 it is approximately -0.025 mV/V. According to the calibration curve in Figure 2 this corresponds to a measured negative pressure of 4-6 MPa. The irreversible behavior is expected to be due to the release of residual hardening stresses in the glue layer when the temperature is increased.

3 EXPERIMENTAL SET-UP

3.1 Measuring technique

The length of the manganin wires used in the sensors is adjusted to a resistance close to 120 Ω. The sensor is placed in a Wheatstone bridge in a quarter bridge configuration. Another quarter consists of a 120 Ω resistor. A potentiometer is used to adjust the two quarters to the same resistance. The temperature coefficient of the resistor and the potentiometer is 2 ppm K^{-1} and 15 ppm K^{-1} respectively. The last half of the bridge is made up of the measuring equipment.

Data are collected each 10 minutes using a datalogger (Campbell, CR10X). The range of the measurements is 2.5 mV resulting in a resolution of 1 μV. The input voltage of 5 V is measured on a voltage divider in the datalogger (resolution: 3.5 mV).

The temperature in the center of the cement paste is measured using a thermocouple placed in the center of a specimen.

3.2 Cement Paste

The cement paste used in the experimental investigation is a high-shrinkage paste containing 20% by weight of microsilica relative to the weight of cement. The cement used is white portland cement, and the water to cement ratio is 0.3 (water/powder is 0.25). This composition of the cement paste is not typical for a normal high performance concrete since 5% to 10% of microsilica usually is considered as maximum content. The reason for using this type of cement paste is that it will ensure high stresses in the sensor as a result of a large autogenous shrinkage deformation.

In the initial hardening process of cement paste the chemical reactions will cause a temperature increase. In order to avoid the influence of the rising temperature on the measurements, the temperature is kept as low as possible by using a small mould for casting, and the whole set-up is placed in a climate chamber at 22°C. The sensors are cast in the center of a cylindrical plastic mould with a diameter of 55 mm and a height of 60 mm. The moulds are closed with a plate and tightened with wax. The moulds are placed in a closed container with water in order to increase the temperature loss during the early hardening, and to obtain a high humidity around the moulds. In case the moulds are not 100% tight it is important to avoid the drying out of the cement paste since this will have a significant effect on the stress state.

4 TEST RESULTS

Two simultaneous measurements have been carried out over a period of 5 weeks. The measurements of eigenstress development inside the sensors are started immediately after casting. Measurements are carried out each 10 minutes. In Figure 4 the results from two simultaneous tests on the same batch of cement paste are shown.

The measurements show that the two sensors behave in the same way. The difference of up to 1 MPa between the measured stresses should be viewed in the light of the sensors being hand-made prototypes, and thus, some variation in the results can be expected. A pressure inside the sensors of up to 8 MPa is observed after approximately 2 weeks. After 2 weeks the stresses decrease monotonically.

The temperature during the initial hardening increased from 22°C to 26.7°C within the first 10 hours after start of mixing the paste. After 1 day the temperature was constant at 22°C.

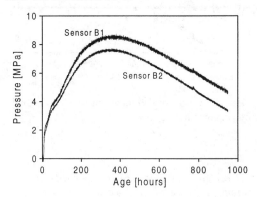

Figure 4. Test results from two simultaneous tests carried out on high shrinkage cement paste

5 DISCUSSION

In order to estimate the magnitude of the stresses to be expected in the sensors embedded in cement paste a simple linear elastic model has been applied. The model considers a homogeneous sphere with stiffness

E_a and Poisson's ratio ν_a embedded in an infinite homogeneous matrix with stiffness E_m and Poisson's ratio ν_m. The matrix undergoes a linear shrinkage ε_s.

The model is derived using the same procedure of superposition as in (Eshelby 1957). Instead of applying a homogeneous stress and strain field on the infinite matrix a stress-free strain of ε_s is assumed to take place in the infinite matrix. A more detailed derivation of the model can be found in (Stang 1996).

For $\nu_a = \nu_m = 0.2$ the hydrostatic pressure σ_p inside, and thus, the magnitude of surface stress on the sphere can be calculated using Equation 1:

$$\sigma_p = \frac{5}{3} \frac{E_m \cdot \varepsilon_s}{\frac{E_m}{E_a} + 1} \tag{1}$$

The stiffness E_a of the sensor is assumed to be 50 GPa, and Poisson's ratio is 0.2 for the sensor as well as for the cement paste. The stiffness E_m and the shrinkage ε_s of the matrix are considered being functions of time. Therefore the time dependent functions $E_m(t)$ and $\varepsilon_s(t)$ are introduced and the hydrostatic pressure $\sigma_p(t)$ in the sensor can be calculated as shown in Equation 2.

$$\sigma_p(t) = \frac{5}{3} \int_0^t \frac{E_m(t)}{\frac{E_m(t)}{E_a} + 1} \frac{\varepsilon_s(t)}{dt} dt \tag{2}$$

The shrinkage $\varepsilon_s(t)$ of the matrix is assumed to be equal to the free autogenous shrinkage of the cement paste used in the test. The stiffness $E_m(t)$ of the matrix is based on a curve fit of test results carried out in tension at 20°C at different ages. The test results as well as the curve fit are shown in Figure 5. In the same figure the autogenous shrinkage measured on the cement paste is shown. The autogenous shrinkage is measured with the use of a dilatometer as described in (Jensen 1995).

The data from the shrinkage is used directly in the model whereas the expression for the curve fit for the modulus of elasticity is used for the stiffness.

The results of the calculations applying Equation 2 and the data presented in Figure 5 are shown in Figure 6 together with the test results. Note that the model presented here is expected to be an upper limit since neither creep nor relaxation has been taken into account.

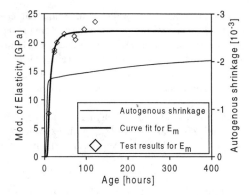

Figure 5. Modulus of elasticity and autogenous shrinkage for cement paste containing 20% microsilica

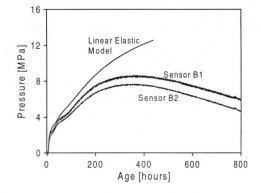

Figure 6. Eigenstresses in sensors as measured and calculated

Comparing the curves in Figure 6 show a good agreement between the model and the test results in the initial hardening stage. Since creep and relaxation are not considered in the model, the calculation is expected to show higher stresses than in the tests. This is confirmed by Figure 6.

The two simultaneous measurements show a difference of approximately 10%. This variation is assumed to be due in part to variations in the hand-made sensors and in part to variations in the cementitious matrix.

A decrease in stresses can be observed after approximately 400 hours. This is assumed to be a result of the rate of creep being higher than the rate of shrinkage after this point in time.

It is remarkable that calculations based on a simple, linear elastic model result in stresses which agree well with the sensor measurements up to an age of about 200 hours.

The highest temperature during the initial hardening of the cement paste was 26.7°C. According to the temperature calibration in Figure 3 and the calibration curve in Figure 2 the error introduced by the temperature dependency is of the order of -1 MPa. The temperature dependency is neglected the test results presented.

It should be noticed that taking the temperature dependency into account will bring the experimental and the theoretical results even closer in the first 400 hours.

6 CONCLUDING REMARKS

The following conclusions can be drawn from the study presented:

- It is possible to measure eigenstresses in an inhomogeneity with the sensors presented here.

- The temperature calibration shows a relatively high irreversible temperature dependency of 1/3 of the total change at temperatures ranging from 22°C to 40°C. This is expected to be due to residual stresses in the glue layer being released when the increase in temperature softens the glue. Further experiments on the temperature dependency remain to be carried out.

- Modeling of eigenstresses using a simplified linear elastic composite model show results comparable with eigenstresses measured with the stress sensors.

- The sensors have been developed aiming specifically at measurements of eigenstresses in concrete. However, it is expected that any stress state can be measured with the use of the sensor, including stresses introduced mechanically and/or thermally.

REFERENCES

Eshelby, J. (1957). The determination of the elastic field of an ellipsoidal inclusion and related problems, *Proc. R. Soc. London* **A241**: 376–396.

Garboczi, E. (1997). Stress, displacement, and expansive cracking around a single spherical aggregate under different expansive conditions, *Cement and Concrete Research* **27**(4): 495–500.

Golterman, P. (1994). Mechanical predictions on concrete detoriation. part 1: Eigenstresses in concrete, *ACI Materials Journal* **91**(6): 543–550.

Jensen, O. (1995). A dilatometer for measuring autogenous deformation in hardening portland cement paste, *Materials and Structures* **28**: 406–409.

Mura, T. (1982). *Micromechanics of defects in solids*, Mechanics of elastic and inelastic solids 3, Martinus Nijhoff Publishers.

Nielsen, K. E. (1971). *Aggregate stresses in Concrete*, PhD thesis, Royal Institute of Technology, Stockholm, Sweden.

Shu, T. T. (1963). Mathematical analysis of shrinkage stresses in a model of hardened concrete, *Journal of the American Concrete Institute* pp. 371–390.

Stang, H. (1996). Significance of shrinkage induced clamping pressure in fiber-matrix bonding in cementitious composite materials, *Advanced Cement Based Materials* **4**: 106–115.

Experimental Mechanics, Allison (ed.)© 1998 Balkema, Rotterdam, ISBN 90 5809 014 0

Viscoelastic viscoplastic characteristics of UHMW-PE

I. Sakuramoto – *Department of Mechanical and Electrical Engineering, Tokuyama college of Technology, Japan*

T. Tsuchida – *Department of Orthopedic Surgery, Chiba University, Japan*

K. Kuramoto – *Department of Medical Business, Nakashima Propeller, Okayama, Japan*

S. Kawano – *Department of Mechanical Engineering, Yamaguchi University, Japan*

ABSTRACT : UHMW-PE has been used as a bearing material in the artificial knee joint. Stress analysis, using FEM, is generally performed to clarify the cause of *in vivo* wear and delamination. A test machine, of the displacement control type, was designed and built and equipped with a measuring system to obtain accurate test results and the constitutive equation of UHMW-PE. The maximum crosshead speed of the machine is 10^4 mm/min. Furthermore, it has the capability of determining stress relaxation and strain recovery behavior immediately after performing a tensile test. Interesting properties of UHMW-PE have been obtained using this machine. Also, the constitutive equations for plasticity and viscoelasticity were derived from the experimental results. The equations have strongly nonlinear properties in the mechanical model described in this work. Stress relaxation simulations based on the equation for viscoelasticity have been performed. From this, it was confirmed that the equation was valid.

1 INTRODUCTION

With the progress of modern society has come an ever increasing life expectancy. As a result, the number of people requiring total knee replacement, because of acute rheumatoid arthritis, for example, is on the increase.

In many cases Ultra high molecular weight polyethylene (UHMW-PE) has been used as a bearing material in the knee and other artificial joints. Investigations on the mechanical and biochemical properties of UHMW-PE have already been undertaken by clinical and biomechanical researchers (Willert et al. 1991 etc.). However, clinical results have shown that the UHMW-PE knee joint inserts are subject to wear, resulting in the delamination and eventual separation of a layer of polyethylene from the component. The failure mechanism has not yet been identified (Ohashi et al. 1996). This has become a very serious clinical problem requiring patients to have a replacement UHMW-PE tibial insert after ten years or so, but in severe cases, after only three or four years.

To investigate the mechanism of delamination, stress analyses on the UHMW-PE tibial insert of the artificial knee joint have been done using Finite Element Methods (FEM) (Fujiki et al. 1993-1995). To apply FEM to the stress analysis of the UHMW-PE, it is necessary to determine first the parameters of the mechanical model of the UHMW-PE, that is, to construct the constitutive

equation from test results (Iwatsubo et al. 1996). However, the stress strain curve of the UHMW-PE is influenced by the strain rate logarithmically in both elastic and plastic stages, i.e. viscoelastic viscoplastic characteristics. Therefore test results under varying strain rates have to be obtained in order to construct the constitutive equation. However, in conventional ballscrew type tensile test machines, maximum crosshead speeds of about 500 mm/min are common. If we need higher loading rates, we can use high speed machines using compressed air, but the use of these machines is not suitable for the lower speed tests and it can not maintain the displacement after completing the tensile test. Furthermore, at the start of loading the effect of the variation of crosshead speed can not be disregarded for conventional ballscrew type tensile test machines. For example, Figure1 shows a Load vs. time curve of a tensile test for the UHMW-PE at a crosshead speed of 200mm/min using a ballscrew type tensile test machine (maximum crosshead speed is 500mm/min). For a crosshead speed of 200mm/min, the actual crosshead speed varies due to inertia at the start of loading.

To obtain accurate stress strain curves and a valid constitutive equation for UHMW-PE, we designed and built a new high speed displacement-controlled test machine (Sakuramoto et al. 1997). The test machine has a crosshead speed range from 1 to 10^4 mm/min and the variation of the crosshead speed at the start of loading is minimized by us-

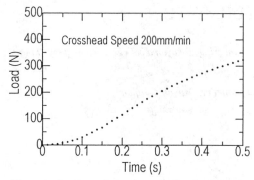

Figure1. Load vs. time curve obtained using a conventional ballscrew tensile test machine.

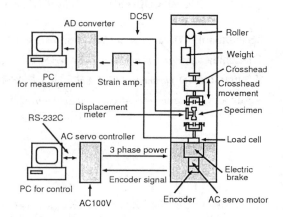

Figure2. A schematic diagram of the system.

ing an improved loading mechanism. Furthermore, our test machine can be used to perform stress relaxation and strain recovery tests directly after performing a loading test without interruption.

Interesting properties of UHMW-PE have been obtained using this machine. Also, the constitutive equations for plasticity and viscoelasticity were derived from the experimental results. The equations have strongly nonlinear properties in the mechanical model described in this work. Stress relaxation simulations based on the equation for viscoelasticity have been performed. From this, it was confirmed that the equation was valid.

2 TEST PROCEDURE AND TEST MACHINE

2.1 Test machine overview

A schematic diagram of the experimental system is shown in Figure2. The AC servo motor (Yaskawa SGM-03B3G24B) is attached to a precision linear guide actuator (THK KR6525B). The motor is driven by an AC servo controller (Yaskawa CACR-HR05BAB11), which is connected to a personal computer through on RS-232C serial interface. The variation of crosshead speed at the start of loading is minimized by the design of the loading system. Loading begins only after the crosshead has reached constant speed. The roller, which contains a bearing, is fixed to the top plate, and the weight of the upper specimen grip is counterbalanced by the weight hung on the roller to enable us to perform a strain recovery test without interruption. The lower specimen grip is fixed to the middle plate through the loadcell (Kyowa LUK-500KBS), which has a maximum load capability of 5000N. The test machine is covered with a box, and its temperature is controlled by a controller. The data are input into a personal computer for measurement through a 12 bit AD converter.

2.2 Specimen

Figure3 shows specimen dimensions conforming to JIS K7113 No.2. Specimens were made from UHMW-PE powder (Hoechst Hostalen GUR1120) by Nakashima Propeller. A molding pressure and temperature of 5.73 MPa and 220 ℃ respectively were maintained for 20 minutes. In addition the specimens were kept at 20 ℃ for an additional 20 minutes under the same pressure. An average molecular weight measured according to ASTM D4020 of 3.35×10^6 was recorded.

2.3 Test procedure

Initially, the displacement meter was mounted on the two gage marks (gage length 25 mm) of the specimen. Next, the specimen was mounted in the test machine and subjected to three tests performed sequentially.

1. (Tensile test) The specimen was loaded at a constant crosshead speed ranging from 10 to 10^4 mm/min (corresponding strain rates:

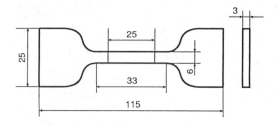

Figure3. Specimen design.

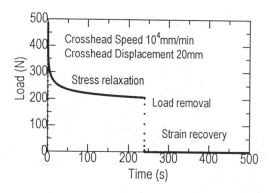

Figure4. Load vs. time curve.

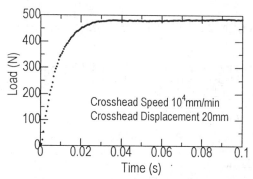

Figure5. The initial part of Figure4.

3.36×10^{-3} to 3.73/s) with a maximum crosshead displacement ranging from 0 to 20mm.

2. (Stress relaxation test) After the tensile test had been completed, the crosshead was maintained at the maximum displacement position and the loadcell output recorded.

3. (Strain recovery test) The applied load was removed by moving back the crosshead after 240 seconds from the start of loading. The weight of the specimen grip was counterbalanced allowing the strain recovery to be measured without removing the specimen.

Loads and displacements were measured for 500 seconds at a constant sampling frequency ranging from 200 to 2000 Hz at each crosshead speed. Finally, the permanent deformation after 24 hours was measured to obtain the plastic strain. Tests were conducted at a temperature of 20 ± 1 ℃.

3 TEST RESULTS

Typical example of Load vs. time curves is shown in Figure4. The curve was generated at a crosshead speed of 10^4 mm/min (corresponding strain rate:

3.73/s) with a maximum displacement of 20mm. It was confirmed that the three kinds of tests, loading, stress relaxation and strain recovery tests, were done continuously. A sampling frequency of 2000Hz was used at this crosshead speed. Figure5 examine in more detail the initial part of the test described in Figure4. At the start of loading the variations in loading and crosshead speed were very small compared with those of a conventional ball-screw type tensile test machine.

Figure6 shows Load vs. time curves concentrating on stress relaxation, which were obtained from three separate tests at a crosshead speed of 10^3 mm/min with maximum crosshead displacements of 10, 15 and 20mm. A sampling frequency of 200Hz was used.

Figure7 shows Displacement vs. time curves concentrating on strain recovery, which were obtained by stress relaxation tests at the same time. They show that elastic recovery ($\overline{AB}$ in the Figure) occurs just after load removal, and that viscoelastic recovery occurs gradually. It is of interest to note that the displacement rate in the strain recovery region decreases with increasing maximum displacement. From the results shown in Figures4-7, it was confirmed that the test machine designed by the authors gave reproducible and accurate test results for UHMW-PE.

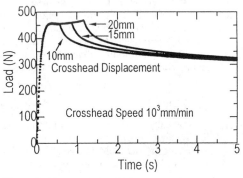

Figure6. Stress relaxation.

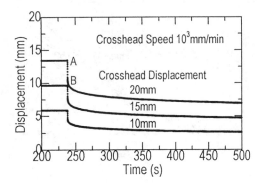

Figure7. Strain recovery.

901

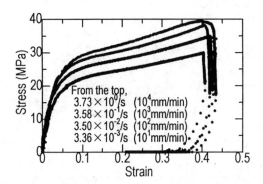

Figure8. Stress vs. strain.

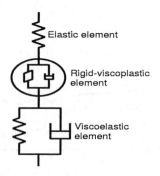

Figure10. Mechanical model of UHMW-PE.

Figure8 shows Stress vs. strain curves at a maximum crosshead displacement of 20mm with strain rates of 3.36×10^{-3}, 3.50×10^{-2}, 3.58×10^{-1} and $3.73/s$ (10, 10^2, 10^3 and 10^4 mm/min). True strain and true stress are used in Figure8, and strain rates were obtained from Displacement vs. time curves. From these results, we were able to see that the effect of strain rate on elastic modulus was very small within the range of our study. Stress was influenced by strain rate and the increase in stress appeared to vary logarithmically with strain rate.

Figure9 shows Plastic strain vs. strain curves obtained from strain recovery tests on 48 specimens. A maximum crosshead displacement ranging from 1 to 20mm and crosshead speed ranging from 10 to 10^4 mm/min were used. In Figure9, solid lines show least-squares approximations based on the quadratic function for each strain rate according to

$$\epsilon_p^{(k)} = a^{(k)} \epsilon^{(k)2} + b^{(k)} \epsilon^{(k)} + c^{(k)} \tag{1}$$

where $k = 0, 1, 2, 3$. There are yield points in each line at low strains (about $\epsilon = 0.01$) and corresponding stresses (about 10MPa). Note that the plastic strains increase with increasing strain rate.

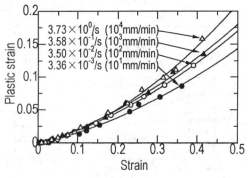

Figure9. Plastic strain vs. strain.

4 CONSTITUTIVE EQUATION

4.1 *Mechanical model*

The mechanical model of UHMW-PE proposed for FEM analysis is shown in Figure10 and can be represented by

$$\epsilon = \epsilon_e + \epsilon_p + \epsilon_c \tag{2}$$

In the model, three elements are connected in series: elastic(ϵ_e), rigid-viscoplastic(ϵ_p), and viscoelastic(ϵ_c).

According to the model, each strain element can be separated from the total strain, which is determined experimentally. Stress vs. plastic strain curves are obtained by applying Eq.(1) to Stress vs. strain curves. From Figure8, the Young's modulus E of the elastic element can be taken as 1000 MPa. Therefore Stress vs. viscoelastic strain curves are obtained by applying the Eq.(2).

4.2 *Constitutive equation for plasticity*

The constitutive equation for plasticity is a sum of the static stress, which is a function of plastic strain, and a dynamic stress, which is a product of the function of plastic strain and the function of plastic strain rate.

$$\sigma(\epsilon_p, \dot{\epsilon}_p) = \sigma_{static}(\epsilon_p) + f_1(\epsilon_p) f_2(\dot{\epsilon}_p) \tag{3}$$

The plastic strain rate is represented as follows.

$$\dot{\epsilon}_p = f(\epsilon_p, \dot{\epsilon}) \tag{4}$$

Taking the time derivative of Eq.(1) gives us the plastic strain rate as

$$\dot{\epsilon}_p^{(k)} = (2a^{(k)} \epsilon^{(k)} + b^{(k)}) \dot{\epsilon}^{(k)} \tag{5}$$

From Eq.(1) and Eq.(5), the plastic strain rate is represented as a function of plastic strain and strain rate as follows.

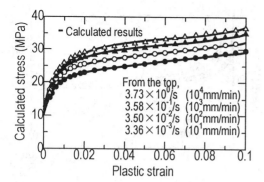

Figure11. Calculated stress vs. plastic strain.

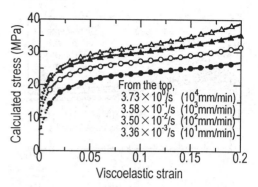

Figure12. Calculated stress vs. viscoelastic strain.

$$\dot{\epsilon}_p^{(k)} = \left(\sqrt{b^{(k)2} - 4a^{(k)}(c^{(k)} - \epsilon_p^{(k)})}\right)\dot{\epsilon}^{(k)} \qquad (6)$$

The constitutive equation for plasticity can be derived as follows.
(1) We assume the stress is represented by a quadratic function of $\ln(1 + \alpha\dot{\epsilon}_p)$ as

$$\sigma = \sigma_0 + A\ln(1 + \alpha\dot{\epsilon}_p) + B(\ln(1 + \alpha\dot{\epsilon}_p))^2 \qquad (7)$$

where α is a constant (2000 was used in our study).
(2) For a given value of plastic strain $\epsilon_{p(j)}$ and at different strain rates $\dot{\epsilon}^{(k)}$, four values of $\ln(1 + \alpha\dot{\epsilon}_{p(j)}^{(k)})$ are plotted against stress. By performing a least-squares approximation on this curve, according to Eq.(7), $\sigma_{0(j)}$, $A_{(j)}$, and $B_{(j)}$ are derived. This process is repeated for $\epsilon_{p(j)}$.
(3) Static stress $\sigma_0(\epsilon_p)$, the coefficients $A(\epsilon_p)$ and $B(\epsilon_p)$ are determined by least-squares approximations of $\sigma_{0(j)}$, $A_{(j)}$ and $B_{(j)}$ respectively.
Figure11 shows the calculated Stress vs. plastic strain curve, which represents the constitutive equation for plasticity of UHMW-PE. It is clear that the calculated values are in good agreement with the experimental results, shown as points in the figure.

4.3 Constitutive equation for viscoelasticity

The constitutive equation for viscoelasticity is a sum of the static stress, which is a function of viscoelastic strain, and the dynamic stress, which is a product of the function of viscoelastic strain and the function of viscoelastic strain rate.

$$\sigma(\epsilon_c, \dot{\epsilon}_c) = \sigma_{static}(\epsilon_c) + f_1(\epsilon_c)f_2(\dot{\epsilon}_c) \qquad (8)$$

The viscoelastic strain rate is represented as follows.

$$\dot{\epsilon}_c = f(\epsilon_c, \dot{\epsilon}) \qquad (9)$$

The concept of the constitutive equation for viscoelasticity is almost the same as that of plasticity, but it is more complex. The viscoelastic strain rate is represented by

$$\dot{\epsilon}_c^{(k)} = \left\{ SQ - \frac{1}{E}\sigma'(\frac{1 - b^{(k)} - SQ}{2a^{(k)}})\right\}\dot{\epsilon}^{(k)} \qquad (10)$$

where $\sigma' = d\sigma/d\epsilon$ and

$$SQ = \sqrt{(b^{(k)} - 1)^2 - 4a^{(k)}(c^{(k)} + \epsilon_c^{(k)} + \frac{\sigma^{(k)}}{E})}$$

Finally, the constitutive equation for viscoelasticity is represented as

$$\sigma = \sigma_0 + A\ln(1 + \beta\dot{\epsilon}_c) + B(\ln(1 + \beta\dot{\epsilon}_c))^2 \qquad (11)$$

where β is a constant (10^5 was used in our study).
Figure12 shows the calculated Stress vs. viscoelastic strain curve, which represents the constitutive equation for viscoelasticity of UHMW-PE.

5 STRESS RELAXATION SIMULATION

Stress relaxation simulations, based on the constitutive equation for viscoelasticity, were performed to confirm that the constitutive equation was valid by comparison with the experimental results.
In the stage of stress relaxation, $\dot{\epsilon} = 0$ and Eq.(2) becomes

$$\epsilon_{max} = \frac{\sigma}{E} + \epsilon_{pmax} + \epsilon_c \qquad (12)$$

Therefore, from Eq.(11), the viscoelastic strain rate is represented as

$$\dot{\epsilon}_c = \frac{e^{f(\epsilon_c)} - 1}{\beta} \qquad (13)$$

where $f(\epsilon_c)$ is

$$\frac{-A + \sqrt{A^2 + 4B\{-\sigma_0 + E(\epsilon_{max} - \epsilon_{pmax} - \epsilon_c)\}}}{2B}$$

Eq.(13), which is given in a form of a nonlinear differential equation, is solved for ϵ_c with an initial condition $\epsilon_c(t_0) = \epsilon_{max} - \epsilon_{pmax} - \sigma(t_0)/E$ when

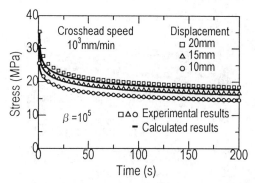

Figure13. Stress relaxation(10^3mm/min,$\beta = 10^5$).

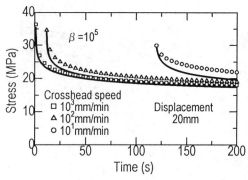

Figure14. Stress relaxation (20mm, $\beta = 10^5$).

$t = t_0$(the start point of stress relaxation). The Euler method was used to obtain the numerical solution for the equation in 0.01s increments.

During stress relaxation, $\sigma(t)$ can be calculated easily if we rewrite Eq.(12) as follows.

$$\sigma(t) = E\{\epsilon_{max} - \epsilon_{pmax} - \epsilon_c(t)\} \qquad (14)$$

The comparisons of the simulations and the experimental results for stress relaxation are shown in Figure13 and Figure14.

Figure13 shows the curves obtained at a crosshead speed of 10^3 mm/min with maximum crosshead displacements of 10, 15 and 20mm.

Figure14 shows the curves obtained at a maximum crosshead displacement of 20mm with crosshead speeds of 10, 10^2, and 10^3 mm/min. It is clear that the calculated values are in good agreement with the experimental results, shown as points in the figure, for $\beta = 10^5$.

For other values of β, the results of stress relaxation simulations do not show such close agreement with experimental results. Therefore, a more valid constitutive equation for viscoelasticity can be obtained by careful selection of the β value from the results of stress relaxation simulations.

6 CONCLUSIONS

1. It was confirmed that the stress of UHMW-PE subjected to high strain rates could be measured reproducibly using the tensile test machine designed by the authors.

2. The effect of strain rate on elastic modulus was very low in our study within the range of $\dot{\epsilon} = 3.36 \times 10^{-3}$ to 3.73/s.

3. A logarithmic increase in the stress was effected by strain rates in both elastic and plastic stages.

4. The plastic strain ϵ_p increased with increasing strain rate.

5. Yield occurred at very low strain (about $\epsilon = 0.01$) within the range of strain rates used in our study. The corresponding stress was approximately 10MPa.

6. The constitutive equation for plasticity and viscoelasticity of UHMW-PE were obtained using four experimental results of Stress vs. strain.

7. It was confirmed that the experimental results of stress relaxation and the simulation results were necessary to obtain a valid constitutive equation.

REFERENCES

Fujiki,H., Ishikawa,H. & Yasuda,K. 1993. Contact mechanisms of ultra high molecular weight polyethylene for artificial knee joint. *Trans. JSME.* 59A: 2800-2807.

Fujiki,H., Ishikawa,H. & Yasuda,K. 1994. The effect of thickness and fixation of ultra high molecular weight polyethylene and of traction on wear behavior in artificial knee joint. *Trans. JSME.* 60A: 1871-1877.

Fujiki,H., Ishikawa,H. & Yasuda,K. 1995. Contact analysis of artificial knee joint during gait movement. *Trans. JSME.* 61A: 262-269.

Iwatsubo,T. et al. 1996. The study for the dynamic response analysis of the artificial knee joint of considering viscoelastic property. *Proc. JSME Bioengineering annual meeting, Osaka.* 96-29: 60-61.

Ohashi,M. et al. 1996. An observation on subsurface defects of ultra high molecular weight polyethylene due to rolling contact. *Bio-Medical Materials and Engineering.*9: 1-11.

Sakuramoto,I. & Kawano,S. 1997. Viscoelastic viscoplastic characterization of ultra-high molecular weight polyethylene. *Proc. The 74th JSME spring annual meeting, Tokyo.* 97-1: 328-329.

Willert H. -G. et al. 1991. Ultra-High Molecular Weight Polyethylene as Biomaterial in Orthopedic Surgery. *Hogrefe & Huber Publishers.*

Experimental Mechanics, Allison (ed.)© 1998 Balkema, Rotterdam, ISBN 90 5809 014 0

On data interpretation of polymers resilience test

L. Goglio & M. Rossetto
Politecnico di Torino, Dipartimento di Meccanica, Italy

ABSTRACT: The paper deals with some problems related to the interpretation of data obtained in instrumented Charpy test of polymers. A dynamic finite element model has been used to investigate on the physical nature of the oscillations. To validate the model, two types of specimen have been instrumented with strain gages and rebound tests have been performed. Some considerations about the interpretation of force oscillation are drawn. The detectability of "first damage force" and "ultimate force" is discussed.

1. INTRODUCTION

The extensive use of plastics for structural applications demands an accurate knowledge of the mechanical properties of such class of materials. The knowledge of the impact behaviour is of specific interest in the evaluation of structural crashworthiness. Typical data used in this field are obtained by resilience testing and drop dart testing.

As far as resilience testing is concerned, the basic datum given by the test is the amount of energy required to fracture the sample under specified testing conditions. A significant improvement is given by the use of an instrumented pendulum, capable to measure the fracturing force during the test and to record its time history. Additional data which can be obtained by such equipment are the ultimate fracturing force and the so called "first damage force", usually referred to as the force level producing the first cracking in the sample. Furthermore it is possible to evaluate the resilience by integration of the force time history.

Obtained data are used by some authors, as Kobayashi et al. (1987), Kishimoto, et al. (1980), Sahraoui & Lataillade (1990) also to evaluate the dynamic fracture toughness.

The main difficulties in data analysis are due to the dynamic response of the system. The typical signal oscillations visible during the impact are partially due to the sequence of force peaks that can be interpreted as the progressive breaking of the sample, but are also affected by the dynamic response of the measuring system. Furthermore, the residual noise in the signal hides the real end of the impact. Digital filtering of the signal may be used to clean the data Cain (1987), but, in order to perform appropriate filtering, it is important to understand the nature of the observed oscillations. These problems have been faced during construction and testing of an instrumented pendulum in the laboratory of our Department.

To understand the nature of the observed oscillations a dynamic model can be used. In the literature some dynamic models are presented which, in our opinion, are unsatisfactory. The model presented by Cain (1987) considers the specimen as a spring-mass system with only one degree of freedom. The model presented by Sahraoui & Lataillade (1990) considers several degrees of freedom of the specimen but only the part between the supports is considered and the pendulum is schematised only with one mass.

In a previous paper by Belingardi et al. (1996), a dynamic finite element model that schematises both the specimen and the pendulum was presented and utilised to draw some considerations about the problems arising in data interpretation and the relationship between stress and force histories.

In this paper this model is refined and validated by means of some rebound tests on specimens instrumented with strain gages to record the strain history during the test.

2. MATERIALS AND METHODS

2.1 *Experimental set-up*

The rebound tests have been performed using the

instrumented Charpy impact machine of our Department. This can be equipped with two different types of impacting pendulum, each one capable of two energy levels (namely 15 and 7.5, 4 and 2 J). In this work the 4 J pendulum was used for all tests.

The striker is instrumented with a piezoelectric load cell (PCB model 201 A04).

Two materials have been considered for testing: polyethylene and steel; the aim of using steel was to have at disposal also a specimen with elastic linear behaviour and Young modulus almost independent from velocity.

The specimens are prismatic, without Charpy notch to allow for strain gauges positioning. The polyethylene specimen dimensions are 15×10×120 mm, as the standard Charpy specimen, the steel specimens dimensions are 6×10×160 mm; the dimension 6 mm for this specimen is chosen to avoid excessive stiffness. The distance between the supports, fixed to 70 mm, is that foreseen by current Italian standards (UNI 6062).

The strains in the samples have been measured by means of Micro Measurement model CEA-06-240-120 strain gauges, arranged in full bridge. Two gauges of the four are applied on the specimen side opposite to that struck by the pendulum, the remaining two are used as compensating gauges applied on an idle specimen. The bridge is wired to a measuring unit SINT SGA-4, with 5 V dc supply voltage and 40 kHz frequency bandwidth.

Force and strain data are sampled and recorded by means of a high speed (0.5 Msample/s on two channels) transient recording board (Bakker BE 490).

2.2 Finite element model

The dynamic finite element model presented by Belingardi et al. (1996) was refined for use in this work. The model, written in the MATLAB environment, schematises both the specimen and the pendulum: half specimen (taking advantage of the symmetry) is schematised with 7 beam elements considering also the part out of the support.

The contact stiffness between knife and specimen, as well as that between supports and specimen, are simulated with spring elements; the stiffness of the load cell is simulated with a spring and the pendulum stem is simulated with 4 beam elements; the block at the end of the stem (where most part of the mass of the pendulum is concentrated) in the first version was assumed to be rigid and its inertial properties are known. The value of contact stiffness k_c is evaluated using an approximate formula found in Ferrari & Romiti (1960):

$$k_c = lE_m / const \qquad (1)$$

where E_m is the Young modulus of the specimen material, l is the contact length and the constant is evaluated empirically (4.622 for steel/steel contact and 11.555 for polyethylene/steel contact).

The stiffness of the piezoelectric load cell, given by the manufacturer, is $1.752 \cdot 10^9$ N/m; the mass of the knife is 11 g.

The response to the impact of the whole system is obtained by means of the modal superposition, using as initial conditions the velocities of the nodes at the initial time.

With respect to the model used in Belingardi et al. (1996) the following improvements have been added:
- an offset has been considered for the nodes of the beam elements corresponding to the supports, to simulate better the kinematics of the part;
- the block at the end of the stem has been simulated by means of two beam elements to account for its deformability;
- modal damping of the system is included.

3. EXPERIMENTAL RESULTS

Rebound tests were carried out on instrumented specimens; the impact energy must be, in general, low to avoid specimens damage. On the basis of some pre-tests on non instrumented specimens the impact velocity has been set to 0.788 m/s (corresponding to a 30° drop angle) for the steel specimens and 0.533 m/s (corresponding to a 20° drop angle) for the polyethylene specimen.

In figure 1 typical time histories of the force measured by the load cell during two rebound tests on the steel specimen are reported (pendulum starting position 30 deg. from the bottom dead centre), whereas figure 2 displays the stress histories. The stress is computed from the strain using a value of Young modulus equal to $2.18 \cdot 10^6$ MPa, determined considering the dynamic behaviour of the specimen as explained in next section.

In figure 3 and figure 4 the analogous results from four rebound tests on polyethylene specimen are reported (pendulum starting position 30 deg.). In this case the Young modulus was set to 2950 MPa.

It is worth of note that the repeatability of the results is excellent. The stress plots are quite smooth, whereas the force plots show complicated oscillations due to high frequency contributions.

In the case of polyethylene specimen the effect of damping on the force time history is clearly more pronounced than in the case of steel.

The fact that the oscillations affect only the force time history (and not the stress) suggest that the force measurement is heavily influenced by the dynamic behaviour of the pendulum.

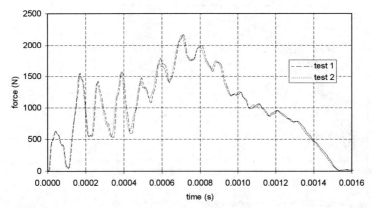

Fig. 1. Force time history recorded during two rebound tests on steel specimen.

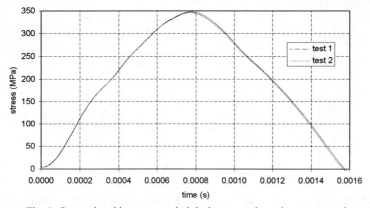

Fig. 2. Stress time history recorded during two rebound tests on steel specimen.

4. MODEL IDENTIFICATION

At a first stage the experimental data have been used to correct the model response, identifying the most influencing parameters. With this purpose, also the separate model for the pendulum and for the specimen were considered.

First, it was checked that the model of the pendulum was adequate to reproduce the dynamic behaviour of the part. Tab. 1 reports the experimental and theoretical values for the first three natural frequencies. The match seems satisfactory, also considering the level of approximation required for our aim.

The apparent Young modulus of the specimen was identified analysing the free vibration frequency of the specimen (Casiraghi 1983). The adopted values are those which give the same fundamental frequencies, respectively for steel and polyethylene,

as obtained experimentally from the instrumented specimens.

The most critical parameters to assign are the damping ratios to be associated at the different modes. Their values were determined by means of a trial and error procedure: assuming as starting value the damping measured for the fundamental mode as logarithmic decrement, the other values were obtained by adding one mode at once into the model response and setting the damping in order to match the amplitude of the experimental response.

Table 1. Lowest eigenfrequencies of the pendulum (Hz)

mode No	experimental	theoretical
1	340	350
2	750	815
3	1520	1480

907

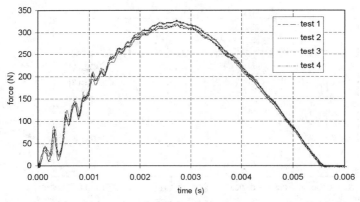

Fig. 3. Force time history recorded during four rebound tests on polyethylene specimen.

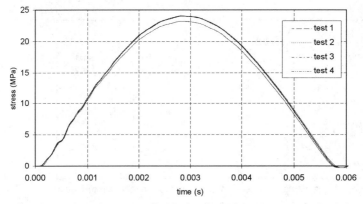

Fig. 4. Stress time history recorded during four rebound tests on polyethylene specimen.

5. MODEL RESULTS

Results obtained from the model considering the steel specimen are reported in figure 5 (force time history) and 6 (stress time history).

The force time history reproduces the typical oscillations which were experimentally observed, due to the contribution of a mode at about 8700 Hz. The peak value of about 2000 N is correct. However, considering the general aspect of the curve, the accordance is more qualitative than quantitative. The individual height of each peak is not the same as in the experiments; moreover the total duration of the simulated impact is lower than the experimental (1.3 instead of 1.5 ms). The latter fact may not be ascribed to a too high Young modulus assumed in the simulation (as it could be argued at a first sight), since this value was identified considering the free vibration of the specimen. More likely, the reason for

such model inadequacy can be sought in some nonlinearities (contact, strain rate dependence, ...) involved in the phenomenon.

Regarding the stress time history, it can be noticed that the simulated curve is approximately "bell shaped" as the experimental one and the same peak value (about 350 MPa) is reached. The stress time history can be reproduced only if the first modes (about twelve in this case) are considered. As for the force, the duration is shorter than that of the experiment.

The same arguments hold for the simulation corresponding to the case of polyethylene specimen. Figure 7 and 8 display respectively the time history for force and stress obtained from the model. The larger discrepancy in the stress peak value is justified by the material behaviour, for which the dependence on strain rate is remarkable.

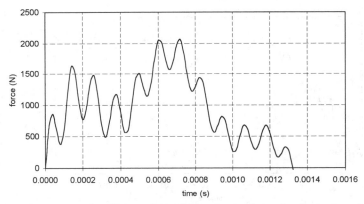

Fig. 5. Force time history obtained from the simulation for the steel specimen.

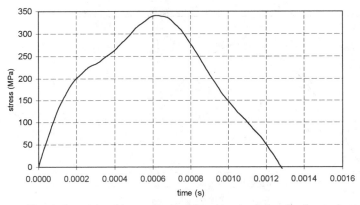

Fig. 6. Stress time history obtained from the simulation for the steel specimen.

6. CONCLUSIONS

Notwithstanding the difficulties encountered in simulating the rebound test on a Charpy pendulum, the results obtained by the model can offer significant help to the interpretation of the experimental data.

Firstly, it comes out that also in a case of elastic rebounding (without damage to the specimen, as proven by the repeatability of the experimental results on the same specimen) the force time history is affected by oscillations. From the simulation it appears clear that such oscillations are due to high order modes involved in the dynamics of the pendulum.

From such arguments it is evident that the force curve is not an intrinsic property of the specimen under test but it depends on the pendulum. The same material tested using machines with different dynamic properties could give different results.

Furthermore, it also follows that observing only the force time history of a destructive Charpy test, it is very difficult to distinguish how much of the early peaks is due to the initial breaking of the sample ("first damage") and how much to the dynamics of the system. Obviously, the first damage could be detected by means of crack detection gages or equivalent techniques.

Secondly, regarding the stress in the specimen, the recorded time history (supported by the simulation, with damping included) suggests that the oscillations of the force are not directly transmitted to the specimen but are filtered by the dynamics of the whole system. It has been found from the simulation that the stress time history can be reproduced only if the first modes (and not only the first) are considered. Conversely, the contribution from high order modes is negligible.

However it has confirmed that the stress may not be calculated by means of static relationships as often found in the literature as already discussed in Belingardi et al. (1996).

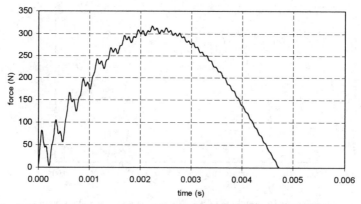

Fig. 7. Force time history obtained from the simulation for the polyethylene specimen.

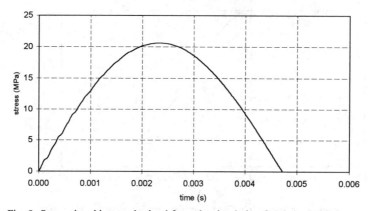

Fig. 8. Stress time history obtained from the simulation for the polyethylene specimen.

REFERENCES

Belingardi G., Goglio L., Rossetto M. 1996. On the force measurement in resilience testing of polymers. *Int. Conf. of Material Eng.* (*XXV AIAS Conf.*). *Gallipoli (I) 4-7 Sept. 1996*:745-752.

Cain P.J. 1987. Digital Filtering of Impact Data. *Instrumented Impact Testing of Plastics and Composite Materials, ASTM STP 936*: 81-102.

Casiraghi T. 1983. Rebound Test to Measure the Strength of Polymeric Materials. *Polymer Engineering and Science* 23(16): 902-906.

Cheresh M.C. & S. McMichael 1987. Instrumented Impact Test Data Interpretation", *Instrumented Impact Testing of Plastics and Composite Materials, ASTM STP 936*: 9-23.

Ferrari C. & A. Romiti 1960. Meccanica Applicata alle Macchine. Torino: UTET.

Hodgkinson J. M. & G. Williams 1987. Analysis of Force and Energy Measurements in Impact testing. *Instrumented Impact Testing of Plastics and Composite Materials, ASTM STP 936*: 337-350.

Ireland D.R. 1974. Procedure and problems associated with instrumented impact testing. *ASTM STP 563*: 3-29.

Jemian W., B.Z. Jang & J.S. Chou 1987. Testing, simulation and interpretation of materials impact characteristics. *Instrumented Impact Testing of Plastics and Composite Materials, ASTM STP 936*: 117-143.

Kalthoff J.F. & G. Wilde 1995. Instrumented impact testing of polymeric materials. *ASTM STP 1248*: 283- 294.

Kishimoto K., S. Aoki & M. Sakata 1980. Simple Formula for Dynamic Stress Intensity Factor of Pre-Cracked Charpy Specimen. *Eng. Fract. Mech.* 13: 501-508.

Kobayashi T. 1984. Analysis of impact properties of A533 steel for nuclear reactor pressure vessel by instrumented Charpy test. *Eng. Fract. Mech.* 19(1): 49-65.

Kobayashi T. et alii 1987. Dynamic fracture toughness testing of epoxy resin filled with SiO_2 particles. *Eng. Fract. Mech.* 28(1): 21-29.

Sahraoui S. & J.L. Lataillade 1990. Dynamic effects during instrumented impact testing; *Eng. Fract. Mech.* 36(6):1013-1019.

Experimental Mechanics, Allison (ed.)© 1998 Balkema, Rotterdam, ISBN 90 5809 014 0

Effect of specimen geometry in dynamic torsion testing

M. N. Bassim

Department of Mechanical and Industrial Engineering, University of Manitoba, Winnipeg, Man., Canada

ABSTRACT: In this study, a dynamic torsion Split Hopkinson bar was developed and implemented to achieve shear strain rates of up to 10^4s^{-1}. This device usually relies on the use of a cylindrical hollow specimen with arbitrary dimensions. The effect of such dimensions on the obtained results (stress-strain curves) was investigated. Steel specimens of AISI 4340 steel were tested at strain rates ranging from 200 s^{-1} to 2,000 s^{-1}. The outside diameter, the inside diameter and the length of the thin walled section of the specimens were changed. It was found that a parameter consisting of the product of the polar moment of inertia and length of specimen has a significant influence on the shear failure mode.

1 INTRODUCTION

The Torsional Split Hopkinson Bar apparatus has been used extensively to study the occurrence of adiabatic shear banding in materials. This process of deformation occurs at very high strain rates and describes a number of complex mechanical and thermal phenomena which are dependent on factors such as the strain rate exponent, strain rate sensitivity, work hardening coefficient, thermal conductivity, material hardness, microstructure and content of imperfections such as inclusions (Hartley 1987, Backman 1973, Costin 1979, Senseny 1978, Rogers 1981).

A Split Hopkinson Bar system was constructed and used to study the phenomena of adiabatic shear bands (ABS) (Cepus 1995). Results on the occurrence of ABS in steels were recently reported (Cepus 1994). It was, however, observed that there are no effective standards for the dimensions of the specimens used in the testing and thus it was rather difficult to compare results on the occurrence of ABS reported from different testing programs. The objective of this present study is to investigate the role of specimen geometry on the results obtained and to establish specific parameters to characterize the specimens, which can then be reproduced in order to insure uniformity of the results obtained.

2 EXPERIMENTAL PROCEDURES

Generally, the materials used in the Torsional Split Hopkinson Bar system are polycrystalline metals. The specimen used consists of a thin-walled tube loaded in torsion. The test section consists of the thin walled section between the flanges which are used for mounting the specimen onto the bars. In the analysis, this thin walled section deforms plastically while the flanges are presumed to remain elastic (Leung 1980). A typical specimen is shown in Fig. 1.

In this study, specimens of AISI 4340 steel were used. All specimens were annealed at 680° C which produced a hardness of HV 114-144.

The specimens were machined in such a way that changes were made to the outside diameter (D), the inside diameter (d) and the length (L) of the thin walled section of the specimens. These changes produce a variation of the polar moment of inertia J which is, for a hollow cylinder, given as

$$J = (\pi/32)(D^4 - d^4) \quad (1)$$

The specimens were then tested using the torsional Split Hopkinson Bar at shear strain rates varying from 200 S^{-1} to 2000 S^{-1}. Stress and strain histories

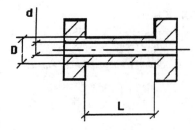

Figure 1. Torsional Hopkinson Bar
Specimen dimensions.

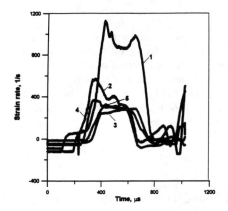

Figure 2. The effect of time on strain
rate under load = 60% of maximum load
for J_L (mm^5) (1) 2405, (2) 3650, (3)
5344, (4) 5550, (5) 6660.

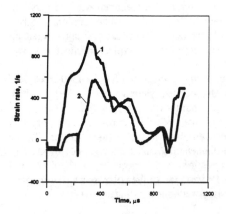

Figure 3. Effect of loading on the strain
rate (1) Load = 70% and (2) Load = 60% of
maximum load. Both specimens have J_L =
3650 mm^5.

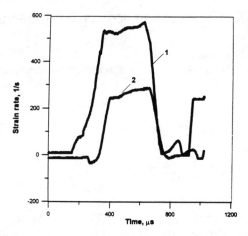

Figure 4. As in Figure 3 but for
specimens with J_L = 5344 mm^5.

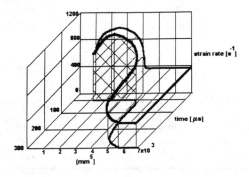

Figure 5. Composite drawing of the
relationship between time (µs), J_L (mm^5)
and strain rate (1/s).

were obtained from the strain gage measurements of
the reflected bar.

3 RESULTS AND DISCUSSIONS

Further to the calculation of J in eq. (1), another
parameter is defined, namely J_L, which is equal to
J x L where L is the length of the specimen shown
in Fig. 1.

The results for all the testing performed are shown
in Table 1, which shows that the parameter J_L has
an important influence on the deformation process
of the specimens. This is given in Fig. 2 which
shows the relationship between strain rate and time
for the various values of J_L tested. It is evident that
as J_L decreases, the shear strain rate increases.

The effect of the applied load for a given value of

Table 1: Summary of testing data obtained

N	D mm	d mm	(D-d)/2 mm	L mm	J mm^4	JL mm^5	strain rate, s^{-1}	stress MPa	strain %	load %	time tγ μs
1	13.34	12.7	0.32	12	555	6660	350	700	8	60	250
2	13.34	12.7	0.32	10	555	5550	350	700	8	60	300
3	13.46	12.7	0.38	8	668	5344	300	350	8	60	300
4	13.46	12.7	0.38	8	668	5344	600	250	16	70	350
5	13.46	12.7	0.38	7	668	4676	630	600	22	60	250
6	13.56	12.7	0.43	5	730	3650	600	700	15	60	20
7	13.56	12.7	0.43	5	730	3650	1000	700	35	70	40
8	13.34	12.7	0.32	4	555	2220	1000	900	25	50	20
9	13.38	12.7	0.34	4	601	2404	1100	900	35	60	20
10	10.41	9.52	0.45	3.5	346	1211	900	900	27	60	10
11	10.41	9.52	0.45	3.5	346	1211	850	650	27	50	40

J_L is shows in Figs. 3 and 4. When the load in the test is decreased by 10% the strain rate increases by 60 to 100% depending on the value of J_L.

The relationship between strain rate, J_L and time to plastic deformation, which is a function of J_L is shown in Fig. 5. This plot shows two distinct regions. Specimens with J_L up to 4000 mm^5 show significant plastic deformation in a very short time period (20-40μs) and high strain rates (up to 1100 s^{-1}) while specimens which have a J_L higher than 4000 mm^5 develop plastic deformation at 200-300 μs and strain rates of 400 s^{-1}.

This value of J_L of 4000 mm^5 appears to be very consistent for all the specimens tested. Considering that J_L depends on the specimen dimensions (D, d, L), a systematic analysis of the role of specimen dimensions has been established which has a value of J_L less than 4000 mm^5.

4 ACKNOWLEDGMENTS

The support of the Natural Sciences and Engineering Research Council of Canada is appreciated. Technical assistance of D. Mardis is also acknowledged.

REFERENCES

Backman, M.E. & S.A. Finnegan 1973. Metallurgical Effects at High Rats of Strain. New York: Plenum Press, 531-543.

Cepus, E., C. D. Liu & M.N. Bassim 1995. Journal de Physique IV, 4(C8):553-558.

Cepus, E. 1994. M. Sc. Thesis. Winnipeg: University of Manitoba.

Costin, L.S., E. E. Crisman, R. H. Hartley & J. Duffy 1979. Inst. Physics Conference, Series 47, 90-100.

Hartley, K.A., J. Duffy & R.H. Hawley 1987. J. Mech. Phys. Solids, 35:283-301.

Leung, E. K. C. 1980. J. Applied Mechanics, 47:278-282.

Rogers, H.C. & C.V. Shastry 1981. Shock Waves and High Strain Rate Phenomena in Metals. New York:221-247.

Senseny, P.E., J. Duffy & R.H. Hawley 1978. J. Applied Mechanics, 45:60-66.

Experimental Mechanics, Allison (ed.) © 1998 Balkema, Rotterdam, ISBN 90 5809 014 0

An approach to the penetration of ceramics

N. L. Rupert & F. I. Grace
US Army Research Laboratory, AMSRL-WM-TA, Aberdeen Proving Ground, Md., USA

ABSTRACT: This paper describes an extension to a previous penetration analysis by Grace and Rupert that identified a dynamic effect mechanism for both metallic and ceramic applique targets. The previous analysis used titanium/rolled homogeneous armor (Ti/RHA), RHA/Ti, and alumina/RHA targets to demonstrate that evaluation of materials using depth of penetration (DOP) testing should take into account physical phenomena regarding material damage to include shock-induced transient and wave reflections from target interfaces. In the current study, titanium diboride/RHA (TiB_2/RHA) and TiB_2/Ti bi-element targets are used to further examine isolation of time-dependent strength effects in ceramic DOP responses and explore a possible approach for normalizing ceramic ballistic data. Particular materials selected allowed sets of density-matched and strength-matched bi-element target combinations to be considered. Experimental data are presented together with an analysis that isolates time-dependent strength effects from other penetration phenomena within these materials.

1 NOTATION

a_N - Normalized thickness of first element in terms of a first element reference material (mm)

a_o - Thickness of first element (mm)

c_{RSW} - Rayleigh wave/shear wave velocity for ceramic damage model (m/s)

DOP   - Depth-of-penetration (mm)

DOP'   - Velocity corrected DOP results (mm)

DOP_N - Normalized DOP results in terms of the two reference materials (mm)

P_{DE} - Penetration shift due to density effect (mm)

$P_{S(Al_2O_3)}$ - Undamaged semi-infinite Al_2O_3 AD-995 CAP 3 penetration value based on theoretical calculations (mm)

$P_{S(RHA)}$ - Semi-infinite RHA penetration value based on experimental data or theoretical calculations (mm)

P_{S1} - Undamaged semi-infinite first element penetration value based on experimental data or theoretical calculations (mm)

P_{S2} - Semi-infinite second element penetration value based on experimental data or theoretical calculations (mm)

v_s - Striking velocity (m/s)

$\acute{\eta}$ - Damage growth rate coefficient (s^{-1})

2 INTRODUCTION

The depth of penetration (DOP) test method has proven to be an inexpensive and valuable tool for comparative testing and ranking of ceramics. However, the analysis used in the DOP tests for determining the performance potential of ceramic materials (or any material) under ballistic impact must go beyond recording the residual penetration in the second target element as a function of the first element's thickness or areal density for a given combination of penetrator and impact velocity. The analysis needs to be more detailed to determine the actual potential of the ceramic eliminating potential bias introduced through such choices as target geometry and the selection of the second target element. The previous analysis by Rupert & Grace (1993) and Grace & Rupert (1993) identified a dynamic effect referred to as a "density" effect mechanism for both metallic and ceramic appliques. In that study, it was concluded that the density of the backup element is responsible for a significant target interaction effect that previously contributed to lack of strength in thin sections (<20 mm), and this can substantially alter perceived performance.

Titanium diboride/rolled homogeneous armor (TiB_2/RHA) and titanium diboride/titanium (TiB_2/Ti) bi-element target designs are used in the current work to further examine the density effect and explore the possibility of normalization of ceramic DOP data. The particular materials selected allowed sets of density-matched and strength-matched bi-element target combinations to be considered. The experimental data are presented together with an analysis of the mechanisms acting in the penetration of these materials.

3 MATERIALS

3.1 Steel

Standard U.S. Army practice measures armor performance in terms of a standard reference steel or RHA steel. Military specifications for the manufacturing process and material properties of RHA are described in MIL-A-12560G (U.S. Army Materials and Mechanics Research Center [AMMRC] 1984). Typical room-temperature property data for RHA were measured from random 100- to 152-mm RHA plates used at the U.S. Army Research Laboratory (ARL) and are listed in Table 1. Large variations in the property data for the RHA are the result of MIL-A-12560G being a performance-based specification. To minimize the effects that these variations could have on the targets ballistic performance, targets were fabricated from a single thickness of plates from the same heat.

Table 1. Property data.

Property	RHA steel[*]	Ti-6Al-4V	TiB$_2$[**]
Density (g/cm^3)	7.85	4.45	4.48
Hardness (Bhn)	241-375	302-340	2,700[***]
Avg. grain size (µm)	NA[****]	NA	15
Comp. strength (GPa)	NA	NA	4.82
Tensile strength (MPa)	793-1,172	896-910	NA
Yield strength (MPa)	655-1,055	827-862	NA
Elongation (%)	8-20	10-12	NA
Young's modulus (GPa)	207	113.8	556
Poisson's ratio	0.29	0.342	0.11
Sonic velocity:			
Longitudinal (m/s)	5,876	6,070	11,285
Transverse (m/s)	3,196	2,974	7,431

[*] MIL-A-12560G
[**] CERCOM, Inc.
[***] Knoop 300 g (kg/mm^2)
[****] Not available.

3.2 Titanium

A low-cost Ti-6Al/4V alloy was selected for this study, since its strength and sound velocity are similar to those of RHA. Property data for armor plates used in this study are listed in Table 1 (Burkins et al. 1994). Targets using Ti were fabricated from a single plate.

3.3 Titanium diboride

Selection of this ceramic was based on two factors. First, the TiB$_2$ has a density similar to that of the Ti alloy. Second, the TiB$_2$ has an acoustic impedance

(product of density and sound velocity) close to that of the RHA. A high-purity TiB$_2$ manufactured by CERCOM Incorporated, Vista, CA, was used in this study. These tiles were nominally 101.6 mm square with thicknesses ranging from 6.35 mm to 35 mm. Additional property data for the ceramic tiles used in this study are listed in Table 1.

4 TEST PROCEDURES

4.1 DOP testing

Performance is generally measured by the DOP of a penetrator into a semi-infinite reference material backplate after passing through a ceramic applique. Ceramic performance comparisons are then made between selected baseline materials using performance maps (as illustrated in Fig. 1). Initially, four performance regions were identified, based on the general appearance of the data when plotted. Later work by Grace & Rupert (1993) provided physical explanations for these four regions. Region I was defined as the region where the dynamic target interaction or density effect is the dominate factor for thin ceramic sections. Region II is the region where time-dependent damage mechanisms, resulting in a mixed solid/granular flow, and the rate of transition from this flow to pure granular flow are dominant. Region III is defined as a region where pure granular flow becomes more dominant in the penetration of the ceramic. Rigid-body penetration of the damaged ceramic may also occur in this region. Region IV is the termination phase, where the unconsumed rod traveling at low velocity abruptly decelerates and stops.

4.2 Projectiles

The tungsten heavy alloy (WHA) projectile used in this study was the 65-g, 93% W-4.91% Ni-/2.11% Fe long-rod penetrator manufactured by Teledyne Firth Sterling. The penetrator was produced using the nominal X-21 process with 25% swaging. The penetrator had a diameter of 7.82 mm, a length-to-diameter (L/D) ratio of 10, and a hemispherical nose. Nominal material properties for these penetrators are as follows: density was 17.6 g/cm^3, hardness was Hardness Rockwell C Scale 40.5-42.6, yield strength was 1.09-1.17 GPa, ultimate strength was 1.13-1.21 GPa, and elongation was 5.8-10.6% (Teledyne Firth Sterling 1991, Farrand 1991).

4.3 Range setup

The penetrators were fired from a laboratory gun consisting of a Bofors' 40-mm-gun breech assembly with a custom-made 40-mm smoothbore barrel. The gun was positioned approximately 3 m in front of the targets. High-speed (flash) radiography was used to

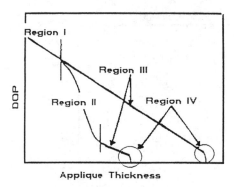

Fig. 1. Generalized performance map.

record and measure projectile pitch and velocity. Two pairs of orthogonal x-ray tubes were positioned in the vertical and horizontal planes along the shot line (as illustrated in Fig. 2). Propellant weight was adjusted for the desired nominal velocity of 1,500 m/s. Projectiles with a striking total yaw in excess of 2° were considered a "no test," and those data were disregarded.

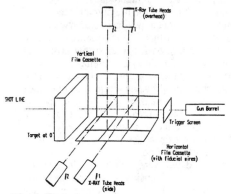

Fig. 2. Test setup.

4.4 *Target construction*

TiB$_2$ DOP target construction followed general features of the standard design (as shown in Fig. 3) (Woolsey et al. 1989, 1990). For the targets used here, no aluminum foil was used, but, rather, the ceramic tile was glued directly to the RHA backing plate. Targets shot in this test series consisted of 101.6-mm (4 in)-square ceramic tile. The ceramic tile was held into a steel lateral confinement frame by EPON 828 and VERSAMID 140, with a mixing ratio of 1:1 and a nominal thickness of 0.5 mm. The frame had a 19-mm (3/4 in) web and a depth equal to or greater than the tile thickness. The frame was mechanically secured to a thick steel backup plate. This steel backup plate was made from a single RHA

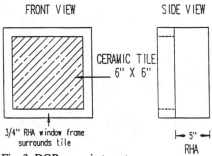

Fig. 3. DOP ceramic target.

steel, MIL-A-12560G, Class 3, 127 mm (5 in) thick, with a nominal hardness of R$_c$ 27. A 102-mm (4 in)-thick Ti-6Al-4V plate was substituted for the 127-mm RHA plate in targets requiring a Ti second element.

5 BALLISTIC RESULTS

5.1 *Penetration into monolithic baseline materials*

Baseline penetration data for RHA and Ti were taken from Woolsey et al. (1990) and Burkins et al. (1994), respectively. For the RHA, a linear empirical fit to the penetration data between 800 m/s and 1,750 m/s was derived. The resulting equation is as follows:

$$DOP_{(RHA)} = 0.0833 \left(\frac{mm \cdot s}{m} \right) V_s - 55.88 \, mm \quad (5.1)$$

where V_s is the striking velocity in meters per second, while DOP and the right-hand constant are in millimeters. To correct for variations in the actual striking velocities, all residual penetration values for ceramic and metallic bi-element targets with RHA second elements were adjusted to a striking velocity of 1,500 m/s by the following correction based on equation (5.1):

$$DOP' = DOP + 0.0833 \left(\frac{mm \cdot s}{m} \right) \left(1,500 \frac{m}{s} - V_s \right) (5.2)$$

where, again, the units are meters per second for velocity and millimeters for depth. This technique is considered uniformly valid for different materials, when a significant amount of the rod length and, therefore, penetration occurs in the RHA steel backplate (Woolsey et al. 1990).

Penetration data for the monolithic Ti-6Al-4V are based on five tests, wherein striking velocity ranged from 1,079 m/s to 1,672 m/s. Over this range, the data are linear; an empirical fit to the data was derived. The resulting equation is as follows:

$$DOP_{(Ti)} = 0.108 \left(\frac{mm \cdot s}{m} \right) V_s - 81.7 \, mm. \qquad (5.3)$$

The Ti alloy second elements targets were normalized to a striking velocity of 1,500 m/s by the following correction, based on equation (5.3):

$$DOP' = DOP + 0.108 \left(\frac{mm \cdot s}{m} \right) \left(1,500 \frac{m}{s} - V_s \right) \quad (5.4)$$

where, again, the units are meters per second for velocity and millimeters depth.

5.2 Penetration into TiB₂/Ti and TiB₂/RHA

DOP' results for TiB₂/RHA and TiB₂/Ti targets (15 each) are shown in Figures 4 and 5. The open diamonds in Figure 4 provide individual data points for the TiB₂/RHA. The open diamond at 0 applique thickness represents a semi-infinite RHA target. The solid diamond represents the semi-infinite DOP datum point for RHA, based on equation (5.1). The open triangles in Figure 5 plot the individual data points for the TiB₂/Ti bi-element targets. The solid triangle represents the semi-infinite DOP for Ti, based on equation (5.3). Corrections for striking velocity variations were generally less than 2 mm for the current data set using equations (5.2) and (5.4).

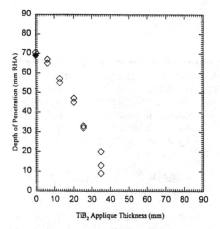

Fig. 4. DOP results for TiB₂/RHA.

6 ANALYSIS AND DISCUSSION

The experimental results show both similarities and contrasting differences to the prior behavior as depicted in DOP performance maps. The TiB₂ ceramic appears to have lower performance associated with Region I when backed by RHA (Fig. 4) in contrast to its relatively high performance (lacking a Region I) when backed by Ti alloy (Fig. 5), even in thin sections (<20 mm). With the density effect analysis, it is believed that the TiB₂ maintains

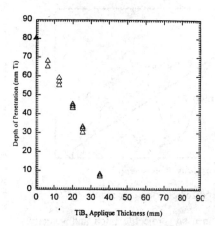

Fig. 5. DOP results for TiB₂/Ti.

its nominal strength in Region I in either case, but becomes subject to time-dependent damage (strength loss) with increased ceramic thickness as the performance extends into Region II. Both data sets show that, with increased ceramic thickness, Region II performance is maintained, as would be expected, based on either the full nominal strength of the ceramic or an initial transition into damage within the ceramic. Further performance losses, characteristic of Region III, wherein the process is dominated by granular flow, are absent.

Previous work (Grace & Rupert 1993) modeled the initial performance shift using the shock-impedance equation, reduced to density only, to describe the density effect. The current data also do not support a shock-impedence effect but rather a density effect. Thus, the previous density effect equation is used here as an empiric relation.

The DOP data were analyzed further by introducing a more extensive normalization of the data by equations (6.1a and b). Reference materials for this normalization were Coors' AD-995 Al₂O₃ CAP 3 ceramic in the first element and RHA in the second element. The first factor in equation (6.1a) subtracts out the target interaction effect contained in Region I. This factor is represented as (DOP' – P_DE), where P_DE is the shift along the ordinate due to the density effect as calculated using the analysis of Grace & Rupert (1993). In addition, the shifted DOP results are multiplied by the ratio of semi-infinite penetration for RHA relative to the second element. The ceramic thickness along the abscissa is adjusted by a factor to reflect differences in semi-infinite penetration into the first element when compared to that of Al₂O₃. Further, to incorporate the time-dependent damage effects within Region II, a damage function is introduced as an additional adjustment to the DOP. This factor reflects any slope changes associated with higher or lower rates of damage in the ceramic material. The preceding factors result in the following equations for a normalized lower performance limit DOP_N:

$$DOP_N = \frac{P_{S(RHA)}}{P_{S2}}(DOP' - P_{DE})$$

$$- \frac{P_{S1} P_{S(RHA)}^2}{P_{S2} P_{S(Al_2O_3)}} \int_0^{a_o} f(c_{RSW}, a_o, \acute{\eta}) da \quad (6.1a)$$

$$a_N = a_o \frac{P_{S(Al_2O_3)}}{P_{S1}}. \quad (6.1b)$$

Transition between penetrator materials may be accomplished by defining DOP', P_{DE}, P_{S1}, and P_{S2} performance in terms of the nonreference penetrator material.

It should be noted that $f(c_{RSW}, a_o, \acute{\eta}) = 0$ for metal-metal targets or ceramic-metal targets where the ceramic tiles exhibit their intact nominal strength throughout the penetration process. Estimates for $P_{S(Al_2O_3)}$ and P_{S1} were obtained using intact strengths of 2.0 GPa (Al_2O_3) and 2.78 GPa (TiB_2) in a one-dimensional penetration model (Grace 1993). The values were $P_{S(Al_2O_3)}$ = 22.71 mm and P_{S1} = 12.73 mm. For material that exhibits time-dependent damage, the $f(c_{RSW}, a_o, \acute{\eta})$ can be determined from the data graphically or calculated by suitable damage models of Cortés et al. (1992) and Curran et al. (1993) or by Grace & Rupert (1993). The normalization in equations (6.1a and b) with the defined $f(c_{RSW}, a_o, \acute{\eta})$ permits a convenient framework through which various factors affecting damage growth rates can be examined. Further, the function should account for all influencing factors to include statistical property variations within the ceramic, as related to the time-dependent damage, such as flaw densities, flaw orientation, etc., as well.

In Figure 6, lower performance limits (upper data points) are presented for TiB_2/RHA, TiB_2/Ti alloy, and Al_2O_3/RHA. The Al_2O_3 thicknesses have been normalized by the ratio of Al_2O_3 density to that of TiB_2 (abscissa). The TiB_2/RHA, TiB_2/Ti alloy, and Al_2O_3/RHA DOP results have been normalized and presented as percent of the semi-infinite penetration for the second element (ordinate). Both TiB_2/RHA and Al_2O_3/RHA systems show the characteristic shift of Region I for the thin ceramic appliques. Further, the slopes of TiB_2/RHA and TiB_2/Ti alloy are similar in Region II, indicating that the time-dependent strength and damage growth rate of the TiB_2 applique may be independent of the backing material so long as sufficient support is provided. The lower slope for the Al_2O_3/RHA may reflect the effects of its lower nominal strength relative to TiB_2 and/or a higher

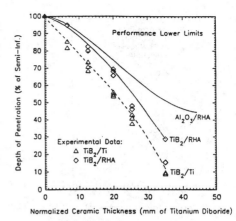

Fig. 6. Normalized data plot.

time-dependent damage growth rate relative to TiB_2 on the penetration process. The nominal intact strength used in the modeling (undamaged ceramic strength) was found to correlate well with the shear component of the compressive strength (Grace & Rupert 1993). Thus, intact strengths used in the modeling here were 2.0 GPa (Grace & Rupert 1993) for Al_2O_3 and 2.78 GPa for TiB_2. The change in curvature of the Al_2O_3 line, representing the lower performance limit at large thickness, suggests a transition toward the higher damage states associated with Region III (pure granular flow).

In Figure 7, ceramic data from the present work are plotted together with the Al_2O_3 lower performance limit curve and the metal-metal response data (Rupert & Grace 1993) using the normalization process of equation (6.1) without the damage function correction. The normalization places all of the metal-metal data on the same curve by subtracting out the shift in DOP response for the Ti/RHA target. Further, when all major factors considered in the penetration process are normalized, except for $f(c_{RSW}, a_o, \acute{\eta})$, the lower performance limits of the two ceramics are not much different. This result suggests that the factors within the function combine so that the overall time-dependent strength loss and its effect on the DOP are approximately equal on a normalized basis for these particular ceramic materials. From Figure 7, the average damage function value for both ceramics is estimated to be 1.47 mm of RHA per millimeter of Al_2O_3. This value was estimated by taking the difference of the slopes between the DOP_N curve and a least-square fit of the ceramic data as plotted in Figure 7. However, the lower performance limits may be more useful for armor designs. Damage function values based on these limits are 1.35- and 1.16-mm of RHA per millimeter of Al_2O_3 for Al_2O_3 and TiB_2, respectively. When the damage function is taken into account in equation (6.1), the ceramic data can be further normalized so that all DOP response

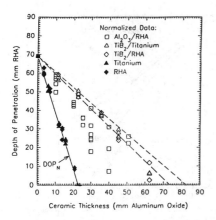

Fig. 7. Partially normalized DOP$_N$ results.

points would lie near the single curve given by the data points for the metal targets. Thus, the damage function has been isolated from the various other contributing factors important to the penetration process. Research in the future can address the specific nature of the damage function.

7 SUMMARY

In this work, DOP responses of TiB$_2$/RHA and TiB$_2$/Ti alloy bi-element targets have been determined and contrasted to the general ceramic behavior in DOP performance maps. The TiB$_2$ ceramic performed very well in thin sections when backed by Ti alloy, demonstrating high ceramic strength. Although the TiB$_2$/RHA targets did not perform as well as the TiB$_2$/Ti alloy systems in thin ceramic sections, the difference was due to a target/target interaction during penetration, or density effect, as a result of the density order of the bi-element targets rather than loss of strength. Lower performance bounds for the TiB$_2$ and baseline Al$_2$O$_3$ ceramic show the density effect is present in these systems. These results are consistent with prior findings regarding Al$_2$O$_3$/RHA responses (Grace & Rupert 1993). The present analysis has been able to separate the overall mechanical response of the penetration process from that which involves time-dependent damage response (strength loss) in the ceramic during penetration. The normalization equation includes the separation of factors and defines a damage function to be determined experimentally or analytically. The model, including the isolated damage function, provides direction for future research efforts.

REFERENCES

Burkins, M.S., J.S. Hansen, & J.I. Paige 1994. Armor applications for Ti-6Al-4V. ARL-MR-1146. U.S. Army Research Laboratory, Aberdeen Proving Ground, MD, USA.

Cortés, R., C. Navarro, M.A. Martínez, J. Rodríguez, & V. Sánchez-Gálvez 1992. Numerical modeling of normal impact on ceramic composite armors. *International journal of impact engineering.* 12:639-651.

Curran, D.R., L. Seaman, T. Copper, & D.A. Shockey 1993. Micromechanical model for communication and grandular flow of brittle material under high strain rate application to penetration of ceramic targets. *International journal of impact engineering.* 13:53-83.

Farrand, T.G. 1991. Unpublished data. U.S. Army Ballistic Research Laboratory, Aberdeen Proving Ground, MD, USA.

Grace, F.I. 1993. Non-steady penetration of long rods into semi-infinite targets. *International journal of impact engineering.* 14:303-314.

Grace, F.I., & N.L. Rupert 1993. Mechanisms for ceramic/metal, semi-infinite, bi-element target response to ballistic impact. *Ballistics '93, 14th international symposium.* TB-16. 2:361-370. Quebec City, Quebec, Canada.

Rupert, N.L., & F.I. Grace 1993. Penetration of long rods into semi-infinite, bi-element targets. *Ballistics '93, 14th international symposium.* TB-27. 2:469-478. Quebec City, Quebec, Canada.

Teledyne Firth Sterling 1991. Material certification sheet. BRL contract number DAAD05-90-C-0431. LaVergne, TN, USA.

U.S. Army Materials and Mechanics Research Center (AMMRC) 1984. Armor plate, steel, wrought, homogeneous (for use in combat vehicles and for ammunition testing). MIL-A-12560G. U.S. Army Materials and Mechanics Research Center. Watertown, MA, USA.

Woolsey, P., S. Mariano, & D. Kokidko 1989. Alternative test methodology for ballistic performance ranking of armor ceramics. *Proceedings of the fifth TACOM armor conference.* Monterey, CA, USA.

Woolsey, P.D., D. Kokidko, & S. Mariano 1990. Progress report on ballistic test methodology for armor ceramics. *Proceedings of the first TACOM combat vehicle survivability symposium.* Monterey, CA, USA.

Experimental Mechanics, Allison (ed.) © 1998 Balkema, Rotterdam, ISBN 90 5809 014 0

Effects of strain rate and temperature on the behaviour of an elastomer

N. Bonnet & T. Thomas
DCE, Centre d'Etudes et de Recherches d'Arcueil, France

J. L. Lataillade
Laboratoire Matériaux Endommagent Fiabilité, ENSAM, France

ABSTRACT: The use of rubber-like materials are considered for applications in impact loading environments where they would be subjected to high strain rates and high pressure levels. The objective of this study was to determine the dynamic behaviour of a particular elastomer. Firstly, the strain rate influence on the hydrostatic part of the behaviour is studied by various experiments (hydrostatic, piston-displacement, split Hopkinson bars and plate impact experiments). Secondly, investigations for studying the deviatoric part of the behaviour were performed. Preliminary technological experiments have shown that in service, pressure levels are higher than the glass transition pressure. Therefore, the deviatoric part of the behaviour corresponds to the glassy state of the material. Several kinds of uniaxial compression tests at low temperature at quasi static and high strain rate, or high pressure level have shown an elastic viscoplastic behaviour. The strain rate and temperature influence on yield stress has been particularly studied.

1 INTRODUCTION

Rubber-like materials are usually used for sound insulation and structure vibrations damping. These uses are restricted to small amplitude and low frequency vibrations. However, their utilisation in impact loading environments is often considered, but a lack of data on these materials in high strain rate and high pressure conditions prevents a convenient behaviour modelling.

The aim of the study is to propose a behaviour law for an elastomer filled by carbon black particles and made of two polymers: a polynorbornene and EPDM. This law will be determined with experimental observations and will have to be valid in working conditions that is for strain rates up to 10^4 s^{-1} and high pressure levels (several GPa).

The in service solicitation leads to a high confining level into the material. It is why we have studied in a first time the influence of strain rate on the response of the material when submitted to an hydrostatic loading.

However, the deviatoric part of the behaviour could play an important part in the real solicitation. After having shown that in service pressure levels combined with high strain rates lead to consider the glassy state of the elastomer, we have studied the influence of strain rate and temperature on the elastic viscoplastic behaviour and more particularly on the yield stress. This work was made for temperatures lower than the glass transition temperature.

2 PRESSURE-VOLUME BEHAVIOUR

Most of experimental tests were conduced in uniaxial deformation condition. We will suppose that the stress field is approximately an hydrostatic stress state. This assumption has been demonstrated by taking into account a deviatoric component and a bulk component in the specific free energy. The deviatoric component has been represented by an hyperelastic neo-Hookean law and the bulk component by a linear law.

Then, for an uniaxial strain test, we have shown that the differences between the axial stress and the hydrostatic stress could be neglected.

Thus, the stress data gathered in these experiments are shown as pressure versus volume curves.

2.1 Experimental tests

The two kinds of quasi static tests (10^{-4} s^{-1}) and the two kinds of dynamic tests (10^3 to 10^4 s^{-1}) that were performed are presented below:

The piston displacement method: experimental data were obtained with the piston-cylinder pressure vessel set up at the "Laboratoire de Recherches et de Contrôle du Caoutchouc et des Plastiques". This apparatus was used at pressures up to 300 MPa. The sample is placed in a steel container whose cavity diameter equals the sample one. The volume strain is determined from the piston displacement from which pressure is generated. After a loading-unloading cycle the elastomer presents an elastic response.

Hydrostatic compression :
experimental data were obtained by means of a high pressure vessel with a maximum hydrostatic pressure of 2 GPa, set up at the Centre d'Etudes de Gramat. Volume strain is obtained from longitudinal and transverse strain measured by two gauges stuck on the sample. The loading-unloading response exhibits a non-linear elastic stress-strain curve. The bulk modulus, K, was deduced from the slope of a plot of P versus $\ln[V/V_0]$ (Lundin 1994). The quantity K(P) showed a change in slope at the glass transition and the low-pressure of this step was taken to define the glass transition pressure $P_g \cong 260$ MPa.

Confined split Hopkinson bar tests (CSHBT):
to ensure a lateral confining, we have modified the standard test by adding a steel ring around the sample (Figure 1).

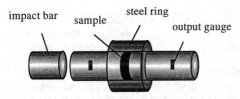

Figure 1: representation of CSHBT

These tests were found to measure the specimen volume change with a non-constant strain rate. But, the pressure-volume data were superimposed with data of piston-displacement tests and these results suggest that the elastomer response is fluid-like and strain rate indépendant (figure 3).

Plate impact tests: five plate impact experiments were performed using the gas gun at the Centre d'Etudes de Gramat. Electromagnetic velocity gauges were used for obtaining the particle velocity histories at different locations in the elastomer (Gupta 1984 & Gupta 1992). A Lagrangian analysis

(Cagnoux et al. 1987) leads to the stress-strain curves (figure 2).

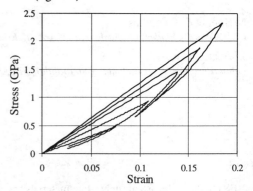

Figure 2 : loading and unloading paths for the five plate impact experiments

Only the end states are considered to be thermodynamic equilibrium states and it is these end states that should be compared with high pressure quasi static data (figure 3). The differences in the impact stress data (locus of the end states from different experiments) and hydrostatic data are found very small.

2.2 Synthesis of the experimental results

The pressure volume data obtained from experiments described above are shown in figure 3.

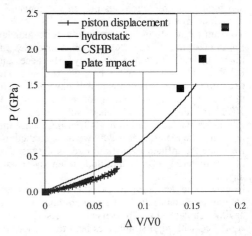

Figure 3: Confined compression at strain rates from 10^{-4} s^{-1} to 10^4 s^{-1}.

The bulk modulus on impact condition is found equal to that determined from hydrostatic loading. The differences observed between results of piston-displacement tests and hydrostatic tests is related to some measure imprecision.

In conclusion, these results show that the pressure volume behaviour may be represented by an strain rate independent relationship in the tested strain rate range, unlike an other work on a Polyvinyl Chloride-based elastomer (Dandekar 1990).

2.3 Behaviour law

For the behaviour description, we use the hydrodynamic framework where equations of balance are relative to the propagation of a shock wave in the material. Rankine-Hugoniot equations provide three equations linking four variables. One additional relationship is needed to complete the determination of all variables. This relation is called Hugoniot curve and reflects the specific behaviour of the material (Davison 1979).

This relation associates values of shock velocity U to the particle velocity u in the form $U = c_0 + su$. It has been found experimentally that some hundreds of materials are described with good approximation over the entire range of plate impact data by the linear relation between U and u (Davison 1979). Results of plate impact experiments provide c_0 and s coefficients.

By combining this relation with equations of balance we obtain equation (1):

$$P = \frac{\rho_0 C_0{}^2 \eta}{(1 - s\eta)^2} \quad \text{with} \quad \eta = 1 - \frac{\rho_0}{\rho} = \frac{V_0 - V}{V_0} \quad (1)$$

2.4 Finite Element simulation

To validate the behaviour law described above, we have tried to simulate the confined split Hopkinson bar test using the explicit version of the ABAQUS finite element code.

The comparison between experimental and simulated results is carried out by comparing the stress levels on the input and output bars at the gauges locations.

We observe in figure 4 a very good agreement between experiments and simulation.

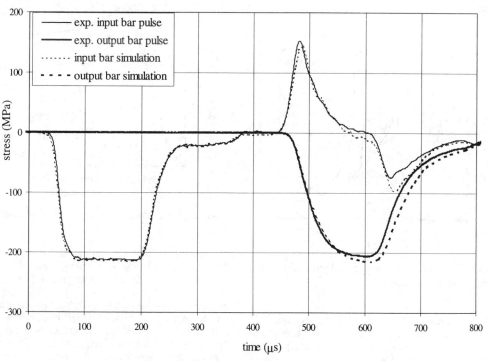

Figure 4: experiments and simulation of CSHBT

3 STUDY OF THE DEVIATORIC PART

3.1 *Preliminary study: real test simulation*

We have seen above that the pressure volume behaviour is strain rate independent, which is not the case for the deviatoric part. According to the strain rate, temperature or pressure levels in the elastomer, the behaviour could be representative of three different states: rubbery, viscoelastic and glassy.

Therefore, it is important to determine the behaviour zone to study.

For this, a technological test which is representative of the in service impact condition is carried out and simulated. The results allow the observation of pressure levels in the elastomer. If the pressure level is higher than the pressure glass transition, which is measured at 260 MPa from the results of hydrostatic tests, the deviatoric behaviour to study will be related to the glassy state of our material.

The technological test consists to carry out a detonation of an explosive placed in between two steel plates. So, the back plate of the system is projected on the tested elastomer. The instrumentation is performed by chronometric gauges and a stress gauge placed in the centre of the rubber plate. The corresponding simulation is performed with the explicit ABAQUS finite element code.

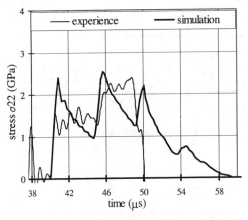

Figure 5: experimental and simulation results for technological impact test.

We have compared on figure 5 the evolution of experimental and simulated stress at the gauge location. The pressure levels are effectively higher than the glass transition pressure. So, at ambient

temperature, pressure effects added to strain rate effects induce a glassy state into the material.

3.2 *Strain rate effects on glassy behaviour for uniaxial compression tests*

The strain rate dependence of the elastomer has been investigated through uniaxial compression tests over the range $\dot{\varepsilon} = 10^{-4}\,\mathrm{s}^{-1}$ to $\dot{\varepsilon} = 10^{3}\,\mathrm{s}^{-1}$ for two temperatures -40°C and -50°C which are below the glass transition temperature (-30°C). At low strain rates we used a static tensile test machine, the strain is calculated from the apparatus displacement. At high strain rate we used the split Hopkinson bars system. The temperature conditioning is produced by using a thermal insulation container (which surrounds the specimen and compression plates or a part of the split Hopkinson bars) in which circulates a liquid nitrogen flow. The compression behaviour presented in figure 6 exhibits an initial elastic response, followed by yield, strain softening and plastic flow. The initial yield stress is found to increase with decreasing temperature or with increasing strain rates.

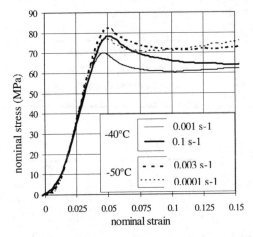

Figure 6: stress strain behaviour of the elastomer

The results are analysed through the Eyring model (Ward 1983) and through the Argon model (Argon 1973) modified by Boyce (Boyce et al. 1988).

Eyring considers that a molecule has to overcome an energy barrier in moving from one position to another in the solid. When the system is stress free, the segments of the polymer jump very infrequently over the barrier and they do so in random directions. But the application of a shear stress modifies the

924

energy barrier height. This analyse conduced Eyring to propose equation 2 to link the yield stress to the strain rate.

$$\frac{\tau_{th}}{T} = \frac{R}{V^*}\left[Ln\frac{\dot{\gamma}}{\dot{\gamma}_0} + \frac{E}{RT}\right] \qquad (2)$$

where V^* = activation volume, E = activation energy, R = gas constant

Argon has proposed a theory based on another concept. The deformation at a molecular level consists in the generation of a pair of molecular kinks. The resistance to double kink formation is considered to arise from intermolecular forces. The intermolecular energy change associated with a double kink is then calculated by modelling these as two wedge disinclination loops. Boyce has modified this theory to introduce the effect of pressure. Then the relation between shear yield stress, temperature, pressure and strain rate is presented in equation 3.

$$\tau_{th} = [s+\alpha P]\left[1 + \frac{T}{A[s+\alpha P]}Ln\left[\frac{\dot{\gamma}}{\dot{\gamma}_0}\right]\right]^{6/5} \qquad (3)$$

where A, s and γ_0 are temperature dependent parameters.

The coefficient of pressure dependence is measured by quasi-static triaxial compression tests using high pressure vessel available at the "Centre d'Etudes de Gramat". Triaxial compression shows a linear augmentation of the yield stress in relation to the confining pressure (figure 7).

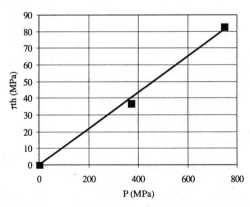

Figure 7: yield stress against pressure at room temperature

The slope α of the yield stress versus pressure is the coefficient of pressure dependence in equation (3) and α =0.108.

All other parameters of these laws are identified from uniaxial compression tests results. We have considered the yield nominal stress because of the dispersion in displacement measurement and the corresponding shear yield stress is calculated from Von-Mises criterion. The results are presented in tables 1 and 2 and in figures 8 and 9.

Table 1: Eyring's model parameters

Temperature °C	V^*/N nm^3	E kJ/mol
-40	3.56	102
-50	3.24	103

Table 2 : modified Argon model parameters

Temperature °C	s MPa	A K/MPa	γ_0 s^{-1}
-40	82.5	288.3	1.06 10^{18}
-50	92	254.5	3.87 10^{18}

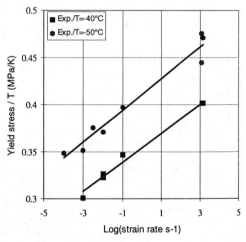

Figure 8: experimental results and corresponding Eyring's law

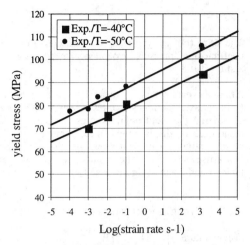

Figure 9: experimental results and corresponding Argon modified law

These results show the importance of yield stress temperature and strain rate dependence. The use of Eyring and Argon model allows a satisfactory description of experimental results in the range of 10^{-4} s^{-1} to 10^3 s^{-1}.

4 CONCLUSION

The study presented in this paper of which the purpose was to determine the dynamic behaviour of an elastomer in the range of 10^{-3} s^{-1} to 10^4 s^{-1} strain rate and up to 2 GPa is divided into two parts. The first one concerns the strain rate influence on the hydrostatic behaviour part. Experimental results analysis shows that this influence is negligible in the strain rate range of 10^{-4} s^{-1} to 10^4 s^{-1}. The bulk behaviour is then correctly represented by a non-linear pressure volume law. This law allowed the finite element simulation of a confined split Hopkinson bars test and the results agree well with experimental data.

The second part concerns the study of the strain rate and temperature influence on the deviatoric part behaviour. Through the finite element simulation of a technological test, we have shown that the combined effects of pressure levels and strain rates borne by the in service material leads the elastomer to its glassy state. So, this glassy state has been investigated over the range of 10^{-4} s^{-1} to 10^3 s^{-1} at two different temperatures (-40°C and -50°C) by means of uniaxial compression tests. The material has an elastic viscoplastic behaviour for which we focused more particularly on strain rate and temperature effects on the yield stress. We show how Eyring's model and Boyce modified Argon model can predict the strain rate evolution of the yield stress.

Acknowledgements: LRCCP, CEG, B. Chartier, J. Clisson and col.

REFERENCES

Argon A.S., "A theory for the low temperature plastic deformation of glassy polymers", Philosophical magazine, 28 (1973) 839-865

Boyce M.C, Parks D.M. & Argon A.S., "Large inelastic deformation of glassy polymers", Mechanics of materials, 7 (1988) 15-33

Cagnoux J., Chartagnac P., Hereil & P. Perez M., "Lagrangian analysis. Modern tool of the dynamics of solids", Annales de physique, 12 (1987) 451-524

Dandekar D.P., "Deformation of a polyvinyl based elastomer subjected to shock loading", Shock compression of condensed matter, (1990), 97-100

Davison, L., "Shock compression of solids", Physics reports 55 n°4, (1979)

Lundin A., "Glass transition in a polymer under pressure: isothermal bulk modulus and relaxation effects", High Temperatures- High Pressures, (26) 1994, 385-391

Gupta S.C. & Gupta Y.M, "High strain rate response of an elastomer", High pressure research, 10 (1992) 785-789

Gupta, Y.M., "High strain rate shear deformation of a polyurethane elastomer subjected to impact loading". Polymer engineering and science. 4 n°11 (1984), 851-860

Ward I.M., "Mechanical properties of solids polymers", second edition, Wiley J.& sons, 1983

Experimental Mechanics, Allison (ed.)© 1998 Balkema, Rotterdam, ISBN 90 5809 014 0

Identification of mechanical properties of component layers in a bimetallic sheet

F. Yoshida
Department of Mechanical Engineering, Hiroshima University, Japan

M. Urabe
Graduate School of Engineering, Hiroshima Universty, Japan

V. V. Toropov
Department of Civil and Environmental Engineering, University of Bradford, UK

ABSTRACT: This paper deals with the identification of the mechanical properties of the individual component layers in a bimetallic sheet. A set of material parameters in a chosen constitutive model with isotropic/kinematic hardening plastic potential was identified using two types of experimental results: the tensile load versus strain curve in the uniaxial tension test and the bending moment versus curvature diagram in the cyclic bending test. An optimization technique based on the iterative multipoint approximation concept was used for the identification of the material parameters. The present paper describes the experimentation, the fundamentals and the technique of the identification, and the verification of this approach.

1 INTRODUCTION

Since the press-formability of bimetallic sheets strongly depends on the mechanical properties of component layers and the layer-thickness ratio, it is important to determine the properties for their individual component layers (Hawkins & Wright 1971, Semiatin & Pihler 1979a, Yoshida et al. 1995, 1997, Kim & Yu 1997). For that purpose, uniaxial tension tests are usually performed on the individual component layers taken from a bimetallic sheet(Semiatin & Pihler 1979b). However, such tension tests have the following drawbacks:

- a time-consuming process of the removal of a layer from a bimetallic sheet by a mechanical or chemical processing is necessary for the preperation of the specimens,
- the stress-strain response during cyclic loading cannot be obtained.

Concerning the latter problem, Yoshida et al. (1998) have recently presented a method of the determination of stress-strain response for monolithic sheet metals from cyclic bending tests. The material parameters in a constitutive model have been identified by minimizing the difference between the test results and the results of the corresponding numerical simulation using an advanced optimization technique (Toropov et al. 1993a).

As an extension of the cyclic bending method, the present paper describes an attempt to identify the mechanical properties of the individual component layers in a bimetallic sheet without the time-consuming removal of a layer. In this research work, a set of material parameters in a chosen constitutive model (Chaboche & Rousselier 1983) is identified using two different types of experimental results for a whole bimetallic sheet. One of them is presented by the tensile load versus strain curve in the uniaxial tension test, and the other by the bending moment versus curvature diagram in the cyclic bending test. The experiments were conducted on a stainless steel clad aluminium sheet.

The identification of the material parameters is based on minimizing the difference between the test results and the corresponding results of numerical simulation using the same optimization technique as employed in the authors' previous work (Toropov et al. 1993a, 1993b, 1997, Yoshida et al. 1998). The above approach has been verified by comparing the simulated stress-strain curves for the layers using the constitutive model incorporating the identified material parameters with the experimental curves determined by the conventional uniaxial tension tests for the individual layers.

2 EXPERIMENTATION AND ANALYTICAL PROCEDURES

2.1 *Experimentation*

In order to determine the stress-strain response for the two individual component layers of a bimetallic sheet, at least two different types of experimental data of the mechanical response are required. In the present work,

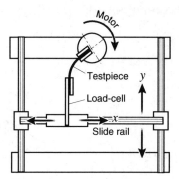

Figure 1. Schematic illustration of experimental setup for cyclic bending test.

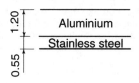

Figure 2. Stainless steel (type-430) clad aluminium (A1100) specimen used in the experiments.

Table 1. Constitutive model

Kinematics: $\quad \dot{\varepsilon} = \dot{\varepsilon}^e + \dot{\varepsilon}^p$

Flow rule:

$$f = \frac{1}{2}(S-\alpha):(S-\alpha) - \frac{1}{3}(Y+R)^2$$

Isotropic hardening rule:

$$\dot{R} = b(Q-R)\dot{\bar{\varepsilon}}, \quad \dot{\bar{\varepsilon}} = \left(\frac{2}{3}\dot{\varepsilon}^p : \dot{\varepsilon}^p\right)^{1/2}$$

Kinematic hardening rule:

$$\dot{\alpha} = \dot{\alpha}_1 + \dot{\alpha}_2$$

$$\dot{\alpha}_1 = \frac{2}{3}H'\dot{\varepsilon}^p$$

[linear kinematic hardening]

$$\dot{\alpha}_2 = C\left(\frac{2}{3}a\dot{\varepsilon}^p - \dot{\bar{\varepsilon}}\alpha_2\right)$$

[non-linear kinematic hardening]

a uniaxial bending test and a cyclic bending test were conducted. In the cyclic bending test, as shown in Fig.1, one end of a specimen was clamped and rotated by a step motor, and the other end was moving freely in x-y directions without rotation. The above condition of the test can be regarded as uniform bending, in which the bending moment is uniformly distributed in the longitudinal direction of the sheet. The bending moment was measured by a load-cell, and the curvature of the specimen was determined from the surface strain measured by strain gauges bonded on the both surfaces of the specimen. A stainless steel (type-430 SS) clad aluminium sheet consisting of 1.20 mm aluminium (A1100) layer and 0.55 mm stainless steel layer, as shown in Fig.2, was employed in the experiments. From this experiment, the bending moment versus curvature diagram was obtained. The second type of experiment (uniaxial tension) produces the tensile load versus strain curve.

In order to verify the identified material parameters, uniaxial tension tests were also performed with the stainless steel specimen which had been taken from the clad sheet, and its stress-strain curve was obtained. The stress-strain curve of the aluminium layer was determined using the rule of mixture by use of the results of the uniaxial tension tests both for the clad sheet and the stainless steel layer.

2.2. Constitutive Model

We used a constitutive model of cyclic plasticity for the infinitesimal deformation (Chaboche & Rousselier 1983) in which the cyclic hardening characteristic is described by the isotropic hardening rule and the Bauschinger effect by the combination of the linear and nonlinear kinematic hardening rules, as given in Table 1. The constitutive model incorporates eight material parameters: two elastic constants E (Young's modulus) and ν (Poisson's ratio); the initial yield stress Y; parameters Q and b in the isotropic hardening rule; and three parameters a, C and H' for the kinematic hardening rules.

2.3 Identification of Material Parameters

The identification of the above material parameters, excluding Poisson's ratio ν which was found to be $\nu=0.3$ for both the stainless steel and the aluminium layer from the conventional measurements, was performed by minimizing the difference between the experiment results (the tensile load versus elongation curve and the bending moment versus curvature diagram) and the corresponding results of numerical simulation.

3 FORMULATION OF THE MATERIAL PARAMETER IDENTIFICATION PROBLEM

3.1 *General Formulation*

Let us consider the phenomenological coefficients of the constitutive model to be identified as components of the vector $x \in R^N$. Then the optimization problem can be formulated as follows:

Find the vector x that minimizes the objective function

$$F(x) = \sum_{\alpha=1}^{L} \theta^{\alpha} F^{\alpha}(x)$$

$$A_i \leq x_i \leq B_i \quad (i = 1, \dots, N) \tag{1}$$

where L is the total number of individual specific response quantities (denoted by α) which can be measured in the course of experiments and then obtained as a result of the numerical simulation. $F^{\alpha}(x)$ is the dimensionless function

$$F^{\alpha}(x) = \left\{ \sum_{s=1}^{S_{\alpha}} \left[R_s^{\alpha} - R^{\alpha}(x, \tau_s^{\alpha}) \right]^2 \right\} / \left\{ \sum_{s=1}^{S_{\alpha}} \left[R_s^{\alpha} \right]^2 \right\} \tag{2}$$

which measures the deviation between the computed α-th individual response and the observed one from the experiment. τ^{α} is a parameter which defines the history of the process in the course of the experiment (e.g. the time or the loading parameter), and the values τ_s^{α} ($\alpha=1,\dots, L$, $s=1,\dots,S$) define the discrete set of S_{α} data points. R_s^{α} is the value of the α-th measured response quantity corresponding to the value of the experiment history parameter τ_s^{α}. $R^{\alpha}(x, \tau_s^{\alpha})$ is the value of the same response quantity obtained from the numerical simulation. θ^{α} is the weight coefficient which determines the relative contribution of information yielded by the α-th set of experimental data. A_i, B_i are side constraints, stipulated by some additional physical considerations, which define the search region in the space R^N of optimization parameters.

The optimization problem (1) has the following characteristic features:

- the objective function is an implicit function of parameters x,
- to calculate values of this function for the specific set of parameters x means to use a nonlinear numerical (e.g. finite element) simulation of the process under consideration, which usually involves a large amount of computer time;
- function values present some level of numerical noise, i.e. can only be estimated with a finite accuracy.

The direct implementation of any of conventional nonlinear mathematical programming techniques would involve too large amount of computer time and, moreover, the convergence of a method cannot be guaranteed due to the presence of numerically induced noise in the objective function values and (or) its derivatives. To solve the problem, the iterative multipoint approximation concept (Toropov et al. 1993a) is used here. The technique is based on the iterative approximation of computationally expensive and noisy functions, $F^{\alpha}(x)$, ($\alpha=1,\dots, L$) by the simplified noiseless functions. To construct such functions, the methods of regression analysis are implemented. Further details of the identification technique can be found in references by Toropov et al. (1993b, 1996, 1997).

3.2 *The Present Identification Problem*

The identification of the material parameters for both the layers of the stainless steel and the aluminium was performed using the tension (tensile load P versus strain ε) curve ($\alpha=1$) and several individual bending/reversed bending (moment M versus curvature κ) curves ($\alpha=2,\dots, L$) which are regarded as individual response quantities. Here in the identification problem,

- the optimization variables $x = \lfloor x_1, x_2, \dots, x_{14} \rfloor$ are the material parameters for the two layers:

$$\lfloor (E, Y, Q, b, H', a, C)_{stainless\ steel},$$
$$(E, Y, Q, b, H', a, C)_{aluminium} \rfloor,$$

- the set of values of R_s^{α} corresponds to the set of values of the tensile load $R_s^1 = P$ (for $\alpha=1$); and experimental bending moment $R_s^{\alpha} = M_s^{\alpha}$ (for $\alpha=2, \dots, L$) both of which are found from the experiment,
- the function $R^{\alpha}(x, \tau_s^{\alpha})$ corresponds to the calculated tensile load $R^1 = \int \sigma_y dA$ (for $\alpha=1$), and bending moment $R^{\alpha} = M^{\alpha} = \int \sigma_y y dA$ (for $\alpha=2, \dots, L$),
- the experiment history parameter τ_s^1 is the strain in uniaxial tension for $\alpha=1$; and τ_s^{α} (for $\alpha=2, \dots, L$) is the curvature κ_s^{α},
- the index α is 1 for the uniaxial tension, 2 for the first monotonic bending, 3 for the subsequent reversed bending, etc. in Eq.(2).

All the response quantities were considered equally weighted ($\theta^1 = \theta^2 = \dots = 1$) in the formulation of the objective function $F(x)$.

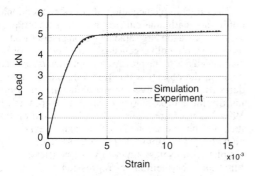

Figure 3. Comparison of experimental curve of tensile load versus strain and the result of simulation with the constitutive model incorporating the identified set of material parameters.

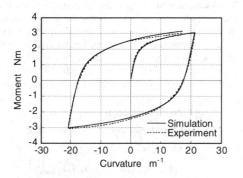

Figure 4. Comparison of experimental diagram of bending moment versus curvature and the result of simulation with the constitutive model incorporating the identified set of material parameters.

4 RESULTS AND DISCUSSIONS

Figure 3 shows the comparison of the experimental results of the tensile load versus strain curve and the corresponding results simulated with the constitutive model incorporating the identified set of material parameters. The results of cyclic bending and the corresponding numerical simulation are shown in Fig.4. These figures show that the load versus strain and bending moment versus curvature diagrams obtained in the simulation fit the experimental results well. Figures 5 shows the calculated stress-strain curves using the identified material parameters for the stainless steel and the aluminium in the bimetallic sheet. Both the component metals exhibit almost no cyclic strain-hardening because they possess large plastic prestrain induced during the cladding by the roll-bonding process.

In order to check the accuracy of this identification method, the stress-strain response for the individual component metals, simulated with the constitutive model incorporating the identified material parameters, were compared with the experimental stress-strain curves, as shown in Fig. 6. Figure 7 shows the simulated surface strain response (strain of the aluminium versus one of the stainless steel) during cyclic bending together with the corresponding experimental results. These results of simulation agree with those obtained in experiments.

Instead of using the set of experimental data of tensile load versus strain in uniaxial tension and bending moment versus curvature in cyclic bending as discussed above, we may use the other experimental data, e.g., the bending moment versus surface strains in cyclic bending for both the stainless and aluminium layers. Figure 8 shows the stress-strain curves for the stainless steel and the aluminium calculated with the constitutive model which incorporates the identified

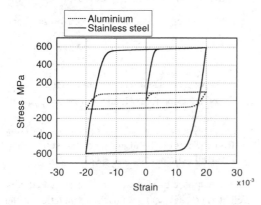

Figure 5. Stress-strain response simulated with the constitutive model incorporating the identified set of material parameters for the stainless steel and aluminium layers in the bimetal sheet.

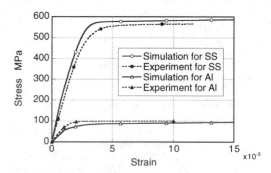

Figure 6. Comparison of the stress-strain curves and the result of simulation with the constitutive model incorporating the identified set of material parameters for the stainless steel (SS)and aluminium (Al)layers in the bimetal sheet.

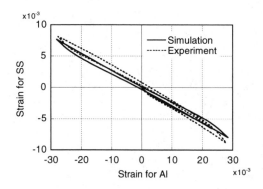

Figure 7. Surface strain of the aluminium(Al) versus that of the stainless steel (SS) during cyclic bending. The results of the experiment and the simulation.

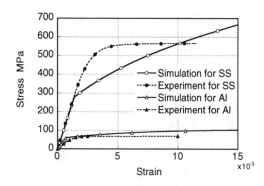

Figure 8. Comparison of the experimental stress-strain curves and the calculated results with the constitutive model incorporating the identified set of material parameters for the stainless steel (SS)and aluminium (Al)layers in the bimetal sheet. The material parameter identification is based on the experimental data of surface strains.

material parameters using only the surface strain data, together with the experimental stress-strain curves. In Fig. 8, a certain discrepancy between the calculated stress-strain curves and the experimental results is found, whereas the results of simulation for the bending moment and curvature agree well with the experimental ones (see Fig. 9). One of the reasons for the discrepancy is that the strain in the cyclic bending is not large enough for the determination of plastic properties. Especially for the bimetallic sheet, the strain in the stainless steel layer is much smaller than in the aluminium layer because of the shift of the neutral surface from the mid-plane of the sheet (Verguts, H. & Sowerby R. 1975). If the curvature in the cyclic bending test had been large enough, the results of the identification with its data would have been better.

5 CONCLUDING REMARKS

A method for the identification of mechanical properties of the individual component layers in a bimetallic sheet by a mixed experimental-numerical approach has been presented. For a stainless steel clad aluminium sheet, a set of material parameters in a complex constitutive model incorporating seven material parameters for each component metal (fourteen material parameters for the two layers) has been successfully identified by using the experimental results of the uniaxial tension and cyclic bending. A special emphasis in the present paper is placed on the following:

- the present method has an advantage of using the experimental results only for a whole bimetallic sheet but not for individual component layers,
- the stress-strain curves not only under monotonic tension but also under the subsequent reversed and

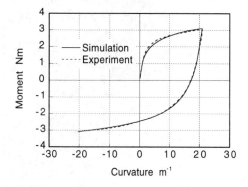

Figure 9. Comparison of experimental diagram of bending moment versus curvature and the calculared result with the constitutive model incorporating the identified set of material parameters. The material parameter identification is based on the experimental data of surface strains.

cyclic loading are used in the identification, so the constitutive model can be used for accurate simulation of complex metal forming processes,
- the identification procedure is based not on the conventional try-and-error approach but on an advanced optimization technique.

Acknowledgement - The present work has been done as a Monbusho (The Ministry of Education, Japan) project of *the Venture Business Laboratory* at Hiroshima University. The authors are grateful to Monbusho for the financial support and also to the British Council for the travel grant offered to V. V.

Experimental Mechanics, Allison (ed.) © 1998 Balkema, Rotterdam, ISBN 90 5809 014 0

A numerical and experimental study on identification of interface surface cracks in bonded dissimilar materials by the electric potential CT method

Shiro Kubo, Takahide Sakagami, Jirasak Sarasinpitak & Hiroyuki Kitaoka
Department of Mechanical Engineering and Systems, Osaka University, Suita, Japan

Kiyotsugu Ohji
Department of Mechanical and System Engineering, Ryukoku University, Ohtsu, Shiga, Japan

ABSTRACT: The present authors proposed the electric potential CT (computed tomography) method for identification of a crack in an electric-conductive body from electric potential distributions measured on surfaces of the body. In the present study the electric potential CT method was applied to the identification of interface surface cracks in three-dimensional bonded dissimilar bodies. As a method of inverse analysis to identify the crack the least residual method was applied: the crack giving the smallest value of residual between the calculated and measured electric potential distributions was taken as the most plausible one. To achieve an efficient estimation a hierarchical inversion scheme was applied, in which a two-dimensional scanning analysis was followed by a three-dimensional inverse analysis. The electric potential distribution for the scanning inverse analysis was approximated by the combined Johnson solution proposed by the present authors. In the three-dimensional analysis, the distribution of electric potential was calculated using the boundary element method. Numerical simulations and experiments were conducted on the identification of interface surface cracks in three-dimensional bodies. It was found that the location and size of interface surface cracks were estimated with good accuracies, demonstrating the applicability of the proposed method.

1 INTRODUCTION

Bonded dissimilar materials have been increasingly used in various engineering applications, since recent severe requirements for materials can be satisfied more easily by using bonded dissimilar materials than by using monolithic materials. In bonded dissimilar materials, however, stress singularity appears at the intersection of the interface and free-edges for usual combination of the material constants of dissimilar materials. Furthermore, the mechanical properties of the interface are usually not so good as those of the constituent materials. Due to these facts the interface of dissimilar materials, in many cases, is supposed to have cracks, small or large. These cracks may develop in manufacturing processes or in service.

The nondestructive inspection for the quantitative evaluation of cracks on the interface of bonded dissimilar materials before and in service is, therefore, very important. For the nondestructive inspection of cracks in monolithic electric-conductive materials, the electric potential method has been applied successfully (Aronson et al. 1979; Gangloff 1981; Hicks et al. 1982; Wilkowski et al. 1983; Murai et al. 1986; Coffin 1988; Abe et al. 1988).

The crack identification can be regarded as an inverse problem (Tikhonov et al. 1977; Romanov 1987; Kubo 1988a, 1992). The present authors proposed the electric potential CT (Computed Tomography) method for the quantitative identification of two- and three-dimensional cracks by using inverse analyses (Ohji et al. 1985; Kubo et al. 1986, 1988b-c, 1991a-d, 1993; Sakagami et al. 1988, 1990a-c). In this method the electric potential distribution observed on the surface of the cracked body was computer-processed to estimate the location, shape and size of cracks. The applicability of the method to the identification of two- and three-dimensional cracks has been shown by many numerical simulations and experiments.

In a previous paper the electric potential CT method was extended to the identification of interface cracks in bonded dissimilar materials (Kubo et al. 1996). In the present paper the method was applied to the identification of an interface surface crack in a three-dimensional body from electric potential distribution measured on the surfaces of cracked body. A hierarchical inversion scheme incorporating a two-dimensional scanning analysis and a three-dimensional analysis was employed. Numerical simulations and experiments were conducted to examine the applicability of the method to the identification of interface surface cracks.

2 INVERSE METHOD FOR CRACK IDENTIFICATION

2.1 Least Residual Method

To identify the crack, measured electric potential distribution $\phi^{(M)}$ on observation surface Γ_{ob} was compared with calculated electric potential $\phi^{(C)}$. Residual R_s between $\phi^{(C)}$ and $\phi^{(M)}$ was defined by the following equation.

$$R_s = \int_{\Gamma_{ob}} (\phi^{(C)} - \phi^{(M)})^2 d\Gamma. \qquad (1)$$

The integration was taken over Γ_{ob}. Based on the least residual method, residual R_s was used as a criterion for the crack identification: the crack giving the smallest value of R_s was taken as the most plausible one. The crack parameters giving the smallest value of R_s were sought by applying the modified Powell optimization method (Powell 1964).

2.2 Hierarchical Inversion Scheme

To speed up the identification the hierarchical inversion scheme (Sakagami 1990a) was employed, in which a two-dimensional scanning analysis was followed by a three-dimensional analysis. In the two-dimensional scanning analysis the least residual method was applied to a cross section shown by dotted area in Fig. 1. The residual R_s was evaluated along line MN on the free surface. By combining the results of identification obtained for several scanning planes a rough estimate of the crack shape and location can be obtained. Then the three-dimensional inverse analysis was made based on the least residual method for a detailed identification. It was assumed that the surface crack was semi-elliptical.

2.3 Boundary Element Analysis of Electric Potential Distribution

The boundary element method (Brebbia & Walker 1980) has been extensively applied to the analyses of electric potential distribution in homogeneous materials. Now consider a dissimilar body, which is consisted of materials 1 and 2 having conductivities of $\sigma^{(1)}$ and $\sigma^{(2)}$, respectively, and has a crack on the interface of two materials. Using Cartesian coordinates $x_1 x_2 x_3$ the governing equation of the distribution of electric potential $\phi^{(m)}$ in material m is written as,

$$\partial \left(\sigma^{(m)} \partial \phi^{(m)} / \partial x_i \right) / \partial x_i = 0 \text{ in material } m. \quad (2)$$

On the outer boundary of the materials, potential $\phi^{(m)}$ or flux $q^{(m)}$ is prescribed, $q^{(m)}$ being defined by the following equation.

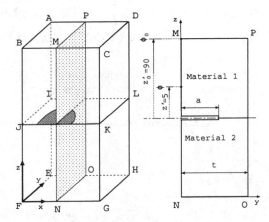

Fig. 1. A three-dimensional body containing an interface surface crack, and a scanning plane.

$$q^{(m)} = \sigma^{(m)} \partial \phi^{(m)} / \partial n = \sigma^{(m)} \left(\partial \phi^{(m)} / \partial x_i \right) n_i. \quad (3)$$

Here n_i denotes the outward unit normal vector on the boundary.

In this study it was assumed that the bonding is complete and the continuity of potential ϕ and flux q across the interface holds on the interface of the dissimilar materials.

$$\phi_{int}^{(1)} = \phi_{int}^{(2)}, \qquad (4)$$

$$q_{int}^{(1)} = -q_{int}^{(2)}, \qquad (5)$$

where $(\bullet)_{int}^{(m)}$ denotes the value on the interface defined in material m. The ratio R of conductivity of materials 1 and 2 is defined as,

$$R = \sigma^{(1)} / \sigma^{(2)}. \qquad (6)$$

The boundary element equation states that the value of electric potential $\phi_A^{(m)}$ at point A on the boundary of material m can be expressed by an integral over its boundary $\Gamma^{(m)}$ as in the followings (Brebbia & Walker 1980):

$$c_A \phi_A^{(m)} + \int_{\Gamma^{(m)}} \left(q^* \phi_B^{(m)} - \phi^* q_B^{(m)} \right) d\Gamma_B^{(m)}$$
$$= 0 \quad (m = 1, 2), \qquad (7)$$

where ϕ^* and q^* denote the fundamental solutions of the Laplace equation, and c_A being a constant dependent on the shape of boundary around point A. The fundamental solutions for the three-dimensional body are expressed in terms of the distance r between points A and B on boundary $\Gamma^{(m)}$ as,

$$\phi^* = \phi^*(r) = 1/(4\pi r \sigma^{(m)}) \qquad (8)$$

934

$$q^* = \sigma^{(m)}\partial\phi^*/\partial n = \sigma^{(m)}\left(\partial\phi^*/\partial x_i\right)n_i. \quad (9)$$

The entire boundary of material m is discretized into boundary elements with N_m nodes. When point A is set to be at the i-th nodal point on boundary Γ_m of material m, Eq. (7) is discretized in the form,

$$c_i\phi_i^{(m)} + \sum_{j=1}^{N_m} q_{ij}^*\phi_j^{(m)} - \sum_{j=1}^{N_m}\phi_{ij}^*q_j^{(m)} = 0 \ (m=1,2),$$
$$(10)$$

where ϕ_{ij}^* and q_{ij}^* are determined by integrals expressed in terms of the fundamental solutions ϕ^* and q^* over elements containing point j. The quadratic element was used in this study.

Equation (10) gives a matrix equation for ϕ and q at nodal points. The distribution of potential ϕ is obtained by solving the matrix equation for materials 1 and 2, together with prescribed boundary conditions on the outer boundary concerning ϕ or q, and the compatibility of ϕ and q written by Eqs. (4) and (5).

2.4 Combined Johnson Solution for Approximate Representation of Electric Potential Distribution

If the cross section on the scanning plane is rectangular, the combined Johnson solution proposed by the present authors (Kubo et al. 1996) may be used for representing approximately the electric potential distribution in it. Consider a rectangular plate of width t made by bonding dissimilar materials 1 and 2 of equal longitudinal length along a line of $a \leq y \leq t, z = 0$ as shown in Fig. 1. Electric potential ϕ on the top end and ϕ on the bottom end of the plate under an application of direct current can be prescribed to be $\phi_0^{(1)}$ and $\phi_0^{(2)}$, respectively.

In the extremes of $R = \infty$ and $R = 0$, ϕ on the interface is constant. In the case of homoge-

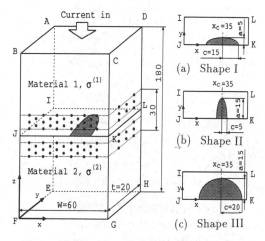

(a) Shape I

(b) Shape II

(c) Shape III

Fig. 2. A model used in numerical simulation of identification of an interface surface crack.

neous material, i.e. $R = 1$, ϕ on the interface is constant due to the symmetry with respect to the interface plane. These may imply that the electric potential on the interface is almost constant for any value of R, and that the distribution of potential ϕ in materials 1 and 2 is approximated by combining Johnson's solution for homogeneous edge-cracked plate, in which ϕ is constant along a line ahead of the crack tip.

From Johnson's solution (1965) potential ϕ in material m $(m = 1, 2)$ at a distance of z from the interface on the front surface $(y = 0)$ is expressed as,

$$\frac{\phi^{(m)} - \phi_{\text{int}}}{\phi_0^{(m)} - \phi_{\text{int}}} = \frac{\cosh^{-1}\left[\cosh\left(\frac{\pi z}{2t}\right)/\cos\left(\frac{\pi a}{2t}\right)\right]}{\cosh^{-1}\left[\cosh\left(\frac{\pi z_0}{2t}\right)/\cos\left(\frac{\pi a}{2t}\right)\right]}, \quad (11)$$

where a and ϕ_{int} denote the crack length and the

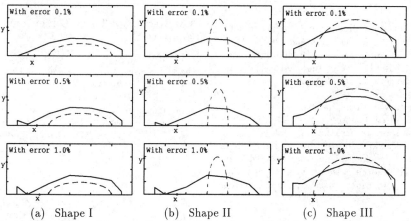

(a) Shape I   (b) Shape II   (c) Shape III

Fig. 3 Cracks estimated by two-dimensional scanning analysis from simulated noisy data.

935

value of ϕ on the interface, with z_0 being the distance between the interface and the top or bottom end where potential ϕ is prescribed to be $\phi_0^{(m)}$.

Due to the continuity of flux across the interface, the ratio of the potential drops in materials 2 and 1 is given by,

$$(\phi_{int} - \phi_0^{(2)})/(\phi_0^{(1)} - \phi_{int}) = \sigma^{(1)}/\sigma^{(2)} = R. \quad (12)$$

This equation gives the value of ϕ_{int}:

$$\phi_{int} = (R\phi_0^{(1)} + \phi_0^{(2)})/(R + 1). \quad (13)$$

The distribution of ϕ on front surface MN is given by combining Eq. (11) with Eq. (13).

3 NUMERICAL SIMULATIONS

3.1 A Model Used in Simulations

Numerical simulations of crack identification

3.2 Results of Two-Dimensional Scanning Analysis

The two-dimensional scanning analysis was applied to seven scanning planes. The calculated electric potential $\phi^{(C)}$ was obtained by applying the combined Johnson solution. The crack depth giving the smallest value of R_s was taken as the most plausible one for each scanning plane.

Figure 3 shows the results of identification for the error level of 0.1%, 0.5% and 1.0%. The cracks estimated by the two-dimensional scanning analysis are shown by solid lines, while thin dashed lines show the actual crack shapes. It is seen from the figure that a rough estimate can be obtained for the shallow crack (shape I) and the big crack (shape III). For the deep crack with small crack length (shape II), however, the estimated crack is much shallower than the actual one.

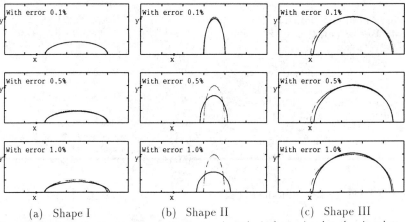

(a) Shape I (b) Shape II (c) Shape III

Fig. 4 Cracks estimated by three-dimensional analysis from simulated noisy data.

were made for three-dimensional surface cracks shown in Fig. 2. A shallow crack (shape I), a deep crack with small crack length (shape II), and a big crack (shape III) were employed in the simulations. The ratio of conductivity R was set at 2.

The boundary element analyses were made to generate measured data of electric potential $\phi^{(M)}$ on the observation surface. The symbol "•" in the figure shows the location of observation points. To investigate the effect of measurement error in $\phi^{(M)}$, artificial random noise of the level of 0.1%, 0.5% and 1.0% was introduced in the calculated potential values.

3.3 Results of Three-Dimensional Analysis

Figure 4 shows the crack shapes estimated by the three-dimensional analysis for the error level of 0.1%, 0.5% and 1.0%. The shallow crack (shape I) and the big crack (shape III) are identified with good accuracies even in the presence of considerable error in $\phi^{(M)}$. The deep crack with small crack length (shape II) was identified with good accuracies only when the error level in $\phi^{(M)}$ was 0.1%.

These results show that good estimation of crack shape can be made when the noise level is of the order of 0.1 %, which can be achieved without difficulties by using commercially available equipment.

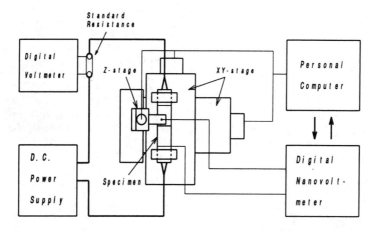

Fig. 6 Experimental set-up for measuring electric potential distribution.

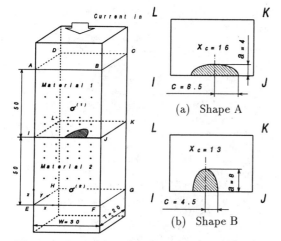

Fig. 5. A model used in experiments of identification of an interface surface crack.

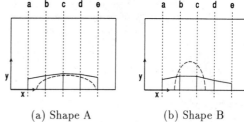

(a) Shape A   (b) Shape B

Fig. 7. Cracks estimated by two-dimensional scanning analysis from measured data.

4 EXPERIMENTS

4.1 A Model Used in Experiments

Experiments of crack identification were made for three-dimensional surface cracks shown in Fig. 5. A shallow crack (shape A) and a deep crack (shape B) were employed in the experiments. The specimens were made by diffusion bonding of a carbon steel with a stainless steel. The cracks were introduced by electro-sparc machining.

The electric potential distribution was measured automatically, using an apparatus composed of a digital nanovoltmeter, a D.C. power supply, an XY stage, a Z stage mounting a pin for potential measurement, and a computer for control, shown in Fig. 6. The electric potential reading was obtained at measurement points on the front surface, shown by " • " in Fig. 5. The ratio of conductivity R was determined from the measured electric potential distribution.

4.2 Results of Two-Dimensional Scanning Analysis

The two-dimensional scanning analysis was applied to five scanning planes. Figure 7 shows the results of identification. It can be seen that a reasonable estimate is obtained for the shallow crack (shape A), while for the deep crack (shape B) the estimated crack shown by a solid line is much shallower than the actual one shown by a dashed line.

4.3 Results of Three-Dimensional Analysis

Figure 8 shows the results of identification using the three-dimensional analysis. It can be seen that the estimated cracks agree reasonably with the actual ones, for the deep crack (shape B) as well as for the shallow crack (shape A).

937

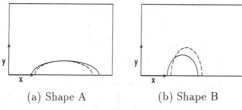

(a) Shape A   (b) Shape B

Fig. 8.   Cracks estimated by three-dimensional analysis from measured data.

5 CONCLUSIONS

The electric potential CT method was applied to the identification of interface surface cracks in bonded dissimilar bodies. As an inverse analysis method for identifying the crack the least residual method was used. The hierarchical inversion scheme incorporating the two-dimensional scanning analysis and the three-dimensional analysis was employed to achieve an efficient identification. Numerical simulations and experiments were conducted on the identification of interface surface cracks. It was found that the location and size of three-dimensional interface surface cracks can be identified by the present method.

ACKNOWLEDGMENTS

The authors would like to express their appreciation to Prof. Yoshiharu Mutoh of Nagaoka University of Technology for his help in preparing the specimens.

This study was partly supported by Grant-in Aid for Scientific Research, Ministry of Education, Culture and Science.

REFERENCES

Abe, H., M. Saka, T. Wachi & Y. Kanoh 1988, Computational Mechanics '88, Vol. 1 (ed. by Atluri, S.N. & G. Yagawa), Springer, Berlin, 12.ii.1-4.

Aronson, G. H. & R. O. Ritchie 1979, J. Testing & Evaluation, 7: 208-215.

Brebbia, C.A. & S. Walker 1980, Boundary Element Techniques in Engineering, Butterworth, London.

Coffin, L. F. 1988, Fracture Mechanics 19th Symposium, ASTM STP 969, Amer. Soc. Testing Mater., 235-259.

Gangloff, R.P. 1981, Fatigue of Eng. Materials & Structures, 4: 15-33.

Hicks, M.A. & A.C. Pickard 1982, Int. J. Fracture, 20: 91-101.

Johnson, H.H. 1965, Materials Research & Standards, 5: 442- 445.

Kubo, S., T. Sakagami & K. Ohji 1986, Computational Mechanics '86, (ed. by Yagawa, G. & S.N. Atluri), Springer-Verlag, Berlin, V-339-344.

Kubo, S. 1988a, JSME Int. J., Ser.I, 31: 157-166.

Kubo, S., T. Sakagami, K. Ohji, T. Hashimoto & Y. Matsumuro 1988b, Trans. Jpn. Soc. Mech. Eng., Ser.A, 54: 218-225.

Kubo, S., T. Sakagami & K. Ohji 1988c, Computational Mechanics '88, Vol. 1 (ed. by Atluri, S.N. & G. Yagawa), Springer-Verlag, Berlin, 12.i.1-5.

Kubo, S., T. Sakagami & K. Ohji 1991a, Fracture Mechanics (Current Japanese Materials Research, Vol. 8) (ed. by Okamura, H. & K. Ogura), Elsevier, London, 235-254.

Kubo, S. 1991b, Inverse Problems in Engineering Science (ed. by Yamaguti, M., K. Hayakawa, Y. Iso, M. Mori, T. Nishida, K. Tomoeda & M. Yamamoto), Springer-Verlag, Tokyo, 52-58.

Kubo, S., K. Ohji & T. Sakagami 1991c, Int. J. Appl. Electromagnetics in Mater., 2: 81-90.

Kubo, S., K. Ohji, K. Kagoshima & T. Imajuku 1991d, Mechanical Behaviour of Materials IV (ed. by Jono, M. & T. Inoue) Vol.4, Pergamon Press, Oxford, 717-722.

Kubo, S. 1992, Inverse Problems, Baifukan, Tokyo.

Kubo, S., K. Ohji & K. Nakatsuka 1993, 1st Int. Conf. on Inverse Problems in Engineering (ed. by Zabaras, N., K.A. Woodbury & M. Raynaud), Amer. Soc. Mech. Engrs., 163-169.

Kubo, S., J. Sarasinpitak & K. Ohji 1996, Mater. Sci. Research Int., 2: 26-32.

Murai, T. & Y. Kagawa 1986, Int. J. Numerical Methods in Eng., 23: 35-47.

Ohji, K., S. Kubo & T. Sakagami 1985, Trans. Jpn. Soc. Mech. Eng., Ser.A, 51: 1818-1825.

Powell, M.J.D. 1964, Computer J., 7: 155.

Romanov, V.G. 1987, Inverse Problems of Mathematical Physics, VNU Sci. Press, Utrecht.

Sakagami, T., S. Kubo, T. Hashimoto, H. Yamawaki & K. Ohji 1988, JSME Int. J., Ser.I, 31: 76-86.

Sakagami, T., S. Kubo & K. Ohji 1990a, Int. J. Pressure Vessels & Piping, 44: 35-47.

Sakagami, T., S. Kubo, K. Ohji, K. Yamamoto & K. Nakatsuka 1990b, Trans. Jpn. Soc. Mech. Eng., Ser.A, 56: 27-32.

Sakagami, T., S. Kubo & K. Ohji 1990c, Engineering Analysis, 7: 59-65.

Tikhonov, A.N. & V.Y. Arsenin 1977, Solutions of Ill-Posed Problems, John Willy & Sons, New York.

Wilkowski, G.M. & W.A. Maxey 1983, Fracture Mechanics 14th Symposium, ASTM STP 791, Amer. Soc. Testing Mater., II-266-294.

Experimental Mechanics, Allison (ed.) © 1998 Balkema, Rotterdam, ISBN 90 5809 014 0

A meso-scopic analysis of fatigue mechanism of TiAl intermetallic alloys

Shinichi Numata & Hideto Suzuki
Department of System Engineering, Faculty of Engineering, Ibaraki University, Japan

ABSTRACT: The fatigue property of Ti34wt.%Al and Ti36wt.%Al alloys was studied in vacuum at 1073K. The crack growth was observed and photographed under high resolution conditions using elevated-temperature loading cyclic stage for the scanning electron microscope. The results revealed influences of microstructure on the fatigue mechanisms and the fatigue strength in these alloys. The fatigue property of Ti34wt.%Al was controlled by the fatigue crack initiation mechanism. The fatigue lives of Ti34wt.%Al can be predicted treating fatigue crack initiation as the predominant fatigue mechanism. In contrast, many microcracks growth stage was the dominant life of Ti36wt.%Al. A many microcracks growth analysis can then be performed to asses the remaining life.

1 INTRODUCTION

Intermetallic alloys based on the TiAl matrix have received considerable interest in recent years because of their potential as elevated-temperature structural materials. This is due to a number of property advantages, including low density, excellent oxidation resistance, and relatively good elevated-temperature mechanical properties. Unfortunately, neither the ductility nor the fracture toughness of these TiAl alloys is particularly high at ambient temperature. With alloy modification and micro-structure control, however, both ductility and toughness have been improved significantly in two phase titanium aluminide alloys, although these alloys are still not adequate for application.

Recent studies have demonstrated that TiAl alloys with in certain composition range can be controlled to exhibit two-phase microstructure containing the following features: (1) equiaxed γ grains and α_2 particles, (2) lamellar colonies of alternating layers of α_2 and γ platelets, and (3) a combination of equiaxed γ grains and lamellar colonies. These microstructure are often referred to as fully lamellar, nearly fully lamellar, duplex, and fully gamma, depending on the relative components of lamellar colonies and equiaxed γ grains. Microstructure plays an important role in the tensile and fracture behaviors of TiAl alloys. However, little is known the effect of microstructure on fatigue property and fatigue mechanism. In studies on fatigue mechanism, it is important to quantify and predict the fatigue lives from an engineering point of view.

The object of this article is to present the results of an investigation aimed at studying the effect of microstructure on the fatigue process in TiAl alloys and to suggest that a life prediction and a life assessment method that involves fatigue mechanisms. In particular, the fatigue mechanism will be examined for both duplex and nearly fully lamellar microstructures at 1073K.

2 EXPERIMENTAL PROCEDURE

The materials used in this investigation are Ti34wt.%Al and Ti36wt.%Al. Both are TiAl alloys of almost similar chemical composition, as shown in Table 1. TiAl alloys were prepared using skull-melting and casting techniques into ingots. Fatigue specimens were machined from Ti34wt.%Al and Ti36wt.%Al casting to the shape shown in Figure 1.

Figures 2 (a) and 2(b) show the microstructure of Ti34wt.%Al and Ti36wt.%Al observed by an optical microscope. The microstructure of Ti34wt.%Al was nearly fully lamellar, consisting of large lamellar colonies and small amount of fine grain boundary gamma grains (Figure 2(a)). The lamellar structure is alternating layers of γ / α_2 plates. Ti36wt%Al contained duplex microstructure consisting equaxed gamma grain and lamellar colonies(Fig.2(b)).

Fatigue tests were conducted in vacuum at 1073K, using a servohydraulic testing machine equipped with scanning electron microscope (SEM-servo testing machine). All tests were carried out in load control at a stress ratio R(minimum to maximum stress) of 0.1

using a frequency of 30Hz.

In-situ observations using SEM-servo testing machine were employed to study the number and follow the initiation and growth of surface fatigue cracks. The surface of failed specimens were examined using scanning electron microscopy.

Table 1 Chemical compositions of the materials in weight percent.

Material	Al	C	N	O	H	Ti
Ti34wt.%Al	33.9	0.01	0.004	0.112	0.0024	Bal.
Ti36wt.%Al	36.3	0.01	0.004	0.112	0.0024	Bal.

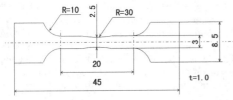

Fig.1 Shape and dimensions of the specimen tested.

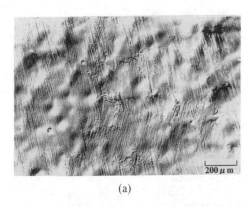

(a)

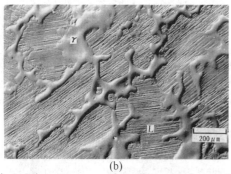

(b)

Fig.2 Microstructures of (a)Ti34wt.%Al and (b) Ti36wt.%Al.

3 RESULTS

The S-N curves of the Ti34wt.%Al and Ti 36wt.%Al are plotted in the form of the stress amplitude(S) versus the number of cycles to failure (Nf) in Figure 3. The fatigue strength of Ti34wt.%Al was higher than that of Ti36wt.%Al. The results indicate that the fatigue strength depends on the microstructure.

The life to fatigue crack initiation exhibited by both microstructures was examined by in-situ observations using a SEM-servo testing machine. The fatigue cracks were initiated at about 10% life for Ti36wt.%Al. This observation indicated fatigue crack growth initiated early during the fatigue tests and that the majority of the sample life was spent in fatigue crack propagation. As in the case of Ti34wt.%Al observations of fatigue damage on a sample interrupted after 80% life didn't reveal a fatigue crack on the surface. It is evident that the fatigue crack growth initiated late in the life.

4 DISCUSSION

4.1 The fatigue mechanism of Ti34wt.%Al

As mentioned early in-situ observations of Ti34wt.%Al at 1073K revealed that most of the fatigue life was spent in crack initiation. The fatigue

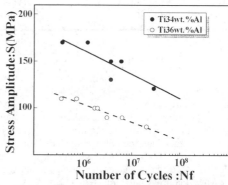

Fig.4 S-N curves of Ti34wt.%Al and Ti36wt.%Al.

Fig.5 A low-magnification fracture surface of Ti34wt.%Al.

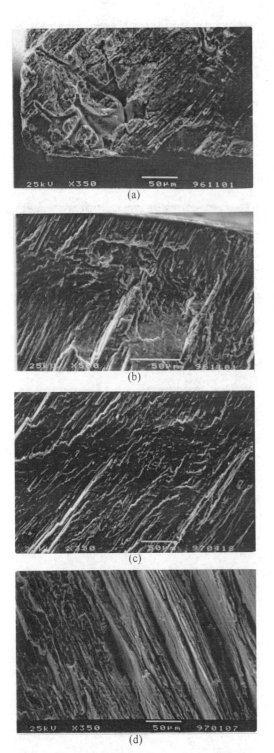

(a)

(b)

(c)

(d)

Fig.6 Detail fractographcs of (a) casting defect, (b) gamma grain, (c) translamellar fracture (d) interface delamination.

crack initiation stage was the dominant life of Ti34wt.%Al. The fracture surfaces of this alloys discuss as follows.

The fracture surfaces of Ti34wt.%Al, tested at 1073K were carefully examined by secondary electron contrast in the SEM. A low-magnification view of the fracture surface is given in Figure 5 which shows a large variety of features, directly related to lamellar microstructure. The separation of a fracture surface can be crack initiation site, or crack propagation site. This observation indicated that the life to final fracture was divided into two stages which were crack initiation and crack propagation.

The detail fractography revealed that the fatigue crack initiation site was observed in a casting defect (Figure 6(a)), or a gamma grain (Figure 6(b)). The site for the initiation of small crack was observed to be the crack origin. The small crack has been defined as mesocrack affected by microstructure. Mesocrack in an early stage of development was seen at edges of preexisting defect. Both the preexisting defect and the mesocrack were related to fatigue crack initiation. The fatigue crack propagation site observed in this microstructure exhibits translamellar fracture (across the lamellar) shown in Figure 6(c) and interface delamination shown in Figure 6 (d).

The crack propagation site was divided into two regions. These two regions were the fatigue crack region and final fracture region. The fracture mechanisms in the fatigue crack region were cracking across the lamellar ,while those in the final fracture region were interface delamination and cracking across the lamellar.

The fatigue process of Ti34wt.%Al are shown schematically in Figure 7. From fractographs and in-situ observations, it may be concluded that the fatigue property of Ti34wt.%Al was controlled by fatigue crack initiation mechanisms. This indicated crack initiation failure criterion will be used in the latter section for the analysis of the data according to the mesoscopic fracture mechanics of Suzuki.

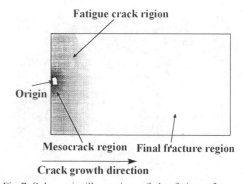

Fig.7 Schematic illustration of the fatigue fracture process for TI34wt.%Al.

941

4.2 The fatigue mechanism of Ti36wt.%Al

The in-situ observation of fatigue damage on a sample interrupted after 40% life at 1073K revealed a substantial number of microcracks on the surface. It was found that many microcracks were initiated on the surface, only a relative few grew to sizes of 100-300 μm while many were often arrested and linked with each other. These microcracks have been defined as mesocracks which were initiated on the surface and several hundred μ m length. These mesocracks could be divided into two cracks. These two cracks are perpendicular crack and slope crack. The perpendicular crack was trying to grow approximately perpendicular to the loading axis. The slope crack ran approximately 45 degrees to the loading axis. The duplex micro-structure affected these mesocracks growth behavior.

The detail mechanisms of mesocrack growth and interaction with microstructure were summarized in Figures 9(a),9(b)and9(c),respectively. In the early stage of the life, many perpendicular cracks were initiated in γ grain, as shown in Figure 9(a). This observation indicated that perpendicular crack in γ grain might have been induced as the result of plastic incompatibility at lamellar grain boundary due to impinging slip developed in the γ grain. At next stage in the life, slope cracks in lamellar microstructure were initiated ahead of the tips of perpendicular cracks, as shown in Figure9(b). And these perpendicular cracks grew into the lamellar microstructure, then stopped, as shown in Figure 9(c). This observation indicated that the orientation of the lamellar relative to the crack growth. The mesocracks began to linked with other mesocracks.

Final fracture eventually occurred in the Ti36wt.%Al when the mesocracks linked to form a macrocrack of a critical length. The fatigue process of Ti36wt.%Al are shown schematically in Figure 10. In the case of the Ti36wt.%Al, fatigue fracture is dominated by mesocrack growth mechanism. The relationship between the mesocracks growth and remaining life will be understood as discussed later.

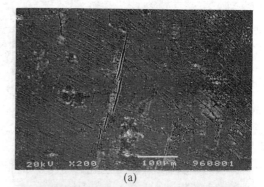
(a)

(b)

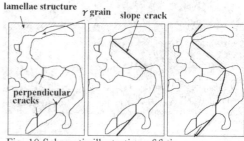
(c)

Fig. 9 Mesocrack growth processes : (a)perpendicular crack in gamma phase (b)slope crack initiation and (c) crack stopped in lamellar.

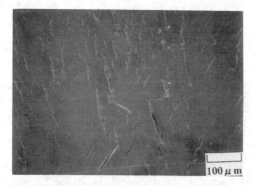

Fig 8 Mesocrack initiation on the surface.

Fig. 10 Schematic illustration of fatigue processes for Ti36wt.%Al.

4.3 Evaluation of mesoscopic factor

The observation that the fatigue fracture of Ti34wt.%Al was controlled by fatigue crack initiation mechanisms, suggests that a life prediction method with the emphasis on mesoscopic fracture behavior may be worthy of consideration. Figure 11 illustrates mesoscopic factor as the predominant fatigue crack initiation mechanism. Measurements of mesoscopic factors within mesocrack region were derived from photographs made in the SEM.

Assuming crack initiation from mesoscopic factor to be the predominant failure mechanism for life calculations, the following relationship;

$$K_{max} = FS\sqrt{(a+c)\pi} \quad \cdots (1)$$

can be used. In this equation K_{max} is the mesoscopic parameter describing fatigue crack initiation, a is the size of defect, c is the length of mesocrack and F is geometry dependent function.

Figure 12 shows K_{max} versus number of cycles to failure relationship for Ti34wt/%Al. It is obvious that linearity is achieved in Figure 12. These data are fit to the correlation

$$K_{max} = A\log Nf + C \cdots (2)$$

where A=-1.54 and C=32.6 for the units express in Figure 12.

In this experiment, the fatigue crack initiation site in Ti34wt.%Al was at a mesoscopic factor of a size between 200 to 420 μ m as observed on the fracture

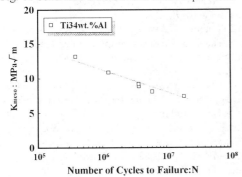

Fig. 11 Schematic illustration of mesoscopic factor

Fig.12 Kmax versus Nf relationship.

surface of individual specimens. From these mesoscopic investigation, 420 μ m was determined as characteristic mesoscopic factor diameter,$(a+c)$,. The value of K_{max} can be determined uniquely by Equation (1). As mention early the K_{max}-N relationship is used in the calculation to determine the stress amplitude versus number of cycles to failure.

It is necessary to emphasize the importance of using the criterion for fatigue crack initiation. In an analysis in this study it was pointed out that using a failure criterion for fatigue crack initiation, the method based on mesoscopic fracture mechanics can be applicable to the fatigue life prediction of Ti34wt.%Al. The present investigation demonstrates the successful application of the method to this alloy using experimentally defined crack initiation failure

4.4 Remaining life assessment for mesocracks

The observation that the fatigue behavior of Ti36wt.%Al was controlled by the mesocracks growth mechanisms, suggests that assessment life method with the emphasis on fractal characteristics may be worthy of consideration. The characteristics of mesocracks was analyzed as a function of the size r of the measuring unit as shown in Figure 13. The estimated numbers N(r) are a function of r according to the following equation:

$$N(r) = N_0 r^{-D} \quad \cdots (2)$$

N_0 is a constant with dimensions of number, D, the fractal dimension, is a constant related to the slope of the linear form of equation (2):.

$$\log N(r) = \log N_0 - D\log r \cdots (3)$$

Characteristic fractal curves are presented in Figure 14 for Ti36wt.%Al. However, instead of plotting log N(r) versus log r, as indicated by equation (2), we use a more useful form of the fractal equation (3). It is obvious that linearity is achieved in Figure 14. A conventional value of D is determined for fractal curves by linear regression.

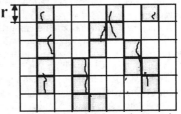

Fig.13 Interaction of mesocracks and measuring box.

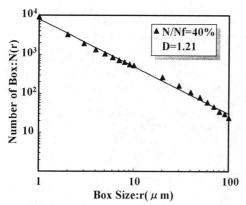

Fig.14 Fractal plot of mesocracks.

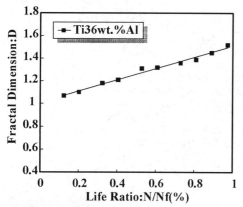

Fig.15 Fractal dimension, D, versus life ratio, N/Nf relationship.

The conventional fractal dimension D values are plotted versus the relative fatigue cycles ratio, N/Nf, in Figure 15. This figure reveals an excellent linear relationship between fractal dimension D and the relative fatigue cycles ratio N/Nf,.

$$D = C_1 (N / Nf) + C_2$$

where C_1=0.5, and C_2=1.01 for the units expressed in Figure 15. The fractal dimension versus life ratio relationship is used in the calculation determined remaining life.

It is necessary to emphasize the importance of using the criterion for mesocrack growth. In an analysis in this study it was pointed out that using a failure criterion for mesocrack growth, the fractal characteristics can be applicable to the remaining life assessment of Ti36wt.%Al. The present investigation demonstrates the successful application of the method to this alloy using experimentally defined mesocracks growth criterion.

5 CONCLUSION

The fatigue property of Ti34wt.%Al and Ti36wht.%Al was investigated in vacuum at 1073K. The test results were analyzed from the mesoscopic view point. The main results obtained are as follows.
(1)The fatigue strength of Ti34wt.%Al was higher than that of Ti36wt.%Al.
(2)The fatigue property of Ti34wt.%Al is controlled by the fatigue crack initiation mechanism while that of Ti36wt.%Al is controlled by the many microcracks growth mechanism.
(3)The application of mesoscopic fracture mechanics to the fatigue life prediction of Ti34wt.%Al was successful. The S-N curves of Ti34wt.%Al could be determined.
(4)The fractal characteristics analysis of the mesocrack growth can be applicable to the remaining life assessment of Ti36wt.%Al.

ACKNOWLEDGMENTS

The authors thank Satellite Venture Business Laboratory, Ibaraki University for technical support. The authors also acknowledgments helpful discussions with the Committee for Hybrid Surface Modification.

REFERENCES

Yoshiaki AKINIWA et al 1997. Propagation of short Fatigue Cracks in Notched Specimens of Inter metallic Compound TiAl. *Journal of the Society of Materials Science, JAPAN.* 46:1261-1267.

D.L.DAVIDSON & J.B.CAMPBELL 1993. Fatigue Crack Growth through the Lamellar Microstructure of an Alloy Based on TiAl. *Metallurgical Transactions A.*24:1555-1573.

K.S.CHAN & Y.W.KIM 1992.Influence of Microstructure on Crack-Tip Micromechanics a Two-Phase TiAl alloy *Metallurgical Transactions A.*23:1663-1677.

.J.J.Pernot et al 1995. Crack Growth Rate Behavior of a Titanium-Aluminide Alloy During Isothermal and Nonisothermal Conditions. *Transactions of the ASME.* 117:1180-126

K. KATAHIRA & H. SUZUKI 1997. Development of a Fatigue Reliability Design Method for Surface Modification Components by a "P-S-N Globe". *Transactions of the Japan Society of Mechanical Engineers A.* 63:1607-1611.

Mandelbrot,B.B.1982. The Fractal Geometry of Nature Freedman New York.

Experimental Mechanics, Allison (ed.) © 1998 Balkema, Rotterdam, ISBN 90 5809 014 0

Experimental characterisation of nonlinear vibration in ultrasonic tools

M. Lucas
Department of Mechanical Engineering, University of Glasgow, UK

J. N. Petzing & G. Graham
Department of Mechanical Engineering, Loughborough University, UK

ABSTRACT: Reliability of tuned components in ultrasonically assisted tooling is often poor due to energy leakage into non-tuned modes. The mechanism of energy leakage can be described by nonlinear vibration behaviour but such behaviour is traditionally difficult to measure. This paper describes an experimental method for characterising nonlinear vibration in ultrasonic tools, including measurement of the in-plane longitudinal response. The normal to surface response can be measured using a laser Doppler vibrometer (LDV). Two in-plane interferometers measure the resultant in-plane surface motion and data is processed to allow the frequency domain response to be obtained. The result is a measurement facility which enables wholefield laser speckle pattern data, measured in three mutually orthogonal axes, to define the full 3D surface motion in a format comparable with conventional contact transducers, allowing the nonlinear response to be characterised in systems tuned to longitudinal modes.

1 INTRODUCTION

Half-wavelength tuned ultrasonic components are used in ultrasonically assisted manufacturing processes to transmit longitudinal vibration from the transducer to the work surface. In many multi-component systems a nonlinear response results from the nonlinear characteristics of the transducers and the interfaces between the bar horns. The result is energy leakage into non-tuned modes and poor operating performance. Often unwanted noisy low frequency modes are excited (Graham & Lucas, 1995). In most high power ultrasonic systems, the required operating amplitude is maintained by a resonance tracking facility in the generator. If the operating response is nonlinear, the generator is unable to lock onto resonance effectively and sufficient vibration amplitude is not sustained for completion of the process. This problem is particularly significant in continuous operations such as cutting and welding applications.

A study of the behaviour of a stack of cylindrical ultrasonic bar horns, shown in Figures 1 and 2, is presented in this paper. Bar horns are often used as transmission elements in multi-component ultrasonic systems and are half-wavelength tuned components which, in this application, operate in a longitudinal mode at a nominal 20 kHz. Mathematical models of the nonlinear mechanism of modal interactions in cylindrical components, resulting in low frequency excitation, have been reported previously (Nayfeh & Nayfeh, 1994). However, measurement of the nonlinear response in cylindrical structures is often difficult due to the highly curved

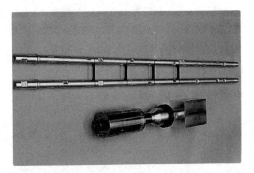

Figure 1: Multi-component stack of bar horns

Figure 2: Individual bar horn components

surface which necessitates the use of noncontact transducers. In this study a laser Doppler vibrometer (LDV) has been used to measure the normal to surface component of the response, but this measurement is insufficient for characterising bar responses which are largely in-plane at the tuned longitudinal mode frequency.

Electronic speckle pattern interferometry (ESPI) has been used successfully to measure wholefield vibration displacement of structures, incorporating two in-plane and one out-of-plane interferometers in order to derive the resultant response from the three mutually orthogonal components (Shellabear & Tyrer, 1989). Recent progress has allowed this technology to be adapted such that the three components of vibration response are extracted in a format which enables the amplitude and phase data to be downloaded into a commercial modal analysis software package for direct comparison with LDV measurement data (Lucas et al. 1996). Using a fine frequency resolution, ESPI can successfully measure the nonlinear response of a bar horn stack, operating in a mode where the response is dominated by the longitudinal in-plane vibration amplitude. This paper discusses the technique and presents the results of both normal to surface measurement, using a LDV, and in-plane response measurement, using ESPI, which identify and characterise the nonlinear response of bar horns in a longitudinal mode.

2 MEASUREMENTS USING 3D ESPI

2.1 Optical configuration

Figures 3 and 4 illustrate the out-of-plane and the in-plane sensitive interferometers (Williams 1993), both of which are used with a frequency doubled, pulsed Nd:YAG laser (wavelength, λ = 532nm).

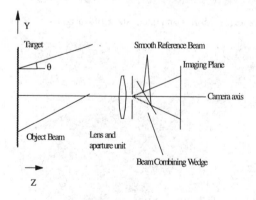

Figure 3: Out-of-plane sensitive interferometer

With respect to the out-of-plane configuration, the reflection of the illuminating beam (object wavefront) from the object surface is collected by the camera lens and directed onto the CCD camera via the beam combining wedge. The reference beam is directed onto the CCD image plane again via the beam combining element, although additional phase stepping optics are included in the reference beam. This arrangement gives a displacement sensitivity described by d=nλ/2, where 'n' is the number of fringes, and this assumes the angle between the reference and object beams is 90°.

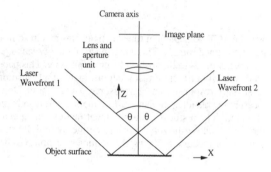

Figure 4: In-plane sensitive interferometer

The in-plane interferometer requires two illuminating beams which are subtended at equal but opposite angles (θ) to the camera/image plane axis, which is typically normal to the object surface   This configuration has an in-plane displacement sensitivity defined by d=nλ/(2sinθ) and can be used for vertical or horizontal in-plane measurements by either rotating the target or the illuminating beams through 90°.

Both configurations have common features. Firstly, the two optical paths should be matched to within 3% of the coherence length of the laser, to produce high visibility fringes. In this case the laser was seeded, giving a coherence length in excess of ten metres, making the matching requirement restrictions insignificant. Secondly, the correlation pattern created by the two interfering beams is collected by the CCD camera. This in turn is connected to a frame store and subsequent images are subtracted from a reference speckle field, results in a correlation fringe pattern representing a contour map of surface displacement.

2.2 Fringe analysis and data manipulation

To identify the response of the ultrasonic bar horns, the horn components are attached to the ultrasonic transducer and the system is mounted separately for each of the optical measurement configurations.

946

In order to obtain vibration amplitude and phase data, a predetermined phase step in one of the optical paths is provided by the piezo mounted mirror. A stepped voltage is supplied to the piezo element which alters the pathlength of the beam by $2\pi/3$ after each step (Robinson & Reid 1993). In this case, two steps were used giving four images: an undisplaced reference image, a displaced image and two phase-stepped displaced images. This procedure is required for each of the three displacement orientations: out-of-plane (oop), horizontal-in-plane (hip) and vertical-in-plane (vip). Ensuring that images are grabbed from the same part of the bar horn's vibration cycle is achieved by synchronising the laser pulses to the vibration response.

Initially, modulo 2π grey scale maps (0 - 255) of wrapped optical phase data are produced, which are then unwrapped using mathematical algorithms based on cosine transforms. The optical phase data requires calibration with respect to the interferometer and the geometry of the object being examined. Specifically, the out-of-plane data requires sensitivity adjustment with respect to the constant curvature of the bar horn.

Wholefield data sets of size 512 by 512 are produced and are subsequently subsampled using routines developed in Matlab in order to reduce the data size and correctly configure the data formats for importing into the modal analysis software (Star Modal). In this study, the frequency response is generated by Matlab to permit illustration of the jump phenomenon associated with bar horn stack nonlinearity.

3. IDENTIFYING THE NONLINEAR RESPONSE

3.1 Out-of-plane response of bar horn set

To measure the resonant response, the bar stack is held at the nodal flange of the transducer and a frequency generator is used to control the excitation output from the transducer. Figure 5 shows the response of a single bar horn stack, excited by a swept-sine input and monitored by a LDV, measured on the side of the bar. The image clearly shows a cubic softening nonlinear response of the horn stack through the operating frequency. Although this result is useful in identifying such a response, it relies on measuring the normal to surface component of vibration response, generated due to Poisson's effect, for a bar which is actually resonant in a longitudinal mode.

The response is characteristic of a Duffing oscillator. When the amplitude is low, the stack behaves as a linear system and does not generate the noise associated with the excitation of untuned modes. At the high amplitudes required in ultrasonic manufacturing operations, which are high power processes, the jump phenomenon is clearly identified. The nonlinearity is highly geometry dependent and

changes significantly if further bar horns are added to the stack. Figure 6 shows the LDV measured out-of-plane response from the side of a stack of three bar horns. Although a softening characteristic is still measured, the stack response exhibits near linear behaviour and is markedly different from the single bar stack.

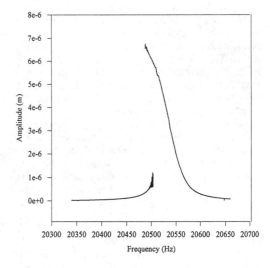

Figure 5: LDV measured oop response for single bar horn stack

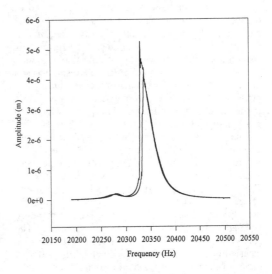

Figure 6: LDV measured oop response for three bar horn stack

947

However, it is difficult to draw conclusions from these measurements since it is not possible to characterise the longitudinal mode by measuring only the normal to surface response.

3.2 *In-plane response of bar horn set*

The two in-plane interferometers are used to measure the in-plane (hip and vip) frequency response of the bar horn stack through resonance. One of the interferometers is used only to confirm that there is no measurable component of in-plane motion in one configuration as the bar is resonant in a longitudinal mode. The second interferometer is subsequently sufficient to measure the response. The interferometers have been configured such that the bar horn stack exhibits only a horizontal in-plane (hip) component of vibration in the longitudinal mode.

Whereas the LDV measurement provides the response at one point on the bar stack surface at a time, ESPI allows wholefield data to be collected from one set of processed fringe patterns and thus the most responsive points on the bar stack surface can be readily identified. However, the LDV has the advantage of producing response curves for a wider range of displacement amplitudes. ESPI is limited to displacements producing a fringe density which can easily distinguish fringes for the processing operations. Therefore in using ESPI, lower vibration amplitudes were excited but were still sufficient to excite and measure nonlinear responses.

Figure 7 shows the in-plane response of the three bar horn stack due to swept-sine excitation through resonance. The resolution of the measurement is merely a function of the number of ESPI measurement sets acquired and can be increased or decreased as necessary to ensure adequate detection of the nonlinearity. In this case, 13 data sets were acquired at 13 one Hz discrete frequency steps. The figure only shows the response at one point on the side of the bar horn stack, but the data sets contain the responses at a grid of points whose number depends on the number of pixels covering the area of the bar stack surface. In this particular case the number of data points which can produce response curves is 200.

The importance of measuring the in-plane response of longitudinal modes is demonstrated by Figure 7. Whereas the LDV measurement for the three bar horn stack showed a near linear softening characteristic, the ESPI measurement clearly shows a hardening spring characteristic. The out-of-plane response is therefore not representative of the longitudinal mode response and cannot be implied from LDV measurements. The ability to measure and characterise the nonlinearity is clearly important in order to understand the different types of behaviour of the ultrasonic stack, which depend on the number and types of components used to transmit vibration to the work surface. It is therefore essential to measure the 3D surface response of components in order to identify nonlinear behaviour accurately.

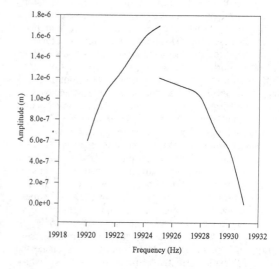

Figure 7: ESPI measured hip response for three bar horn stack

4. CONCLUSIONS

Combining LDV and ESPI measurements can provide successful measurements of the nonlinear behaviour of ultrasonic tools. ESPI has been demonstrated to be a viable and valuable technique in identifying and characterising the nonlinear in-plane response of multi-component ultrasonic systems. For the example of a multiple bar horn stack, both cubic hardening and cubic softening responses are identified and the characteristic of the jump phenomenon measured at the tuned operating frequency depends on the number of bars in the stack. The advantage of ESPI, which offers a wholefield measurement technique rather than conventional pointless measurements, is demonstrated. Critically, ESPI allows measurement of the in-plane response through resonance, which is essential for components exhibiting longitudinal modal response at the ultrasonic operating frequency.

5. REFERENCES

Graham, G. & M. Lucas 1995. Off-resonance system behaviour in multi-component ultrasonic tools. *BSSM Conf., Sheffield, Sept. 1995.*

Lucas, M., G. Graham & J. N. Petzing 1996. Modal analysis of tuned longitudinal mode ultrasonic bar horns. *Porch. ISMA21, Leuven, Sept. 1996*, 893-901.

Lucas, M., G. Graham & A. C. Smith 1996. Enhanced vibration control of an ultrasonic cutting process. *Ultrasonics*, 34:205-211.

Nayfeh, S. A. & A. H. Nayfeh 1994. Energy transfer from high to low modes in a flexible structure via modulation. *ASME Journal of Vibration and Acoustics*, 116:203-207.

Robinson, D. W. & G. T. Reid 1993. *Interferogram analysis; digital fringe pattern measurement techniques*. Bristol, UK: Institute of Physics Publishing.

Shellabear, M. C. & J. R. Tyrer 1989. Three-dimensional analysis of volume vibrations by electronic speckle pattern interferometry. *Proc. SPIE*, 1084.

Williams, D. C. 1993. *Optical methods in engineering metrology*. London, UK: Chapman & Hall.

Experimental Mechanics, Allison (ed.)© 1998 Balkema, Rotterdam, ISBN 90 5809 014 0

Assessment of local effects of notches in wall of welded pipeline

J. Jíra
Faculty of Transportation Sciences, CTU, Prague, Czech Republic

J. Jírová & M. Micka
Institute of Theoretical and Applied Mechanics, Prague, Czech Republic

ABSTRACT: The local effects of notches in the form of long cuttings on the development of deformations in the wall of a spiral welded pipeline were modelled in the course of the hydraulic strength tests of hp gas pipelines. The influence of notches and the spiral weld on the pipe state of stress were investigated experimentally together with the numerical simulation of the behaviour of pipes with defects by means of the application of the ANSYS programme.

1 INTRODUCTION

The high-pressure (hp) gas pipelines of the Czech national gas supply network were constructed in about 1960 and their operation life was nearly 30 years. Such pipelines represent in the Czech Republic more than 20% of the total length. The coal gasification plants, using pressure gasification technology, were the reason for constructing main gas lines with nominal pressure 2.5 MPa (PN 25). But the typical level of the gas pressure for inland natural gas transmission is 4 MPa (PN 40). As the delivery of the natural gas became much greater than of the town gas, the whole regions had to be changed to natural gas gradually. The transmission and distribution of gas was to secured through newly constructed gas pipelines as well as through main gas lines that originally were used for the town gas. The present hp gas pipeline system used older pipelines with PN 25 as well as the new pipelines PN 40. Their interconnecting often means that the new pipelines have to be operated at PN 25 because they are connected with older gas pipelines. During the last ten or more years were realised several research programmes that had to find out and describe faults caused by corrosion, fatigue or a less suitable foregoing construction technology. The objective of our research was to find crucial data about strength and residual life of hp pipelines of the national gas supply network after long-term operation and to enable an output increase by changing the operational pressure from 2,5 MPa to 3,5 MPa together with achieving a necessary amount of operational reliability.

The ageing of the gas pipelines manifests itself by degradation processes of the steel of the pipes, the origin of corrosion defects and the generation of local fields of very high stress levels in the pipe walls, the origin of which can be traced back frequently to the processes of manufacture and construction. The assessment of the reliability of operating pipelines necessitates the knowledge of the actual stress and strain magnitudes in the pipe wall. The real magnitudes of these quantities often differ significantly from theoretical values computed in the design of a hp pipeline of ideal dimensions and properties. Plenty hp pipelines after a long-term of operation shows different faults. These faults cause higher concentrations of stress in such a way that at specific local areas the plastic strain limit is reached and often exceeded. Corrosion damage, the deviations from circularity, local variations of wall thickness, geometric imperfections in welded joints and other inaccuracies of the pipeline form due to its production and placing in the ground play their role in the stress state of the pipeline. The local effective values of total stresses in pipeline operation in sensitive wall places of the real pipe may lie even above yield point. Therefore the strength computation based on the values of stresses produced in the wall of an ideal pipeline by the internal operating pressure is not conclusive.

The hp gas pipelines were assembled of welded, mainly spirally, pipes. The manufacturing and assembly welded joints of high-pressure pipelines of

DEFORMATION OF SPIRAL PIPE 426/6 MM

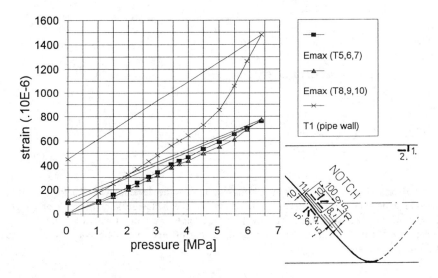

Figure 1. The notch parallel with the spiral weld

the national gas distribution network represent an enormous length and may influence considerably the stress state of the pipes and their deformation processes in the course of operation. The material of the pipes of the national gas supply network is ferritic perlitic carbon steel of different strength. In most cases the steel is of good plastic qualities, considerable yield point and the ability to hardening by plastic deformations. The purpose of the research of hp pipelines was to find decisive information on their serviceability and strength after long-term operation, when they gradually age and on their sensitivity to overloading. According to experience defects of various types originate of the pipelines. One of the defect types may be a notch in the form longitudinal cutting which originates either by an external physical interference in the course of construction or by corrosion in the course of pipeline operation. Because the environs of the manufacturing spiral weld of the hp pipeline often becomes the decisive for the origin of the pipe degradation, the influence of cracks in the form of notches on the deformation development in the wall of a spiral welded pipeline was modelled in the course of experimental research of boundary conditions of hp gas pipelines.

2 EXPERIMENTAL RESEARCH

The hydraulic strength tests of pressure vessels (Jíra & Jírová 1994) manufactured from specific types of hp gas pipeline were conducted on: (i) spiral welded pipes, (ii) lengthwise welded pipes. Artificial defects was created in the wall of the pressure vessel which simulated different location of notches. Before making the defects, a defect check was carried out so as to expose any defects already present in the pipes. The components of the prepared strength research were complex tensile tests of materials of all steel pipes used for experiments.

2.1 Hydraulic stress test

Experimental research was made with spiral welded pipes 426/6 mm and 324/6 mm from ferrous-pearlite carbon steel of class 11 365.1. The longitudinal notches were 100 mm long and 4 mm deep which corresponded with 2/3 of the initial wall thickness. They were cut so as to ensure a smooth and continuous transition between the notch bottom and edges in the longitudinal direction. Before making the artificial defects, a defect check was carried out so as to expose any natural defects already present in

DEFORMATION OF SPIRAL PIPE 426/6 MM

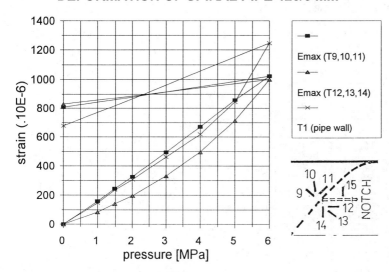

Figure 2. The longitudinal notch close to the weld

pipes. In the pressure vessel made of the pipe the notches were usually situated in the direction of the longitudinal axis of the vessel and in the close proximity of the spiral weld, parallelly with it (Fig.1) and longitudinally close to the spiral weld (Fig.2).

The hydraulic strength test was carried out in stages, thus making it possible to measure the deformation processes in the walls of pipes in the course of the pressurisation of the pressure vessel. The deformations were measured by strain gauges. The purpose of strain gauge measurements was to obtain information on the conditions of the deformation origin and strain development in the near proximity of the spiral weld and the notch-shaped artificial defects. The strain-gauge roses were fastened with adhesive in the immediate proximity of the spiral weld. The same rose was fastened also near the cutting to enable a comparison of the measurements of the deformation process in the defect region with deformation development in the undisturbed part of the spiral weld on the opposite side. The strain-gauge roses enabled the analysis of the development of principal strains and of the changes of their angles. The ascertained deformation values were compared also with the circumferential deformation of the pipe wall.

2.2 Results of strain gauge measurements.

The strain gauge measurements have proved that in the course of manufacture local deviations from the ideal circular shape appear in the proximity of the spiral weld. These imperfections of a shape produce in these regions change of the stress state and the measured strain values differ from the values ascertained for the ideal circular cross section of a pipe. Apart from that the residual deformations due to other loading cycles and the pressure-strain diagrams show that an equalisation of both internal stress due to manufacture and imperfections of form takes place.

The spiral weld of a pipe has a reinforcing effect which arises from the strain gauge measuring (Fig.1 and Fig.2) and the comparison of circumferential deformations in the pipe wall (T1) with the maximum deformations Emax computed from the strain-gauge rose measurements (T5,6,7 and T9,10,11). The maximum deformations in the environs of the spiral weld develop linearly and subsequently slightly non-linearly in dependence on the inner pressure and are inferior to those in the pipe wall (T1), as a rule. In the course of loading also the angles of principal deformations rotate in dependence on the plastic equalisation of imperfections. The

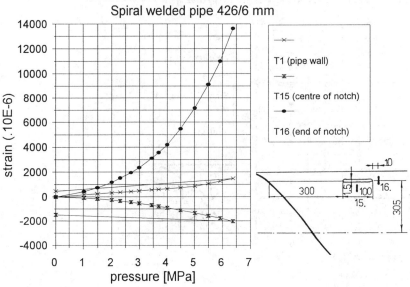

Figure 3. The longitudinal notch between the welds

parallel cutting (T8,9,10) and the longitudinal cutting (T12,13,14) do not influence substantially the stiffening effect of the spiral weld. The strain development in the course of loading is similar to that in the weld without a notch. The deformations, however, are slightly lower than those in the place not weakened by the notch. This is more distinct in the case of the longitudinal notch (T12,13,14), where it looks as though the deep cut has loosened the near environs. In actual fact, however, a more distinct strain take place in the notch ligament, where the deformation process practically takes place. The process of notch opening is testified to also by the history of the strain gauge measurements with the tensometer T15 (Fig.2) situated in the centre on the notch periphery in the circumferential direction, where a significant strain development begins only from 3 MPa upwards and heralds the buckling of the edge.

3 COMPUTATIONAL MODELLING

The local defect is the greatest danger for failure of the pipe wall. The notches can originate in the places of damaged pipeline insulation and are concentrated on limited pipe surface, but attains considerable depth

and length. We used the programme ANSYS (Jíra & Micka 1996) for the numerical analysis of the stress field and deformation process inside the notch and in the vicinity of it and the spiral weld. The environs of the cutting was modelled by 3d section in the shape of the cylindrical parallelogram. The notch had a constant width and its bottom was elliptical curved. For node definition an automatic meshing with smaller node distance in the vicinity of points located on the edge of a cutting was used. The boundary conditions were defined with the use of symmetry. An isoparametric spatial element SOLID95 consisting of 20 nodes was used for the solution of the given problem. The material was considered as being in elastic-plastic region with linear hardening.

The computation proved the stiffening effect of the spiral weld on the deformation of the pipe wall (Fig. 4). The fields of stress or strain and directions of principal stresses in the vicinity of the notch were the result of the computational modelling. The elastic-plastic model of notch enabled to discover the local process of the plastic deformation of notch. For the determination of the reparative overpressure (Jíra et al. 1996) during the piping rehabilitation it is important to know the characteristic sizes of the notches and the corresponding coefficients K of the stress concentration. The task was solved experimentally together with the numerical simulation of the notch behaviour. The outputs can be plotted graphically or it is possible to make a print of the displacement and stress in all points of the FEM network. The values of the local maximal stresses in the bottom or at the periphery of the notch can be determined   and the coefficients K of the stress concentration can be determined by comparison with the average membrane stresses.

4 CONCLUSION

Under repeated loading an increase of the threshold of origin of plastic deformations or, in other words, strengthening, takes place even in case of notches. The failure took place in the cuttings by their opening and the rupture of the weakened cutting ligament. In case longitudinal notch the crack width was 4 times as large as the original width   and in case of the notch parallel with the spiral weld only twice as large, a fact due to the stiffening effect of the weld along the whole cutting length.

The experiments have shown that the defect in the form of a cutting is more dangerous, if located in the direction of the lengthwise axis of the pipeline. Its strength is approximately 8-10% lower than the strength of the notch situated parallelly with the spiral weld, but 15% higher than the strength of a notch situated longitudinally in the pipe wall in the area between the spiral welds. A parallel notch, in spite of its great depth, is less dangerous than expected, as it attains as many as 90% of the theoretical strength of the pipe.

We have studied the reliability of hp pipelines (jíra et al. 1997) weakened by the notches. The research has demonstrated that if the material is not too damaged by fatigue and is not considerably heterogeneous and weakened by microcracks in the structure, it is possible to accept very deep weakening of the pipe wall. According to the size of the local wall weakening the limit depths of the cuttings have been determined, so that a reliable residual service life can be determined for different pipeline sizes and pressures, provided we know the speed of wall corrosion loss.

REFERENCES

Jíra, J., Jírová, J. 1994. Strain analysis of spiral welded pipe 426/6 mm with defects. *Research Report ITAM* (in Czech).

Jíra, J., Micka, M. 1996. Stresses in Walls of National High-Pressure Gas Pipelines in the Region of Plastic Deformations. *Proc. of 5th International Colloquium "Operational Reliability of High-Pressure Gas Pipelines", Prague, March 1996*: 25-34.

Jíra, J., Jírová, J., Kolář, F. 1996. Research of behaviour of corrosion pits with reflection photoelastic coatings. *Proc. of 34.Conference EAN 96, Plzeň, June 1996*: 89-94 (in Czech).

Jíra, J., Jírová, J., Kastlová, O., Micka, M. 1996. Local elastic and plastic deformations of high-pressure pipeline with imperfections. *Proc. of International Conference on MATERIAL ENGINEERING, Gallipoli-Lecce, Italy, 4-7 September 1996*: 915-922.

Jíra, J., Jírová, J., Micka, M., Němec, J. 1997. Complex method for assessment of acceptance of corrosive defects in wall of high-pressure pipeline. *Proc. of the International Conference on Carrying Capacity of Steel Shell Structures (Fracture, Stability, Fatigue, Life-time), under the auspices of European Convention for Constructional Steelwork, Brno, October 1997*: 277-283.

The research has been sponsored by the project no.103/97/0729 of the Grant Agency of the Czech Republic.

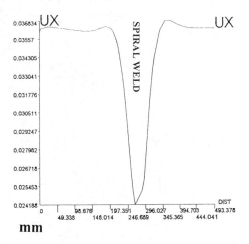

Figure 4. Radial displacement in the vicinity of the spiral weld

Microelectronics

Experimental Mechanics, Allison (ed.) © 1998 Balkema, Rotterdam, ISBN 90 5809 014 0

Measurement of mechanical properties of MEMS materials

W.N.Sharpe, Jr
Department of Mechanical Engineering, The Johns Hopkins University, USA

ABSTRACT: Techniques and procedures have been developed that permit one to measure the tensile stress-strain curve of thin-film polysilicon and electrodeposited nickel used in Microelectromechanical Systems (MEMS). Polysilicon tensile specimens (3.5 μm thick) surrounded by a supporting frame are deposited on a single crystal silicon chip. The central portion of the chip is etched which leaves the tensile specimen stretched across the frame. After gluing the grip ends into a small test machine, the supporting strips are cut away. The nickel specimens (200 μm thick) are deposited in the shape of 3 millimeter long 'dog biscuits' which are placed into matching grip inserts. In both cases, friction is eliminated in the load train by a linear air bearing. Strain is measured directly on both specimens by laser interferometry from two gold lines or reflective indentations on the specimens. Techniques and procedures are described and typical results are presented.

1 INTRODUCTION

Thin-film polysilicon is the predominant material in current commercial devices; it is only a few micrometers thick. A candidate material for larger MEMS is electrodeposited nickel which is a few hundred micrometers thick. To measure the tensile stress-strain curve of such materials, one must address three main issues: specimen preparation and handling, friction in the loading mechanism, and accurate strain measurement. Novel approaches have been developed at Hopkins over the past few years, and they are described in this paper with particular emphasis on the strain measurements.

2 THIN-FILM POLYSILICON

The techniques and procedures described in this section, while perhaps viewed as elaborate, have proven to be manageable and reliable (Sharpe, Yuan, et al. 1997). In fact, much of the testing is now done by undergraduate students. Though used only for polysilicon, this approach can be applied to the mechanical testing of other thin films.

2.1 *Specimen preparation and handling*

It is actually possible to release a 3.5 μm thick polysilicon film from a silicon substrate by etching. One cannot pick up the specimen, but must slide it off the substrate onto a glass slide while it is in the etchant and then slide it onto the grips. Attempts to grasp it with tweezers or even a vacuum pipette fail because the specimen is so fragile.

Figure 1. SEM photograph of a polysilicon tensile specimen after it has been released by back etching.

The approach here is to deposit the thin film onto the substrate in the shape of a tensile specimen with integral gripping ends and supporting side bars. One then etches away only the substrate under the tensile specimen portion leaving it supported between the two grip ends that are held together by the side bars. Figure 1 is a SEM photograph of such a specimen after the rectangular area under it has been etched away. This is a robust structure that can be handled easily and placed into a test machine.

2.2 Loading mechanism

Once the grip ends of the support structure have been aligned and glued to the fixed and movable ends of the test machine, the two side bars are cut with a thin diamond saw. Figure 2 shows a specimen after this step. The left grip is fixed, and the right one is the slider portion of a linear air bearing. Friction in the load train is eliminated by this approach. The other end of the slider bar is attached to a 1-lb load cell which is attached to a piezoelectric actuator with 0.9 µm least-count displacement.

Figure 2. A mounted polysilicon specimen after the side bars have been cut. The silicon chip is 1 cm square.

2.3 Strain measurement

The strain on the thin tensile specimen is measured using a variant of the optical technique presented at the BSSM conference in 1987. The principle is interference between two closely spaced reflective surfaces on the specimen— Young's two-slit experiment in reflection. At that time, pyramidal indentations were placed astride a small crack in a metal specimen and used to measure crack opening displacement. Obviously one cannot indent a thin, brittle film so some other means of generating a suitable reflecting surface must be found. Figure 3 shows a biaxial strain gage on a 3.5 µm thick tensile specimen.

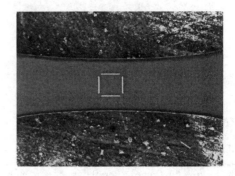

Figure 3. SEM Photograph of the biaxial strain gage. The gold lines are 250 µm apart.

The gold lines are patterned onto the specimen with photoresist and deposited by sputtering. Each line is 200 µm long, 10 µm wide, and 0.5 µm thick. The edges of the lines are rounded as a natural consequence of the sputtering process, so interference fringes are generated at large angles from the vertical. The interference fringe patterns are monitored by four linear photodiode arrays located at approximately 45° from the vertical and along the tensile direction and transverse to it. This arrangement enables the measurement of axial and lateral strain so that one can determine Poisson's ratio.

2.4 Test system

A schematic of the test system is given in Figure 4. The base is attached to a plate with tapped holes placed on a table; this enables easy positioning of the four fringe detectors which are mounted with orthogonal translation stages for easy positioning. These fringe detectors are linear diode arrays – each diode has an aperture 13 µm high by 2.5 mm wide and there are 512 of them spaced on 25 µm centers for a total length of 12.5 mm. The rest of the strain measurement consists of a 10 mw laser and a microcomputer with the associated software. The piezoelectric actuator is controlled by the microcomputer, and the test is run under displacement control.

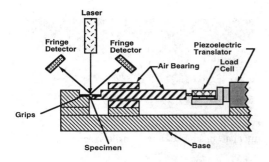

Figure 4. Schematic of the polysilicon test machine.

2.5 Results

A representative result is shown in Figure 5. Polysilicon is a very linear and brittle material so the strain range is not very large. A very slight preload is applied before the test is started; this is to straighten the specimen so that the gold lines are properly aligned to generate fringe patterns. It is a usual practice to preload tensile specimens in the testing of larger specimens. There are 353 data points in each of the strain plots; the resolution is conservatively estimated at 5 microstrain. The overall relative uncertainty is estimated at ± 3 %.

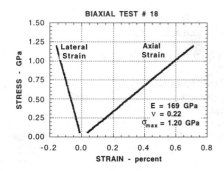

Figure 5. A representative stress versus biaxial strain plot for 3.5 μm thick polysilicon.

Forty seven tensile tests have been conducted on specimens from various production runs at the Microelectronics Center of North Carolina, and the Young's modulus is 169 ± 6.2 Gpa. The tensile strength is 1.20 ± 0.15 Gpa. The coefficient of variation is 3.6% for the modulus, which is not as good as would be expected in testing large metal specimens, but is quite acceptable. The strength varies more widely at 12.5%, but that is to be expected. In 23 of the tensile tests, the Poisson's ratio was measured to be 0.22 ± 0.01; that technique

is an extension of the first approach which used only two gold lines. Table 1. presents a summary in some detail of the results from five production runs.

Table 1. Mechanical property data for polysilicon.

MUMPs	Young's Modulus (GPa)	Poisson's Ratio	Tensile Strength (GPa)
6	163 ± 4.78	-	1.23 ± .200
8	173 ± 6.94	-	1.27 ± .107
10	173 ± 2.00	-	1.14 ± .155
11	168 ± 7.15	0.22 ± .015	1.19 ± .155
12	168 ± 1.59	0.22 ± .003	1.17 ± .100
All Data	169 ± 6.15	0.22 ± .011	1.20 ± .150

These techniques and procedures have yielded consistent and accurate data on the elastic and strength values of this important MEMS material. Others have measured the modulus indirectly, but these are the first measurements in which the strain is measured directly on the specimen in the time-honored manner. No measurements of Poisson's ratio existed when these tests were conducted; a series of indirect measurements have since been made on polysilicon membranes. The strength is an important although at this stage of MEMS technology it is customary to over-design components by a large factor. But, mechanical properties are always important, and this is a way of measuring them in a new, thin material.

3. ELECTRODEPOSITED NICKEL

One cannot expect to fabricate MEMS with very large load-carrying capacity from thin materials. Thicker MEMS are under development, and a leading candidate material is electrodeposited nickel. It can be patterned in a manner similar to that used to make polysilicon devices, but the thicker photoresist requires special processing. Testing is considerably easier in that it is easier to handle the specimen, but direct strain measurement is still below the limit of traditional techniques (Sharpe, LaVan, and Edwards 1997).

3.1 Specimen preparation and handling

Specimens are deposited in their final shape – that of a 'dog biscuit'. The wedge-shaped ends can be inserted into matching inserts in the grips, which obviates the need for mounting screws, adhesives, etc. A set of six specimens is shown in Figure 6 as they are received. The substrate is silicon, just as it is for the polysilicon specimens, and they are released by etching away the layer between the specimens and the substrate. Tensile specimens this large (3 mm long overall) can be easily handled with tweezers.

Figure 6. Six nickel microtensile specimens on a silicon substrate.

3.2 Loading mechanism

The loading mechanism is similar in concept to that used for polysilicon except that the piezoelectric actuator is replaced with a screw-driven translation stage because it does not have a large enough displacement range. This stage is not controlled by the microcomputer, but is simply turned on after the strain measurement program is activated. The test is therefore also run in displacement control.

3.3 Strain measurement

measured in exactly the same fashion as for the polysilicon except that the reflecting gage markers are Vickers microhardness indentations instead of gold lines (reflective lines could be applied, but indentations are easier). Figure 7 is a photomicrograph of a specimen with two tiny indents (biaxial strains can be measured with three indents, but that is not done here). These indents are only approximately 4 μm deep, while the specimen is 200 μm thick. A special feature of this testing is the placing of indents on both sides of the specimen so that the two strains can be averaged to eliminate bending effects.

3.4 Results

Figure 8 shows the results from 17 tests on electrodeposited nickel. The variation in results is considerable, but the electrodeposition process is known to vary spatially because of nonuniformity of the plating solution. This nickel is quite strong compared to ordinary cast nickel. The yield strength of 323 ± 30 MPa (50 ksi) is much higher than the 60 MPa of cast nickel. Also, as is obvious from the

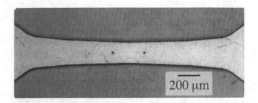

Figure 7. Vickers microhardness indentations 300 μm apart in a nickel microtensile specimen.

plots, it is a very ductile material. It is not only suitable for larger MEMS undergoing elastic deformation, but for those devices where plastic collapse is a key to their function.

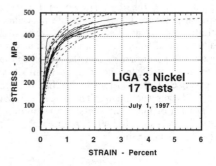

Figure 8. Tensile stress-strain curves of 17 nickel microspecimens.

4. CONCLUDING REMARKS

The two materials tested in this work have come into prominence with the origination of MEMS technology. However, there are other situations where it is desirable to test small and/or thin specimens. Prototype alloys are often available in only very small quantities and mechanical properties are needed to guide the development of the new materials. Inhomogeneities in properties of structures such as weldments can be studied by excising microspecimens. Variation of properties such as brittleness over long periods of exposure to radiation can also be tracked via tiny specimens removed from the operating component.

In all of these cases, the key need is strain measurement directly on the specimen. It is not too difficult to handle a small specimen and measure its breaking strength (although specialized procedures must be used), but it is much more challenging to measure an accurate stress-strain curve. The techniques and procedures described herein offer a solution to this important challenge in strain measurement.

ACKNOWLEDGMENTS

The research program under which this work has been conducted is supported by the National Science Foundation, the Army Research Laboratory, and the Naval Surface Warfare Center. The contributions of colleagues R. L. Edwards and R. Vaidyanathan and students B. Yuan, J. Fox, D. LaVan, and A. McAleavey are very gratefully acknowledged.

REFERENCES

Sharpe, W. N., Jr., B. Yuan, R. L. Edwards, and R. Vaidyanathan, ''Measurements of Young's modulus, Poisson's ratio, and Tensile Strength of Polysilicon'', *Proceedings of the Tenth IEEE Inter'l Workshop on Microelectromechanical Systems*, Nagoya, Japan pp. 529-534 (1997).

Sharpe, W. N., Jr., D. A. LaVan, and R. L. Edwards ''Mechanical Properties of LIGA-Deposited Nickel for MEMS Transducers'', *Proceedings Transducers '97,* Chicago, IL, pp. 607-610, [1997].

Experimental Mechanics, Allison (ed.)© 1998 Balkema, Rotterdam, ISBN 90 5809 014 0

Thermo-mechanical analysis of microelectronic components and chipcards

B. Michel & D. Vogel
Department of Mechanical Reliability and Micro Materials, Fraunhofer Institute for Reliability and Microintegration (IZM) Berlin, Germany

ABSTRACT: This paper presents the results of a series of experiments and combined numerical simulations for advanced electronic packaging structures. With growing miniaturization the „local" material properties and local temperature gradients exert a greater influence on the reliability of microcomponents and microsystems than in any macroscopic component. The authors apply such experimental techniques as acousto-microscopy, laser scanning microscopy, thermography, and the microDAC method to characterize the material behaviour of microelectronic packaging components. The experiments have been combined directly with FE simulations. This finally leads to an improved reliability assessment of the microcomponents (e.g. chip cards, airbag sensor components). Special attention is also given to the experimental analysis of thermal fatigue behaviour of solder bumps in microsolder interconnects of flip chip assemblies and chip size packages. These problems are very important for applications in automotive electronics and telecommunication as well.

1 INTRODUCTION

In the recent years thermo-mechanical reliability analysis of microelectronic components has become very important due to new chip interconnection technologies (e.g. flip chip, ball grid array, chip size package etc.), increasing miniaturization and higher packaging densities. An advanced microdeformation analysis of the critical regions of microelectronic packages like flip chip assemblies, in multichip modules or chipcard components has become more and more important for the breakthrough of low cost interconnection technologies in microelectronics. Many problems of defective microsystems can be related to the mechanical and thermal stresses (e.g. thermal „misfit" stresses) that develop at various stages of microsystem processing (see e.g. Michel et al. 1994).

Field measuring techniques performed in a laser scanning microscope and in a scanning electron microscope allow an effective determination of strongly localized deformation fields, e.g. within the critical regions of flip chip solder bumps or in the chip-board interconnection layers. Until now optical methods, namely Moiré techniques, have been widely used by different authors (Han and Gou 1995) to measure component deformation on the

package object surface. Therefore, FE results can be compared with experimental data from real components in terms of displacement or strain fields. Successful measurements were carried out for flip chip assemblies and ball grid arrays. Measurements of the overall assembly deformation during thermal loading as well as of local deformations in medium sized structures (e.g. inside bigger BGA solder balls), have been reported (Han and Gou 1995). Nevertheless, a lot of open questions is remaining with concern to the mechanical behaviour smaller sized structures like solder bumps, different kind of thin layers, etc.

This work applies a method recently established at the Fraunhofer Institute IZM in Berlin - the microDAC deformation measurement tool (see e.g. Vogel et al. 1997). It possesses particular advantages compared to existing conventional methods. Contactfree measurement, downscaling capability and simplicity are essential features of microDAC. In the following its application to micro solder joint strains in chip scale packages and flip chip assemblies is outlined with some emphasis on the comparison between finite element results and those obtained by measurements.

2 THE MICRODAC CONCEPT

MicroDAC (Vogel et al. 1997) stands for deformation field measurement inside microscopically small objects (DAC - **d**eformation **a**nalysis by **c**orrelation method). The method's basic items are mathematical algorithms and software codes developed in cooperation between the Fraunhofer Institute IZM in Berlin, the CWM GmbH and K&T Measuring Systems in Chemnitz.

The computer codes can be applied to microscopic object images of different mechanical and/or thermal load states. Computational comparison of these images results in deformation fields over the object surface. The basic idea of the underlying mathematical algorithms proceeds from the fact that microscopes of different kinds commonly allow to record tiny local and unique object patterns. They can function as a local image area marker. Commonly, these microstructures even remain stable during severe thermal and/or mechanical component load and can be recognized after loading. A correlation based image processing algorithm is applied to determine a set of local pattern displacements between two object states, and finally whole displacement and strain fields are measured.

The developed computer codes are able to track local structures with a subpixel accuracy of approximately 0.1. Subpixel accuracy of 0.1 means that local displacements with a resolution of $\delta U = 0.1 \, L/M$ can be measured (see Fig. 1). Here, L indicates the length of the field of view for the imaging equipment. M is the number of pixels along the image edge. For the example of a typical low resolution SEM image of microsolder bumps we used the values L = 100 μm, M = 1024 and n = 15. Then, a measurement resolution of $\delta U \approx 10$ nm can be achieved. This corresponds to a lateral structure displacement resolution of about 1.5 μm. This is a very good approximation for many practical applications. Measurement resolution in terms of strain equals $1 \cdot 10^{-3}$, independently on magnification.

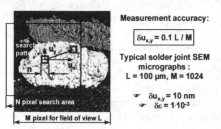

Fig. 1 MicroDAC structure tracking by digital correlation analysis

Commonly, measurement points are set for a regular grid overlaying the object images with selectable node distances. If desired, measurement points can also be defined from finite element input files. In this case finite element meshes have to be generated for the real component making use of component micrographs. A special software tool has been developed for this purpose. Relevant FE mesh nodes within the measurement area of a later microDAC analysis are red by the microDAC program code from the standard finite element code. The authors used the ANSYS, ABAQUS and ASTOR codes for the applications.

The microDAC measurement tool has the following major advantages:

- no special surface preparation for measurement has to be performed,
- the application of different imaging equipment allows to change the measurement resolution in a wide scope,
- the experimental set up is rather simple, especially in comparison with optical interferometric and moiré methods.

3 DEFORMATION STUDIES OF MICRO-SOLDER JOINTS

Aiming at low cost assemblies attempts have been made at flip chip bonding on organic substrates like FR-4 (Lau 1996). A main problem occurs for this approach with regard to the large thermal mismatch between epoxy and the silicon die. Underfill encapsulants are used to improve the mechanical integrity of the assemblies. They have to compensate mechanical and thermal stresses between substrate and chip. Their material characteristics as thermal expansion coefficients, Young's modules, creep parameters and others are important inputs to be optimized for particular configurations. Furthermore, a good adhesion is desired to avoid cracks between underfill, substrate and chip.

In order to obtain information upon the influence of material properties, assembly geometry and manufacturing imperfections on the strain/stress behaviour and component life times, finite element simulations for thermal cycling have been performed. The results are compared for particular cases with the microDAC measurement to make sure, that adequate mechanical modeling has been applied.

FE simulation has been carried out for temperature intervals beginning from room temperature to match measurement conditions. The flip chip specimen discussed here is characterized by

- a silicon chip of 5 mm × 5 mm × 0.5 mm,
- an FR-4 substrate stripe of 35 mm × 5 mm × 1 mm,
- Cu(8 μm)-Ni(3μm) top-surface metallurgy,
- Cu(5μm)-CuSn(2μm) under bump metallization,
- eutectic PbSn63 solder bumps, peripherally arranged with a bump height of 70 μm and a pitch of 600 μm,
- 70 % SiO$_2$ filled epoxy underfill,
- 14 μm solder mask.

For further details of the theoretical model including the constitutive behaviour of the flip chip assembly see the paper of the authors (Michel 1997b).

In the microDAC measurement of the same package the specimens were cross-sectioned along a bump row of the package bumps and thermally loaded inside a scanning electron microscope. Measurements were done during heating of the specimens from room temperature to levels above 100 °C.

Comparison between FE calculations and the experimental investigations show a measured and simulated strain field of the outermost package bumps, respectively (Fig. 2, 3). In both cases the maximum strain in the direction perpendicular to the board was found in the inner part of the bumps. Nevertheless, the comparison of the theoretical and experimental data reveals an essential difference for the strain values. More detailed investigations using improved constitutive parameters for the polymeric underfiller showed that the anisotropy in the CTE has a strong influence on the results of the FE model. Taking into account the anisotropic CTE values an improved simulation was carried out. Then, the gap between measurement and theory has dropped down significantly (see Fig. 4).

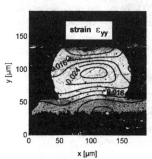

Fig. 2 PbSn solder bump strain field for assembly heating from 31 °C to 118 °C, vertical strain component, micro-DAC measurement inside SEM

Summarizing all measurements and simulation results obtained for the investigated assembly with eutectic PbSn solder bumps, it could be established

- that the used underfill with 70 % SiO$_2$ is essentially stiffening the whole assembly, solder shear is rather moderate for outside bumps,
- that underfill and solder strain perpendicular to the board direction are unexpectedly high, their amount exceeds the value of unrestricted thermal strain between a half and one order of magnitude.

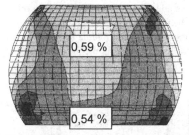

Fig. 3 PbSn solder bump strain field for assembly heating from 31 °C to 118 °C, vertical strain component, FE simulation for *isotropic* underfill CTE

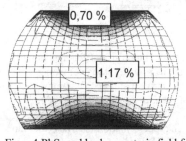

Fig. 4 PbSn solder bump strain field for assembly heating from 31 °C to 118 °C, vertical strain component, FE simulation for *anisotropic* underfill CTE

Lateron the experiments were extended taking into account also component imperfections like e.g. underfill voids, material property gradients and asymmetric assembling as well (Fig. 5, 6).

The delamination of the polymeric materials at the chip edges was studied by means of acousto-microscopy and laser scanning experiments. Finally, the investigations lead to a better understanding of the failure mechanisms. The quantitative evaluation of the local deformation were taken as input data for the reliability analysis. Crack quantities (K, J, C*, ΔT* etc.) were calculated for different assemblies and material combinations also taking into account creep deformation. The procedure resulted in high reliable optimized package structures.

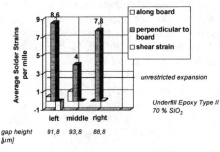

Fig. 5 Average solder bump strains for flip chip on FR-4 measured by microDAC, strain components for the left, right outermost and middle bumps, *symmetric* die-to-board gap along the die edge

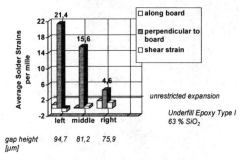

Fig. 6 Average solder bump strains for flip chip on FR-4 measured by microDAC, strain components for the left, right outermost and middle bumps, *asymmetric* die-to-board gap along the die edge

Besides the microdeformation analysis for flip chip assemblies also ball grid arrays and chip size package structure have been investigated. Last not least the approach described above has been successfully applied to chipcard package assemblies and airbag sensor components. The microdeformation analysis (microDAC etc.) combined with FEA also in these applications was a very good base for an advanced reliability analysis of the electronic assemblies. An improved damage model based on the local deformaion analysis revealed very good results also for life-time predictions of the thermal fatigue behaviour of the investigated microcomponents. The method, taking into account the local deformation fields within the critical regions, in most cases was in a better correspondence with real failure behaviour of the components than the widely used Manson/Coffin approach.

4 CONCLUSIONS

The microdeformation field measurement method microDAC has been successfully applied to microscopic structures in microelectronic packaging. Its usefulness has been demonstrated with regard to flip chip and chip size package technology. A comparison between measured and simulated strain fields allows to refine mechanical models for finite element simulations. The combination of microDAC and FEA leads to improved fatigue and fracture concepts which have been successfully applied to microelectronic reliability analysis („Fracture Electronics").

REFERENCES

Han, B. & Y. Gou 1995. Thermal deformation analysis of various electronic packaging products by moiré interferometry. *J. of Electronic Packaging* 117(9): 185-191.

Lau, J. H. 1996. *Flip Chip Technologies,* New York, Springer Verlag.

Michel, B.; Schubert, A.; Dudek, R. & V. Grosser 1994. Experimental and numerical investigations of thermo-mechanically stressed micro-components. *Micro System Technologies* 1(1): 14-22.

Michel, B.; Winkler, T. & H. Reichl 1997a. Fracture Electronics - Application of fracture mechanics to microelectronic systems and chip packages. *Proc. 9th Conference Fracture (ICF9),* Sydney, 1-5 April 1997: 3107-3112, Amsterdam, Pergamon.

Michel, B.; Vogel, D.; Schubert, A.; Auersperg, J. & H. Reichl 1997b. The microDAC method - a powerful means for microdeformations analysis in electronic packaging. *Proc. ASME Congress* Dallas, USA, Nov. 1997, AMD-Vol. 226, EEP-Vol. 22: 117-123

Vogel, D.; Auersperg, J.; Schubert, A.; Michel, B. & H. Reichl 1997. Deformation analysis on flip chip solder interconnnects by microDAC. *Proc. Reliability of Solders and Solder Joints Symposium* at 126th TMS Annual Meeting, Orlando, USA, In: *TMS Publication* (ed. R.K. Mahidhara) No. 96-80433: 429-438.

ACKNOWLEDGEMENT

The authors want to thank Dr. R. Kuehnert, K&T Measuring Systems Chemnitz, for his support of the microDAC experiments.

Experimental Mechanics, Allison (ed.)© 1998 Balkema, Rotterdam, ISBN 90 5809 014 0

Investigation of vibration characteristics of printed circuit boards

W.C.Wang & C.B.Cheng

Department of Power Mechanical Engineering, National Tsing Hua University, People's Republic of China

ABSTRACT: In this paper, the electronic shearography and modal analysis technique were used to investigate the vibration characteristics of copper film epoxy/glass fiber laminated printed circuit boards (PCBs). Both flat-plate and chip-welded PCBs were tested and their full-field displacement derivatives were compared. Finally, the location of maximum stress was obtained based on the approximation that the PCB is treated as an isotropic material.

1 INTRODUCTION

Electronic instruments and devices may be mal-functioned due to peeling off of the solder on the printed circuit boards (PCBs) under environmental or accidental vibration. Because of their complicated material properties and boundary conditions, analytical and numerical solutions of the dynamic characteristics of copper film epoxy/glass fiber laminated PCBs are difficult to obtain. Experimental method is the best alternative for investigating the vibration characteristics of PCBs. In this paper, the electronic shearography and modal analysis technique were employed to compare the full-field displacement derivatives of flat-plate and chip-welded PCBs.

Shearography is a full-field and non-contact optical method. It measures the derivatives of the displacement. With proper optical arrangement, it can measure directly the gradients of the out-of-plane displacement component (i.e. $\partial w/\partial x$, $\partial w/\partial y$, etc.) for structural vibration, warpage, etc. Since only derivatives of the displacement are measured by the shearography, rigid body motion will not produce additional strain. Essentially, the method itself is insensitive to the environmental vibration.

Mohan et al. (1996) applied the electronic shearography (i.e. an integration of the digital image processing technique and the shearography) to investigate the static and vibration characteristics of a clamped aluminum plate. Yang et al. (1996) applied dual beam digital shearography to measure the 3-axis strains of a diametrically loaded circular disk. Note that digital shearography is the synonym of the electronic shearography.

In the literature, limited experiments have been done on the printed circuit boards (PCBs). Tossell & Ashbee(1989) applied the shadow moire method to investigate the deformation due to thermal expansion mismatch around thickness-through holes in multilayer circuit boards. Similar work was performed by Watt et al. (1991) as well as Taniguchi & Takagi(1994),however,holographic interferometry was used instead. Experimental modal analysis on PCBs with surface mount electronic components was investigated by Wong, et al. (1991). In the above-mentioned research work on PCBs, only displacement field was measured. While the peel off of the solder on the PCBs or the fracture of the PCBs is mainly due to the stress or strain, it is therefore important to perform experimental investigation. In this paper, the stress or strain produced by the environmental or accidental vibration was investigated by the electronic shearography and modal analysis technique. The natural frequencies and associated vibration modes of the test specimens were measured by the modal analysis technique. Besides, the frequency of the exciting force for the electronic shearography was set at the natural frequency for the corresponding mode.

2 THEORY

The intensity of light recorded before the specimen was deformed can be expressed as (Hung & Liang, 1979)

$$I = 2a^2(1 + \cos\phi) \qquad (1)$$

where I is the light intensity ; a is the amplitude ; ϕ is the phase difference between two neighboring points of the test specimen.

If the CCD camera accumulates the light intensity from t_1 to $t_1+\tau$ and δ approaches zero, the output voltage V_{CCD} relates to the light intensity is (Cheng, 1997)

$$V_{CCD} = 4a^2\alpha\left(\frac{2N\pi}{\omega}\right)\left[1 + J_0(K)\cos\phi\right] \qquad (2)$$

where $\delta = \tau - \dfrac{2N\pi}{\omega}$; δ is the time difference between vibration period and shutter opening time interval; ω is the vibration frequency; τ is the shutter opening time interval; N is the number of cycle; α is a proportionality constant; $J_0(K)$ is the zero-order Bessel function and $K=f(\lambda,\delta x,H(x,y,z))$, where $H(x,y,z)$ is the vibration amplitude of the point $P(x,y,z)$ to be measured; λ is the wavelength of the light source.

3 EXPERIMENTAL SETUP

3.1 Test Specimen

Two test specimens were used in this paper. The material of both specimens is glass fiber reinforced copper film with epoxy additive. The dimensions of both specimens are 222mm×255mm and about 1.6mm in thickness. One of the two specimens is a flat-plate PCB, i.e. without any chips and wires on it. The other specimen is a chip-welded PCB used in a typical personal computer (Figure 1). Note that the PCB is attached to a steel frame by using four plastic joints and two screws. Thus the boundary conditions are rather complicated. A special loading frame was designed and manufactured so that the mounting conditions of the PCB in the experiments performed are exactly the same as the real applications.

3.2 Vibration system

In this paper, the vibration system (Figure 2) was used to evaluate the natural frequencies and associated modes of the specimens. Besides, the vibration system was also used to provide the sinusoidal exciting force for the electronic shearography. The devices and components of the vibration system are a shaker (LDS Co., UK), an accelerometer (PCB Co., USA), an amplifier (LDS Co., UK) and a PC-based spectrum analyzer (Prowave Co, ROC). The maximum output of the shaker is limited to 17.8N. The weight of the accelerometer is 0.5g. The amplifier was used to magnify the signal output from the spectrum analyzer in order to push the shaker to excite the specimens. For modal testing, the spectrum analyzer generated the pseudo random signal to the amplifier and evaluated the frequency response functions (FRFs) of each measuring point on the specimens.

3.3 Real-time Electronic Shearography System

The real-time electronic shearography system consists of the optical setup and the real-time image-processing system with associated software.

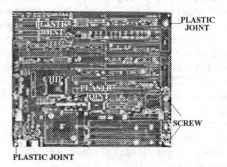

Figure 1. The specimen of chip-welded PCB.

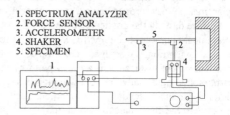

1. SPECTRUM ANALYZER
2. FORCE SENSOR
3. ACCELEROMETER
4. SHAKER
5. SPECIMEN

Figure 2. The vibration system.

The optical setup is depicted in Figure 3. A 75mW He-Ne laser (NEC GLG5-900) of 632.8 nm wavelength is used as the light source. The beam steering device is used to elevate and change the direction of the light beam. A spatial filter is used to filter the optical noise and expand the laser beam. An unbalanced Michelson interferometer is used as the shearing device which is composed of one 50/50 beam splitter and two reflective mirrors.

The real-time image-processing system consists of the advanced frame grabber (AFG) (Image Technology Co. USA) image-processing board with associated software, a standard RS-170 CCD camera(Pulnix Co. USA) and a PC.

4 EXPERIMENTAL PROCEDURE

4.1 *Modal testing*

1. The sampling points were first selected on the specimens as shown in Figure 4. The locations of the sampling points of both PCBs are the same. Therefore, only the location of the chip-welded PCB is shown.
2. The force sensor was attached at point 64 on the surface of the two specimens by the quick set epoxy adhesive. The force sensor was connected to the shaker by a stick.
3. The maximum output voltage of the pseudo random noise was first set. In this study, the frequency range and frequency resolution were selected as 200 Hz and 0.25 Hz, respectively. The accelerometer was mounted on one sampling point and then moved to other points to obtain FRFs of all sampling points. The obtained FRFs were analyzed by the Star software (1990).

4.2 *Electronic Shearography*

1. Adjusting the optical setup.
2. A white powder coating was sprayed on the surface of the specimen in order to increase the light-reflection capability of the specimen.
3. Frequencies of the sinusoidal exciting force were set to the natural frequencies obtained from the modal testing.
4. Adjusting the shearing direction and the amount of shear.
5. The image-signal-subtraction method was employed in this study. The reference image was taken without any driving force and stored in the memory of image-processing board at first. Then the power amplifier was turned on and the reference image was subtracted by the image under load. The results are displayed on the screen.

5 RESULTS AND DISCUSSION

The natural frequencies of the two PCBs are listed in Table 1. All frequencies of the flat-plate PCB are lower than those of the chip-welded PCB for the same mode. The weights of the flat-plate PCB and the chip-welded PCB are 189.1g and 194.2g, respectively. Generally speaking, due to its larger weight, the natural frequency of the chip-welded PCB seems to be lower than that of the flat-plate PCB. However, the modal testing shows the opposite results. It is very possible that the effect of increase of stiffness due to the mounted chip on the chip-welded PCB is stronger than the effect of increase of weight of the chip.

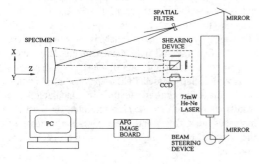

Figure 3. Schematic of optical setup of electronic shearography.

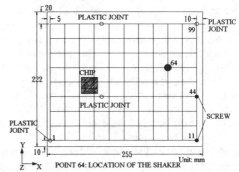

Figure 4. The sampling points of the chip-welded PCB.

Mode	Frequencies		
	Flat-plate, F_f (Hz)	Chip-welded, F_c (Hz)	Difference
1	46.55	51.14	9.86%
	58.83	65.58	11.47%
3	84.74	91.38	7.84%
4	102.41	114.39	11.70%
5	117.68	121.58	3.31%
6	136.11	145.02	6.55%

$$\text{Difference} = \frac{F_c - F_f}{F_f} \times 100\%$$

Table 1. Natural frequencies of the flat-plate and chip-welded PCB.

It is noteworthy that the neighboring natural frequencies of each specimen are very close. Besides, the magnitudes of all natural frequencies are quite low. The maximum one is only 145.02 Hz! Similar magnitudes of natural frequencies for the first five modes (<152Hz) were also obtained for a two slot-ended PCB (Pitarresi et al.,1991).

Although the displacement data obtained from the modal testing are relative magnitude, it is still instructive to utilize them to obtain the approximate location of the maximum stress. With the help of Mathematica software (1993), those displacement data were fitted by a 4th degree surface equation of x and y. Then, the partial derivative contours of the fitted displacement equation w(x,y), i.e. $\partial w/\partial x$ and $\partial w/\partial y$ are plotted as shown in Figure 5.

The shearographic fringe patterns of the two PCBs in the x and y directions are shown in Figure 6. It is obvious that the fringe patterns of the chip-welded PCB nearby the chip are significantly different from those of the flat-plate. By comparing Figures 5 and 6, it is clear that the tendencies of the fringes in both figures are similar. However, the shearographic fringes are slightly inclined from the contours shown in Figure 5. This phenomenon is mainly due to the image-doubling effect (Waldner, 1996) and the error in the fitting process. The good agreement between the results from two different experimental techniques strengthen the confidence for further calculation of the data obtained from the modal analysis technique.

According to the plate theory, the stress components σ_x and σ_y are related to the curvatures $\frac{\partial^2 w}{\partial x^2}$ and $\frac{\partial^2 w}{\partial y^2}$ as follows (Timoshenko & Woinowsky-Krieger, 1959):

$$\sigma_x = K\left(\frac{\partial^2 w}{\partial x^2} + v\frac{\partial^2 w}{\partial y^2}\right)$$

$$\sigma_y = K\left(v\frac{\partial^2 w}{\partial x^2} + \frac{\partial^2 w}{\partial y^2}\right)$$

(3)

where $K = -\dfrac{Et}{1-v^2}$; E, t, and v are the Young's modulus, thickness and Poisson's ratio of the plate respectively. Assume that the PCB as an isotropic material and let

$$\overline{\sigma}_x = \left(\frac{\partial^2 w}{\partial x^2} + v\frac{\partial^2 w}{\partial y^2}\right)$$

$$\overline{\sigma}_x = \left(v\frac{\partial^2 w}{\partial x^2} + \frac{\partial^2 w}{\partial y^2}\right)$$

(4)

If v is given as 0.23, the relative magnitudes of the stress components σ_x and σ_y, i.e. $\overline{\sigma}_x$ and $\overline{\sigma}_y$, can be calculated by differentiating $\partial w/\partial x$ and $\partial w/\partial y$ with respect to x and y, respectively. The results of eqn(3) are plotted in Figure 7. Note that point M in Figure 7 indicates the approximate location of the maximum stress.

6 CONCLUSION

Owing to its complicated boundary conditions and material properties, the analytical and numerical solutions for the vibrations of the PCBs are rather difficult to obtain. In this paper, the electronic shearography and modal analysis technique were used to find the full-field displacement derivatives, natural frequencies and mode shapes of a flat-plate PCB and a chip-welded PCB. Approximate location of the maximum stress can then be calculated. The results also show that the chip mounted on the PCB increases the stiffness and overweighs the increase of the weight. Hence, higher natural frequencies are obtained.

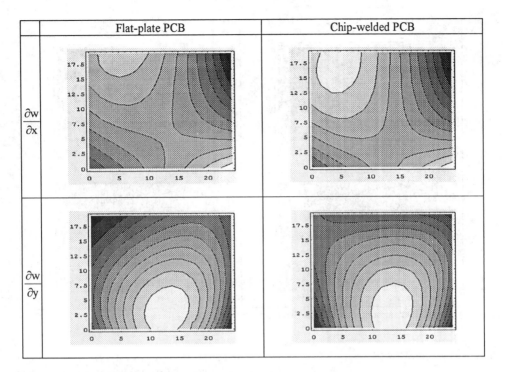

Figure 5. The partial derivative contours of the fitting function.

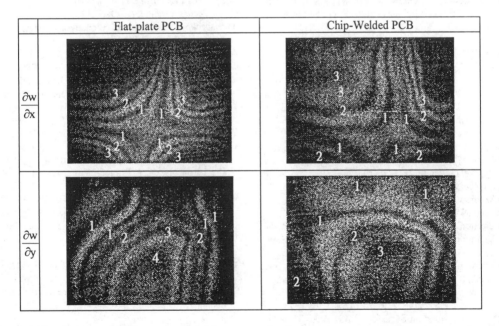

Figure 6. The shearographic fringe patterns of the two specimens.

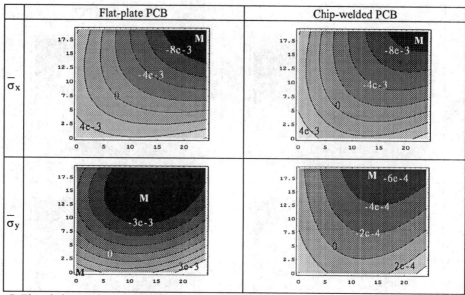

Figure 7. The relative maximum stress of the 3rd mode of the two specimens.

ACKNOWLEDGEMENTS

This research was supported in part by the National Science Council (Grand No.'s: NSC-83-0404-E007-076 and NSC-84-2212-E007-031) of the Republic of China.

REFERENCES

Cheng, C. B. 1997. Investigation of Vibration of Printed Circuit Boards Using Electronic Shearography, *Master Thesis*, Department of Power Mechanical Engineering, National Tsing Hua University, Taiwan, Republic of China.

Hung, Y. Y. & Y. Y. Liang 1979. Image-Shearing Camera for Direct Measurement of Surface Strains, *Applied Optics,18(7)* :1046-1051.

Mathematica For Windows 1993. Enhanced Version 2.2, Wolfram Research Inc., Illinois, USA.

Mohan, N. K., H. O. Saldner and N. E. Molin 1994. Electronic Shearography Applied to Static and Vibrating Objects, *Optics Communications, 108*:197-202.

Pitarresi, J. M., D. V. Caletka, R. Caldwell and D. E. Smith 1991. The Smeared Property Technique for the FE Vibration Analysis of Printed Circuit Cards, *J. of Electronic Packaging, 113* : 250-257.

Structral Measurement Systems (SMS) 1990. *The STAR System*, A Gen Red Product Line, Canada

SURFER 1994, Version 5.01, Surface Mapping System, Golden Software, Inc., Golden, Colorado, USA.

Taniguchi, M. & T. Takagi 1994. Holographic Pattern Measuring System and Its Application to Deformation Analysis of Printed Circuit Board Due to Heat of Mounted Parts, *IEEE Transactions on Instrumentation and Measurement,43(2)* :326-331.

Timoshenko, S. P. & S. Woinowsky-Krieger 1959. *Theory of Plates and Shells, Second Edition*, McGraw-Hill Book Co., USA.

Tossell, D. A. and K. H. G. Ashbee 1989. High Resolution Optical Interference Investigation of Deformation Due to Thermal Expansion Mismatch Around Plates Through Holes in Multilayer Circuit Boards, *J. of Electronic Materials,18(2)* :275-286.

Watt, D. W., T. S. Gross & S. D. Hening 1991. Three-illumination-beam Phase-shifted Holographic Interferometry Study of Thermally Induced Displacements on a Printed Wiring Board, *Applied Optics, 30(13)* :1617-1623.

Wong,T. L. , K. K. Stevens & G. Wang 1991. Experimental Modal Analysis and Dynamic Response Prediction of PC Boards with Surface Mount Electronic Components, *Journal of Electronic Packaging,113* :244-249.

Yang, L. X., W. Steinchen & G. Kupfer 1996. Digital Shearography for In-plane Strain Measurement of the Object Surface under Three-Dimensional Strain Conditions, *SPIE 2860*:175-183.

Experimental Mechanics, Allison (ed.) © 1998 Balkema, Rotterdam, ISBN 90 5809 014 0

Fatigue life prediction of microelectronics solder joints

K. Kaminishi & M. Iino
Department of Mechanical Engineering, Yamaguchi University, Japan

M. Taneda
Department of Mechanical Engineering, Fukuyama University, Japan

ABSTRACT: Fatigue tests on solder joints in product size electronic package models were carried out in order to examine crack initiation life Nc, crack extension path, crack extension rate da/dN and fatigue life N_f. Torsion fatigue tests on bulk solder specimen were performed in parallel, and prediction formulae for Nc of the bulk solder were proposed as a function of an equivalent inelastic strain range. Further, Nc of microelectronics solder joints were estimated in terms of maximum equivalent inelastic strain range calculated by a 3D elasto-inelastic FEM. As to crack extension behaviors, 2D-FEM were performed using quadtree automatic mesh generation technique. The FEM results showed that crack extension path and rate were controlled by a maximum opening stress range at the crack tip, $\triangle \sigma_{\theta\,max}$. The crack extension rate was experimentally found to be related to $\triangle \sigma_{\theta\,max}$ as da/dN = β $[\triangle \sigma_{\theta\,max}]$ $^\alpha$, with α = 2.0 and β =2.65 · 10^{-10} mm^5/N^2 determined independently of the test conditions. The calculated values of crack initiation and extension lives by the FEM using the above equations were in good agreement with the experimental ones.

1 INTRODUCTION

Microelectronics solder joints are constantly subjected to fatigue induced by thermal expansion mismatch between the IC package and the substrate. From a viewpoint of reliability assessment, it would be important to predict fatigue life of the solder joints, which should help in improvement of the package design accuracy and efficiency.

The fatigue life of solder joints may be divided into crack initiation life and extension life. As to crack initiation life, many formulae based on the Coffin-Manson's law or the modified Coffin-Manson's law have been proposed (Mukai et al. 1991, Lau et al. 1992, Busso et al. 1994, Shiratori et al. 1996). By applying the strain range partitioning approach and a linear cumulative damage concept, and using material-dependent parameters, the authors also have proposed a fatigue life prediction formula on eutectic (63Sn - 37Pb) and low melting-point (37Sn - 45Pb - 18Bi) solders, expressing number of cycles to failure as a function of cyclic frequency and equivalent inelastic strain range, which can be understood to successfully incorporate a creep damage effect (Taneda et al. 1992). However, little is known about the prediction method of the crack extension life.

In this study, fatigue tests on 53Sn-2Bi-45Pb solder joints in product size electronic package models were carried out at different temperatures, cycling frequencies and controlled displacement amplitudes in order to examine the crack initiation life Nc, crack extension path, crack extension rate da/dN and fatigue life N_f. Torsion fatigue tests on bulk specimen of 53Sn-2Bi-45Pb solder were performed in parallel to obtain basic fatigue data of 53Sn-2Bi-45Pb solder material such as cyclic stress-strain curves for various combinations of temperature and frequency. Thus, prediction formulae for Nc of the bulk solder were proposed again for the new material as a function of an equivalent inelastic strain range. Further, Nc of microelectronics solder joints were estimated in terms of maximum equivalent inelastic strain range calculated by a 3D elasto-inelastic FEM using a cyclic stress strain curve as a constitutive equation which incorporates both plasticity and creep effect. Finally, 2D-FEM were performed using quadtree automatic mesh generation technique in order to predict the crack extension path and life.

2 TESTING PROCEDURES

The configuration and the dimension of the test model manufactured by cutting the quad flat type LSI-package mounted on a substrate are shown in Figure 1. The test model consists of lead wire, solder joint and sliced package with substrate of Gullwing type. The shape and the size of the solder joints are shown magnified in Figure 2. The lead

Table 1. Chemical compositions of solder material.

(Wt%)

Sn	Pb	Sb	Cu	Bi	Fe	Al	Zn	As	Cd
52.93	45.13	0.004	0	1.93	trace	trace	0.0003	0.002	0.0002

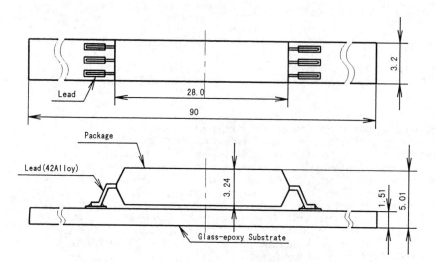

Figure1. Geometry of test model.

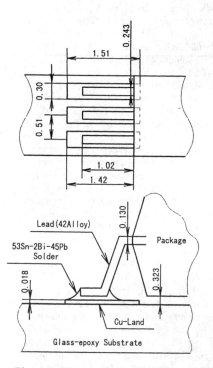

Figure 2. Shape and size of solder joint.

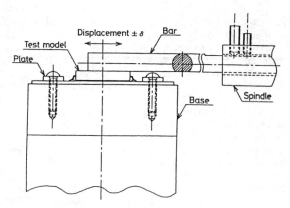

Figure 3. Fitting of test model to fatigue testing system.

wire is made of 42 Alloy, and the substrate of glass epoxy resin. Chemical composition of the solder material is given in Table 1. The top surface of the test model is adhered to a bar, which is in turn clamped to a displacement-controlled reciprocating spindle, as shown in Figure 3. Motion of the spindle, and the rod, can be controlled to reciprocating displacement of $\pm 30 \mu$ m or $\pm 60 \mu$ m by combined use of a variable speed motor and a cam.

976

The fatigue tests on solder joins were carried out at temperatures controlled to 303 ±2K and 318 ±2K, the reciprocating frequencies 0.1 and 1.0Hz and the displacement amplitudes being 30 and 60 μ m. In addition, torsion fatigue tests on bulk specimen of 53Sn-2Bi-45Pb solder were performed in the same way as in the previous paper (Taneda et al. 1992) to obtain basic fatigue data using FEM analysis for life prediction of the solder joints.

3 EXPERIMENTAL RESULTS

3.1 Fatigue crack initiation

Fatigue crack initiation life, N_c, is defined by the value of N, number of cycles, at which a microcrack has grown to approximately 20 μ m, long enough to be detectable as a surface crack. The crack initiation, as observed microscopically takes place as follows: With increase in number of cycles, a surface of the solder fillet in the comparatively upper position loses metallic luster, followed by appearance of a fissured pattern, which grows into cracks. Experimentally determined life data, N_c, are listed

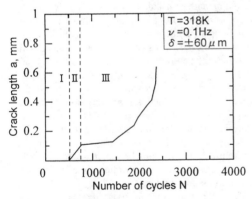

Figure 4. Experimental results of crack length against number of cycles.

and compared with FEM analysis results, [N_c]cal, in Table 3.

3.2 Fatigue crack extension

Typical plots of crack length versus number of cycles and experimentally observed fatigue crack extension path are presented in Figures 4 and 5, respectively. As shown in theses figures, fatigue life can be divided into three domains, namely life to crack initiation (I), period from crack initiation to crack growth up to the vicinity of the interface with lead wire (II) and period for crack propagation along the vicinity of the solder-lead interface to final failure (III).

4 FATIGUE LIFE PREDICTION

4.1 Crack initiation life

For prediction of the crack initiation life Nc three dimensional elasto-inelastic FEM analyses were performed, employing cyclic stress-strain curves obtained from the fatigue tests using bulk specimen at various combinations of temperature and frequency. Typical cyclic stress-strain relationship and mechanical properties used for FEM analysis are shown in Figure 6 and Table 2, respectively.

Equivalent inelastic strain range, $\triangle \varepsilon^{in}_{eq}$, distributions were calculated by FEM, performed for all the test conditions. The FEM analysis results show satisfactory agreement with experimental observations in that the location of experimentally observed crack initiation generally accorded with the zone of elevated $\triangle \varepsilon^{in}_{eq}$.

On the other hand, based on the experimental results of torsion fatigue test using 53Sn-2Bi-45Pb bulk specimen, prediction formulae for Nc of the bulk solder were obtained, as a function of cyclic frequency and equivalent inelastic strain range, by applying the strain range partitioning approach and liner cumulative damage concept.

Figure 5. Experimentally observed fatigue crack extension path.

In the case of T=303K

$$N_c = 332(\triangle \varepsilon^{in}_{eq})^{-1.46} \quad \text{for } \nu \geq 1.0\text{Hz}$$

$$N_c^{-1} = 3.01 \times 10^{-3}\{f_{303}(\nu)\triangle \varepsilon^{in}_{eq}\}^{1.46} \quad (1)$$
$$+ 6.92 \times 10^{-3}\{(1 - f_{303}(\nu))\triangle \varepsilon^{in}_{eq}\}^{1.67}$$
$$\text{for } \nu < 1.0\text{Hz},$$

where $f_{303}(\nu)$ is a function of cycling frequency and $f_{303}(0.1) = 0.384$.

In the case of T=318K

$$N_c = 294(\triangle \varepsilon^{in}_{eq})^{-1.87} \quad \text{for } \nu \geq 1.0\text{Hz}$$

$$N_c^{-1} = 3.41 \times 10^{-3}\{f_{318}(\nu)\triangle \varepsilon^{in}_{eq}\}^{1.87} \quad (2)$$
$$+ 2.98 \times 10^{-3}\{(1 - f_{318}(\nu))\triangle \varepsilon^{in}_{eq}\}^{1.65}$$
$$\text{for } \nu < 1.0\text{Hz},$$

where $f_{318}(0.1) = 0.155$.

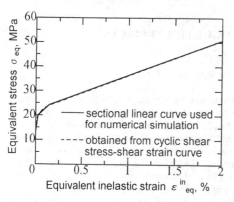

Figure 6. Cyclic stress-strain curve.

Table 2. Mechanical properties used for numerical simulation.

Material		Solder		42Alloy
Young's	T=303K	ν=0.1Hz	24.2	
modulus		ν=1.0Hz	25.0	147
(GPa)	T=318K	ν=0.1Hz	21.2	
		ν=1.0Hz	23.4	
Yield	T=303K	ν=0.1Hz	13.3	
stress		ν=1.0Hz	24.1	
(MPa)	T=318K	ν=0.1Hz	13.0	
		ν=1.0Hz	17.7	

Table 3. Crack initiation lives obtained

Test condition	Temperature T, K	Cycling frequency ν, Hz	Displacement δ, mm	Nc	$\triangle \varepsilon^{in}_{eq}$ %	(Nc)cal	Nc/(Nc)cal
1		1.0	±30	3287	0.932	335	9.8
2	318		±60	337	2.794	43	7.8
3			±30	1511	0.790	626	2.4
4		0.1	±60	648	2.258	109	5.9
5	303		±60	468	2.132	76	6.2

By substituting into the above equations the maximum equivalent inelastic strain range calculated by FEM, fatigue crack initiation lives in a product size solder joint can be obtained. Table 3 shows the calculated crack initiation life, [Nc]cal, compared with experimental results Nc. It is reasonable to suppose that experimental Nc can be predicted fairly well by using proposed formulae and FEM since the ratios Nc/[Nc]cal are found to be 2.4 to 9.8. Difference in coupon size and geometry between the torsion fatigue specimen and gullwing type model would be a possible cause of the discrepancy.

4.2 Crack propagation life

Numerical techniques in the application of the finite element method to crack extension analyses in microelectronics solder joint were proposed in order to predict the crack extension path and life. The FEM program consists of the subroutines for the automatic element re-generation using quadtree automatic mesh generation technique, the element configuration in the near-tip region being provided by a super-element demonstrated in Figure 7, elasto-inelastic stress analyses, prediction of crack extension path and calculation of fatigue life. In Figure 7 OP indicates a crack having proceeded from the left up to a current crack tip denoted by P, at which crack can shift its azimuth in accordance with the adopted fracture criterion.

In the process of crack extension it is found that general yielding proceeds in solder fillets. Here the linear fracture mechanics representation would be inappropriate to describe stress fields at crack tip to determine crack extension path. It is assumed in this instance that crack extends in the direction of maximum $\triangle \sigma_\theta$ at a small radial distance of $r = d$, where d is chosen to be a grain diameter's distance, 3.5μ m, in solder material.

The FEM analysis results using the above criteria are demonstrated in Figure 8, showing that the simulated crack extension path is in good agreement

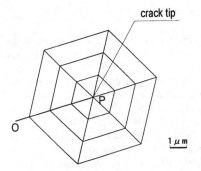

Figure 7. Super-element embedded in the near crack-tip region.

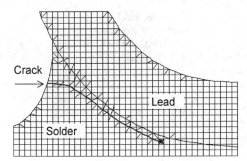

Figure 8. Simulated crack extension path (T=318K, ν =0.1Hz, δ =±60 μ m).

with experimentally observed fatigue crack extension path.

The maximum $\Delta \sigma_\theta$ at r =3.5 μ m is denoted by $(\Delta \sigma_\theta)_{CT}$ in Figure 9, where an example of plot of $(\Delta \sigma_\theta)_{CT}$ versus crack length is presented. It is characteristic of this plot that $(\Delta \sigma_\theta)_{CT}$, after an initial rise, rapidly decreases as the crack approaches a length of about 0.1 mm, followed by a gradual rise and a final rapid increase, as crack length exceeds 0.2 mm as shown in the figure.

In Figure 10 crack extension rate, da/dN, is plotted against $(\Delta \sigma_\theta)_{CT}$, the maximum $\Delta \sigma_\theta$ at r= 3.5 μ m, which describes crack extension rate as a function of $(\Delta \sigma_\theta)_{CT}$ of the form

$$da/dN = \beta [(\Delta \sigma_\theta)_{CT}]^{\alpha}, \qquad (3)$$

where numerical analysis showed that exponent α and factor β takes a constant value of

$$\alpha =2.0 \text{ and } \beta =2.65 \times 10^{-10} \text{ mm}^5/\text{N}^2 \qquad (4)$$

independently of the test conditions, which encourages prediction of fatigue crack propagation life by integrating eq.(3). The square element generated by

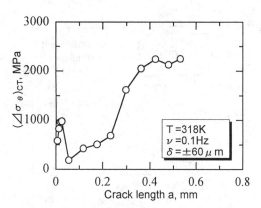

Figure 9. Plot of computed maximum tangential stress range against crack length.

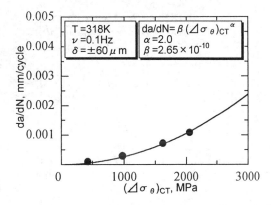

Figure 10. Plot of experimentally determined crack extension rate against maximum tangential stress range.

the quadtree mesh generating method is 10 μ m, which must be greater in size than the super-element shown in Figure 7, but is chosen to be sufficiently small compared with specimen geometry, i.e., diameter of the lead wire and curvature of the solder fillet for the present FEM analyses to be sufficiently reliable. The parameters α and β are determined under these conditions. It may be assumed that the parameters α and β are material constants, being possibly environment-dependent, provided that the same super-element is employed in the analyses.

Crack length-to-number of cycles relation obtained by numerical integration of eq.(3) is compared with experimental plots in Figure 11 for test conditions indicated in the figure. This comparison would suggest validity of the present FEM analyses for the prediction of fatigue crack extension life, and total life, of the solder joint.

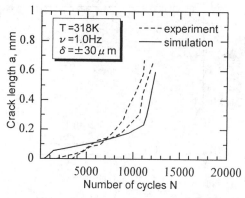

(a) T=318K, ν =1.0Hz, δ =$\pm$30 μ m

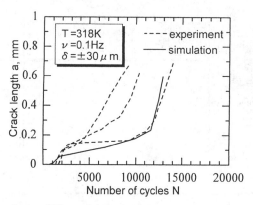

(b) T=318K, ν =0.1Hz, δ =$\pm$30 μ m

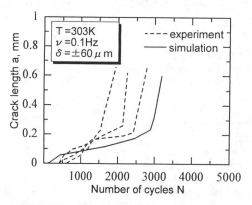

(c) T=303K, ν =0.1Hz, δ =$\pm$60 μ m

Figure 11. Comparison of calculation with experiment.

5. CONCLUSIONS

To predict the fatigue life of solder joints in an electronic package, experimental and numerical analyses were performed. The FEM analysis results show a satisfactory agreement with experimental observations in that the location of experimentally observed crack initiation accorded with the zone of elevated equivalent inelastic strain range, $\Delta \varepsilon^{in}_{eq}$.

On the other hand, prediction formulae for crack initiation life, Nc, of the bulk solder were proposed as a function of $\Delta \varepsilon^{in}_{eq}$. By substituting the maximum $\Delta \varepsilon^{in}_{eq}$ calculated by FEM into the proposed equations, Nc of a product size solder joint were estimated. It is reasonable to suppose that experimental Nc can be predicted fairly well by using proposed formulae and FEM since the ratios Nc/[Nc]cal are found to be 2.4 to 9.8.

Experimentally obtained crack extension rate is found to be related to a maximum tangential stress range, $\Delta \sigma_{\theta max}$, at crack tip in FEM analysis by
$$da/dN = \beta [\Delta \sigma_{\theta max}]^\alpha,$$
where $\alpha = 2.0$ and $\beta = 2.65 \cdot 10^{-10}$ mm^5/N^2 are determined independently of the test conditions. The FEM analysis result is in agreement with experimentally observed fatigue crack extension path and life.

REFERENCES

Busso, E. P., Kitano, M. & Kumazawa, T. 1994. Modeling complex inelastic deformation processes in IC packages solder joints. ASME Journal of Electronic Packaging, Vol.116:6-15.

Lau, J.H., Rice,D.W. & Erasmus, S. 1992. Thermal fatigue life of 256-Pin, 0.4mm pitch plastic quad flat pack(QFP) solder joint. ASME Advances in Electronic Packaging 1992, EEP-Vol- 2:855-865.

Mukai, M., Kawakami,T., Endo, T. & Takahashi, K. 1991. Elastic-creep behavior and fatigue life in an IC package solder joint. 4th Annual Meeting on Computational Mechanics, JSME, No.910-79:223-224.

Shiratori,M. & Qiang,Y. 1996. Fatigue-strength prediction of microelectronics solder joints under thermal cyclic loading. Fifth Intersociety Conference on Thermal Phenomena in Electronic Systems:151-157.

Taneda,M. & Kaminishi,K. 1992. Effect of cycling frequency on fatigue life of solder. ASME Advances in Electronic Packaging 1992, EEP-Vol-1:337-342.

Experimental Mechanics, Allison (ed.)© 1998 Balkema, Rotterdam, ISBN 90 5809 014 0

Measurement of residual stresses in electronic chips

C.A.Sciammarella & F.M.Sciammarella
MMAE, Illinois Institute of Technology, Chicago, Ill., USA

B.Trentadue
Dipartimento di Progettazione e Produzione Industriale, Politecnico di Bari, Italy

ABSTRACT: Thermally induced residual stresses play a very important role in the failure of electronic packages. In this paper a non-destructive methodology to measure residual stresses in an electronic package is presented. Optical measurements are combined with numerical computations to obtain the residual stresses. An excellent agreement is obtained between experimental and predicted values

1 INTRODUCTION

Residual stresses in electronic packages can cause the failure of the package by inducing the fracture of the supporting strata of the package. These stresses are generated during the manufacturing process. Figure 1 shows the type of package that is the subject of this paper. The packages' material is alumina. The alumina is sintered with several metallic inserts (solder pads), contact pads and a brazed seal ring. The metallic parts are made of iron nickel alloys of the invar type. The present study deals with the experimental determination of the residual stresses generated by the fabrication process, when to the ceramic body is added a metallic ring. The difference of thermal expansion coefficients between the metal and the ceramic produces residual stresses.

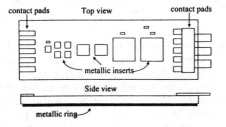

Figure 1. Sketch of analyzed hybrid module.

2 ANALYSIS OF THE SOURCE OF RESIDUAL STRESSES

The hybrid module under study can be analyzed as a composite beam. Figure 2 shows the cross section of the module. The coefficient of expansion of the ceramic is larger than the coefficient of expansion of the metal. The equilibrium diagram of the cross section of the module at room temperature is shown in Figure 2. The module is subjected to the simultaneous action of a bending moment and an axial load. Since the coefficient of expansion of the ceramic is larger than the coefficient of expansion of the metal upon cooling the ceramic wants to contract more than the metallic ring does and therefore is subjected to tension and the metal ring to compression. The bending moment created by the difference of the coefficients of expansion, produces compression in the ceramic above the neutral axis and tension in the metal. The bending moment predominates over the axial force and the upper face of the module is subjected to compression. Using the bending theory of composite beams subjected to thermal changes one can obtain the strains of the ceramic from the curvature of the module through the equation:

$$\varepsilon_c = C\kappa \tag{2.1}$$

where :

$$C = \frac{I_T}{(y_m - y_{cl})A_c} - y \tag{2.2}$$

In equation (2.2): I_T is the moment of inertia of the composite beam, y_{cl} is the distance between the centroid of the ceramic to the uppermost fiber of the beam, y_m is the distance of the centroid of the metal to the same fiber and y is the distance of the fiber under analysis to the neutral axis, A_c is the area of the ceramic. It is also possible to derive the following equation for the change of curvature due to a temperature change:

$$\Delta k = \frac{n(\alpha_c - \alpha_m)\Delta T A_c A_m (y_m - y_{cl})}{I_T(A_c + nA_m)} \quad (2.3)$$

In equation (2.3) α_c, α_m are respectively the coefficients of expansion of the ceramic and of the metal, n is the ratio of the modulus of elasticity of the metal to the modulus of elasticity of the ceramic. It is possible to see that if the curvatures of the surface are known, equation (2.1) can be used to convert curvatures to strains in the module.

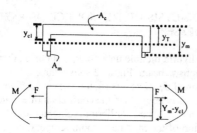

Figure 2. Analysis of the hybrid module as a bi-material beam.

3 DETERMINATION OF THE CURVATURES OF THE MODULE

The curvatures of the module can be obtained by measuring the shape of the module. To perform the measurement of the shape of the module the optical technique introduced in Sciammarella (1982) for a single illumination beam and extended in Sciammarella & Ahmadshahi (1987) to double illumination is used in this work. Modifications to the above technique are introduced in this paper to adapt it to the use of the Holo-Strain system (Sciammarella et al. 1993). The corresponding optical setup is shown in Figure 3. The specimen is inside a chamber to make it possible to control the temperature during the performance of the experiments. The specimen is subjected to double illumination.

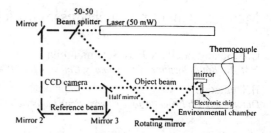

Figure 3. Optical set-up.

The equations of the fringes produced by each one of the illumination beams are:

$$I_1(x) = I_0 + I_1 \cos\phi_1(x) \quad (3.1)$$

$$I_2(x) = I_0 + I_1 \cos\phi_2(x) \quad (3.2)$$

The arguments of the functions $\phi_1(x)$ and $\phi_2(x)$ are given by the following equations (Sciammarella, 1982):

$$\phi_1(x) = \frac{2\pi}{\lambda}[x(\cos\alpha\Delta\alpha) - h(x)\sin\alpha\Delta\alpha] + \phi_{1n}(x) \quad (3.3)$$

$$\phi_2(x) = \frac{2\pi}{\lambda}[x(\cos\alpha\Delta\alpha) + h(x)\sin\alpha\Delta\alpha] + \phi_{2n}(x) \quad (3.4)$$

In equations (3.3) and (3.4), x is the coordinate point, α is the angle of illumination of the beam with respect to the normal of the plane of the module, $\Delta\alpha$ is the angle of rotation introduced to the illumination direction between to exposures, h (x) is the change of height with respect to the reference plane. The terms $\phi_{1n}(x)$ and $\phi_{2n}(x)$ are noise terms. From the subtraction of the arguments given in equation (3.3) and (3.4), calling $\beta(x)$ this difference one obtains:

$$h(x) = \frac{\beta(x)\lambda}{4\pi\sin\alpha\Delta\alpha} \quad (3.5)$$

The term h (x) can be considered the addition of two terms, one corresponding to the trend of the surface h*(x), and another h' (x) giving local variations. The term h*(x) represents the deflections of the neutral plane of the module caused by the constant moment introduced by the cooling of the module from brazing temperature to the room temperature. The term h' (x) represents the local variations of the upper surface of the module introduced during manufacturing. These local variations are assumed to change more rapidly than the deflections and its ensemble average $\langle h'(x)\rangle$ is assumed to be zero:

$$h(x) = h^*(x) + h'(x) \qquad (3.6)$$

Using the theory of cylindrical bending of long rectangular plates (Timoshenko, 1940), and since the applied moment is constant, h*(x) can be obtained by fitting a second order cylinder to the measured shape of the module. This cylindrical surface provides an average curvature, $\kappa_{avg}(x)$. However this is an ideal model, the actual deflected shape may have differences with the average values. The properties of the cross sections of the module represent averages that have some local variations due to the presence of metallic inserts that are not symmetrically located and connecting pads at the end of the module. To correct the average curvature the following procedure is followed. The local curvature of a point at a given temperature can be expressed as:

$$\kappa_{LT} = \kappa_{avgT} + \Delta\kappa_{LT} \qquad (3.7)$$

In equation (3.7) κ_{LT} represents the local curvature at a given point, at a temperature T, κ_{avgT} is the average curvature at T, $\Delta\kappa_{LT}$ is the local correction of the curvature at T. Deflected shape measurements are carried out at three different temperatures. At the same time the strains on the surface of the specimen can be determined by applying the double illumination technique described by Sciammarella and Bhat (1992). The strains caused by temperature changes: $\Delta T_1 = 55°C$, $\Delta T_2 = 50°C$, are measured $(\Delta T_1 = 21°C-(-34°C))$, $\Delta T_2 = 71°C-21°C$. The thermally induced strains are separated from the thermal dilatation using the equation:

$$\varepsilon_x = \varepsilon_{xm} - \alpha_c \Delta T \qquad (3.8)$$

where ε_{xm} is the measured strain. Applying equation (2.1) the strain changes are converted to changes in curvature. The surface strain measurements provide the differences of the curvatures for the corresponding temperatures. The sum of the curvatures for the same temperatures can be obtained by twice differentiating the deflected shape. The differentiation is carried out after the values h'(x) are removed by numerical filtering. One can write the following equations with the

indices i, j=1,2,3:

$$\kappa_{LTi} - \kappa_{LTj} + \kappa_{avgTj} - \kappa_{avgTi} = \Delta\kappa_{LTi} - \Delta\kappa_{LTj} \qquad (3.9)$$

$$\kappa_{LTi} + \kappa_{LTj} - \kappa_{avgTi} - \kappa_{avgTj} = \Delta\kappa_{LTi} + \Delta\kappa_{LTj} \qquad (3.10)$$

In equation (3-9) the left most term from the equal sign is obtained from measurements of the surface strains. The next term is obtained by fitting a cylindrical surface to the deflected shape of the module. In equation (3.10) the left most term from the equal sign is obtained from the second derivative of the deflected shape. Using equation's (3.9) and (3.10) it is possible to write a redundant system of six equations with three unknowns, the corrections of the local curvatures for the three temperatures $T_1 = 21° C$, $T_2 = -34°$, $T_3 = 71° C$. The redundant system is solved by the least squares method. The surface of the module is covered with a grid and the system of equations is solved for each node of the grid. The residual stresses were determined by using the theory of cylindrical bending of long rectangular plates:

$$\sigma_x = \frac{\varepsilon_x E}{1 - v^2} \qquad (3.11)$$

$$\sigma_y = v\sigma_y \qquad (3.12)$$

4 EXPERIMENT

The hybrid module was inside a chamber that has a temperature control system. A thermocouple was attached to one point of the module to determine its temperature. This was done in order to know the exact temperature of the hybrid module since there was a difference between the chamber indicator and what the thermocouple displayed. Measurements were carried out when the module reached a stable pre-selected temperature. A vacuum pump was connected to the chamber and the air was removed keeping a very low pressure. The angle α was equal to 45° and the rotation applied to the illumination beam was $\Delta\alpha$ equal to 10 minutes of arc. The wavelength of light was $\lambda = 0.6328$ microns.

5 RESULTS

Figure 4 shows the measured deflections of the hybrid modules at the indicated temperatures. It is very difficult to put the module perpendicular to

983

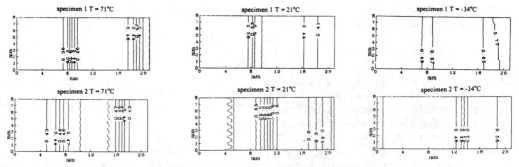

Figure 4. Deflections of the hybrid modules 1 and 2 (deflections in mm).

the direction of observation. Consequently the plotted deflections include the inclination of the neutral plane of the module with respect to the reference plane. Figure 5 shows the strains of the upper surface of the module. To get these strains the reference exposure was taken at T= 21° C and the second images at T= 71°C and T= –34°C respectively. The deflections plotted in Figure 4 were fitted to a cylindrical surface as shown in Figure 6 that represents a section of the deflected hybrid module 1 at temperature T= 21°C. The deflection surfaces shown in Figure 4 were differentiated twice to get approximated values of the curvatures at the corresponding temperatures. From the strains the difference of the curvatures were computed. The approximate values of the sum of the curvatures are obtained by the double differentiation of the deflection curves plotted in Figure 4. The differences of the average curvatures obtained as trends of the deflected surfaces are introduced in equations 3-9 and 3-10. A system of six equations with three unknowns $\Delta\kappa_{L21}$, $\Delta\kappa_{L71}$ and $\Delta\kappa_{L-34}$ is obtained. The system is solved by using the least squares method. The stresses at the three temperatures of interest are plotted in Figure 7.

6 CONCLUSIONS

The applied procedure is more involved than the usual procedure of twice differentiating the deflected shape. This procedure is applicable when the studied surface is originally flat. This is not the case in the present application. The upper surface of the hybrid module is not flat to the required degree to get accurate curvatures of the deflected shape by double differentiation. The proposed technique provides additional information and thus improves the values of the obtained curvatures. A check of the validity of the model has been done by computing the changes of curvatures using equation (2.3) and by taking the average stresses computed for the different temperatures, reducing the stresses to curvatures and then computing the differences of curvatures at the corresponding temperatures. Table I shows the results for module 1. An excellent agreement is obtained indicating that the model is correct on the average.

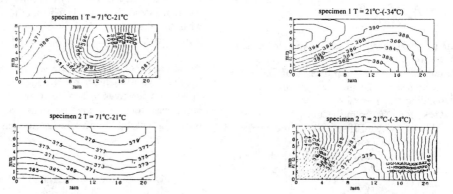

Figure 5. Strains of the hybrid modules 1 and 2 (values in microstrain).

Table 1. Curvature changes predicted by theory and experimentally determined

ΔT	$\Delta\kappa_{th.}$	$\Delta\kappa_{exp}$
55^{0}C	8.50x10^{-6}mm	8.41x10^{-6}mm
50^{0}C	8.01x10^{-6}mm	8.32x10^{-6}mm
105^{0}C	16.50x10^{-6}mm	16.10x10^{-6}mm

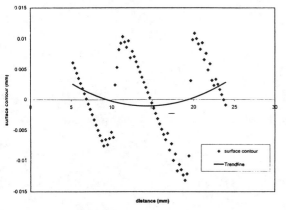

Figure 6. Trend of Specimen 1 at T=21°C.

REFERENCES

Sciammarella, C.A. 1982. Holographic moire, an optical tool for the determination of displacements, strains, contours, and slopes of surfaces. Optical Engineering:21(3) 447-457.
Sciammarella, C.A. & M. Ahmadshahi 1987. Holographic interferometry method for assessment of stone surface recession and roughening caused by weathering and acid rain. Proceedings of IMEKO: (2) 486-494. Plzen, Czechoslovakia.
Sciammarella, C.A. & G. Bhat 1992. Two dimensional Fourier transform methods for fringe pattern analysis. Proceedings of VII International Congress of Experimental Mechanics (SEM): 1530-1537.
Sciammarella, C.A., G. Bhat & P. Bayeux 1993. A portable electro-optical interferometer. Proceedings on advanced technology in experimental mechanics (JSEM):155-160.
Timoshenko, S. 1940. Theory of plates and shells. New York:McGraw-Hill.

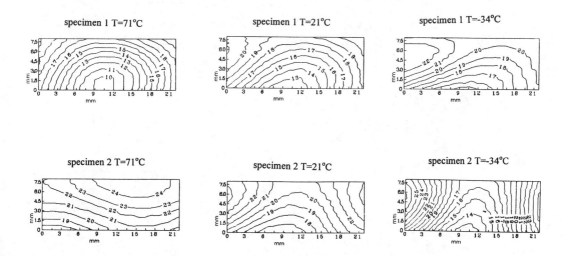

Figure 7. Compressive stresses in the upper face of hybrid modules 1 and 2 (stresses in Mpa).

Experimental Mechanics, Allison (ed.)© 1998 Balkema, Rotterdam, ISBN 90 5809 014 0

A new method of residual stress measurement on electronic packaging

Zhu Wu & Jian Lu
LASMIS – Mechanical Systems and Engineering, Universite de Technologie de Troyes, France

Yifan Guo
Packaging Mechanics, Semiconductor Products Sector, Motorola, Tempe, Ariz., USA

ABSTRACT: A new method, combining moiré interferometry and hole-drilling method is used in determining the process induced residual stresses in electronic packaging. The PBGA package consists of four layers of different materials. The upper surface of the silicon chip was molded by the plastic compound at a high temperature. Considerable residual stresses in the package is found and is ascribed to the coefficient of thermal expansion (CTE) mismatch between the plastic and the silicon chip when the package is cooled down from the curing temperature of the plastic compound to room temperature. The grating with a frequency of 1200 line/mm is replicated onto the upper surface of the plastic compound at room temperature. A blind hole with a diameter of 2.0 mm is drilled from this side. By using moiré interferometry, after each hole-drilling procedure, the surface in-plane displacement fields around the hole due to the relaxation of residual stresses are obtained. The relationship between the in-plane surface displacements and the corresponding residual stress is established by introducing a set of calibration coefficients. These coefficients are determined by establishing a multi-layered 3-D FEM model. The tensile residual stress in the plastic layer and the compressive residual stress in the silicon chip layer are determined successfully. The method proposed in the paper can be applied practically to many other types of electronic packaging.

1 INTRODUCTION

Plastic Ball Grid Array packages, PBGAs are widely used in the electronic industry. As the top layer of the package, plastic compound is cured at a relatively high temperature, due to the coefficient of thermal expansion (CTE) mismatch between the silicon chip and the plastic compound, considerable residual stresses are developed in the chip assembly after the package is cooled down. The thermal residual stresses of the chip assembly are becoming of concern in many aspects. Residual stresses in PBGAs may lead to mold compound and chip cracking, delamination and other failures when they are superposed on external loads during their service cycle (Wu, 1995; Han, 1995). They can also cause problems in assembly processes such as the inplanary and dimension instability (Guo, 1995). It is no doubt that the accurate determination of residual stresses in a real PBGA package is extremely important. The theoretical analyses and the numerical simulations were used in the prediction of thermal residual stresses (Tummala, 1989). However, due to the uncertainties of material properties of the assembly and the manufacturing processes, the actual residual stress in any individual packages may deviate from each other. Moiré interferometry was recently applied to the PBGA packaging (Zhu, 1996; Han and Guo, 1996). In these studies, the thermal shearing strains between molding compound and silicon chip or between silicon chip and PCB board were obtained from the polished cross section surface of packaging by using *in-situ* moiré interferometry. The specimen grating was replicated at the elevating temperature of the assembly by a relatively sophisticated technique. The shearing residual strains located in the specimen grating was assumed equivalent to the actual global residual strain of the assembly. Obviously, this procedure would be inaccurate and impractical. Moreover, in the method, it is not the entire package was studied, it seemed to be difficult to establish the qualitative relationship between the residual stress in each layer and the shearing strains detected by the method. Thus, the combined numerical analysis was inevitable (Guo, 1996).

The traditional strain gage hole drilling method has matured for many years. It was widely applied in residual stress measurement for many engineering materials (Lu, 1996). However, the size of a standard strain rosette is too large to be adopted for many types of electronic packaging. X-ray diffraction method and neutron diffraction method which were widely used for multi-cristal materials are also invalid to the plastic compound and PCB board. A newly developed technique, combining moiré interferometry and incremental hole drilling device has sufficient accuracy for many residual stress problems (Wu, 1997). As the U_x and U_y displacement fields due to the relaxation of residual stresses were generally localized in a small region around the hole and featured high signal-to-noise ratio even in the regions very close to hole boundary, It may be suitable to detect residual stresses in electronic packaging.

2. BACKGROUND: MOIRÉ INTERFEROMETRY

Moiré interferometry is an optical method, providing real time and whole field contour maps of in-plane displacements. The high displacement measurement sensitivity and high spatial resolution made it suitable for a broad range of problems in solid mechanics (Post, Han and Ifju, 1994). The in-plane surface displacements U_x and U_y can be determined from fringe orders by the following relationships:

$$U_x = \frac{1}{2f_s} N_x, \qquad U_y = \frac{1}{2f_s} N_y \qquad (1)$$

where N_x and N_y are fringe orders in U_x and U_y fields, respectively. In routine practice, a frequency of 1200 lines/mm is used, providing a contour interval of 0.417 µm/fringe order.

3. RELATIONSHIP BETWEEN DISPLACEMENT AND RESIDUAL STRESS

When residual stresses exist in a isotropic homogenous material, a blind hole solution was proposed to consider the three dimensional effect (Schajer, 1981). Schajer's solution was adopted in the holographic interferometry hole drilling method (Nelson, 1994; Makino, 1994) and was expressed as:

$$u_r(r,\theta) = A(\sigma_{xx} + \sigma_{yy}) + B[(\sigma_{xx} - \sigma_{yy})\cos 2\theta$$
$$+ 2\tau_{xy}\sin 2\theta] \qquad (2)$$

where, A, B are calibration coefficients; $u_r(r,\theta)$ is surface radial displacement in the cylindrical coordinate system; σ_{xx}, σ_{yy} and τ_{xy} are in-plane residual stress components in the Cartesian coordinate system.

For in-plane biaxial residual stress problems, In moiré interferometry, the relationship between the fringe orders of U_x and U_y displacement fields, and the corresponding residual stress was established in the following expression:

$$\left[N_x^i(x_k, y_k) \quad N_y^i(x_k, y_k) \right] \begin{bmatrix} \cos\theta_k \\ \sin\theta_k \end{bmatrix} = \sum_{j=1}^{i} 2f \cdot$$

$$\left[A^{ij} + B^{ij}\cos 2\theta_k \quad A^{ij} - B^{ij}\cos 2\theta_k \quad 2B^{ij}\sin 2\theta_k \right] \begin{bmatrix} \sigma_{xx}^j \\ \sigma_{yy}^j \\ \tau_{xy}^j \end{bmatrix},$$

$$k=1,2,3; \; i=1,2,..., n \qquad (3)$$

where, n is the total number of incremental steps; A^{ij}, B^{ij} are the coefficients of jth layer after ith incremental step is drilled; $N_x^i(x_k, y_k)$ and $N_y^i(x_k, y_k)$ are the total fringe numbers of U_x-field and U_y-field obtained from moiré interferometry respectively after ith incremental step is drilled; f_s is the frequency of the specimen grating; σ_{xx}^j, σ_{yy}^j and τ_{xy}^j are the residual stress components of ith current layer respectively.

In Eq. (3), coefficients A^{ij}, B^{ij} were calibrated by 3-D FEM analysis, in which two specific residual stress fields were considered: (1) $\sigma_{xx} = \sigma_{yy} = \sigma$, $\tau_{xy} = 0$, an equibiaxial residual stress field, which is equivalent to a uniform pressure p=σ acting on the hole side surface; (2) $\sigma_{xx} = -\sigma_{yy} = \sigma$, $\tau_{xy} = 0$, an pure shearing stress field, which is equivalent to the harmonic distributions of the normal stress $\sigma_{rr} = \sigma\cos 2\theta$ and the shearing stress $\tau_{r\theta} = -\sigma\sin 2\theta$ acting on the hole side surface. A multi-layered 3-D finite element model was established, where the geometrical parameters and the material properties of the elements were taken the same as that for the tested specimen. Generally, for different packaging problems, the calibration coefficients are also different and need to be calibrated.

4. TEST SPECIMEN

The dimension of the PBGA package is drawn schematically in Fig. 1. The material properties of the chip assembly are listed in Table 1.

Table 1 Material properties packaging

Materials	Elastic Modulus (GPa)	Poisson's Ratio	CTE (ppm/°C)
Plastic	14.3	0.33	15
Silicon Chip	112	0.28	2.7
PCB	21	0.3	18

5. EXPERIMENTAL PROCEDURE

The newly developed system combining moiré interferometry and blind hole drilling device (Wu, 1997) was used in the experiment.

The holographic phase type cross line specimen grating with a frequency of 1200 lines/mm was replicated on the top surface of the package. The optical set up of moiré interferometry was pre-adjusted to produce an initial null field condition, which was devoid of fringes. The drilling machine subassembly was locked into the U-shaped base and the XY translation stage was adjusted to locate the drill bit to the center of the package. The diameter of the drill bit used in the experiment was 2.0 mm. As the residual stress level in the PBGA package would be not very high, the hole was drilled only with four incremental steps. The first step and the second step had equal drilling increment of 0.2mm, which just reached the upper surface of the silicon chip. The third step and the fourth step also with an identical increment of 0.2mm but for the silicon chip layer.

After each hole-drilling step, the drilling machine subassembly was then removed and the U_x displacement field in the region around the hole was recorded by moiré interferometry. The experiment continued with rotating the specimen by 90° and the corresponding U_y displacement field was obtained. The specimen can be repositioned to its original U_x field precisely for the next step hole drilling procedure.

The moiré fringes of U_x and U_y displacement fields within the local region of the hole were recorded by a high resolution CCD camera and counted by using a post processing software at some desired points. They are presented in Figs. 2(a)~(e), where a contour interval was 0.417 μm per fringe order.

Fig. 2(a) The U_x field when the hole was drilled just throughout the specimen grating, depth h=10 μm

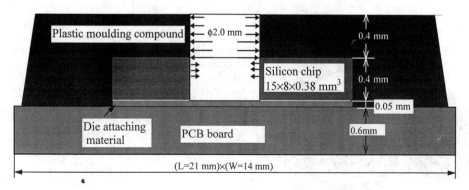

Fig. 1 A cross section diagram for the glob tope encapsulated packaging

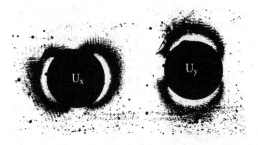

Fig. 2(b) The U_x and U_y fields after the first step hole drilling in the plastic, hole depth h = 200 μm

Fig. 2(c) The U_x and U_y fields after the second hole drilling step in the plastic, hole depth h = 400 μm

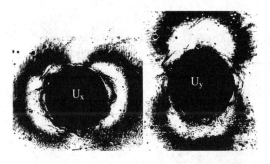

Fig. 2(d) The U_x and U_y fields after the third hole drilling step in the chip layer, hole depth h = 600 μm

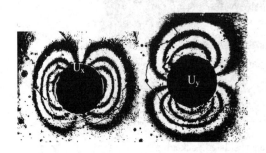

Fig. 2(e) The U_x and U_y fields after the fourth hole drilling step in the chip layer, hole depth h = 800 μm

6. ANALYSIS AND RESULTS

In order to get more accurate result, the hole depth of each step was controlled precisely. Fig. 2(a) showed the original position on the top surface of the package. It indicated that no residual stress was released in the specimen grating layer. Fig. 2(b) and Fig. 2(c) represented the result of residual stress relaxation in the plastic layer. Fig. 2(d) and Fig. 2(e) was the result in the silicon chip layer. The silicon chip was very hard, as a result, the quality of fringe patterns was bad in the close to hole boundary region, as shown in Fig. 2(d). More seriously, a circular crack was initiated near the interface between the plastic compound and the silicon chip on the chip side during the fourth step hole drilling procedure. With the relatively weak constraint, more complete relaxation of residual stresses in the plastic surface was found in Fig. 2(e).

The plastic compound and the silicon chip can be treated approximately as isotropic and homogeneous materials in the macro scale, Eq. (2) was believed valid and used directly in the calculation of residual stresses. Although fringe orders of three points, used in Eq. (2), can be chosen relatively arbitrary, in order to improve the fringe counting accuracy and to simplify coefficients calibrating process, points on the concentric circle of the hole with a radius of $1.2r_0$ was suggested. Moreover, some characteristic points such as: $(1.2r_0, 45°)$, $(1.2r_0, 0°)$ and $(1.2r_0, -45°)$; $(1.2r_0, 0°)$, $(1.2r_0, 45°)$ and $(1.2r_0, 90°)$; $(1.2r_0, 135°)$, $(1.2r_0, 180°)$ and $(1.2r_0, 225°)$; $(1.2r_0, 90°)$, $(1.2r_0, 135°)$ and $(1.2r_0, 180°)$... were routinely chosen in the analyses. In this experiment, by using a powerful image processing software, with the aid of intensity distribution analysis, fringe orders at three points $(1.2r_0, 0°)$, $(1.2r_0, 45°)$ and $(1.2r_0, 90°)$ were counted accurately, where an accuracy of 0.05 frictional fringe orders was ensured. Finally, the average levels of residual stresses for the three steps were determined, respectively. They are plotted in Fig. 3.

The result indicated: (1) residual stresses were tensile stress in the plastic layer and changed its sign in the silicon chip layer; (2) The residual stress level in the silicon chip was higher than that in the plastic compound.

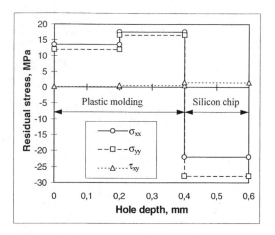

Fig. 3 The distribution of residual stresses for the glob-top encapsulated packaging

7. CONCLUSIONS

A combined method of moiré interferometry and incremental hole-drilling method was used successfully in determining residual stress for the Plastic Ball Grid Array packaging. The residual stresses detected are relatively large and it would be an important influencing factor to reliability and the optimal design of PBGA packaging. The method proposed in this study is anticipated to be suitable to determine the real residual stresses of various electronic packaging.

REFERENCES

Dai, X., Kim, C. and Willecke, *In-situ Moiré Interferometry Study of Thermomechanical Deformation in Glob-top Chip-on-board Packaging*, Experimental/Numerical Mechanics in Electronic Packaging, Vol. 1, Edited by Han. B., Barker, D. and Mahajan, R., SEM, 1996, PP. 15-21.

Guo, Y., *Applications of Shadow Moiré Method in Determinations of Thermal Deformations in Electronic Packaging*, Proc. of 1995 SEM Spring Conf. on Exp. Mech., Grand Rapids, MI June 1995.

Guo, Y., Li, L., *Hybride Method for Local Strain Determinations in PBGA Solder Joints*, Experimental/Numerical Mechanics in Electronic Packaging, Vol. 1, Edited by Han. B., Barker, D. and Mahajan, R., SEM, 1996, PP. 15-21.

Han, B. and Guo, Y., *Thermal Deformation Analysis of Various Electronic Packaging Products by Moiré and Microscopic Moiré Interferometry*, J. Electronic Packaging, Vol. 117, 1995, PP. 185.

Han, B. and Guo, Y., *Photomechanics Tools as Applied to Electronic Packaging Product Development*, Experimental/Numerical Mechanics in Electronic Packaging, Vol. 1, Edited by Han. B., Barker, D. and Mahajan, R., SEM, 1996, PP. 1-14.

Edited by Lu, J., Editorial Board, James, M., Lu, J. and Roy, G., *Handbook of Measurement of Residual Stresses*, Published by The Fairmont Press, INC, 1996.

Makino, A. and Nelson, D., *Residual stress Determination by Single Axis Holographic Interferometry and HoleDrilling, Part I: Theory*, Exp. Mech., 3, 1994, PP. 66-78.

Miura, H., Kumazawa, T. and Nishimura, A., *Effect of Delamination at Chip/encapsulant Interface on Chip Stress and Transistor Characteristics*, Applications of Experimental Mechanics to Electronic Packaging, ASME-EEP, Vol. 13, 1995, NY, USA, PP. 73-78.

Nelson, D, Fuchs, E., Makino, A. and Williams, D., *Residual stress Determination by Single Axis Holographic Interferometry and Hole Drilling, Part II: Experiments*, Exp. Mech., 3, 1994, PP. 79-88.

Post, D., Han, B. and Ifju, P., *High Sensitivity Moiré (Experimental Analysis for Mechanics and Materials)*, Published by Springer-Verlag Inc, 1994.

Schajer, G. S., *Application of Finite Element Calculations to Residual Stress Measurements*, J. Eng. Mater. Tech., 4, 1981, Vol. 103, PP. 157-163.

Tummala, R. and Rymaszewski, E., Editors, *Microelectronics Packaging Handbook*, Van Nostrand Reinhold, New York, 1989.

Wu, T., *Process-induced Residual Stresses in a Glob Top Encapsulated Silicon Chip*, Applications of Fracture Mechanics to Electronic Packaging and Materials, EEP, Vol. 11, MD-Vol. 64, ASME, 1995, PP. 123-131.

Wu, Z., Lu, J. and Joulaud, P., *Study of Residual Stress Distribution by Moiré Interferometry Incremental Hole Drilling Method*, The Fifth Int. Conf. on Residual Stresses, Linkoping, Sweden, June 1997.

Wu, Z. and Lu, J., *A New Optical Method for Shot Peening Residual Stress Measurement*, MAT-TEC 97, 'Analysis of Residual Stresses from Materials to Bio-Materials', Nov. 1997.

Strength of joints

Experimental Mechanics, Allison (ed.)© 1998 Balkema, Rotterdam, ISBN 90 5809 014 0

Determination of tensile and shear strengths of adhesive joints under impact loading

T. Yokoyama

Department of Mechanical Engineering, Okayama University of Science, Japan

ABSTRACT : The tensile and shear strengths of adhesive joints under impact loading are determined with a modified split Hopkinson bar using two kinds of specimen geometry. A cyanoacrylate adhesive (commonly termed the instantaneous adhesive) is used as an adhesive material. The effects of loading mode, loading rate, adherend material and thickness of adhesive layer on the joint strength of the cyanoacrylate adhesive are examined.

1 INTRODUCTION

The joint (or bond) strength of structural adhesives under impact loading has recently attracted a great deal of attention with their widespread use in industrial applications. Several test techniques for assessing the mechanical performance of adhesive joints under uniaxial tension, tensile-shear, compressive-shear loading, *etc.* have been specified in the ASTM Standards (ASTM 1995). Nevertheless, there do not exist the standard test methods for determining the impact strength of adhesive joints, except for the block impact test method (ASTM D950-94). In this test, the kinetic energy required to break a bonded specimen is in the usual way measured with a pendulum-type impact machine without considering the effect of stress wave loading. This method does not provide the strength data on adhesive joints available for engineering design purposes (note that ASTM regard this test as a comparative test). Adams and Harris (1996) made a critical assessment of this impact test using the finite element analysis. To data, however, little work (e.g., Harris & Adams 1985; Cayssials & Lataillade 1996) has been done in the area of impact testing of adhesive joints.

The purpose of the current work is to determine the tensile and shear strengths of adhesive joints under impact loading. A cyanoacrylate adhesive (hereafter referred to as the CA adhesive) is chosen in this study. The impact tensile and shear strengths of the CA adhesive joints are determined with a modified split Hopkinson bar using two kinds of specimen geometry. The corresponding static strengths are obtained in an Instron testing machine (Instron, Model 4505) with specimens of the same geometry as those used in the impact tests. The influences of loading mode, loading rate, adherend material and thickness of adhesive layer on the strength of the CA adhesive joints are investigated in detail.

2 PREPARATION OF ADHESIVE JOINTS

2.1 Adhesive and adherend materials

A commercial cyanoacrylate adhesive (Toagosei, #201) which belongs to one-component adhesives was chosen because of its availability, high-speed of bonding and easy handling. The physical properties of the bulk CA adhesive are given in Table 1. Two adherend materials were used: bearing steel (JIS SUJ2) and high-strength aluminum alloy (Al 7075-T6). Their nominal tensile properties are listed in Table 2.

Table 1. Physical properties of bulk CA adhesive.

Young's modulus, E_a	588 MPa
Linear expansion coefficient, β	1.1×10^{-4} 1/℃
Glass transition temperature, T_g	140 ℃
Mass density, ρ	1.2483×10^3 kg/m³

Table 2. Tensile properties of adherend materials.

Adherend material	Young's modulus E (GPa)	Yield strength σ_Y (MPa)	Tensile strength σ_B (MPa)	Elongation at fracture ε_f (%)
Bearing steel (JIS SUJ2)	209	360	639	26
Al 7075-T6	72	489	554	21

2.2 Specimen design

Figures 1 and 2, respectively, depict the form and dimensions of a butt joint tensile specimen (ASTM 2094-91) with threaded (M8 x 1.0) grip ends and a pin-and-collar shear specimen (ASTM D4562-90). The pin-and-collar specimen is designed to have a uniform shear stress distribution along the overlap length. For the pin and the collar with equal cross-sectional area, a theoretical analysis based on the theory of axi-symmetric elasticity shows that the shear stress distribution in the pin-and-collar specimen under static compressive-shear loading is fully symmetric and almost uniform along the overlap length. As seen from Figure 3, the theoretical prediction agrees well with the experimental shear strength of the CA adhesive joint. In an attempt to eliminate stress concentration induced near the adhesive edges under compressive-shear loading, the upper and lower internal edges of the collar corresponding to the adhesive edges were slightly rounded (see, Figure 2).

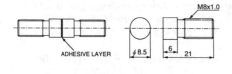

ADHESIVE LAYER $\phi\,8.5$ 6   21

(DIMENSIONS IN MM)

Figure 1. A butt joint tensile specimen.

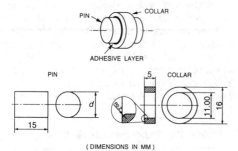

(DIMENSIONS IN MM)

Figure 2. A pin-and-collar shear specimen.

2.3 Assembly of specimens

All the specimens were degreased and cleaned in an acetone bathe before bonding. The CA adhesive was then applied at room temperature by hand with simple alignment fixtures. The thickness of the CA adhesive layer above 50 µm in the butt joint tensile specimen was controlled using glass beads, while that in the pin-and-collar specimen was varied from about 20 µm to 100 µm by adjusting accurately the clearance between the pin and the collar. In practice, the diameter of the pin was slightly varied, while the internal diameter of the collar was held constant. The

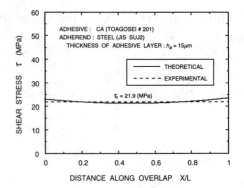

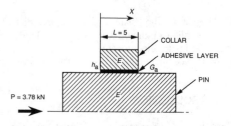

Figure 3. Shear stress distribution in CA adhesive bond-line of pin-and-collar steel specimen.

machining accuracy of the pin and the collar was confirmed with two kinds of digital micrometers (Mitutoyo, GMA-25DM and IMP-30DM) with an accuracy of ±5 µm. The surface profiles of the specimens for a 3-mm travel distance were measured with a surface roughness tester (Rank Taylor Hobson, Talysurf plus). In this work, the surface roughness on the specimens was held constant. All the specimens bonded with the CA adhesive were cured for nearly 30 hours at room temperature prior to testing.

3 IMPACT TEST TECHNIQUES

3.1 Impact tensile tests

Figure 4 indicates a schematic drawing of the tensile Hopkinson bar apparatus (Nicholas 1981) used in the tests. The apparatus consists principally of a striker bar, a gun barrel, two Hopkinson bars (bar No.1 and bar No.2) and associated recording system (not shown). The striker bar, 350 mm long, is made of a 15.8-mm diameter carbon tool steel rod (JIS SK5). The Hopkinson bars supported on V-shaped blocks are made of 16-mm diameter bearing steel rods (JIS SUJ2), which remain elastic during the test. The assembled butt joint specimen is firmly attached to the two Hopkinson bars. After the specimen has been screwed into one of the Hopkinson bars, a split collar with a length of 12 mm is placed over the specimen and the specimen is then

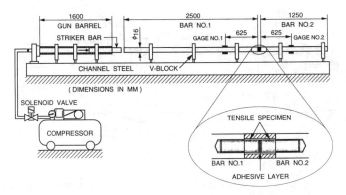

Figure 4. Schematic drawing of tensile Hopkinson bar apparatus.

screwed in until the Hopkinson bars fit tightly against the collar. The split collar made of the same material as the Hopkinson bars has an outer diameter of 16 mm and an inner diameter of 9 mm. The ratio of the cross-sectional area of the collar to that of the Hopkinson bars is 7: 10, whereas the ratio of the area of the collar to the net cross-sectional area of the specimen is about 5: 2. Figure 5 displays a Lagrangian x-t diagram which illustrates the details of the longitudinal wave propagation in the Hopkinson bars. A compressive pulse is generated in bar No.1 by axial impact with the ·striker bar fired through the barrel by compressed air released from a high-pressure reservoir, and travels through the assembled specimen. The compressive pulse continues to propagate until it reaches the free end of bar No.2. There, it reflects and propagates back as a tensile pulse (ε_i) and passes strain gage No.2. Upon reaching the assembled specimen at point A as depicted in Figure 5, the tensile pulse is partially transmitted through the assembled specimen (ε_t) and partially reflected back into bar No.2 (ε_r). The entire loading of the assembled specimen takes places between points A and B in the Lagrangian x-t diagram. The incident, reflected and transmitted strain pulses were measured by two sets of semiconductor strain gages with a gage length of 2 mm (Kyowa, Type KSP-2-E4) mounted on the Hopkinson bars. The output signals from the strain gages were fed through a bridge circuit into a 10-bit digital storage oscilloscope, where the signals were digitized and stored at a sampling rate of 1 μs/word. The digitized data were then passed to a 32-bit microcomputer for data processing.

The theory and analysis of the tensile split Hopkinson bar is almost identical to that of the compressive one (Lindholm 1964), except for the sign of the strain pulses. By application of the elementary theory of the one-dimensional elastic wave propagation, the tensile stress history applied in

the CA adhesive bond-line is obtained from

$$\sigma(t) = \frac{AE}{A_s}\varepsilon_t(t) \qquad (1)$$

where E = Young's modulus; A = the cross-sectional area of the Hopkinson bars; ε_t = the transmitted strain pulse; A_s = the bonded area of the butt joint specimen. By use of Eq.(1), the impact tensile strength of the CA adhesive joint is determined from a maximum value of the tensile stress history applied in the CA adhesive bond-line.

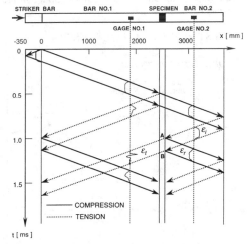

Figure 5. Lagrangian x-t diagram for tensile Hopkinson bar.

3.2 Impact shear tests

Impact shear tests were performed using a modified compressive Hopkinson bar apparatus adapted to the geometry of the pin-and-collar specimen as described in detail elsewhere (Yokoyama 1998) and shown in Figure 6. The fixing of the specimen between the

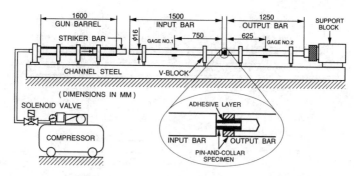

Figure 6. Schematic drawing of modified compressive Hopkinson bar apparatus.

Hopkinson bars is given in the inset in Figure 6. As in the analysis of the tensile Hopkinson bar, the shear stress history applied in the CA adhesive bond-line is given by

$$\tau(t) = \frac{AE}{A_s}\varepsilon_t(t) \tag{2}$$

where E = Young's modulus; A = the cross-sectional area of the Hopkinson bars; ε_t = the transmitted strain pulse.; A_s = the bonded area of the pin-and-collar specimen. With the help of Eq. (2), the impact shear strength of the CA adhesive joint is determined from a peak value of the shear stress history applied in the CA adhesive bond-line.

4   TEST RESULTS AND DISCUSSION

4.1   *Impact tensile test results*

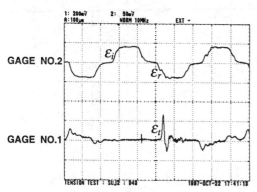

Sweep rate: 100μ s/div.
Vertical sensitivity
      Upper trace: 200 mV/div. (456 $\mu\varepsilon$/div.)
      Lower trace: 50 mV/div. (114$\mu\varepsilon$/div.)

Figure 7. Typical oscilloscope records from tensile Hopkinson bar test on CA adhesive joint.

Figure 7 shows a set of typical strain-gage records from the tensile Hopkinson bar test on the CA adhesive joint using the butt joint steel specimen. The three strain pulses are recorded as predicted from the Lagrangian diagram of Figure 5. The upper trace gives the incident and reflected strain pulses, while the lower trace gives the transmitted strain pulse. The incident strain pulse is nearly flat with a rise time of nearly 30 μs and a duration of about 180 μs.

Figure 8 represents the tensile stress history applied in the CA adhesive bond-line from Eq.(1). After the tensile peak, the tensile wave is distorted as release (or unloading) waves emanating from new fracture surfaces within the CA adhesive joint interact. In this test, the impact tensile strength of the CA adhesive joint was determined to be $\sigma_f = 106.6$ MPa and the time to fracture was determined to be $t_f = 16$ μs.

The tensile strength data for the CA adhesive joints from the steel and Al alloy specimens are shown in Figure 9 as a function of the loading rate defined by

$$\dot{\sigma} = \frac{\sigma_f}{t_f} \quad [\text{MPa/s}] \tag{3}$$

It is observed that the tensile strength of the CA adhesive joints increases with increasing loading rate, independent of the adherend materials. This increase is entirely attributed to the inherent dynamic viscoelastic properties of the CA adhesive itself which is very sensitive to strain rate. Note that the tensile strength of the CA adhesive joints from the Al alloy specimens is much lower than that from the steel specimens. This is because an adhesive failure (fracture within the interfacial region) occurred dominantly on the surface of the Al alloy specimen due to the presence of the oxidized layer. To achieve the inherent shear strength of the CA adhesive joints, a particular preparation of the Al surface for adhesive bonding is needed as specified in the ASTM Standards (ASTM D2651-90).

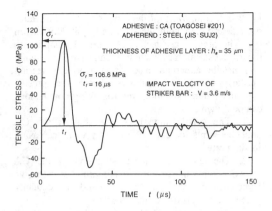

Figure 8. Tensile stress history applied in CA adhesive bond-line of butt joint steel specimen.

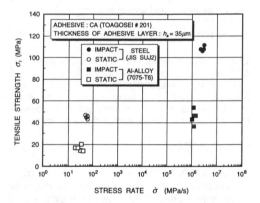

Figure 9. Effect of loading rate on tensile strength of CA adhesive joints.

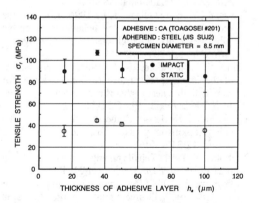

Figure 10. Effect of thickness of adhesive layer on tensile strength of CA adhesive joints.

In an effort to study the effect of the adhesive layer thickness, the static and impact tensile strengths of the CA adhesive joints are plotted versus the thickness of adhesive layer in Figure 10. The data points and the vertical bars denote, respectively, the mean value and the standard deviation of each five test data. It is seen that the tensile strength of the CA adhesive joints increases to a maximum at an adhesive layer thickness of about 35 μm and subsequently decreases. The decrease in the tensile strength of the CA adhesive joints with increasing adhesive layer thickness may be due to the presence of the residual internal stresses or the statistical probability of larger flaws (or defects) in the adhesive bond-line.

4.2 Impact shear test results

Figure 11 presents a set of typical strain-gage records from the compressive Hopkinson bar test on the CA adhesive joint using the pin-and-collar steel specimen. The top and bottom traces exhibit, respectively, the output signal from strain gages No.1 and No.2. Figure 12 indicates the shear stress history applied in the CA adhesive bond-line from Eq.(2). As in the tensile Hopkinson bar tests, the impact shear strength was evaluated from a peak value of the shear stress history. The shear strength data for the CA adhesive joints from the steel and Al alloy specimens are plotted in Figure 13 as a function of the loading rate. The effect of the adhesive layer thickness on the shear strength of the CA adhesive joints is illustrated in Figure 14. As a whole, the shear test results show the same general trends as found in the tensile test results.

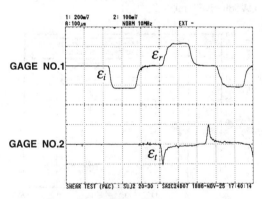

Sweep rate: 100 μ s/div.
Vertical sensitivity
Upper trace: 200 mV/div. (456 με /div.)
Lower trace: 100 mV/div. (228με /div.)

Figure 11. Typical oscilloscope records from compressive Hopkinson bar test on CA adhesive joint.

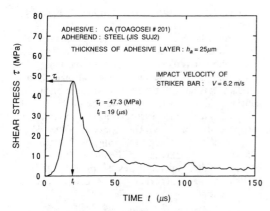

Figure 12. Shear stress history applied in CA adhesive bond-line of pin-and-collar steel specimen.

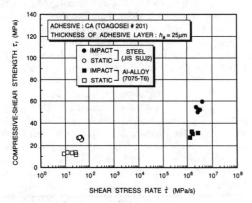

Figure 13. Effect of loading rate on shear strength of CA adhesive joints.

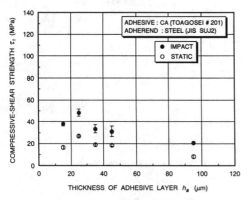

Figure 14. Effect of thickness of adhesive layer on shear strength of CA adhesive joints.

5   CONCLUSIONS

The tensile and shear strengths of the CA adhesive

joints at loading rates of the order of 10^6 MPa/s have been determined with the modified split Hopkinson bar. The present test technique cannot be applied to the impact strength evaluation of adhesive joints using non-metallic adherend materials such as engineering plastics, woods and rubbers. From the present experimental investigation, the following conclusions can be drawn :

1. The tensile and shear strengths of the CA adhesive joints increase significantly with increasing loading rate.

2. The tensile and shear strengths of the CA adhesive joints increase to a maximum at an adhesive layer thickness of about 25-35 μm and subsequently decrease, irrespective of loading rate.

3. The joint strength of the CA adhesive from the Al specimens is always lower than that from the steel specimens under quasi-static and impact loading.

4. The shear strength of the CA adhesive joints is much lower than the corresponding tensile strength under quasi-static and impact loading.

ACKNOWLEDGMENTS

This research program was supported in part by the Light Metal Educational Foundation Inc., Osaka. The supply of the CA adhesives by TOAGOSEI Co. Ltd., Nagoya, is gratefully acknowledged. Sincere thanks are also extended to H. Shimizu and K. Ohkubo for their assistance in the conduct of the experiments.

REFERENCES

Adams, R. D. & Harris, J. A. 1996. A critical assessment of the block impact test for measuring the impact strength of adhesive bonds. *Int. J. Adhesion and Adhesives.* 16 : 67-71.

Annual Book of ASTM Standards, 1995. Vol. 15.06, *Adhesives.* Philadelphia: ASTM.

Cayssials, F. & Lataillade, J. L. 1996. Effect of the secondary transition on the behaviour of epoxy adhesive joints at high rates of loading. *J. Adhesion.* 58: 281-298.

Harris, J. A. & Adams, R. D. 1985. An assessment of the impact performance of bonded joints for use in high energy absorbing structures. *Proc. Inst. Mech. Engrs.* 199-C2 :121-131.

Lindholm, U.S. 1964. Some experiments with the split Hopkinson pressure bar. *J. Mech. Phys. Solids.* 12:317-335.

Nicholas, T. 1981. Tensile testing of materials at high rates of strain. *Exp. Mech.* 21(5): 177-185.

Yokoyama, T. 1998. Determination of impact shear strength of adhesive joints with the split Hopkinson bar. *Key Engineering Materials.* 145/149 : 317-322.

Experimental Mechanics, Allison (ed.)© 1998 Balkema, Rotterdam, ISBN 90 5809 014 0

Stress analysis and strength of adhesive lap joint of shafts subjected to tensile loads

Y. Nakano & M. Kawawaki
Department of Mechanical Engineering, Shonan Institute of Technology, Fujisawa, Japan

T. Sawa
Department of Mechanical Engineering, Yamanashi University, Kofu, Japan

ABSTRACT: The stress distributions in adhesive lap joints have been analyzed using an elastoplastic finite element method where two hollow shafts of different materials were bonded by an adhesive at an overlap length and subjected to tensile loads. The effects of Young's modulus of the hollow shafts and the overlap length on the stress distribution at the interfaces between an inner hollow shaft and an adhesive, and an outer hollow shaft and the adhesive in the axial direction have been clarified by FEM analyses. Then a method for predicting joint strength has been proposed using the interface stress distributions. Moreover the tensile strengths of the adhesive lap joints of the hollow shafts composed of different materials, carbon steel and aluminum alloy were measured and compared with the values obtained from the prediction method. As a result, the joint strengths can be evaluated using the prediction method within an error of -15% of the test results in all combinations of the shafts.

1 INTRODUCTION

Recently, hollow shafts and shafts made of composite materials or light alloy metals are widely used for the purpose of lightening and increasing the specific strength, i.e. strength to density ratio, of the shafts for power transmission in mechanical structures. There are several serious problems in connecting the shafts made of different materials, by conventional ways such as using a key and key way, a flange attached at the ends of the shaft and welding. Therefore, making the shaft joints by using structural adhesives could be a more effective way, especially in the case where the shafts are made of different materials.

However, the strength and reliability of the adhesively bonded joints are still quite low as compared with those of the joints connected by the conventional ways and the basic data for designing the adhesive joints is not sufficiently established until now. Hence, there are many difficulties to convert the joint making procedure from the conventional ways to adhesive bonding. From the viewpoint of designing the adhesive joints of the shafts, a lot of analysis has been done by using the theory of elasticity (Lubkin & Reissner 1956, Chon 1982, Sawa et al. 1987, Nakano et al. 1989, Nakano et al., in press) and the finite element method (Adams & Peppiat 1977, Adams et al. 1978, Hipol 1984, Reedy JR. & Guess 1993, Lee et al. 1995, Choi & Lee 1996) concerning the stress distribution and strength evaluation of the joints subjected to various kinds of loadings such as tension, torsion, bending

and so on. However, most research has been limited within an elastic region even in finite element analysis and the results of the elastic analysis are no longer sufficiently related to the strength design and the strength evaluation of the adhesive joints with shafts when the adhesive and the shafts possess the elastoplastic mechanical properties.

In this study, the adhesive lap joints with hollow shafts subjected to tensile loads are analyzed by an elastoplastic finite element method taking the elastoplastic behaviors of the shafts and the adhesive into consideration and a new prediction method for the joint strength is proposed based on the stress-strain behavior of the adhesive and the frictional resistance between the shafts and the adhesive generated after rupture of small pieces of the adhesive. Furthermore, tensile tests were carried out for three kinds of adhesive lap joints with similar and dissimilar hollow shafts and the tensile strengths were compared with the values evaluated by the prediction method.

2 ANALYTICAL PROCEDURE

Figure 1 shows the model for FEM analysis of an adhesive lap joint where an inner hollow shaft I and an outer hollow shaft III are bonded by an adhesive II at an overlap length of l_2. In the figure, the upper half of the joint is shown because of symmetry with respect to the axial z direction. The joint length is denoted by l and the inner and outer diameters and the length of the

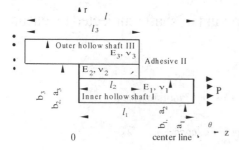

Fig 1 A model for FEM analysis of the adhesive lap joint of hollow shafts

Table 1 Material properties of the shafts and the adhesive

Property	Shaft (A5056)	Shaft (S45C)	Adhesive (1838B/A)
Young's modulus E	71.5GPa	205.8GPa	2.63GPa
Poisson's ratio ν	0.33	0.3	0.37
Yield stress σ_Y	261MPa	425.3MPa	19.7MPa
Work hardening C	9.0GPa	19.8GPa	—

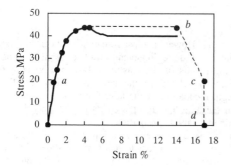

Fig 2 Stress-strain curve of the epoxy adhesive

inner hollow shaft I are denoted by $2a_1$, $2b_1$ and l_1 and those of the outer hollow shaft III are by $2a_3$, $2b_3$ and l_3, respectively. Young's modulus and Poisson's ratio are denoted by E and ν and the subscripts 1, 2 and 3 designate the inner hollow shaft I, the adhesive II and the outer hollow shaft III, respectively. The boundary conditions of the joint are assumed that an axial tensile load P is applied at the right end of the inner hollow shaft I and the opposite left end of the outer hollow shaft III is fixed in the z direction. A coordinate system used here is such that r represents the radial direction, θ the hoop direction and z the axial direction of the joint and the origin of the z direction is fixed at the left end of the inner hollow shaft I.

The joint model shown in Figure 1 is divided by using a four-node axisymmetric element and analyzed by a general purpose structural analysis finite element code, MARC. In the FEM analysis, materials of the

hollow shafts are taken as steel for structural use (S45C, JIS) and aluminum alloy (A5056, JIS) and an epoxy adhesive (Sumitomo 3M, 1838B/A) is as the adhesive, for coincidence with the joints made and used in the tensile strength tests. The dimensions of the joint model (the inner and outer diameters and the overlap length) are also the same as those of the joints used in the experiments. In the calculations, the adhesive lap joints composed of shafts of the same material, i.e. steel/steel and aluminum alloy/aluminum alloy combinations, called St-St joint and Al-Al joint hereinafter, respectively and of different materials, i.e. steel/aluminum alloy and aluminum alloy/steel combinations, the former is the inner shaft and the latter is the outer one, called St-Al joint and Al-St joint, respectively are analyzed.

Table 1 lists the material properties of the shafts and the adhesive used in the elastoplastic finite element analysis. The values of the epoxy adhesive were measured in advance by a uni-axial tensile test of its bulk specimen.

Figure 2 shows the stress-strain curve of the epoxy adhesive. The elastoplastic characteristic of the adhesive, shown by the solid line in the figure, is approximated as the dotted line a - b (rupture strength is 43.6 MPa and maximum strain is 14 %) in the FEM stress analysis. The meaning of the followed dotted line b - c - d will be mentioned later.

3 EXPERIMENTAL METHOD

Figure 3 shows the dimensions of each hollow shaft and the adhesive lap joint used in the tensile test. Figure 3a shows the inner hollow shaft machined by drilling so that the inner and outer diameters are 13.8 and 17.8 mm, respectively, i.e. the wall thickness is 2 mm and the depth of bore is 50 mm. Figure 3b shows the outer hollow shaft, the inner and outer diameters of which are 18.0 and 22.0 mm, respectively, and the wall thickness and the depth of bore are the same as those of the inner hollow shaft. Figure 3c shows the adhesive lap joint where the overlap length is 15.0 mm and the thickness of the adhesive layer is 0.1 mm. Bonding procedure was as follows: Grease was removed by acetone from the bonding surfaces of the inner and outer hollow shafts. These shafts were then dried in an electric chamber whose temperature was kept at 40 °C and bonded at an overlap length of 15 mm by an epoxy adhesive (1838B/A). In order to adjust the overlap length and to prevent oozing the adhesive into the inner free surface of the outer hollow shaft, a Teflon rod of 18 mm diameter was inserted prior to bonding, as shown in Figure 3c. After bonding, the joint was kept at room temperature for 1 hour and then cured at 80 °C in an electric chamber for 4 hours.

The adhesive lap joints of hollow shafts of three different combinations were made, i.e. the St-St and

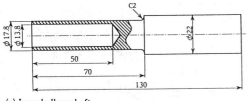

$\phi17.8$ $\phi13.8$ C2 $\phi22$

50
70
130

(a) Inner hollow shaft

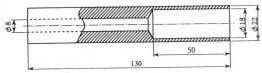

$\phi8$ $\phi18$ $\phi22$

50
130

(b) Outer hollow shaft

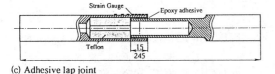

Strain Gauge   Epoxy adhesive

Teflon   15
245

(c) Adhesive lap joint

Fig 3 Dimensions of the hollow shafts and the adhesive lap joint used in the experiments [mm]

Al-Al combinations of shafts of the same material and the St-Al combination of shafts of different materials. The tensile strength of the joint of each combination was measured using a 200 kN capacity Instron testing machine 4507 under monotonic tension with a constant crosshead rate of 0.5 mm/min.

4 TENSILE STRENGTH MEASUREMENTS METHOD

Figure 4 shows an example of the tensile load-crosshead displacement curve in the case of the St-Al joint. As shown in the figure, after the applied tensile load reached maximum of about 24.5 kN, it dropped rapidly. However, the joint did not rupture immediately and a constant amount of resistance force F_s, of about 9 kN in the figure, remained successively until complete debonding of the adhesive.

Figure 5 shows the mean tensile stress-axial strain curve which was measured by strain gauges attached to the outer surface of the outer aluminum alloy shaft during the application of the tensile load to the St-Al joint. The ordinate in the figure denotes the mean axial tensile stress which is calculated by dividing the tensile load by cross sectional area of the inner hollow shaft. It can be seen from the figure that the curve shows linear behaviour, i.e. the mean tensile stress increases directly with the axial strain within smaller stress and strain ranges and it shows non-linear behaviour over the mean stress of about 143 MPa. As the

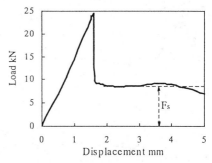

Fig 4 Tensile load – crosshead displacement curve of the adhesive lap joint (St-Al joint)

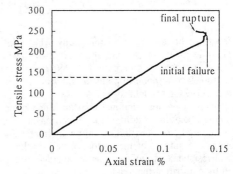

Fig 5 Mean tensile stress-axial strain curve at the outer surface of the joint (St-Al joint)

tensile load increases more, the axial strain changes to a decrease around the mean stress of 244 MPa and such transition is thought to be caused by partial debonding at the interface between the adhesive and the outer hollow shaft and is called the initial failure point of the joint in this study. The joint, however, does not rupture immediately at the initial failure point and bears some more increase of the tensile load and finally ruptures at the mean stress of about 250 MPa, which is called the final rupture point of the joint in this study.

From these experimental results shown in Figures 4 and 5, the adhesive reaches its yield stress state gradually from the edge of the overlapped bonded region as the tensile load increases and finally reaches its rupture strength. However, in the adhesive lap joint, even if small pieces of the adhesive between the hollow shafts reach the rupture strength, the joint does not lose the load carrying capacity immediately but still possesses it owing to some kind of frictional resistance generated after partial debonding at the interface between the adhesive and the shaft or in the adhesive layer: therefore, in consideration of such behavior of the adhesive, the following assumption is made in order to predict the tensile strength of the adhesive lap

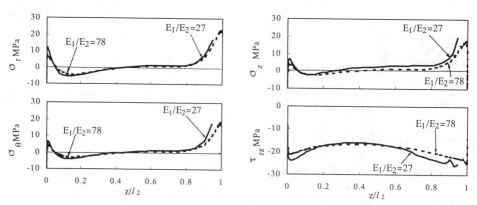

Fig 6 Effect of Young's modulus ratio on the stress distribution at the inner interface (P=16kN) (inner and outer hollow shaft are made of the same material, $E_1=E_3$, $v_1=v_3$)

joints with hollow shafts in this study. The point at which the mean stress-axial strain curve of the joint becomes non-linear and the initial failure point shown in Figure 5 correspond to points a and b in the stress-strain curve of the adhesive shown in Figure 2, respectively. Then the frictional resistance generated at the interface or in the adhesive layer carries the tensile load applied from the initial failure to final rupture points. The amount of the region carried by the frictional resistance is evaluated as an average strain of 3 % by comparing the axial strain increment between the non-linear and the initial failure point with that between the initial failure and final rupture points in Figure 5. Hence, as shown in Figure 2 by the dotted line b - c - d, the adhesive is thought to have the following stress-strain characteristic, i.e. after the tensile strength of 43.6 MPa and the maximum strain of 14 % (point b), the stress-strain curve decreases gradually to point c which corresponds to the mean shear strength which is calculated by dividing the remaining resistance force F_s shown in Figure 4 by an overlapped bonded area and the strain of 17 %, then finally drops to point d where the adhesive loses its load carrying capacity completely. In the finite element analyses, when whole elements of one adhesive layer in the axial direction reach point d, the joint is said to be ruptured and the tensile load applied at the time is regarded as the joint tensile strength.

The mean shear strengths obtained from the measurements for each combination of hollow shafts expressed as the Mises equivalent stress were 19.6 MPa for the St-St joint, 13.8 MPa for the Al-Al joint and 18.6 MPa for the St-Al joint. Furthermore, a joint fracture begins and propagates not in the hollow shafts but in the adhesive or at the interface under the conditions of the materials and dimensions of the joint components used in this study and a fracture mode was almost interfacial from the observation of fractured surface of the joint.

5 NUMERICAL RESULTS AND DISCUSSIONS

5.1 Numerical results of stress distribution

Stress distributions given below are in the adhesive.

5.1.1 Effects of Young's modulus ratio

Figure 6 shows the effect of Young's modulus ratio E_1/E_2 on the distributions of the stress components σ_r, σ_θ, σ_z and τ_{rz} in the axial direction at the inner interface (r = 8.9 mm) between the inner hollow shaft and the adhesive for the case where the hollow shafts are made of the same material, i.e. $E_1=E_3$ and $v_1=v_3$ and the joint is subjected to a tensile load of 16 kN. In the figures, the abscissa denotes the axial position at the interface which is normalized by the overlap length l_2 and the ratio E_1/E_2 of 27 (solid line) corresponds to the Al-Al joint and that of 78 (dotted line) to the St-St joint. From the result, the normal stresses σ_r and σ_θ become tensile and maximum at the loaded end of the interface (z/l_2 = 1.0) and σ_z is maximum near the loaded end of the interface. These stresses are also tensile near the free end of the interface ($z/l_2 = 0.0$) and are smaller than those at the loaded end of the interface; however, they are almost zero in the middle of the interface. On the other hand, the shear stress τ_{rz} distributes more uniformly at the interface except near both ends of it. Moreover, the adhesive within a range of about 5 % inward from the loaded end of the interface reaches its tensile strength (point b shown in Fig. 2) by applying the tensile load of 16 kN to the joint of smaller Young's modulus ratio, E_1/E_2 = 27, as shown by the solid line in the figure. In the joint of larger Young's modulus ratio, E_1/E_2 = 78, shown by the dotted line, however, whole adhesive still remains under the tensile strength.

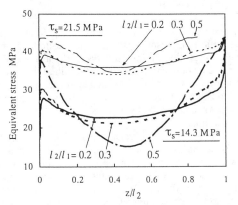

Fig 7 Effect of the overlap length on the Mises equivalent stress distribution at the inner interface (St-St joint, l_1=50mm, E_1/E_2=78)

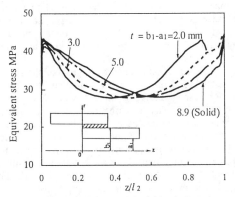

Fig 8 Effect of the thickness of the inner shaft on the Mises equivalent stress distribution at the inner interface (Al-Al joint, P=16kN)

5.1.2 Effects of overlap length

Figure 7 shows the effect of the ratio of the overlap length to the shaft length, l_2/l_1, on the Mises equivalent stress distribution at the inner interface under the condition that the mean shear stress τ_s, calculated by dividing the tensile load by a bonded surface area, is constant. In the figure, the joint is composed of hollow shafts of the same material, the St-St joint, and the abscissa is the normalized axial position z divided by each overlap length l_2. From the result, in the case of smaller τ_s (= 14.3 MPa), as the ratio l_2/l_1 increases, the Mises equivalent stress increases steeply at both ends of the interface, especially at the loaded end (z/l_2 = 1.0), however, it is rather small around the middle of the interface (z/l_2 = 0.5). When the ratio l_2/l_1 is small as 0.2 or 0.3, the equivalent stress distributes more uniformly at the interface and that means the external tensile load is carried evenly by the adhesive in the

whole overlap length. In the case of greater τ_s (= 21.5 MPa), the adhesive in the range between z/l_2 of 1.0 and 0.85 reaches its tensile strength when the ratio l_2/l_1 is large as 0.5 and hence the equivalent stress increases around the middle of the interface. When the ratio l_2/l_1 is small as 0.2 or 0.3, the equivalent stress becomes larger at the whole interface; however, the adhesive is still under its tensile strength. As seen above, the fracture initiating stress tends to decrease relatively as the overlap length ratio l_2/l_1 increases; therefore, it can not be expected that the tensile strength simply increases directly with the overlap length or the bonded area.

5.1.3 Effects of the wall thickness of the inner shaft

Figure 8 shows the effect of the wall thickness of the inner hollow shaft on the Mises stress distributions at the inner interface in the case where the outer diameter $2b_1$ of the inner hollow shaft is kept constant (17.8 mm) while the inner diameter $2a_1$ is varied. In the figure, the joint is composed of hollow shafts of the same material, the Al-Al joint, and is subjected to the tensile load of 16 kN and the overlap length ratio l_2/l_1 is 0.3. The wall thickness t (= b_1-a_1) of 8.9 mm in the figure is the case where the inner shaft is a solid one. From the result, the Mises equivalent stress decreases at the loaded end of the interface (z/l_2 = 1.0) and increases slightly at the free end of it (z/l_2 = 0.0) as the inner diameter $2a_1$ decreases, i.e. the wall thickness t of the inner hollow shaft becomes thick. Moreover, the adhesive in about 10 % inward from the loaded end of the interface reaches its tensile strength by applying the tensile load of 16 kN when the wall thickness is thin as t = 2.0 mm; however, such range does not appear as the wall thickness is large as t = 3.0, 5.0 and 8.9 mm (a solid shaft). Therefore, it can be concluded that the tensile strength will increase with an increase of the wall thickness of the inner hollow shaft which results in relatively higher rigidity of the inner shaft.

5.2 Joint tensile strength

Figure 9 is an example of the measured tensile strength distribution of the St-St joint plotted on a normal probability chart. In this study, a non-ruptured value over 95 % of more than 25 joints is evaluated as the tensile strength for each joint combination of inner and outer hollow shafts taking a relatively wide range of scatter of measured values shown in the figure into consideration.

Table 2 shows the tensile strengths (kN) of each combination of inner and outer hollow shafts. The values shown in [Exp.] in the table are the experimental results and those in [FEM]′ are numerically predicted ones. The overlap length of the adhesive lap joints was 15.0 mm and only in the case of the St-St joint, the joints with the overlap length of 25 mm were

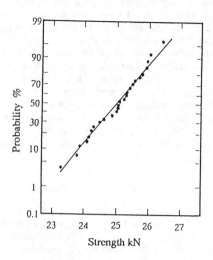

Fig 9 Tensile strength distribution (St-St joint)

Table 2 Comparison of the joint strengths with the experimentally predicted results

Joint Type		Strength [kN]		Error[%]
				$\dfrac{FEM-Exp.}{Exp.}$
Inner Shaft I	Outer Shaft III	Exp.	FEM.	
Steel	Steel (l_2=15)	23.6	20.7	-12
Steel	Steel (l_2=25)	34.5	31.4	-9
Al-alloy	Al-alloy	21.2	17.8	-16
Steel	Al-alloy	23.7	20.0	-16

additionally made and tested. The lower values shown in the row of the St-St joint are the results of such joints with larger overlap length. From these results, when the joints are composed of hollow shafts of the same material, the tensile strength (23.6 kN) of the St-St joint of relatively large Young's modulus is larger than the tensile strength (21.2 kN) of the Al-Al joint of relatively small Young's modulus. The tensile strength (23.7 kN) of the St-Al joint composed of hollow shafts of different materials is almost the same as that of the St-St joint. The scatters of the measured tensile strength expressed by the standard deviation are 0.8 kN for the St-St joint, 1.4 kN for the Al-Al joint and 1.5 kN for the St-Al joint.

From the comparison of the predicted values of the tensile strength with the experimental ones, the former values are 9 ~16 % smaller than the latter ones in all combinations of inner and outer hollow shafts; however, the predicted tensile strengths show a similar tendency with the experimental results that means the tensile strengths of the adhesive lap joints can be evaluated in the safety side by this prediction method.

6 CONCLUSIONS

The stress distributions of adhesive lap joints with dissimilar hollow shafts subjected to tensile loads are analyzed by the elastoplastic finite element method taking the nonlinear behaviors of the adhesive and the hollow shafts into consideration and the prediction method of the tensile strength is proposed based on the FEM analysis. The results obtained in this study are as follows:

1. When the adhesive lap joint of hollow shafts made of the same material is subjected to a tensile load, each stress component becomes maximum at the loaded end of the interface between the inner hollow shaft and the adhesive. When the joints are composed of hollow shafts of larger Young's modulus, a rupture of the adhesive layer is hard to occur at the end of the interface and the tensile strength is expected to become large.

2. In the case of the hollow shafts made of different materials, the tensile strength becomes larger when the inner hollow shaft of larger Young's modulus is bonded to the outer one of smaller Young's modulus than that of the reverse combination of inner and outer hollow shafts.

3. The prediction method of the tensile strength is proposed based on the Mises equivalent stress distribution in the adhesive calculated by the elastoplastic finite element analysis taking a frictional resistance generated after rupture of small pieces of the adhesive into consideration. By comparing the predicted values of the tensile strength with the experimental results for both joints with hollow shafts made of the same material and of different materials, the proposed prediction method can evaluate the tensile strength of the adhesive lap joints within an error of about –15 % on the side of safety.

REFERENCES

Adams, R.D. and Peppiat, N.A.1977 J.Adhesion, 9: 1-18.
Adams, R.D. Coppendale, J. and Peppiat, N.A. 1978 J. Strain Analysis, 13(1): 1-10.
Choi, J.H. and Lee, D.G. 1996 J. Adhesion, 55: 245-260.
Chon, C.T. 1982 J.Composite Mater., 16: 268-284.
Hipol, P.J. 1984 J. Composite Mater., 18: 298-311.
Lee, D.G., Jeong, K.S. and Choi, J.H. 1995 J. Adhesion, 49: 37-56.
Lubkin, J.L. and Reissner, E. 1956 Trans., ASME, 78: 1213-1221.
Nakano, Y., Sawa, T. and Arai, S. 1989 Intl. J.Adhesion Adhesives, 9 (2): 83-87.
Nakano, Y., Kawawaki, M. and Sawa, T. in press J.Adhesion Science Technol.
Reedy,JR.E.D.and Guess, T.R. 1993 Intl. J. Fracture, 63: 351-367.
Sawa, T. Nakano, Y. and Temma, K. 1987 J.Adhesion, 24: 245-258.

Experimental Mechanics, Allison (ed.)© 1998 Balkema, Rotterdam, ISBN 90 5809 014 0

Strength of single-lap adhesive joints under static and repeated loads

Toshiyuki Sawa & Hidenori Fujimura
Department of Mechanical Engineering, University of Yamanashi, Japan

Izumi Higuchi
Kofu High School of Technology, Japan

Yasuaki Suzuki
Nippon Sharyo Limited, Nagoya, Japan

ABSTRACT: Experiments to measure the strength of single-lap adhesive joints subjected to tensile loads and external bending moments were carried out. The effects of Young's modulus, the yield stress of adherends on the joint strength were examined experimentally. In addition, fatigue tests for single-lap adhesive joints were carried out. Stress distributions of single-lap adhesive joints under tensile loads and external bending moments are analyzed using an elasto-plastic finite element method in order to predict the joint strength. It is seen that the peel stress at the interfaces increases as the rigidity of the adherends decreases. A fairly good agreement is seen between the analytical and the measured results of the joint strength. In addition, it is observed that the joint strength increases as the yield strength, Young's modulus and the rigidity of the adherends increase. It is also found that the pulsating fatigue strength of the joints at 2×10^6 cycles under tensile loads is the one-fourth of the joint strength under static tensile loads.

1. INTRODUCTION

Single-lap adhesive joints have been used in many kinds of industries and many researches have been carried out on the stresses in elastic ranges and the strength of single-lap adhesive joints[1],[3]~[7]. However, it is well known empirically that plastic deformation is occurred in thinner adherends of single-lap adhesive joints as tensile loads increase. In addition, it is predicted that the rigidity and the yield stress of adherends will influence the joint strength. Furthermore, single-lap adhesive joints are subjected to external bending moments as well as tensile loads. Thus, it is necessary to know the effects of Young's modulus and the yield stress of adherends on the joint strength experimentally when single-lap adhesive joints are subjected to tensile loads and external bending moments. A few researches[2] have investigated on the stress in plastic deformation of single-lap adhesive joints. In practice, a crack due to the peel stress is occurred at the interfaces of a single-lap adhesive joint under tensile load and bending moments, and the interface stress state will be changed. In order to predict the joint strength, it is necessary to know the crack growth at the interfaces of single-lap adhesive joints under tensile loads and external bending moments.

In this paper, experiments to measure the strength of single-lap adhesive joints subjected to tensile loads and bending moments were carried out and the effects of Young's modulus, the adherend thickness and the yield stress of the adherends on the joint strength were examined. The stress and the deformation of single-lap adhesive joints are analyzed in elasto-plastic state by using finite element method. When the maximum principal stress at the edge of the interfaces reaches a fracture stress of the adhesive under certain tensile load or bending moment, the nodal point is released and the computation is repeated. Thus, the crack growth at the interfaces is examined. In addition, the effects of Young's modulus, the adherend thickness, the lap length and the yield stress of the adherends are shown on the stress state of joints and the crack growth at the interfaces. The comparisons of the joint strength are made between the analytical and the experimental results.

2. EXPERIMENTAL METHOD

2.1 *Strength measurement of joints subjected to static tensile loads*

Figure 1 shows the dimensions of adherends used in the experiments. The material of the adherends is chosen as aluminum alloy (A5052, A7075) and mild steel (SS400, JIS). Two similar adherends are bonded by an epoxy adhesive. Figure 2 shows a schematic experimental setup. A tensile load is applied by a material testing machine. Table 1 shows the material properties of the adherends and the adhesive.

2.1.1 *Effect of Young's modulus ratio E_1/E_2 between adherends and an adhesive on joint strength*

In order to examine the effect of Young's modulus ratio E_1/E_2 on joint strength, the material of adherends is chosen as aluminum (E_1=69.69GPa) and mild steel (E_1=200.4 GPa) while Young's modulus E_2 of the adhesive employed in the experiments is 1.77GPa as shown in Table 1. Strength measurements were carried out 30 times for each joint specimen.

2.1.2 *Effect of yield stress of adherends σ_{y1}*

In order to examine the effect of the yield stress of adherends on the joint strength, the material of adherends is chosen as aluminum A5052 (σ_{y1}=261.7MPa) and A7075 (σ_{y1}=779.1MPa) while Young's modulus of the adherends is assumed to be approximately the same.

2.1.3 *Effect of lap length*

In order to examine the effect of lap length l_2, the lap length ratio (l_2/l_1) is chosen as 0.1(l_2=19mm), 0.125(l_2=24mm) and

0.25(l_2=47mm), where l_1 denotes the adherend length and l_1=188mm as shown in Figure 1.

2.2 *Strength measurement of joints subjected to static bending moments*

Figure 3 shows a schematic experimental setup. Four-point bending tests were carried out. The load 2W is applied using a material testing machine. A bending moment M is obtained as M=60W.

2.2.1 *Effect of Young's modulus ratio E_1/E_2 on joint strength*

In order to examine the effect of Young's modulus ratio E_1/E_2, the material of adherends is chosen as aluminum and mild steel while Young's modulus E_2 of the adhesive is shown in Table 1. Strength measurement tests were carried out 30 times for each specimen.

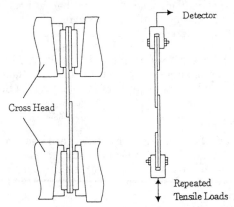

Fig.2 Schematic apparatus in static tensile tests

Fig.4 Schematic apparatus in fatigue tests

Table.1 Material properties of adherends and an adhesive used in the tests

	Adherend A5052	Adherend A7075	Adherend SS400	Adhesive EA9430
Young's modulus E GPa	69.69	77.82	200.4	1.77
Poison's ratio ν	0.314	0.308	0.291	0.370
Yield stress σ_Y MPa	261.7	779.1	424.7	30.38
c GPa	8.992	18.00	19.83	0.03038

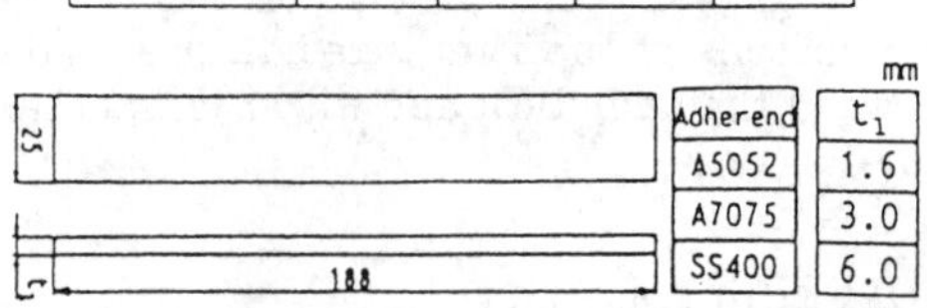

Adherend	t_1
A5052	1.6
A7075	3.0
SS400	6.0

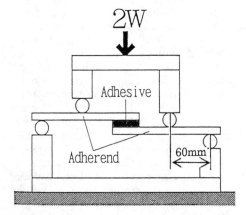

Fig.1 Dimensions of adherends used in tensile and bending tests

Fig.3 Schematic apparatus of static bending moments tests

2.2.2 *Effect of yield stress of adherends* σ_{yl}

In order to examine the yield stress of adherends, the material of adherends is chosen as aluminum alloy A5052 and A7075.

2.2.3 *Effect of lap length*

In order to examine the effect of lap length l_2, lap length ratio (l_2/l_1) is chosen as 0.074 $(l_2=9.4mm)$ and $0.148(l_2=18.8mm)$.

2.3 *Fatigue strength measurement of joints under pulsating loads*

Figure 4 shows a schematic experimental apparatus for fatigue tests under pulsating loads. In this experiment, pulsating fatigue limit at 2×10^6 cycles was examined using staircase method.

3. FINITE ELEMENT STRESS ANALYSIS AND ESTIMATION OF JOINT STRENGTH UNDER STATIC TENSILE LOADS AND BENDING MOMENTS

3.1 *FEM analysis of joints under static tensile loads*

Figure 5 shows an example of mesh divisions in the elasto-plastic finite element analysis of single-lap adhesive joints under tensile loads.

Taking into consideration the symmetry of the joints, the center of the adhesive is fixed and the nodal points over the length of 56.5mm in the upper adherends are fixed as shown in Fig. 5. Figure 6 shows a crack growth process of joints under static tensile loads. Under a static tensile load W, when the maximum principal stress at the edge of the interfaces reaches a fracture stress of the adhesive, the nodal point at the interface is released. Furthermore, when a tensile load is applied from zero to a certain load W_1 and the stress at the edge of the interfaces reaches a fracture stress, nodal points are also released. This procedure is continued until the adhesive ruptures. Thus, comparing among the values of W_1, W_2, W_3 and W_4 shown in Fig.6, the largest value is determined as the joint strength.

3.2 *FEM analysis of joints under external bending moments*

Figure 7 shows an example of mesh divisions for analyzing joints subjected to bending moments. Figure 8 shows an illustration of a crack growth process of joints under external bending moment. In the same way described in

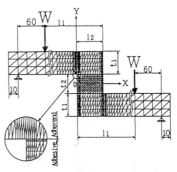

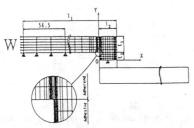

Fig.5 Boundary conditions and an example of mesh divisions for FEM analysis under static tensile loads

Fig.7 Boundary conditions and an example of mesh divisions for FEM analysis under static bending moments

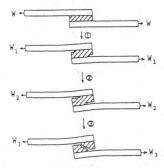

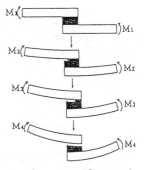

Fig.6 Assumption on crack growth process of joints under static tensile loads

Fig.8 Assumption on crack growth process of joints under static bending moments

1009

Fig.6, comparing among the values M_1, M_2, M_3 and M_4 shown in Fig.8, the largest value is determined as the joint strength for external bending moments. Figure 9 shows an example of stress-strain diagram obtained from the tests (dotted line). In the elasto-plastic finite element analysis, the diagram is approximated by a bi-linear line (solid line) shown in Fig.9. After yielding, the slope of the stress-strain diagram is denoted as c and the values are shown in Table 1.

4. EXPERIMENTAL RESULTS OF JOINT STRENGTH AND COMPARISONS WITH NUMERICAL RESULTS

4.1 *Joints strength under tensile loads*

4.1.1 *Effect of Young's modulus ratio E_1/E_2*

The experimental results of joint strength were treated by statistic method and it was found that they were in the normal distribution. The experimental result of the joint strength was determined using 90% non-destructive value taking into account the scatter of the joint strength. Table 2 shows the effect of Young's

modulus ratio on the join strength. The joint strength was obtained as 10.5kN for aluminum adherends, 14.4kN for mild steel adherends as shown in Table 2, where the adherend thickness t_1= 3 mm. From the results, it can be concluded that the joint strength of single-lap adhesive joints subjected to tensile loads increases as Young's modulus of adherends increases. A fairy good agreement is observed between the experimental and the numerical results.In addition, joint strength for aluminum adherends of t_1=1.6mm was measured as 9.05kN and 16.8kN for aluminum adherends of t_1=6.0mm. It is found that the joint strength increases as the adherend thickness t_1 increases.

4.1.2 *Effect of the yield stress of adherends*

Figure 10 shows the numerical results of the yielded region(black color) which is occurred as a tensile load increases. Figure 10(a) is the case where adherends are A5052 and Fig.10(b) is the case of A7075. In the case where adherends are A5052, a part of the adherends is yielded because the yield stress of he adherends is smaller than that of A7075. Thus, the maximum principal stress at the interface in Fig.10(a) is greater and the joint strength is

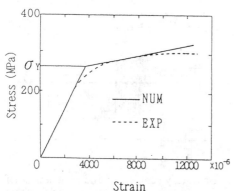

Fig.9 An example of stress-strain diagram (aluminum A5052)

Table.2 Comparison of experimental and numerical results under static tensile loads (Effect of Young's modulus ratio E_1/E_2 on joint strength)

E_1/E_2	Num.(kN)	Exp. (kN)
39.37(A5052)	10.62	10.5
133.2(SS400)	13.07	14.4

Table.3 Effect of the yield stress of adherends on joint strength

σ_{y1}(MPa)	Num.(kN)	Exp. (kN)
261.7(A5052)	10.62	10.5
779.1(A7075)	12.63	12.9

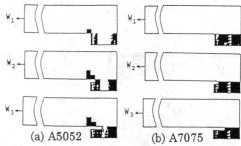

(a) A5052　　　　(b) A7075

Fig.10 The numerical results of the yielded region (black color) ocurred as a tensile load increases

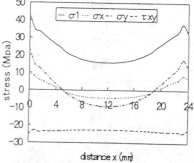

Fig.11 The interface stress distribution under static tensile loads (A5052 W=14.0kN t_1=6mm)

smaller than that of the case in Fig.10(b).

Figure 11 shows the interface stress distribution of joints, where adherends are A5052, the thickness is t_1=6 mm and the tensile load W=14.0kN. The maximum principal stress σ_1 is about 43.2MPa at the edge of the interface when the load W=14.0kN is applied. When W=14.0kN is applied, a fracture initiates from the edge of the interface because the fracture stress of the adhesive was measured as 43.2MPa. It is seen that the peel stress σ_y among the stress components is the largest. Figure 12 shows the interface stress distributions, where the adherends are A5052, the thickness t_1=1.6mm and the load is W=5.6kN. It is seen that the maximum principal stress is about 43.2MPa at the edge of the interface when the load W is 5.6kN. From Figs.11 and 12, it is seen that the joint strength increases as the adherend thickness increases while the material of adherends is the same.

Table 3 shows the comparison of joint strength. The joint strength was measured as 10.5kN for A5052 and 12.9kN for A7075 as shown in Table 3. It is seen that the joint strength increases as the yield stress of

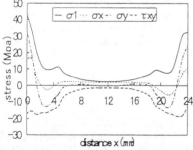

Fig.12 The interface stress distribution under static tensile loads (A5052 W=5.6kN t_1=1.6mm)

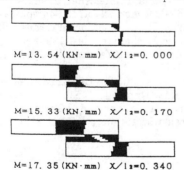

M=13. 54 (KN·mm) X/l₂=0. 000

M=15. 33 (KN·mm) X/l₂=0. 170

M=17. 35 (KN·mm) X/l₂=0. 340

Fig.13 The numerical results of the yielded region (black color) as a bending moment increases

adherends increases. The experimental results are in a fairly good agreement with the numerical results.

4.1.3 *Effect of lap length*

Table 4 shows the effect of lap length on the joint strength. The joint strength was measured as 9.05kN for the length ratio l_2/l_1=0.1, 10.5kN for l_2/l_1=0.125 and 16.8kN for l_2/l_1=0.25. From the results, it can be concluded that the joint strength increases as the lap length increases. The experimental results are fairly coincided with the numerical results.

4.2 *Joint strength under bending moments*

4.2.1 *Effect of Young's modulus ratio E_1/E_2*

Table 5 shows the comparison of joint strength under bending moments. The joint strength was measured as 18.73kN·mm for aluminum adherends, 23.12kN·mm for mild steel adherends as shown in Table 5. From the results, it can be concluded that the joint strength of single-lap adhesive joints subjected to bending moments increases as Young's modulus of adherends increases. A fairly good agreement is seen between the experimental and the numerical results.

4.2.2 *Effect of the yield stress of adherends*

Figure 11 shows the numerical results of the

Table.4 Comparison of experimental and numerical results under static tensile loads (Effect of lap length on joint strength)

l_2/l_1	Num. (kN)	Exp. (kN)
0.1 (l_2=19mm)	9.102	9.05
0.125 (24mm)	10.62	10.5
0.25 (47mm)	16.75	16.8

Table.5 Effect of Young's modulus ratio E_1/E_2 on joint strength under static bending moments

E_1/E_2	Num.(kN·mm)	Exp. (kN·mm)
39.37(A5052)	17.35	18,73
133.2(SS400)	21.22	23.12

Table.6 Effect of the yield stress of adherends on joint strength under static bending moments

σ_{y1}(MPa)	Num.(kN·mm)	Exp. (kN·mm)
261.7(A5052)	17.35	18.73
779.1(A7075)	18.16	20.25

Table.7 Effect of lap length on joint strength under static bending moments

l_2/l_1	Num. (kN·mm)	Exp. (kN·mm)
0.074(l_2=9.4 mm)	10.61	11.84
0.148 (18.8 mm)	17.35	18.73

yielded region (black color) as a bending moment is increased. It is observed that the adherends are yielded. Table 6 shows the comparisons of joint strength. The joint strength was obtained as 18.73kN·mm for A5052 and 20.25kN·mm for A7075 as shown in Table 6. It is seen that the joint strength increases as the yield stress of adherends increases. The experimental results are fairy coincided with the numerical results.

4.2.3 *Effect of lap length*

Table 7 shows the comparison of joint strength. The joint strength was measured as 11.84kN·mm for the lap length ratio l_2/l_1=0.074 and 18.73 kN·mm for the ratio l_2/l_1=0.148. It is seen that as the lap length increases, the joint strength increases. A fairly good agreement is seen between the experimental and the numerical results.

4.3 *Fatigue strength measurement of joints under repeated tensile loads*

Figure 14 shows an example of staircase method for A5052 adherends (t_1=6mm), where the ordinate is pulsating stress and the abscissa is the number of the tests. The joint pulsating fatigue strength at 2×10^6 cycles was obtained as 3.058GPa or adherend thickness t_1=6mm, 1.588GPa for t_1=3mm, and 1.490GPa for t_1=1.6mm. From these results it is found that the pulsating fatigue limit at 2×10^6 cycles is about 1/4 of the joint strength under static tensile loads which are described in 4.1.1.

5.Conclusions

In this paper, joint strength of single-lap adhesive joints under static tensile and bending moments was measured. In addition, fatigue strength was also measured. The results obtained are as follows.

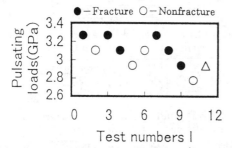

Fig.14 Experimental results under repeated tensile loads (A5052 t_1=6mm)

(1) The effects of Young's modulus, the yield stress of adherends and the lap length on the joint strength of single-lap adhesive joints under tensile loads and bending moments were examined experimentally. From the results, it was confirmed that the joint strength increases as Young's modulus and the yield strength of adherends increase and the lap-length increases. In addition, it was seen that the fracture initiated from the edges of the interfaces.
(2) Fatigue strength of single-lap adhesive joints under repeated tensile loads was measured by using staircase method. It is found that the pulsating fatigue strength of joints at 2×10^6 cycles is about one-fourth of the joint strength of single-lap adhesive joints under static tensile loads.
(3) Elasto-plastic stress states of single-lap adhesive joints under static tensile and bending moments are analyzed using finite element method. In addition, crack growth of the interfaces is analyzed and joint strength is also estimated. From the analytical results, it is shown that the joint strength increases as Young's modulus and the yield stress of adherends, the adherend thickness and the lap length increases. The experimental results of the joint strength are a fairy good agreement with the numerical results. In other word, the joint strength increases as the rigidity of adherends increases because the joint strength depends on the maximum principal stress at the interfaces.

REFERENCES
(1) Chen,D., and Sheng,S., Trans.ASME, J.Applied Mechanics,Vol.50(1983)pp.109-115.
(2) Harris,J.A., Adams,R.D., Int.J.Adhesion and Adhesives, Vol.4, No.2(1984)pp.65-78.
(3) M.Goland and E.Reissner, Trans.ASME, J.Applied Mechanics,Vol11(1944)pp.A17-A27
(4) Ming-Yi Tsai and J.Morton, Trans.ASME, J.Applied Mechanics,Vol.61(1994)pp.712-715
(5) O.Volkersen, *Luftfahrtforschung*,15(1983) pp.41-47.
(6) Sawxer,J.W., and P.A.Cooper AIAA.J, Vol.19 No.11(1981)pp.1443-1451
(7) T.Sawa, K.Nakano and H.Toratani, J.Adhesion Sci.Technol., Vol11,No8(1997)1039

Experimental Mechanics, Allison (ed.)© 1998 Balkema, Rotterdam, ISBN 90 5809 014 0

Photoelastic thermal stress measurement of scarf adhesive joint under uniform temperature changes

F. Nakagawa
Department of Mechanical Engineering, Tokyo Metropolitan College of Aeronautical Engineering, Japan

T. Sawa
Department of Mechanical Engineering, Yamanashi University, Japan

Y. Nakano & M. Katsuo
Department of Mechanical Engineering, Shonan Institute of Technology, Japan

ABSTRACT: This study dealt with thermal stresses and delamination growth in scarf joints under uniform temperature change using photoelasticity and a two-dimensional finite element method. Adherends were aluminum plates and an adhesive layer was modeled and manufactured from an epoxide resin plate. The adherends and the epoxide resin plate were bonded using a heat setting and one component type adhesive. It was confirmed that the initiating position of the delamination growth from the edge of the interface was not the obtuse angle side but the acute angle side. When the scarf angle was 90 degree, the strength of the joint for the delamination was minimum. It was expected that the thermal strength increased with a decrease of the adhesive layer thickness. It was recognized that the stress singularity due to the thermal loadings was quite different from that due to tensile loads.

1 INTRODUCTION

Heat setting adhesives are used for mechanical and structural use, because they are excellent in thermal resistance and mechanical property. In adhesive joints, the linear thermal expansion coefficient of adhesives is generally quite different from that of adherends. When a heat setting adhesive is used, the adhesive is cured at high temperature and cooled down to ambient temperature in the curing process. Thus, thermal stress maybe occurs in the adhesive joints. It is supposed that the thermal stresses occurred in the adhesive joints may affect on the joint strength.

The singular stresses near the edge of the interface of joined dissimilar materials have been studied analytically (Bogy 1971). Also, the thermal stress near the edge of the interface has been investigated. But, the adhesive layer is very thin and sandwiched between adherends. Therefore, it is expected that the size of the stress concentration area near the edge of the interface in the adhesive layer of the adhesive joints and the effect of the stress concentration are small.

The corner angles of the adherends at the edge of the interface in single lap joints and in butt joints are generally 90 or 180 degree. The stress concentrations near the edge of the interface are large in these joints. The scarf joints have been used to prevent the stress concentrations. Many studies on the stress distribution of the scarf adhesive joints subjected to static loads have been done (Lubkin 1957, Wah

1976, Wang 1981, Adkins et al. 1985, Kyogoku et al. 1987, Suzuki 1987, Chen et al. 1990, Ikegami et al. 1990, Imanaka et al. 1994). However, few researches on the stress distribution of scarf adhesive joints subjected to thermal loads have been done.

This study deals with the thermal stresses and delamination growth in scarf adhesive joints under uniform temperature change using photoelasticity and a two-dimensional finite element method. Aluminum plates are used as adherends and an epoxide resin plate as an adhesive. The adherends and the epoxide resin plate were bonded using a heat setting and one component type adhesive. The adhesive joint was cured at a high temperature and cooled down to room temperature. The residual thermal stress then occurred in the scarf adhesive joints. Photoelastic experiments were carried out at room temperature. Then the scarf joints were cooled stepwise and delamination growth at the edge of the interface was measured. The thermal stresses in the scarf joints with actual and thin adhesive layer were also analyzed. Concerning the joints of the several scarf angles, the stress singularities near the edge of the interface were examined.

2 EXPERIMENTAL METHOD AND EXPERIMENTAL RESULTS

Figure 1 shows a scarf joint with the scarf angle θ' under uniform temperature change. The material of the adherends (I) and (III) is aluminum plate and that

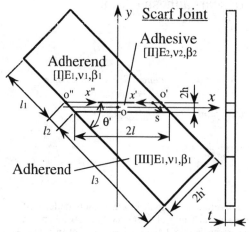

Scarf Joint

Figure 1. Scarf joint bonded with epoxide resin adhesive under uniform temperature change (2h=20mm(θ'=90°, 60°, 52°, 45°,30°,2h'=50mm), 2h=5mm,10mm(θ'=90°, 75°, 60°, 52°, 40°, 30°, 2h'=43mm), $l_1+l_2+l_3$=160mm(2h=20mm), $l_1+l_2+l_3$=150mm(2h=10mm),t =6mm,ΔT=-50°C, E_1=69.6GPa, v_1=0.33, β_1=23.5x10^{-6}/°C, E_2=3.43GPa, v_2=0.38, β_2=60x10^{-6}/°C)

of adhesive is modeled and manufactured from the epoxide resin plate in photoelastic experiments. The epoxide resin plate specimens were manufactured by using an abrasive waterjet cutting machine. The thicknesses 2h of the epoxide resin plate are 20, 10 and 5mm in the photoelastic experiments. The width of the epoxide resin plate is denoted by 2h'. The scarf angles θ' are chosen as 90, 60, 52, 45 and 30 degree, respectively, when 2h is 20mm and 2h' is 50mm. In addition, the scarf angles θ' are 90, 75, 60, 52, 40 and 30 degree, respectively, when 2h is 10 and 5mm and 2h' is 43mm. In Figure 1, the length $l_1+l_2+l_3$ is 160mm when 2h is 10mm (2h'= 43mm) and is 150mm when 2h is 20mm (2h'= 50mm). The adherends and the epoxide resin plate were bonded using a heat setting adhesive (EP128, CEMEDINE) and cured at the temperature of 85°C. After the curing process, the joints were cooled down to room temperature and the photoelastic experiments were carried out. Furthermore, the joints were cooled stepwise to -70°C and the delamination near the edge of the interface was observed.

Figure 2 shows the examples of the photoelastic experiments where the thickness 2h of the epoxide resin plate is 20mm and the angles θ are 52 and 30 degree, respectively.

Figure 3 shows examples of the photoelastic experiments after the delamination occurred. It is seen that the delamination initiates at the edge of the interface with the acute angle. From the delamination observation, it was found that the delamination

initiated from the interface with the acute angle in all scarf joints.

Figure 4 shows the results of the photoelastic experiments where the thickness 2h of the epoxide resin plate is 10 mm. It is seen that the stress concentrations occur near the edge of the interface.

Figure 5 shows the results of the photoelastic experiments where the thickness 2h is 5mm.

Near the edge of the interface, the distributions of the isochromatic fringes in the epoxide resin plates are similar in these models of Figures 4 and 5. When the thickness of the epoxide resin plate was 10 or 5mm, it was observed that the delamination initiated from the edges of the interface of the acute angle in the same manner as the joint where the thickness of the epoxide resin plate was 20mm.

From Figures 2-5, it is expected that the thermal stress distributions near the edge of the interface are similar although the thickness of the epoxide resin plate changes.

3 ANALYTICAL METHOD AND ANALYTICAL RESULTS

The finite element analyses of the scarf joints subjected uniform temperature change are carried out using the model shown in Figure 1. The quarter model is made when the scarf angle θ' is 90 degree and the whole model of the scarf joint is made when θ' is other than 90 degree. The corner angle of epoxide resin plate side is denoted by θ. The acute angle θ is equal to the scarf angle θ' and the obtuse angle θ is 180°-θ' in this study. When the thickness 2h is 0.05mm, the epoxide resin plate isn't used and the two adherends are bonded directory using the heat setting adhesive. Young's modulus, Poisson's ratio and a linear thermal expansion coefficient (LTEC) of the adherend (I) and (III) are denoted by E_1, v_1 and β_1, respectively. Also, those of the adhesive (II) are designated using a subscript 2. In this paper, it is assumed that these material properties are held constant and independent of temperature change and the epoxide resin plate and the epoxide resin adhesive have the same properties.

MARC and FINAS are employed as solvers (FEM). MENTAT and CADAS/W are employed as pre- and post- processors. In FEM analysis of MARC and FINAS, the dimensions of the scarf joints are the same but the mesh divisions are different.

The model shown in Figure 1 is divided by using a four-node and plain stress element when the thicknesses of the epoxide resin plates are 20 and 10mm and is divided by using a eight-node and plain strain element when the thickness of the adhesive layer is 0.05mm. The numbers of the elements in FEM models are 2592 to 3024 (2h=10mm, whole model) and 960 (2h = 0.05mm). In numerical

(a) θ'=52°   (b) θ'=30°

Figure 2. Isochromatic fringe patterns from photoelastic measurements (2h=20mm, 2h'=50mm)

(a) θ'=52°   (b) θ'=30°

Figure 3. Isochromatic fringe patterns from photoelastic measurements after delamination growth occurred (2h=20mm, 2h'=50mm)

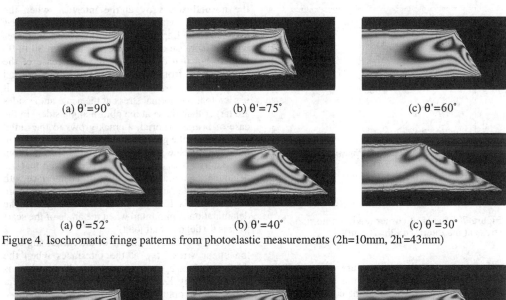

(a) θ'=90°   (b) θ'=75°   (c) θ'=60°

(a) θ'=52°   (b) θ'=40°   (c) θ'=30°

Figure 4. Isochromatic fringe patterns from photoelastic measurements (2h=10mm, 2h'=43mm)

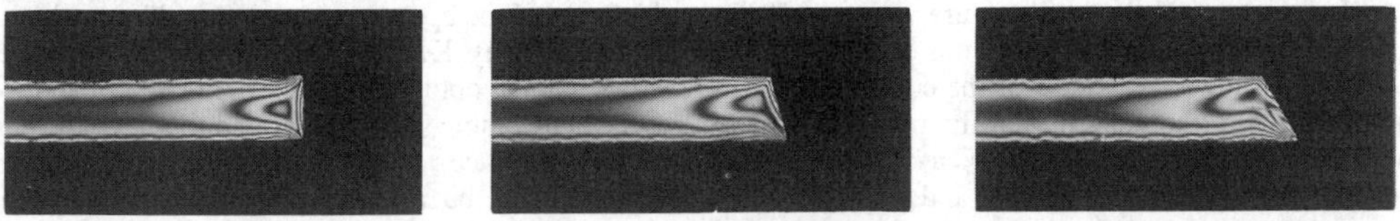

(a) θ'=90°   (b) θ'=75°   (c) θ'=60°

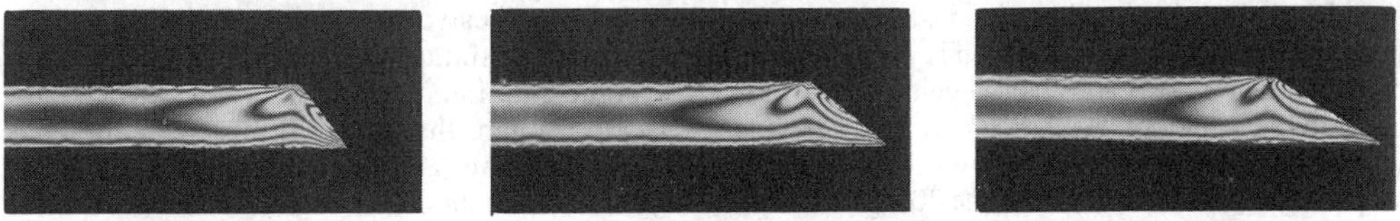

(a) θ'=52°   (b) θ'=40°   (c) θ'=30°

Figure 5. Isochromatic fringe patterns from photoelastic measurements (2h=5mm, 2h'=43mm)

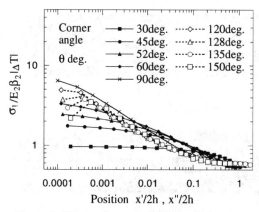

Figure 6. Effects of corner angle on maximum principal stress at the interface (2h=20mm)

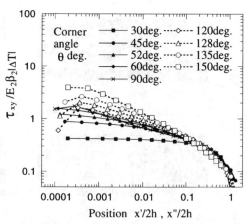

Figure 8. Effects of corner angle on shear stress at the interface (2h=20mm)

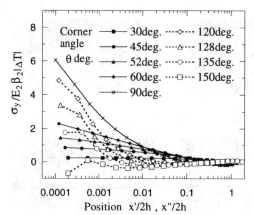

Figure 7. Effects of corner angle on normal stress at the interface (2h=20mm)

calculations, a uniform temperature change -50°C (=ΔT) of the scarf adhesive joints was chosen.

Figure 6 shows the effects of the corner angle on the maximum principal stress at the interface when the thickness of the epoxide resin plate is 20mm. The ordinate shows the positions (x' and x") from the edges of the interface (o' and o") in logarithmic scale which are normalized by the thickness 2h of the epoxide resin plate. The abscissa shows the maximum principal stress σ_1 which is normalized by $E_2\beta_2|\Delta T|$ where $|\Delta T|$ is absolute value of the uniform temperature change ΔT. At the edge of the interface, the maximum principal stress shows the singularity except the joint with the corner angle 30degree. At the edge of the interface and the very small region, the maximum principal stresses at the obtuse angle sides are larger than those at the acute angle sides.

Figure 7 shows the effects of the corner angle on

the normal stress σ_y at the interface when the thickness of the epoxide resin plate is 20mm which is normalized by $E_2\beta_2|\Delta T|$. The normal stress increases with an increase of the corner angle from 30 to 90 degree, but that decreases with an increase of the corner angle from 90 to 150 degree. The normal stresses show tensile near the edge of the interface. It is seen that the normal stress at the acute angle sides are larger than those at the obtuse angle sides. In the case of brittle material, a crack grows to the radial direction and the growth direction is perpendicular to maximum tensile stress (Erdogan et al. 1963). From this theory ($\sigma_{\theta max}$ criterion), it is expected that the delamination initiates at the edge of the interface with the acute angle but that with the obtuse angle. From Figure 7, it is forecasted that the strength to the delamination is minimum when the angle of the scarf joint is 90 degree (butt joint).

Figure 8 shows the effects of the corner angle on the shear stress τ_{xy} at the interfaces when the thickness of the epoxide resin plate is 20mm which is normalized by $E_2\beta_2|\Delta T|$. In Figure 8, the shear stresses at the obtuse angle sides are larger than those at the acute angle sides. It is found that near the edge of the interface, the normal stress is tensile and dominant at the acute angle sides and the shear stresses are dominant at the obtuse angle sides, respectively.

In the adhesive scarf joints subjected to tensile loads, the optimum angle exist where the stresses become minimum in the adhesive layer (Lubkin 1957). When the material of the adherend is aluminum plate and the adhesive is epoxy resin type, the optimum scarf angle θ' is 51.8 degree ($\fallingdotseq$ 52 degree). The stress distributions in the scarf joints subjected to tensile loads were calculated by FEM using the same models shown in Figure 1. From the results, it is confirmed that the order of stress

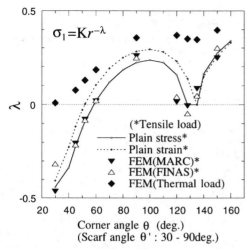

Figure 9. Effects of corner angle of epoxide
resin plate on the order of stress singularity λ
($2h=20$mm)

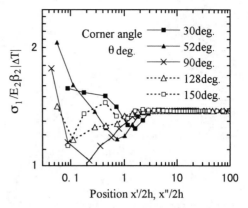

Figure 10. Effects of corner angle on maximum
principal stress at the interface ($2h=0.05$mm)

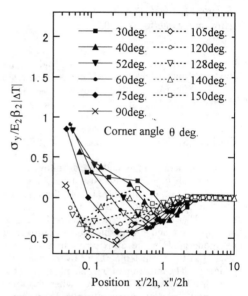

Figure 11. Effects of corner angle on normal
stress at the interface ($2h=0.05$mm)

singularity vanishes when the corner angles are 52
and 128 degree. The order of stress singularity λ is
calculated using the maximum principal stress
distributions at the interface of the epoxide resin plate
under uniform temperature changes, shown in Figure
6.

Figure 9 shows the effects of the corner angle of
the epoxide resin plate on the order of stress
singularity λ. The ordinate shows the corner angle θ
and the abscissa shows the order of stress singularity
λ. The dotted and solid lines show analytical results
of the order of stress singularity (Bogy 1971) in the
plane stress and the plane strain state. The symbols ▼
(FEM (MARC)) and △ (FEM(FINAS)) show the
order of stress singularity λ obtained from FEM
analysis for the scarf joints subjected to a tensile load.
The symbol ◆ (FEM (Thermal load)) shows the
order of stress singularity λ of the scarf joints
subjected to thermal load. From Figure 9, it is seen
that the order of stress singularity λ for the joints
subjected to thermal loadings become zero when the
angle is about 30 degree while the value λ shows
positive. It can not be found that the optimum scarf
angle of the joints is under thermal loadings.

Figure 10 shows the effects of the corner angle on
the maximum principal stress distributions at the
interface when the thickness $2h$ of the adhesive layer
is 0.05mm. The ordinate shows the position
normalized by $2h$ (0.05mm) from the edge of the
interface. It is shown that the value of the normalized
maximum principal stress is about 1.4 (constant) in
the region where is three times far from the adhesive
thickness from the edge of the interface. The
maximum principal stresses become small near the

edge of the interface but show the stress singularity at
the very small region. It is thought that the effect of
the stress singularity on the strength of the scarf
joints is small.

Figure 11 shows the effects of the corner angle on
the normal stress at the interface. The normal stress
shows compression near the edge of the interface.
The region where the stress singularity occurs is less
than several μm.

Figure 12 shows the effects of the corner angle on
the shear stress at the interface. It is seen that the
distributions of the stress components shown in
Figures 10-12 are similar to those shown in Figures

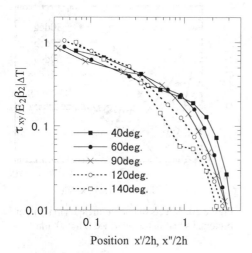

$\tau_{xy}/E_2\beta_2|\Delta T|$

Position x'/2h, x"/2h

Fugure 12. Effects of corner angle on shear
stress at the interface (2h=0.05mm)

6-8, respectively, when the positions (x' and x") are
normalized by the thicknesses of the epoxide resin
plate (2h=20mm) and the adhesive layer (2h=
0.05mm).

4 CONCLUSIONS

The results obtained in this study are as follows;
1. The scarf joints, the adhesive layer of which
was modeled and manufactured from the epoxide
resin plate, were made using the heat setting
adhesive. The scarf joint was heated and cured at
high temperature and after curing process the thermal
stress generated in the epoxide resin plate was
measured at the room temperature by photoelastic
experiments. The initiation of the delamination at the
edge of the interface was measured when the scarf
joints were cooled stepwise from the room
temperature to -70°C. It was found that the
delamination initiated at the edge of the interface with
the acute angle of the epoxide resin plate.
2. The orders of stress singularity under thermal
load and tensile load are examined. Under tensile
load, the optimum scarf angle exists, but under
thermal load, it doesn't exist.
3. The thermal stress distribution in the thin
adhesive layer (2h = 0.05mm) of the scarf joints is
similar to those in the thick adhesive layer (2h=20,10
and 5mm).
4. It is expected that the effect of the stress
singularity on the strength of the scarf joint is small
when the thickness of adhesive layer is thin.

REFERENCES

Adkins, D.W. & Pipes, R.B. 1985. End Effects in
Scarf Joints. *Composites Science and Technology.*
22:209-221.
Bogy, D.B. 1971. Two Edge-Bonded Elastic Wedges
of Different Materials and Wedge Angles Under
Surface Tractions. *J. Applied Mechanics*, 38:377-
386.
Chen D. & Cheng S. 1990. Stress Distribution in
Plane Scarf and Butt Joints. *Transactions of the
ASME Journal of Applied Mechanics.* 57(3):78-83.
Erdogan, F. & Sih, G.C. 1963. On the Crack
Extension in Plates Loading and Transverse Shear.
*Transactions of the ASME. Journal of Basic
Engineering.* 85:519-527.
Ikegami, K., Takeshita, T., Matsuo, K. &
Sugibayashi, T. 1990. Strength of Adhesively
Bonded Scarf Joints Between Glass Fibre-Reinforced
Plastics and Metals. *Int. J. Adhesion and
Adhesives.* 10(3):199-206.
Imanaka, M. & Iwata, T. 1994. Fatigue Failure
Criterion of Adhesively-Bonded Joints Under
Combined Stress Conditions. *Adhesion Science and
Technology. Proceeding of the International
Adhesion Symposium.* :329-344. Yokohama, Japan.
Kyogoku, H., Sugibayashi, T. & Ikegami, K. 1987.
Strength Evaluation of Scarf Joints Bonded with
Adhesive Resin. *JSME International Journal.*
30(265):1052-1059.
Lubkin, J.L. 1957. A Theory of Adhesive Scarf
Joints. *J. Appl. Mech.* 24:255-260.
Suzuki, Y. 1987. Adhesive Tensile Strengths of Scarf
and Butt Joints of Steel Plates (Relation Between
Adhesive Layer Thicknesses and Adhesive Strengths
of Joints). *JSME International Journal.*
30(265):1042-1051.
Wah, T. 1976. The Adhesive Scarf Joint in Pure
Bending. *Int. J. Mech. Sci.* 18:223-228.
Wah. T. 1976. Plane Stress Analysis of a Scarf Joint.
Int. J. Solids Structures. 12:491-500.
Wang, S.S. & Yau, J.F. 1981. Interfacial Cracks in
Adhesive Bonded Scarf Joints. *AIAA J.* 19(10):
1350-1356.

Metal forming

Experimental Mechanics, Allison (ed.)© 1998 Balkema, Rotterdam, ISBN 90 5809 014 0

Influence of thickness variation on the limit strains of sheet metals

Fernando Ferreira Fernandez
Instituto de Aeronáutica e Espaço, São José dos Campos, SP, Brazil

Hazim Ali Al-Qureshi
Instituto Tecnológico de Aeronáutica, São José dos Campos, SP, Brazil

Abstract: This paper examines the influence of surface roughness at the onset of necking. A mathematical model is presented to predict the necking point, which is considered here as an upper bound limit of deformation in the process of stretch forming anisotropic thin sheet metals. The theoretical model assumes that the roughness can be represented by sinusoidal distribution. Stretching operations were performed with a spherical punch, where circular grids were imprinted on the specimens by photographic technique, enabling measurement of strains up to the necking point. The experiments were performed on aluminium 2024-O sheets. The results of the present work are analysed together with theoretical data from other authors and compared and found to be in good agreement.

1 INTRODUCTION

Classical instability theories do not take into account neither geometrical imperfections nor microstructural imperfections, considering the sheet surface geometrically perfect. Marciniak and Kuczynski (MK, 1967) acknowledge the significance of such defects calling non-homogeneity the lower thickness point of the whole sheet. Such a defect develops until the necking point, a fact not foreseen by the classic theory on the stretch zone (ε_1, $\varepsilon_2 > 0$).

The present work studies the relationship between these defects and the surface roughness of the sheet. It is well known that a metallic sheet subjected to plastic strains has its roughness increased, obeying Fukuda's law:

$$r = r_0 + k.d_0.\varepsilon \tag{1}$$

where r is the surface roughness at an instantaneous effective strain ε, r_0 is the initial roughness, d_0 is the initial grain size diameter, and k is the roughness to deformation ratio. Yamagushi et al (1976) studied the plastic instability and limit strain on isotropic metallic sheet, considering the roughness as high and low plane regions. Their model assumed independent stretching of thin and thick portions of the sheet, the solution being incompatible in the strains. Parmar et al (1977) refined the Yamagushi's model, characterising the roughness as pyramids and assuming the dependency in the different thickness regions on the balance of forces.

2 THEORETICAL ANALYSIS

The main concept is quite simple and has firstly the purpose to describe the behaviour and influence of roughness on the limit strains and secondly to provide data for experimental comparisons.

In this work, roughness is represented in a sinusoidal form, in both directions of the sheet plane, producing peaks, and valleys. The measured roughness is the peak to valley roughness, which is the larger height of a peak and an adjacent valley. It is also admitted that the roughness on both sides of the sheet are out of phase π rads, this being a critical condition because the loads are supported by the lower thickness (t) section. The material is assumed to be incompressible, to obey the hardening law $\sigma = c.(\varepsilon_0 + \varepsilon)^n$. Also there are only plane stress, normal anisotropy and the Bauschinger effect does not occur.

The present model describes a system where material flows from the volume that represents the loaded section to increase the volume of the surface roughness, which cannot support any load.

The strains are monitored in two steps: First until the diffuse necking (plastic instability), where a maximum load is achieved and second until the necking point (limit strain), when a localized groove appears and the material outside this groove stops deforming.

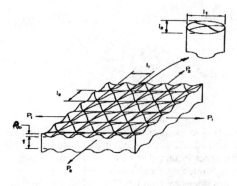

Figure 1- Proposed model for roughness.

2.1 Plastic Instability

The effective stress, for a plane stress case, considering the Levy-Mises flow law, can be written as

$$\overline{\sigma} = \sigma_1 \left[1 - \frac{2 \cdot R \cdot x}{(R+1)} + x^2 \right]^{1/2} \tag{2}$$

where $x = \sigma_2/\sigma_1$, is admitted constant during all the process and R is the normal anisotropy.

Approaching the volume of an element with the sinusoidal profile on directions 1 and 2, like figure 1, without deformation, by:

$$V_0 = \frac{4}{3} \pi \, l_0^2 r_0 + \pi \, l_0^2 t_0 \tag{3}$$

and the volume at any moment of strain, by

$$V = \frac{4}{3} \pi \, l_1 l_2 r + \pi \, l_1 l_2 t \tag{4}$$

By equating equations (3) and (4) and considering initially $r_0 \ll t_0$, then:

$$\frac{dt}{t} = \frac{\left\{ \dfrac{-(1+x)}{2.A} \cdot \text{Exp} \left[\dfrac{-(1+x)\overline{\varepsilon}}{2.A} \right] - \dfrac{4}{3} \dfrac{kd_0}{t_0} \right\}}{\left\{ \text{Exp} \left[\dfrac{-(1+x)\overline{\varepsilon}}{2.A} \right] - \dfrac{4}{3} \dfrac{\left(kd_0 \overline{\varepsilon} + r_0 \right)}{t_0} \right\}} \cdot d\overline{\varepsilon} \tag{5}$$

where $A = \left[x^2 - \dfrac{2Rx}{(R+1)} + 1 \right]^{1/2}$. The effective strain at plastic instability is obtained by Swift's analysis (1952). By manipulating equations (2),(3),(4) and (5) and remembering the hardening law $\overline{\sigma} = C(\overline{\varepsilon})^n$, then

$$\overline{\varepsilon}_d = \frac{-2.A^3.n}{\left[B.C.(R+1).(x+1) + FF \right]} \tag{6}$$

where $B = \left[1 - \dfrac{Rx}{(R+1)} \right]$, $C = \left[x - \dfrac{R}{(R+1)} \right]$ and FF is called form factor, given by:

$$FF = 2 A^3 \left[\frac{ - \dfrac{(1+x)}{2A} \text{Exp} \left[\dfrac{-(1+x)\overline{\varepsilon}_d}{2A} \right] - \dfrac{4}{3} \dfrac{kd_0}{t_0} }{ \text{Exp} \left[\dfrac{-(1+x)\overline{\varepsilon}_d}{2A} \right] - \dfrac{4}{3} \dfrac{\left(kd_0 \overline{\varepsilon}_d + r_0 \right)}{t_0} } \right] \tag{7}$$

The present work differs from the classical theory due to the coefficient FF, which includes the initial roughness and the roughness to deformation ratio. If FF=0 and R=1 (isotropic material), equation (6) reduces to Swift's result. The increase of roughness during the process reduces the thickness t of the sheet to a lower value than that expected from Swift's analysis and instability occurs at a lower level of strain.

2.2 Limit Strain

MK proved that even under biaxial tension on the stretch region, failure might occur with the appearance of a localized groove. Such failure mode was explained assuming non-homogeneity in the sheet, represented by a lower thickness region, which forms the groove as the plastic strains proceed. It is assumed here that the depth of the groove is the measured peak to valley roughness, and that the necking is generated due to this thickness difference between the regions A and B, with thickness given, respectively by (t+2.r) and t.

Basically the analysis is similar to the MK model, however with the strains starting from the instability, with different hardening states for the regions A and B. It is assumed that the load ratio $(x = \sigma_2/\sigma_1)$ remains constant until instability, providing that the hardening degree experienced by regions A and B depends only on the stress state in each region. Tadros et al (1975) assumed the same hardening degree for A and B. The limit strain in this work will be the sum of strain up to the plastic instability with the strains between the instability and the necking point.

Considering the volume of an element (figure 1), it can be seen from geometry that at plastic instability:

$$V_d = \pi \, t_d l_{1d} l_{2d} + \frac{4}{3} \pi \, r_d l_{1d} l_{2d} \qquad (8)$$

Similarly, the volume after instability, assuming that the element maintains its sinusoidal profile on both directions 1 and 2, then

$$V^* = \pi \, t^* l_1^* l_2^* + \frac{4}{3} \pi \, r^* l_1^* l_2^* \qquad (9)$$

where the suffix "d" refers to an instability condition and the "*" notation refers to a posterior condition. From equations (8) and (9),

$$\frac{t^*}{t_d} = \left[1 + \frac{4\left(kd_0 \bar{\varepsilon}_d + r_0\right)}{3 \, t_d} \right] \cdot \mathrm{Exp}\left(-\varepsilon_1^* - \varepsilon_2^*\right) - \frac{4\left(kd_0\left(\bar{\varepsilon}_d + \bar{\varepsilon}^*\right) + r_0\right)}{3 \, t_d} \qquad (10)$$

The stress and strain relation, is, after the instability, written as:

$$\bar{\sigma}_A = C\left(\bar{\varepsilon}_{Ad} + \bar{\varepsilon}_A^*\right)^n, \text{ for A, and}$$

$$\bar{\sigma}_B = C\left(\bar{\varepsilon}_{Bd} + \bar{\varepsilon}_B^*\right)^n, \text{ for B.} \qquad (11)$$

From the equilibrium of forces in the direction 1, then

$$\sigma_{1A} \cdot t_A \cdot N \cdot l_2 = \sigma_{1B} \cdot t_B \cdot N \cdot l_2 \qquad (12)$$

Considering equations (11) and (12), hence

$$\frac{\sigma_{1A} / \sigma_A}{\sigma_{1B} / \sigma_B} = \left(\frac{\bar{\varepsilon}_{Bd} + \bar{\varepsilon}_B^*}{\bar{\varepsilon}_{Ad} + \bar{\varepsilon}_A^*} \right)^n \cdot \frac{t_B^*}{t_A^*} \qquad (13)$$

The stress ratio σ_{1A} / σ_A is obtained through equation (2). The stress ratio for the B region is not constant with the strains, and can be determined assuming that the strain transverse to the groove is the same for both regions A and B. So,

$$\varepsilon_{2A}^* = \varepsilon_{2B}^*, \text{ and } d\varepsilon_{2A}^* = d\varepsilon_{2B}^* \qquad (14)$$

If $t_B^* < t_A^*$, then $d\varepsilon_{1B}^* > d\varepsilon_{1A}^*$, and consequently $\dfrac{d\varepsilon_{2B}^*}{d\varepsilon_{1B}^*} < \dfrac{d\varepsilon_{2A}^*}{d\varepsilon_{1A}^*}$.

When the groove appears, the B region is under a plane strain state. This can be represented by:

$$\frac{d\varepsilon_{2B}^*}{d\varepsilon_{1B}^*} = 0, \text{ ou } \frac{d\varepsilon_A^*}{d\varepsilon_B^*} = 0 \qquad (15)$$

Taking into account equations (2) for the B region, and (14), equation (13) can be rewritten as

$$\frac{\sigma_{1B}}{\sigma_B} = \frac{(R+1)}{(2R+1)^{1/2}} \cdot \left[1 - \frac{[x(R+1)-R]^2}{(R+1)^2\left(x^2 - \frac{2Rx}{(R+1)} + 1\right)} \left(\frac{d\bar{\varepsilon}_A^*}{d\bar{\varepsilon}_B^*}\right)^2 \right]^{1/2} \qquad (16)$$

Substituting equations (2) and (16) into (13), then

$$\frac{d\varepsilon_{1B}^*}{d\varepsilon_B^*} = \frac{(2R+1)^{1/2}}{2} \cdot \left[1 - \frac{[x(R+1)-R]^2}{(R+1)^2\left(x^2 - \frac{2Rx}{(R+1)} + 1\right)} \left(\frac{d\bar{\varepsilon}_A^*}{d\bar{\varepsilon}_B^*}\right)^2 \right]^{1/2} +$$
$$- \frac{R[x(R+1)-R]}{2(R+1)\left(x^2 - \frac{2Rx}{(R+1)} + 1\right)^{1/2}} \left(\frac{d\bar{\varepsilon}_A^*}{d\bar{\varepsilon}_B^*}\right) \qquad (17)$$

To obtain the second differential equation, it is necessary to substitute equations (2), (14), and (16) into (13):

$$\frac{d\bar{\varepsilon}_A^*}{d\bar{\varepsilon}_B^*} = \left\{ \frac{(R+1)^2\left(x^2 - \frac{2Rx}{(R+1)} + 1\right)}{[x(R+1)-R]^2} - \frac{(2R+1)}{[x(R+1)-R]^2} \left[\left(\frac{\varepsilon_{Ad} + \bar{\varepsilon}_A^*}{\varepsilon_{Bd} + \bar{\varepsilon}_B^*}\right)^n \frac{t_A^*}{t_B^*} \right]^2 \right\}^{1/2} \qquad (18)$$

The simultaneous equations (17) and (18) were solved numerically, using the third order Runge-Kutta method. The solution allows determination of $\bar{\varepsilon}_A^*$ at the limit condition $\dfrac{d\bar{\varepsilon}_A^*}{d\bar{\varepsilon}_B^*} = 0$, in the whole stretch-forming region, i.e., in the range $\dfrac{R}{R+1} \le x \le 1$. So, $\bar{\varepsilon}_A^*$ is considered as the limit strain, and once determined for every x, it is possible to plot the entire Forming

Limit Diagram (FLD), i.e., $\left(\varepsilon_{1A}^{*} \; x \; \varepsilon_{2A}^{*}\right)$. Figure 2 shows the influence of initial roughness on the FLD:

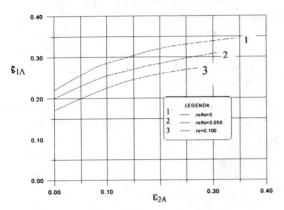

Figure 2 – Influence of initial roughness on the limit strains of a material with R=0,5; n=0,2; k=0,5 and d_0/t_0 =0,020.

3 EXPERIMENTAL PROCEDURE

The material used was an aluminium 2024-O alloy sheet with nominal thickness of 0,813mm, and the main properties are shown in table 1.

Table 1 - Mechanical Properties of 2024-O Alloy

Yield Strength (MPa)		78.48
Tensile Resistance (MPA)		231.52
Young's Modulus (MPA)		72397.8
Strain Hardening Exponent		0.191
Resistance Coefficient (MPA)		328.64
Anisotropy	Planar	0.072
	Normal	0.507
Thickness (mm)		0.813

The data above were obtained by tensile test (ASTM-E8-79/84), micrographies (ASTM-E112-85) and with a roughness meter machine (Hommel T 2000). To plot the entire stretch region of the FLD, the Erichsen test was used with blanks of 240 mm diameter and with different lubrication conditions.

At one side of the specimens, circular grids were imprinted by photographic technique, enabling the measurement of the limit strains with a profile projector.

4 DATA ANALYSIS

For comparison purposes, the specimens for the Erichsen test were polished and classified into three classes:
1. High roughness: $r_0 > 25\mu m$,
2. Medium roughness: $10 < r_o \leq 25\mu m$ e
3. Low roughness: $r_o \leq 10\mu m$.

All specimens as received, would be classified as a low roughness class, but in a fabrication system, the others classes should also be considered.

Figure 3 shows the experimental data and the theoretical curves of Parmar [4], MK [1], and the present model for the material used.

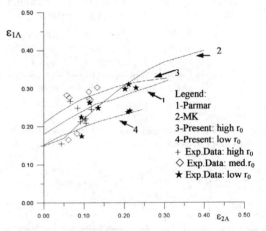

Figure 3 – Limit strains for the aluminium 2024-O, with $R = 0,507$, $n = 0,191$, $k = 0,48$ e $\dfrac{d_0}{t_0} = 0,020$

From figure 3, it can be seen that all the models represented the data well. From the present model two curves were generated: with high initial roughness ($r_0 = 25$ μm) and low initial roughness ($r_0 = 10$ μm). At the beginning region of the FLD, the proposed theory occurs as an upper bound and the MK theory as a lower bound, and at the region near the balanced biaxial state, MK occurs as an upper bound and the present work as a lower bound.

The present model revealed behaviour sensitive to roughness, and adequate to other models, differing in the roughness pattern adopted and the normal anisotropy.

As the initial roughness increases, the strains up to the necking point decrease. A suitable justification for this fact is that the roughness performs as a starting point to fracture, and also that the increase of roughness during the process contributes to reduce, in a faster way, the sheet's thickness.

The present work shows the importance of roughness on the limit strain of a material. As shown before, the parameter r_0 is the maximum measured peak to valley roughness, and represents an initial defect in the material. It becomes evident from the present analysis, that the limit strain is sensitive to the sheet metals roughness. For example, a sheet with a nominal thickness of 0,813 mm, and with an initial roughness of 5μm, this roughness representing 1,2% of the sheet's thickness.

5 REFERENCES

Marciniak, Z. & K. Kuczynski 1967. Limit strain in the processes of stretch forming sheet metal. Int. J. Mech. Sci, pg. 609.

Yamagushi, K. & P. B. Mellor 1976. Thickness and grain size dependence of limit strains in sheet metal stretching. Int. J. Mech. Sci., pg. 85.

Parmar, A., P. B. Mellor & J. Chakrabarty 1977. A new model for the prediction of instability and limit strains in thin sheet metal. Int. J. Mech. Sci., pg. 389.

Tadros, A. K. & P. B. Mellor 1975. Some comments on the limit strains in sheet metal stretching. Int. J. Mech. Sci., pg. 203.

Fernandez, F. F. 1994. Análise da influência da rugosidade nas deformações limites de chapas metálicas finas e anisotrópicas na região de estiramento, Master thesis, Instituto Tecnológico de Aeronáutica, Brasil.

Experimental Mechanics, Allison (ed.) © 1998 Balkema, Rotterdam, ISBN 90 5809 014 0

The influence of non-symmetrical rolling on the mechanical properties of final plates

P. Korczak & H. Dyja

Faculty of Metallurgy and Material Engineering, Częstochowa Technical University, Poland

J. Wachniak

Steel Plant Częstochowa, Poland

ABSTRACT: Parameters influencing stability distribution mechanical properties of plates after hot rolling are described. Various combinations of deformation temperatures, reduction, inter-pass times and cooling rates were tested. The experimental rolling was made in reversing hot mill, for low carbon steel. The samples were quenched at various stages of the process and the microstructure was investigated. The main advantage of actual work was construction of a mathematical model for prediction the influence of the spread of retained strain after hot rolling on stability distribution of the ferrite grain size and on level of mechanical properties.

1. INTRODUCTION

The basic metallurgical theories for recrystallization, transformation and precipitation and the technology for computer simulations have contributed to promoting the studies on the predictions of the recrystallization behaviour of austenite in hot rolling, of the transformation behaviour in subsequent cooling and, consequently, of the final microstructure and mechanical properties of hot-rolled steels (Dutta & Sellars 1987). In recent years, stricter requirements relating to homogeneous mechanical properties in the through-length, width and thickness directions have been demanded in hot-rolled steels. Technology for the prediction of mechanical properties using mathematical models can be applied to the selection of optimum manufacturing conditions and the control of mecha- nical properties. Finite element method programme was used to predict a distribution strain rate stress and temperature during multi-pass rolling processes. The microstructure and mechanical properties in the thickness and width directions can be predicted by combining the mathematical semi-empirical models for the microstructure and mechanical properties (Gibbs et al. 1993) and FEM (Lee & Kobayashi 1982, Pietrzyk & Lenard 1991) for strain and temperature. Based on previous works by Dyja & Korczak (1996) and Dyja et al. (1997) in this paper influences of non- symmetrical conditions of rolling and the spread of retained strain (the result of partial

recrystallization) on stability of distribution of structural parameters which described the mechanical properties of final plates are presented.

2. FEM MODEL

To predict the distribution of strain, strain rate and temperature changes, during hot-rolling and next cooling in air a model of non- symmetrical rolling was used. The finite-element programme Elroll (Pietrzyk, 1994) has been used in the investigation. The mathematical model is composed of three parts: mechanical, thermal and microstructural components. The mechanical component is based on the fundamental work of Lee and Kobayashi (1982). The detailed description of the coupled thermal-mechanical solution is given in Pietrzyk & Lenard (1991). The velocity field in the deformation zone is calculated by minimisation of the functional:

$$J = \int_V \left(\sigma_i \, \dot{\varepsilon}_i + \lambda \, \dot{\varepsilon}_V \right) dV - \int_S \{f\}^T \{v\} dS \qquad (1)$$

where V - volume; S - contact surface; σ_i - effective stress; $\dot{\varepsilon}_i$ - effective strain rate; λ - Lagrange multi- plier; $\dot{\varepsilon}_V$ - volumetric strain rate; $\{f\}$ - vector of boundary tractions; $\{v\}$ - vector of velocities.
The boundary conditions include velocity compo- nent $v_y = 0$ along the horizontal axis of symmetry,

$v_x = -v_y \tan\alpha$ at the contact with the roll, where α is an angle between the tangent to the roll and the x axis and x, y are the co-ordinates. The temperature distribution in the deformation zone is calculated by a numerical solution of the general convective-diffusion equation:

$$\nabla^T(k\nabla T) - \rho c_p \{v\}^T \nabla T + Q = 0 \qquad (2)$$

where k - conductivity; T - temperature; ρ - density; c_p - specific heat; Q - heat generated due to plastic work.

The boundary conditions include heat generated due to friction and heat exchange with the roll. The solution is performed in a typical finite-element manner. Thermal equations are solved during each iteration of the minimisation of the power functional (1). Current local change of temperatures are used to determine the flow stress of the material while the local stresses and strains are used to calculate the heat generated due to plastic work and due to friction.

For description of the rheology values of the material in present work equation proposed by Hodgson-Collinson was used:

$$\sigma_p = K\varepsilon^m \sinh^{-1}\left(\frac{Z}{A}\right)^p \qquad (3)$$

where the coefficient $K = 100$ $p = 0.2$, $A = 5\ 10^{12}$ and $m = 0.2$ for HSLA steel rolling during a experimental investigations, Z is Zener-Hollomon parameter and ε represents strain.

The microstructural part of model is based on empirical equations describing the processes of recrystallization, precipitation of carbo-nitrides Nb or V and austenite grain growth. The local values of temperature, strains and strain rates are used to describe a value of austenite grain size and amount of recrystallization. For calculation of degree of recrystallization, X the equation governed by Avrami rule was used:.

$$X = 1 - \exp\left[-0.693\left(\frac{t}{t_{0.5}}\right)^2\right] \qquad (4)$$

where t - time, $t_{0.5x}$ - time for 50% recrystallization The amount of static recrystallization depends on chemical composition of rolled steel, parameters which describe a plastic deformation process,

distribution of temperature and time between passes. In present work for calculation of a time for 5% precipitation $t_{0.05p}$, time for 5% recrystallization $t_{0.05x}$ and time for 50% recrystallization $t_{0.5x}$ was used equations developed by Hodgson (Sellars & Whiteman 1979, Hodgson 1993).

$$t_{0.5x} = (-5.24 + 550[Nb]) \cdot 10^{-18} D^2$$
$$\varepsilon^{-4+77[Nb]} \exp\frac{330000}{RT} \qquad (5)$$

$$t_{0.05p} = 6 \cdot 10^{-6}[Nb]^{-1}\varepsilon^{-1}Z^{-0.5}$$
$$\exp\frac{270000}{RT}\exp\frac{2.5 \cdot 10^{10}}{T^3(\ln k_s)^2} \qquad (6)$$

k_s is the supersaturation ratio which is given by:

$$k_s = \frac{[Nb]\left([C] + \frac{12}{14}[N]\right)}{10^{2.26-\frac{6770}{T}}} \qquad (7)$$

where $[C], [Nb], [N]$ - carbon, niobium and nitrogen contents, respectively.

During multi-pass rolling there are possible two cases:
- if $t_{0.05x} < t_{0.05p} < t_{0.95x}$ then recrystallization is retarded by precipitation
- if $t_{0.05x} > t_{0.05p}$ then recrystallization does not take place.

In some cases partial recrystallization can occur between passes and the microstructure entering the ntext pass will be a mixture of recrystallized and unrecrystallized austenic grains. There is then the question how to model the subsequent recrystallization of this mixed microstructure. From studies of C-Mn (Sellars & Whiteman 1979, Sellars 1983) and Nb (Hodgson & Collinson 1990, Beynon & Sellars 1991) steels it was shown that this situation could be modelled by assuming that the partial recrystallization resulted in a uniform loss of driving force. If the mixed microstructure entering the next pass was divided into to separate volume fractions: one fully recrystallized and the other unrecrystallized with retained strain ε_1 and then the kinetics of the two volume fractions were modelled individually after applying the next strain ε_2, then the model prediction does not fit the actual measurements.

If however, uniform softening is assumed then an effective residual strain ε_r is calculated and the subsequent kinetics after the second pass are modelled using a total strain ε_{TOT} and ε_2.

$$\varepsilon_r = K\varepsilon_1(1 - X_s) \qquad (5)$$

$$\varepsilon_{TOT} = K\varepsilon_r + \varepsilon_2 \qquad (6)$$

where X_s - fractional softening after 1st pass; K - constant.

For C-Mn steels (Sellars & Whiteman 1979, Sellars 1983) and certain conditions in the Nb steels (Hodgson & Collinson 1990, Beynon & Sellars 1991) K was equal to 1.0, while under other conditions (Hodgson & Collinson 1990) and work elsewhere (Gibbs & Hodgson 1990) K was found to be 0.5. When full recrystallization occurs, the recrystallized austenite grain size D_γ and grain after growth D can be calculated.

$$D_\gamma = AD^{0.67}\varepsilon^{-0.67} \qquad (7)$$

$$D^m = D_\gamma^m + k_s \exp\left(\frac{-Q_g}{RT}\right) \qquad (8)$$

where for the steel used coefficients are (Hodgson & Collinson 1990): $m = 4.5$, $k_g = 4.4 \; 10^{23}$, $Q_g = 435$ kJ mol^{-1}

The simulation of the austenite to ferrite transformation is included into the model. It is based on (Hodgson & Collinson 1990) where it is shown that the amount of the retained strain ε_r has an essential influence on the nucleation of the ferrite grains. The empirical equation developed in (Gibbs et. al 1990) and describing the dependence of the ferrite grain size on the austenite grain size, cooling rate c_r and retained strain ε_r is used here.

$$D_a = \left(1 - 0.8\varepsilon_r^{0.15}\right)\left\{29 - 5c_r^{0.5} + 20\left[1 - \exp(-0.15D)\right]\right\} \quad (9)$$

3. SYMMETRY AND NON -SYMMETRIC HOT PLATE ROLLING

Non- controlled asymmetry between the upper and lower surfaces of rolled material can often be found in industrial rolling processes which may lead to cooling of the rolled stock and determination of its dimensional accuracy. Main factors influencing

destabilisation of symmetry of plate rolling are:
- non uniform heating of slabs,
- non overlapping of neutral axis of the plate with a neutral axis of the roll gap,
- entry of plate under some angle to a roll gap.
Typical mesh and thermal conditions for non-symmetric kind of rolling is presented in Figure 1.

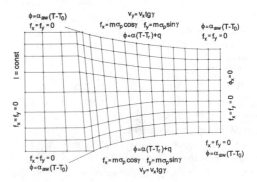

Figure 1. Conditions of non--symmetrical rolling

4. EXPERIMENTAL PROCEDURE

The objective of the work was to perform a computer simulation in conditions similar to those observed during last pass industrial rolling. As a test material in this study steel with chemical composition given in Table 1 was used.

Table 1. Chemical composition of the tested steel.

%C	%Mn	%Nb	%N$_2$	%V
0.09	1.37	0.034	0.0048	0.002

The investigation was made for symmetric and asymmetric kind of rolling process. Parameters of symmetric process are given: initial austenite grain size 30μm, initial thickness of plate 20mm, rotational speed of rolls 60 rpm, heat transfer coefficient between rolls and hot plate 50,000 Wm^{-2}K^{-1} and friction coefficient 0.25. For asymmetric hot rolling condition the following parameters were used : heat transfer coefficient between up roll and hot plate: 50,000 Wm^{-2}K^{-1}, between bottom roll and hot plate: 70,000 Wm^{-2}K^{-1} and two different friction coefficient: upper - 0.2, lower - 0.25. In Table 2 are shown schedules exploited in numerical simulation. For asymmetric case of rolling names of plane of passes addicted on asymmetric ratio and presented: PI1aA - a_v = 1.05, PI1bA - a_v = 1.15, PI1cA - a_v = 1.25.

Table 2. Rolling schedules used for computer simulation

Tempe-rature	$T_I = 950°C$			$T_{II} = 900°C$			$T_{III} = 912°C$; not-uniform distribution of temperature		
Redu-ction	$\varepsilon_1 = 10\%$	$\varepsilon_2 = 25\%$	$\varepsilon_3 = 35\%$	$\varepsilon_1 = 10\%$	$\varepsilon_2 = 25\%$	$\varepsilon_3 = 35\%$	$\varepsilon_1 = 10\%$	$\varepsilon_2 = 25\%$	$\varepsilon_3 = 35\%$
Name of plane	PI1s	PI2s	PI3s	PII1s	PII2s	PII3s	PIII1s	PIII2s	PIII3s

Changes of amount of recrystallization in function pass reduction for start rolling temperature is shown in Figure 2.

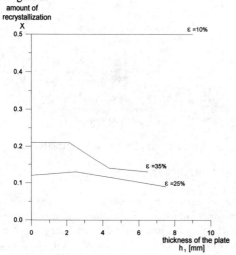

Figure 2. Amount of static recrystallization in function pass reduction for start temperature T = 900°C

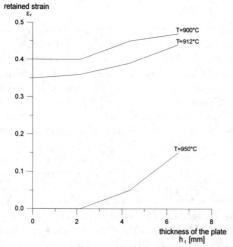

Figure 3. Distribution of retained strain on thickness of the plate for different start rolling temperature and reduction in pass 35%.

In Figure 3 is shown influence of initial rolling temperature and reduction in pass (35%) on distribution retained strain ε_r, on thickness of the final plate.

Distribution of ferrite grain size through the final plate thickness for reduction in pass 25% and different start rolling temperatures is shown in Figure 4.

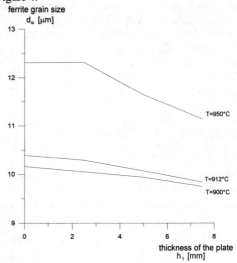

Figure 4. Ferrite grain size in function start rolling temperatures and reduction 25%.

In present work are presented influence of start rolling temperature, reduction in pass on distribution amount of static recrystallization and retained strain. The value of nodes temperature, retained strain and cooling rate describe ferrite grain size. In Figure 5, 6, 7 are presented distribution of ferrite grain size on final plate thickness in function temperature, reduction and values of retained strain.

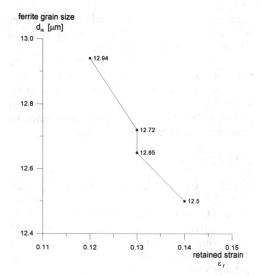

Figure 5. Distribution of ferrite grain size on plate thickness for T = 900°C, ε = 10%

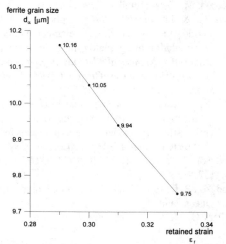

Figure 6. Distribution of ferrite grain size on plate thickness for T = 900°C, ε = 25%

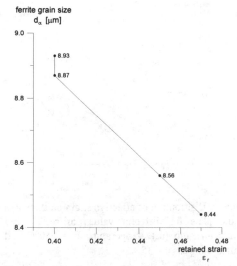

Figure 7. Distribution of ferrite grain size on plate thickness for T = 900°C, ε = 35%

For demonstrate a influence of parameters describing a non-symmetry of rolling condition on stability properties of final plates in Figure 8-10 are shown distribution of effective strain (Figure 8, 9) and temperature immediately after pass (Figure 10).

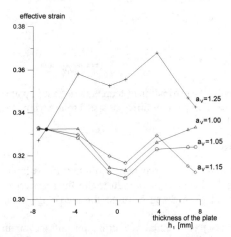

Figure 8. Distribution of effective strain on thickness of the plate for different values of rolls speed asymmetry ratio (a_v), temperature T = 950°C and reduction in pass ε = 35%.

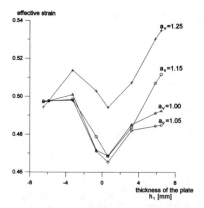

Figure 9. Distribution of effective strain on thickness of the plate for different values of rolls speed asymmetry ratio (a_v), temperature T = 950°C and reduction in pass ε = 35%.

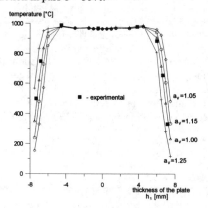

Figure. 10. Distribution of temperature field in plate at once after pass for different speed rolls asymmetry ratio

CONCLUSIONS

The finite-element program Elroll gave a possibility of the real simulation of industrial rolling process of thick plates and can be used for preparing of the optimal rolling schedule.

The program simulates distributions of the strain rates, strains, temperatures and austenite grain size in the deformation zone. The ferrite grain size after experimental rolling was compared which microstructure parameters after simulation. On distribution of mechanical properties have an influence not only chemical composition of rolled in hot condition steel but also factors influencing on non-symmetry of rolling conditions.

The value of retained strain accumulated in austenité structure effects on the end recrystallization tempe

rature, the spread of retained strain after hot rolling in finishing mill have an influence on stability of the distribution of the ferrite grain size.

Present work demonstrate an influence of nonsymmetry condition rolling process on stability of distribution temperature field and effective strain in hot rolled plate, allow to state that using a purposely asymmetry speed of rolls affect on control structural parameters on thickness of final plate.

REFERENCES

Dutta B. & C. M. Sellars 1987. *Mat. Sci. Techn.*, 3, 197-206.

Gibbs R.K. B.A. Parker & P. D. Hodgson 1993. *Proc. Symp. Low-Carbon Steels for 90's*: 173-179. Pittsburgh: TMS.

Lee, G. & S. Kobayashi 1982. *ASME, J. Eng. Ind.*, 104, 55-64.

Pietrzyk M. & J. G. Lenard 1991. *Thermal-Mechanical Modelling of the Flat Rolling Proces.* Berlin: Springer-Verlag.

Pietrzyk M. 1994. Hutnik - *Wiad. Hutn.* 61.

Dyja H. & P. Korczak 1996. Application of the thermo- mechanical finite - element model to the simulation of hot plate rolling. *International Conference Numerical Methods in Engineering and Applied Sciences CIMENICS'96, Merida, Venezuela, 25-29 March 1996.*

Dyja H., P. Korczak & M. Głowacki 1997 Validation of Thermo - mechanical and microstructural models for nobium steels in hot plate rolling, *ASTM'97, Australia 1997.*

Hodgson P. D. & R. K. Gibbs 1992. *ISIJ Int.*, 32: 1329-1338.

Hodgson P. D. 1993. *PhD dissertation*, University of Queensland.

Sellars C. M. & A. Whiteman 1979. *Met. Sci*: 13, 178.

Sellars C. M. 1983. Hot working and Forming Processes, In C. M. Sellars, G. J. Ratz & P. J. Wary (eds), *The Metall. Soc. AIME.*

Beynon J. H. & C. M. Sellars 1992 Modelling microstructure and its effects during multi-pass hot rolling, *ISIJ International*: 32, 3: 359-367.

Hodgson P. D. & D. C. Collinson 1991 *Proc. Symp. Mathematical Modelling of Hot Rolling of Steel*, In S. Yue (ed.), 239-250. Hamilton.

Gibbs R. K., P. D. Hodgson & B. A. Parker 1990. Effect of deformation history on the recrystallization behaviour of Nb microalloyed steels. In T. Chandra (ed.). *Recrystallization'90*. The Minerals 585.

Experimental Mechanics, Allison (ed.)© 1998 Balkema, Rotterdam, ISBN 90 5809 014 0

Experimental evaluation of strain paths driving to failure in sheet metal forming

S. K. Kourkoulis & N. P. Andrianopoulos
National Technical University of Athens, Department of Engineering Science, Section of Mechanics, Greece

ABSTRACT: The experimental and theoretical construction of the Forming Limit Diagrams (FLDs), used as limiting guides during sheet-metal forming operations is revisited in the present work, from the point of view of Fracture Mechanics using concepts of classical Strength of Materials. An experimental procedure is described, the main feature of which is that it permits the continuous record of the strain path followed during the loading procedure up to the final failure. A number of critical points concerning the validity and reliability of the conventional methods adopted for the construction of FLDs are discussed and new evidence is presented regarding some factors whose influence appears to be important. The parallel theoretical construction of FLDs with the aid of a generalized version of the T-criterion of failure helps rationalizing a number of apparently conflicting observations.

1 INTRODUCTION

In a number of important metal forming operations the deformation is predominantly stretching. It is known that when a sheet is progressively thinned two modes of plastic instability are possible (Dodd & Bai 1987): Diffuse and localized necking. However, in cold-working processes localized necking dominates, the diffused mode being undetectable. The localized necking modes are basically two: If one of the in-plane strains is compressive then the angle of localized necking depends on the imposed stress ratio and the angle of the neck with respect to the pulling axis is a characteristic angle of the problem. In the second case, if both in-plane strains are tensile the localized neck forms usually in a direction perpendicular to the direction of the larger stress.

In most practical problems the combination of major and minor in-plane strains beyond which failure occurs in the form of localized necking are presented with the aid of the Forming Limit Diagrams (FLDs), following the original method introduced by Keeler & Backofen (1963). According to this pioneering work the ductility of metal sheets subjected to stretching is studied by measuring the deformation of a suitable grid, previously marked on the surface of the test sheets, in order to obtain the limiting strain combinations for various strain paths.

Typically, an experimentally obtained FLD appears as it is being consisted of two distinct curves: One including the limiting strain combinations in the Left Hand Side (LHS) of the $(\varepsilon_1, \varepsilon_2)$ plane which is

usually simulated by a straight line with slope $d\varepsilon_1/d\varepsilon_2 \cong -1/2$ and a second one including the limiting combinations in the Right Hand Side (RHS) of the $(\varepsilon_1, \varepsilon_2)$ plane. This second curve appears to have an initial small positive inclination which is gradually reduced tending to zero as the minor in-plane strain increases (Hecker 1978).

Concerning, the theoretical construction of FLDs various criteria have been developed, varying from simple empirical or semi-empirical ones based on curve fitting procedures of experimental data (critically reviewed by Atkins and Mai 1985) to extremely complicated ones based on local necking theories (Hill 1952), bifurcation analysis (Hutchinson & Tvergaard 1980), and imperfection assumptions (Marciniak & Kuczynski 1967). However, besides the intensive research work devoted to the subject and the fact that the majority of these criteria gave satisfactory results for many practical applications, the lack of a sound and unified theoretical basis is a disturbing point and the answers to a number of crucial questions are pending. The strain path followed during stretching, the detection of the onset of strain localization marking the end of the forming process and the role of the third strain (the out-of-plane one) are among them.

In the present work an attempt is made to answer these questions taking advantage of both the results of a series of stretching experiments executed with the aid of a modified procedure compared to the one described by Keeler & Backofen (1963) and a generalized version of the T-criterion of failure (Andrianopoulos & Boulougouris 1990).

2 EXPERIMENTAL PROCEDURE

2.1 *The stretching device*

For the execution of the stretching experiments a simple, "portable" stretching device was constructed, consisting of the motor, the gear-box, the load cell, the internally threaded punch that transforms the rotational motion to linear, the blankholder, the die, a data acquisition system and a personal computer with the suitable software. A draft sketch of the loading device is shown in Figure 1. The terminal part of the punch is replaceable so that both in-plane and out-of-plane stretching techniques can be realized, by simply substituting the flat-nosed head with the circular cutout, (Fig.1), by a compact hemispherical one. This is unavoidable since it has been proved by Ghosh and Hecker (1974) that the instability conditions for in-plane deformation can differ significantly from those for out-of-plane deformation. The strain combination required is obtained by adjusting the width of the specimen and the lubrication conditions.

As it can be seen, the set-up described here, beyond its horizontal arrangement, induced by practical reasons, is more or less similar to the one used by Hecker (1978), concerning both the gripping of the specimen and the loading procedure. The main difference lies at the strain measurement system.

Hecker (1978), as well as many other researchers, follow the initial method of Keeler & Backofen (1963) and measure the deformation of a suitable grid of circles of the smallest possible diameter, previously marked on the surface of the test sheets. However, in this case "...the proper placement of the FLD in a diagram ...requires some subjective judgment" since "...the onset of localized necking is determined by the appearance of a distinct, localized

thickness trough as seen by the unaided eye" (Hecker 1978). On the other hand, it is practically impossible, using this method, to record the strain path followed during loading up to the failure point. It has been pointed out, however, by Azrin & Backofen (1970), Korhonen (1981), Ghosh and Hecker (1975), that at least for punch stretching the strain path followed is not linear. After a relatively small linear portion it becomes curved upwards showing a definite drift off to plane strain. In other words the strain path tends to become vertical with $d\varepsilon_2=0$. This trend of the strain paths can cause significant problems if it is ignored, since Korhonen (1981) proved that the position of the limit curves is very sensitive to strain path. If the strain paths are curved upward then this decreases the level of the whole forming limit curves. It seems, thus, that the knowledge of the exact strain path realized during loading is indispensable demand if one intents to predict theoretically the FLDs. Such information can only be obtained experimentally.

Towards this direction, a different strain measurement method is adopted in the present work, based on a system of 5 or 9 orthogonal strain rosettes of the Kyowa KFG-1-120-D16 type, suitably glued at various strategic points of the specimens. The extremely small dimensions of these strain gauges (their active diameter is less than 1.5 mm) was the decisive factor for their selection.

The main advantage of the above described system, beyond the fact that it permits the continuous record of the in-plane strains combination during the loading procedure, is that it permits the detection of the onset of strain localization without any subjectivity, since characteristic changes of the strain paths at points far away from the finally visible neck are recorded by the respective strain gauges due to elastic unloading at these points, as will be seen later.

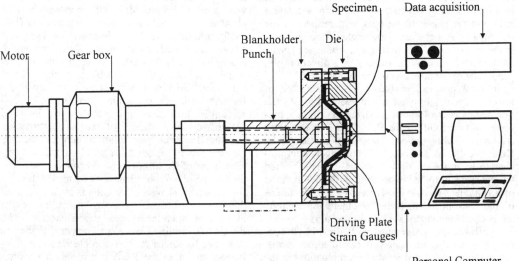

Figure 1. The experimental set-up (not scaled).

2.2 The material and the specimens

Carefully polished sheets of commercially pure aluminum (99.8%), prepared by the Aluminium Pechiney company under the commercial name H19-IV, were used for the construction of the specimens. The mechanical properties of the material were determined with the aid of a preliminary series of simple uniaxial tension experiments and are summarized in Table 1. The axial stress - axial strain diagram is shown in Figure 2. The thickness of the sheet was equal to 1.1×10^{-3} m and it was constant with maximum variation less than 0.1%. This is very important since the out-of-plane strain is calculated by post-mortem measuring the change of the thickness of the specimens.

Table 1. Mechanical properties of the material.

Property	Value	Unit
Young's modulus	82	GPa
Linearity limit	134	MPa
Yield stress	168	MPa
Ultimate tensile strength	218	MPa
Ductility	1.1	%
Poisson's ratio	0.35	

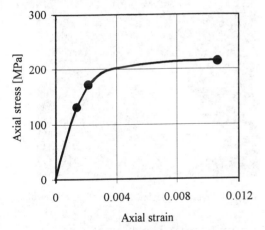

Figure 2. Axial stress versus axial strain.

Three different types of specimens were constructed: The first one, the familiar strip specimen, standardized according to ASTM E 8-79 for the execution of tension tests, was used to obtain the uniaxial tension point of the FLD. The second one, of the form of circular plate, was used for the execution of equi-biaxial tension tests, in order to obtain points in the RHS of the FLD. The third type is the one shown in Figure 3. It was used to obtain the plane-strain point of the FLD.

After preliminary stretching tests it was concluded that for the specific material and for the in-plane case 5 strain rosettes should be used, while for the out-of-plane case this number should be increased to 9. In all cases one gauge was glued at the geometric center of the specimens and the remaining ones at the

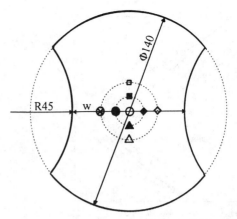

Figure 3. The plane-strain specimens and the position of the gauges: in-plane: w=65 mm, out-of-plane: w=60 mm.

end points of two mutually perpendicular diameters, either on a single circle of diameter 12 mm (in-plane case) or on two successive concentric circles of diameters 12 mm and 24 mm (out-of-plane case), as it is shown in Figure 3, where all dimensions are in mm.

All tests were carried out at a stretching speed of 2 mm/min. In the out-of-plane case the clamped specimens were stretched directly over the hemispherical punch. However to obtain the equi-biaxial point a piece of polytetrafluorothene (PTFE) film (0.5 mm thickness) was placed between the punch and the specimen. On the contrary, during the in-plane stretching process the punch did not touch the specimen directly. A driving plate with a hole in its center was placed between the punch and the specimen, causing different forming speeds between the two plates. The difference in speeds and the friction between the two plates produced the stretching deformation in the central part of the specimen. For the aluminium studied it was found that ductile steel plates of thickness 1.5 mm with a 20 mm diameter hole, gave the most successful results.

2.3 Experimental results

In Figures 4, 5 the strain paths have been plotted as they have been measured during two representative experiments, corresponding to in-plane and out-of-plane stretching, respectively. It is seen from these figures that especially for the case of out-of-plane stretching the strain paths followed by all four perimetric rosettes are not linear. Indeed, after a relatively small linear portion the strain paths are curved upwards. It seems that at the moment of strain localization the strain path of the point closest to the neck fulfills the condition $d\varepsilon_2 \to 0$. It should be mentioned, here that in all plane strain experiments the neck was formed in a direction perpendicular to the major strain for both stretching cases. The fractured specimens corresponding to the two experiments described

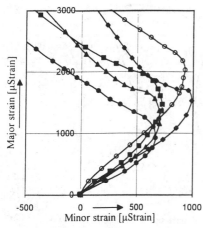

Figure 4. Strain paths during in-plane stretching. The symbols on the curves correspond to those of Fig.3.

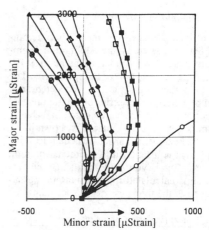

Figure 5. Strain paths during out-of-plane stretching. The symbols on the curves correspond to those of Fig.3.

Figure 6. Fractured specimen by in-plane stretching.

Figure 7. Fractured specimen by out-of-plane stretching.

above are shown in Figures 6 and 7. For the first case the deformed driving plate is, also, shown.

The limiting strain combinations obtained from the present series of experiments are shown in Figure 8 of the next paragraph, together with the respective predictions of the T-criterion of failure. It is seen that these points form two well-separated groups that belong to the first and second quadrants, respectively. Each group contains points from both stretching techniques. The limiting combinations corresponding to in-plane deformation conditions are systematically lower than the respective ones of the out-of-plane stretching, in the first quadrant. Such a trend was not detected in the second quadrant.

Concerning the third strain, ε_3, it was found that its experimental values almost coincided with the ones obtained from the incompressibility condition.

3 THEORETICAL CONSIDERATIONS

For the theoretical construction of FLDs the T-criterion of failure was adopted, since beyond its sound natural basis it is also free from the assumption of linearity of the strain paths. According to it the limiting strains are obtained when either the dilatational, T_V, or the elastic distortional strain energy, T_D, exceed their critical thresholds, $T_{V,0}$ and $T_{D,0}$, respectively. The quantities T_V and T_D were computed by means of numerical integration of the following equations:

$$dT_v = p d\theta \tag{1}$$

$$dT_D = \bar{\sigma} d\bar{\varepsilon}^e \tag{2}$$

where $p = \sigma_{ii}/3$, $\theta = \varepsilon_{ii}$, with σ_{ii} and ε_{ii}, i=1,2,3 being the principal stresses and strains, respectively, calculated by means of the theory of plastic flow, along a given load path. $\bar{\sigma}$ and $\bar{\varepsilon}^e$ are the equivalent stress and the elastic part of the equivalent strain according to von Mises, respectively (Andrianopoulos et al. 1996).

Concerning the critical thresholds $T_{V,0}$ and $T_{D,0}$, they are material constants determined with the aid of axial tension and torsion experiments as described by Andrianopoulos & Kourkoulis (1994).

For the needs of the present study the material is considered as macroscopic continuum without any kind of discrete structure or structural imperfections. Its equivalent stress-equivalent strain diagrams consists of a linear part, followed by a non-linear elastic portion and finally by a region where the material hardens isotropically (Fig.2).

The stresses and the strains are related through the generalized Hooke's Law and the incremental theory of plasticity with the von Mises flow rule. For the calculation of the limiting strain combinations a flexible and fast convergent numerical code has been developed. Under the plane-stress conditions that are assumed here, the iterative procedure starts with an arbitrary initial stress biaxiality ratio and stops either when $T_V=T_{V,0}$ or $T_D=T_{D,0}$ is fulfilled, along a predefined strain path, exactly the one obtained by numerically simulating the experimental paths described in the previous paragraph for each type of loading.

The predictions of the T-criterion, obtained by ap-plying the above described procedure have been plotted in Figure 8, together with the experimentally obtained critical strain combinations. It is seen from this figure that according to this criterion the FLD consists of two separate curves. The first one is an almost straight line, slightly distorted at its lowest portion that lies mainly in the second quadrant and corresponds to fulfillment of the condition $T_V=T_{V,0}$, i.e. it is a locus of constant dilatational strain energy density. The second curve is part of an ellipse, slightly distorted, and corresponds to the fulfillment of the condition $T_D=T_{D,0}$, i.e. it is a locus of constant distortional strain energy density.

The first line shows very good agreement with the experimental results as far as it concerns the plane-strain and uniaxial-tension points. At a first glance, the same conclusion can not be drawn for the second curve and the experimental points that correspond to equi-biaxial tension. Really, these points appear to lie out side the failure locus, since in the first quadrant this locus is limited by the $T_V=T_{V,0}$, line, and apparently points like A and B violate the prediction of the T-criterion. However, if extreme strains are considered and plotted (by using the equivoluminal assumption for plastic strains) all these points are transformed to the second quadrant, as it shown in Figure 8 for points A and B, and they lie again very close to the elliptical curve (points A′ and B′). On the

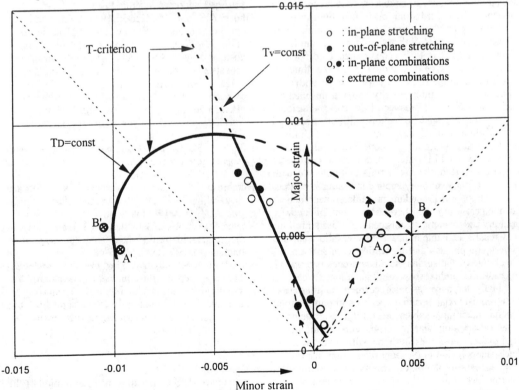

Figure 8. Experimental results and theoretical predictions.

other hand the points close to the $T_V=T_{V,0}$ line move only slightly to the left when they are replaced by their respective extreme values. Thus, in terms of extreme strains, a complete FLD, consists of an almost straight line that starts at the first quadrant and intersects with an elliptic curve at the second quadrant, as it shown by the heavily drawn line of Figure 8.

4 DISCUSSION AND CONCLUSIONS

An experimental procedure was described in the present work that permits the continuous recording of the strain paths at various strategic points of stretched specimens. It was concluded, that at points close to the final strain localization, the strain paths followed were far from linear. Clear differences between the in-plane and the out-of-plane stretching procedures were also observed, concerning both the relative position of the limiting strain combinations and the strain paths realized during each one of them. Taking advantage of these experimental data the T-criterion of failure was employed for the theoretical construction of FLDs. The agreement between the theoretical predictions and the experimental evidence was very good for the whole range of experimental points, especially if the extreme strains were taken into account.

However, in spite of this satisfactory agreement a crucial point remains to be discussed, concerning the influence of the third strain ε_3. In fact, both in-plane and out-of-plane stretching processes exhibit more or less strain-triaxiality that should not be ignored if a complete FLD is to be constructed theoretically. An FLD of this type is unavoidably a three dimensional surface in the (ε_1, ε_2, ε_3) space, even under the plane-stress assumption usually adopted. In practice, such a surface can be reduced in a two-dimensional curve by making suitable sections with other surfaces on which one of the three strains is constant. Unfortunately, such a procedure is complicated and is not followed when FLDs are constructed, for practical purposes. Instead, the two strains that coincide with the axes of the externally applied stresses are plotted. Although such a procedure presents some practical advantages, from the theoretical point of view is arbitrary and, sometimes, misleading. Indeed, preliminary results available indicate that the failure surface is very complicated and its section with surfaces on which one of the strains is constant does not always resemble the familiar shape of FLDs.

Finally, the observation concerning the translation of the equi-biaxial tension experimental points to the second quadrant resolves the, intuitively, inconvenient observation that materials cope with higher equi-biaxial tensile strains than with uniaxial ones. Such a conclusion in the light of the present discussion, appears to be just the result of considering the intermediate strain in place of the minimum one.

Summarizing, the most attractive conclusion of the present work is that metal forming obeys to general laws containing physically interpretable parameters which are part of the identity of the material and that can be determined with the aid of simple experiments of the classical Strength of Materials.

REFERENCES

Andrianopoulos, N.P. & B.C. Boulougouris, 1990. A generalization of the T-criterion of failure in case of isotropically hardening materials. *Int. J. of Fracture*. 44:R3-R6.

Andrianopoulos, N.P., B. Dodd, S.K. Kourkoulis & Luo Limin, 1996. FLDs for Al 2124 and Al-Li 8090 through fracture mechanics and perturbation analysis. *J. Mat. Proc. Techn.* 62:275-282.

Andrianopoulos, N.P. & S.K. Kourkoulis, 1994. FLDs for modern Al and Al-Li alloys and MMCs. J.F. Silva Gomez et al. (eds), *Proc. 10th Int. Conf. Exp. Mech., Lisbon, July 1994*:995-1000. Rotterdam: Balkema.

Atkins, A.G. & Y.M. Mai 1985. *Elastic and Plastic Fracture*. Chichester: Ellis Horwood and Wiley.

Azrin, M. & W.A. Backofen 1970. The deformation and Failure of a Biaxially Stretched Sheet. *Metallurgical Transactions*. 1:2857-2865.

Dodd, B & Y. Bai 1987. *Ductile Fracture and Ductility*. London: Academic Press.

Ghosh, A.K. and S.S. Hecker 1974. Stretching limits in sheet metals: In-plane vs out-of-plane deformation. *Metallurgical Transactions*. 5:2161-2164.

Ghosh, A.K. and S.S. Hecker 1975. Failure in thin sheets stretched over rigid punches. *Metallurgical Transactions*. 6A:1065-1074.

Hecker, S.S. 1978. Sheet stretching experiments. In H. Armen & R.F. Jones (eds), *Applications of numerical methods to forming processes*: 85-94. ASME.

Hill, R. 1952. On discontinuous plastic states, with special reference to localized necking in sheets. *J. Mech. Phys. Solids* 1:19-30.

Hutchinson, J.W and V. Tvergaard 1980. Surface instabilities on statically strained plastic solids. *Int. J. Mech. Sci.* 22:339-354.

Keeler, S.P. & W.A. Backofen 1963. Plastic instability and fracture in sheet stretched over rigid punches. *Trans. ASM*. 56:25-48.

Korhonen, A.S. 1981. *Necking, fracture and localization of plastic flow in metals*. Doctor of Technology Thesis, Helsinki Univ. of Technology.

Marciniak, Z. & K. Kuczynski 1967. Limit strains in the processes of stretch-forming sheet metal. *Int. J. Mech. Sci.* 9:609-620.

ACKNOWLEDGEMENTS
The help of Mr S.G. Doumas in the execution of the experiments, is gratefully acknowledged.

Experimental Mechanics, Allison (ed.) © 1998 Balkema, Rotterdam, ISBN 90 5809 014 0

The effect of fabrication processes on strength properties of GMM

T. Nakamura & T. Noguchi
Department of Mechanical Science, Hokkaido University, Sapporo, Japan

ABSTRACT: The influence of fabrication processes on static and fatigue strength properties of Giant Magnetostrictive Materials were investigated. Test results with Bridgman materials (BM) and powder metallurgy materials (PM) showed: (1) Compression strength of PM was about 2/3 of BM, and the strength scatter of PM was much smaller than BM. The smaller compression strength can be explained by the porosity of PM, and the smaller strength scatter by the microstructure anisotropy of BM. (2) Radial compression strength was about 1/20 of the compression strength. (3) Both kinds of materials showed stress dependence on Young's modulus especially at low stresses, and this may be considered a ΔE effect. At high stresses Young's modulus of PM showed about 3/4 of BM because of its porosity. (4) There was an apparent fatigue limit at 60% of the compression strength. This ratio is almost the same in both kinds of materials so fatigue limits can be evaluated by static strengths.

1 INTRODUCTION

The application of a magnetic field is known to change the dimensions of magnetic substances but magnetostriction is about $10^{-5} \sim 10^{-6}$ and generally considered too small for industrial applications. In 1970s, however, A. E. Clark discovered rare-earth-iron compounds demonstrating $100 \sim 1000$ times larger magnetostriction at room temperature (Clark & Belson 1972). Such materials showing enormous magnetostriction are called as "Giant Magnetostrictive Materials" and have been under dynamic research and development.

The production of giant magnetostrictive materials requires that the easy axis of magnetization be aligned in the same direction. In recent years, powder metallurgy using press forming under a magnetic field as well as Bridgman method using the process of unidirecional solidification have been applied. Powder metallurgy benefits from good linearity of magnetostriction at low magnetic fields, and ease of formation of complex shapes; whereas the Bridgman method has a large magnetostriction constant. Much research and development of devices applying giant magnetostrictive materials has been undertaken, but there are few studies discussing the mechanical properties of the materials, and no method to evaluate the strength of such materials has been established. In particular, there is no fatigue strength data of giant magnetostrictive materials manufactured by powder metallurgy. Compression tests, radial compression tests, and compression fatigue tests were conducted to determine the basic mechanical and fatigue properties of giant magnetostrictive materials and the effect of fabrication methods.

2 MATERIALS

The experiments used laves-phase rare-earth-iron compounds $Tb_{0.3}Dy_{0.7}Fe_{1.95}$ produced by the Bridgman and

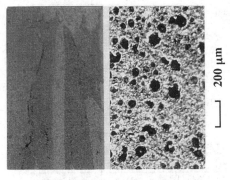

(a) Sp:Brigman   (b) Sp:Powder metallurgy

Figure 1. Microstructures of the materials.

by the Powder metallurgy method. In the Bridgman method, the crystallographic direction is aligned by a unidirectional solidification technique and the primary growth direction lies along the <112> (Verhoeven et al. 1987) which is inclined 19 degrees to the easy axis of magnetization <111>. In the Powder metallurgy method, crushed particles of R_2T and RT_2, are sintered after press forming under a magnetic field. This method theoretically aligns the crystallographic direction to <111>.

Microstructures of both kinds of materials are shown in Figure 1. The Bridgman material (BM) is composed of platelike thin aligned grains of widely varying thickness. The Powder metallurgy material (PM) has many pores, and grain boundaries like in BM are not observed.

3 MECHANICAL PROPERTIES

3.1 Compression strength

Specimens for the compression tests were solid cylinders, 8 mm in diameter and 12 mm long. Fractures occurred explosively with light flashes. Test results and weibull plots of compression strengths are shown in Table 1 and Figure 2 respectively. Average compression strength of BM is 894MPa, almost the same as the results in other research (Kondo & Shinozaki 1992). The average strength of PM is 596MPa and about 2/3 of BM.

To describe the effect of pores on the strength of brittle materials, the following equation is frequently used (Duckworth 1953).

Table 1. Results of static strength tests.

	Bridgman	Powder metallurgy
Average compression strength (MPa)	894	596
Weibull shape parameter of compression strength	8.0	23.6
Average radial compression strength (MPa)	31	24

Table 2. Strength at different porosities

		(p = 0.141)
σ/σ_0	Equation	Notes
0.86	(1- p)	——
0.49- 0.72	exp(- Bp)	B=2.3-5
0.62- 0.68	$(1- p)^l$	l =2.5-3.1

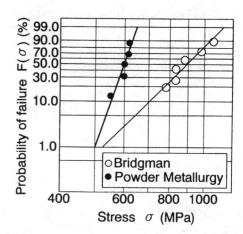

Figure 2. Weibull plots of compression strength for Bridgman and Powder metallurgy materials

$$\sigma(p) = \sigma_0 \exp(-Bp) \qquad (1)$$

Here, σ is the strength of a pore free material, $\sigma(p)$ is the strength with porosity of p, B is a constant which is usually 2.3~5.

The next semi empirical equation is also used (Dutta et al. 1988, Wagh et al. 1993):

$$\sigma(p) = \sigma_0 (1-p)^l \qquad (2)$$

where l is a constant.

Wagh applied equation (2) to the strength data of ceramics without a glassy second phase and obtained l≅2.5~3.1. Table 2 shows calculated results with equation (1) and (2) considering the porosity of the material (14.1%). This table also notes the effect of the pores, which reduce the cross-sectional area. The strength ratio of PM to BM in this experiment was 0.67, which is well explained by the above equations. This indicates that pores influence fracture strength more than reductions in cross-sectional area, with the result that the strength of PM is lower than BM.

As shown in Figure 2, the compression strength of both materials can be described by weibull distributions. The scatter in the strength of PM was much smaller than in BM considering that the weibull shape parameter of PM was about 3 times BM.

Both kinds of materials exhibited some exfoliation (chipping) at the edges of specimens during loading, while there were differences in the chipping behavior of PM and BM. Here PM fractured soon with small increase in load after chipping was observed. Chipping of BM started at 500~600MPa and continued intermittently until bulk fracture. The BM is composed of platelike grains as shown in Figure 1, and the bonding force in the direction perpendicular to the rod axis appears to be smaller than in the longitudinal direction. The thickness of platelike grains vary widely, and the location of grains especially near the edges of specimens where stress is concentrated must affect the scatter behavior and chipping phenomena. Different from BM, PM has a relatively isotropic microstructure, and pores are distributed uniformly in the material. These factors may cause the smaller strength scatter in PM.

3.2 Radial compression strength

The specimens were solid cylinders of 8mm diameter (d) and 4 mm length (l). Radial compression tests were conducted to load the side of the cylinders in the radial direction. In these tests, the nominal tensile stress, $\sigma = 2W/(\pi dl)$, is generated perpendicular to the loading direction (load = W).

Fracture occurred with crack propagation in the loading direction, and the specimens separated into two pieces. Tests results shown in Table 1 express that the radial compression strength is smaller than 1/20 of the compression strength of the two kinds of materials.

3.3 Stress dependence of Young's modulus

Typical stress-strain curves of compression strength tests are shown in Figure 3. The symbols $\bigcirc$, $\bullet$ and $\diamondsuit$, $\blacklozenge$ shows pairs of strains on the same BM and PM specimens respectively. The slopes of the stress strain curve is smaller at low stresses than at high stresses. The stress dependence of Young's modulus is expressed in Figure 4 using the polynomial approximation of Figure 3. Where the applied stress is near zero, Young's modulus of both materials is about 30 GPa, this value is close to data measured by the resonance

Table 3. Young's modulus at different porosities

(p = 0.141)

E(p)/E$_0$	Equation	Notes
0.57-0.75	exp(-Bp)	B = 2 - 4
0.75-0.76	(1-p)/(1+Ap)	A = (1+v)(13-15v)/2/(7-5v), v=0.2-0.4
0.68-0.74	(1-p)^m	m = 2-2.5

method. Young's modulus rapidly increases with stress in the low stress region. The increase below 200 ~ 300 MPa is due to a ΔE effect and is considered to be caused by the rotation of the magnetization. In the high stress region, the Young's modulus of PM is smaller than that of BM, and the PM to BM ratio, E_P/E_B is about 0.75 at 400 MPa.

The next equations are used to describe the effect of pores on Young's modulus.

$$E(p) = E_0\exp(-Bp) \tag{3}$$

$$E(p) = E_0(1-p)(1+Ap) \tag{4}$$

$$E(p) = E_0\exp(1-p)^m \tag{5}$$

E(p) is the Young's modulus of a material with porosity p, and E_0 is the Young's modulus of the pore free material. Equation (3) has been used for a long time (Spriggs 1961), and usual values of the constant B are 2~4. Constant A of equation (4) (Hasselman 1962) relates to the Poisson's ratio v when spherical pores are randomly distributed in the material, as follows:

$$A = (1+v)(13-15v)/2/(7-5v) \tag{6}$$

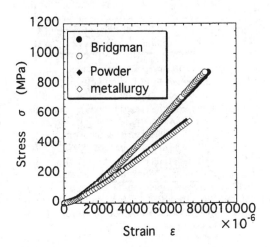

Figure 3. Stress-strain curves for the compression tests

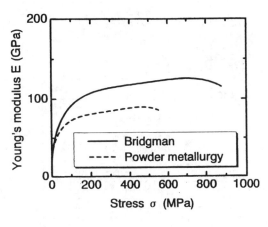

Figure 4. Young's modulus versus applied compression stress

Table 4. Results of compression strength tests of powder metallurgy materials

Form of test pieces	Prism (S) 4.50×4.50×8.00	Prism (L) 7.38×7.38×11.07	Rod φ 8.00×12.00
Aspect ratio	1.8	1.5	1.5
Volume (mm²)	162	603	603
Average porosity (%)	16.4	14.6	14.1
COV of porosity	0.09	0.01	0.07
Average compression strength (MPa)	556	558	596
Weibull shape parameter	15.6	25.3	23.6

Equation (5) was proposed by Wagh using an open-pore model, and for usual ceramics m is nearly equal to 2, while 2~2.5 is obtained for rare earth compounds (Wagh et al. 1991). $E(p)/E_0$ calculated by the above equations are shown in Table 3. These values are almost the same as the E_p/E_B obtained in this experiment, showing that differences in the Young's modulus of PM and BM are largely explained by the porosity.

4 Strength properties of powder metallurgy materials

4.1 *Effect of volume and shape of specimens*

Materials made with powder metallurgy are easily formed into desired shapes, and shapes other than rods are often used. This section details compression strength tests using PM with different configurations and volumes to investigate how differences affect material strength.

Table 4 shows the forms of test pieces and test results. Test results with rods of PM were shown in the previous chapter. The average compression strength of prisms (L) were approximately 7% lower than that of rods. Weibull shape parameters of both materials are very similar and there are no differences in the strength scatter. Because volume and aspect ratios of rods and prisms(L) are equal and the average porosity are very similar, the differences in the shapes of rods and prisms are considered a reason for the different material strengths.

Stress analysis of compression strength tests on rods by Kondo (Kondo & Shinozaki 1992) indicated that stress concentration occurred near the edge of test pieces. This stress concentration is based on differences in compression plates and partial displacement in the test pieces. Thus compression strength depends on configuration, modulus of elasticity, and friction rates of compression plates and test pieces. The lower compression strength of prisms in this experiment may result from larger stress concentrations on the square parts of the test pieces.

Prism(S) and prism (L) showed almost equal average compression strength, and there were no significant differences due to volume differences. However, they demonstrated differences in Weibull shape parameters. In particular, a very large pore of around 500μm in diameter was observed in one of the prism (S) in this experiment which showed especially low compression strength (430MPa). Table 4 excludes this test data, but differences in the Weibull shape parameter of the two prisms is considered to be an effect of pore distribution, particularly when pore size is large.

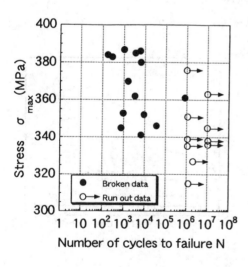

Figure 5. S-N diagram of compression fatigue tests for the powder metallurgy materials

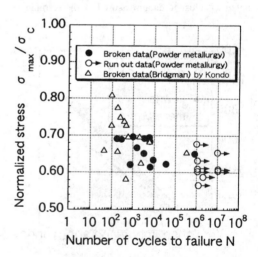

Figure 6. Normalized S-N diagram of compression fatigue tests with powder metallurgy and bridgman materials

4.2 Compression fatigue strength

The test pieces here were prism(S) in Section 4.1. This experiment used a closed loop servo fatigue testing machine and was conducted at a stress ratio R=0.1, and a cyclic frequency of 110Hz.

When test pieces were fractured by fatigue, the situation is the same as static compression tests, fracturing explosively into small pieces with flashes. Figure 5 shows the S-N diagram of compression fatigue tests. Data demonstrated an apparent fatigue limit of ca. 340MPa, about 2/3 of BM (ca. 500MPa). Different from brittle materials such as ceramics, the scatter of life is short, at most 2 decades.

Figure 6 shows the S-N diagram normalized by static compression strength, with test results of BM (Kondo& Shinozaki 1992) also included. Fatigue data of PM and BM distribute in nearly the same region, and the ratio fatigue strength to static compression strength is approximately 60% for both materials. In conclusion, the absolute values of the compression strengths showed some differences but the fatigue strengths of the two materials can be approximately evaluated by the static strengths.

5 CONCLUSIONS

1. Compression strength of powder metallurgy material (PM) was about 2/3 of Bridgman material (BM). The strength scatter of PM was much smaller than BM, considering that the weibull shape parameter of PM was about 3 times that of BM. The smaller compression strength can be explained by the porosity of PM, and the smaller strength scatter and larger weibull shape parameter by the microstructure anisotropy of BM and the uniformly distributed pores in PM.

2. Radial compression strength was almost always less than 1/20 of the compression strength, and both kinds of materials showed very brittle fracture features.

3. Both kinds of materials showed stress dependence on Young's modulus especially in the low stress region, which may be considered a ΔE effect. In the high stress region the stress dependence was small and the Young's modulus of PM was about 3/4 of BM because of its pores.

4. The compression strength of PM differed by configuration: prisms showed lower compression strengths than rods. Within the limits of these experiments, there were no effects of volume on the compression strength.

5. There were apparent fatigue limits for the two materials, and those may be approximately evaluated by the static strengths.

The authors wish to thank Dr. Teruo Mori of TDK Corporation for his valuable advice on giant magnetostrictive materials as well as his generous offer of materials.

REFERENCES

Cark A.E. & H.S.Belson 1972. Giant Room-Temperature Magnetostrictions in TbFe2 and DyFe2. *Phys. Rev. B5*, 3642-3644.

Duckworth W. 1953. *J. Am. Ceram. Soc.* 36-2:68.

Dutta S.K. et al. 1988. Assessment of Strength by Young's Modulus and Porosity. *J. Am. Ceram. Soc.* 71-11:942-947.

Hasselman D.P.H.1962. *J. Am. Ceram. Soc.* 45:452-453.

Kondo K. & K. Shinozaki 1992. Strength of Standard Giant Magnetostrictive Materials. *Proc. of Int. Sympo. of Giant Magnetostrictive Materials and Their Applications*: 63-68.

Spriggs R.M. 1961. *J. Am. Ceram. Soc.* 44:628-629.

Verhoeven J.D. et al. 1987. The Growth of Single Crystal Terfenol-D Crystals. *Metallurgical, Trans*. 18A: 223-231.

Wagh A.S. et al. 1991. Open Pore Description of Mechanical Properties of Ceramics. *J. Mater. Sci.* 26:3862-3868.

Wagh A.S. et al. 1993. Dependence of ceramic fracture properties on porosity. *J. Mater. Sci.* 28:3589-3593.

Fatigue test methods

Experimental Mechanics, Allison (ed.) © 1998 Balkema, Rotterdam, ISBN 90 5809 014 0

Evaluation of fatigue damage by surface SH wave in carbon steel

Yutaka Nagashima
Graduate School, University of Electro-Communications, Chofu & Research and Development Division, TOMOE Company Limited, Chuo, Tokyo, Japan

Yasuo Ochi & Takashi Matsumura
Mechanical and Control Engineering, University of Electro-Communications, Chofu, Tokyo, Japan

ABSTRACT: The majority of the accidents for machines and structures is the cause of the fatigue fracture, so many investigations on the nondestructive inspection have been carried out about the fatigue fracture. As one of the methods, the investigation by using a diffracted surface SH ultrasonic wave (a horizontally shear wave) technique have been started recently. In the present study, the received wave forms during the fatigue process with rotating bending tests of low carbon steel have been measured by using a diffracted surface SH wave technique under three kinds of stress amplitude and two kinds of ultrasonic probe. It was found that changing of the wave area and of the propagation time of the received wave forms depend on number cycles of fatigue, a stress amplitude, fatigue properties and the kind of probe. From these results, the fatigue damage was evaluated in the subsurface of low carbon steel.

1 INTRODUCTION

In order to prevent the fracture of materials destructively, it is important to test deterioration and defects of materials. Therefore, the nondestructive inspection has been used various physical energies or physical phenomena. For examples, physical energies such as a mechanical vibration (a vibration frequency, an acoustics, an ultrasonic wave), an electromagnetic wave, a magnetism and an electricity, and physical phenomena such as an electromagnetic induction, a penetration and a leak. The surface wave technique and the creeping wave technique are taken to test deterioration and defects of materials (H.Fukuoka 1993). These techniques, however, have the following problems. In the surface wave technique, received wave forms tend to have parasitic echoes, increase attenuation by the surface conditions. In the creeping wave technique, the analysis of received wave forms is usually difficult because of transverse waves propagate into the material at the same time (J.D.Achenbach 1991). The feature of the surface SH wave technique includes no influence of the surface conditions and no need to transform the longitudinal wave mode to the transversal wave mode (K.Mano 1962, M.Takahashi et al 1989). Previously, this technique had been required a couplant of high viscosity. The improvement of couplants have lately resulted in brief scanning. For this reason, it is thought that the surface SH wave technique is effective to evaluate the fatigue damage in the subsurface.

Several investigations have been conducted on the use of the surface SH wave. Relatively, the themes of these investigations are the evaluation of property of surface SH wave such as the effect of pulse amplitude on artificial cracks of the material surface, the experimental study of phase shift and comparison between the surface SH wave and the SV wave. The paucity of report on evaluation of fatigue damage using the surface SH wave technique during the rotating bending fatigue tests prompted us to investigate it.

In the present study, received wave forms by using the surface SH wave technique were measured to examine the influence of its wave form modulations and fatigue properties such as propagation curves of the maximum crack length and the variations of the crack density. And these measurement were carried out under a higher, a middle and a lower stress amplitude and two kinds of ultrasonic probe.

2 EXPERIMENTAL PROCEDURES

2.1 Surface SH wave technique

The surface SH wave technique is the inspecting method by using horizontal shear waves propagate along the material surface. As the material changes during the fatigue process influence the pulse amplitude, the wave area, the wave velocity and the propagation depth of the surface SH wave. The major transformation are that surface SH wave forms during the fatigue process extend to the longer time because of as fatigue degree increase, the incidence wave was received as the diffracted wave which is refracted and detoured in direction to depth of the medium. That is, the wave velocity of surface SH wave and the propagation time are very important to evaluate the fatigue damage. In generally, the wave velocity can be expressed as the equation (1) is about 90% of one of a shear wave, determined by a Poisson's ratio and a modulus of rigidity of the medium.

$$Vs = 0.9 \cdot \sqrt{\frac{G}{\rho} \frac{1}{2(1+\upsilon)}} \tag{1}$$

where Vs = a wave velocity of surface SH wave; G = a modulus of rigidity; ρ = a density of medium and ν = a Poisson's ratio.

The surface SH wave technique propagates horizontally shear waves generated by a crystal in direction of 90° through a polymer wedge. The oscillations of particle propagate as transverse waves of horizontal direction. Therefore, the surface SH wave has no influence of condition of test surface and the other modes of ultrasonic wave do not propagate into the medium. For this reason, the surface SH wave is very convenient for the quantitative analysis of microstructual changes just under the materials.

2.2 Experimental apparatus

The experimental apparatus used for generating the diffracted surface SH wave was SONIC SCOPE, for analyzing received wave forms was SONIC ANALYZER made by Toshiba Tungaloy Co. Ltd. The personal computer had built-in these apparatus. Figure 1 shows the schematic drawing of the experimental equipment, Figure 2 shows the flow chart of surface SH wave (Catalog 1995). Water-free glycerin (Sonicoat SH) was used as a couplant.

Two shape types of ultrasonic probes which are receiving diffracted waves were used. The type-A probe is designed to probe the test surface which is the parallel direction to long dimension of the specimen. And the type-B probe is designed to probe the test surface which is the perpendicular direction to long dimension of the specimen. These probes have two crystals with shape of $5 \times 5mm^2$ in the bridge, are used for a transmitter and a receiver. The two crystals are partitioned off by a ditch of 10mm width. The surface SH waves are generated by refraction from a transverse wave PZT [Pb (ZrxTi1-x) O3] transducer to a polymer wedge. The angle of incident and receiver are fixed at 0.332 rad. In order to prevent propagation loss due to high frequency and signal broadening due to low frequency, the frequency of each probes of 1.6MHz was selected.

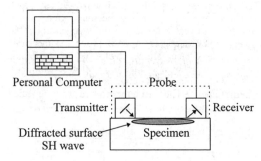

Figure 1. Schematic drawing of experimental equipment.

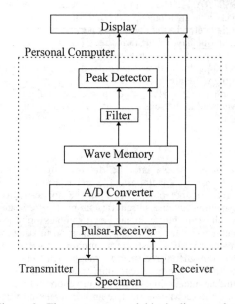

Figure 2. The apparatus and block diagram for measurement.

2.3 Materials and experimental procedures

The material used was JIS SS400 steel, and the chemical composition is shown in Table 1. The specimen of geometry and dimensions shown in Figure 3, was machined. The specimen have a shape of solid rod which circumferential shallow cut introduced in center as shown in Figure 3. The test surface covers within the shallow notch where fatigue crack initiate naturally. After the machining, all specimen were polished by emery paper of #400-1500, which were annealed at 930°C for 1hr in vacuum condition. Then, the center part of test specimens were polished to 1 μ m finish by polishing cloth.

During rotating bending tests, the surface SH wave was measured in area of whole circumference of the center part of the specimen by using each probes. For this purpose, the specimens were rotated on its axis every a 60° using the type-A probe and every a 120° using the type-B probe. At the same time, the surface fatigue cracks of the specimens were observed by a surface replication technique in the same place as the surface SH waves measurement. The same procedure was repeated until the specimen fractured.

Table 1. Chemical composition (mass %).

C	Mn	P	S
0.20	0.10	0.040	0.040

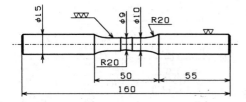

Figure 3. Shape of fatigue test specimen (in mm).

3. RESULTS AND DISCUSSION

3.1 *Behavior of surface fatigue cracks.*

Rotating bending fatigue tests were carried out at room temperature in air. Figure 4 shows as S-N curve. The fatigue limit of the material used was about 255MPa. The stress amplitude used for the surface SH wave measurement was set to as follows; the high stress amplitude = 330MPa, the middle one = 300MPa and the lower one = 270MPa.

Crack behaviors on the surface of the specimen were observed by a surface replication technique. The propagation curves of the maximum crack length in each stress amplitude are shown in Figure 5, and the variations of the crack density are shown in Figure 6. Where, the crack length of more 10 μ m was recognized as the crack in the experiment. And the crack density was defined as the number of cracks growing in unit area (in mm²). The crack initiation is early in order of the higher stress amplitude. All maximum crack length curves show a rapid propagation of increase after the crack initiation, and the specimen resulted to final fracture. All maximum crack length curves just before the fracture are growing about 4000 μ m. For the results, the stress amplitude does not affect the growth of maximum crack length.

On the other hand, the stress amplitude affects the growth of crack density. The crack density is higher in order of the higher stress amplitude at same experiment N/N_f. The results obtained agreed

approximately with those expected. The effect its on the received wave forms of surface SH wave is described in later chapter.

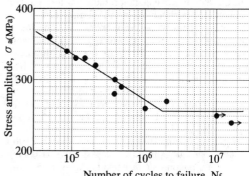

Figure 4. S-N curve.

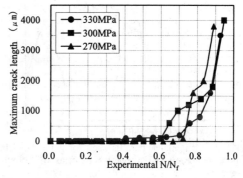

Figure 5. Maximum crack length during fatigue process.

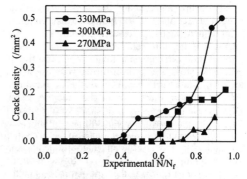

Figure 6. Crack density during fatigue process.

3.2 *Changes of area ratio of received wave forms*

As above description, the wave area which is one of

important element of received wave forms was measured. It is known that beyond the forth wavelets of received wave forms during the fatigue process extend its in direction to the longer time. Therefore, the wave area of the wavelets passing through the specimen during the fatigue process for the parallel and the perpendicular directions to the long dimension was measured from the four wavelets, and the relationship between the wave area and number cycles of fatigue was researched. For example, Figure 7 shows a part of received wave form patterns of output signal measured from the experiment. The wave area is obtained by integral of received wave form curves of output signal, illustrated the area oblique line in Figure 7. The plots are expressed by value of the area ratio which is the quotient the measured wave area (W_i) and the initial one (W_0), considering difference in the specimen.

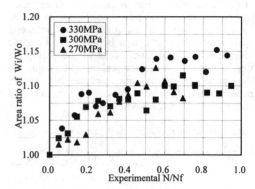

Figure 8. Effect of area ratio of received wave forms for parallel direction during fatigue process.

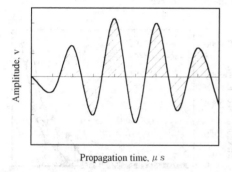

Figure 7. Received wave form patterns of output signal.

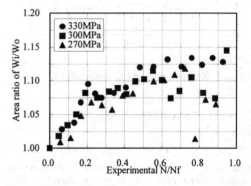

Figure 9. Effect of area ratio of received wave forms for perpendicular direction during fatigue process.

Showing Figures 8-9, the area ratios are nearly straight up to experimental N/N_f :0.40 (330MPa), 0.45 (300MPa) and 0.55 (270MPa), above which they almost are saturated and they are scattered. The increase of area ratio of the higher stress amplitude is bigger than one of the lower stress amplitude. It is thought that the influence of the stress amplitude appeared in the change of the wave area sensitively. Next, researching the relationship between the wave ratios and the surface crack behavior, all wave ratios tend to saturate after the surface fatigue crack initiation. This means that the wave area of received wave forms passing through the specimen with surface fatigue cracks do not increase by a certain ratio, and the product of the pulse amplitude and the propa-gation time almost unchanged. Showing kinds of probes, the result using the type-A probe is better sensitive for the change of area of wave form than the type-B probe, but it can be considered that there is no problem of kind of probe to evaluate fatigue damage. After all, it is difficult to evaluate the fatigue damage only by the change of the wave area.

3.3 Analysis of propagation time

It has been suggested that the propagation time was analyzed by the multiple regression analysis to estimate the degree of fatigue during the fatigue process destructively. In the other words, the received wave forms by using the diffracted surface SH wave techique have the relationship between the propagation time and number cycles of fatigue, but this suggestion had not been considered difference in the stress amplitude and used a kind of ultrasonic probe (M.Fukuhara et al 1996).

In a similar way of method, the received wave forms of the diffracted surface SH wave were measured under the same stress amplitude as the surface fatigue crack behavior and the wave area measurement. The method used for measuring the propagation time during the fatigue process was as follows. The time deviation of the wavelets passing through the specimen during the fatigue process for parallel and perpendicular directions to the long dimension was measured from each peak positions of four wavelets.

The number of measurement and the place of measurement are the same as measurements of the surface fatigue crack behavior and the area ratio. As measurers do not ascertain the positions of surface fatigue cracks, the diffracted surface SH wave in the area of whole circumference of the center part of the specimen was measured, but the propagation time expressed in figures used the average value of the same wavelet of different place of measurement for the parallel and the perpendicular direction at the same number cycles of fatigue. The results of measurement which were obtained in the way described above are shown in Figures 10-11.

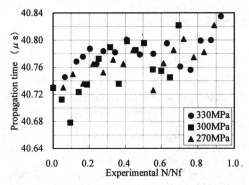

Figure 10. Propagation time calculated from second wavelet using type-A probe.

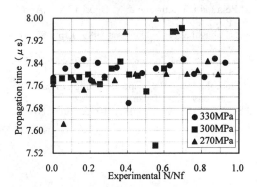

Figure 11. Propagation time calculated from second wavelet using type-B probe.

The figures as the experimental N/N_f increased, the numbers of the propagation time increased further. But, they had no clear relationship because of the propagation time had the large scattering against the number cycles. In order to solve these problems, the propagation time of four wavelets for both directions were analyzed with an aid of multiple regression analysis. The analysis used in the experiment can be expressed by the equation as equation (2).

$$Y = \begin{bmatrix} Y_1 \\ \vdots \\ Y_j \end{bmatrix} \quad X = \begin{bmatrix} X_{1,1} & \cdots & X_{1,j} \\ \vdots & \vdots & \vdots \\ X_{i,1} & \cdots & X_{i,j} \end{bmatrix} \quad and \quad \beta = \begin{bmatrix} \beta_0 \\ \vdots \\ \beta_j \end{bmatrix} \quad (2)$$

where Y_i = predictor variables; X_i = a average of propagation time of peak position; β_j = variables; i =20; and j =4.

The calculated N/N_f from the equation (2) is plotted against the experimental N/N_f in Figures 12-13. These figures are divided by using the kind of probe. The correlation coefficients of all multiple regression analysis are shown in Table 2.

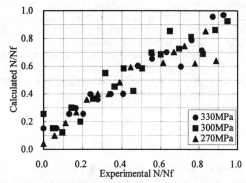

Figure 12. Relation between calculated N/N_f by mult- iple regression analysis and experimental N/N_f using type-A probe.

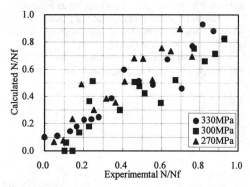

Figure 13. Relation between calculated N/N_f by mult- iple regression analysis and experimental N/N_f using type-B probe.

Table 2. Correlation coefficient of the multiple regression analysis.

Stress amplitude	Type-A probe	Type-B probe
330MPa	0.889	0.698
300MPa	0.838	0.621
270MPa	0.700	0.618

The results of both figures in common are as follows. The calculated N/N_f are roughly straight changing against the experimental N/N_f, and the calculated N/N_f is higher than experimental ones. In the other words, the calculated N/N_f is the safe values for the fracture of materials. In the detail, the scattering of the calculated N/N_f in range of the experimental N/N_f:0-0.4 are less than beyond the experimental N/N_f:0.4. These results were the same tendency as the area ratio shown in Figures 7-8. Thus, after the surface fatigue crack initiation, the calculated N/N_f scattered. These analytical results were affected remarkably by the surface fatigue crack behaviors. Specifically, the scattering of the calculated N/N_f depends on the stress amplitude. There were the relationships between the result of propagation time by the multiple regression analysis and the crack density. The maximum crack lengths were not affected very much. As can be seen from Table 2, using the type-A probe can good analyze in comparison with the type-B probe. It was considered that the good scanning method was that the surface SH wave was propagated in perpendicular direction to surface fatigue cracks.

4 CONCLUSIONS

The nondestructive inspection by using diffracted surface SH wave technique were carried out during the fatigue process in a low carbon steel. From viewpoint of the propagation time and the area of wave form on received wave forms, the fatigue damage was evaluated in the subsurface of low carbon steel. The results were summarized as follow.
1. Of the major transformation of the surface SH wave, it was confirmed that the propagation time extends to the longer time experimentally.
2.The scanning method of the test surface which was parallel direction to long dimension of the specimen was more advantageous than one for perpendicular direction.
3. The area ratio of received wave forms changed during the fatigue process, and tendered to increase with increasing the stress amplitude.
4.The wave area of received wave forms by the multiple regression analysis were increased rough straightly against the experimental N/N_f:0-0.4, and after that it almost saturated .
5.The propagation time of received wave forms by the multiple regression analysis were depended on the stress amplitude and the surface fatigue crack density.
6.The calculated N/N_f of propagation time by the multiple regression analysis increased straightly in proportion to the experimental N/N_f.

5 REFERENCES

A.J.Dobson 1988, An Introduction to Statistical Modelling, Chapman and Hall, London.
Catalog of Ultrasonic SH Wave Diagnosis and Analyzing System for Deterioration 1995, Hardening and Fatigue of Materials Surface Depth, by Toshiba Tungaloy Co. Ltd, Japan.
K.Mano 1962, Railway Technical Research Report, 276:44, Japan.
M.Fukuhara & A.Sanpei 1994, Phys.Rev.B., 49: 13099.
M.Fukuhara, Y.Kuwano & M.Oguri 1996, J.Appl. Phys., 35:418.
N.Takahashi & N.Hatakeyama 1989, J.Jpn. Soc.Nondest Inspect, 38:823.
R.B.King & C.M.Fortunko 1983, J.Appl.Phys., 54, 54(6):3027.
H.Fukuoka 1993, Basis and application of aacoustoelasticity, ohm, Japan.
J.D.Achenbach & I.Komsky 1991, Mater.Eval., Aug., 977.

Experimental Mechanics, Allison (ed.) © 1998 Balkema, Rotterdam, ISBN 90 5809 014 0

Measurements of step height at grain boundary under cyclic loadings by means of AFM

J. Ohgi, K. Hatanaka & T. Zenge
Yamaguchi University, Japan

ABSTRACT : The push-pull low cycle fatigue tests were performed and the step height formed at grain boundary, which is precursor of crack initiation, was measured in annealed copper, using Atomic Force Microscope (AFM). According to the AFM observations, the steps were easily generated at grain boundaries inclined at angles ranged from $50°$ to $80°$ with respect to the loading axis at first loading. Their heights increasingly grew with progress in the cyclic loading process. The ranges of variation in step height during one load cycle increased with increase in number of loading cycles. The growth behavior of the step was successfully evaluated by the cumulative effective plastic range $\sum \Delta \varepsilon_{pe}$. The critical step height at fatigue crack initiation was successfully assessed by a parameter σ_c^* which is concerned with the stress concentration at the step-root.

1 INTRODUCTION

Fatigue crack initiation at grain boundary has been investigated and several mechanisms of the crack initiation at the grain boundary has been proposed through observations of crack initiation by means of optical and/or scanning electron microscopes in early papers[1],[2]. The requirements for the crack initiation at grain boundary, however, is not quantitatively clarified yet.

Kim et al.[1],[2] investigated the step-growth at grain boundary which leads to crack initiation in high strain fatigue of OFHC copper, using optical microscopy and interferometry. They illustrated that a step was gradually built up at an active grain boundary up to be approximately $1.5 \mu m$ high, at which the stress concentration is sufficient to propagate a crack, and that the step always appeared to be higher on tension side than on compression side in one cycle. However, the nano-order load cycle dependent variations in specimen surface geometry could not be measured precisely at the crack initiation sites, using the interferometry. Additionally, such quantitative measurements of the step height were not correlated with the external loading conditions.

In the present study, the nano-order development of step height with progress in the cyclic loading process was measured by means of Atomic Force Microscope (AFM). Then the quantitative requirements for the crack initiation at the grain boundary were discussed by quantifying the development of the step as functions of nominal stress, plastic strain range and a parameter being considered to express fatigue damage.

2 EXPERIMENT

Polycrystalline 99% purity copper was chosen as the test material. Specimen was machined into the geometry shown in Figure 1. The specimens were heat-treated at $940°C$ for 9 hour in a vacuum after straining of 3%. The resultant grain size was about $320 \mu m$ in average. After the specimen was electropolished in the gage portion, grids were scribed on its surface (Figure 2) to facilitate locating successfully the same portion in replicas taken at various stages in fatigue process. The push-pull low cycle fatigue tests were performed at nominal stress amplitudes $\Delta \sigma_n / 2 = 80$, 100 and 120MPa under load-controlled condition of load ratio $R_\sigma = -1.0$, where the stress rate was $\dot{\sigma} = 80$MPa/sec. The axial strain was measured with the extensometer of 12mm in gage length.

Acetyl-cellulose film of 0.034mm thickness was used for the replication. The replica films were removed from the specimen surface at proper intervals during the fatigue tests. The fatigue tests were continued until the main crack reached about $1000 \mu m$ in length.

The nano-order step height formed at grain boundary was measured by scanning the surface of the replica films taken from specimen by means of AFM (Nanoscope- II : Digital Instrument Co.) of contact mode type. The maximum size of scanning area was $15 \mu m \times 15 \mu m$.

3 RESULTS AND DISCUSSIONS

3-1 *AFM observations*

The step height was measured at grain boundary with increasing number of cycles. A typical variation of surface geometry with progress of stress cycling at $\Delta\sigma_n / 2 = 80$MPa is shown in Figure 3. According to these figures, a small step was formed at N=10 cycles (Figure 3(b)), and its height at grain boundary increased gradually with stress cycling. At N=50000 cycles, neighboring two grains were separated at the grain boundary in the vicinity of the specimen surface due to crack initiation (Figure 3(d)). The surface profile at A-A' section shown in Figure 3 was illustrated in Figure 4, where the step height at grain boundary is defined by H. The growth of step height with progress in cyclic loading process also can be confirmed from this figure. Thus crack initiation was preceded by growth of step at an active grain boundary.

3-2 *Requirement for crack initiation*

The fatigue cracks generated at grain boundary and the adjoining grains were observed using SEM. According to SEM observations, the intense slip trace was developed in both or either of two grains adjacent to the grain boundary at which step develops well and crack originates. The inclined angles of such a grain boundary with respect to the loading axis and slip trace formed in adjoining grains, α and β, and the angle of the slip trace inclined with respect to the direction of maximum shear stress, which has inclination angle of $45°$ to loading axis, γ were measured. Such angles were illustrated in Figure 5. The distributions of the number of intergranular cracks are shown as functions of α, β and γ in Figures 6, 7 and 8. According to these figures, crack initiates much easier at grain boundaries inclined at angles ranged from $50°$ to $80°$ with respect to loading axis. It is noted that cracks originate concentrically at grain boundaries inclined at around $90°$ for β. This suggests that the formation of step at grain boundary might be closely associated with the slip inclined at high angle to the grain boundary.

The specimens were longitudinally sectioned to investigate the mode of fatigue crack propagation in inner direction. The section was ground step by step up to the initiation site. Figure 9 is the optical microscope observation in the longitudinal section, where crack growing along grain boundary into the inside is shown. It was confirmed from such observations that crack tends to originate at angles from $70°$ to $90°$ with respect to specimen surface.

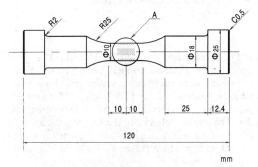

Figure 1 Shape and dimensions of specimen.

Detail of A in Figuer 1

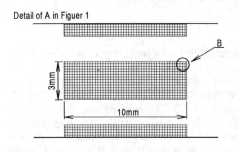

Detail of B

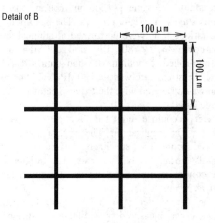

Figure 2 Grid pattern scribed in gage portion of specimen.

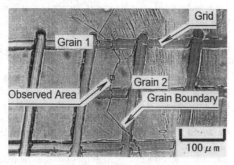

(a) General view of AFM observation area.

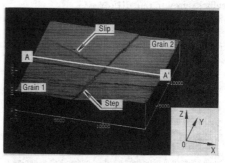

(b) At N=10 cycles.

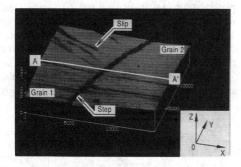

(c) At N=10000 cycles.

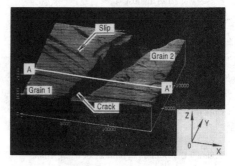

(d) At N=50000 cycles.

Figure 3 Development of step at grain boundary at which crack initiated at $\Delta\sigma_n / 2 = 80$MPa.

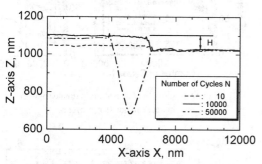

Figure 4 Surface profile at A-A' section in Figure 3.

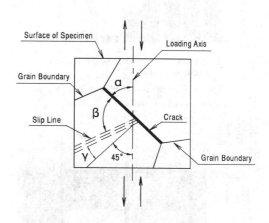

Figure 5 Definition of angles α, β and γ.

3-3 Growth behaviors of steps at grain boundary

AFM observations of replica film taken from the specimen surface were performed at proper intervals in cyclic loading process. The growth of step at grain boundary in the cyclic loading process was shown in Figure 10, which was measured at the point A on the hysteresis loop inserted in the figure. Crack initiation point is shown by an arrow in the figure. According to this figure, the step grows rapidly at early stage with increase in number of cycles, decreasing

1055

gradually its growth rate dH/dN.

The variations of the step height during one load

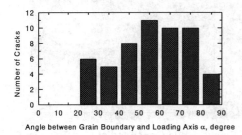

Figure 6 Distribution of number of intergranular cracks against α.

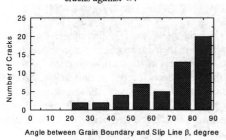

Figure 7 Distribution of number of intergranular cracks against β.

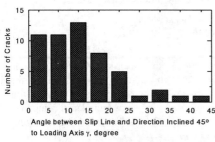

Figure 8 Distribution of number of intergranular cracks against γ.

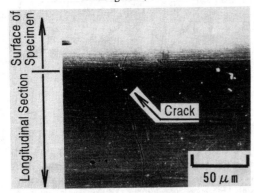

Figure 9 Intergranular crack in the longitudinal section.

cycle were measured at proper number of cycles and illustrated in Figure 11. The step height range Δ H which is defined as the difference between step heights at the points B and D on the hysteresis loop increases gradually with stress cycling.

It is well known that the reversibility of slip induced by cyclic stressing increases with progress in fatigue process. The gradual decrease in dH/dN and increase in Δ H with increase in number of cycles shown in Figs.10 and 11 have a tendency similar to such a cyclic slip behavior. Thus grain boundary step formation is closely related to slip deformation in grains adjacent to the grain boundary.

One of the authors proposed a component of plastic strain dominating fatigue damage[3]. This is defined as effective plastic strain range $\Delta\varepsilon_{pe}$ in Figure 12, where unloading and loading profile from the compressive tip in the hysteresis loop is presented as the relationship between stress σ' and plastic strain ε'_p on double logarithmic scales. The cumulative effective plastic strain range $\sum\Delta\varepsilon_{pe}$ was estimated

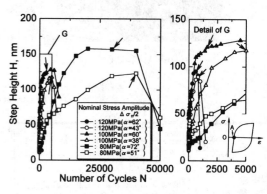

Figure 10 Growth of step at grain boundary with increase in number of cycles.

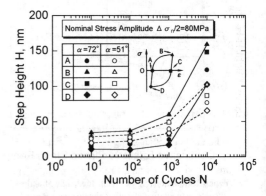

Figure 11 Variation of step height during one loading cycle.

1056

from the stress-strain hysteresis loops obtained in cyclic loading process. Development of the grain boundary step was successfully plotted against this parameter in Figure 13.

3-4 Step height generated by first tension

The step height generated at first tensile loading, H_1 was plotted against the plastic strain yielded during that loading excursion in Figure 14. Although data has quite large scatter, the average value of the step height tends to increase with increase in the plastic strain. Incidentally, H_1 is independent of α, β and γ under all stress conditions.

3-5 Step height at crack initiation

The critical step height at crack initiation period, H_c was shown as a function of applied stress amplitude in Figure 15. The step heights decreased with

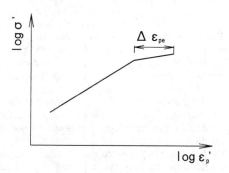

Figure 12 Definition of effective plastic strain range, where σ' and ε'_p are stress and plastic strain on the upper hysteresis loop profile estimated from its compressive tip.

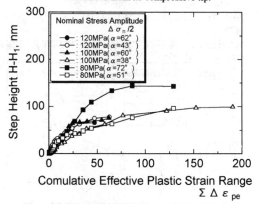

Figure 13 Relationship between step height H and cumulative effective plastic strain range $\sum \Delta\varepsilon_{pe}$.

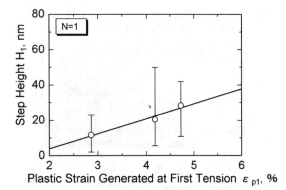

Figure 14 Step height formed by first tensile load excursion.

increase in stress level in the figure. Kim et al.[1] presented in their earlier paper that crack initiation is caused by stress concentration generated at the root of the step which reached critical height. Since the critical stress at the root of the step for crack initiation is the same under all applied stress conditions, the critical step height might be in inverse proportion to the applied stress.

Marsh[4] proposed that the stress concentration factor K_T at the step-root is given by the following equation in the infinite plate,

$$K_T = 1 + f(\theta)\sqrt{H/\rho} , \qquad (1)$$

where $f(\theta)$ is non-dimensional function of θ, and ρ is the step root-radius. Geometrical correlation among them is presented in Figure 16. According to this paper, provided ρ is quite small compared with H, then $f(\theta)\sqrt{H/\rho} >> 1$. The following equation,

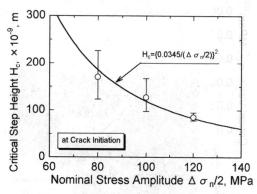

Figure 15 Stress amplitude dependency of step height at crack initiation.

1057

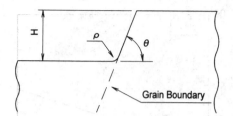

Figure 16 Definition of quantities associated with
stress concentration at step-root.

$$K_T = f(\theta)\sqrt{H/\rho}$$ (2)

is derived at that time. The angle θ is approximately
$90°$ as mentioned in the section 3-1, and ρ seems to
be constant, independent of H and applied stress level.
Then the following equation,

$$f(90°)/\sqrt{\rho} = C = constant$$ (3)

holds.
Consequently, the stress concentration factor and the
concentrated stress are given at the step-root by

$$K_T = C\sqrt{H}$$ (4)

and

$$\sigma_c = K_T\sigma_n = \sigma_n C\sqrt{H_c} .$$ (5)

Since the critical stress value at the step-root σ_c is
considered constant at crack initiation under all
applied stress conditions, the equation

$$\sigma_c^* = \sigma_c / C = \sigma_n\sqrt{H_c} = constant$$ (6)

is derived.

The values of σ_c^* were calculated from eq.(6) and
plotted against nominal stress amplitude in Figure 17.
It is found from the figure that σ_c^* was nearly
constant, independent of applied stress amplitude,
showing successful prediction of the critical step
height by eq.(6). The validity of eq.(6) is also
confirmed from the solid line expressing the
relationship between H_c and $\Delta\sigma_n / 2$, which is
calculated through this equation, in Figure 15.

4 CONCLUSIONS

The growth of step was examined at grain boundary
under load controlled-low cycle fatigue by means of
AFM in annealed copper. Then the development of
grain boundary-step was evaluated in terms of stress,
strain and effective component of plastic strain. The
main results obtained are as follows.

(1) The intense slip was developed in both or either
    of two grains adjacent to the grain boundary at
    which the step grows greatly and crack initiates.
    Such a grain boundary is inclined at angles ranged
    from $50°$ to $80°$ with respect to the loading axis.

(2) The step heights H increased with increase in
    number of cycles, reducing gradually its growth
    rate dH/dN. Meanwhile, the range of variation of
    step height during one stress cycle Δ H increases
    with the progress in cyclic loading process.

(3) The growth of grain boundary-step was
    successfully evaluated by the cumulated effective
    plastic strain range $\sum \Delta\varepsilon_{pe}$ proposed by one of
    the authors.

(4) The parameter σ_c^* was proposed for assessing the
    critical step height at crack initiation, based on
    the stress concentration at the step-root.

REFERENCES
(1) W.H.Kim, C.Laird : Acta Metallurgica, Vol.26, p.777,
    1978.
(2) W.H.Kim, C.Laird : Acta Metallurgica, Vol.26, p.789,
    1978.
(3) K.Hatanaka, T.Yamada, Y.Hirose :Bull. of JSME, Vol.23
    No.180 p.791, 1980.
(4) D.M.Marsh : Fracture of Solids, Interscience, New York,
    1962.

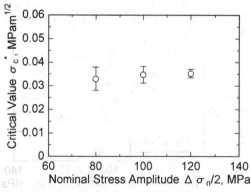

Figure 17 Parameter σ_c^* plotted against applied
stress amplitude.

Experimental Mechanics, Allison (ed.)© 1998 Balkema, Rotterdam, ISBN 90 5809 014 0

Luminance decline and microfracture during fatigue loading in notched FRP plates

H. Hyakutake & T. Yamamoto

Department of Mechanical Engineering, Fukuoka University, Japan

ABSTRACT: The validity of a fatigue failure criterion based on the idea of severity near the notch root of notched FRP plates is investigated experimentally. This is accomplished by obtaining experimental data of fatigue tests of pulsating tension on a glass fiber-reinforced polycarbonate containing notches for a wide range of notch-root radii and notch depths. To evaluate the fatigue damage near the notch root, we measured the luminance distributions by means of the luminance-measuring technique using a CCD camera. Closer observation of the microfracture in the fatigue-damage zone revealed the process of the luminance decline in the damaged zone along with the increase of loading cycles. The experimental result shows that the initiation of the fatigue damage was governed predominantly by both the notch-root radius and the maximum elastic stress at the notch root and it is independent of notch depth.

1 INTRODUCTION

There is considerable literature on the fatigue fracture for notched specimens of fiber-reinforced composite materials. The attention in these studies is mostly on revealing the mechanism of fatigue damage development at the notch root. Our goal is to elucidate the fracture behavior of fiber-reinforced plastics (FRP) containing stress concentrations in various notch geometries and to develop a limiting condition for predicting the fatigue strength of notched bars of FRP.

Studying stress distribution near the notch root, we have obtained a fracture criterion for notched bar of engineering plastics having intermediate notch-root radii (Nisitani & Hyakutake 1985). The criterion is based on the idea of severity near the notch root. Several experiments have shown that the criterion is applicable to not only notched bars of engineering plastics (Hyakutake et al. 1987) but notched FRP plates (Hyakutake et al. 1989, 1990) over a wide range of notch geometries and dimensions of specimens.

The aims of the present research are to provide experimental evidence of the validity of the fracture criterion based on the severity near the notch root for the fatigue failure of notched FRP plates. This is accomplished by obtaining experimental data of fatigue tests of pulsating tension on a glass fiber-reinforced polycarbonate containing notches for a wide range of notch-root radii and notch depths.

To evaluate the fatigue damage near the notch root of notched FRP plates, we measured the luminance distributions with CCD camera. Closer observation of the microfracture in the fatigue-damage zone was made by a microscope with the transmitted mode to reveal the decline of luminance in the damaged zone near the notch root.

2 THEORY

The severity near the notch root is determined by both the maximum elastic stress and the notch-root radius. It is suggested that the elastoplastic stress distributions near the notch root after small-scale yielding are the same in all specimens, for which both the maximum elastic stress and the notch-root radius are equal in all cases.

On the basis of the evidence mentioned above, the fracture criterion for a notched bar under static load is expressed as (Hyakutake et al. 1989)

$$\sigma_{max} = \sigma_{max,c}(\rho),\qquad(1)$$

where σ_{max} is the maximum elastic stress at fracture and is determined as the product of the nominal stress σ_n and the geometrical stress concentration factor K_t. The parameter $\sigma_{max,c}$ on the right-hand side of eq. (1) is the material constant, which is governed by the notch-root radius ρ only and is independent of notch geometry and specimen size.

To verify the validity of the idea of severity near the notch root for composite laminates, stress distributions near the notch root of a thin, orthotropic, notched plate in tension were examined, and an example of the results is shown in Fig. 1. This figure is drawn from the result of an analysis using the finite element method. The coordinates 1 and 2 are principal axes of the lamina. Material constants are as follows: Young's modulus $E_1 = 5.71$ GPa, $E_2 = 2.15$ GPa, Poisson's ratio $\nu_{12} = 0.33$, and the shear modulus $G_{12} = 1.31$ GPa (Hayashi 1959).

Figure 1 shows that the relative elastic stress distribution near the notch root is governed predominantly by the notch-root radius ρ, and it is independent of the notch depth a in the range of $a / \rho > 1$. We obtained the same result as Fig. 1 for an isotropic notched plate.

In view of the idea of severity near the notch root mentioned above, it is appropriate to discuss the fatigue failure criterion for notched plates in terms of a combination of the maximum elastic stress σ_{max}, the notch-root radius ρ and the number of loading cycles to fatigue failure N_d. It is reasonable to assume that the fatigue failure criterion for a notched FRP plate in cyclic loading is expressed as (Hyakutake et al. 1993)

$$\sigma_{max} \cdot (N_d)^m = C(\rho), \tag{2}$$

where m is the material constant. The parameter C on the right-hand side of eq. (2) is the material constant, which is governed by the notch-root radius ρ only. It should be noted that N_d in eq. (2) is not the cycles to fracture.

3 EXPERIMENTAL PROCEDURE

The material used was a glass fiber-reinforced polycarbonate (GF/PC). Idemitsu "Toughlon" PC was applied to the matrix. The plates of GF/PC were made by injection molding. The dimensions of the plates were 70 mm width, 270 mm long, and 3 mm thick. The plate contains 30 % E-glass fiber (about 0.013 mm in diameter) by weight.

Photomicrograph of a transverse section of the

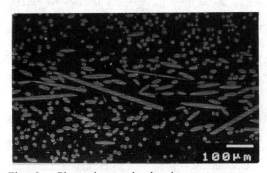

Fig. 2. Photomicrograph showing a transverse section of the GF/PC plate.

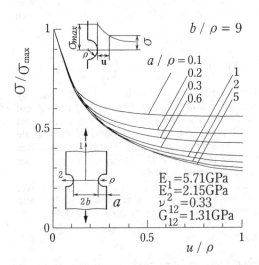

Fig. 1. Relative elastic stress distribution near the notch root for an orthotropic lamina in tension.

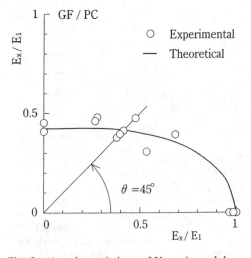

Fig. 3. Angular variations of Young's modulus.

GF/PC plate is shown in Fig. 2. Angular variations of Young's modulus of the GF/PC plates are represented in Fig. 3, where θ is the angle between the principal direction of the specimen and the longitudinal direction of the plate. The theoretical curve in Fig. 3 is determined by the primary theory of an orthotropic lamina. The principal axis 1 coincides with the longitudinal direction of the plate.

Unnotched specimen of the GF/PC plate fails in a brittle manner in tension test. The tensile strength was about 120 MPa for which $\theta = 0°$ and 76 MPa for which $\theta = 90°$.

All specimens were cut from the plate, so that the principal direction of the specimen coincided with the longitudinal direction of the plate. The shapes and dimensions of the notched specimens are shown in Fig. 4. The specimens were notched in a U-shape on both sides at the midpoint of their length. The notch-root radius ρ had the following four values: 0.2, 0.5, 1 and 2 mm. The notch depth a had two values: 4 and 8 mm. The width of the minimum section of notched plate $2b$ is 12 mm in all specimens. The geometrical stress concentration factor K_t is 2.21

to 6.41 for tension.

Fatigue tests of pulsating tension were made on a servohydraulic material test system at frequency 2 Hz with an R value of 0.1 at room temperature.

To evaluate the fatigue damage, we measured successively the luminance distributions near the notch root during fatigue test. The luminance-measuring system with CCD camera is shown in Fig. 5.

4 RESULTS AND DISCUSSION

Figure 6(a) shows the S-N curve for notched FRP plates in fatigue tests of pulsating tension. The notch-root radii were varied from 0.2 mm to 2 mm while the notch depth a was 4 mm in all specimens. The coordinate σ_n is the maximum nominal stress and N_f is the number of cycles to fracture. It can be seen from Fig. 6(a) that there is little effect of the notch-root radius ρ on the

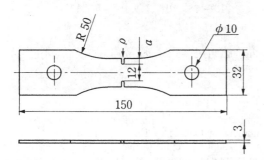

Fig. 4. Test-specimen dimensions (mm).

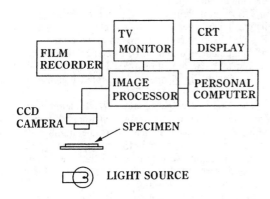

Fig. 5. Luminance-measuring system.

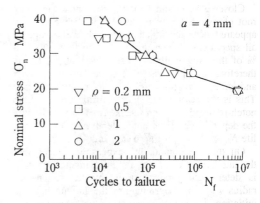

(a) Notch depth $a = 4$ mm.

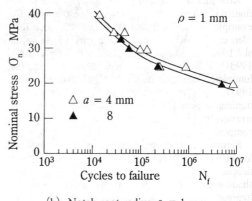

(b) Notch-root radius $\rho = 1$ mm.

Fig. 6. S-N curve for notched GF/PC plates.

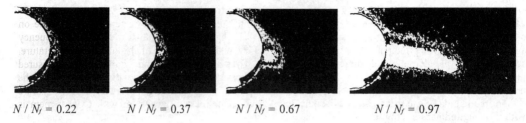

| $N / N_f = 0.22$ | $N / N_f = 0.37$ | $N / N_f = 0.67$ | $N / N_f = 0.97$ |

Fig. 7. Luminance distributions near the notch root ($\rho = 1$ mm, $\sigma_n = 39.2$ MPa, $N_f = 1.34 \times 10^4$).

fatigue life N_f. It seems likely that the notch sensitivity decreases in fatigue tests of notched FRP plates (Maier et al. 1987).

Fig. 6(b) shows another S-N curve for notched FRP plates in fatigue tests. The notch depths were varied from 4 mm to 8 mm while all specimens had a constant notch-root radius ($\rho = 1$ mm). It can be seen from Fig. 6(b) that there is the effect of the notch depth a on the fatigue life N_f.

Closer observation on the surface near the notch root of the specimen revealed that fatigue damage appeared near the notch root in an early stage in all specimens; it occurs within the first 10 to 20 % of the fatigue life. Most of the fatigue life is therefore the process of fatigue-damage growth and crack growth (Fig. 11) as described later. This is the reason why there is little effect of the notch-root radius ρ on the fatigue life and there is the dependence of the notch depth a on the fatigue life N_f, as shown in Fig. 6(a) and (b).

On the other hand, the experiment shows that the number of cycles to fatigue damage initiation is determined predominantly by the notch-root radius ρ. For determination the fatigue damage initiation, it is necessary to evaluate the damage near the notch root quantitatively.

Attempts to determine the fatigue damage for fiber-reinforced composite materials have used, for example, the delaminated area near the notch root (Jen et al. 1993), X-ray radiographs (Spearing et al. 1991) and ultrasonic detection (Kaczmarek 1995). We observed the decrease of the luminance near the notch root (Hyakutake et al. 1993). It is evident that the decrease of luminance near the notch root was associated with irreversible damage and microfracture of composites.

Figure 7 shows an example of the change of luminance distributions near the notch root during fatigue test. The striped patterns with light and shade correspond to the value of luminance. The luminance distributions in Fig. 7 are made up by four values of the relative luminance (R.L.): 60 %, 65 %, 70 % and 75 %. The value of relative

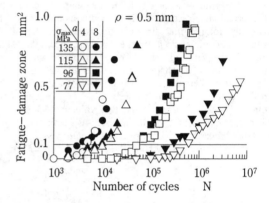

Fig. 8. Growth of the area of fatigue-damage zone.

luminance is the ratio of the luminance at a number of loading cycles to the luminance before testing. It is evident that the fatigue damage accumulated severely at the region where the value of R.L. is small. As described later, there is a main fatigue crack in the damaged zone near the notch root (Fig. 11).

The area of fatigue-damage zone increased with increasing number of loading cycles, as shown in Fig. 7. Examples of the growth curve of the area of fatigue-damage zone of the relative luminance, R.L. = 60 % are shown in Fig. 8. For all specimens in Fig. 8, there is the rapid increase of the area of fatigue-damage zone.

The fatigue-damage initiation N_d was determined from the experimental results mentioned above (Fig. 8). We assumed that N_d is the number of cycles at the area of fatigue-damage zone = 0.1 mm². It can be seen from Fig. 8 that the value of N_d depends on the maximum stress σ_{max} for a constant notch-root radius.

Fatigue-damage initiation N_d is relatively small compared to the fatigue life. Most of fatigue life

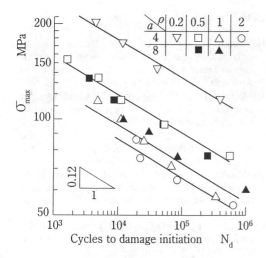

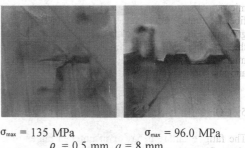

$\sigma_{max} = 135$ MPa $\sigma_{max} = 96.0$ MPa
$\rho = 0.5$ mm, $a = 8$ mm

0.02 mm

$\sigma_{max} = 103$ MPa $\sigma_{max} = 72.5$ MPa
$\rho = 1$ mm, $a = 4$ mm

Fig. 9. Effect of the notch-root radius ρ on the relationship between the maximum elastic stress σ_{max} and the number of loading cycles to fatigue damage initiation N_d.

Fig. 10. Microcracks in the damaged zone near the notch root.

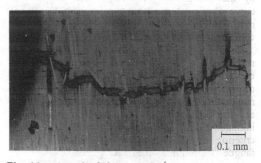

0.1 mm

Fig. 11. A main fatigue crack ($\rho = 0.5$ mm, $\sigma_{max} = 96.0$ MPa, $N = 6 \times 10^5$).

N_f is therefore the process of fatigue-damage growth. This is the reason why there is little effect of the notch-root radius ρ on the fatigue life N_f as shown in Fig. 6(a). On the other hand, it seems likely that the number of loading cycles to fatigue-damage initiation N_d is determined predominantly by the notch-root radius ρ.

Figure 9 shows the relationship between σ_{max} and N_d for a constant notch-root radius ρ. There are four kinds of ρ. As seen in Fig. 9, each experimental point fell in close proximity to a characteristic straight line for which ρ is constant and the four characteristic straight lines are parallel. It should be emphasized that the characteristic straight line is determined by notch-root radius ρ and is independent of notch depth a.

From the experimental results mentioned above, we will confirm the validity of the fatigue failure criterion of eq. (2) derived from the extension of eq. (1), which is based on the idea of severity near the notch root. The value of material constant m is 0.12 for the FRP used in our research. It is evident from the present studies that for any specimens, the number of loading cycles to fatigue-damage initiation N_d can be determined. The value of N_d is determined by both the maximum elastic stress σ_{max} ($= K_t \cdot \sigma_n$) and the notch-root radius ρ.

It is likely that the parameter C on the right-hand side of eq. (2) corresponds to the area of fatigue-damage zone near the notch root. It seems likely the area of fatigue-damage zone at N_d is governed predominantly by the notch-root radius ρ only and is independent of the stress and notch depth, as shown in Fig. 8.

To reveal the growth of microfracture along with the luminance decline, we observed the transverse section of the damaged zone near the notch root. Typical examples of microcracks in the damaged zone are shown in Fig. 10. The microcracks initiated the end corner of a glass fiber. Several microcracks joined together and grew up a main crack (Fig. 11). It is likely that the process of the initiation and growth of microcracks is independent of notch geometries and stress as shown in Fig. 10.

5 CONCLUSIONS

The fatigue failure behavior of notched plates of a glass fiber-reinforced polycarbonate was studied over a wide range of notch-root radii and stress amplitudes. The luminance distribution near the notch root was measured successively during fatigue test to evaluate the fatigue damage. It is evident that the decrease of luminance near the notch root was associated with irreversible damage and microfracture.

Closer observation on the transverse section of the fatigue-damage zone near the notch root revealed that the microcracks initiated the end corner of a glass fiber. Several microcracks joined together and grew up a crack. The process of initiation and growth of the microcracks produced the decline of luminance near the notch root.

We determined the fatigue damage initiation N_d by the luminance distributions and the growth curve of the fatigue-damage zone near the notch root. It was found that N_d was relatively small compared to the fatigue life.

On the basis of the idea of severity near the notch root, the experimental results can be explained and a fatigue failure criterion is determined in terms of a combination of the maximum elastic stress σ_{max}, notch-root radius ρ and the number of cycles to the fatigue-damage initiation N_d. Applying the fatigue failure criterion derived here, we can estimate of the fatigue failure for notched FRP plates.

REFERENCES

Hayashi, T. 1959. On the tension in an orthogonally aeolotropic strip with a circular hole. *Trans. Jpn. Soc. Mech. Eng.* (*in Japanese*), 25-159: 1125-1133.

Hyakutake, H. & H. Nisitani 1987. Conditions for Ductile and brittle fracture in notched polycarbonate bars. *JSME Int. J.*, 30-259: 29-36.

Hyakutake, H., H. Nisitani & T. Hagio 1989. Fracture criterion of notched plates of FRP. *JSME Int. J.*, Ser. I 32-2: 300-306.

Hyakutake, H., T. Hagio & H. Nisitani 1990. Fracture of FRP plates containing notches or a circular hole under tension. *Int. J. Pres. Ves. & Piping* 44-3: 277-290.

Hyakutake, H., T. Hagio & T. Yamamoto 1993. Fatigue failure criterion for notched FRP plates. *JSME Int. J.*, Ser. A 36-2: 215-219.

Jen, M.H.R., Y.S. Kau & J.M. Hsu 1993. Initiation and propagation of delamination in a centrally notched composite laminate. *J. Compos. Mater.* 27-3: 272-302.

Kaczmarek, H. 1995. Ultrasonic detection of damage in CFRPs. *J. Compos. Mater.* 29-1: 59-95.

Maier, G., H. Ott, A. Protzner & B. Protz 1987. Notch sensitivity of multidirectional carbon fibre-reinforced polyimides in fatigue loading as a function of stress ratio. *Composites.* 18-5: 375-380.

Nisitani, H. & H. Hyakutake 1985. Condition for determining the static yield and fracture of a polycarbonate plate specimen with notches. *Eng. Fract. Mech.* 22-3: 359-368.

Spearing, M., P.W.R. Beaumont & M.F. Ashby 1991. Fatigue damage mechanics of notched graphite-epoxy laminate. *Composite Materials: Fatigue and Fracture.* T.K. O'Brien (ed.). ASTM STP 1110, p.617-637. Philadelphia, ASTM.

Experimental Mechanics, Allison (ed.) © 1998 Balkema, Rotterdam, ISBN 90 5809 014 0

Evaluation of low cycle fatigue damage using laser speckle sensor

Akira Kato
Department of Mechanical Engineering, Chubu University, Kasugai, Aichi, Japan

Mitsue Hayashi
Graduate School, Chubu University, Kasugai, Aichi, Japan

ABSTRACT: We investigated a method to evaluate low cycle fatigue damage of metal materials without contact using laser speckle sensor. One of the objectives of this study is that we develop a method to evaluate fatigue damage or maximum stress subjected to the material so far when several times of very large magnitude of loading has been applied like an earthquake. Slipbands are produced on metal surface after this kind of low cycle cyclic loading. Surface property change invoked by slipbands is detected by observing change of laser speckle pattern in this method. We observed change of laser speckle pattern depending on maximum stress and number of loading cycles subjected on steel specimens. We investigated a possibility of estimating maximum stress and fatigue life at the early stage of low cycle fatigue based on this experimental result.

1 INTRODUCTION

We have been working on evaluating plastic strain and high cycle fatigue damage in steels using laser speckle sensor. Result of our previous work for static tensile test showed that light intensity distribution of the speckle pattern relates closely to frequency distribution of the surface profile (Kato et al., 1993). The intensity distribution of the laser speckle broadens with increase of the plastic strain. This is because that density of slipbands increases with the increase of the plastic strain and thus ratio of high frequency component in surface profile diagram increases.

When steel specimens are subjected to cyclic loading, slipbands are produced on the specimen surface by fatigue. Density of the slipbands increases with progress of fatigue damage and then initial cracks are produced in the persistent slipbands. This means that slipband density will increase and ratio of high frequency component in surface profile will increase corresponding to increase of loading cycles and thus the light intensity distribution of the laser speckle pattern will expand with the progress of fatigue damage (Kato et al., 1996).

In this study, we investigated a possibility of evaluating fatigue damage in low cycle case based on the above experimental results. We observed the laser speckle pattern and surface property under cyclic loading with constant maximum stress which exceeds yield stress using steel specimens. Relation between the speckle pattern and the surface property change

occurred by slipbands was investigated. Speckle pattern change corresponding to progress of fatigue damage was observed. Influence of the maximum stress on this relation was investigated and relation between fatigue life and speckle pattern was observed. Possibility of estimation of maximum stress and fatigue life using this relation was investigated.

2 EXPERIMENTAL PROCEDURE

Material used in this experiment is SM490A (JIS) steel. The specimen is a strip with width of 20 mm and thickness of 5mm. Surface of the specimen was polished finally with Aluminum dioxide powder in random direction so that the surface properties became macroscopically isotropic and uniform. Initial surface roughness was about 0.05μm Ra. Mechanical properties of the material are shown in Table 1. The specimens were loaded by uni-axial cyclic tension with the minimum load of 0 N. The maximum stress subjected to the specimen was set to the value exceeds yield stress of the material (Table 1). Cycle speed of the loading was 0.01Hz.

Figure 1 shows layout of the experimental system. He-Ne laser is illuminated on the specimen surface. Diameter of the laser beam is about 1mm. Laser speckle pattern is formed on the ground glass placed in front of the specimen as a screen. Image of the speckle pattern is input into the image-processing system through a CCD camera. Resolution of the image used is 512×512 pixels and a pixel is expressed with 256 gray levels. The laser speckle

Table 1 Mechanical properties of the material

Yield Strength (MPa)	Tensile Strength (MPa)	Elongation (%)
455	559	35

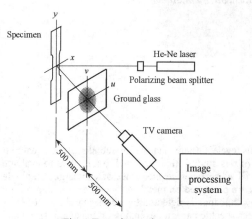

Fig. 1 Experimental system

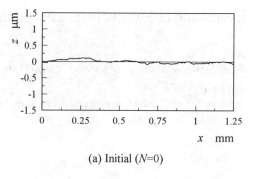

(a) Initial ($N=0$)

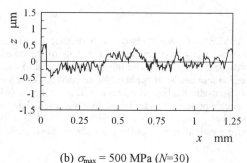

(b) $\sigma_{max} = 500$ MPa ($N=30$)

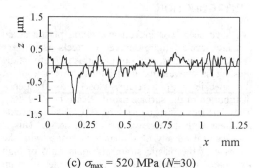

(c) $\sigma_{max} = 520$ MPa ($N=30$)

Fig. 2 Surface profile diagrams

pattern can be analyzed automatically and quantitatively using this image-processing system.

3 LASER SPECKLE PATTERN AND SURFACE PROPERTY

Figure 2 shows surface profile diagrams of the specimens after fatigue. Figure 2 (a) is one for the initial surface before loading and Fig. 2 (b) and (c) are ones for maximum stress of $\sigma_{max} = 500$ MPa and $\sigma_{max} = 520$ MPa after 30 loading cycles. It is found that surface roughness is larger for larger maximum stress. This is because that slipband on the specimen surface occurs more densely for larger stress.

Figure 3 shows spatial frequency distribution of the surface profile diagrams shown in Fig. 2. These figures show autocorrelation functions obtained from magnitude of Fourier transform of the diagrams of Fig. 2 (a), (b) and (c), respectively. It is found that frequency distribution expands to the higher frequency region with increase of maximum stress. This is caused by the increase of slipband density with the increase of the maximum stress.

Figure 4 shows speckle patterns for the same maximum stress as Fig. 2 and Fig. 3. The pictures show that speckle pattern broadens with the increase of maximum stress. Figure 5 shows the autocorrelation functions of the light intensity distribution of the speckle patterns of Fig. 4 on the radial direction from centroid of the image for each speckle pattern. The figure shows that light intensity distribution expands with the increase of the maximum stress and changes corresponding to the frequency distribution of the surface profile diagrams

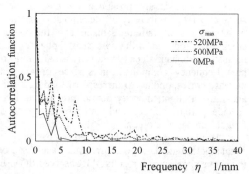

Fig. 3 Autocorrelation functions of frequency distribution of the surface profile diagrams

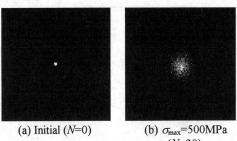

(a) Initial (N=0)   (b) σ_{max}=500MPa   (c) σ_{max}=520MPa
                         (N=30)   (N=30)

Fig. 4 Speckle patterns

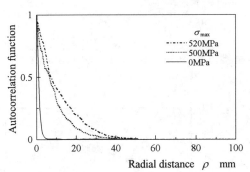

Fig. 5 Autocorrelation functions of light intensity distributions

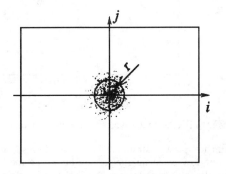

Fig. 6 Image of speckle pattern

correlation function was calculated between gray level data of whole image plane of the speckle pattern and inside the circular window of radius r = 20 pixels at the centroid of the image (Fig. 6). Distribution of the normalized correlation function on the radial direction is shown in Fig. 7. This distribution can be approximated with an exponential function. The radius at which the exponential function takes the value of 0.25 is considered and we use this value as a parameter to express the width of the speckle pattern. We denote this parameter as B_s.

4 CHANGE OF THE LASER SPECKLE PATTERN UNDER LOW CYCLE FATIGUE

Fatigue tests were made under seven different maximum stresses from 457 to 540MPa and laser speckle pattern was observed stopping the testing machine during the test. Figure 8 shows an example of load-displacement curve during the test. Large plastic deformation occurs by the first cycle of loading and increase of plastic displacement d_p is very small after second cycle of loading. Figure 9 shows relation between the parameter B_s and number of loading cycles N for different maximum stresses. Initial value of the parameter B_s before loading (N=0) was about 3 for every specimen. The parameter B_s increases at the first cycle and change of B_s is very small after second cycle up to about 30 cycles for all the maximum stresses except the case of σ_{max} = 540MPa. The parameter B_s is larger for larger maximum stress when the number of loading cycles is the same. Relation between B_s and maximum stress is shown in Fig. 10 for N = 1 to 10. Difference of B_s is very small for different N except σ_{max} = 540MPa. There is almost a fixed relationship between B_s and

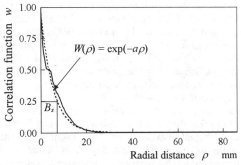

Fig. 7 Evaluation of width of speckle pattern

shown in Fig. 3. This means that we can detect change of surface property caused by the occurrence of slipbands under low cycle fatigue by observing change of the speckle pattern. In the figure, speckle patterns at number of loading cycles N=30 are shown for σ_{max} = 500MPa and 520MPa, but speckle pattern was almost similar for N=1 and N=30 as will be mentioned later. Thus surface properties are presumed to be similar for N=1 and N=30.

The method to evaluate width of the speckle pattern has been already reported by the authors (Kato et al., 1995). The method is summarized as the following. First, to smooth the gray level distribution and make clear the characteristics of the distribution,

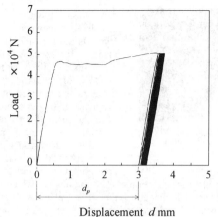

Fig. 8 Load-displacement curve

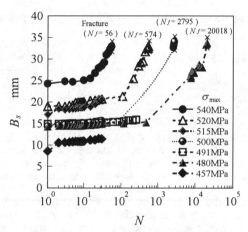

Fig. 9 Relation between width of speckle pattern B_s
and number of loading cycles N

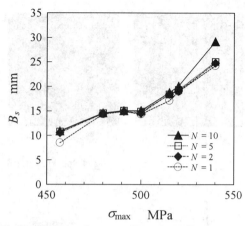

Fig. 10 Relation between B_s and maximum stress
σ_{max}

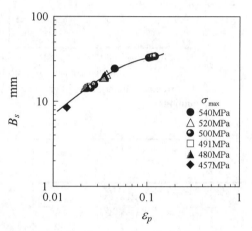

Fig. 11 Relation between B_s and plastic strain ε_p

σ_{max}, if σ_{max} is smaller than 540MPa. Thus it is possible to estimate maximum stress which was subjected to the specimen so far by observing the speckle pattern.

Increase rate of the parameter B_s to the number of loading cycles is very small at the earlier stage of fatigue, but B_s increases rapidly at the later stage and take similar value just before the final fracture of the specimens for all the maximum stresses as shown in Fig. 9. The final value of B_s was about 34mm. Plastic strain ε_p subjected in the specimen during the test can be obtained from plastic displacement d_p in the load-displacement curves as shown in Fig. 8 and relation between B_s and ε_p is shown in Fig. 11. The figure shows that there is a fixed relation between B_s and ε_p for different maximum stresses. From this result, width of the speckle pattern B_s is found to be dominated by plastic strain ε_p. This shows that we can

estimate plastic strain subjected to the specimen during low cycle fatigue test by observing the laser speckle pattern. From Fig. 9, B_s takes almost similar value at the final stage of fracture and this means that ε_p is similar when the specimens fracture. There is a possibility to evaluate fatigue damage by knowing plastic strain from laser speckle pattern.

From the experimental result when the fatigue test was made until the specimen broke among the data shown in Fig. 9, relation between the parameter B_s at $N = 1$, 2, 5 and 10 cycles and fatigue life N_f for each specimen is shown in Fig. 12. We have only 4 data for different maximum stresses in the figure and number of the experimental data is not sufficient to make a clear decision that there is a fixed relation between B_s and N_f. But we can conclude that there is a tendency as shown in Fig. 12 between them. If we obtain this kind of relation in advance, we can

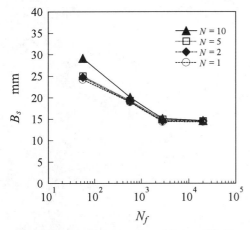

Fig. 12 Relation between B_s and fatigue life N_f

estimate fatigue life of the material by observing B_s at N=1 to 10. There is a possibility of fatigue life estimation in low cycle case at very early stage of fatigue using this relation.

5 CONCLUSIONS

In this study, we investigated a method to evaluate low cycle fatigue damage using laser speckle sensor. We observed change of speckle pattern during fatigue of steel specimen and we derived relation between the parameter B_s which expresses width of the speckle pattern and plastic strain and also a relation between B_s and fatigue life. It was found that B_s is dominated by plastic strain and we can evaluate plastic strain subjected in the specimen during low cycle fatigue test. And we obtained a relation between B_s at N=1 to 10 and fatigue life and it was found that there is a possibility to estimate fatigue life based on change of speckle pattern at the early stage of fatigue using this

relation. It is very useful for practical applications if we can estimate fatigue life at very early stage using this kind of non-contact method.

The work reported has been supported by The Science Research Promotion Fund (1996 and 1997) by Japan Private School Promotion Foundation.

REFERENCES

Kato, A. & Kawamura, M. 1993, Damage Monitoring of Metal Materials by the Laser Speckle Method (Measurement of Plastic Strain), Proc. of Asia Pacific Conference on Fracture and Strength 93, 379-383.

Kato, A. & Kawamura, M. 1993, Measurement of Plastic Strain in Steel Specimens by Means of Intensity Distribution of Laser Speckle, J. of the Society of Materials Science Japan (in Japanese), **43**(489), 696-702.

Kato, A., Kawamura, M. & Nakaya, I. 1995, Damage Monitoring of Metal Materials by Laser Speckle Assisted by Image Processing Techniques (Relationship between Distribution of Laser Speckle and Surface Properties), JSME Int. J., **38**(2), 249-257.

Kato, A., Kawamura, M. & Ito K. 1995, Measurement of Plastic Strain in Steel Materials by Laser Speckle (Influence Due to the Difference of Grain Size and Materials), Journal of JSNDI (in Japanese), **44**(7), 529-535.

Kato, A. & Ito K. 1995, Damage Evaluation Using Laser Speckle Sensor, Proc. of the International Symposium on Advanced Technology in Experimental Mechanics, 189-194.

Kato, A. & Ito K. 1996, Damage Evaluation of Fatigue Using Laser Speckle Sensor, Proc. of the VIII International Congress on Experimental Mechanics, 58-59.

Experimental Mechanics, Allison (ed.) © 1998 Balkema, Rotterdam, ISBN 90 5809 014 0

Combined fatigue rig and moiré interferometer to measure crack closure

S.Güngör & L.Fellows
Department of Engineering Science, University of Oxford, UK

ABSTRACT: High sensitivity moiré interferometry was used in the study of crack closure to investigate the R-ratio effect on the fatigue crack behaviour of Ti-6Al-4V. For this purpose, a moiré interferometer was built and attached to a servo-hydraulic fatigue testing machine. The stability of the system was such that the automated fringe analysis by temporal phase stepping could successfully be applied to obtain both in-plane displacement maps. The paper describes the system and addresses the problems associated with the use of moiré interferometry in crack closure measurements, such as, the influence of convection currents and vibrations on the fringe stability and the durability of diffraction gratings during fatigue cycling.

1 INTRODUCTION

In 1970 Elber discovered that cracks remain closed for part of even a tensile fatigue cycle. It is widely accepted that this closure could account for the R-ratio effect and the retardation of crack growth following a stress overload. The idea was put forward that the crack growth rate is simply dependent upon the difference between the stress intensity factor at maximum load and the stress intensity factor at crack tip opening point. The difficulty with this relationship is that the instant at which a crack is open or closed to the crack tip is not known. Therefore in order to develop improved predictions of fatigue life accurate crack closure measurements are necessary.

Crack opening and closure are not easy to measure because they are not discrete events and the displacements are small. Cracks peel open gradually from the mouth to the tip and from the mid-section to the surface. Consequently, different measurement techniques, such as those based on the compliance of either bulk of the specimen or locally near the crack tip, yield different estimates of crack closure. It is therefore believed that this could be the reason for so much uncertainty and controversy when considering crack closure (Kemp, 1990).

Gray and MacKenzie (1990) have shown that the crack closure phenomena can be investigated more effectively by the analysis of in-plane displacements around cracks. Moiré interferometry, which is a well established experimental technique used in stress analysis, produces fringes that represent contours of constant in-plane displacements upon deformation of the sample under observation. They detected crack closure by observing the shape of the moiré fringes close to the crack. They considered that if the crack is open the fringes will have an abrupt change of direction close to the point where the fringes intersect the crack.

Moiré technique has many advantages over other techniques:

i)   it is less time consuming and more direct than the stereo-imaging or replica techniques;

ii)   back-face strain gauges and potential drop methods give descriptions of the bulk behaviour of the crack that make correlations with a model difficult to interpret;

iii) gauges close to the crack give only the displacements of discrete points.

The disadvantage of moiré interferometry is that only surface displacements are known. However, some authors believe that closure at the surface governs crack growth rates, the increased level of closure acting as an anchor on each side of the crack. Also if the crack front is growing with a constant shape then all points through the thickness are progressing at a constant rate. The assumption that the surface will give information from which the crack propagation rate can be determined therefore seems reasonable.

The use of moiré interferometry in the study of crack closure is not a straightforward task as measurements using moiré interferometry are usually conducted in laboratory environments where the optical elements of the interferometer and the specimen are located on an isolated optical table in order to minimize vibrations and to obtain stable fringe patterns. This is particularly crucial when phase stepping methods are used to analyse the fringe patterns. The phase stepping method is based on acquiring a series of fringe patterns while introducing a constant phase shift between each fringe pattern. The resulting interferograms are then processed to obtain accurate full-field phase distribution with increased displacement resolution (Poon et al. 1993). Any disturbance, however, in the fringe stability during the acquisition of the fringe patterns may lead to errors in the calculated phase maps.

It is, of course, possible to grow a fatigue crack in a fatigue machine first, and then transfer the specimen to a moiré interferometer for the measurements. However, the study of non-zero R-ratio loading histories requires the moiré experiments to be done without unloading the specimen. This is only possible by conducting the measurements *in-situ*, unless a practical way is found to transfer the specimen to the interferometer without unloading it, for example using a compact tension specimen and a jacking screw. For the present study, it was decided to build a moiré interferometer into the fatigue rig. This paper reports the design of this interferometer and describes how the vibration and other problems associated with the fringe stability were overcome.

Another difficulty associated with the use of moiré interferometry in crack closure studies is the durability of diffraction gratings attached to the specimen surface, which are required for in-plane displacement measurements. The conventional replicated aluminium gratings tend to flake off around the crack tip during cyclic loading. This precluded their use in the present study as displacement information in the vicinity of both the tip and the wake of the crack was required. Renewing the grating before each measurement is tedious and impractical when information is required at several cycles after an event, such as an overload. Therefore, a method has been developed to write gratings directly onto the specimens using positive photoresists. These gratings proved to be resistant to cyclic loading. The processes involved to make these gratings are also outlined in this paper.

2 MOIRÉ INTERFEROMETER

The in-plane moiré interferometry is a well established technique and a comprehensive treatment of its principle and practice is given elsewhere (Post, 1987). Numerous different optical schemes can be arranged for moiré interferometry (Post, 1987, Epstein & Dadkah, 1993). The system chosen for the present work is based on the three-mirror design (Post, 1987). It is a low-cost system which is relatively easy to design and operate. Figure 1 shows an overview of the system. A crossed-line diffraction grating of frequency 1200 lines/mm is replicated on the specimen and it deforms together with the specimen. A 10mW helium/neon laser provides the light beam which is expanded through a pin hole and collimated using a collimating lens, illuminating the specimen grating and the interferometer consisting of a light frame and three adjustable mirrors, M1, M2 and M3. To obtain the u-displacement field, half the incident beam B1 in Figure 1b impinges directly on the specimen surface while the other half B2 impinges on the mirror M3 and then reflects to the specimen in a symmetrical direction to form a virtual (reference) grating. This virtual grating of frequency twice of the specimen grating, interacts with the vertical set of lines of the specimen grating to form a moiré fringe pattern which is recorded using a CCD camera. Similarly, two incident beams, A1 and A2, impinge on the mirrors M1 and M2 and then illuminate the set of grating lines perpendicular to the y-axis to produce the v-displacement fields. In this four-beam optical arrangement, the u- and v-displacement fields are obtained by using beam pairs A1-A2 and B1-B2 simultaneously. This is achieved by selectively filtering the beam with an opaque object with a slot or slots allowing only the required beam pairs to illuminate the interferometer, as shown in the inset of Figure 1. The resulting interferograms were then collected by a CCD camera and digitised, using a frame grabber plugged in a personal computer, in an array of 256 by 256 data points in 8 bit (i.e. 256 gray levels) resolution. Displacement sensitivity of moiré fringes are determined by the frequency of the reference grating. For the system described here the frequency of the virtual grating is 2400 lines/mm and therefore the corresponding displacement sensitivity is 0.417 µm per fringe order.

The traditional way of analysing moiré fringe patterns involved manually locating the fringe centres and

1072

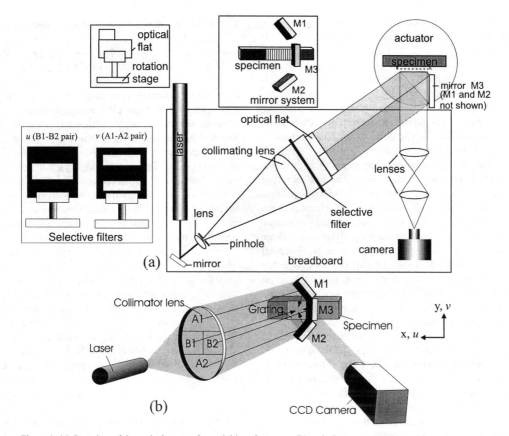

Figure 1. (a) Overview of the optical system for moiré interferometry, (b) optical arrangement for u and v displacement measurement.

determining the order and the gradient of the fringes. The use of phase stepping methods eliminates these tedious and labourious procedures as computers can be used to analyse the fringe patterns and produce phase maps in a matter of minutes (Greivenkamp & Bruning, 1992). Furthermore, since these methods reveal the grey scale information between fringe extremas (that is, the fractional fringe orders), the displacement resolution of the system is also greatly enhanced by the use of these methods. The automation of the fringe pattern analysis here was done using temporal phase stepping (TPS) method which is explained in detail by Poon et al. (1993). The method is based on acquiring at least three interferograms with a constant phase shift and determining the phase (corresponding to the displacement value) at each pixel. The intensity of an interferogram at a point (x,y) is given by

$$I_i(x, y) = a(x, y) + b(x, y)\cos[\phi(x, y) + \delta_i] \quad (1)$$

where i denotes each phase shifted interferogram. $a(x,y)$ is the background intensity, $b(x,y)$ is the local contrast, $\phi(x,y)$ is the phase to be determined and δ is the phase shift. By using the three phase-shifted interferograms, the three unknowns in equation 1, namely a, b and ϕ can be determined on a pixel by pixel basis. The phase can also be obtained by acquiring more than three interferograms which leads to an over-deterministic system that affords errors in the phase shift setting to a certain extent and therefore results in more accurate phase maps. In general, the phase shift is introduced by changing the path length of one of the interfering beams. This is accomplished in the present system by placing an optical flat into the path of the collimated beam (Poon et al. 1993). The optical flat is an 'L' shape, as shown in the inset of Figure 1, such that the path length of one half of the beam is changed for both u- and v-displacement measurements. The calculated values of $\phi(x,y)$ vary from $-\pi$ to π between two subsequent fringes, therefore an unwrapping process (Takeda, et al., 1982) was used to transform the phase fringes into a continuous function (Poon et al. 1993).

3 FRINGE STABILITY

In the literature, the main problem effecting the stability of fringes has been reported as mechanical vibrations (Guo, 1991, Perry, 1996). In general, the sensitivity of a moiré interferometer to vibrations is directly related to the displacement sensitivity of the system and the direction of the vibrations. The vibration sensitivity is equal to the displacement sensitivity when the direction of the vibration is the same as the direction of the displacement measured (that is, when vibrations are perpendicular to the grating lines used for the particular displacement measurement). Although the main effect of vibrations is to introduce fringe motions, the actual manifestation of the vibration on the fringe pattern depends on the frequency of the vibrations. When the frequency of the vibrations are much higher than the data acquisition rate of the device used, the contrast of the fringe pattern will be reduced. Whereas, if the vibrations are more subtle, the effect will be seen as movements in the fringes. Nevertheless, as long as the vibrations are not severe, it is usually possible to obtain accurate fringe patterns, especially, where the analysis is done using only one fringe pattern. In this study, the measurements using moiré interferometry were performed while the high pressure of the testing machine was off; the load on the specimen was maintained by the low pressure facility of the machine. With this arrangement, no vibrational effects were observed on the stability of moiré fringes. While the specimen is undergoing fatigue cycling the mirror system is removed for protection.

The main problem associated with the fringe stability in the present system, however, was fringe motion caused by convection currents due to temperature fluctuations in the laboratory. Since the refractive index of a medium varies with temperature, the temperature variations in air and the material of the optical flat caused the phase difference between the beam pairs A1-A2 and B1-B2 in Figure 1 vary with time. The effect of this was observed as a slow movement of fringes which adversely affected the accuracy of the phase stepping procedures. A frame was built around the interferometer to overcome this problem and curtains were attached to the frame using velcro strips. All the heat sources were removed from the enclosure. These included the laser for the interferometer and a light source for a microscope that

was used to monitor the crack length. The laser was mounted on top of the enclosure and the beam was directed through a hole. The only heat source that could not be removed was the actuator and this caused significant problems as it reached a temperature of about 40°C. Water cooling by coiling a plastic pipe around the actuator was used to reduce its temperature to ambient conditions. Furthermore, significant pressure differences between the corridor and the fatigue laboratory causing large draughts. So, an area of the fatigue laboratory was sectioned off using a heavy blackout curtain and draught proofing was used around the door to the laboratory. With all these above measures taken, stable fringes were obtained.

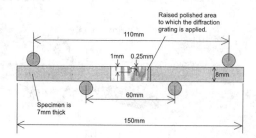

Figure 2. Geometry of the four-point-bend fatigue specimen

4 PRODUCTION OF PHOTORESIST GRATINGS

Polymers exist that, having been exposed to uv light, can be developed and washed away in the exposed areas. These polymers are called positive photoresists and can be used in the production of diffraction gratings directly on the specimen surface. The processes involved to make such gratings are outlined below:

1. The specimens were cleaned thoroughly then baked for thirty minutes in an oven at 115°C.
2. The specimens were put in a container overnight along with a petri dish of Hexamethyldisilazane ($C_6H_{19}NSi_2$) which improves the adherence of photoresist to the specimen surface.
3. Hunts 514 photoresist was applied to the specimen by dipping, using a withdrawal rate of three centimetres per minute.
4. The specimens were baked in an oven at 90°C for three minutes.
5. The photoresist was exposed for 80 seconds using

The specimens were made from a titanium alloy, Ti-6Al-4V, chosen for its lack of work hardening. The specimen geometry is shown in Figure 2. The cracks were grown by applying loads using four point bending. The surface of the specimen was polished to enable the cracks to be measured with a long range microscope as they grew. Polishing the whole of a specimen is difficult, so when the specimen was machined a small raised area was made on both sides, and these areas were polished. A notch 1.5mm in length was machined into the specimen using wire erosion to accelerate and localize crack initiation. After a photoresist diffraction grating was printed on its surface, as described above, the specimen was cycled at the load ratio of 0 for the particular example given here. Then, measurements using moiré interferometry were conducted at varying crack lengths. Figures 3a and 3b show two fringe patterns of u-displacements (i.e. displacements normal to the crack at the surface) taken at a phase difference of $\pi/2$. The crack length is 0.8 mm and the load applied to the specimen is 3 kN. Figure 3c gives the corresponding phase map computed using five phase-shifted interferograms (two of which are shown in Figures 3a and 3b). This phase map was then unwrapped as described earlier and a continuous displacement map was obtained (Figure 4).

6 CONCLUSIONS

A combined fatigue rig and moiré interferometer has been designed and commissioned. The rig had fringe stability problems that have now been identified and solved. This has enabled the required sensitivity of measurement to be achieved by phase stepping.

Applying photoresist diffraction gratings to titanium beam specimens before fatigue cracking is possible. The specimen has to be machined oversize as a bead of photoresist along the edges of the specimen needs to be removed after the grating is applied. High contrast moiré fringes could be obtained for the photoresist gratings. However, after the removal of the bead, defects in the grating were observed. The photoresist gratings remain adhered to the specimen close to the crack faces after fatigue crack growth. Hence, the gratings may be used for measuring crack closure accurately at various R-ratios and after overload cycles. The photoresist gratings may extend the use of moiré interferometry to new applications.

Additional benefits may be obtained by etching the

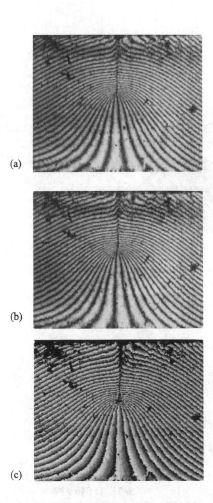

(a)

(b)

(c)

Figure 3. (a) and (b) Fringe patterns of u-displacements around a fatigue crack at a phase shift of 0 and $\pi/2$ respectively, and (c) the computed phase map of the same area.

an Argon laser ($\lambda = 488$nm). The grating frequency was set to 1200 lines per mm.

6. The specimens were developed using PLSI developer for 60 seconds at 21°C, rinsed with distilled water and dried using compressed air.

7. The specimens were baked in an oven at 90°C for 10 minutes.

8. Gold was evaporated onto the specimens to a depth of 0.05μm.

9. Notches were wire eroded into the specimens to a depth of 1.5mm

10. Cracks were grown to a combined notch and crack length of 2.5mm

11. The beads of photoresist were removed by milling 2.0mm off the cracked side of the specimens.

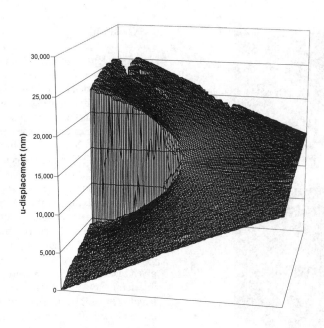

Figure 4. Surface plot of u-field displacement map computed from the phase map of Figure 3.

grating into the metal. In order to do this the grating would need to be fully resolved, i.e. have no photoresist in the grating troughs. An ion beam milling machine could then be employed that would erode both the photoresist and the bare metal in the troughs. Some preliminary work has been carried out in this area but problems of the photoresist melting were encountered.

REFERENCES

Brownell, J.B. & Parker, R.J. 1991. Automated fringe analysis for moiré interferometry. *Proc. SPIE*. 1554b:481-492.

Elber, W. 1970. Fatigue crack closure under cyclic tension. *Eng. Fract. Mech*. 2:37-45.

Epstein, J.S. & Dadkah, MS. 1993. Moiré interferometry in fracture research. In *Experimental Techniques in Fracture*, Chap.11, 427-508. Ed. J.S. Epstein. VCH Publishers, Inc.

Gray, T.G.F. & MacKenzie, P.M. 1990. Fatigue crack closure investigation using moiré interferometry. *Int. J. Fatigue*.12(5): 417-423.

Greivenkamp, J.E. & Bruning, J.H. 1992. Phase shifting interferometry. In *Optical Shop Testing, Second Edition*. Ed. D. Malacara.501-598. John Wiley & Sons, Inc.

Guo, Y. 1991. A vibration insensitive moiré interferometry system for off-table applications. *Proc. SPIE*. 1554b:412-419.

Kemp, R.M.J. 1990. Fatigue Crack Closure - A Review. Royal Aerospace Establishment, Technical Report 90046 ICAF Document 1776.

Perry, Jr. K.E. 1996. Delamination and damage studies of composite materials using phase-shifting interferometry. *Optics and Lasers in Eng*. 24:467-483.

Poon, C.Y., Kujawinska, M. & Ruiz, C. 1993. Automated fringe pattern analysis for moiré interferometry. *Exp. Mech*. 33(3):234-241.

Post, D. 1987. Moiré interferometry. In *SEM Handbook on Experimental Mechanics*, Chap. 7, 314-387. Ed. A.S. Kobayashi, Prentice Hall Inc.

Takeda, M., Ina, H. & Kobayashi, S. 1982. Fourier transform method of fringe pattern analysis for computer based topography and interferometry. *J. Opt. Soc. Amer*. 72:156-160.

Fatigue crack growth

Experimental Mechanics, Allison (ed.) © 1998 Balkema, Rotterdam, ISBN 90 5809 014 0

Fatigue, creep and creep/fatigue crack growth behaviour of a nickel base superalloy at 700°C

T.H.Hyde, L.Xia & A.A.Becker
Department of Mechanical Engineering, University of Nottingham, UK

ABSTRACT: This paper presents the results of a series of experimental tests on the high temperature behaviour of a Ni-base superalloy at 700° C; the material is used in the manufacture of aeroengine turbine discs. It is shown that the fatigue crack growth data correlates on the basis of the stress-intensity-factor range and creep crack growth rates appear to correlate on the basis of the C* parameter. Constant load dwell periods introduce significant scatter in fatigue crack growth data. However, there is an indication that the threshold stress-intensity-factor range reduces when constant load dwell periods are introduced and that crack growth rates can be approximately correlated on the basis of stress-intensity-factor range.

NOTATION

A, n, B, χ, φ	constants in creep continuum damage equation (Equ. 2)
a	crack length
f	loading frequency (Hz)
K, ΔK, ΔK_{th}	stress-intensity factor, stress-intensity factor range and threshold stress-intensity factor range
N	load cycle number
P	loads applied to compact-tension specimens
R	ratio of minimum to maximum load during a load cycle
S_{ij}	deviatoric stress tensor ($= \sigma_{ij} - \frac{1}{3} \delta_{ij} \sigma_{kk}$)
t	time
W	width of compact-tension specimen
δ_{ij}	Kronecker delta ($= 1$ for $i = j$ and $= 0$ for $i \neq j$)
$\dot{\Delta}$	load-line displacement rate
ϵ^c, $\dot{\epsilon}^c_{min}$	creep strain and minimum creep strain-rate
$\sigma, \hat{\sigma}, \sigma_a, \sigma_{eq}, \sigma_u, \sigma_y, \sigma_{ref}$	stress, maximum principal stress, alternating stress, equivalent stress, ultimate tensile stress, yield stress and reference stress
$\omega, \dot{\omega}$	damage and damage-rate ($d\omega/dt$)

1. INTRODUCTION

Waspaloy, a Ni-based superalloy, is used in the manufacture of aeroengine turbine discs. The high stress and temperature ranges in which these components operate make it necessary to consider creep, fatigue and the possibility of creep/fatigue interaction when designing and assessing the life of the components. Although the discs would be initially "defect-free" and the design would be such that no cracks should be generated during the design life, the susceptibility of the material to creep and fatigue crack growth is of obvious interest to designers.

In this paper, the results of a series of crack growth tests performed on Waspaloy specimens, at 700° C, which were all machined from the same forging blank of a gas-turbine disc, are presented. These include creep, fatigue and creep/fatigue interaction tests of compact-tension (CT) and corner-crack (CC) specimens.

2. TEST SPECIMENS, EQUIPMENT AND INSTRUMENTATION

The specimens were machined from a forging blank of a gas-turbine engine disc; the composition of the material in % weight was as follows: Ni 56.57, Cr 19.99, Co 3.82, Mo 4.19, Ti 3.35, Al 1.51 and Fe 0.57.

Compact-tension (Fig. 1(a)) and corner-crack (Fig. 1(b)) specimens were used to investigate fatigue, creep and creep/fatigue crack growth behaviour of the material. Prior to testing at 700° C, pre-fatigue cracks were produced from the sharp notches, at room temperature, using an Amsler Vibrophore, resonant test machine, at a frequency of about 100 Hz. Maximum load levels used to produce the pre-fatigue cracks were significantly lower than any of the loads used in the subsequent tests at 700° C.

Chromel-Alumel thermocouples (TC.C70, Type K) were attached to each specimen; all tests were carried out at 700° ± 1.0° C. Crack growth was monitored using a DC potential drop system. The change in DC potential drop reading obtained during the test was used in conjunction with the measured crack growth (after post-test, fatigue crack growth of the specimens) to self-calibrate each test. Load-line displacements were obtained using linear voltage displacement transducers (LVDT's) and high temperature extensometry attached to the surfaces of the specimens.

Thermocouple, extensometer LVDT voltages and DC potential drop voltages were recorded at predetermined times using a Solartron Orion data logger.

3. UNIAXIAL BEHAVIOUR

The uniaxial creep properties for the material, at, 700°C, have been described in detail elsewhere [Hyde et al 1997].

The Norton power-law creep representation of the uniaxial data in the stress range 350-850 MPa exhibits a two-stage behaviour, i.e.,

$$\dot{\epsilon}^c_{min} = 2.355 \times 10^{-43} \, \sigma^{14.0} \qquad (1a)$$

for $\sigma > 520$ MPa, and

$$\dot{\epsilon}^c_{min} = 9.226 \times 10^{-34} \, \sigma^{10.65} \qquad (1b)$$

for $\sigma < 520$ MPa, where $\dot{\epsilon}^c_{min}$ and σ have units of h^{-1} and MPa, respectively.

The two-stage damage mechanics, creep constitutive equation [Hayhurst et al 1984] fit to the uniaxial creep data was as follows:-

$$\dot{\epsilon}^c = A \left(\frac{\sigma}{1 - \omega} \right)^n \qquad (2a)$$

and

$$\dot{\omega} = B \frac{\sigma^\chi}{(1 - \omega)^\phi} \qquad (2b)$$

where A = 2.355 x 10^{-43}, n = 14.0, B = 1.907 x 10^{-41}, χ = 13.49 and ϕ = 13 for $\sigma > 520$ MPa and A = 9.226 x 10^{-34}, n = 10.65, B = 8.46 x 10^{-27}, χ = 8.133 and ϕ = 13 for $\sigma < 520$ MPa, with σ and $\dot{\epsilon}^c$ having units of MPa and h^{-1}, respectively.

The material was found to be "creep ductile", having average failure strains, at the end of the creep tests, of 24% (the lowest measured value was 10%). The failure strain results were subjected to a wide scatter band and did not indicate any significant stress (or time to failure) dependence.

4. CRACK GROWTH TEST RESULTS

4.1 Fatigue crack growth

Compact-tension specimens, Fig. 1(a), were pre-fatigue cracked, at room temperature, to produce initial a/W values of 0.52 to 0.54. Three of these specimens were subsequently subjected to fatigue loading, with R = 0.1 and f = 0.5 Hz; test details are given in Table 1. From the crack length versus cycle number data, obtained from the tests, the log (da/dN) vs log (ΔK) behaviour shown in Fig. 2 was obtained. The formulae used to determine ΔK are given in the Appendix. Also shown in Fig. 2 are the corresponding results obtained from a test of a corner-crack specimen, Fig. 1(b); the initial crack which was approximately quarter-circular with a radius of 1.0 mm remained approximately quarter-circular and the average final crack radius at the end of the test was 1.6 mm. There is good correlation between the compact-tension and corner-crack test data.

The data, Fig. 2, exhibits the usual behaviour with a power-law fit [Parker 1981] to the stage II crack growth, ie.

$$\frac{da}{dN} = 6.792 \times 10^{-8} \, (\Delta K)^{2.70} \qquad (6)$$

where da/dN and ΔK have units of mm/cycle and MPa $m^{½}$, respectively. Also, the results indicate that there is a threshold stress-intensity factor range, ΔK_{th}, below which fatigue crack growth does not occur, ie. ΔK_{th} = 16 MPa $m^{½}$.

4.2 Creep crack growth

Five creep crack growth tests were performed on compact-tension specimens (Fig. 1(a)), with initial a/W values in the range 0.51 to 0.55; and on 2 corner-crack specimens (Fig. 1(b)) with initial crack radii to side length ratios of 0.2 and 0.27; the test details are given in Table 2. After completion of the creep tests, post-fatigue loading, at relatively low load levels and at room temperature, was used to expose the creep crack growth surfaces; typical surfaces for the compact-tension and corner-crack specimens are shown in Figs. 3(a) and 3(b), respectively. It can be seen that the creep crack growth in the compact-tension specimens is significantly higher at the middle of the crack than at the surface. However, at distances greater than about 2 mm from the surfaces, the creep crack growth is practically constant. The numerical creep crack growth data presented for the compact-tension specimens is based on the average of measurements taken at eight equally spaced positions on the crack front; it should be noted that the initial fatigue cracks are also slightly curved and the same procedure was used to determine these initial crack lengths. The tunnelling creep crack growth in the Ni-base superalloy compact-tension specimens is less than that obtained in 316 stainless steel specimens at 600° C [Hyde 1988]. The initial shape of the fatigue cracks in the corner-crack specimens can be seen to be quarter-circular (Fig. 3(b)). However, the creep crack growth exhibits tunnelling with very small creep crack growth at the surfaces. The numerical creep crack growth data presented for the corner-crack specimens is based on the measurements taken at the centres of the cracks, ie. they are the maximum creep crack growth data.

The C* parameter [Kanninen and Popelar 1985, Hyde 1986, Ainsworth and Haigh 1987] is commonly used to characterise creep crack growth behaviour. The creep crack growth rates obtained from the experimental tests were used together with the loads (for compact-tension specimens) or reference stress [Ainsworth and Haigh 1987] (for the corner-crack specimens) and the instantaneous crack length to produce the data shown in Fig 4. The formulae used to determine C* are given in the Appendix. Although the C* values for the corner-crack specimens increase due to the increase in crack length, the creep crack growth rates (as indicated by the DC potential drop equipment) were practically constant. Therefore, the data points plotted for the corner crack specimens in Fig. 4 are represented by horizontal lines indicating a constant da/dt while C* is slightly increasing during the tests.

In a previous paper [Xia, Becker and Hyde, to appear] it has been shown that the C* values, presented in Fig. 4 for the compact-tension specimens, are not steady-state values. However, Fig. 4 demonstrates that the C* value obtained from load-line displacement-rate measurements correlates the compact-tension data and is also in reasonable agreement with data obtained from corner crack creep crack growth tests.

4.3 Creep/fatigue crack growth

In order to determine the effect of constant load dwells on fatigue crack growth behaviour, four creep/fatigue interaction tests were performed on compact tension specimens. The specimens, which had initial fatigue cracks with a/W = 0.49 to 0.53, were subjected to load histories of the types shown in Fig. 5, with dwell periods in the range 1 to 18 seconds; test details are given in Table 3. It should be noted that the maximum loads are much less than the loads required to cause significant creep crack growth (Table 2). The log (da/dN) versus log (ΔK) data derived from the results of these tests are given in Fig. 6, together with the fatigue crack growth data shown in Fig. 2. The results indicate that the one-second dwell period has no significant effect on the crack growth rates obtained, whereas there seems to be a significant increase in initial crack growth rates, when dwell periods of 8 and 18 seconds are applied. However, in general, subsequent fatigue crack growth rates correlate closely with the stage II crack growth data from the fatigue crack growth tests (Fig. 2); one of the tests produced some very low crack growth rates (symbol ◊ in Fig. 6), in the stage II region, compared with all of the other data.

5. CONCLUSIONS

Fatigue crack growth data was found to correlate on the basis of stress-intensity-factor range and creep crack growth rates appear to correlate on the basis of the C* parameter.

Although constant load dwell periods introduced significant scatter in the fatigue crack growth data (Fig. 6) there is an indication that the threshold stress-

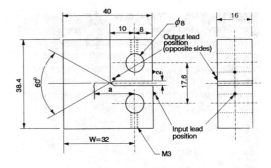

(a)   Compact-tension specimen

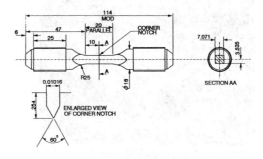

(b)   Corner-cracked specimen

Fig. 1   Test Specimens

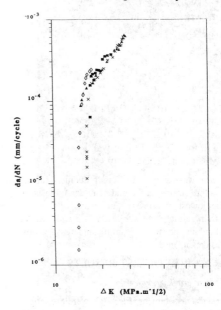

Fig. 2   Compact-tension (x, ▲, □) and corner-crack (◊) fatigue crack growth data, log (da/dN) vs log (ΔK)

1081

(a)   Compact-tension

(b)   Corner-crack

Fig. 3 Typical failure surfaces

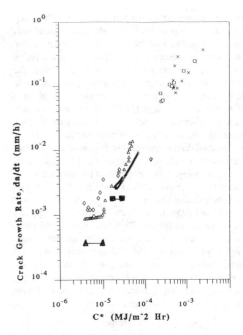

Fig. 4 Compact-tension (x, □, +, △, ◇) and corner-crack (■, ▲) creep crack growth data, da/dt vs C*

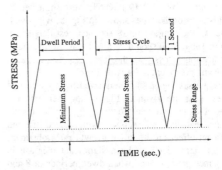

Fig. 5 Load cycles used in the creep/fatigue interaction tests

intensity- factor range reduces when constant load dwell periods are introduced and that the crack growth rate (with respect to cycle number) can be approximately correlated on the basis of stress-intensity factor range.

ACKNOWLEDGEMENTS

The authors wish to acknowledge the financial support of the Science and Engineering Research Council and Rolls-Royce plc through a collaborative research grant, and the help and advice of Dr A C Chambers of Rolls-Royce plc.

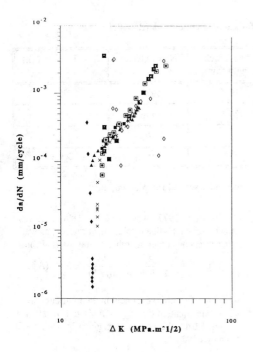

da/dN (mm/cycle)

ΔK (MPa.m^1/2)

Fig. 6 Compact-tension fatigue (x, ▲, ■) and creep/fatigue interaction (□ for 1-1-1), (■, ◇ for 1-8-1), and (♦ for 1-18-1) data with R=0.1

REFERENCES

Ainsworth, R.A. and Haigh, J.R., 1987, CEGB High temperature assessment procedure for defects under steady loading, Conf. on High Temperature Crack Growth, IMechE, 1987, Mechanical Engineering Publications Ltd, London 9-13.

BS-5447, 1977, Methods of test for plane strain fracture toughness (K_{IC}) of metallic materials.

Haigh, J.R. and Richards, C.E., 1974, Yield point load and compliance functions of fracture specimens, CERL RD/L/M461.

Harper, M.P. and Ellison, E.G., 1977, The use of the C* parameter in predicting creep crack propagation rates, J. Strain Analysis, 12, 167-179 (1977).

Hayhurst, D.R., Brown, P.R. and Morrison, C.J., 1984, The role of continuum damage in creep crack growth, Phil. Trans. R. Soc. London, A311, 131-158.

Hyde, T.H., 1986, Estimates of the creep parameter C* in terms of reference stress, J Strain Analysis, 21, 117-119 (1986).

Hyde, T.H., 1988, Creep crack growth in 316 stainless steel at 600° C, J. High Temperature Technology, 6 (2), 51-61.

Table 1 - Compact-tension (CT) and corner-crack (CC) specimen fatigue test conditions (R= 0.1 and f = 0.5 Hz)

Specimen Type	Maximum Load (kN)	Initial Crack Length (mm)	Final Crack Length (mm)	No of Cycles	Test Duration (h)
CT	4.3	16.8	22.2	25500	14.7
CT	4.5	17.2	21.7	23320	14.0
CT	5.0	16.7	19.7	12140	6.5
CC	20.0	1.0	1.6	20060	11.0

Table 2 - Compact-tension (CT) and corner-crack (CC) specimen creep test conditions

Specimen Type	Load (kN)	Initial Crack Length (mm)	Final Crack Length (mm)	Test Duration (h)
CT	8	16.8	19.6	655.5
CT	8	16.8	18.9	928
CT	8.5	17.6	21.1	36.1
CT	9.0	17.1	22.0	30.8
CT	9.0	16.4	18.3	966
CC	20.0	1.90	2.5	1610
CC	22.5	1.41	2.05	300.3

Table 3 - Compact-tension specimen creep/fatigue test conditions (R = 0.1)

Maximum Load (kN)	Loading Wave-form (*)	Initial Crack Length (mm)	Final Crack Length (mm)	No of Cycles	Test Duration (h)
4.5	1-18-1	15.7	16.5	49865	308
5.0	1-1-1	16.9	24.1	14960	12.7
5.0	1-8-1	17.0	22.7	18930	53.1
5.5	1-8-1	17.0	22.7	15770	47.4

(*)1-8-1 indicates 1 second ramp-up, 8 seconds hold and 1 second ramp-down, etc.

Hyde, T. H., Xia, L., Becker, A. A. and Sun, W., 1997, "Fatigue, creep and creep/fatigue behaviour of a Ni-base superalloy at 700°C", Fatigue and Fracture of Engineering Materials and Structures, Vol 20 (9), 1295-1303

Kanninen, M.F. and Popelar, C.H., 1985, Advanced fracture mechanics, Oxford University Press, Oxford (1985).

Parker, A.P., 1981 The mechanics of fracture and fatigue, E & FN Spon Ltd, London.

Pickard, A.C., 1986, The application of 3-dimensional finite element methods to fracture mechanics and fatigue life prediction, 1986, Engng. Materials Advisory Services Ltd, ISBN 0 947817 22 0, Chamelon Press Ltd.

Xia, L., Becker, A.A. and Hyde, T.H., An assessment of the C* and K_I parameters for predicting creep crack growth in a Ni-base superalloy at 700° C, to appear in International Journal of Fracture

APPENDIX

(a) Stress Intensity Factors

For compact tension specimens,

$$K_I = \frac{P}{BW^{\frac{1}{2}}} \, Y \, (a/W) \qquad (A1)$$

where Y (a/W) is the compliance function obtained from BS5447 [1977].

For semi-circular corner cracks, radius = a, in square cross-section bars, side-length = W,

$$K_I = \frac{2\sigma \sqrt{\pi a}}{\pi} \, f \, (a/W) \qquad (A2)$$

where numerical data from which f (a/W) can be determined has been published by Pickard [10].

(b) C* values

In order to determine the C* values for compact tension specimens, the method proposed by Harper and Ellison 1977 was used. This requires instantaneous load-line displacement rates, $\dot{\Delta}$, as well as instantaneous crack lengths, a, and loads, P, ie.

$$C^{\cdot} = -\frac{n}{n+1} \, \frac{P\dot{\Delta}}{BW} \, \frac{1}{n} \, \frac{dm}{d(a/W)} \qquad (A3)$$

where n is the stress exponent in the minimum creep strain-rate equation (ie. $\dot{\epsilon}^c_{min} = A\sigma^n$) and m (a/W) is obtained from the relationship given by Haigh and Richards 1974.

For the semi-circular corner-cracks, C* can be estimated [Ainsworth and Haigh 1987] using

$$C^{\cdot} = \sigma_{ref} \, \dot{\epsilon}^c_{min} \, (\sigma_{ref}) \, R \qquad (A4)$$

where σ_{ref} is the reference stress (= σ_{net} in this case) and

$$R = (K_I / \sigma_{ref})^2 \qquad (A5)$$

Experimental Mechanics, Allison (ed.) © 1998 Balkema, Rotterdam, ISBN 90 5809 014 0

Control of fatigue crack propagation in the TiNi shape memory fiber reinforced smart composite

Akira Shimamoto
Saitama Institute of Technology, Okabe, Japan

Yasubumi Furuya & Hiroyuki Abé
Tohoku University, Sendai, Japan

ABSTRACT : Shape memory alloy(SMA) TiNi fiber reinforced /epoxy matrix composite was fabricated to demonstrate the active enhancement of mechanical properties, especially, fatigue *resistance* in our proposed new type of smart composite. Fatigue crack propagation behaviors were investigated by using the originally developed fine-grids method. Fatigue crack propagation was arrested or retarded immediately after heating the specimen above Af temperature. It depended on the increase of the pre-strain value for embedded TiNi fibers. Therefore, it seems to be attributed mainly to the introduced compressive stress field in the matrix due to the shrinkage of the pre-strained TiNi fibers .The local retardation phenomena of a fatigue crack approaching to the embedded fiber was also recognized especially very near the fiber. From a measurement of the crack-tip opening displacement (CTOD) in one loading cycle, the macroscopic compressive stress induced in the matrix and following crack closure phenomena play a key role to suppress the crack-tip stress intensity and fatigue crack propagation. These experimental results with suppression of fatigue crack propagation in this smart composite are discussed by comparing with theoretical analysis which are based on the elastically equivalent inclusion modeling and linear fracture mechanics approach.

1. INTRODUCTION

Active enhancement of the mechanical material strength (stiffness, yield strength and fracture toughness) and suppression of the degradation damages(crack, delamination) during in service time are becoming very important subject in the development of advanced engineering composite systems. Recently, the authors developed a new type of "smart" composite where shape memory TiNi fiber was used as a reinforcement and actuator to improve the mechanical properties of the composite at higher use temperatures above the inverse phase transformation temperature (T>Af) of TiNi alloy (Furuya, 1994). The design concept of enhancing the mechanical properties of the SMA smart composite is described in the following paragraph 2. using Fig.1. Up to the present, the enhancing effects of the tensile yield strength as well as the increase of the fracture toughness (K-value) in a TiNi/epoxy composite have been experimentally confirmed because of the large compressive stresses in the matrix and following suppression effect of crack-tip stress concentration

by the shape memory shrinkages of embedded fibers (Shimamoto,1995). In the present paper,

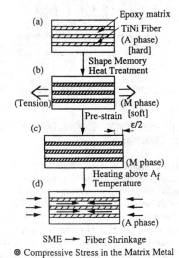

SME → Fiber Shrinkage
⊚ Compressive Stress in the Matrix Metal

Fig.1.Design cosign concept of smart composite

suppression effect of fatigue crack propagations in TiNi fiber reinforced epoxy matrix (TiNi/Epoxy) composite are mainly described. In order to verify these controlling factors and mechanism for fatigue crack retardation in the TiNi shape memory fiber reinforced smart composite, crack-tip deformation and following propagation behaviors near the embedded fibers have been experimentally investigated by using the fine-grid method that was originally developed for local strain measurement with 10 μ m order gage length. Fatigue crack propagation rate (da/dN) gradually decreased as a crack-tip approached to embedded fibers. The increase of the crack-closure stress clearly was observed from the hysteresis curve of crack-tip displacement vs applied load. Fatigue crack retardation , crack-tip deformation and crack closure behaviors will be discussed by correlating with the theoretical analysis(Fujin,1984) for elastic stress distribution in the domain around the embedded inhomogeneous fibers in the composite as shown schematically in Fig.2.

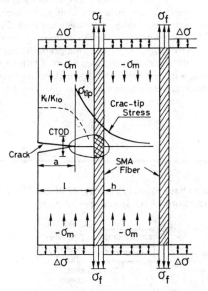

Fig.2. Control factors of fatigue crack propagation in smart shape memory fiber

2. DESIGN CONCEPT FOR SHAPE MEMORY SMART COMPOSITE

Thermoelastic shape memory effect (i.e.shape memory and recovery phenomenon) takes place during martensite(M) to austenite(A)phase transformation in SMA with increasing temperature. Therefore, material functional properties of SMA changes clearly depending on the changes of temperature {Y.Furuya,1994}. It should be noticed as a unique property that SMA shows more higher stiffness (2-3 times) and large recovery stress at the higher temperature region due to inversely thermoelastic phase transformation in opposition to weakening of those properties in the general metals. In consequence, SMA natively has the smart functions ,i.e., (1)sensor (thermal), (2)actuator (shape memory deformation) and (3) memory and shape recovery (namely, processor function) . These unique properties natively with SMA can be utilized to strengthen the composite. The design concept of enhancing the mechanical properties of the SMA composite is schematically shown in Fig.1. TiNi fibers are heat-treated to shape-memorize their initial length at higher temperatures (>Af), then quenched to room temperature(nearly, martensite start temperature =Ms),given tensile prestrain ε (>0) and embedded in the matrix material to form a composite. The composite is then heated to temperature(>Af) at which the TiNi fibers tend to shrink back to their initial length by the amount of prestrain ε , then the matrix is subjected to compressive stress. It is this compressive stress in the matrix that contributes to the enhancement of the tensile properties, fracture toughness and fatigue resistance of the composite.

3. EXPERIMENT
3.1 Specimen preparation

Experimental processing and mechanical testing of TiNi/epoxy composite are described. The shape memorized TiNi fibers(Ti-50.2at%Ni) of 400μ m are diameter supplied by Kantoc Ltd, Fujisawa, Japan are arranged in a mold to which phtoelastic epoxy and specified amount of hardener were poured and then kept at 130°C for 2hours for curing. The photo-elastic sensitivity (α) of the epoxy resin was 0.116 mm/N. After curing, as-molded composite was cooled to room temperature. During the process, TiNi fibers were kept in tension with four different prestrains of 0,1,3 and 5%. A side notch of 2 mm length ,0.3 width with a notch-tip angle θ =60° was then cut into the as molded composite specimen. The geometry of the composite specimen is shown in Fig.3. In order to determine the transformation temperatures and stress-strain relations of TiNi fiber, tensile test of TiNi fiber at different temperatures (T) was conducted. The stress vs. strain curves at constant temperature: T=20, 40, 60, 80 and 100°C were obtained as shown in Fig.4. From another relationship between strain vs. T at constant stress of 94MPa, four transformation temperatures of TiNi fiber (500°C, 30min, water quenched(WQ)) were determined: martensitic start Ms=31°C, martensite finish Mf=15 °C, austenitic start As=57 °C, and austenitic finish temperature Af=63°C. Pre-fatigue crack of about 254μ m (10 grids) length was made by cyclically constant applied stress, σmax=8.17MPa (R= σ min/ σ max=0) and then, the fatigue crack propagation and crack-tip deformation behaviors

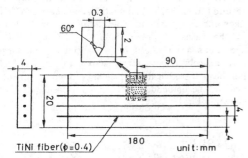

Fig.3.Geometry of TiNi/epoxy composite specimen

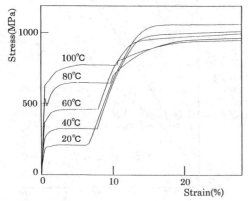

Fig.4.Stress-straincurves of *TiNi fiber* at different temperatures

were investigated. The fine dotted grids of 25.4μm pitch were engraved on the epoxy specimen surface by photo-etching technique for local strain measurement at the crack-tip. By measuring the distortion of the fine-dotted grid, crack-tip deformation and following propagation behaviors near the embedded fibers could be experimentally investigated in detail.

3.2 Fatigue test

Fatigue tests were conducted by using a uni-axially hydraulically, digital-controlled fatigue machine which was equipped with an elevated temperature chamber. The outside view of the elevated temperature fatigue test system is shown in Fig.5. Fatigue crack propagation test was performed under a constant cyclic stress range ($\Delta \sigma$) with sinusoidal pull-pull loading type(R= σ min/ σ max=0, σ max=5.454MPa) of a frequency of 8Hz. The behavior of a fatigue crack propagation was investigated continuously by watching a fatigue crack-tip in an enlarged view through the optical microscope with relay lens. After a fatigue crack propagates to a certain length (ao) from a notch at room temperature(20℃),the temperature of the specimen was heated up to 80℃ above the inverse phase

transformation temperature (Af) of the embedded TiNi fiber, and then, the fatigue crack propagation behavior was investigated until final failure of the specimen. The photographs of the deformed fine grid at the crack-tip were taken continuously during one cyclic load at each certain crack length by using full-automatic camera which was set into optical microscope(x 100 magnification) as shown in Fig.5. Using the enlarging projection equipment, local crack-tip strain was calculated by measuring the distortion of the grid space along the loading direction on the basis of the virgin state of the grid spacing before fatigue test . The gage length for local crack-tip strain measurement is 76.2μm (i.e. three grids length).The accurate crack length was determined by combining X-Y micro stage with its accuracy of 0.001mm and the photo-negafilm of the fatigue crack-tip. Fatigue crack propagation rate (da/dN) was calculated by averaging the two measuring points of the crack lengths. Stress intensity factor range (ΔK) is obtained by using the following formula. $\Delta K = Y(a/W) \Delta \sigma \sqrt{\pi a}$.

Fig.5.Experimental apparatus

4.RESULTS AND DISCUSSION

4.1 Control of fatigue crack propagation by shape memory effect

An typical example of the fatigue crack propagation behavior in the TiNi/epoxy smart composite is shown in Fig.6 where the relationship between crack length(a) and cyclic loading number (N) in the pre-strained 1% for TiNi fibers is shown. Fatigue crack arrest(i.e. point A in the figure) can be clearly recognized immediately after the increase of the environmental temperature of the chamber from 20℃(R.T.) into 80℃(>Af). In order to continue to investigate the following fatigue crack propagation behavior, the cyclic stress range ($\Delta \sigma$) was increased up to 15% for 1% prestrain specimen, up to 20% for

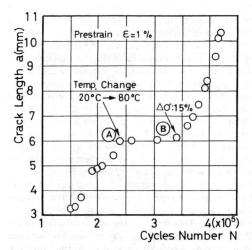

Fig.6.Crack length(a) vs cycles (N) curve

3% prestrain specimen, respectively(i.e. point B in the figure). The relationships between fatigue crack propagation rate (da/dN) and stress intensity factor range(ΔK) in three different pre-strained , ε =0,1 and 3% specimens are shown in Fig.7. As a result, it

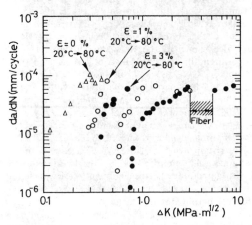

Fig.7.da/dN vs ΔKin TiNi/epoxy composite in three cases of ε =0%, ε =1% and ε =3%

could be experimentally confirmed that the suppression effect of fatigue crack propagation was attained to about 12 times for 1% prestrain case and 14 times for 3% prestrain case respectively in comparison with the case of 0% prestrain. In this figure, the initial difference of starting crack propagation level would be resulted from the initially introduced residual compressive stresses during the fabrication process(i.e. heating, solidification and following curing for polymer etc.) of the TiNi/epoxy composite. Besides, a slight drop of da/dN after temperature

change in the prestrain ε =0% specimen would be resulted from thermal mismatch strain between TiNi fiber and epoxy matrix. The distinct drops of da/dN in the cases of pre-strained, ε =1% and 3% , specimens are thought to be caused by the large residual compressive stresses(-σm in Fig.2) which has been introduced by the shape memory shrinkages of the embedded TiNi fibers as schematically shown in Fig.1.

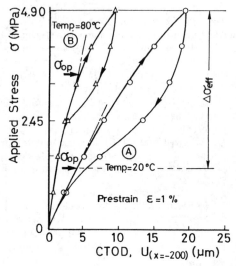

(a)prestrain, ε =1%

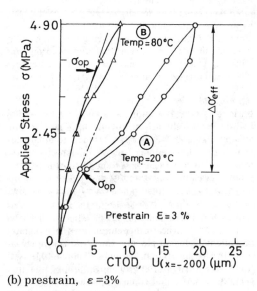

(b) prestrain, ε =3%

Fig.8.Applied stress(σ) vs CTOD curves before and after heating above Af temperature

4.2 Local fatigue crack-tip deformation behavior

The large residual compressive stresses($-\sigma$m in Fig.2) associated with the shape memory shrinkages of the embedded TiNi fibers mentioned above will decrease the stress concentration at the crack-tip ,and moreover, it also will make an important role to change the fatigue crack closure phenomena. In order to confirm this effect, it becomes necessary to investigate the crack opening and closing behavior in one loading cycle in detail. By using real-time fine grid strain measuring technique (Shimamoto, Yokota, Furuya and Takahashi,1988), the hysteresis loops of the relationship between applied external stress(σ) and crack-tip opening displacement U (CTOD) were investigated. The value of U(CTOD) was measured at the point (X=-200 μm) behind a propagating crack-tip. The variations of the $\sigma \sim$ U (CTOD) hysteresis loops before(temp.=20 ℃) and after(temp.=80℃) heating above Af temperature for two kinds of prestrained, ε =1% and ε =3% specimens are shown in Fig.8(a)(b) respectively. It can be clearly recognized that CTOD decreased more than 50% , and crack-opening stress (σ op) which is defined the crooked point (i.e. arrow point) in a loading half cycle
increased toward higher loading level as shown in Fig.8. In consequence, it was found that the effective cyclic stress range($\Delta \sigma_{eff}$ =σmaxσop) decreased as much as over 50% after heating temperature (80℃) above Af for two cases of pre-strains, ε =1% and ε =3%. The decreasing degree of ε =3% seems to be slightly more larger than that of ε =1% that will play a key Role to active suppression of fatigue crack propagation in the smart shape memory TiNi /epoxy composite.

4.3 Comparison between theoretical analysis and experimental result

1) Fatigue crack propagation retardation

AS we can find from the $\sigma \sim$U(CTOD) hysteresis loops in Fig.8, the difference between crack opening stress levels before and after heating, i.e. 20℃ and 80 ℃ respectively, means Increasing stress(σup = σop(80℃)- σop(20℃)) which becomes necessary for propagating fatigue crack-tip opening. The values of σup in these two cases become about 2.2 MPa for prestrain ε =1% and 2.4 MPa for prestrain ε =3% specimen respectively might correspond to the introduced compressive stress in the matrix associated with shape memory shrinkage effect of the TiNi fibers. By the theoretical prediction using Eshelby's equivalent inclusion model, the compressive stress value in the matrix, σm becomes about -6.3MPa (4) (Taya and Furuya ,1993). This difference, about a twice value for σm estimation ,between theory and experiment might arise mainly from the incomplete bonding state at the interface between fibers and

matrix. Anyhow, it can be concluded that the large compressive residual stresses σm which can be actively introduced using shape memory shrinkages of TiNi fibers in this type of "smart" composite will be able to act very effectively for decreasing the stress intensity at a crack-tip, and then for suppressing fatigue damage, especially, fatigue crack propagation.

2)Crack propagation behavior within a domain very near a embedded TiNi fiber

As for the stress intensity factor .K value of the approaching single crack into the interface between inhomogeneous filler and matrix, the theoretical analysis for the variations of K has been studied based on elastic fracture mechanics theory (Fujino, Sekine and Abe, 1984). From their analysis for the semi-infinite composite model with a long reinforced phase with an edge crack, the normalized K value, KI/σo$\sqrt{\pi a}$. is apt to decrease exponentially depending on the normalized distance (a/l) between the crack-tip and the reinforced inhomogeneous phase , although K also depends a non-dimensional para-meter (Γ) that is related with volume function ratio(h/l) as well as stiffness ratio(G*/G). In order to compare the experiment with this theory, the fatigue crack propagation and local fatigue crack-tip deformation behaviors studied in detail ,especially, in the domain D very near the fibers. The arrest orretardation of fatigue crack propagation near the fibers are also shown in the right part in Fig7. In this experimentally used specimen of finite width (Wo), the actual K value should be inevitably affected by another increasing factor from a finite width correction factor Y(a/Wo),therefore, the real K value in the testing specimen will not be changed so much because of the cancellation of the above mentioned two factors, inhomogenity(-) and finite correlation(+) factors as shown in Fig.9. The CTOD behaviors of the fatigue crack approaching to TiNi fiber were also investigated to verify this domain effect. It was found that the width of the hysteresis loop , i.e. the displacement (X-axis) range of σ vs U curve as similarly shown in Fig8, gradually decreased as a crack-tip approached to the domain near the embedded TiNi fiber and finally it was saturated at a constant small value when a fatigue crack began to arrest at the domain region (D) very near the fiber. This typical fatigue crack arrest phenomena would be correlated with the theoretically predicted decrease of K value at the domain area very near the inhomogeneous reinforced fiber. Such a retardations of a fatigue crack propagation near the fibers are experimentally confirmed by the fracture surface traces of propagating crack by fractography observation using SEM as shown in Fig.10. The quantitative discussions will be expected in the next study by correlating the crack-tip stress/strain behaviors with theoretical analysis using more ideally wider composite specimen.

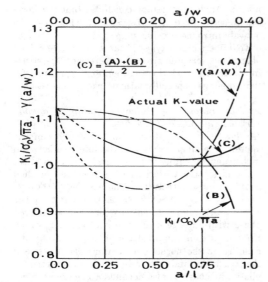

Fig.9.Estimation of the variation of actual of KI value with increasing the crack length(a) of the test specimen with finite width(W). The KI value can be calculated by adding the two factors,(1)inhomogeneity of fiber for decreaing K value near the fiber and (2) finite width correlation factor Y(a/W)for decreaing K value with increasing the crack length. Each value is shown as a normalized value in reference with its the initial state. Each paramater relates with the specimen geometry and stress intensity factor in Fig.2.

5. CONCLUSION

Shape memory TiNi fiber reinforced /epoxy matrix composite was fabricated to demonstrate the active enhancement of mechanical properties, especially, fatigue resistance in our proposed new type of smart composite. Fatigue crack propagation behaviors were investigated by using the originally developed fine-grids which were engraved on the specimen surface by photo-etching. Fatigue crack propagation was arrested or retarded immediately after heating the specimen above Af temperature and it depended on the increase of the pre-strained value for embedded TiNi fibers. Therefore, it seems to be caused mainly by the introduced compressive stress field in the matrix due to the shrinkage of the pre-strained TiNi fibers. The local retardation phenomena of a fatigue crack approaching to the embedded fiber was also recognized especially in the local domain region very near the fiber. From measurement of the crack-tip opening displacement (CTOD) through the distortion of grid at a crack-tip in one loading cycle ,the macroscopic compressive stress induced in the matrix as well as following crack closure phenomena play a key

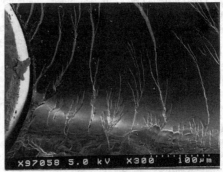

Fig.10. An example of a SEM fractography ashowis the fracture surface near the embedded TiNi fiber in the prestrain 1% specimen. It was found that the striation spacing near the fiber decreased in correspondence with the retardation phenomena of fatigue crack propagation at the domain region around the fiber

role to suppress the crack-tip stress intensity and fatigue crack retardation. From the comparisons between theoretical analysis and experiment, the analytical model which is based on the elastically equivalent inclusion modeling combining with fracture mechanics approach can explain the same trend as the experimental results of the suppression effects for a fatigue crack propagation in this shape memory fiber reinforced composite.

REFERENCES

1.)Y.Furuya,A.Sasaki and M.Taya, "Enhanced Mechanical Properties of TiNi Shape Memory Fiber/Al Matrix Composite" Mater. Trans. JIM, Vol.34(3), pp224-227,1993

2)A.Shimamto and M.Taya, "Reduction in KI by shape memory effect in a TiNi shape memory fiber reinforced epoxy matrix composite"J.JSME,63-605.A,pp.26-31.1997

3.)K.Fujino, H.Sekine and H.Abe, "Analysis of an edge crack in a semi-infinite composite with a long rein-forced phase",Inter.J.Fracture,25(1984),81-94

4.)A.shimamoto, A.Yokota, Y.Furuya and S. Takahashi, "Fundamental study on rupture by low cycle fatigue of polymers applying fine grid method",Proc.5th Inter. Conf. Exp. Mech., (Portland) pp.538-543.1988

5.)M.Taya,Y.Furuya,Y.yamada,R.Watanabe,S.Shibata and T.Mori,"Strengthening mechanism of TiNi shape memory fiber /Al matrix composite" ,SPIE, 1916(1993),373-383

High cycle fatigue

Experimental Mechanics, Allison (ed.)© 1998 Balkema, Rotterdam, ISBN 90 5809 014 0

Failure of metallic wires under fatigue cycles induced through point contact

R.Ciuffi
Dipartimento di Meccanica e Tecnologie Industriali, Università di Firenze, Italy

ABSTRACT: A new approach has been explored to get fatigue tests on wires which could have a better significance than the standard ones for the preview of the behaviour of the wire when assembled in a wire rope. At this end a simple and inexpensive machine has been designed and built and a series of tests were performed on standard harmonic steel wire. This paper exposes the principles of the testing procedure and the results of the first performed.tests

1 INTRODUCTION

In wire ropes axially loaded and running over sheaves or pulleys the fatigue life depends on the variation of the load on the single wires due to the change of curvature of the rope and on the local hertzian load at the points of contact with the other wires or with the pulley.

The importance of the influence of the point stresses is attested by the fact that, all other conditions standing, fatigue life of ropes on very hard pulleys is largely shorter than on soft ones and in addiction the wire breaking sequence is generally different, with wire fractures starting under the external layers and not at the surface of the rope as usual with hard pulleys. The influence of contact stresses is obviously larger when accompanied by fretting.

At present for wire ropes acceptance tests wire tests are performed too (traction, torsion, flexion) but the importance of these tests is mainly for the control of the uniformity of the product and not for a well defined influence on the fatigue behaviour of the rope. The testing procedure which will be exposed in this paper and the related testing machine, have been designed just to correlate, if possible, the fatigue behaviour of the wire rope with that of the wires it is composed with.

2 THE MACHINE

The design of the machine is very simple and its construction almost inexpensive. The scheme is illustrated in Figure 1.

A needle bearing with the external ring removed is fitted on a shaft connected with an electric motor. A

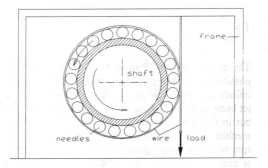

Figure 1. Scheme of the machine

loop of the wire to be tested is wound around the bearing, one of the ends being fixed to a load frame and the other loaded with the test load (for instance a deadweight). If, as in Figure 1, the straight sections of the wire, which are tangent to the bearing.external circumference are also on the same line, no trans-tverse load is applied to the shaft which can be designed accordingly. The wire spans in the loop around the bearing are loaded in traction by the external load and in shear and flexion by the needles which are pushed against the inner ring by the wire (Figure 2).

When the shaft is put into rotation by the motor the wire acts as the fixed external gear of a planetary gear so that the cage rotates and the point of contact of the needles with the wire moves along the wire loop which, in each section, is subjected to a periodically variable (i.e. fatigue) bending and hertzian stress while the traction, neglecting the influence of the (low) friction and secondary periodical variations, remains constant.

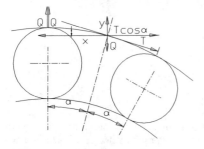

Figure 2. Geometry and equilibrium of wire and needles.

The number of fatigue cycles per unit of time f is proportional to cage speed and to the number of needles n in the bearing. Let the rpm of the shaft be N_f, D the external diameter of the inner race and d' the diameter of the needles

$$f = \frac{N_f \times D \times n}{2 \times (D \times d')}$$ (1)

The motor was feed through an inverter to have the possibility of changing the operating speed. The influence of the frequency on the tests results has not jet been explored and the velocity has been kept slow but in any case the testing frequency is high for a mechanical machine as f.i. at a velocity of only 120 rpm with a 40 needles bearing the testing frequency is almost 40 cycles per second. In any case the temperature of the wire remained low even at the highest frequencies attained as the points of maximum stress move along the wire which is fully immersed in air, neglecting the lubricated points of contact with the needles.

In Figure 3 the picture of a detail of one of the testing machines(a modified rotating bending one) is shown. One dedicated machine has been designed, manufactured and used for the tests.
Similar arrangements were utilized in France and Belgium many years ago in wire ropes traction-bending testing equipment with four or more idle pulleys hinged at the periphery of rotating big wheels to increase the testing frequency avoiding the problems of alternating motion of standard machines but the cost of manufacturing many expensive pulleys for a single rope diameter and the traction variations induced in the rope by the changes of

geometry of the pulleys-rope system discouraged the use of this kind of machines.
In wire testing the availability of standard, cheap bearings with a large number of rolling elements for at least small diameter wires makes this kind of application very promising, as our test have demonstrated.

Figure 3. A 90mm nominal diameter bearing inner ring and cage with test wire wound around shown installed on a modified rotating bending machine.

3 THE TESTS

To initiate a test a length of wire is cut and one end is fixed with a simple screw clamp to the frame through a lever and microswitch device to interrupt power when the wire breaks down. The wire specimen is then wound around the bearing and the load T is applied to the other end. In our tests the bearing axis was horizontal and the wire section small so that simple metric weights where used to apply the load.
In most tests.fixing points had to be added to keep the wire within the needle track.
When switching on the motor the test begins and the number of shaft revolutions is recorded.

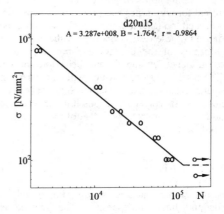

Figure 4. Fatigue tests results, d20n15 bearing.

Failure occurred mostly at the tangential lead in or lead out sections of the wire.

The test results are presented here simply as plots of the number N of cycles to failure vs nominal traction stress σ (σ =T/wire section area) as is common for fatigue reports for wire ropes. The continuous straight lines in the log-log plots represent the regression lines according with the formula

$$N = A\sigma^B \tag{2}$$

In Figures 4÷7 the results of the tests on four different bearings of a d=0.5mm diameter music steel wire are reported.

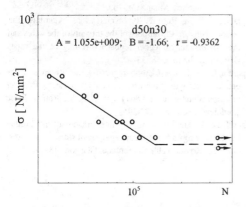

Figure 5, Fatigue tests results, d50n30 bearing.

The letters and numbers within the plots represent: d the nominal bearing diameter, n.the number of needles, A and B the coefficients of (2), r the cor-relation coefficient. For instance d50n41 means a bearing having 50mm nominal (inner) diameter with 41 needles. The dotted horizontal lines represent the fatigue limit σ_L assumed to be 90% of the maximum stress for which a life of more than 2 million cycles was attained and the test interrupted.

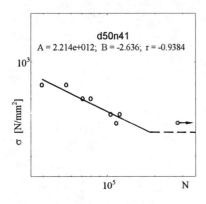

Fig. 6. Fatigue tests results, d50n41 bearing.

In Table 1 the main characteristics of the bearings and the corresponding values of σ_L are summarized. R is the radius of the circle.circumscribed about the needles, α is half the angle between two adiacent needles (Fig.2)

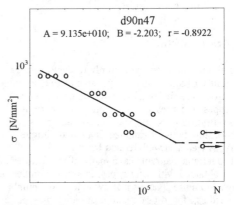

Figure7. Fatigue tests results, d90n47 bearing.

Table 1 Geometrical characteristics of the test bearings and fatigue limit σ_L.

Bearing	d' [mm]	R [mm]	α [rad]	σ_L [N/mm²]
1 d20n15	3.5	16	.20944	91.67
2 d50n30	4	34	.10472	229.2
3 d50n41	2.5	30	.07662	504.2
4 d90n47	4	54	.06684	458.4

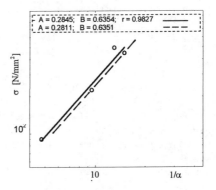

Figure 8. Fatigue limit σ_L vs 1/α.

The values of σ_L of Table 1 have been plotted vs 1/α in Figure 8. The lines in the figure represents the regression curve

$$\sigma_L = A(1/\alpha)^B \tag{3}$$

The solid line takes into account all the four results while the dotted one takes into account only the bearings 2 and 4 having d'=4. The corresponding constants are not very different so it is probable that the influence of the number of needles is by far more important than their diameter. A more accurate analysis of the results taking into account the bending and hertzian stresses is being performed and will be presented in the near future.

4-CONCLUSIONS

The aim of this paper was simply to demonstrate the feasibility of a new and inexpensive testing machine and procedure for metallic wires, mainly steel wires for ropes and tires construction and the results of these first tests seem very encouraging. A clear relationship between applied load and fatigue life exists as well as a well defined fatigue limit.

The fatigue diagrams have many of the features of those obtained with wire ropes. Here the number of cycles to failure depends mainly on load and number of needles, for ropes mainly on load and pulley diameter.

A peculiar characteristic of the results is that the number of fatigue cycles near the fatigue limit is not in the order of one million or more as in standard fatigue tests on steel specimens but a lot lower, suggesting a peculiar type of fatigue is applied.

Even if a deeper analysis (again, as for ropes) and much more experiments are needed to asses the influence of the many factors affecting fatigue life of wires, at least for comparative tests and quality control the proposed testing procedure seems to have a consistent value.

REFERENCES

Verwilst, J. 1964. Recherches sur les fils. OIPEEC Bulletin n.4: 11-12.

Gousseau, G. 1964. Essais de fils pour cables. OIPEEC Bulletin n.4: 17-18.

Luboz, G. 1973. Essais comparatif sur fils. OIPEEC Round Table Conference, Milan: 158-165.

Oplatka, G & M. Roth 1980. Investigations regarding the relation between the number of the torsions of the wires and the fatigue strength under repeated bending of the ropes. OIPEEC Bulletin n. 38: 33-40.

Oplatka, G. 1981. Investigation regarding the influence of the breaking strength of the wires on the fatigue strength under repeated bending of stranded ropes. OIPEEC Round Table Conference, Krakow: 278-285.

Hansel, J & W. Olesky 1985. The determination of the fatigue life of wire ropes based on an analysis of stresses in wires. OIPEEC Round Table Conference, Glasgow: 13-24.

Experimental Mechanics, Allison (ed.)© 1998 Balkema, Rotterdam, ISBN 90 5809 014 0

High-cycle fatigue properties of austenitic stainless steels

Nobusuke Hattori & Shin-ichi Nishida
Faculty of Science and Engineering, Saga University, Japan

ABSTRACT: Fatigue tests had been performed using rotating bending fatigue testing machine about three kinds of austenitic stainless steels (SUS304, SUS304N and YUS170) observing by the successive-taken replica method. The main results obtained in this test are as follows; (1) The fatigue limit exists distinctly for the specimens of SUS304 and SUS304N, respectively but not for those of YUS170. In addition, the fatigue strength of these steels by 10^7 cycles does not necessarily depend on their tensile strength. (2) The reason why the fatigue strength of YUS170 decreases as compared with those of SUS304 and SUS304N is that the effects of strain aging and martensitic transformation induced by deformation of YUS170 are smaller than those of SUS304 and SUS304N. (3) The fatigue limits of SUS304 and SUS304N increase by 5 and 30 MPa, respectively by the annealing at 300°C for 2 hours in vacuum during fatigue test by the specified cyclic ratio under the constant higher stress amplitude than their original fatigue limit. (4) The non-propagating micro crack does not exist in the specimens of SUS304 and SUS304N suffered by 10^7 cycles under the stress amplitude of fatigue limit.

1 INTRODUCTION

According to the social demand toward high quality of materials, stainless steels have been widely used because of their superior properties such as their high corrosion resistance, tensile strength in high temperature, especially in view of maintenance-free even in pretty severe environment. Therefore, the stainless steels have been utilized in the various fields such as chemical plants, automobile, electric, building, ship, subway trains industries etc. As most of past experimental studies have only focused to corrosion properties or creep strength, there are few reports about the fatigue properties of these materials, which become one of the most important one for structural materials. In addition, the high strength stainless steels have been recently requested in the industries. Then, one of the effective methods to improve the tensile strength of austenitic stainless steels is to make use of solid solution strengthening by adding carbon or nitrogen content in these materials. The authors have reported on fatigue strength and fatigue crack initiation behavior of various kinds of stainless steels until now (NISHIDA 1995,1995&1998) and have mainly tried to investigate the effect of nitrogen content on the above subjects in this study.

2. EXPERIMENTAL PROCEDURE

2.1 *Testing materials*

The materials used in this test are three kinds of austenitic stainless steels, i.e., the most representative austenitic stainless steel SUS304 , nitrogen-increased SUS304 (SUS304N) and alloy elements-increased SUS304N (YUS170). Table 1 and 2 list the chemical composition and mechanical properties of these three kinds of materials. Their nitrogen content were 0.042% (SUS304), 0.21% (SUS304N) and 0.335% (YUS170), respectively and the ratio of nitrogen content becomes about 1 to 5 to 8 in this order.

Table 1. Chemical composition.   mass %

Materials	C	Si	Mn	P	S	Ni	Cr	Nb	N
SUS304	0.04	0.51	0.86	0.027	0.05	8.81	18.86		0.042
SUS304N	0.05	0.64	0.83	0.029	0.001	7.64	19.02	0.1	0.21
YUS170	0.03	0.78	0.58	0.027	0.001	13.22	24.35	—	0.335

Table 2. Mechanical properties.

Materials	Proof stress MPa	Tensile strength MPa	Elongation %
SUS304	296	663	71
SUS304N	469	793	49
YUS170	445	808	55

Figure 1 shows the shape and dimensions of fatigue specimen. All of the specimens were cut out by coinciding the rolling direction with the specimen's axis and by making a partial shallow notch at its rolling surface side. This shallow notch exists for limiting fatigue damaged part and does not affect for its fatigue strength at all.(NISITANI 1978) All of the specimens were annealed in vacuum at 500°C for 2 hours after polishing with fine emery paper and thereafter electro-polished to the depth of about 40 μm in diameter.

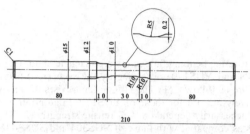

Figure 1. Shape and dimensions of fatigue specimen.

2.2 Testing method

The main fatigue tests have been performed to investigate the effect of nitrogen content on the fatigue strength are as follows; Condition (A); conventional fatigue tests to compare fatigue properties among the three kinds of materials. Condition (B); fatigue tests using the specimen, which was annealed at 300 °C for 2 hours in vacuum during test by the specified cycle ratio (just before of fatigue micro-crack initiation) and suffered under the same stress amplitude as before. Condition (C); fatigue tests using the specimen, which was suffered by the higher stress amplitude than the fatigue limit until just after of fatigue micro-crack initiation and loaded successively by the original stress level of fatigue limit.

The fatigue tests had been performed using Ono-type rotating bending fatigue testing machine with 98 N-m load capacity. In the case of fatigue tests of these kinds of materials, testing part of each specimen was cooled by pouring fresh water for prevention of exothermic fracture. It is confirmed that there is no difference between the result under low frequency in air and that under high frequency with cooled by distilled water. (NISITANI 1990) Fatigue crack initiation behavior of each specimen was observed by the successive taken replica method in the circumferential direction at the specimen's surface.

3. RESULTS AND DISCUSSION

3.1 Fatigue strength

Figure 2 shows the S-N curves, which are determined from the fatigue tests under the above condition (A). The fatigue limits of SUS304 and SUS304N distinctly exist and they are 280MPa and 320MPa, respectively. The former fatigue limit is smaller than the latter one by 40MPa. In addition, the fatigue strength of SUS304N becomes considerably higher than that of SUS304. It should be noted that the nitrogen content improved the fatigue strength of austenitic stainless steels. The high strength of SUS304N would be attributed in the solid solution strengthening effect, (MURATA 1992) strain hardening, (NISHIDA 1995) etc., by nitrogen in matrix. On the other hand, though the fatigue strength of YUS170 becomes nearly equal to that of SUS304N in high stress level, the specimen is broken under the lower stress level than not only the fatigue limit of SUS304N but also that of SUS304. That is, the fatigue limit does not distinctly exist in the case of YUS170. This fact would be unexpected result for most of the metallurgists related to stainless steels.

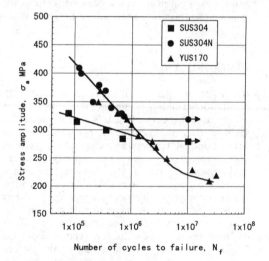

Figure 2. S-N curves.

Figure 3 shows the representative successive observation results for fatigue crack initiation behavior of SUS304, SUS304N and YUS170, respectively. The fatigue cracks of these materials initiate in grain boundary or in the neighborhood by the cyclic ratio between 5 to 15%.

Figure 4 shows the surface state of the specimens suffered under the stress amplitude of each fatigue limit or fatigue strength by 10^7 cycles. Though slip bands exist in the specimens of

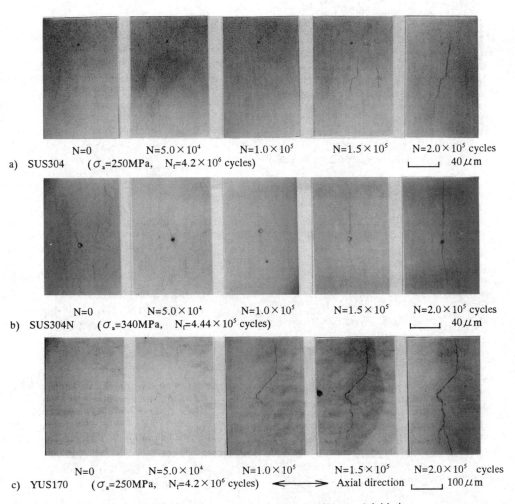

N=0   N=5.0×10⁴   N=1.0×10⁵   N=1.5×10⁵   N=2.0×10⁵ cycles

a) SUS304   (σ_a=250MPa,   N$_f$=4.2×10⁶ cycles)   40 μm

N=0   N=5.0×10⁴   N=1.0×10⁵   N=1.5×10⁵   N=2.0×10⁵ cycles

b) SUS304N   (σ_a=340MPa,   N$_f$=4.44×10⁵ cycles)   40 μm

N=0   N=5.0×10⁴   N=1.0×10⁵   N=1.5×10⁵   N=2.0×10⁵ cycles

c) YUS170   (σ_a=250MPa,   N$_f$=4.2×10⁶ cycles)   ←——→ Axial direction   100 μm

Figure 3. Successive surface observation of fatigue crack initiation.

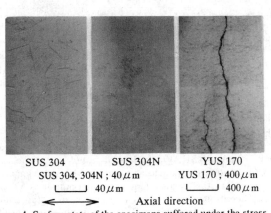

SUS 304   SUS 304N   YUS 170

SUS 304, 304N ; 40 μm   YUS 170 ; 400 μm

└——┘ 40 μm   └———┘ 400 μm

←————————→ Axial direction

Figure 4. Surface state of the specimens suffered under the stress
amplitude of each fatigue limit or fatigue strength by 10⁷ cycles.

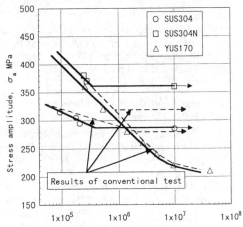

Figure 5. S-N curves obtained from the
fatigue tests under the . test condition (B).

1099

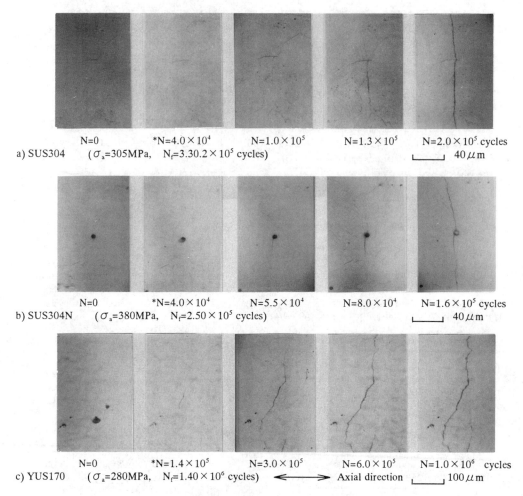

	N=0	*N=4.0×10⁴	N=1.0×10⁵	N=1.3×10⁵	N=2.0×10⁵ cycles

a) SUS304   (σ_a=305MPa, N_f=3.30.2×10⁵ cycles)   ⌊————⌋ 40μm

	N=0	*N=4.0×10⁴	N=5.5×10⁴	N=8.0×10⁴	N=1.6×10⁵ cycles

b) SUS304N   (σ_a=380MPa, N_f=2.50×10⁵ cycles)   ⌊————⌋ 40μm

	N=0	*N=1.4×10⁵	N=3.0×10⁵	N=6.0×10⁵	N=1.0×10⁶ cycles

c) YUS170   (σ_a=280MPa, N_f=1.40×10⁶ cycles) ◄————► Axial direction ⌊————⌋100μm

Figure 6   Successive surface observation of fatigue crack initiation

SUS304 and SUS304N, there do not exit non-propagating micro-cracks. In addition, as there is no distinct fatigue limit in YUS170, pretty long fatigue cracks are observed in this specimen.

3.2 Change of fatigue strength by annealing at low temperature

Figure 5 shows S-N curves obtained from the fatigue tests under the test condition (B). In the case of SUS304, after cycling to N/N_f = 0.075 at the specified stress amplitude over the fatigue limit of this material obtained under the condition (A), and annealed at 300℃ for 2 hours, the fatigue tests had been continued using the specimens under the same stress amplitude until failure of specimen or by 1×10⁷ cycles. On the other hand, the specimens in the case of SUS304N were annealed at the same low temperature as the above for 2 hours after the cycling to N/N_f = 0.15 and the fatigue tests were continued. Though the fatigue tests for SUS304 had been also performed using the specimen annealed after cycling to N/N_f = 0.15, the improvement of fatigue strength was not observed. Thus, the cycling number of annealing of the specimen of SUS304 is different from that of SUS304N. As shown in Fig. 5, the fatigue limits of SUS304 and SUS304N under this condition were 285 MPa and 350MPa, respectively. The fatigue limit of SUS304 had been tested under the test condition (B) increases only 5 MPa as compared with the results under the condition (A). In contrast, that of SUS304N had been tested under the test condition (B) is larger than the results under the condition (A) by 30MPa and there is scarcely difference on fatigue strength between condition (A) and (B). In particular, it should be noted that the fatigue strength of SUS304N increased using the specimens annealed at low temperature during tests. It is considered that the effect of strain aging by nitrogen in SUS304N increases its fatigue limit. On the other hand, in the case of YUS170, the fatigue strength becomes decrease under the test condition (B) than that of condition (A). Then, it is considered that there does not exist the aging effect in YUS170.

Figure 6 shows the representative successive observation results for fatigue crack initiation behavior of SUS304, SUS304N and YUS170, respectively. Each asterisk mark means the cyclic number when the specimen is annealed in vacuum at 300°C × 2hr. There is no difference between the surface state of these specimens and those as shown in Fig. 4.

Table 3 lists the micro-Vickers hardness number of grains before and after fatigue test under the same stress amplitude of 350MPa by fatigue crack initiation in the length of 1～2 grains. The hardness of the specimen of after test becomes increased for all of the materials than those of before test and this hardness increase would be mainly due to work hardening effect during fatigue test. When focusing to the hardness increase of the crack tip, there is much difference among three kinds of material. The hardness increase becomes smaller in the order of SUS304, SUS304N and YUS170 and would be mainly due to aging effect and strain induced transformation (NISHIDA 1995).

Table 3. Micro-Vickers hardness number of grain before and after test, Hv(0.029N).

Materials	Before test	After test*		
		No-slip bands	Slip bands	Crack tip
SUS304	230	246	267	318
SUS304N	246	273	301	311
YUS170	266	286	286	289

*: σ_a=350MPa, cycle ratio ; just after fatigue crack initiation in the length of 1～2grains.

3.3 Fatigue micro-cracks

According to the results of section 3.1, the non-propagating fatigue micro-cracks are not observed in the specimens of SUS304 and SUS304N, which are suffered under the stress amplitude of fatigue limit by 10^7 cycles.

Figure 7 shows the testing results under the condition (C). That is, the specimens, which are suffered under the higher stress amplitude than its fatigue limit by 20 or 50MPa, respectively, and in which fatigue micro-cracks initiate in the length of 1～2 grains, are loaded under stress amplitude of its original fatigue limit. As shown in Fig. 7, all of these specimens were broken by about 3.0 × 10^4 cycles. That is, the non-propagating micro-cracks

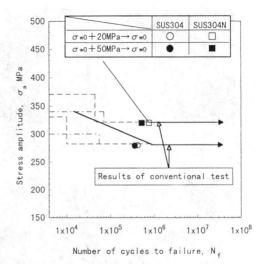

Figure 7. S-N curves obtained from the fatigue tests under the test condition (C).

do not exist in the case of SUS304 and SUS304N and their fatigue limit is controlled by the condition of fatigue crack initiation.

Figure 8 shows the representative successive observation results for fatigue crack initiation behavior of SUS304 and SUS304N, respectively. Each asterisk mark means the cyclic number when the stress amplitude is changed from over stress to the original stress level of fatigue limit. There is also no difference between the surface state of these specimens and those as shown in the above.

4 CONCLUSION

The main results obtained in this study are as follows;
(1) The fatigue limits of SUS304, SUS304N and YUS170 are 280, 320 and 230MPa (fatigue strength by 10^7 cycles), respectively. Though there distinctly exist fatigue limits in SUS304 and SUS304N, but not in YUS170. The high strength of SUS304N is considered to be due to strain aging and solid solution strengthening by nitrogen in matrix. On the other hand, in the case of YUS170 there is not distinct fatigue limit and the fatigue strength by 10^7 cycles is smaller than fatigue limit of SUS304, because YUS170 has very stable structure and does not show strain-induced transformation and strain aging.
(2) The fatigue strength of SUS304 'and SUS304N are both improved using the specimens which was annealed during fatigue test. This phenomena is mainly considered due to the effect of strain-aging based on the difference of nitrogen content. While,

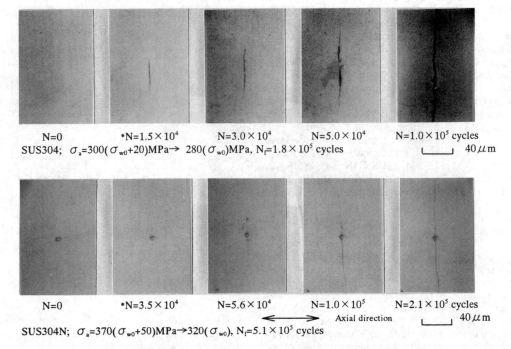

N=0   *N=1.5×10⁴   N=3.0×10⁴   N=5.0×10⁴   N=1.0×10⁵ cycles

SUS304; σ_a=300(σ_{w0}+20)MPa→ 280(σ_{w0})MPa, N_f=1.8×10⁵ cycles   ⌐___⌐ 40μm

N=0   *N=3.5×10⁴   N=5.6×10⁴   N=1.0×10⁵   N=2.1×10⁵ cycles

←——→ Axial direction   ⌐___⌐ 40μm

SUS304N; σ_a=370(σ_{w0}+50)MPa→320(σ_{w0}), N_f=5.1×10⁵ cycles

Figure 8. Successive surface observation of fatigue crack initiation.

fatigue strength of YUS170 are slightly decreased under the same condition as the above and the strain-aging effect was not observed in YUS170.

(3) The non-propagating micro-cracks have not been observed in SUS304 and SUS304N specimens subjected to the stress amplitude of fatigue limit by $1×10^7$ cycles under the condition not only of (A) but also of (B).

(4) When the micro-cracks in the length of 1∼2 grains initiated in the specimen suffered by the specified cycles under the higher stress amplitude than the fatigue limit of SUS304 and SUS304N,respectively, the specimens have broken before 1 ×10^7 cycles under the same original stress amplitude of fatigue limit.

REFERENCES

MURATA. Y et al 1992, Tetsu-to-Hagane Trans. of JSME (in Japanese), Vol.78, No.3: 346-353.
NISHIDA. S et al 1995, Proc. of ATEM'95: 177-182.
NISHIDA. S et al 1995, Proc. of JSME(in Japanese), No.95-1: 356-357
NISHIDA. S et al 1995 Trans. of JSME(in Japanese), Vol.61, No.582: 211-216.
NISHIDA. S et al 1996, Proc. of JSME(in Japanese), No.96-1: 364-365
NISITANI. H et al 1978, Trans. of JSME (in Japanese), Vol.44,No.377: 1-7.
NISITANI. H et al, 1990, Trans. of JSME (in Japanese), Vol.56, No.525: 1067-1073.

Experimental Mechanics, Allison (ed.) © 1998 Balkema, Rotterdam, ISBN 90 5809 014 0

Fatigue properties of notched austenitic stainless steel

N. Hattori & S. Nishida
Faculty of Science and Engineering, Saga University, Japan

Y. Yano
Faculty of Science and Engineering, Saga University (Graduate School), Japan

A. Yamamoto
Nippon Steel Corporation, Japan

ABSTRACT: The effect of stress concentration factor on the fatigue properties have been investigated using the circumferentially notched specimens of typical austenitic stainless steel SUS304, whose notch radii only were changed among $\rho = \infty$ (i.e. plain specimen) to 0.1 mm. Though the fatigue cracks in the specimens with a blunt notch initiate at one point, those in the specimens with a sharp notch initiate at several points. There exist the slip bands in the surface of the specimen under the stress amplitude of fatigue limit by 1×10^7 cycles, and do not exist the non-propagating micro-cracks in all kinds of the specimens. Furthermore, it has been found that notch sensitivity of austenitic stainless steels is higher than that of a typical plain carbon steels such as 0.16%, 0.36%, or 0.46% carbon steels under the higher stress concentration factor region.

1 INTRODUCTION

Stainless steels have been widely used because of their superior properties such as corrosion resistance, and heat resistance, especially, in view point of maintenance-free for corrosive environment. Thus, these materials are used in the many parts of automobile, electric, structure, subway, ship, etc.. As most of the previous experimental studies have been focused on corrosion properties and creep strength, there are few papers about fatigue properties about these materials. However, the fatigue properties of these materials are important when using in structural materials.

With regard to the classified results of failures according to cause, one of the authors (Nishida 1992) has reported that more than 90 % of failures are caused by fatigue directly or indirectly. Furthermore, it has been found that more than 90 % of fatigue crack initiated from stress concentrated parts.

Therefore, it is very important to investigate the fatigue strength of notched parts. There are a few reports on the fatigue strength of stainless steel of notched specimen from the viewpoints of the fundamental mechanism or the theoretical analysis.

The authors mainly try to investigate the fatigue strength and fatigue crack initiation behavior using the typical austenitic stainless steel with a notch.

2 EXPERIMENTAL PROCEDURE

2.1 Testing materials

The materials used in this test is a typical austenitic stainless steel, i.e. JIS type SUS304. Table 1 and 2 list the chemical composition and the mechanical properties, respectively.

Figure 1 shows the shape and dimensions of fatigue specimen, which was cut out by coinciding the

Table 1. Chemical composition. mass %

	C	Si	Mn	P	S	Ni	Cr	Nb	N
SUS304	0.04	0.51	0.86	0.027	0.05	8.81	18.36	-	0.042

Table 2. Mechanical properties.

	Proof stress $\sigma_{0.2}$ MPa	Tensile strength σ_B MPa	Elongation %	Fatigue limit σ_{w0} MPa
SUS304	245	590	68	280

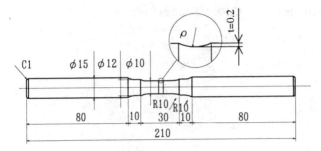

Figure 1. Shape and dimension of specimen.

rolling direction with the specimen axis and by making a circumferential notch. The notch of specimens has six kinds of radii, i.e. $\rho=\infty$(i.e. plain specimen), 2.0, 1.0, 0.6, 0.3, and 0.1 mm with constant notch depth (t=0.2mm). All specimens were annealed in vacuum at 500°C for 2 hours after polishing with fine emery paper, and thereafter electropolished to the depth of about $40\,\mu$m in diameter.

2.2 Testing method

The fatigue tests have been performed using Ono-type rotating bending fatigue testing machine with 98 N·m load capacity. In the case of fatigue tests of these materials, testing part of each specimen was cooled by pouring fresh water for prevention of exothermic fracture. It is confirmed that there is no difference between the results under low frequency in air and that of high frequency with cooled by distilled water. Fatigue crack initiation behavior of each specimen was observed by the successive taken replica method in the circumferential direction at the specimen's surface. Furthermore, the stress concentration factor at the root of notch is estimated with Neuber's triangular law.

3 RESULTS AND DISCUSSION

3.1 S-N curves

Figure 2 shows the S-N curves. In this figure, the ordinate shows the nominal stress at minimum cross section. Considering the fatigue limit is the maximum stress in which each specimen can stand by 1 × 10^7 cycles, the fatigue limit of all specimens are listed in Table 3. As shown in Figure 2, the gradients of these curves in all notched specimens become larger than that of the plain specimen. Furthermore, the knee point of S-N curve shifts to the direction of the long life region with decreasing notch radius.

Figure 3 shows the relation between $K_t\sigma_a$ and N_f, where K_t is stress concentration factor. Consid-

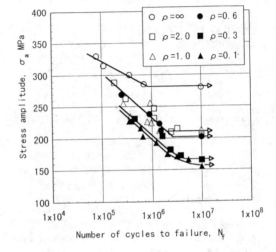

Figure 2. S-N curves.

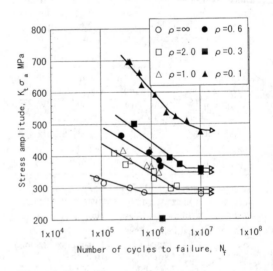

Figure 3. $K_t\sigma_a$–N curves.

ering the effect of stress concentration at the root of notch on fatigue strength, the maximum fatigue stress $K_t \sigma_w$ increases with decreasing the notch radius. In addition, the knee point of each curve shifts in the direction of the short life region with increasing the fatigue limit.

3.2 Comparison with other materials on notch sensitivity

Figure 4 shows the relation between $1/K_f$, i.e. the ratio of fatigue limit of notched specimen to that of smooth specimen; σ_w/σ_{w0}, and stress concentration factor K_t for SUS304. In this figure, the results of 0.16% carbon steel and 0.46% carbon steel (NODA et al. 1994) are also plotted. In general, the fatigue limit σ_w of notched steel specimen is defined by the crack initiation limit σ_{w1} for blunt notches and by the crack propagation limit σ_{w2} for sharp notches (ISIBASI 1967). In this test, it has been found that the fatigue limit of SUS304 for sharp notches is also decided by fatigue crack initiation limit because the non-propagating micro-cracks do not exist in the specimen. Under this test condition for SUS304, therefore, it is said that σ_{w2} does not exist. Consequently, the notch sensitivity of SUS304 is higher than that of carbon steel under the lower stress concentration factor region, and is lower under the higher stress concentration factor region.

Figure 5 shows the relation between stress concentration factor K_t and notch sensitivity factor q. Notch sensitivity factor q is defined as

$$q = \frac{K_f - 1}{K_t - 1}$$

Notch sensitivity factor q varies from zero for no notch effect to unity for the full effect predicted by the theory of elasticity. The notch sensitivity of all specimens is listed in Table 3 and the value of q in SUS304 is roughly the same as those of S15C and S45C under the stress concentration factor region lower than 2, while the notch sensitivity factor q in YUS170 is larger than those of plain carbon steels under the stress concentration factor region higher than 2. Therefore, it is said that the sensitivity of SUS304 for notch is sensitive as compared with those of plain carbon steels. Particularly, the notch sensitivity factor of SUS304 does not decrease rapidly with an increase of stress concentration factor as compared with those of carbon steel.

3.3 Fatigue crack initiation and propagation behavior

Figures 6 and 7 show the successive observation results of fatigue crack initiation and propagation behaviors at the root of notch. Fatigue crack in

Table 3. Notch radius and main experimental results.

ρ mm	K_t	K_f	σ_w MPa	q
∞	1.00	1.00	280	−
2.0	1.42	1.37	204	0.88
1.0	1.64	1.35	207	0.55
0.6	1.73	1.38	202	0.52
0.3	2.17	1.69	166	0.59
0.1	3.07	1.91	146	0.44

where ρ : Notch radius
      K_t : Stress concentration factor
      K_f : Fatigue notch factor
      σ_w : Fatigue limit
      q : Notch sensitivity factor

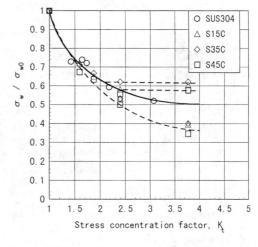

Figure 4. Relation between σ_w/σ_{w0} and K_t.

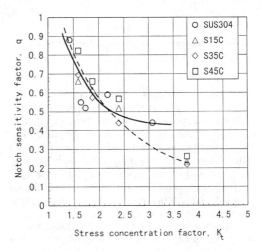

Figure 5. Relation between notch sensitivity factor q and stress concentration factor K_t.

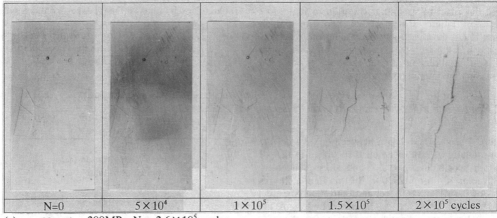

(a) $\rho = \infty$, $\sigma_a = 300\text{MPa}$, $N_f = 3.6 \times 10^5$ cycles

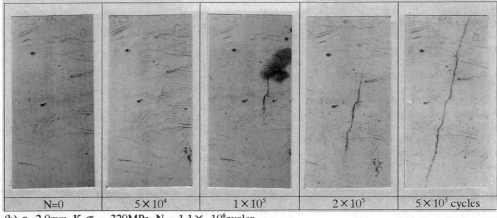

(b) $\rho = 2.0\text{mm}$, $K_t \sigma_a = 320\text{MPa}$, $N_f = 1.1 \times 10^6$ cycles

Axial direction   100 μm

Figure 6. Successive observation results of fatigue crack initiation and propagation behaviors at the root of blunt notch.

plain specimen initiates at about 10 to 15 % of total fatigue life. On the other hand, fatigue cracks in each notched specimen initiate at about 5 to 10 % of total fatigue life. From this result, it is found that the fatigue cracks of notched specimens initiate at a little smaller cyclic ratio N/N_f (where N_f is the number of cycles to failure) to that of plain specimen, and the ratio of crack propagation life to total life for each notched specimen is larger than that for plain specimen. The fatigue cracks of all specimens initiate at grain boundary or its neighborhood and inclusions. In addition, the fatigue cracks of the specimens whose notch radius is larger than 1 mm (i.e. $\rho = \infty$, 2.0 and 1.0 mm) initiate at one point, then the

crack propagates. On the other hand, the fatigue crack of specimens whose notch radius is smaller than 1mm (i.e. $\rho = 0.6$, 0.3 and 0.1 mm) simultaneously initiate at several points, then the cracks combine and propagate. The reason why there are different between fatigue phenomena in the blunt notch specimen and that in the sharp notch one is considered that the sharper notch makes higher stress concentration region at the root of notch. In the case of more higher stress concentration factor, the limit of fatigue crack initiation, i.e. σ_{w1}, in SUS304 increases due to the generation of strain hardening and strain-induced martensitic transformation. On the contrary, the effect of strain aging is insufficient in

1106

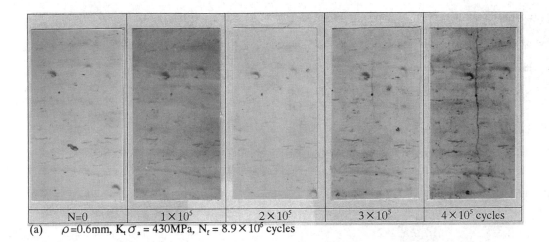

| N=0 | 1×10^5 | 2×10^5 | 3×10^5 | 4×10^5 cycles |

(a) ρ=0.6mm, $K_t \sigma_a$ = 430MPa, N_f = 8.9 $\times$ 10^5 cycles

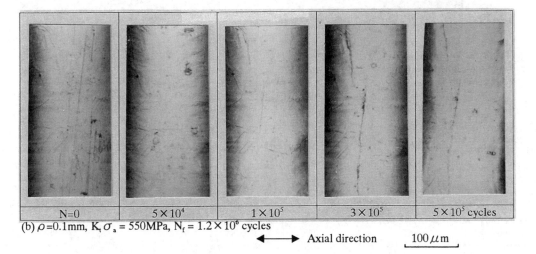

| N=0 | 5×10^4 | 1×10^5 | 3×10^5 | 5×10^5 cycles |

(b) ρ=0.1mm, $K_t \sigma_a$ = 550MPa, N_f = 1.2 $\times$ 10^6 cycles

◄──────► Axial direction   └─ 100μm ─┘

Figure 7. Successive observation results of fatigue crack initiation and propagation behaviors at the root of sharp notch.

this material. Therefore, it becomes very difficult that the non-propagating micro-cracks exist in this material. In general, it is confirmed that non-propagating micro-cracks exist in plain carbon steels at the region of higher stress concentration factor. On the other hand, from the above reason, it is considered that non-propagating micro-cracks do not exist in the specimen of SUS304 in this test.

4. CONCLUSIONS

Fatigue tests had been accomplished using rotating bending fatigue testing machine about SUS304 with several different notch radius. The main results

obtained in this study are as follows;

1. The knee point in S-N curve shifts in the direction of long life region with increasing the notch radius.

2. The notch sensitivity of SUS304 is more sensitive as compared with that of carbon steel under the lower concentration factor region, while is dull under the higher concentration factor region.

3. The fatigue cracks in plain specimen initiate at about 10 to 15 % of total fatigue life and those in notched specimens initiate at about 5 to 10 % of total fatigue life.

4. The fatigue limit in SUS304 for all notches in this test decides by fatigue crack initiation limit because the non-propagating micro-cracks does not

exist in the specimen.

REFERENCES

ISIBASI, T. 1967. *Prevention of fatigue and fracture of metals.* Tokyo: Yokendo.

NISHIDA, S. 1992. *Failure analysis in engineering applications.* UK: Butterworth Heineman Co. LTD

NODA,N. et al. 1994. Trans. of JSME (in Japanese), Vol.60, No.575; 1517-1523

Thermal fatigue

Experimental Mechanics, Allison (ed.)© 1998 Balkema, Rotterdam, ISBN 90 5809 014 0

Fatigue design method of aluminum bonding wires subjected to thermal cycles

M. Yano, J. Yano, D. Inoue & K. Koibuchi
Department of Precision Machinery Engineering, Graduate School of Engineering, The University of Tokyo, Japan

ABSTRACT: Fatigue design method of aluminum bonding wires where fatigue failures originated at the highest part of the loop or the bond in thermal cycles, was studied. A mechanical fatigue testing system which can obtain displacement/loads induced to the specimen, was developed to simulate the displacement of the bonding wire by thermal deformation of the devices. By the mechanical fatigue tests applied to the specimens with several profiles, it was proved that the wire showed typical load/displacement hysteresis loops indicating low cycle fatigue. And it was found that the different profile specimens with the same fatigue life had almost the same stabilized bending stress range applied to the failure points. In regard to the fatigue behavior of the bond, a stress singularity analysis using three-dimensional finite element method was performed and the effects of the cross-sectional dimensions of the bonded wire on the fatigue strength were made clear by parametric analyses.

1 INTRODUCTION

Aluminum is used as the bonding wire materials in a power electronic device such as IC igniter; an electronic device of an automobile for engine ignition. The aluminum bonding wire is from 200 to 300 μm in diameter and forms a loop. And both its ends are bonded to the terminals of power modules for electrical connection. The configuration of the wire is restricted by the package size and is automatically determined from 5 to 10 mm in width between its ends, from 1 to 4 mm in height of the loop.

When these devices are set up near the engine, they are subjected to thermal cycle due to ambient temperature changes and heat dissipation from the modules. Thermal stress is generated at various parts in electronic devices by thermal deformation because of the difference in coefficients of thermal expansion of component materials. As the result, fatigue failures are sometimes observed at the aluminum bonding wires; the highest part of the loop or the bonded surface area shown in Figure 1. The neck part of the bond could be also considered to be a critical point, but wasn't dealt with as the object in this study because there were few reports of the fatigue fracture at that point in the field. As far as the authors are aware, little work, if any, has been reported to date on the analyses of the fatigue behaviors of the bonding wires. This paper is concerned with the fatigue design method of aluminum bonding wires in electronic devices subjected to thermal cycle, where the highest part of the loop and the bonded surface area are dealt with as the objects in particular.

Since the fatigue failures are due to the cyclic thermal deformations of the component materials constructing the electronic devices, thermal fatigue tests of aluminum bonding wires built in the devices should be performed by applying such as a thermostatic chamber in order to simulate the real thermal fatigue behaviors. Thermal fatigue tests, however, have various weak points:

1. It needs much testing time;
2. Fatigue life can't be measured precisely, because the tests can do nothing but judge whether the wire breaks before a certain number of cycles or not;
3. The measurement of the various data of the wire in continuous aspect is difficult.

These weak points mean that it is impossible to apply the thermal fatigue tests on the analyses of the fatigue behaviors by reason of obtaining poor data such as the relation between the number of cycles to failure and the temperature range. Hence the analysis methods of the fatigue behaviors on the highest part of the loop and the bonded surface area respectively by not applying the thermal fatigue tests, were proposed in this study.

By cyclic thermal deformation, countless creases is produced on the surface of the highest part of the loop,

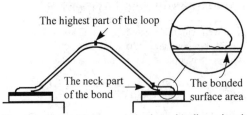

Figure 1. Thermal fatigue of aluminum bonding wires in electronic devices.

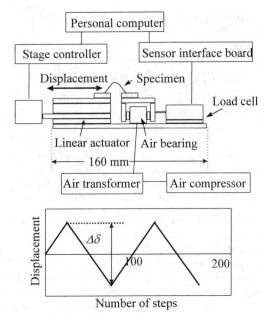

Figure 2. Mechanical fatigue testing system.

which comes to be down. That fatigue behavior is considered to be due to strain concentration on that part which results from working bending force caused by relative displacement between the ends of the wire. In this paper, a mechanical fatigue testing system was developed to simulate cyclic deformation of the wire. And fatigue tests using that system were carried out to analyze the fatigue mechanism of the bonding wire. Further more, a life prediction method was proposed based upon the results of the fatigue tests.

At the bond, large shear strain is induced on the interface between the wire and the terminal due to the thermal-expansion mismatch of utilized materials. A fatigue crack initiates at the edge of the bonded area and propagates into aluminum region parallel to the interface. And it has been definitely shown by thermal fatigue tests that the fatigue life of crack initiation and crack propagation combined was dependent on the cross-sectional shape of the wire at the bond. In this paper, the stress field of the bonded surface area was calculated elastically by three-dimensional finite element method (FEM). Then parametric analyses of the effects of the cross-sectional dimensions were carried out using a stress singularity parameter to assess the results of thermal fatigue tests.

2 FATIGUE BEHAVIOR OF THE HIGHEST PART OF THE LOOP

2.1 *Mechanical fatigue testing system*

It is possible to carry out the mechanical fatigue tests equivalent to thermal cycle by giving the displacement to the end of the wire compulsorily since cyclic thermal deformations of each components of the devices produce relative displacement between the ends of the wire. Figure 2 shows the mechanical fatigue testing system developed in this study. This system can measure a load working on a specimen through the air bearing in case of giving displacement to the specimen by a linear actuator controlled by a personal computer.

Because the displacement range of the bonding wire caused by thermal cycle of $\Delta T = 230$ degrees (from -50 to 180 °C) was considered to be from 100 to 300 μm, the resolution of the displacement of the linear actuator was decided to be 1 μm. The accuracy of the displacement was confirmed by optical measurement. The air bearing was developed to transmit the load to the load cell without the friction loss. And as the result, the minimum load which could be measured by the load cell was 0.1 gf.

2.2 *Application to aluminum bonding wires*

In order to carry out the fatigue tests using the mechanical fatigue testing system, the aluminum bonding wire was modeled for the specimen. Figure 3 shows the model and the profile can be determined by

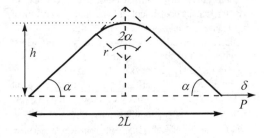

Figure 3. Aluminum bonding wire model.

three parameters L, r, and α. Because it was necessary for the mechanical fatigue testing system to verify the application to aluminum bonding wire, the relation between the displacement and the load given at the end of the wire was obtained by elastic calculation on the model shown in Figure 3.

To begin with, elastic simulation on the model applying finite element method was carried out and the result revealed that bending stress was dominant over the cross-section of the highest part of the loop. Elastic strain energy U in case of acting the horizontal load P on the end of the wire was calculated and the horizontal displacement δ was found from Castigliano's theorem. Consequently the relation between P and δ can be expressed as

$$\delta = \frac{\partial U}{\partial P} = \frac{2}{EI}\left[\frac{1}{3}\sin^2\alpha\left(\frac{L}{\cos\alpha} - r\tan\alpha\right)^3\right.$$
$$+ r\left\{\frac{r^2\alpha}{2} + \frac{r^2\sin 2\alpha}{4} + 2r\sin\alpha\left(L\tan\alpha - \frac{r}{\cos\alpha}\right)\right.$$
$$\left.\left. + \left(L\tan\alpha - \frac{r}{\cos\alpha}\right)^2\alpha\right\}\right]P \qquad (1)$$

where E is Young's modulus and I is moment of inertia of area.

Then the experiment using the mechanical fatigue testing system was performed and the loads were measured on the occasion of giving the displacement of $\delta = 100$ μm on the end of the wire shown in Figure 3. The wire used in the experiment was 300 μm in diameter and had the profiles of $2L = 6$mm, $r = 2.15$ mm and $\alpha = 45°$. Figure 4 shows the comparison between the experimental data and the elastic calculation applying the equation(1). It could be recognized from the experimental result that plastic

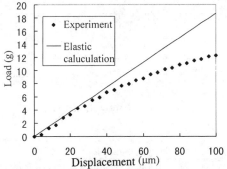

Figure 4. Comparison of results between experiment and elastic calculation.

deformation occurred at the displacement of over 40 μm. And below the linear limits, the experimental data agreed well with the elastic calculation. For that reason, it could be confirmed that the mechanical fatigue testing system could apply to aluminum bonding wires.

2.3 Mechanical fatigue tests

Fatigue tests of the highest parts of the loops of aluminum bonding wires were carried out using the mechanical fatigue testing system. In order to make clear the mechanism of the fatigue behavior of the wire, test conditions which could compare profiles, diameters, and processing methods of the wires were determined. Table 1 gives the fatigue test conditions. In the fatigue tests, load ranges of every hundred cycles and load/displacement hysteresis loops of every hundred cycles were recorded in the personal computer. The load range means the peak to peak value of the measured loads in a cycle.

Table 1. Fatigue test conditions.

Specimen	(a)	(b)	(c)	(d)	(e)
$2L$ (mm)	6	6	9	6	6
r (mm)	2.15	2.15	2.15	2.15	2.15
α (°)	45	30	30	45	45
Diameter (μm)	300	300	300	250	300
Processing Method	SOFT	SOFT	SOFT	SOFT	HARD
$\Delta\delta$ (μm)	150	150	200	200	200
	200	200	300	250	250
	250	250	400		
	300	300	500		

*HARD is the only wire drawing process and SOFT is the process of annealing after wire drawing.

According to the results of the fatigue tests, the fatigue failures occurred at the highest parts of the loops and the numbers of cycles to failure were from about 3000 to 25,000 cycles. Those behaviors indicate low cycle fatigue, which is the same as the behavior of the actual thermal fatigue fracture. Figure 5 shows the changes of the load ranges of the specimen(a) and it can be confirmed that cyclic softening occurs in the wire. As the common features to all specimens, it can be found that the load ranges decreased rapidly in the first hundreds cycles and that the wires came to be down through the long steady states where the load ranges made little change. Figure 6 shows the comparison of load/displacement hysteresis loops between the first cycle and $N_f/2$ cycles in the steady state, where N_f is the number of cycles to failure. It is well shown that plastic

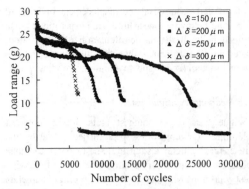

Figure 5. Load ranges of different displacements.

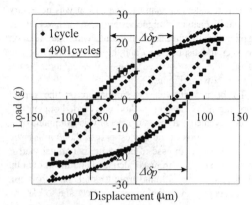

Figure 6. Load/displacement hysteresis loops.

deformation occurs in the wire and that the plastic deformation range $\Delta\delta_p$ enlarges as the number of cycles increases. Figure 7 shows the comparison of the changes of the load ranges of different processing methods, *SOFT* and *HARD*. It can be confirmed that it makes little difference between the numbers of cycles to failure but makes clear difference between the inclinations of the steady states limits. Since work hardening of the *HARD* wire is heavier than that of the *SOFT* wire, it is considered that there is difference between the intensities of the strain concentrations of the highest parts of the loops.

2.4 Prediction of the fatigue life

In this paper, fatigue life prediction method was developed by making use of the data of mechanical fatigue tests. Because the bending force was dominant at the highest part of the loop, the bending stress of that part was decided as the life evaluating parameter. Bending moment can be calculated by multiplying the load P by the loop's height h shown in Figure 3 and the bending stress σ_M can be expressed elastically as

$$\sigma_M = \frac{Ph}{Z} \tag{2}$$

where Z is section modulus.

Because it could be considered that the load range in the steady state shown in Figure 5 had influence on the fatigue life, the stabilized bending stress range $\Delta\sigma_{MC}$ was calculated from the equation(2) using the load range in $N_f/2$ cycles ΔP_C. Figure 8 shows the relation between $\Delta\sigma_{MC}$ and the number of cycles to failure about all specimens. It was shown that the relation could be represented by a single band, which indicated that uniform life assessment of the highest parts of the

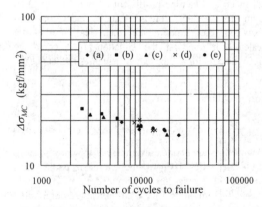

Figure 7. Load ranges of different processing methods.

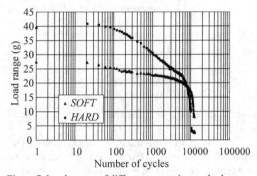

Figure 8. $\Delta\sigma_{MC}$ versus number of cycles to failure N_f.

loops was possible by making use of $\Delta\sigma_{MC}$.

Further more, in order to estimate stabilized bending stress range $\Delta\sigma_{MC}$ without performing fatigue tests, the

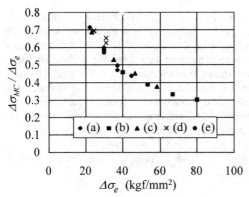

Figure 9. $\Delta\sigma_{MC}/\Delta\sigma_e$ versus $\Delta\sigma_e$.

other parameter $\Delta\sigma_e$ was proposed in this study. $\Delta\sigma_e$ is an elastic bending stress range which is found from the load range ΔP_e calculated by the equation(1) in elastic condition. Then the relation between $\Delta\sigma_e$ and $\Delta\sigma_{MC}/\Delta\sigma_e$ is shown in Figure 9. The reduction rate of the stress range $\Delta\sigma_{MC}/\Delta\sigma_e$ means the damage of the wire and it is shown that the relation is expressed by a single curve. As the result, it is shown that the number of cycles to failure of aluminum bonding wire can be predicted independently of profiles, diameters, and processing methods of the wires, by $\Delta\sigma_e$ found from elastic calculation without the fatigue test.

3 STRESS SINGULARITY ANALYSIS AT THE BOND

Aluminum bonding wires are usually bonded to terminals by supersonic bonding process. Fatigue cracks are sometimes observed at the aluminum parts near the bonded interface, which initiate at the bonded edge caused by cyclic plastic strain due to the thermal misfit of bonded materials.

3.1 Thermal fatigue tests

Figure 10 shows typical examples of the cross-sections of bonded wires. Thermal fatigue tests on condition that it took 30 minutes for a thermal cycle of $\Delta T = 170$ degrees (from -40 to 130 ℃), were carried out in order to compare fatigue lives at the bonds of two cross-sectional shapes. Since it was hard to observe the crack propagation directly, shear strengths of every thousand cycles were measured by a load cell as the evaluation basis of crack propagation. Shear strength after crack propagation are dealt with conventionally to compare the uncracked parts of the bonded surface areas. The

results of fatigue tests showed that the fatigue life of model(b) was about one and a half times as long as that of model(a), where the fatigue life was defined as the number of cycles that the shear strength was reduced to a half of the initial strength.

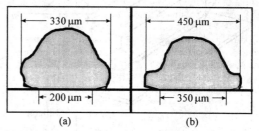

(a)   (b)

Figure 10. Cross-sections of bonded wires.

3.2 Elastic simulation by finite element method

Elastic simulations using three-dimensional finite element method (FEM) were carried out to get the stress fields of the wire-bonded surface areas in case of giving thermal changes of $\Delta T = 170$ degrees. Figure 11 shows an analytical model of the bond used in FEM simulation. Because the shear force working at the bond by thermal deformation of the wire was about 10 to 20 gf and was small enough to be neglected, only the end of the wire was modeled to take notice of shear stress on the interface caused by thermal-expansion mismatch of utilized materials. And the symmetry was considered in the model. Then the results of elastic simulations showed that the shear stress was concentrating at the edge of the bonded surface area. Further more, the stress distribution showed a singularity where the bonded edge was the infinite point of the shear stress. It was watched that the stress distribution near the bonded edge could be approximately denoted by

$$\tau = \frac{K}{r^\lambda} \qquad (3)$$

where τ is the shear stress, r is the distance from the stress singularity point shown in Figure 11, K is the intensity of the stress singularity field, and λ is the singularity exponent.

The fatigue life of the bond can be considered to be evaluated at the initial condition though the fatigue life indicates the process of crack propagation. Since the stress distribution near the bonded edge can be determined by K and λ, it is possible to evaluate the strength of the bond relatively by using that two parameters.

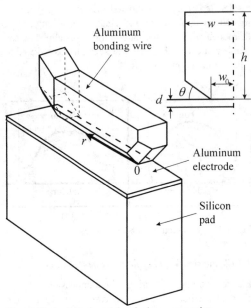

Figure 11. Analytical model and wire cross-section.

w, and d had little effects on K. However, as the width of the bonded surface area w_0 and the angle of the bonded edge θ increased, K decreased. Consequently it was confirmed that the shadowed part shown in Figure 12 affected the fatigue strength at the bond. The ratios of cross-sections A_1/A_2 of model(a) and model(b) were 0.65 and 0.35 respectively. Therefore the lifetime of the bond becomes longer as that part becomes smaller. The intensities of stress singularity K of model(a) and model (b) were 23.2 and 16.6 respectively, which indicated that K was inversely proportional to the fatigue life of the bond.

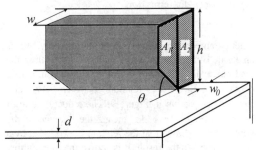

Figure 12. Effects of cross-sectional dimensions.

3.3 Parametric analyses of the effects of the cross-sectional dimensions

In order to assess the results of thermal fatigue tests, the effects of the cross-sectional dimensions of the bonded wire on the stress distribution of the bonded surface area was investigated in this paper. Five shaping parameters (h, w, w_0, d, θ) shown in Figure 11 were selected for parametric analyses. And elastic simulations applying FEM were carried out by changing one parameter while keeping the rest constant to compare the effects between five parameters. Table 2 gives changeable ranges of five shaping parameters. Further more, singularity parameters K and λ were obtained from the stress distribution, the result of FEM simulation.

Table 2. The ranges of shaping parameters.

Parameter	Minimum value	Maximum value
h (μm)	200	300
w (μm)	165	250
w_0 (μm)	50	250
d (μm)	4	8
θ(°)	0	90

As the result, the singularity exponent λ kept almost constant of 0.22 through all simulations. In regard to the intensity of the stress singularity K, the parameters h,

4 CONCLUSION

In order to study the fatigue behaviors of the highest parts of the loops of aluminum bonding wires subjected to thermal cycle, a mechanical fatigue testing system which could obtain displacement/loads induced to the specimens was developed and applied to the bonding wires. And the mechanical fatigue tests were carried out on the various conditions which could compare profiles, diameters, and processing methods of the wires. The experimental results revealed the fatigue mechanism of the bonding wire; plastic deformation, cyclic softening, and low cycle fatigue. Further more, based upon the results of the fatigue tests, the fatigue life prediction method applying the bending stress ranges was proposed. This method made the life assessments of the highest parts of the loops possible without performing fatigue tests.

In regard to the fatigue behavior of the bond, elastic simulations applying three-dimensional finite element method were carried out to get the stress fields of the wire-bonded surface areas. Based upon the FEM results, the effects of the cross-sectional dimensions of the bonded wires were analyzed by the stress singularity parameters.

Experimental Mechanics, Allison (ed.)© 1998 Balkema, Rotterdam, ISBN 90 5809 014 0

Fatigue life characteristics of Ti-6Al-4V in very high cycle region

M. Kaneko
Hokkaido University, Sapporo, Japan

T. Nakamura & T. Noguchi
Department of Mechanical Science, Hokkaido University, Sapporo, Japan

ABSTRACT: Fatigue tests of Ti-6Al-4V alloy up to a very high numbers of cycles were performed. The S-N curve showed a change from gentle to steep near 10^7 cycles. This feature is different from ordinary non-ferrous alloys. In the short life, high stress region fractures originated from the surface of specimens while in the long life, low stress region they originated from the interior. The SEM observations of fracture surfaces indicated that the changes in the S-N slope and fracture orgin site resulted from a competition between the two fracture modes, surface and interior originated fractures. The fatigue life distribution in two modes can be described by the competing-risk model.

1 INTRODUCTION

Titanium alloys are known for their high strength particularly used in the aerospace industry. Today the field of application of Titanium alloys is expanding and becoming more diversified and includes general industrial applications. Their fatigue characteristics in the very high cycle region, over 10^7 cycles, are not entirely clear. Here fractures originating from the interior are observed like in high strength steels (Neal & Blenkinsop 1976, Atrens et.al. 1983).

This research reports fatigue tests of Ti-6Al-4V alloy with very high cycle numbers. The fracture surfaces were observed, and the fatigue life distribution is discussed.

2 MATERIAL AND EXPERIMENT

The experiments used Ti-6Al-4V alloy annealed at 730℃ for 1 hour after rolling at 950℃, with a chemical composition of Al:6.04, V:4.01, N:0.01,

C:0.02, Fe:0.20, and H:0.0008 (mass%). The manufacturing route is: after heat treatment, lathe turning to suit the configuration of the test piece shown in Figure 1, and then finished by polishing with emery paper from #120 to #2000. The mechanical properties are: tensile strength:988MPa; yield stress:899MPa; and Vickers hardness:309.

Tension-tension fatigue tests with a stress ratio of 0.1 were conducted in a laboratory environment. The load was a sine wave of 120 Hz.

3 EXPERIMENTAL RESULTS

Figure 2 shows the S-N diagram. It consists of two parts, a gentle slope in the high stress region and a

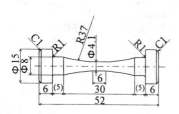

Figure 1. Configuration of test piece.

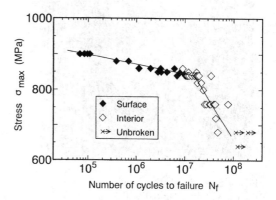

Figure 2. S-N diagram.

Figure 3. A surface originating crack (σ_{max}=840MPa, N_f=6.83x10^6).

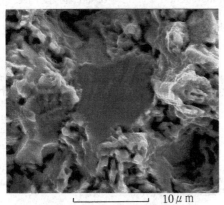

Figure 4. An interior originating crack (σ_{max}=840MPa, N_f=3.55x10^7).

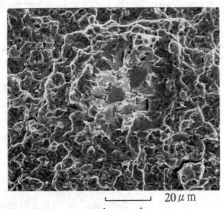

Figure 5. A propagating crack (σ_{max}=860MPa, N_f=6.07x10^6).

steep incline in the low stress region. The border of these two is around 10^7 cycles.

In non–ferrous alloys, the slope of the S–N curve generally becomes gentler as applied cyclic stress

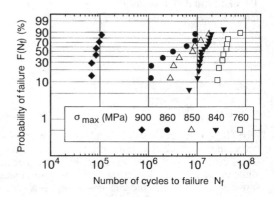

Figure 6. Fatigue life distributions.

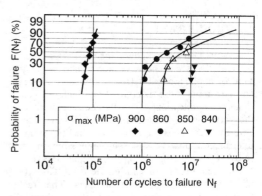

Figure 7. Fatigue life distributions of surface originating mode extracted from whole data.

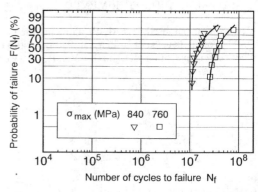

Figure 8. Fatigue life distributions of interior originating mode extracted from whole data.

1118

decreases and the number of cycles to failure increases. The fatigue strength at 10^7 cycles is refered as a basis of the fatigue strength. This particular S–N diagram will be discussed further below.

4 OBSERVATION OF FRACTURE SURFACE

4.1 Fracture origins

To investigate the reasons for this feature, the fracture origins were observed by SEM. The observations showed that the fracture origins were located at the surface of the specimens in the gentle slope region, while they were in the interior of the specimen in the steep region.

Figure 3 shows the surface origin type fracture and Figure 4 the interior type. The observations also indicate that the origins are facets and do not show glide lines regardless of the initiation site. The initiation of fatigue cracks in this experiment is not caused by the separation of slip planes.

The surroundings of the facets, the stage IIa fracture surface, show parallel glide lines in Figure 3. In Figure 4, however these lines are not clear. And this fuzziness suggests a recombination of fracture surfaces from lack of the atmosphere in interior crack. It is infered that the growth rate of interior cracks is very slow and that this is a cause of the long fatigue life of fractures originating in the interior.

4.2 Propagating cracks

Apparently, the break in S–N slope may be considered to result from the change between the two fracture modes, from the surface and from the interior. On the fracture surface, propagating cracks are observed as shown in Fig.5. These cracks are frequently observed in fractures both from the surface and from the interior. This indicates that surface and interior cracks are co-existing and competing. Both the changes in the S–N slope and the fracture orgins are infered to result from competition between the two fracture modes.

5 FATIGUE LIFE DISTRIBUTION

5.1 General characteristic

Figure 6 shows the weibull plots. With decrease in stress, the fatigue life distribution is first dispersed, and then there is a reduction in the scatter again. This development is unusual, and is thought to be caused by the two competing fatigue modes, originating from the surface and the interior.

5.2 Competing–risk model

5.2.1 Description of fatigue life distribution

The characteristics of the two fatigue modes may be described by the competing–risk model (Ichikawa 1984, Kaneko et.al. in press). When two fatigue modes are competing, the fatigue life distribution is made by the two competing fatigue life distributions. Accordingly fatigue life distribution $F(N_f)$ can be described in formula (1) using $F_A(N_f)$ and $F_B(N_f)$ for mode A and B respectively.

$$F(N_f)=1-(1-F_A(N_f))(1-F_B(N_f)) \tag{1}$$

5.2.2 Calculation of failure probability

Fatigue life data may be classified into two groups as surface–originating (Mode A) or interior–originating (Mode B) according to observations of fracture surfaces. Then when the fatigue life data of Mode A are realigned to calculate the failure probability, the data of Mode B may be treated as random–cut data according to Johnson's method. Johnson's rank method is expressed in formulas (2) and (3).

$$j=j_0+\frac{(n+1)-j_0}{(n+1)-(i-1)} \tag{2}$$

$$F_A(N_f)=\frac{j}{n+1} \tag{3}$$

where j = rank of Mode A; j_0 = rank of preceding Mode A data; i = rank in whole data; and n = total number of a whole data.

The interior–originating data, Mode B, are described in the same way.

5.3 Extraction of two fatigue modes

Figures 7 and 8 show the fatigue life distributions as described by the above surface and interior type considerations, and next a Weibull distribution was applied to the data of each mode. The estimation of the Weibull parameters is according to SAKAI and TANAKA (1981). The fitted Weibull distributions are shown as solid lines.

The Weibull distribution is:

$$F(N_f)=1-\exp\{-(\frac{N_f-\gamma}{\beta})^\alpha\} \tag{4}$$

where α = shape parameter; β = scale parameter; and γ = locate parameter.

In Figure 7, the data fit the Weibull distribution well except at 840MPa. The data at 840MPa do not fit as they are all below 30%.

Stress is lower in surface–originating fractures, and

the distributions shift to longer life with dispersion and without saturation of failure probability. This is the general situation in the high cycle region of ordinary non–ferrous alloys. In the high strength steels, the fatigue life distribution of surface type fractures show a clear saturation of failure probability (Kaneko et.al. in press).

For the interior–originating fractures, figure 8 shows that data are not scattered even in the very long life region above 10^7 cycles. This is similar to the interior type of high strength steels (Kaneko et.al. in press).

6 CONCLUSIONS

1. The S–N diagram consists of two parts, a gentle slope in the high stress, short life region; and a steep incline in the low stress, long life region.

2. In the high stress region, fractures originate from the surface of the specimens, and in the low stress region from the interior.

3. Observation of the fracture surface suggests that the S–N relation results from competition between surface and interior mode fractures.

4. The fatigue life distributions of the two modes can be described by the competing–risk model.

REFERENCES

Neal, D.F. & P.A.Blenkinsop 1976. Internal fatigue origins in $\alpha - \beta$ titanium alloys, *Acta Metallurgica*. 24:59–63.

Atrens, A., W.Hoffelner, T.W.Duerig & J.E.Allison 1983. Subsurface crack initiation in high cycle fatigue in Ti6Al4V and in a typical martensitic stainless steel, *Scripta METALLURGICA*. 17:601–606.

Ichikawa, M. 1984. Possible models for quasi–composite distributions, *Reliability Engineering*. 8:117–128.

Kaneko, M., T.Nakamura & T.Noguchi, in press. A study of method utilizing fatigue life data on two competimg fatigue modes, under contribution to *JOURNAL OF THE SOCIETY OF MATERIALS SCIENCE, JAPAN*.

SAKAI, T. & T.TANAKA 1981. Estimation of Three Parameters of Weibull Distribution in Relation to Parameter Estimation of Fatigue Life Distribution [Continued Report], *24th Japan Congress Material Research*. 141–147.

Experimental Mechanics, Allison (ed.) © 1998 Balkema, Rotterdam, ISBN 90 5809 014 0

Fatigue life of TiNiCu shape memory alloy under thermo-mechanical conditions

T. Sakuma & U. Iwata
Central Research Institute of Electric Power Industry, Japan

N. Kariya & Y. Ochi
Department of Mechanical and Control Engineering, University of Electro-Communications, Japan

ABSTRACT: In this paper, the fatigue lives of Ti-41.7at%Ni-8.5at%Cu alloy are investigated experimentally. Experiments are carried out by repeating the combination of the thermal cycles and the loading-unloading cycles under various heating temperatures and the constant cooling temperature. Result shows that the fatigue lives are related to the dispersion strain energy obtained by the amount of decrease of recovery strain energy per cycle.

1 INTRODUCTION

Shape memory alloy has the property of the pseudo elasticity that it returns to original shape with unloading and of the shape memory effect that it returns to original shape with heating after unloading. Therefore, the shape memory alloy with peculiar properties is applied to various fields such as heat engines and actuators.

Authors have proposed a reciprocating heat engine incorporating, as an energy conversion element, TiNiCu shape memory alloy (Sakuma & Iwata 1996). And, operating characteristics of the heat engine have been examined (Sakuma et al. 1997).

From the viewpoints of the design of heat engines or actuators for the practical application, it is necessary to optimize the heating/cooling temperature and the deformation strain. Researches on a shape memory alloy have demonstrated that the recovery stress decreases and the irrecoverable strain increases with the increase in the number of thrmo-mechanical cycles (Tobushi et al. 1989,1991, Sakuma & Iwata 1997).

Therefore, it is important to grasp the variations of these characteristics for determining the basic specifications of heat engines and actuators such as the external load and the deformation strain.

Furthermore, from the viewpoints of maintenance of the heat engine and the actuator, it is important to estimate the fatigue life of the alloy against the operation conditions. The fatigue life of TiNi alloy or of Cu group alloy are investigated in relation to the strain amplitude, the plastic strain and the stress (Melton et al. 1979, McNichols et al. 1981, Miyazaki et al. 1989, Sakamoto & Shimizu 1986).

The fatigue life becomes shorter as the strain and the stress become larger. However, when the thermal and mechanical cycles are considered simultaneously, a comprehensive understanding of the fatigue life is still lacking.

This paper discusses the fatigue life of TiNiCu shape memory alloy. In order to achieve these purposes, the variations of the recovery stress and the irrecoverable strain with the number of cycles are investigated for various heating temperatures and strains.

Furthermore, the fatigue lives for various heating temperatures and strains are investigated. And the estimation using the dispersion energy of the fatigue life for a shape memory alloy is proposed.

2 EXPERIMENTS

2.1 *Experimental samples*

The shape memory alloy used is composed of Ti-41.7 at%Ni-8.5 at%Cu and is a smooth wire rod without notch of 2mm in diameter and 200mm in length.

The transformation temperatures of the TiNiCu alloy are A_f=333K, A_s=331K, M_s=324K and M_f=321K respectively at a stress free stage.

2.2 *Experimental procedure*

The experimental apparatus is shown in Fig.1.

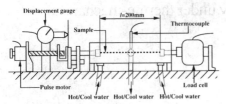

Figure 1. Schematic drawing of experimental apparatus.

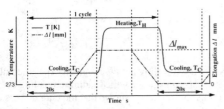

Figure 2. Schematic diagrams of variations of temperature and displacement with time.

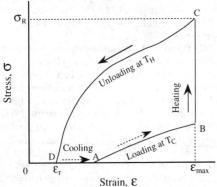

Figure 3. Schematic drawing of the stress-strain curve of a cyclic test.

The sample is elongated by a pulse motor. The elongation and the stress are measured with a displacement gauge and a load cell. The temperature is measured with a thermocouple. The cyclic tests are that the alloy which is cooled at temperature T_c is elongated to a given elongation Δl_{max} by a pulse motor. And the alloy is heated at the elongation Δl_{max} up to a given temperature T_H by hot water, thereafter the alloy is recovered by a pulse motor maintaining a given temperature T_H. Figure 2 shows the variations of temperature and elongation with time. Such a thermo-mechanical cycle recurs until the alloy breaks. Stress-strain diagram of cyclic tests is shown in Fig.3. The strain ε_{max} in this figure is a value of $\Delta l_{max}/l$ (l is the initial distance between the grips). And the stress σ_R and the strain ε_r express the recovery stress and the irrecoverable strain.

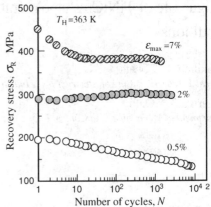

Figure 4 Variations of recovery stress σ_R with the number of cycles N for various strains ε_{max}.

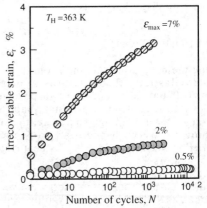

Figure 5 Variations of irrecoverable strain ε_r with the number of cycles N for various strain ε_{max}.

The cooling temperature T_c is 293K constant, and the heating temperature T_H is 348K - 363K. The variations of strain ε_{max} are 0.5 - 7%.

3 RESULTS AND DISCUSSIONS

3.1 Cyclic behaviors

Figure 4 shows the relationship between the recovery stress σ_R of the alloy and the number of cycles N at the condition of T_H=363 K.

The recovery stress σ_R varies with the number of cycles. When the strain is more than 5%, the recovery stresses rapidly decrease at the early stage of the number of cycles N. On the other hand, when the strain is smaller than 4%, the recovery stresses are almost same and do not vary with the cyclic numbers. Therefore, it seems to obtain the recovery stress which was

1122

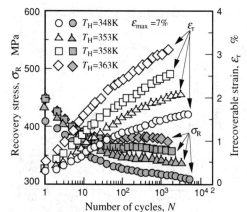

Figure 6 Variations of recovery stress σ_R and irrecoverable strain ε_r with the number of cycles N for

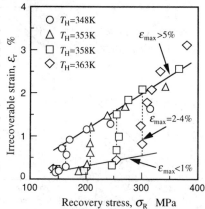

Figure 7 Variations of irrecoverable strain ε_r with recovery stress σ_R for various heating temperatures T_H

stabilized for the number of cycles by conducting several tens of training in advance.

Figure 5 shows the variations of the irrecoverable strain ε_r with the number of cycles N under the same condition of Figure 4. The irrecoverable strains increase with the increase in the cyclic numbers. And as the strain ε_{max} becomes large, the irrecoverable strains ε_r also increase.

The general tendency of the recovery stress and irrecoverable strain are also recognized in the case which changed the condition of the heating temperature. Figure 6 shows the variations of the recovery stress σ_R and the irrecoverable strain ε_r with the cyclic numbers N for various heating temperatures T_H.

The recovery stress and the irrecoverable strain are largely dependant upon the heating temperature. The recovery stress increases with the increase in heating temperature T_H. And the

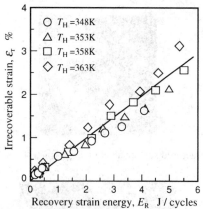

Figure 8 Variations of irrecoverable strain ε_r with recovery strain energy E_R.

irrecoverable strain also increases with the increase in T_H.

Figure 7 shows the variations of the irrecoverable strain ε_r with the recovery stress σ_R for various heating temperatures T_H. The values of the recovery stress and the irrecoverable strain are the average values from $N=1$ to N_f (the number of cycles to failure).

When the strain ε_{max} is less than 1% and larger than 5%, the irrecoverable strains increase with the increase in the recovery stresses, that is, they increase with the heating temperature. On the contrary, in the strain range of 2-4%, the irrecoverable strain increase with the increase in the heating temperature although the recovery stresses rarely vary. These increases of irrecoverable strain are considered to be due to the rearray of the orthorhombic phase in structures. Moreover, when the strain ε_{max} is below 1% and 5% or more, the increasing of the irrecoverable strain is caused by irreversible deformation due to the dislocation in structures.

Figure 8 shows the relationship between the irrecoverable strain ε_r and the recovery strain energy E_R for various heating temperatures. The recovery strain energy E_R is defined as the area surrounded by the unloading C-D curve and the strain axis in Figure 3. It shows the energy per unit volume when the alloy recovers shape. It is defined by the following equation.

$$E_R = \frac{1}{N_f} \sum_{i=1}^{N_f} \left\{ \int_{\varepsilon_r}^{\varepsilon_{max}} \sigma_{R,i}(\varepsilon, T_H)\, d\varepsilon \right\} \quad (1)$$

The irrecoverable strain ε_r increases approximately proportionally to the recovery

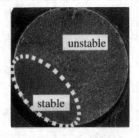

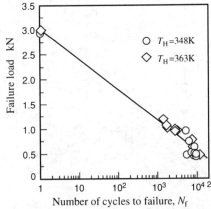

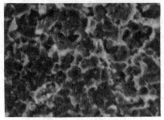

(a) Fractured surface at strain ε_{max}=2%,

200μm

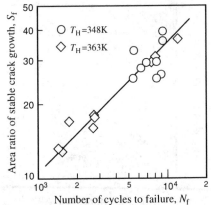

10μm

(b) area of the unstable crack growth,

Figure 9 Observations of fatigue fractured surface by SEM.

strain energy E_R. Therefore, the irrecoverable strains which vary with the heating temperature and the strain ε_{max} can be arranged by the recovery strain energy E_R.

3.2 *Fatigue life*

Figure 9 shows observations of fatigue fractured surface by SEM at strain ε_{max}=0.7%. It is observed on the fractured surface that the fatigue fractured surface is distinguish between the area of the crack grown up stably and that of the crack grown up unstably (Fig.9(a)). Furthermore, it is found that the area of the unstable crack growth on the fractured surface is covered with a dimple pattern (Fig.9(b)). It is conjectured that this is due to the ductile fracture.

As a result of observations of the fractured surface at other strains, the larger the strain, the area of the stable crack growth becomes smaller. Additionally, when the strain ε_{max} is small, the generation of plural crack initiation sites are observed on the fractured surface.

Figure 10 shows the relationship between the failure load and the number of cycles to failure N_f. The failure loads decrease with the increase in the number of cycles to failure N_f. Thus, the damaged area due to fatigue increases with the decrease in the failure load because the stress

Figure 10 Relationship between the failure load and the number of cycles to failure N_f.

Figure 11 Relationship between the area ratio of stable crack growth S_f and the number of cycles to failure N_f.

value when an alloy is fractured are almost equal in all alloys.

Figure 11 shows the relationship between the area ratio of the stable crack growth and the number of cycles to failure N_f. It can be stated that the area ratio is almost proportionately to the number of cycles to failure. As a result, it is conjectured that the form of failure in the alloy is fatigue fracture which the crack grows up stably with the thermo-mechanical cycles.

Fatigue life of metallic materials is generally estimated according to the stress amplitude and the strain amplitude given to the materials. Then, the relationship between the recovery stress, the irrecoverable strain and the fatigue life N_f for various heating temperatures are examined by experiments.

Figure 12 shows the relationship between the stress amplitude and the fatigue life N_f for various heating temperatures. The fatigue life

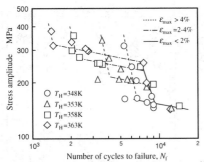

Figure 12 Relationship between the stress amplitude and the number of cycles to failure N_f for various heating temperatures T_H.

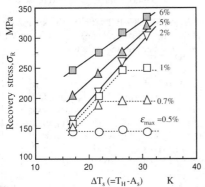

Figure 13 Variations of the recovery stress σ_R with temperature ΔT_s for various strains ε_{max}.

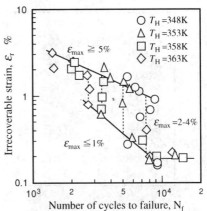

Figure 14 Relationship between the irrecoverable strain ε_r and the number of cycles to failure N_f for various heating temperature T_H.

N_f becomes longer as the stress amplitude becomes smaller overall. However, when the strain ε_{max} is larger than 4%, the fatigue life is influenced by heating temperature. And the lower the heating temperature, the fatigue life becomes longer. When the strain ε_{max} is 2-4%, the fatigue life becomes longer even though the stress amplitude does not almost decrease. Furthermore, when the strain ε_{max} is smaller than 1%, the fatigue life is not influenced by the heating temperature. As the cause for, it is considered that the austenite transformation is not finished because the larger the strain ε_{max}, the austenite transformation finish temperature A_f becomes higher.

Figure 13 shows the relationship between the recovery stress σ_R and the temperature difference $\Delta T_s (T_H - A_s)$. When the strain ε_{max} is 0.5%, the recovery stress σ_R is constant regardless of increase in heating temperature because the austenite transformation is finished in the test temperature range. However, when the austenite transformation is not finished, the recovery stress σ_R increases with temperature. Therefore, when the strain ε_{max} is larger than

2%, the austenite transformation is not finished under heating temperature conditions of this test, and the recovery stress σ_R increases with the increase in the heating temperature.

Figure 14 shows the relationship between the irrecoverable strain ε_r and the fatigue life N_f. When the fatigue life N_f is estimated using the irrecoverable strain ε_r corresponds to the residual strain, the fatigue life becomes longer with the decrease in the irrecoverable strain. However, when examining more in detail, the region of the strain range is divided into the region of 5% or more, of 2%-4% and of 1% or less.

The inclination of the straight line in the region of 5% or more is about 0.5 and this inclination is almost equal to that of low cycle fatigue in general metallic materials. Therefore, it is considered that the main cause of fatigue fracture in this region is due to the slip. Furthermore, in the region of 1% or less, the alloy is within the limit of elasticity of the austenite phase, and the fatigue limit is estimated to be about 0.2%. In addition, in the region of 2%-3%, the fatigue life does not almost vary with the irrecoverable strain because this is due to the rearrangement of martensite phase.

As mentioned above, the fatigue life is influenced by the heating temperature(or stress) and the strain range. Then, the fatigue life N_f is estimated using the dispersion energy. The dispersion energy can be obtained using a following equation.

$$\Delta E_R = \frac{1}{N_f} \sum_{i=1}^{N_f} \Delta \sigma_{R,i} \qquad (2)$$

1125

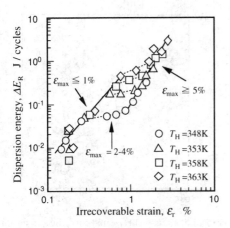

Figure 15 Relationship between the dispersion energy ΔE_R and the irrecoverable strain ε_r for various heating temperatures T_H.

where,

$$\Delta \sigma_{R,i} = \int_{\varepsilon_r}^{\varepsilon_{max}} \sigma_{R,i}(\varepsilon, T_H) d\varepsilon - \int_{\varepsilon_r}^{\varepsilon_{max}} \sigma_{R,i+1}(\varepsilon, T_H) d\varepsilon \tag{3}$$

Figure 15 shows the relationship between the dispersion energy and the irrecoverable strain. The dispersion energy ΔE_R increases with the increase in the irrecoverable strain ε_r. However, when the strain ε_{max} is 2%-4%, the dispersion energy is almost constant, thought the irrecoverable strain increases. Therefore, the fatigue life N_f can be estimated by the dispersion energy ΔE_R including the strain and the heating temperature.

Figure 16 shows the relationship between the dispersion energy ΔE_R and the fatigue life N_f. It is clear that the fatigue life N_f of shape memory alloy can be estimated by the dispersion energy including the effect of strain and heating temperature.

4 CONCLUSIONS

The fatigue life for TiNiCu shape memory alloy with the thermo-mechanical cycles N are investigated by experiments. The results are summarized as follows.
(1) The recovery stresses rapidly decrease at an early stage of the number of cycles N, and thereafter, these gradually decrease. On the other hand, the irrecoverable strain increases with the increase in the cyclic numbers.
(2) The irrecoverable strains which vary with

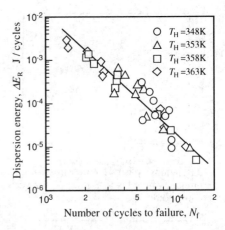

Figure 16 Relationship between the dispersion energy ΔE_R and the number of cycles to failure N_f for various strains ε_{max} and heating temperatures T_H.

the strain and the heating temperature can be estimated by the recovery strain energy.
(3) It is clarified that there are two regions where the crack grows up stably and unstably by SEM observations of the fractured surface.
(4) The smaller the strain and the lower the heating temperature, the fatigue life becomes longer. However, the fatigue life N_f can be estimated by the dispersion energy, including the influence of heating temperature and strain.

REFERENCES

McNichols Jr. J. L., P. C. Brookes & J. S. Cory 1981. *J. Appl. Phys.* 52 : 7442-7447.
Melton, K. N. & O. Mercier 1979. *Mater. Sci. Engg.* 40 :81-86.
Miyazaki, S., Y. Sugaya & K. Otsuka 1989. *Proc. MRS Int. Mtng. Adv. Mater.* 9 : 251-256.
Sakamoto, H. & K. Shimizu 1986. *Trans. JIM.* 27 : 601-606.
Sakuma, T. & U. Iwata 1996. *Trans. Jpn. Soc. Mech. Eng.* 62-597, B : 2086-2091.
Sakuma, T. & U. Iwata 1997. *Trans. Jpn. Soc. Mech. Eng.* 63-610, A : 1320-1326.
Sakuma, T. U. Iwata & M. Arai 1997. *JSME Int. J.* Series B, 40-4 : 599-606.
Tobushi, H., H. Iwanaga, A. Inaba & M. Kawaguchi 1989. *Trans. Jpn. Soc. Mech. Eng.* 55-515, B : 1663-1668.
Tobushi, H., H. Iwanaga, K. Tanaka, T. Hori & T. Sawada 1991. *Trans. Jpn. Soc. Mech. Eng.* 57-543, B : 2747-2752.

Experimental Mechanics, Allison (ed.) © 1998 Balkema, Rotterdam, ISBN 90 5809 014 0

Thermal mechanical cyclic stress strain testing of a type 316 steel

M.W. Spindler
Structural Integrity Branch, Nuclear Electric Limited, UK

ABSTRACT: Procedures for assessing creep-fatigue crack initiation, in components which are subject to thermal cycles with creep dwells, commonly use isothermal data to determine the appropriate input parameters. To validate the use of isothermal data it is necessary to investigate material behaviour under thermal mechanical loading. In order to do this a low cycle fatigue machine has been modified to simultaneously control both the temperature cycle and the mechanical strain cycle. Details of the modifications are given along with the results of some thermal mechanical tests on a Type 316 austenitic stainless steel under combined strain and temperature cycling. A simple method of describing the cyclic stress strain data obtained is also presented.

1 INTRODUCTION

Within the UK, a comprehensive procedure, called R5 (Ainsworth et al. 1995), is often used to assess the high temperature response of structures. One part of R5 deals with creep-fatigue initiation, and requires values for the cyclic strain range, the stress at the start of the creep dwell and the elastic follow-up factor. These values are then used to predict the number of cycles to initiate a creep-fatigue crack.

The cyclic stress strain response of a component is determined by the magnitude of the thermal cycle acting on the constrained body and its material properties under non-isothermal conditions. However, the majority of the materials data used during R5 assessments, and indeed the data used to validate R5 were obtained from isothermal testing. In order to validate the current approach of using isothermal data it is necessary to investigate material behaviour under thermal mechanical loading. This can be achieved in a uniaxial test specimen by simultaneously controlling both the temperature cycle and the mechanical strain cycle.

This paper presents results of thermal mechanical tests conducted on a Type 316 austenitic stainless steel under combined strain and temperature cycling. These data are intended for use in validating procedures for predicting thermal mechanical cyclic stress strain behaviour from isothermal data.

2 EXPERIMENTAL METHODS

2.1 Material and Specimens

The tests were carried out on material from a 50mm plate of Type 316 austenitic stainless steel with composition shown in Table 1. The plate had been solution treated at 1070-1100°C and then water quenched. Button headed low cycle fatigue (LCF) specimens with a diameter of 12.7mm were manufactured to the design shown in Figure 1.

Table 1. Composition of type 316 plate in weight %.

C	Si	Mn	Ni	Cr	Mo	S	P
0.05	0.37	1.86	11.7	18.0	2.25	0.009	0.023

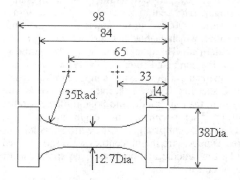

Figure 1. LCF specimen dimensions in mm.

2.2 Testing

Seven push-pull thermal mechanical tests were conducted during which the temperature was cycled between 400 and 650°C simultaneously with the mechanical strain. Five tests were conducted with mechanical strain ranges of ±0.15, ±0.25, ±0.4, ±0.8 and ±1.5% applied in phase with the temperature cycle. Two further tests were conducted where a ±0.8% strain range was applied 180° and 90° out of phase with the temperature cycle. The 90° out of phase cycle was constructed such that the peak temperature occurred at zero mechanical strain, whilst the strain was increasing. The test conditions are summarised in Table 2. In the absence of a suitable standard for thermal mechanical testing the code of practice for LCF testing (Thomas et al. 1989) was followed as closely as possible.

Table 2. Conditions of the thermal mechanical tests.

Test No.	Mechanical strain range %	Phase angle °	Strain at peak Temperature %	No. of cycles
#015	±0.15	0	0.15	350
#025	±0.25	0	0.25	200
#040	±0.4	0	0.4	110
#080	±0.8	0	0.8	100
#1808	±0.8	180	-0.8	75
#908	±0.8	90	0	100
#15	±1.5	0	1.5	50

2.3 Thermal Mechanical Test Machine

The thermal mechanical tests were conducted using a 100kN Instron 8562 servo-electric test machine with an 8500 digital controller. Axial strain was controlled over a 25mm gauge length with a side contacting high temperature capacitance extensometer to an in-house design (Hales & Walters 1982). The specimen was heated by a three zone resistance furnace, each zone of which was controlled by a Eurotherm 902 temperature controller. The central zone was equipped with a Eurotherm 902P programmable temperature controller, which enabled heating and cooling ramps to be defined. The outer zones were configured to follow the command signal of the central zone and thus maintain a low axial temperature gradient. Three sheathed Type K thermocouples were wire locked to the top, middle and bottom of the parallel portion of the specimen gauge length and were used to control the temperature of the zones. In addition, three Type K wire thermocouples were spot-welded to the top, middle and bottom of the specimen for measurement.

Control of the thermal mechanical cycle and logging of the data was achieved using software written in TransERA[*] HT Basic which runs on a Personal Computer. The computer was interfaced with both the Instron 8500 digital controller, via an IEEE parallel interface, and the central zone Eurotherm 902P, via an RS-232 serial interface. In general, the Eurotherm 902P was used to control the heating and cooling rates (0.033 and 0.025°C/s respectively) giving a total cycle time of 4.9 hours for each test. This cycle time was dictated by thermal properties of the furnace and test rig and represented the shortest time possible. During temperature cycling the program continually monitored the temperature and used a simple equation to determine the total strain that should be applied to the specimen to achieve the desired thermal mechanical cycle. The desired total strain was controlled by the Instron 8500 controller, the value of which was continually updated by the computer. Since the heating and cooling rates were the same for all tests the mechanical strain rates increased as the strain range increased.

2.4 Mechanical Strain Control

To achieve the desired mechanical strain it was necessary to apply a total strain (ε_T) to the specimen, where the total strain was the sum of the mechanical strain (ε_M), the thermal strain arising from the temperature (ε_α) and an apparent strain due to differential expansion of the extensometer and specimen (ε_{ext}), such that

$$\varepsilon_T = \varepsilon_M + \varepsilon_\alpha + \varepsilon_{ext} \qquad (1)$$

Prior to each test the value of $\varepsilon_\alpha + \varepsilon_{ext}$ at a given temperature was determined by cycling the temperature between 400 and 650°C with the specimen maintained at zero load. It was found that the value of $\varepsilon_\alpha + \varepsilon_{ext}$ at a given temperature was different during the heating and cooling ramps (Fig. 2). In addition, it was found that the centroid of the $\varepsilon_\alpha + \varepsilon_{ext}$ versus temperature loops initially shifted with cycling before a stable loop was achieved. Taking these considerations into account the procedure for determining $\varepsilon_\alpha + \varepsilon_{ext}$ was to cycle the temperature between 400 and 650°C at zero load until the loops stabilised (typically 12 cycles) and to record the data at 1°C intervals. The $\varepsilon_\alpha + \varepsilon_{ext}$ versus temperature data during heating and cooling from the stabilised cycle were used to fit two polynomials of the form

for heating $\varepsilon_\alpha + \varepsilon_{ext} = a_h + b_h T + c_h T^2 + d_h T^3 + e_h T^4$

$$(2)$$

for cooling $\varepsilon_\alpha + \varepsilon_{ext} = a_c + b_c T + c_c T^2 + d_c T^3 + e_c T^4$

where a_h to e_h and a_c to e_c are constants for the heating and cooling polynomials and T is the temperature in °C. An example of a stabilised cycle with the fitted polynomials is shown in Figure 2.

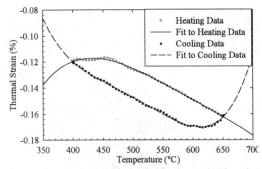

Figure 2. Thermal strain in the final zero load cycle.

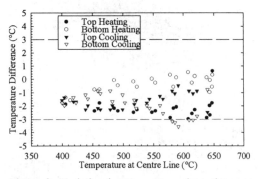

Figure 3. Variation in temperature along the gauge length.

In LCF testing the errors in the mechanical strain are usually determined by a room temperature calibration of the extensometer to Grade C as defined in BS3846:1970 (1985). However, in thermal mechanical testing the total strain is the sum of the mechanical and thermal strains, see equation (1). Thus, for these tests to comply with the requirements of the code of practice (Thomas et al. 1989) it was necessary to show that the total strain met the requirements of Grade C. This was achieved by showing that both the errors in the mechanical and thermal strains met the requirements of Grade B as defined in BS3846:1970 (1985). Since Grade B only permits half the error of Grade C, the error in the total strain (mechanical plus thermal strain) must meet the requirements of Grade C. Prior to each test the extensometer was calibrated at room temperature to Grade B. In addition, the differences between the measured and predicted values of $\varepsilon_\alpha + \varepsilon_{ext}$ were checked against the requirements of Grade B.

2.5 Temperature Control

The code of practice for LCF testing (Thomas et al. 1989) requires the temperature to be maintained to within $\pm 3\,^\circ\text{C}$ of the desired temperature. In isothermal tests this is usually achieved following a soak of about 1 hour to ensure an even temperature distribution. To comply the $\pm 3\,^\circ\text{C}$ criterion it was necessary to demonstrate that the temperature along the gauge length and through the thickness at any instant varied by less than $\pm 3\,^\circ\text{C}$.

The temperatures at the centre and the ends of the gauge length were measured using the spot welded thermocouples (Fig. 3). The variation between these was least at the lower temperatures and increased as the temperature increased (Fig. 3). The $\pm 3\,^\circ\text{C}$ criterion was usually achieved but could be exceeded briefly for the bottom thermocouple during cooling at around 600°C (Fig. 3). A comparison of the results from the in phase and out of phase tests shows that the consequences of exceeding this limit are negligible (see Section 4.1). The temperature at

the core of the specimen was measured by inserting sheathed Type K thermocouples into holes drilled into a dummy specimen. The difference between the core and the surface of the specimen was less than $\pm 3\,^\circ\text{C}$ (Fig. 4).

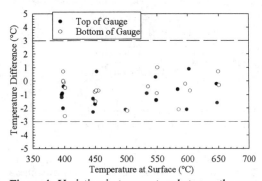

Figure 4. Variation in temperature between the core and the surface.

3 RESULTS

In isothermal tests it is sufficient to report the maximum and minimum stress for the cycle. However, for thermal mechanical data it is necessary to report data corresponding to the maximum and minimum stress and strain since these are not necessarily coincident. Therefore, to summarise the thermal mechanical stress strain data eight points on the hysteresis loops have been chosen which characterise the loop. These points include the turning points, ie where the stress or strain reach maxima or minima. In addition, the crossing points where either the stress or strain are zero have been reported. The locations of these points and the nomenclature are shown in Figure 5.

The thermal mechanical data from each test after saturation have been summarised as above and are given in Table 3. Examples of the hysteresis loops

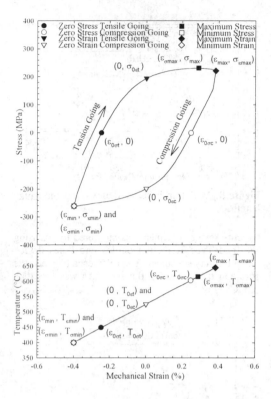

Figure 5. Points defining the thermal mechanical hysteresis loop for an in-phase cycle.

from tests at ±0.8% strain range are plotted in Figures 6-8 and to show the cyclic hardening behaviour of Type 316 under thermal mechanical fatigue the values of σ_{max} and σ_{min} have been plotted against cycle number in Figures 9, 10.

4 DISCUSSION

4.1 *Performance of Test Rig*

The thermal mechanical test machine and control software worked well for the in phase and 180° out of phase tests (Figs 6, 7). However, the performance was not as good for the 90° out of phase test (Fig. 8). At zero mechanical strain, during the tensile going ramp of test #908, the software occasionally unloaded the specimen by a few tens of MPa (Fig. 8). Also not all of the cycles experienced the full mechanical strain range. Both of these faults resulted from the software failing to recognise when heating had finished and cooling had begun. This caused the wrong total strain to be applied to the specimen. Minor software

Table 3. Thermal mechanical cyclic stress strain at saturation.

Test	#015	#025	#040	#080	#1808	#908	#15
ε_{0ot}	-0.035	-0.113	-0.245	-0.587	-0.633	-0.589	-1.268
T_{0ot}	496	469	449	433	624	558	420
ε_{0oc}	0.056	0.138	0.264	0.636	0.588	0.600	1.287
T_{0oc}	572	594	607	624	433	494	632
$\sigma_{0\varepsilon t}$	52	126	194	260	274	224	335
$T_{0\varepsilon t}$	525	525	525	525	525	646	525
$\sigma_{0\varepsilon c}$	-84	-156	-204	-274	-266	-281	-323
$T_{0\varepsilon c}$	525	525	525	525	525	403	525
$\varepsilon_{\sigma max}$	0.144	0.221	0.288	0.353	0.801	0.787	0.302
σ_{max}	154	189	232	277	311	301	341
$T_{\sigma max}$	645	635	615	580	400	527	550
$\varepsilon_{\sigma min}$	-0.15	-0.25	-0.40	-0.799	-0.351	-0.780	-1.497
σ_{min}	-198	-233	-262	-318	-283	-323	-370
$T_{\sigma min}$	400	400	400	400	580	522	400
ε_{max}	0.15	0.25	0.384	0.769	0.801	0.787	1.442
$\sigma_{\varepsilon max}$	152	179	223	251	311	301	286
$T_{\varepsilon max}$	650	650	645	645	400	527	645
ε_{min}	-0.15	-0.25	-0.40	-0.799	-0.798	-0.788	-1.497
$\sigma_{\varepsilon min}$	-198	-233	-262	-318	-233	-301	-370
$T_{\varepsilon min}$	400	400	400	400	650	527	400

Units of strain are %, stress are MPa and temperature are °C.

modifications are required to correct this fault for future tests. Due to these difficulties only the complete cycles of #908 have been reported. It should be noted that the presence of cycles which did not experience the full mechanical strain range would affect the cyclic hardening behaviour of test #908 (Fig. 10) and the results of this test should be treated with caution.

Most of the calibration and test control requirements described in Section 2 were met by the

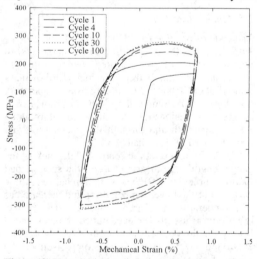

Figure 6. Hysteresis loops from test #080.

1130

test rig. However, the ±3°C on temperature criterion could be exceeded briefly for the bottom thermocouple during cooling at around 600°C (Fig. 3). The consequences of exceeding this limit should be negligible because during the cooling ramp of in phase and 180° out of phase tests temperatures of around 600°C correspond to elastic unloading and can not affect the plastic strain.

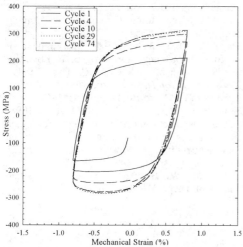

Figure 7. Hysteresis loops from test #1808.

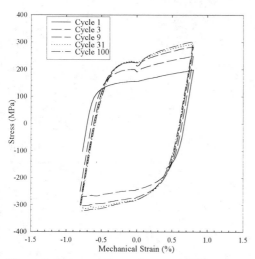

Figure 8. Hysteresis loops from test #908.

4.2 *Thermal Mechanical Stress Strain*

This paper is not intended to provide an analysis of the thermal mechanical cyclic stress strain data. However, it is worth noting some of the features of

the data which may provide an insight into the mechanical properties of Type 316 under thermal mechanical cycling.

Saturation in the thermal mechanical tests was reached after relatively few cycles (Figs 9, 10) when compared with the isothermal tests on the same material and at the same strain range (Spindler, unpubl.). It is not clear if this is due to differences between the strain rates of the thermal mechanical and isothermal tests or if it is a feature of material properties under thermal mechanical cycling.

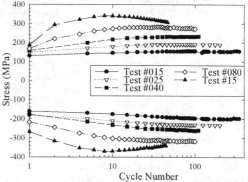

Figure 9. Cyclic hardening during in-phase tests.

During in phase testing the maximum stress did not occur at the maximum strain and temperature (Table 3). Also when the strain range was increased the temperature at which the maximum stress occurred decreased. After the maximum stress is reached the stresses fall even though the mechanical strain is increasing (Fig. 6). This is probably due to the reductions in the proof stress of Type 316 with increasing temperature, although stress relaxation which accelerates at higher temperatures may also contribute to the reductions in stress, particularly in view of the slow strain rates which were used in these tests.

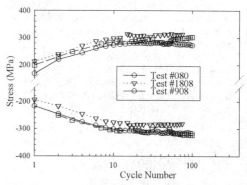

Figure 10. Comparison of cyclic hardening during in phase and out of phase tests.

Due to the varying temperature, the plastic strain range of the thermal mechanical tests is difficult to define since the Young's modulus is varying. However, as the elastic strain is small in comparison with the plastic strain a single value of modulus can be used to estimate elastic strain without giving large errors in the plastic strain. The plastic strain ranges at saturation in the thermal mechanical tests are less than the plastic strain ranges of the isothermal tests at 650°C (Spindler, un-publ.). At mechanical strain ranges of ±0.15, ±0.25 and $\pm0.4\%$ the plastic strain ranges were between those of the 600 and 650°C isothermal tests. For mechanical strain ranges of ±0.8 and $\pm1.5\%$ the plastic strain ranges were between those of the 550 and 600°C isothermal tests. This observation is relevant to understanding how thermal mechanical cycling affects fatigue endurance since plastic strain range is often thought to control endurance, although it should be noted that hysteresis area or energy might be important also (Skelton et al. 1996). The plastic strain range of the 180° out of phase test at $\pm0.8\%$ was identical to the in phase test at $\pm0.8\%$ (Table 3). This is because these hysteresis loops are the mirror image of one another (Figs 6, 7). However, the plastic strain range of the 90° out of phase test at $\pm0.8\%$ (Table 3) was less than the in phase and 180° out of phase tests. These observations can be used to test the theories of whether plastic strain range or hysteresis area control endurance.

5 CONCLUSIONS

1. A test machine has been developed and commissioned which can simultaneously impose thermal and mechanical cycles to a fatigue specimen either in phase or 180° out of phase. Minor modifications to the control software are required to improve the 90° out of phase test.

2. Seven thermal mechanical cyclic stress strain tests have been conducted on a Type 316 steel. The tests cover strain ranges from ±0.15 to $\pm1.5\%$ and include in phase, 90° out of phase and 180° out of phase tests. The temperature was cycled between 400 and 650°C.

3. A method for summarising the cyclic stress strain data from thermal mechanical tests has been described which reports the turning points, ie where the stress or strain reach maxima or minima and the crossing points where either the stress or strain are zero.

ACKNOWLEDGEMENT

This paper is published with permission of Nuclear Electric Ltd.

REFERENCES

Ainsworth, R. A., Hales, R., Budden, P. J. & D C Martin 1995. (eds), Assessment procedure for the high temperature response of structures, Nuclear Electric Report R5 Issue 2.
BS3846:1970, 1985. Methods for calibration and grading of extensometers for testing of metals.
Thomas, G. B., Hales, R., Ramsdale,J., Suhr, R. W. & G Summer 1989. A code of practice for constant amplitude low cycle fatigue testing at elevated temperature. *Fatigue Fract. Engng. Mater. Struct.* 12:135-153.
Hales, R. & D. J. Walters 1982. Measurement of strain in high temperature fatigue. In Loveday, M. S., Day, M. F. & B. F. Dyson (eds), *Measurement of High Temperature Mechanical Properties of Materials.* London: HMSO.
Skelton, R. P., Rees, C. J. & G. A. Webster 1996. Energy damage summation methods for crack initiation and growth during block loading in high-temperature low-cycle fatigue. *Fatigue Fract. Engng. Mater. Struct.* 19:287-297.
Spindler, M. W. Isothermal cyclic stress strain testing of a type 316 steel. un-published.

Experimental Mechanics, Allison (ed.) © 1998 Balkema, Rotterdam, ISBN 90 5809 014 0

Fatigue strength evaluation of large extruded aluminium alloy shapes

H.J. Lee, S.W. Han & S.R. Lee
Korea Institute of Machinery and Materials, Korea

B.S. Kang
Busan National University, Korea

ABSTRACT : One of the effective ways to reduce the weight of the carbody without sacrificing the safety of railway vehicles is to change the materials of the carbody structure with light-weight materials. Stainless steels and high strength steels are commonly used for the present carbody of railway vehicles. These materials are however more or less inadequate from viewpoint of weight reduction. Aluminium alloys are widely used in the high speed automobiles and aircraft structures because of their high specific strength and high damping capability. For the fatigue strength evaluation of the large scale aluminium extruded shapes, stress analysis, static and fatigue tests, and strength evaluation were carried out. Finite element analysis has been performed to obtain stress distribution in the large scale extruded aluminium shapes. Also fatigue tests have been carried out until fracture occurs. It has been observed that the location of fatigue crack initiation coincided with the location of the maximum principal stress predicted by finite element analysis. In addition, the fatigue life of both specimen and the extruded shape has been obtained.

1. INTRODUCTION

Recently the adoption of the light weight aluminium alloy in railroad vehicle industries is increasing remarkably. It is due to the fact that the aluminium alloys have many advantages compared with the traditional steel and stainless steel.

By reducing the total weight of railroad vehicles by use of aluminium alloy, it is well known that the energy consumption rate and the maintenance cost could be reduced considerably.

The adoption of aluminium alloys for light weight design in railroad cars could be possible due to the development of high strength aluminium alloys, development of large capacity extruder, advancement of extrusion technology for large extruded shapes, and the automatic welding technology for aluminium alloys, etc.

In Japan, it is reported that the Nozomi bullet train reduced its weight about 25% by adoption of aluminium alloy instead of steel. As a result, the speed of the train could be increased significantly (Froes 1994).

In Europe, the aluminium alloys have been widely employed in commuter and subway lines until 1988 (Railway Gazette Int. 1991). Recently the application of aluminium alloys is increasing in German high speed trains, ICE. Also it is reported that next generation French high speed train, TGV NG, will be made of aluminium alloys soon.

The trend is very similar in America too. More than 90% of the freight cars manufactured these days in U.S.A are made of aluminium alloys rather than steel counterparts (Pennington 1994).

The increase of aluminium alloys in railroad vehicles can be attributed to both the advancement of extrusion technology for large scale structural components and the advancement of related automatic welding technology. Both technologies made it possible that aluminium vehicles have competitive edge in total cost along with steel/stainless steel counterparts.

In this study, the structural aluminium extruded shape, shown in Fig. 1, is considered. It could be used as a main structural part of the vehicle floor. The material is Al 6005A-T6, and its width and height are 545.5 mm and 80.5 mm, respectively. The length of 95 mm in the direction of extrusion was employed for static and fatigue tests. The thickness in the cross section of the shape varies between 2.7 and 5 mm as shown in Fig. 1.

The loading condition expected in railroad car floor is variable and complicate due to the weight of passengers and carbody itself along with various attachments in carbody. In other words, the aluminium structural parts will be under complicate loading conditions with varying mean load and amplitude.

Thus it is very important to estimate the strength of the structural components under complex loading conditions before they are welded to complete the car body. That is because in railroad vehicles the safety of passengers and carbody itself is very critical.

In this study, both structural analysis and the fatigue strength analysis for the extruded shapes have been performed. Finite element analysis has been employed to estimate the static stress distributions in the component. Also static and fatigue tests for the component have been fulfilled for the component.

2. FINITE ELEMENT ANALYSIS

Finite element analysis for the extruded shape (refer to Fig. 2) has been performed to obtain the stress distribution induced. The commercially available NASTRAN has been used (MSC 1996).

Fig. 2 shows the finite element model used for stress analysis along with the loading and boundary conditions. Quarter model has been employed considering the symmetry conditions. The concentrated vertical load of 16.84 kg per unit length (mm) is applied for the model and the vertical displacement at the roller support is constrained in y direction. In z direction, there is no constraint applied, i.e., it is stress free in that direction.

The first principal stress distributions are shown in Fig. 3(a). The maximum principal stress of 11.8 kg/mm^2 near the roller support is obtained. Fig. 3(b) shows the enlarged view around the roller support where the maximum principal stress occurred.

3. CHARACTERIZATION OF MECHANICAL PROPERTIES

To obtain the mechanical properties of the extruded shapes, both tensile and fatigue tests have been performed by use of Instron hydraulic material tester (10 TON, model No. 8516).

The specimens were taken from the extruded shapes as shown in Fig. 4. Both parallel and perpendicular specimens to the direction of extrusion were considered. They are machined according to ASTM standards (Annual Book 1996). The dimension of the specimen is shown in Fig. 5.

From tensile test for the specimens, the tensile strengths of the parallel and perpendicular specimens to the extrusion direction were obtained to be 26.4 and 28.9 kg/mm^2 , respectively. In addition, the 0.2% offset yield stresses for the parallel and perpendicular specimens were obtained to be 21.4 and 23.0 kg/mm^2 , respectively.

Both tensile and 0.2% offset yield strengths of the perpendicular specimens to the extrusion direction were observed to be higher compared with those for parallel ones.

Also fatigue tests have been performed varying the applied stress. During the test, the load ratio has been maintained constant to be zero. The experiments were done at room temperature atmosphere with applied frequency between 10 and 20 hertz. Sine wave repeated loading was applied.

The results from fatigue tests are summarized in Fig. 6. Failure was defined as the number of cycles reached, when the specimen was completely fractured. As shown in Fig. 6, the maximum stress applied for the perpendicular specimens to the extrusion direction are found to be higher than those for the parallel ones.

In addition, fatigue limit stresses are estimated using the fatigue test results. In this case, the fatigue limit stress is defined to be the fatigue strength when the number of cycles at fracture reaches one million cycles. Fatigue limits for both perpendicular and parallel specimens to the extrusion direction are estimated to be 11.4 and 10.6kg/mm^2, respectively.

4. FATIGUE TESTS FOR EXTRUDED SHAPES

Both static and fatigue experiments were employed in order to characterize the mechanical properties of the extruded shapes shown in Fig 1. 4-point bending tests have been performed as shown in Fig. 7.

From static 4-point bending tests, the relation between load and displacement is obtained as shown in Fig. 8. The maximum load applied was 4,828 kg, when the displacement rate was maintained constant to be 0.01 mm/sec.

Fatigue tests are also included using hydraulic material tester (Instron 8500 series) at room temperature atmosphere.
The experiment conditions are as follows :
、 Load type : 4-point bending test
、 Control : closed loop load control
、 Applied load : sine wave , above 200 kg
、 Applied frequency : 1 - 5 hertz
The fatigue test results are summarized in Fig. 9. Failure was defined as the number of

cycles reached, when the displacement at load line point has been observed to be increased about 10-20 percent compared with the initial displacement. The fatigue cracks were found to be initiated and propagated near the roller support as shown in Fig. 10. The location of the crack initiation from fatigue tests was found to be coincided with the that of the maximum principal stress predicted by finite element analysis mentioned in Chapter 2.

5. STRENGTH EVALUATION FOR EXTRUDED SHAPES

In the previous chapters, the fatigue strengths obtained from specimen tests and from extruded shapes(components) were mentioned. In this chapter, we would like to compare those results and get some relationships between them, if possible.

The test results for specimens taken normal to the extrusion direction will be compared with those for extruded shapes. In Fig. 11, the relations between equivalent stress amplitude and fatigue life are compared. For the specimen test results, the data in Fig. 6 were replotted through the modified Goodman equation with mean stress of zero and load ratio(R) of -1. The results obtained from specimen tests were represented in Fig. 11 by triangle (△).

For the extruded shapes, the data shown in Fig. 9 were employed. The load obtained from the extruded shapes tests were used as input data for applied load in order to get the stress distributions. Thus from finite element analysis for 4-point bending test, the maximum load of 16.84 kg per unit length (mm) induced principal stress of 11.8 kg/mm^2 at the crack initiation site. The data for extruded shapes were represented by square (□) in Fig. 11.

As we see in Fig. 11, the fatigue strengths obtained from specimen tests was observed to be much higher than those obtained from components. It means that if we use the data from specimen tests in the real structure design, it could be dangerous. For conservative design, the fatigue strength data obtained from the extruded shapes, not specimen, are recommended.

The gaps in fatigue strengths shown Fig. 11 could be attributed to the following reasons. First, the size effects and the difference in thickness might bring the gaps. Second, different material tests were involved, i. e., tensile tests for specimen and 4-point bending tests for extruded shapes. Third, the surface roughness might cause the gap. Last, the surface hardening effects induced in machining of specimens can be considered.

6. CONCLUSIONS

Through static and fatigue tests, finite element analysis, and comparison of fatigue strengths obtained from specimen and component itself for the extruded shapes, we obtained the following conclusions.

1. Finite element analysis for the extruded shapes under 4-point bending conditions predicted the weakest area to be near the roller support. That area coincided with the location of crack initiation observed from 4-point bending fatigue test for the extruded shapes.

2. From tensile and fatigue tests for specimens taken from the extruded shapes, the tensile and fatigue strengths depend on direction relative to the extrusion direction.

3. 4-point bending fatigue tests for the extruded shapes were performed to get the relation between maximum load and fatigue life. In addition, it was observed that the crack initiated and propagated near the roller contact region.

4. The fatigue strengths obtained by use of specimen test data were found to be higher than those obtained using component test data. Thus for conservative design, the latter is recommended.

REFERENCES

Froes, F. H., 1994, "Aluminum : Key to advancing the performance of the bullet train," Light Metal Age.

Railway Gazette International, 1991, "Aluminum invades Europe's main line coach market".

Pennington, J. N., 1994, Modern Metals, "Aluminum : Big gains coming from rail cars, bridges".

MSC, 1996, MSC/NASTRAN User Manual.

Annual Book of ASTM Standards, 1996, E 8M - 96, Vol. 03.01.

ACKNOWLEDGEMENTS: This study was fulfilled under the auspices of Ministry of Science and Technology and Dongyang Gangchul Co., Korea

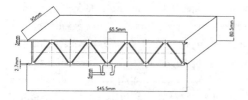

Fig. 1 Extruded aluminium shape

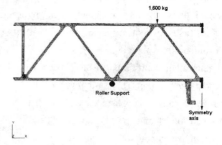

Fig. 2 Finite element model & boundary conditions

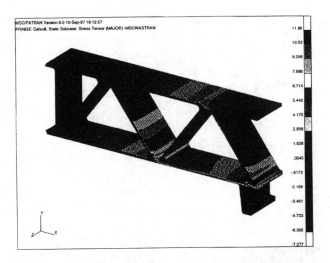

(a) 1st Principal Stress Distribution

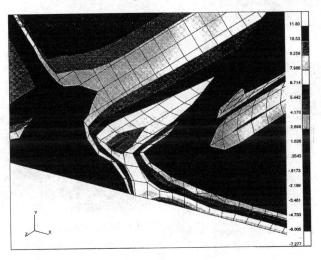

(b) Zoom-in of max. 1st principal stress area

Fig. 3   Principal stress distribution

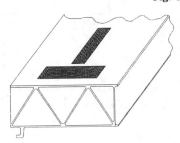

Fig. 4   Specimen Orientation

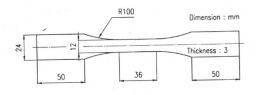

Fig. 5   Tensile & fatigue testing Specimen

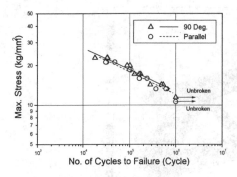

Fig. 6 Fatigue test result

Fig. 7 4-Point bending test

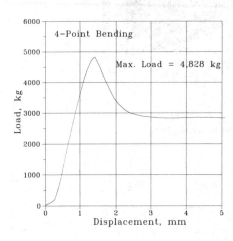

Fig. 8 Static load – displacement of extruded
aluminium shape

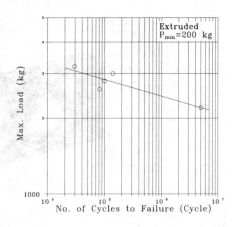

Fig. 9 Fatigue test results of extruded
aluminium shape

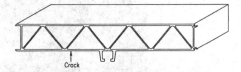

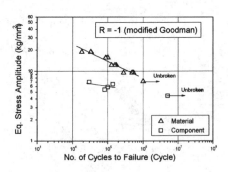

Fig. 10 Fatigue crack of extruded aluminum

Fig. 11 Equivalent stress amplitude – fatigue
life relation

Fracture mechanics

Experimental Mechanics, Allison (ed.) © 1998 Balkema, Rotterdam, ISBN 90 5809 014 0

Crack growth mechanism below T_g in an epoxy strip

K. Ogawa & M. Takashi
Department of Mechanical Engineering, Aoyama Gakuin University, Japan

A. Misawa
Department of Mechanical Engineering, Kanagawa Institute of Technology, Japan

ABSTRACT: The authors discuss crack growth behavior below the glass temperature T_g in a viscoelastic epoxy strip. Below T_g, not only stable and unstable crack growth of a single main crack but also, sometimes, branching to multicracks are observed. Several types of characteristic mark on the fracture surface are observed below T_g. These types of characteristic mark appear when a crack changes its velocity drastically from stable to unstable. Characteristic tongue mark developed from the initial crack front shows remarkable temperature dependence even below T_g. Several aspects of transient behaviors, such as shape, size and location of tongue mark reveal several features as follows; 1] A temperature and time dependent transient process exists at the transition from stable to unstable crack growth, 2] The tongue mark is outlined by plastic deformation near outer sides under plane stress state, and 3] Both the upper and lower fracture surface leaves a trace of convexity or concavity on fracture surface poking each other.

1 INTRODUCTION

It is well known that the micro-structure of polymeric materials which consists of 3D crosslinks and entanglements of long molecular chain is quite different from that of metallic and crystalline materials. On this account, not only the mechanical but fracture behavior of polymer such as epoxy resin show remarkable dependence on time and temperature (Knauss 1974). As a result of the mechanical behavior and microscopic structural features(Newman & Wolock 1958), the fracture behavior and mechanism can be very complicated, so that various difficulties exists when it is expected to establish a unified understanding of the phenomena. Particularly in amorphous polymer, it is difficult to identify some sort of intermediate structures comparable to grains in ordinary metallic or crystalline materials. Thus, the quantitative evaluation and description of fracture behavior and microscopic mechanisms are also difficult because of divergent information from various bases of different sciences with wide varieties of observation scale.

In the previous papers (Ogawa & Takashi 1990, Ogawa et. al. 1991), slow and stable crack growth behaviors above T_g was carefully observed using a wide strip specimen having a long (semi-infinite) crack under constant rate displacement loadings and several temperatures. The possiblity of time and temperature independent crack growth resistance was

discussed by taking precise and reproducible crack growth behaviors into consideration and utilizing the time dependent $J'(t)$-integral for a linearly viscoelastic materials.

In this study, the authors will discuss crack growth behavior below T_g in the same material. Below T_g, not only stable and unstable crack growth of a single main crack, but sometimes branching to multicracks occur. A characteristic tongue mark is observed on fracture surface obtained below the glass temperature. This type of characteristic tongue mark appears during stable growth of a crack before it changes into unstable (rapid) growth drastically. Such marks show remarkable temperature dependence even below the glass temperature. The configuration of the tongue marks on fracture surface gives us important information on the transient behavior from stable to unstable crack growth. In addition, careful observation of fracture surface obtained in the transient process gives several important clues for better understanding fracture mechanisms in this type of material on the basis of relationship between crack growth behavior and the configuration of characteristics on the fracture surface.

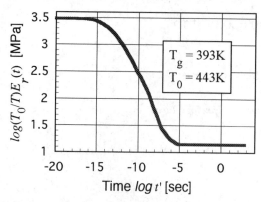

Figure 1. Master curves of the relaxation function in tension, $E_r(t')$.

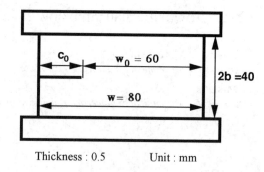

Figure 2. Geometry of specimen.

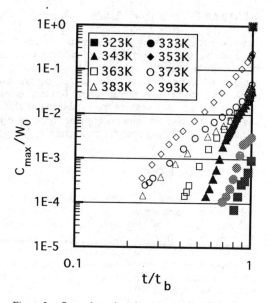

Figure 3. Several crack extension curves below T_g.

2 EXPERIMENTS

2.1 Mechanical properties of the material

The materials adopted in this experiment was a type of hard epoxy plate prepared by mixing Bisphenol-A type resin (Epikote 828, Shell Chem.) with an amine type hardener (Triethylene-Tetramine). Figure1 shows the master curves of the relaxation functions, $E_r(t')$, in tension measured under several constant strain rates and temperatures, then composed using the W.L.F. shift factor. It is obviously seem that the $E_r(t')$ of the material show remarkable dependence on the reduced time t'. The material is, then, considered a typical example of linearly viscoelastic materials. The values of $E_r(t')$ vary more than two hundred times between rubbery and glassy state over a wide range of the reduced time more than ten decades in a logarithmic scale. The glass temperature T_g is measured as 393K. Also, the master curve of $E_r(t')$ is approximated by use of Prony series to be applied to the computation of $J'(t)$-integral.

2.2 Geometry of specimen and loading condition

The specimen adopted is a strip with w = 80 mm in width, 2b = 40 mm in height and t = 0.5 mm in thickness having an initial edge crack c_0 = 20 mm as shown in Figure 2. The ratio of initial crack length to the other dimensions of specimen is chosen to satisfy approximately the condition of a semi-infinite crack. The constant rates of displacement loading, 8.33×10^{-6} m/s is applied at the lower grip fixed rigidly not to change loading angle with increase of crack length. Eight different constant temperatures are adopted for tests below T_g.

3 RESULTS AND DISCUSSIONS

3.1 Crack growth behavior

Figure 3 shows examples of crack growth curves obtained at different temperature conditions. The normalized time, i.e. t/t_b: where t_b is the rupture time, is the abscissa and the normalized crack growth increment, i.e. C/W_0; where W_0 is the expected total length of crack path, is the ordinate. A crack growth curve obtained under 413 K above T_g is also shown for comparison; it shows no evidence of transient behavior from stable to unstable crack growth with a very good straight line relation. On the

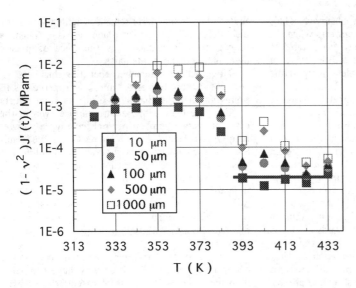

Figure 4. Temperature dependence of the values of $J'(t)$.

other hand, below T_g, not only stable but unstable crack growth is obviously seen, and crack growth curves are remarkably dependent on temperature. Upon lowering the temperature, each normalized slow growth curve shows fairly good straight line behavior on a double logarithmic scale, and is shifted downward to the right-bottom corner of the figure. A remarkable temperature effect is also seen in the slope of the slow crack growth curve. The duration of slow and stable up to final unstable crack growth becomes shorter with a decrease of temperature. In general, it is considered that macroscopic mechanical behavior such as the relaxation modulus below T_g does not show any temperature dependence. It should be emphasized, however, that not only the crack growth behavior but also the strength itself shows very remarkable dependence even below T_g. In other words, crack growth mechanisms below T_g could not be so simple as the case of macroscopic mechanical behavior such as, the relaxation function in tension, $E_r(t)$.

3.2 Extended J'-integral for viscoelastic material

Rice's well known J-integral (Rice1968) has already been modified into the extended $J'(t)$-integral for a linearly viscoelastic thin plate in one of the previous papers (Misawa et al. 1990). The authors have already pointed out in one of the papers (Ogawa 1991) that there could be a time and temperature independent resistance J'_{Ic} for crack growth threshold over a wide range of test conditions. Selecting an incubation time corresponding to the crack growth increment of 10 μm, a material

property unique to crack growth and independent on time and temperature is successfully obtained from accurate experiments on crack growth and by use of $J'(t)$-integral for a linearly viscoelastic material under monotonically increasing loading. Figure 4 shows the temperature dependence of the apparent $J'(t)$-integral value for each apparent incubation time t^*, i.e. for each increment of crack length, 10 to 1000 μm. It shows results obtained above T_g for comparison, together. When calculating $J'(t)$-integral below T_g (= 393 K), it is assumed that the W.L.F. equation can be valid till 373 K, while $E_r(t')$ keeps constant values below 363 K. In spite that the difference of crack length at the middle line and the side of specimen thickness is large or not, the value of $J'(t)$-integral was calculated by using the incremented length of crack at the middle line. Even below T_g, the temperature dependence of $J'(t)$ is obviously recognized between 323 K and 393 K, independent on the increment of crack length. $J'(t)$ increases gradually with increasing temperature till 353 K and decreases considerably until 393 K although the mechanical behavior does not depend on temperature below T_g. The value of $J'(t)$-integral calculated for the smallest crack length increment, 10μm, on which $J'(t)$ is regarded as constant above T_g, shows temperature dependence. The value of $J'(t)$ at below T_g is very large as compared with those above T_g.

3.3 Morphological characteristics of fracture surface

Figure5 shows a typical example of fracture surface obtained at a temperature below T_g. The characteristic tongue mark is clearly observed. This type of characteristic tongue mark appears when the crack changes its velocity drastically from stable to unstable. The shape and length of this type of tongue mark show remarkable temperature dependence even under the conditions below T_g.

Next let us consider the streamline marks indicating the flow path trace of crack growth. The fracture surface inside tongue mark formed by stable crack growth shows clear evidence that the core of the crack develops at the middle part of specimen thickness and that the front of the crack moves towards both side edges symmetrically and perpendicularly to the side of specimen. At the instant of critical state for unstable crack growth, crack bursts and extends in all directions, thus in certain portions crack goes backwards. It is marvelous to point out that even in a very thin strip of only 0.5 mm thickness, the formation of clear tongue marks on the fracture surface is affected not only by the complicated stress and strain but also by their changing rate at the curved crack front.

Figure 6 shows a typical example set of three-dimensional shape matching of the upper and lower fracture surface of the specimen which was obtained with a laser microscope. As for the characteristic feature of fracture surface with tongue mark, the central part and both side parts show similar height of wall formed by plastic flow in the early stage. On the other hand, the central part becomes lowered flat into the final stage. Thus, they face each other, convex to convex and concave to concave. The central part grows progressively forming a flat tongue shaped fracture surface under plane strain

first, and both side parts remain as some wall by the high deformability under plane stress. Then, a tongue mark is remained as the trace of plastic deformation on both side parts.

Now, let us remember that the thickness of specimen is very thin as 0.5 mm. It is marvelous to point out that the triaxiality of stress state affects strongly the feature of fracture surface even in the case of very thin strip, particularly at lower temperature below T_g. With regard to the semi-microscopic view point, cold drawing and pulling out of molecular chains from their entanglement, which is closely related to plastic flow, can be easily generated under only plane stress condition. Once, this type of molecular movement and plastic flow occur, the deformed wall formed by plastic flow prevents crack growth with a sharp tip under triaxial stress state.

The value of $J'(t)$-integral at each temperature between 323 K and 373 K is very large compared with the value above T_g in the case of the smallest crack length increment, 10 μm. These phenomena are caused by a unique crack growth behavior, making a tongue mark on fracture surface and bringing out slow crack growth. Therefore the use of $J'(t)$-integral as a measure for the critical condition of crack growth might be also possible at the temperature below 323 K. To explain crack growth mechanism forming the characteristic fracture surface, it would be inevitable to make clear the mechanical behavior between T_g and $T_g - 60K$ and develop a new approach to the treatment of complicated three dimensional stresses at crack front and plastic flow under plane stress condition.

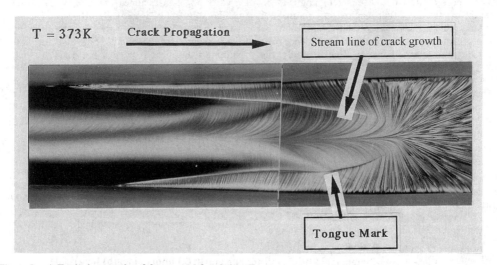

Figure 5. A Typical example of fracture surface below T_g.

T = 373K

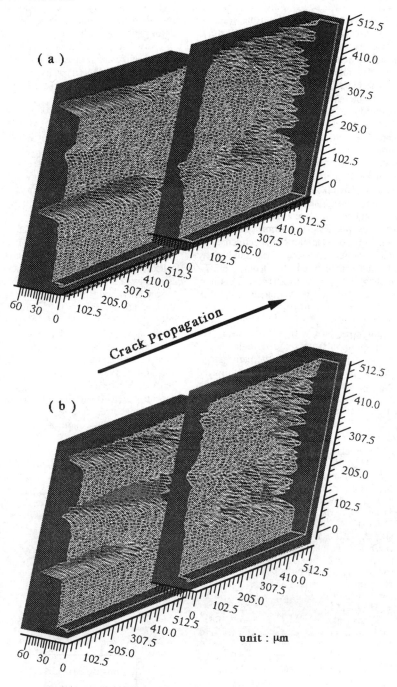

Fig.6 Three-dimensional display of tongue mark, (a) upper side, (b) lower side

4 CONCLUDING REMARKS

In order to investigate crack growth mechanism and behavior of a viscoelastic epoxy strip below T_g, careful observations on the characteristic feature of fracture surface and crack growth behaviors were performed. The results obtained are briefly summarized as follows:

(1) Temperature dependent crack growth behavior was observed in the region of temperature between T_g and $T_g - 60$ K. To explain this phenomena mechanical behavior including time-temperature shift effect have to be investigated more precisely in conjunction with material structure and molecular mobility.

(2) The tongue mark formed during the stable crack growth before the drastic unstable growth is observed on fracture surface obtained below T_g. It is formed by plastic deformation under plane stress condition near side parts of specimen.

(3) The direct application of the extended $J'(t)$-integral to stable crack growth between T_g and $T_g - 60$ K is difficult or impossible because of very complicated crack tip configuration in this stage.

ACKNOWLEDGMENT

The authors appreciate a partial support of the Center for Science and Engineering Research Institute of Aoyama Gakuin University.

REFERENCES

Knauss, W.G. 1974. On the steady propagation of a crack in a viscoelastic sheet: Experiments and Analysis. In H.H. Kausch (ed), *Deformation and fracture of high polymers*: 501-541. Plenum Press.

Misawa, A., M. Takashi & T.Kunio. 1990. Application of the conservation law to threshold of crack growth in viscoelastic strip. *Proc. the 9th Int. Conf. on Exp. Mech.* 5: 1786-1794.

Newman, S.B. & K.I. Wolock 1958. Fracture phenomena and molecular weight in polymethyl methaacrylate. *J. Appl. Phys.* 29: 49-52.

Ogawa, K. and M. Takashi 1990. The quantitative evaluation of fracture surface roughness and crack propagation resistance in epoxy resin. *Trans. JSME. Ser.* A, **56**, 1133-1139.

Ogawa, K., A. Misawa, M. Takashi & T. Kunio 1991. Evaluation of crack growth resistance and fracture surface charcteristics in several epoxy resins, *Proc. 6th Int. Conf. Mech. Behav. of Materials* 6: 105-110.

Rice, J.R. 1968. A path independent integral and the approximate analysis of strain concentration by notches and cracks. *J. Appl. Mech.* 35: 379-386.

Experimental Mechanics, Allison (ed.)© 1998 Balkema, Rotterdam, ISBN 90 5809 014 0

Multi-term elastic strain fields around the crack tip

D. M. Constantinescu

Department of Strength of Materials, 'Politehnica' University of Bucharest, Romania

ABSTRACT: A new proposed solution outlines that the Sanford's solution is much too general, and a three or four terms series extension based on the Eftis and Liebowitz solution is useful in order to evaluate the stress, strain and displacement fields at greater distances from the crack tip. Very good agreement between the experimental and theoretical strain fields is obtained up to distances $r / a = 1$ where the influence of non-singular terms is important. The proposed solution for strain fields, being in agreement with the developments of Dally and Sanford, has multi-term known coefficients and, as experiments show, can be used with confidence in order to establish the elastic strain fields around the crack tip.

1 INTRODUCTION

Westergaard (1939) showed that a limited range of Mode I crack problems can be solved by using a complex stress function technique. His solution played an important role in the development of linear elastic fracture mechanics because it connected the remote loading applied on the boundaries of the cracked body to the local crack tip stress and displacement fields. This solution is valid only for a through crack in an infinite plate subjected to biaxial uniform remote tension. Irwin (1958) noted that photoelastic fringe patterns observed by Wells & Post (1958) on the centre cracked panels did not match Westergaard's solution. Irwin achieved good agreement between theory and experiment by subtracting a uniform horizontal stress that depends on the remote stress. When a centre cracked panel is loaded in uniaxial tension, a transverse compressive stress develops in the plate. Sih (1966) provided theoretical basis for Irwin's modification. Sih generalized the Westergaard approach by applying a complex potential formulation for the Airy stress function. He imposed the condition of zero shear stress along the crack axis (Mode I) and found that a constant term should be subtracted from the horizontal normal stress, this being in agreement with Irwin's modification. Later, Eftis & Liebowitz (1972) and Eftis et al. (1977) showed that for infinite plates biaxially loaded in tension-tension or tension-compression, a constant term depending on the loading can be added to the

Westergaard's stress function. This constant is subtracted from the horizontal normal stress and can be established by imposing the boundary conditions; it is zero for the Westergaard solution for uniform biaxial tension problem.

Sanford (1979) showed that the previous approaches are too restrictive. A complex function can be added to Westergaard's stress function. By considering series expansion for both of them, a Williams (1957) type solution can be obtained. If the complex function is a real constant the problem is reduced to Sih's approach, or, if it is zero, to the original Westergaard solution. For a reasonable solution one should consider only several terms of such a series expansion and establish the series coefficients. Following the theoretical developments (Sanford 1979) a three term series (six parameter and six coefficient) representation of the strain field was adopted by Dally & Berger (1986), and Dally & Sanford (1987) for determining the Mode I or Mode II stress intensity factors by using strain gauges. For a proper inclination of the strain gauge direction of measurement that depends on Poisson's ratio, the Mode I stress intensity factor can be easily established. It seems that for Mode II the results are not so reliable. The validity of the three term crack tip strain fields was analysed (Berger et al. 1991) for various crack length to width of the plate ratios, for single edge notched specimens, in comparison with the twelve term solution that was considered to be exact. Photoelasticity was also used to determine the coefficients of up to twelve term series expansion.

2 THEORETICAL SOLUTION

Eftis & Liebowitz (1972) and Eftis et al. (1977) considered the infinite plate with crack length 2a, biaxially loaded. Generalizing Westergaard's problem, the remote tractions are considered to be $\sigma_x = \alpha\sigma$ and $\sigma_y = \sigma$. The positive sign for the parameter α signifies tension - tension loading, while a negative value means tension - compression. If $\alpha = 0$, the common case of uniaxial tension is obtained.

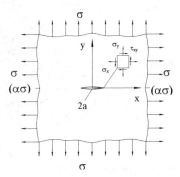

Figure 1. Biaxially loaded plate with central crack.

It has been shown (Eftis et al. 1977) that for an infinite plate loaded in Mode I, the following equation is satisfied on the horizontal symmetry axis of a centre cracked plate:

$$z \cdot \phi''(z) + \psi'(z) + B = 0 \qquad (1)$$

where $\phi''(z)$ and $\psi'(z)$ are the second and respectively the first derivatives of the Goursat-Kolosov holomorphic functions. The real constant B can be determined from the Mode I traction boundary condition on the outer surface of the cracked body as $B = -(1-\alpha)/2$.

The stresses close to the crack tip are obtained as:

$$\sigma_x = \text{Re}[2\phi'(z)] + y \cdot \text{Im}[2\phi''(z)] + B$$
$$\sigma_y = \text{Re}[2\phi'(z)] + y \cdot \text{Im}[2\phi''(z)] - B \qquad (2)$$
$$\tau_{xy} = -y \cdot \text{Re}[2\phi''(z)]$$

Confining attention only to the crack tip region and using the notation $z^* = z - a = r \cdot e^{i\theta}$, the following expressions are derived:

$$2\phi'(r,\theta) \cong \frac{\sigma\sqrt{\pi a}}{\sqrt{2\pi r}}\left(\cos\frac{\theta}{2} - i\sin\frac{\theta}{2}\right) - \frac{(1-\alpha)\sigma}{2}$$

$$\phi(r,\theta) \cong \sigma\sqrt{\pi a}\sqrt{\frac{r}{2\pi}}\left(\cos\frac{\theta}{2} + i\sin\frac{\theta}{2}\right) - \frac{(1-\alpha)\sigma a}{4}\left(\frac{re^{i\theta}}{a} + 1\right) \qquad (3)$$

Relations (3) were obtained for series expansions in which the terms of order $(z^*/a)^{1/2}$ and above were neglected in the first relation, respectively terms of order $(z^*/a)^{3/2}$ and above were neglected in the second relation. This means in fact that the relations are valid only for $0 < r/a \ll 1$.

In the above relations, more terms of the series expansion can be considered in order to improve the solution and extend the limit of validity as a function of r/a. The proposed solution (Constantinescu & Voloshin, in prep.) reconsiders the series expansion of relations (3) in the following form:

$$2\phi'(r,\theta) = \frac{\sigma}{\sqrt{2}}\left[\left(\frac{re^{i\theta}}{a}\right)^{-\frac{1}{2}} - \frac{(1-\alpha)}{\sqrt{2}}\left(\frac{re^{i\theta}}{a}\right)^{0} + \frac{3}{4}\left(\frac{re^{i\theta}}{a}\right)^{\frac{1}{2}} - \frac{5}{32}\left(\frac{re^{i\theta}}{a}\right)^{\frac{3}{2}}\right]$$

$$\phi(r,\theta) = \frac{\sigma a}{\sqrt{2}}\left[\left(\frac{re^{i\theta}}{a}\right)^{\frac{1}{2}} - \frac{(1-\alpha)}{2\sqrt{2}}\left(\frac{re^{i\theta}}{a}\right)^{1} + \frac{1}{4}\left(\frac{re^{i\theta}}{a}\right)^{\frac{3}{2}} - \frac{1}{32}\left(\frac{re^{i\theta}}{a}\right)^{\frac{5}{2}}\right] - \frac{(1-\alpha)\sigma a}{4}$$

With $e^{in\theta} = \cos(n\theta) + i\sin(n\theta)$ and $K_I = \sigma\sqrt{\pi a}$ it results:

$$2\phi'(r,\theta) = \frac{K_I}{\sqrt{2\pi r}}\left(\cos\frac{\theta}{2} - i\sin\frac{\theta}{2}\right) - \frac{(1-\alpha)\sigma}{2} + \frac{3\cdot\sigma}{4\sqrt{2}}\left(\frac{r}{a}\right)^{\frac{1}{2}}\left(\cos\frac{\theta}{2} + i\sin\frac{\theta}{2}\right) - \frac{5\cdot\sigma}{32\sqrt{2}}\left(\frac{r}{a}\right)^{\frac{3}{2}}\left(\cos\frac{3\theta}{2} + i\sin\frac{3\theta}{2}\right) \qquad (4)$$

$$2\phi''(r,\theta) = \frac{K_I}{\sqrt{2\pi r}}\left[-\frac{1}{2r}\left(\cos\frac{3\theta}{2} - i\sin\frac{\theta}{2}\right)\right] +$$

$$+\frac{3\cdot\sigma}{8\sqrt{2}}\frac{1}{\sqrt{a}\sqrt{r}}\left(\cos\frac{\theta}{2} - i\sin\frac{\theta}{2}\right) - \qquad (5)$$

$$-\frac{15\cdot\sigma}{64\sqrt{2}}\frac{1}{a}\left(\frac{r}{a}\right)^{\frac{1}{2}}\left(\cos\frac{\theta}{2} + i\sin\frac{\theta}{2}\right)$$

and

$$\phi(r,\theta) = K_I\sqrt{\frac{r}{2\pi}}\left(\cos\frac{\theta}{2} + i\sin\frac{\theta}{2}\right) -$$

$$-\frac{(1-\alpha)\cdot\sigma\cdot a}{4}\left[1 + \left(\frac{r}{a}\right)(\cos\theta + i\sin\theta)\right] +$$

$$+\frac{\sigma\cdot a}{4\sqrt{2}}\left(\frac{r}{a}\right)^{\frac{3}{2}}\left(\cos\frac{3\theta}{2} + i\sin\frac{3\theta}{2}\right) - \qquad (6)$$

$$-\frac{\sigma\cdot a}{32\sqrt{2}}\left(\frac{r}{a}\right)^{\frac{5}{2}}\left(\cos\frac{5\theta}{2} + i\sin\frac{5\theta}{2}\right)$$

Using eqns (4), (5), and (2), the stress field is obtained as:

$$\sigma_x = \frac{K_I}{\sqrt{2\pi r}}\cos\frac{\theta}{2}\left(1 - \sin\frac{\theta}{2}\sin\frac{3\theta}{2}\right) - (1-\alpha)\cdot\sigma +$$

$$+\frac{3}{4\sqrt{2}}\cdot\sigma\cdot\left(\frac{r}{a}\right)^{\frac{1}{2}}\cos\frac{\theta}{2}\left(1 + \sin^2\frac{\theta}{2}\right) -$$

$$-\frac{5}{32\sqrt{2}}\cdot\sigma\cdot\left(\frac{r}{a}\right)^{\frac{3}{2}}\left(\cos\frac{3\theta}{2} - \frac{3}{2}\sin\frac{\theta}{2}\sin\theta\right)$$

$$\sigma_y = \frac{K_I}{\sqrt{2\pi r}}\cos\frac{\theta}{2}\left(1 + \sin\frac{\theta}{2}\sin\frac{3\theta}{2}\right) +$$

$$+\frac{3}{4\sqrt{2}}\cdot\sigma\cdot\left(\frac{r}{a}\right)^{\frac{1}{2}}\cos\frac{\theta}{2}\left(1 - \sin^2\frac{\theta}{2}\right) -$$

$$-\frac{5}{32\sqrt{2}}\cdot\sigma\cdot\left(\frac{r}{a}\right)^{\frac{3}{2}}\left(\cos\frac{3\theta}{2} + \frac{3}{2}\sin\frac{\theta}{2}\sin\theta\right)$$

$$\tau_{xy} = \frac{K_I}{\sqrt{2\pi r}}\cos\frac{\theta}{2}\sin\frac{\theta}{2}\cos\frac{3\theta}{2} -$$

$$-\frac{3}{4\sqrt{2}}\cdot\sigma\cdot\left(\frac{r}{a}\right)^{\frac{1}{2}}\sin\frac{\theta}{2}\cos^2\frac{\theta}{2} + \qquad (7)$$

$$+\frac{15}{32\sqrt{2}}\cdot\sigma\cdot\left(\frac{r}{a}\right)^{\frac{3}{2}}\sin\frac{\theta}{2}\cos^2\frac{\theta}{2}$$

Thus, the proposed solution completes the well known Westergaard solution (first term of each of the stress relations) and the Eftis and Liebowitz solution. The last two terms of each of the stress relations are added in this solution.

With the stresses established by relations (7) and in the conditions of plane strain, the in plane strains are:

$$\varepsilon_x = \frac{1}{E}\left[(1-v^2)\sigma_x - v(1+v)\sigma_y\right]$$
$$\qquad (8)$$
$$\varepsilon_y = \frac{1}{E}\left[(1-v^2)\sigma_y - v(1+v)\sigma_x\right]$$

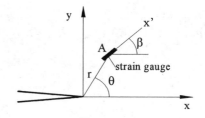

Figure 2. Strain gauge around the crack tip.

If the strain is measured on the direction x' established in Figure 2, it becomes:

$$4G\varepsilon_{x'} = \frac{K_I}{\sqrt{2\pi r}}\left[(\kappa - 1)\cos\frac{\theta}{2} - \sin\frac{3\theta}{2}\sin\theta\cos 2\beta +\right.$$

$$\left. + \cos\frac{3\theta}{2}\sin\theta\sin 2\beta\right] -$$

$$-(1-\alpha)\sigma\left[\frac{(\kappa-1)}{2} + \cos 2\beta\right] +$$

$$+\frac{3\sigma}{4\sqrt{2}}\left(\frac{r}{a}\right)^{\frac{1}{2}}\left[(\kappa-1)\cos\frac{\theta}{2} + \sin\frac{\theta}{2}\sin\theta\cos 2\beta -\right.$$

$$\left. - \cos\frac{\theta}{2}\sin\theta\sin 2\beta\right] -$$

$$-\frac{5\sigma}{32\sqrt{2}}\left(\frac{r}{a}\right)^{\frac{3}{2}}\left[(\kappa-1)\cos\frac{3\theta}{2} - 3\sin\frac{\theta}{2}\sin\theta\cos 2\beta -\right.$$

$$\left. - 3\cos\frac{\theta}{2}\sin\theta\sin 2\beta\right]$$
$$\qquad (9)$$

For plane strain $1 - 2v = (\kappa - 1)/2$; the constant κ is a parameter of the material and is known to be considered as: $\kappa = 3 - 4v$ for plane strain, and $\kappa = (3 - v)/(1 + v)$ for plane stress.

The second term becomes zero if:

$$\cos 2\beta = -\frac{\kappa - 1}{2} \tag{10}$$

- for plane strain: $\cos 2\beta = -(1 - 2\nu)$ (11)

- for plane stress: $\cos 2\beta = -\frac{1-\nu}{1+\nu}$ (12)

Relation (12) is identical with a relation that is based on Sanford's solution (Dally & Sanford 1987), but obtained on different theoretical bases.

The third term of eqn (9) is zero if:

$$(\kappa - 1)\cos\frac{\theta}{2} + \sin\theta\cos 2\beta\left(\sin\frac{\theta}{2} - \cos\frac{\theta}{2}\,\mathrm{tg}2\beta\right) = 0$$

It results:

$$\mathrm{tg}\frac{\theta}{2} = -\mathrm{ctg}2\beta \tag{13}$$

that is the same as the corresponding relation from the same reference.

The use of relation (9) makes possible the calculation of the strain in a point defined by the polar coordinates (r,θ), and along a direction inclined with the angle β measured from the horizontal axis (Fig. 2). The above mentioned relation considers 4 terms of the series expansion used in order to calculate the stress field around the crack tip.

For the 3 terms solution, if the angle β is taken as in relation (12) and the angle θ as a function of β from relation (13), the strain is calculated as:

$$4G\varepsilon_{x'} = \frac{K_1}{\sqrt{2\pi r}}(\kappa - 1)\cos\frac{\theta}{2}\left[1 + \sin\frac{\theta}{2}\sin\frac{3\theta}{2}\left(1 + \mathrm{ctg}\frac{\theta}{2}\,\mathrm{ctg}\frac{3\theta}{2}\right)\right] \tag{14}$$

For the 4 terms solution, the strain becomes:

$$4G\varepsilon_{x'} = \frac{K_1}{\sqrt{2\pi r}}\left[(\kappa - 1)\cos\frac{\theta}{2} - \sin\frac{3\theta}{2}\sin\theta\cos 2\beta + \cos\frac{3\theta}{2}\sin\theta\sin 2\beta\right]$$
$$- \frac{5\sigma}{32\sqrt{2}}\left(\frac{r}{a}\right)^{\frac{3}{2}}(\kappa - 1)\cos\frac{\theta}{2}\left(3 - 8\cos^2\frac{\theta}{2}\right) \tag{15}$$

For the particular case when angles $\beta = \theta$, relation (9) has the particular form:

$$4G\varepsilon_{x'} = \frac{K_1}{\sqrt{2\pi r}}\left[(\kappa - 1)\cos\frac{\theta}{2} - \sin\frac{3\theta}{2}\sin\theta\cos 2\theta + \cos\frac{3\theta}{2}\sin\theta\sin 2\theta\right] -$$
$$- (1-\alpha)\sigma\left[\frac{(\kappa - 1)}{2} + \cos 2\theta\right] +$$
$$+ \frac{3\sigma}{4\sqrt{2}}\left(\frac{r}{a}\right)^{\frac{1}{2}}\left[(\kappa - 1)\cos\frac{\theta}{2} + \sin\frac{\theta}{2}\sin\theta\cos 2\theta -\right.$$
$$\left. - \cos\frac{\theta}{2}\sin\theta\sin 2\theta\right] -$$
$$- \frac{5\sigma}{32\sqrt{2}}\left(\frac{r}{a}\right)^{\frac{3}{2}}\left[(\kappa - 1)\cos\frac{3\theta}{2} - 3\sin\frac{\theta}{2}\sin\theta\cos 2\theta -\right.$$
$$\left. - 3\cos\frac{\theta}{2}\sin\theta\sin 2\theta\right] \tag{16}$$

Equation (9) - respectively (16) - is based on the proposed solution with 4 terms of the series expansion, and proves to be in agreement with Sanford's solution. This time the solution is complete, as the coefficients of each term are known.

For a centre cracked plate loaded in uniaxial tension ($\alpha = 0$) by the stress σ, with G and ν known, the theoretical strain of the 4 terms solution is:

$$\varepsilon_{x'} = \frac{1}{4\sqrt{2}}\frac{\sigma}{G}\left(\frac{r}{a}\right)^{-\frac{1}{2}}\left[(\kappa - 1)\cos\frac{\theta}{2} - \sin\frac{3\theta}{2}\sin\theta\cos 2\theta + \cos\frac{3\theta}{2}\sin\theta\sin 2\theta\right]$$
$$- \frac{1}{4}\frac{\sigma}{G}\left(\frac{r}{a}\right)^{0}\left[\frac{(\kappa - 1)}{2} + \cos 2\theta\right] +$$
$$+ \frac{3}{16\sqrt{2}}\frac{\sigma}{G}\left(\frac{r}{a}\right)^{\frac{1}{2}}\left[(\kappa - 1)\cos\frac{\theta}{2} + \sin\frac{\theta}{2}\sin\theta\cos 2\theta -\right.$$
$$\left. - \cos\frac{\theta}{2}\sin\theta\sin 2\theta\right]$$
$$- \frac{5}{128\sqrt{2}}\frac{\sigma}{G}\left(\frac{r}{a}\right)^{\frac{3}{2}}\left[(\kappa - 1)\cos\frac{3\theta}{2} - 3\sin\frac{\theta}{2}\sin\theta\cos 2\theta -\right.$$
$$\left. - 3\cos\frac{\theta}{2}\sin\theta\sin 2\theta\right] \tag{17}$$

3 EXPERIMENTAL PROCEDURE

The experimental measurements are done on a plate with a central crack of length 2a, loaded in uniaxial tension by a uniform remote stress σ.

The plate is made from aluminum with the values of the elastic constants established by us as: E = 61,000 MPa , and ν = 0.3. The plate has the following dimensions: height H = 580 mm, width W = 240 mm and thickness t = 2.5 mm. The through thickness crack has a width of 0.2 mm. The length of the central crack is the same 2a = 48 mm, and the dimensions of the width of the plate are changed as to obtain: 2a / W = 0.2 ; 2a / W = 0.3 and 2a / W = 0.4. A dead load first order lever device is used to apply the traction force and two fixtures with joints are distributing it as a tensile stress to the plate; the upper fixture is fixed to a rigid frame and the lower one to the lever.

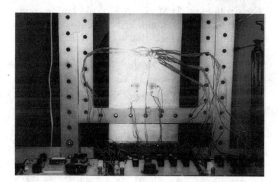

Figure 3. Experimental setup for measurements.

The value of the tensile stress is established in a section in which there is no boundary effect and that is far from the crack plane; at a distance of 90 mm from the crack plane and at 25 mm from the vertical axis (Fig. 3) are symmetrically fixed four Höttinger strain gauges, type 10 / 120 XA 21 with the constant k = 2.04 ± 1 %. With this strain gauges one can establish the value of the tensile stress and Poisson's ratio.

Around the crack tip, along radial lines defined by the angles of θ = 0° ; θ = 45° ; θ = 90° and θ = 135°, the strains are measured on each direction in four points with Höttinger gauges 0.6 / 120 LY 13, compensated for aluminum, a gauge length of 0.6 mm and the constant k = 1.65 ± 1.5 % . A blow up of the crack tip region is shown in Figure 4. It is tried to avoid the immediate vicinity if the crack tip, where the influences of the process zone could be

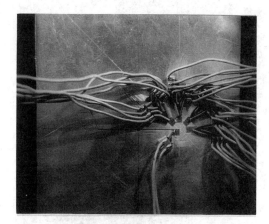

Figure 4. Blow up of the crack tip region.

important. With another strain gauge is measured the strain close to the lower crack flank for θ = 180°.

Thus, one can obtain experimentally the strains around the crack tip for different polar angles θ and, on each radial line, in four measuring points defined by r / a ratios between 0.25 and 0.87. This values can be compared to the 4 terms or 3 terms proposed solution – equation (17), for the specified polar angles and distances from the crack tip, taking also into account the different crack length to width ratios.

4 COMPARISON BETWEEN THEORETICAL AND EXPERIMENTAL RESULTS FOR STRAIN FIELDS

As an example, for a crack length to width of plate ratio of 2a / W = 0.3 and for all considered polar angles, the experimental strains are very close to the 3 terms and 4 terms theoretical strains of the proposed solution. In Figures 5-6 the theoretical and experimental results are compared. A nondimensional representation is chosen in order to obtain the general nature of the strain field around the crack tip. For each stress level three measurements are done, and all resulting experimental values are represented in same graphs; many points do overlap. If θ = 0° (Fig. 5) the experimental strains are much different from the classical solution and close to the proposed solution. For θ = 90° (Fig. 6) strains have the greatest values. The experimental strains are greater than those of the classical solution, and this results are confirmed by the proposed 3 terms or 4 terms solutions. The complete experimental results (Constantinescu & Voloshin, in prep.) are going to be presented elsewhere.

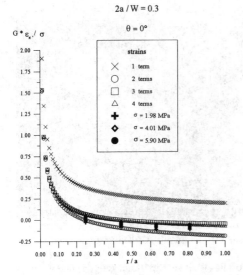

Figure 5. Theoretical and experimental strains
for 2a / W = 0.3 and θ = 0°.

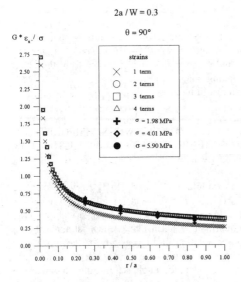

Figure 6. Theoretical and experimental strains
for 2a / W = 0.3 and θ = 90°.

5 CONCLUSIONS

Experimental evidence of the proposed solution is
given by strain gauge measurements in order to
establish the strain field around the crack tip. For
different crack length to width ratios (2a / W = 0.2 ;
2a / W = 0.3 ; 2a / W = 0.4) it is proven that the
experimental strains are similar to the theoretical
ones obtained from the 3 terms and 4 terms
solutions. As one can notice, the established theoretical relations can be trustfully used in order to
determine the strain fields around the crack tip even
for greater crack lengths to width of plate ratios and,
what is also important, up to r / a = 1 which is far
enough from the crack tip from an engineering point
of view.

REFERENCES

Berger, J.R., J.W. Dally & R.J. Sanford 1991. Extend of validity of three-parameter crack-tip
strain fields. *Proc. SEM Spring Conf. on Exp.
Mech., Milwaukee, 10-13 June 1991*: 572-578.
Constantinescu, D.M. & A.S. Voloshin in prep.
Elastic stress, strain and displacement fields
around the crack tip - I. Theoretical developments.
Constantinescu, D.M. & A.S. Voloshin in prep.
Elastic stress, strain and displacement fields
around the crack tip - II. Numerical and
experimental evidence.
Dally, J.W. & J.R. Berger 1986. A strain gage
method for determining K_I and K_{II} in a mixed
mode stress field. *Proc. SEM Spring Conf. on
Exp. Mech., June 1986*: 603-612.
Dally, J.W. & R.J. Sanford 1987. Strain gage
methods for measuring the opening-mode stress
intensity factor "K_I". *Exp. Mech.* 27: 391-388.
Eftis, J. & H. Liebowitz 1972. On the modified
Westergaard equations for certain plane crack
problems. *J. of Fracture Mech.* 8: 383-391.
Eftis, J., N. Subramonian & H. Liebowitz 1977.
Crack border stress and displacement equations
revisited. *Engng Fracture Mech.* 9: 189-210.
Irwin, G.R. 1958. Discussion of article by Wells &
Post. *Proc. SESA* 16: 93-96.
Sanford, R.J. 1979. A critical re-examination of the
Westergaard method for solving opening mode
problems. *Mechanics Research Communica-
tions* 6(5): 289-294.
Sih, G.C. 1966. On the Westergaard method of crack
analysis. *Int. J. Fracture Mech.* 2: 628-631.
Wells, A.A. & D. Post 1958. The dynamic stress
distributions surrounding a running crack - a
photoelastic analysis. *Proc. SESA* 16: 69-92.
Westergaard, H.M. 1939. Bearing pressures on
cracks. *J. Applied Mech.* 6: A49-A53.
Williams, M.L. 1957. On the stress distribution at
the base of a stationary crack. *J. Applied Mech.*
24: 109-114.

Experimental Mechanics, Allison (ed.) © 1998 Balkema, Rotterdam, ISBN 90 5809 014 0

Strain gauge methods for determining stress intensity factor of the edge cracks in the strip subjected to transverse bending*

Shigeru Kurosaki
Tokyo National College of Technology, Japan

ABSTRACT: This paper describes the strain gauge method for determining the stress intensity factor of an edge crack in a strip subjected to transverse bending. This paper proposes two methods of stress intensity factor analysis. The strain gauges used in this paper are extremely small strain gauges consisting of 5 grids, with gauge length of 0.15 mm and gauge pitch of 0.5 mm. The four-point bending tests of a strip were carried out using a universal testing machine. The accuracy of the stress intensity factors determined by the experiments are within 10 percent compared to the analytical values, except for the data of a short crack length specimen.

1. INTRODUCTION

It is important in the field of engineering to determine stress intensity factor by using strain gauges experimentally. For the case of tensile testing, some studies are reported, of using strain gauges for determining stress intensity factor.(1)-(5) However there are no reports for the case of a bending plate with an out-of-plane crack using the strain gauge. In this paper, we propose that strain gauges can be effectively employed to measure the stress intensity factor for the case of plate bending with an out-of-plane crack (Fig.1).

The stress singular region for the crack tip is small compared to the crack length.

Hence the strain for the vicinity of the crack tip must be measured precisely, if we are to determine the stress intensity factor with accurately. Recently, extremely small strain gauges consisting of 5 measuring grids (Fig.2) were developed by a company which manufactures strain gauges.(6) In this study we propose two methods to determine Mode I stress intensity factor K_I with the use of these extremely small strain gauges.

The experiments for the bending tests of the cracked plate are carried out with two kinds of specimens to examine the accuracy. These specimens are strips with single-edge and double-edge cracks.

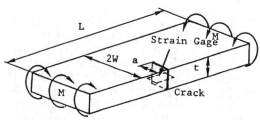

Fig.1 A strip with a edge crack subjected to transverse bending .

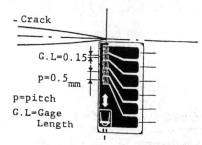

Fig.2 Position of the strain gauges.

2. STRESS COMPONENTS OF CRACK TIP SUBJECTED TO OUT-OF-PLANE BENDING

Sih et al. (7) and Williams(8) introduced the stress components $(\sigma_r, \sigma_\theta, \tau_{r\theta})$ of the crack tip for the plate subjected to out-of-plane bending, as shown in Fig.1. The stress components of the crack tip are expressed in series form with respect to the distance "r" from the crack tip, when the coordinate of the crack tip is as shown as Fig.3. The first terms of stress components for the crack tip are expressed as

$$\sigma_r = \frac{K_I}{\sqrt{\pi r}} \left\{ \cos\frac{3\theta}{2} - \frac{(3+5\nu)}{(7+\nu)} \cdot \cos\frac{\theta}{2} \right\} \cdot \frac{3bGZ}{2}$$

$$\sigma_\theta = \frac{K_I}{\sqrt{\pi r}} \left\{ -\cos\frac{3\theta}{2} - \frac{(7+3\nu)}{(7+\nu)} \cdot \cos\frac{\theta}{2} \right\} \cdot \frac{3bGZ}{2}$$

$$\tau_{r\theta} = \frac{2K_I}{\sqrt{\pi r}} \left\{ -\sin\frac{3\theta}{2} - \frac{(1-\nu)}{(7+\nu)} \cdot \sin\frac{\theta}{2} \right\} \cdot \frac{3bGZ}{2}$$

(1)

where

$$b = -\frac{(7+\nu)}{3\sqrt{2} \cdot (3+\nu) G t}$$

(2)

The strain component ε_r used in this study is obtained by substituting Eq.(1) into the following plane stress-strain relation.

$$E \varepsilon_r = \sigma_r - \nu \sigma_\theta$$

(3)

The strain component ε_r expressed in generalized form is given by

$$E \varepsilon_r = A r^{-1/2} + A_0 + A_1 r^{1/2} + A_2 r^{2/2} +$$

(4)

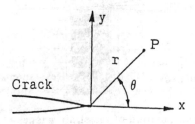

Fig.3 Coordinate of the crack tip.

The coefficient "A" of the first term of Eq.(4) is expressed as

$$A = \frac{K_I}{\sqrt{\pi}} \cdot S \cdot \left\{ (7+\nu) \cdot \cos\frac{3\theta}{2} + 3(\nu-1) \cdot \cos\frac{\theta}{2} \right\}$$

(5)

where

$$S = \frac{(1+\nu) \cdot Z}{2\sqrt{2}(3+\nu) t},$$

(6)

K_I: stress intensity factor of mode I,
E: modulus of longitudinal elasticity,
G: modulus of shearing elasticity,
ν: Poisson's ratio, $C_i (i=0,1,2,\cdots)$: unknown coefficients, t: thickness of plate and z: coordinate along the z axis.

The coordinate along the z axis is in the direction of the thickness of the plate. The origin of the z axis is at the centerline through the thickness of the plate. The positive direction of z axis is toward the lower part of the plate. If the thickness is "t", the lower surface of the plate is at z=+t/2 and the upper surface is at z=-t/2.

3. DETERMINATION OF STRESS INTENSITY FACTOR USING THE STRAIN GAUGE DATA

In this study, the position to put a strain-gauge on is the angle $\theta = \pi/2$ for the x axis (Fig.2) and z=+t/2. The position of the gauge is close for the crack tip. Equation(4) is obtained to substitute $\theta = \pi/2$ and z=+t/2 into Eq.(5) Thus,

$$E \cdot \varepsilon_r = \frac{(5-\nu)(1+\nu)}{4(3+\nu)} \cdot \frac{K_I}{\sqrt{\pi r}} + A_0 + A_1 r^{1/2} + A_2 r^{2/2} + \cdots$$

(7)

where the only first term of the Eq.(7) is represented in detail. The two methods are proposed in this study. One of them is the method to use the first term only ("one point gauge method"), and another one is the method to use first and second term in the series of Eq.(7). ("the five point gauges method")

CASE1:
"THE ONE POINT GAUGE METHOD"

This method treats the only one datum of the most closest strain gauge from crack tip. The analytical equation of this method is the first term of Eq.(7). Thus

$$K_I = \frac{4E(3+\nu)}{(5-\nu)(1+\nu)} \cdot \varepsilon_r \sqrt{\pi r} \qquad (8)$$

CASE 2:
"THE FIVE POINTS GAUGE METHOD"

This method treats strain data of five gauges consisting five measuring grids. The first and second term of Eq.(7) are expressed as

$$\frac{4E(3+\nu)}{(5-\nu)(1+\nu)} \cdot \varepsilon_r \cdot \sqrt{\pi r} = K_I + A_0' \sqrt{r} \qquad (9)$$

By representing $\sqrt{r} = \xi$, $A_0' = C$ and the left part of Eq.(9) as F, Equation(9) becomes a linear equation for ξ. Thus,

$$F = K_I + C \cdot \xi \qquad (10)$$

Equation(10) expresses a linear equation for ξ with inclination C and the cross point (K_I) at the F axis(Fig. 4). The strain gauges to be used are consisted with five measuring grids. The strain ε_y at five point are measured. The F value of Eq.(9) are calculated by these strain and the stress intensity factors are determined by the extrapolation method.(Fig.4)

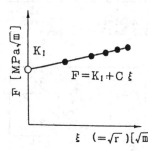

Fig.4 Extrapolation of stress intensity factor

4.EXPERIMENTAL VERIFICATION

The test specimens are two kinds of type in this study, One of them is the single edge cracked plate bending specimens (SBND) (Fig.5), another is double edge cracked plate bending specimen (WBND) (Fig.6). In this paper , the name of a test specimen is called as ommissive name (SBND or WBND) and crack length. For example, it is shown as WBND20 for crack length 20mm of double edge cracked plate bending specimen .The material of this specimens is mild steel ss41. The modulus of longitudinal elasticity is 218GPa and poisson's ratio is 0.28. The notch of the test specimen are cut by the electrical wire arc cutting machine. The wire of the machine is 0.2 mm diameter. The radius of a notch tip is 0.1 mm. The notch length of the single edge cracked plate bending specimens are 5, 10, 15 and 20 mm, and the notch length of double edge cracked plate bending specimen are 7.5, 10, 12.5 and 15 mm. The strain gauges to be used in this study are extremely small sized strain gauges (KFR-015-120-D9-11 made by KYOWA Corp) as shown fig.1. It is measured by using microscope in the accuracy 1/100 mm for the distance between notch tip and the position of the most closest strain gauge at a notch tip. The experiments are carried out as follow. Firstly the preload applies and simultaneously sets up the strains zero. Next, the each load applies and the strains are measured. The four-point bending tests are carried out by the universal testing machine with the capacity 100KN and used at the minimum range 12kN. Figure 7 shows the experimental setup.

5.RESULTS OF EXPERIMENTS

5.1 Load - strain relation near the crack tip

As the example, The load-strain relation of SBND20 is shown as Fig.8. Figure 8 is a straight line with every line as the elastic field. The strain of the most nearest gauge to the crack tip is maximum strain.

5.2 Comparison between experimental and theoretical value

This section makes a comparison between experimental and analytical values which is solved to use the body-force method by Nishitani and Mori(9).The stress intensity factor of the analytical Eq.(7) is expressed as

$$K_I = F \cdot \sigma_b \sqrt{\pi a} \qquad (11)$$

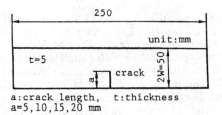

a:crack length, t:thickness
a=5,10,15,20 mm

Fig.5 Single edge cracked plate bending
specimen(SBND).

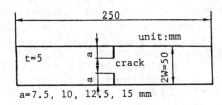

a=7.5, 10, 12.5, 15 mm

Fig.6 Double edge cracked plate bending
specimen(WBND).

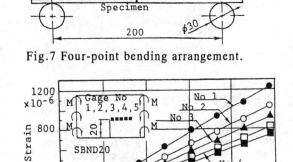

Fig.7 Four-point bending arrangement.

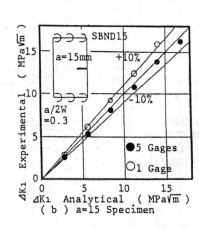

Fig.8 Load-strain relation (SBND20).

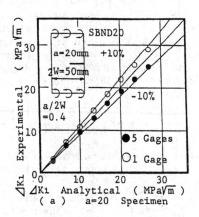

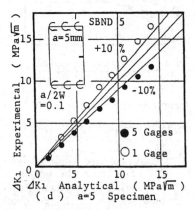

Fig.9 Accuracy of the specimen (SBND).

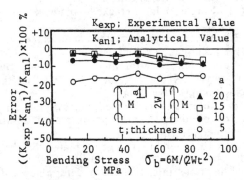

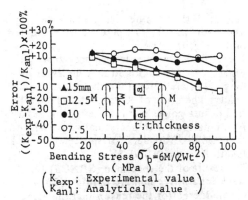

Fig.10 Error for the crack length
of the specimen SBND.

Fig.12 Error for the crack length
of the specimen WBND.

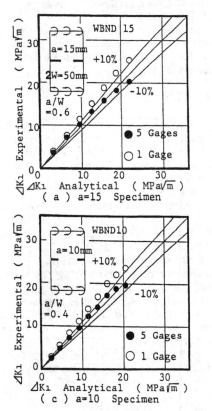

(a) a=15 Specimen

(c) a=10 Specimen

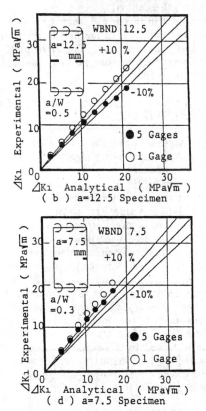

(b) a=12.5 Specimen

(d) a=7.5 Specimen

Fig.11   Accuracy of   the specimen (WBND).

When the width of a plate is W, The bending stress σ_b of Eq.(11) is as follow

$$\sigma_b = 6M/(W \cdot t^2) \qquad (12)$$

The value F is difference with single edge cracked plate bending specimen(SECB) and double edge cracked plate bending specimens (WBND).

The stress intensity factor K is expressed as ΔK,   because the K value of preload is basis.

5.3 Single edge cracked plate bending specimen (SECB)

The comparative figure between experimental

1157

and analytical value are shown in Fig.9. The error line of ±10 percent are presented in the Fig.9.

The error for a crack length are shown in Fig.10. It is recogniged that the error of a long crack are better than the one of a short crack. The error of K value determined by five points gauge method are within ±10 % in comparison with analytical value with the exceptions of the data for specimen a crack length five mm(SBND5).

The result of the specimen (SBND5) for the crack length five mm are within ±20 %. It is considered for the reason that the effect of the radius at the notch tip of the crack length five mm (SBND5) appeared sensitivity as the crack length is short.

It is evident in this Fig.9 that the accuracy of the five point gauge method is better than one point gauge method. The error of K value determined the one point gauge method are within ±15 %.

5.4 Double edge cracked plate bending specimen(WBND)

The analytical formula to be used with comparison to experimental value are Eqs.(11) and (12). In this analytical formula, the width of a plate is 50mm as 2W. The comparative figure between experimental and analytical value are shown in Fig.11. The error for a crack length are shown in Fig.12. The errors of double edge cracked plate bending specimen are more than the one of single edge cracked plate bending specimen. This reason is thought that the load are applied unequal to the double edges and the accuracy of stress intensity factor are influenced by a position to applied bending load. The error of stress intensity factor obtained by five point gauge method are within ±16% of all specimens and the error by one point gauge method are within maximum error +30%.

6.CONCLUSION

This report propose the methods with extremely small sized strain gauges to determine the stress intensity factor of the edge cracks in a strip subjected to transverse bending. The experiments are carried out to examine the accuracy of the stress intensity factor to be determined with these methods by using the two kinds of test specimens. The conclusion obtained by these

experiments are as follow

(1) Single edge cracked plate bending specimen

The stress intensity factors determined with five point gauge method are within ±10% errors compare with analytical value. But these results are excepted the data for the crack length five mm (a/2W=0.1). For the one point gauge method, stress intensity factor obtained by the experiments are within maximum error ±15%.

(2)Double edge cracked plate bending specimen

The stress intensity factor determined with five point gauge method are within ±16% errors compare with analytical value. For one point gauge method, the maximum error are +30%.

(3) For all specimen, the value of the stress intensity factor obtained by one point gauge method are larger than the value by five point gauge method.

REFERENCES

(1).H.kitagawa and H.Ishikawa, Determination of Stress Intensity of Through-Crack in a Plate Structure under Uncertain Boundary Condition by Means of Strain Gauges, ASTM Spec.Tech.Publ. 631, (1976), p232

(2). Y.Kondo, "Development of K-Gauge for Stress Intensity Factor Measurement", Trans.Jpn.Soc.Mech.Eng., (in Japanese), vol.53, No.495, A (1987), p1977

(3). J.W.Dally and R.J.Sanford, Strain-gauge methods for measuring the opening-mode stress intensity factor KI,, Exp. Mech., vol.27, No.4, (1987), p381

(4). J.W.Dally and Riley, Experimental stress analysis (Third Edition) Mcgraw-Hill, (1991), p104

(5). S.Kurosaki, H.Nozaki and S.Fukuda, Mode I Stress Intensity Factor Measurements by the Electrical Strain Gauge, Trans.Jpn. Soc.Mech.Eng., (in Japanese), vol.56, No524, A, (1990), p875

(6). KYOWA DENGYOU COMPANY, The Catalog of Electronic Instruments 1990, (in Japanese), Cat.No.300V/1990 , (1990), p31

(7). G.C.Sih, P.C.Paris, F.Erdogan, Trans. ASME, Series E, J.Appl.Mech, Vol.29, June(1962), p306

(8). M.L.Williams, Trans. ASME, Series E, J.Appl. Mech, Vol.28, March(1961), p78

(9). H.Nishitani and K.Mori, Transverse Bending of Rectangular Plate with a Single-Edge Crack or Double-Edge Cracks, Trans.Jpn.Soc.Mech.Eng.(in Japanese), vol.55, No.511, A, (1989), p508

Experimental Mechanics, Allison (ed.) © 1998 Balkema, Rotterdam, ISBN 90 5809 014 0

Evaluation of static strength by the application of stress intensity factor for a V-shaped notch

D.H.Chen

Science University of Tokyo, Japan

ABSTRACT: A method of evaluating the static strength of a V-shaped notch based on stress intensity factors of the notch is studied. The singular stress field at the notch tip is defined by two parameters, K_{I,λ_1} and K_{II,λ_2}, which correspond to the intensities of the symmetric stress field and the skew-symmetric field, respectively. Fracture tests are carried out on plane specimens of acrylic resin having single or double sharp notches. It is found that the fracture values K_{I,λ_1} and K_{II,λ_2} are almost constant independent of the notch depth and the loading conduction. Four kinds of fracture criteria are considered; two of them are based on the tensile strength σ_B and the other two are based on the fracture toughness K_{IC}. The usefulness of the criteria is investigated through the experimental results.

1 INTRODUCTION

A sharp notch problem can be seen as an extension of a crack problem. For the crack problem, the strength is evaluated by the stress intensity factors K_I and K_{II}, which dominate stresses near the crack tip. For the sharp notch problem, the singular stress field at the notch tip can also be defined in terms of two parameters K_{I,λ_1} and K_{II,λ_2}, analogous to the crack problem. Therefore, it is worth investigating the application of the parameters K_{I,λ_1} and K_{II,λ_2} to evaluation of the notch strength. In this study, fracture tests are carried out on plane specimens of acrylic resin having either single or double sharp notches. The usefulness of K_{I,λ_1} and K_{II,λ_2} for the notch fracture problem is confirmed through the experiment. Also, four kinds of fracture criteria are presented to predict critical values of K_{I,λ_1} and K_{II,λ_2} necessary to the notch fracture.

2 STRESSES NEAR THE NOTCH TIP

A local configuration around the notch tip is shown in Figure 1. The singular stress field in Figure 1 can be expressed as a sum of the symmetric state with a stress singularity $1/r^{1-\lambda_1}$ and the skew-symmetric state with a stress singularity $1/r^{1-\lambda_2}$,

as follows (Chen 1995):

$$\sigma_{ij}(r,\theta) = \frac{K_{I,\lambda_1}}{r^{1-\lambda_1}} f_{ij}^I(\theta) + \frac{K_{II,\lambda_2}}{r^{1-\lambda_2}} f_{ij}^{II}(\theta)$$

$$(ij = r,\theta,r\theta) \tag{1}$$

where

$$f_r^I(\theta) = \frac{1}{\sqrt{2\pi D_I}} [C_I \cos\{(\lambda_1+1)\theta\} - (\lambda_1-3)\cos\{(\lambda_1-1)\theta\}]$$

$$f_r^{II}(\theta) = \frac{1}{\sqrt{2\pi D_{II}}} [-C_{II} \sin\{(\lambda_2+1)\theta\} + (\lambda_2-3)\sin\{(\lambda_2-1)\theta\}]$$

$$f_\theta^I(\theta) = \frac{1}{\sqrt{2\pi D_I}} [-C_I \cos\{(\lambda_1+1)\theta\} + (\lambda_1+1)\cos\{(\lambda_1-1)\theta\}]$$

$$f_\theta^{II}(\theta) = \frac{1}{\sqrt{2\pi D_{II}}} [C_{II} \sin\{(\lambda_2+1)\theta\} - (\lambda_2+1)\sin\{(\lambda_2-1)\theta\}]$$

$$f_{r\theta}^I(\theta) = \frac{1}{\sqrt{2\pi D_I}} [-C_I \sin\{(\lambda_1+1)\theta\} + (\lambda_1-1)\sin\{(\lambda_1-1)\theta\}]$$

$$f_{r\theta}^{II}(\theta) = \frac{1}{\sqrt{2\pi D_{II}}} [-C_{II} \cos\{(\lambda_2+1)\theta\} + (\lambda_2-1)\cos\{(\lambda_2-1)\theta\}]$$

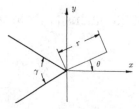

Figure 1: Local configuration near the notch tip

$$
\begin{aligned}
C_I &= \lambda_1 \cos\gamma + \cos\{\lambda_1(2\pi - \gamma)\} \\
C_{II} &= \lambda_2 \cos\gamma - \cos\{\lambda_2(2\pi - \gamma)\} \\
D_I &= \lambda_1 + 1 - C_I \\
D_{II} &= \lambda_2 - 1 - C_{II}
\end{aligned}
$$

λ_1 and λ_2 are roots, in range of $0 < \lambda < 1$, of the following eigenequations:

$$
\begin{aligned}
\sin\{\lambda_1(2\pi - \gamma)\} - \lambda_1 \sin\gamma = 0 \quad &\text{(for mode } I) \\
\sin\{\lambda_2(2\pi - \gamma)\} + \lambda_2 \sin\gamma = 0 \quad &\text{(for mode } II)
\end{aligned}
$$
(2)

In eq.(1), the singular stress field at the notch tip is defined in terms of two parameters, K_{I,λ_1} and K_{II,λ_2}, which are related with the stress components σ_θ and $\tau_{r\theta}$ at $\theta = 0$ as follows:

$$
\begin{aligned}
K_{I,\lambda_1} &= \lim_{r\to 0}\left[\sqrt{2\pi}r^{1-\lambda_1}\sigma_\theta(r,0)\right] \\
K_{II,\lambda_2} &= \lim_{r\to 0}\left[\sqrt{2\pi}r^{1-\lambda_2}\tau_{r\theta}(r,0)\right]
\end{aligned}
$$
(3)

3 TESTED MATERIAL AND SPECIMENS

The material used in this investigation is an acrylic resin, whose Young modulus E, ultimate tensile strength σ_B and the fracture toughness K_{IC} are 3230 MPa, 72.57 MPa and 37.10 $MPa\sqrt{mm}$, respectively.

Specimens, 5 mm thick and 40 mm in width, are shown in Figure 2. The centers of two holes, through which a pulling load is applied, may be shifted from the center line of the specimen by d, $d = 0, 3, 5$ or 8 mm, to yield a bending moment. A single or double V-shaped artificial notches are introduced in the specimen. Geometry of the notch is defined by its opening angle γ, inclined angle β and the notch depth a. In this investigation, γ is taken to be $\gamma = 30^o, 60^o, 90^o$, β is taken to be $\beta = 0^o, 15^o, 30^o$, and the depth of the notch is varied from $a/W = 0.1$ to $a/W = 0.4$.

The tension-test was carried out in a testing machine, Shimadzu AUTOGRAPHAG 10-TC, controlled by a computer, which exactly measures load value for specimen fracture.

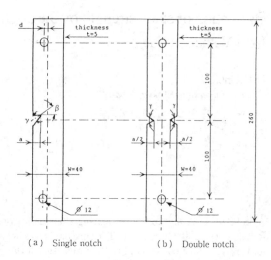

(a) Single notch   (b) Double notch

Figure 2: Geometry of specimens tested

4 EXPERIMENTAL RESULTS

Figures 3 and 4 show experimental results of critical values of K_{I,λ_1} necessary to fracture of the specimens with identical opening angle of $\gamma = 60^o$ and identical inclined angle of $\beta = 0$.

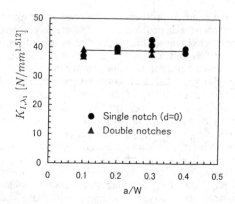

Figure 3: Experimental results of K_{I,λ_1} for notches of $\gamma = 60^o$ with various notch depth a

When the inclined angle $\beta = 0$, created at the notch tip are only mode I singular stresses, whose intensity is defined by one parameter K_{I,λ_1}. The values of K_{I,λ_1} necessary to specimen fracture, shown in Figures 3 and 4, are obtained from the load value corresponding to the specimen fracture, P, by the following equation:

$$
K_{I,\lambda_1} = F_{I,\lambda_1}|_{\sigma_T^\infty}\sigma_T^\infty\sqrt{\pi}a^{1-\lambda_1} + F_{I,\lambda_1}|_{\sigma_B^\infty}\sigma_B^\infty\sqrt{\pi}a^{1-\lambda_1}
$$
(4)

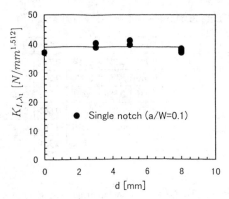

Figure 4: Experimental results of K_{I,λ_1} for notches of $\gamma = 60^o$ with various bias d

Table 1: Values of $F_{I,\lambda_1}|_{\sigma_T^\infty}$ and $F_{I,\lambda_1}|_{\sigma_B^\infty}$

		Single notch		Double notches
γ	a/W	$F_{I,\lambda_1}|_{\sigma_T^\infty}$	$F_{I,\lambda_1}|_{\sigma_B^\infty}$	$F_{I,\lambda_1}|_{\sigma_T^\infty}$
	0.0	1.133	1.133	1.133
30°	0.1	1.201	1.056	1.123
	0.2	1.381	1.063	1.145
$\lambda_1 = 0.501$	0.3	1.675	1.130	1.251
	0.4	2.129	1.266	1.604
	0.0	1.186	1.186	1.186
60°	0.1	1.258	1.102	1.176
	0.2	1.445	1.105	1.199
$\lambda_1 = 0.512$	0.3	1.751	1.172	1.309
	0.4	2.223	1.309	1.672
	0.0	1.308	1.308	1.308
90°	0.1	1.388	1.209	1.298
	0.2	1.597	1.207	1.323
$\lambda_1 = 0.544$	0.3	1.939	1.278	1.445
	0.4	2.472	1.433	1.864

where

$$\sigma_T^\infty = P/tW, \quad \sigma_B^\infty = 6Pd/tW^2$$

In eq.(4), $F_{I,\lambda_1}|_{\sigma_T^\infty}$ and $F_{I,\lambda_1}|_{\sigma_B^\infty}$ are dimensionless stress intensity factors for a V-notched strip subjected to a tension load with a uniform tension stress $\sigma_T^\infty = 1$ and an in-plane bending load with the maximum tension stress $\sigma_B^\infty = 1$, respectively. Values of $F_{I,\lambda_1}|_{\sigma_T^\infty}$ and $F_{I,\lambda_1}|_{\sigma_B^\infty}$ related to this study are shown in Table 1, which are obtained by the author in a previous study (Chen 1995) by numerical analysis.

In Figure 3, the fracture values K_{I,λ_1} are shown for specimens having either a single notch or double notches with different notch depth a, but with d fixed as $d = 0$. In Figure 4, the results are shown for single notch specimens with d varied from $d = 0$ to $d = 8$mm, but with the notch depth

Table 2: Experimental results

γ	β	K_{I,λ_1} $(N/mm^{1+\lambda_1})$	K_{II,λ_2} $(N/mm^{1+\lambda_2})$	θ_c $(^o)$
	0°	37.49	0.0	0.0
30°	5°	36.56	2.17	6.1
	15°	36.78	7.16	16.8
	30°	34.60	13.57	30.0
	0°	38.93	0.0	0.0
60°	5°	39.82	2.89	3.2
	15°	38.35	8.32	12.8
	30°	36.41	15.84	25.7
	0°	41.99	0.0	0.0
90°	5°	45.26	3.08	1.4
	15°	40.45	8.38	6.9
	30°	40.32	16.17	17.0

fixed as $a/W = 0.1$. It is seen from Figures 3 and 4 that the fracture values K_{I,λ_1} are almost constant, $K_{I,\lambda_1} \cong 38.93N/mm^{1+\lambda_1}$, independent of the notch depth a, the amount of bias d and whether the notch is single or double.

For the other opening angles and incline angles, experimental results show a similar characteristic. That is, the notch fracture is mainly controlled by the notch stress intensity factors, K_{I,λ_1} and K_{II,λ_2}. The experimental results are listed in Table 2, which shows the obtained critical values of K_{I,λ_1} and K_{II,λ_2} necessary to notch fracture.

5 PREDICTION OF FRACTURE VALUE K_{I,λ_1} AND K_{II,λ_2}

5.1 Fracture criterion

The notch fracture problem is different from the crack fracture problem in that the critical values of stress intensity factors necessary to notch fracture are a function of the notch opening angle γ. This is because different notch opening angles have different stress singularities. Thus, even for mode I loading, many data of K_{I,λ_1} are necessary for various notches with different opening angles, unlike in the crack problem, in which only one data of K_{IC} applies to all the crack problems. Therefore, a fracture criterion, by which the critical values of stress intensity factors can be estimated from the mechanical properties of the considered material such as the tensile strength σ_B or the fracture toughness K_{IC}, would be useful in application.

Figure 5 shows four kinds of available fracture criteria. In Figures 5(a) and (b), the criteria are based on the tensile strength σ_B, and in Figures 5(c) and (d) the criteria are based on the fracture

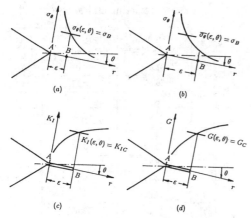

(a)

(b)

(c)

(d)

Figure 5: Criteria on fracture of notch

toughness K_{IC}.

In the criterion of Figure 5(a), occurrence of fracture depends on stress $\sigma_\theta(\varepsilon, \theta)$ at a point B at a distance of ε from the notch tip. That is, fracture will occur when the stress $\sigma_\theta(\varepsilon, \theta)$ at the point B, which are a function of K_{I,λ_1} and K_{II,λ_2} and can be readily obtained from eq.(1), reaches the tensile strength σ_B of the considered material.

$$\sigma_\theta(\varepsilon, \theta) \geq \sigma_B \qquad (5)$$

In the fracture criterion of Figure 5(b), the average stress of σ_θ over the range of AB, $\overline{\sigma_\theta}(\varepsilon, \theta)$, is used to replace the stress at the point B. Thus, the criterion condition is as follows:

$$\overline{\sigma_\theta}(\varepsilon, \theta) \geq \sigma_B \qquad (6)$$

where

$$\overline{\sigma_\theta}(\varepsilon, \theta) = \frac{\int_0^\varepsilon \sigma_\theta(r, \theta) dr}{\varepsilon}$$

In Figure 5(c), fracture at the notch tip is simulated by propagation of a small crack, with a length of ε, imagined at the notch tip. Fracture occurs when the stress intensity factor at the crack tip $K_I(\varepsilon, \theta)$ is larger then the critical value K_{IC}.

$$K_I(\varepsilon, \theta) \geq K_{IC} \qquad (7)$$

In Figure 5(d), using the energy balance criterion the fracture condition is given by

$$G(\varepsilon, \theta) \geq G_C \qquad (8)$$

where G_C is related to K_{IC} as

$$G_C = K_{IC}^2 / E'$$
$$E' = \begin{cases} E & \text{(plane stress)} \\ E/(1 - \nu^2) & \text{(plane strain)} \end{cases}$$

Table 3: Dimensionless SIF $F_I^I(\theta)$ and $F_I^{II}(\theta)$

γ	θ	0°	15°	30°	45°	60°
30°	$F_I^I(\theta)$	0.991	0.967	0.896	0.790	0.659
	$F_I^{II}(\theta)$	0.000	0.367	0.690	0.934	1.075
60°	$F_I^I(\theta)$	0.984	0.960	0.891	0.785	0.655
	$F_I^{II}(\theta)$	0.000	0.356	0.663	0.883	0.992
90°	$F_I^I(\theta)$	0.966	0.941	0.870	0.761	0.627
	$F_I^{II}(\theta)$	0.000	0.349	0.642	0.834	0.902

where ν is Poisson's ratio.

In eq.(8), $G(\varepsilon, \theta)$ is the average energy release rate for creating a crack with length of ε at the notch tip and is defined as

$$G(\varepsilon, \theta) = \frac{\frac{1}{2} \int_0^\varepsilon [\sigma_\theta(r, \theta)\Delta u_\theta(r) + \tau_{r\theta}(r, \theta)\Delta u_r(r)] dr}{\varepsilon} \qquad (9)$$

where $\sigma_\theta(r, \theta)$ and $\tau_{r\theta}(r, \theta)$ are the stresses around the notch tip before the crack initiation, and $\Delta u_\theta(r)$ and $\Delta u_r(r)$ are the opening and sliding displacements of the created crack surfaces.

In the criteria of Figures 5(c) and (d), the assumed crack is so small that the $K_I(\varepsilon, \theta)$ in eq.(7) and $G(\varepsilon, \theta)$ in eq.(8) can be expressed by a function of the stress intensity factors for the notch tip, K_{I,λ_1} and K_{II,λ_2}, as follows:

$$K_I(\varepsilon, \theta) = F_I^I(\theta)\varepsilon^{\lambda_1 - 1/2}K_{I,\lambda_1} + F_I^{II}(\theta)\varepsilon^{\lambda_2 - 1/2}K_{II,\lambda_2} \qquad (10)$$

$$G(\varepsilon, \theta) = \left[g_{11}(\theta)\varepsilon^{2\lambda_1 - 1}K_{I,\lambda_1}^2 + g_{22}(\theta)\varepsilon^{2\lambda_2 - 1}K_{II,\lambda_2}^2 \right.$$
$$\left. + g_{12}(\theta)\varepsilon^{\lambda_1 + \lambda_2 - 1}K_{I,\lambda_1}K_{II,\lambda_2} \right] / E' \qquad (11)$$

where $F_I^I(\theta)$, $F_I^{II}(\theta)$, $g_{11}(\theta)$, etc. should be determined by numerical calculation. Table 3 shows values of $F_I^I(\theta)$ and $F_I^{II}(\theta)$.

5.2 Application to mode I loading

When the inclined angle $\beta = 0$, only mode I stresses occur around the notch tip. Thus, fracture will occur on direction of $\theta = 0$ and is controlled by only one parameter K_{I,λ_1}. Here we try using the criteria in Figure 5 to predict the critical values of K_{I,λ_1} for notch fracture.

In the criteria shown in Figure 5, the fracture condition depends on the assumed value ε, which is corresponding to the position of the point B in Figures 5(a) and (b) and to the assumed crack length in Figures 5(c) and (d). Figures 6, 7, 8 and 9 show relations between the predicted critical values of K_{I,λ_1} and the assumed value of ε,

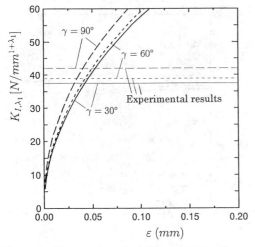

Figure 6: K_{I,λ_1} vs ε predicted by eq.(5) for $\beta = 0$

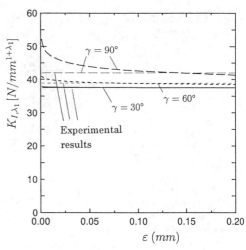

Figure 8: K_{I,λ_1} vs ε predicted by eq.(7) for $\beta = 0$

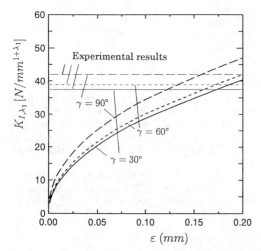

Figure 7: K_{I,λ_1} vs ε predicted by eq.(6) for $\beta = 0$

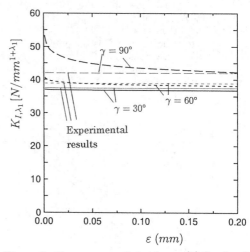

Figure 9: K_{I,λ_1} vs ε predicted by eq.(8) for $\beta = 0$

obtained by eqs.(5), (6), (7) and (8), respectively. It is seen from Figures 6 and 7 that the value of ε has a significant effect on the predicted values of K_{I,λ_1}. In order to give a satisfactory prediction, one has first to select an adequate value of ε. By supposing that the fracture criteria shown in Figure 5 also apply to the crack problem, the value of ε can be obtained from the σ_B and K_{IC} of the tested material as

$$\varepsilon = \frac{1}{2\pi} \left(\frac{K_{IC}}{\sigma_B} \right)^2 = 0.042\,mm \qquad (12)$$

for the criterion shown in Figure 5(a) and

$$\varepsilon = \frac{1}{2\pi} \left(\frac{2K_{IC}}{\sigma_B} \right)^2 = 0.166\,mm \qquad (13)$$

for the criterion shown in Figure 5(b).

For the criteria shown in Figures 5(c) and (d), however, the effect of ε isn't so significant, as shown in Figs.7 and 8, which show that the predicted values of K_{I,λ_1} are almost constant over a quite broad range of ε. Therefore, for the criteria of Figures 5(c) and (d), one can use either the value of $\varepsilon = 0.042mm$ obtained in eq.(12) or the value of $\varepsilon = 0.166mm$ obtained in eq.(13) to give a satisfactory prediction, as shown in Table 4.

5.3 *Application to mixed mode loading*

When $\beta \neq 0$, a mixture of mode I and II is created at the notch tip, and then the fracture will depend

Table 4: Critical values of K_{I,λ_1} $\left(N/mm^{1+\lambda_1}\right)$ obtained by eqs.(5), (6), (7) and (8)

γ	with $\varepsilon = 0.042mm$			with $\varepsilon = 0.166mm$		
	30°	60°	90°	30°	60°	90°
experimental results	37.49	38.93	41.99	37.49	38.93	41.99
results from eq.(5)	37.22	38.52	42.69	73.49	74.94	79.47
results from eq.(6)	18.67	19.73	23.24	36.85	38.39	43.27
results from eq.(7)	37.61	39.21	44.26	37.54	38.56	41.65
results from eq.(8)	36.86	38.83	45.21	36.78	38.19	42.55

on the combination of K_{I,λ_1} and K_{II,λ_2}. Here, we use the criterion of Figure 5(c) with $\varepsilon = 0.166mm$ to estimate the critical values of K_{I,λ_1} and K_{II,λ_2} for various notches.

First, we have to determine the direction of the fracture, namely, the angle θ in eq.(7). For any given value of θ, the load P necessary to specimen fracture can be calculated from the critical values of K_{I,λ_1} and K_{II,λ_2} obtained by eq.(7). Therefore, the direction of the fracture θ is determined as such that the corresponding load P is the minimum for all the possible directions. The fracture directions θ so obtained are shown in Figure 9 which agree very well with the experimental results.

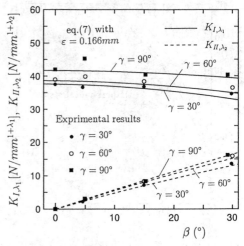

Figure 11: K_{I,λ_1} and K_{II,λ_2} vs ε predicted by eq.(8) with $\varepsilon = 0.166mm$

results.

6 CONCLUSIONS

1. Fracture of a notch is mainly controlled by the notch stress intensity factors K_{I,λ_1} and K_{II,λ_2}.

2. Four kinds of fracture criteria are presented and their usefulness in predicting critical values of K_{I,λ_1} and K_{II,λ_2} is confirmed through the experiment.

3. The criteria of eq.(7) and eq.(8) are better than the criteria of eq.(5) and eq.(6) because they don't depend on the parameter ε significantly.

REFERENCES

Chen, D.H. 1995. Stress intensity factors for V-notched strip under tension or in-plane bending. *International Journal of Fracture*. 70:81-97.

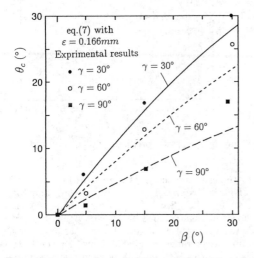

Figure 10: θ_c vs β predicted by eq.(7) with $\varepsilon = 0.166mm$

After determining the fracture direction θ, the critical value of K_{I,λ_1} and K_{II,λ_2} can be readily obtained from eq.(7). Figure 10 shows the predicted values of K_{I,λ_1} and K_{II,λ_2}, which appear to show reasonable agreement with the experimental

Experimental Mechanics, Allison (ed.) © 1998 Balkema, Rotterdam, ISBN 90 5809 014 0

Analysis of automatically monitored three-dimensional crack growth in thermally loaded compound models

K. Linnenbrock, F. Ferber & K. P. Herrmann
University of Paderborn, Germany

ABSTRACT: The development and expansion of cracks in three-dimensional thermally self-stressed two-phase compounds was measured by means of a special image processing technique. In order to get detailed information on crack driving forces intensive additional numerical examinations have to be performed. Following the well-known technique of the crack-closure-integral the fracture mechanical parameter G as the energy-release-rate was calculated by an analysis of the stress and strain fields with a newly developed post processor.

1 INTRODUCTION

The use of composite structures in aviation, astronautics, and in other high-tech areas, for example in microelectronics, can lead to certain thermal stress problems. Very often the materials are exposed not only to mechanical but also to thermal loads. In most cases materials with different thermal expansion coefficients are combined such that cooling or heating generates thermal stresses which eventually destroy the composite structure.

Figure 1a. Three-dimensional self-stressed structure in a soldered microchip; crack initiation and crack growth by thermal failure.

For example during the manufacturing of microchips semiconductor materials are used which, by initiation of thermal stresses, can fail at the solder joints or at the various interfaces of the special component.

Figure 1b. Shell-shaped crack-face of a destroyed soldered microchip.

The failure at the interface of the composite often shows a short interface crack followed by a kinking of the crack into the brittle material like a shell as shown in the Figures 1a and 1b. Because of these problems it seems reasonable to analyze damage in compound materials by means of state-of-the-art continuum mechanics.

2 EXPERIMENTAL EXAMINATION

For a first systematical examination of the development and expansion of cracks in three-dimensional thermally self-stressed two-phase compounds a model was built in order to reproduce the crack growth experimentally and to observe the crack faces.

Figure 2. Model consisting of two segments of glass and aluminum with a 3-D crack due to thermal loading.

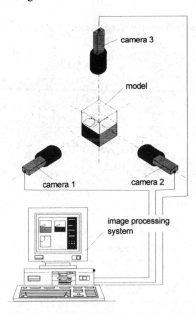

Figure 3. Three-camera system for optical observation of the crack growth.

Therefore a specimen was chosen, which consists of two block-like segments made of aluminum and glass as presented in Figure 2. It allows an optical

measurement of the expected three-dimensional crack geometry in the glassy segment.

The crack growth was monitored by means of three digital cameras in combination with a frame grabber board as displayed in Figure 3.

The model was cooled down step by step from room-temperature (289 K) to at least 253 K at a speed of 7 K/h. For the cooling process a special chamber was built to enable optical measurements.

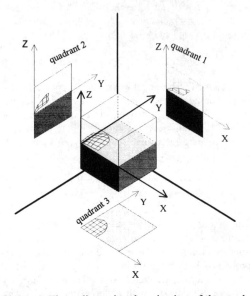

Figure 4. Three-dimensional projection of the crack front.

Figure 5. Experimentally taken crack front in the three view points.

At several times photos were taken from the different projections to get the whole three-dimensional information of the crack growth.

These digital gray scaled pictures, as shown in Figure 5, were used to reconstruct the crack front by optical measurement methods (Figure 6).

In Figures 7a to 7c the reconstructed crack front is shown for three different time-steps. The crack is growing up from an interface crack into the brittle material glass. It forms a shell-like surface which is the typical crack front behavior for compound materials as presented in Figures 1a and 1b.

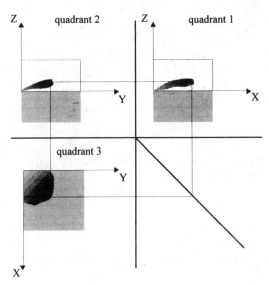

Figure 6. Determination of the crack front coordinates.

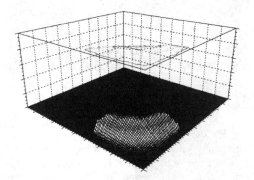

Figure 7a. Reconstructed crack front at the time t_2.

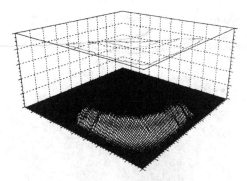

Figure 7b. Reconstructed crack front at the time t_4.

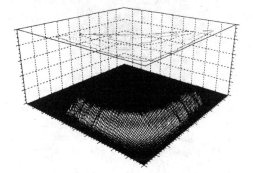

Figure 7c. Reconstructed crack front at the time t_6.

Table 1. Material properties

Material	Optical Glass	Aluminum
Young's Modulus E [MPa]	$66.0 \cdot 10^3$	$72.0 \cdot 10^3$
Thermal Expansion α [K^{-1}]	$6.1 \cdot 10^{-6}$	$2.39 \cdot 10^{-5}$
Poisson's ratio ν [1]	0.237	0.34

A classification of the crack growth cannot be done by experimental methods. Therefore, the finite element method is chosen to calculate stresses and strains and especially a fracture mechanical parameter like the energy release rate or the J-integral.

3 NUMERICAL ANALYSIS

The finite element meshing is done by the CAD program IDEAS by using 20-node elements.

For the first calculation of the stresses (Figure 8) and strains in the model the real crack front is idealized by a nearly spherical topology with a typical crack-tip mesh as shown in Figure 9. The FE-processing has been done by the FE-program ABAQUS.

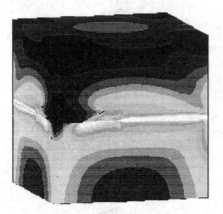

Figure 8. v. Mises stress distribution for a cracked model

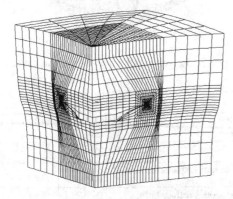

Figure 9. Deformed finite element mesh for a cracked model with a typical crack-tip mesh.

For the calculation of fracture mechanical parameters in complex three-dimensional bodies in combination with finite element simulations mostly two different energy methods are used. Alternatively the J-integral as proposed by Rice and Eshelby as well as the virtual crack closure integral (3D VCCI) technique based on Irwin's energy release rate are used (Irwin 1960, Rice 1968, Eshelby 1975).
In this work fracture mechanical parameters like the energy release rate are computed by the crack closure technique. Consequently the work necessary to close the crack virtually at a small plane is

computed. The small area of the crack plane is given by $\Delta t \Delta a$ where Δt is the length between the nodes for the forces F_2 and F_4 as shown in Figure 10a and 10b. Δa is the finite element length of a crack front element behind and ahead of the crack front which are almost equal.

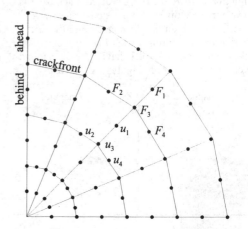

Figure 10a. Topology for calculating the energy release rate at the corner nodes.

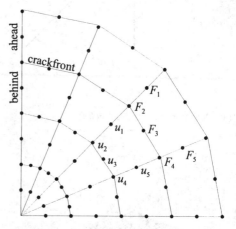

Figure 10b. Topology for calculating the energy release rate at the midnodes

The 3D VCCI method needs two finite elements simulations with two different crack lengths a and $a+\Delta a$ where a is the original crack length and Δa is a small new crack opening. With the modified virtual crack closure integral (3D MVCCI) it is possible to avoid these two steps. Therefore, one assumes that within a small Δa the forces $F(a)$ are nearly the same as $F(a+\Delta a)$ as shown by Rybicki and Kaninnen (Rybicki et al. 1977).

The energy release rate for the nodes at the corners (Figure 10a) is computed with the 3D MVCCI technique by

$$G_i = \frac{1}{\Delta t \Delta a}\left(\frac{1}{4}F_{2i}u_{2i} + \frac{1}{2}(F_{1i}u_{1i} + F_{3i}u_{3i}) + \frac{1}{4}F_{4i}u_{4i}\right) \quad (1)$$

where i = I, II, III and F_{ji} and u_{ji} are the forces and the total crack opening displacements which are responsible for the separated energy release rate G_i (Buchholz et al. 1993).

For the midnodes (Figure 10b) another calculation is used (Shivakumar et al. 1988).

Therefore, the energy-release-rate is computed by

$$G_i = \frac{1}{\Delta t \Delta a}\sum_{j=1}^{5}F_{j,i}u_{j,i} \quad (2)$$

where the notation of i, j, F and u is the same as in equation (1).

In the Figures 11a – 14 the energy release rates for six different crack fronts are presented. The plots 11a and 11b show the total energy release rate:

$$G = \sum_{i=1}^{3}G_i. \quad (3)$$

The diagrams 12a and 12b display the separated mode I part and diagram 13 the mode II part. In Figure 14 the separated mode III part, which only exists for the interface crack, is shown.

The first step is always the interface crack between the two materials. In the next steps (2-6) the crack is growing into the glassy segment. For the interface crack the energy release rates are higher than for the other steps as displayed in the diagrams. Also the shell-structure of the slipping crack can be explained from the diagrams. The energy release rate is increasing from the edge to the middle of the specimen and then decreasing to the second edge. This behavior is responsible for the shell-shaped growing crack. Furthermore, the decrease of the slope of the growing crack can be seen in the reduction of the energy release rate from step 2 until step 6.

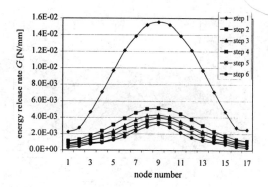

Figure 11a. Energy release rate G for all steps.

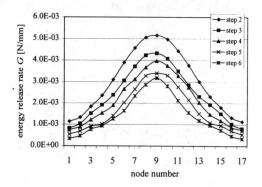

Figure 11b. Energy release rate G without representation of the interface crack.

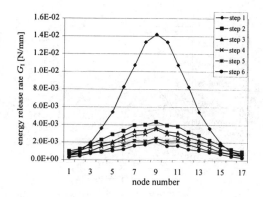

Figure 12a. Energy release rate G_I for all steps.

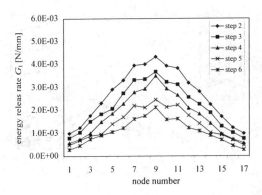

Figure 12b. Energy release rate G_I without representation of the interface crack.

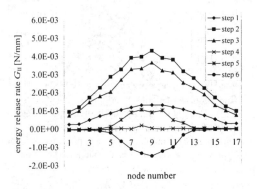

Figure 13. Energy release rate G_{II} for all steps.

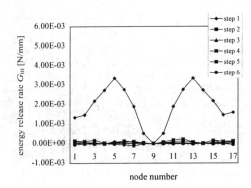

Figure 14. Energy release rate G_{III} for all steps.

4 CONCLUSION

The initiation and expansion of three-dimensional cracks in thermally self-stressed two-phase compound models were measured by image processing. In order to get detailed information about the fracture mechanical parameters for the description of the behavior of the crack front the total and the separated energy release rates were computed by the three dimensional modified crack closure integral. With this technique it is possible to explain the shell-like growing crack. Furthermore, it will be possible to simulate the crack growth by an appropriate fracture criterion which is under investigation.

REFERENCES

Irwin, G.R. 1960. Fracture mechanics. Structural mechanics, Proc. 1-th Symposium on Naval Structural Mech., 557-591.

Eshelby, J.D. 1975. The elastic energy-momentum tensor. Journal of Elasticity, 5, 321-335.

Rice, J.R. 1968. A path independent integral and the approximate analysis of strain concentrations by notches and cracks. Journal of Applied Mechanics, 35, 379-386.

Rybicki, E.F.; Kanninen, M.F. 1977. A finite element calculation of stress intensity factors by a modified crack closure integral. Engng. Fracture Mech. 9, 931-938.

Buchholz, F. G., Umlauf, B., Dietrich, S. 1993. Mixed-Mode- und Mehrachsigkeitseinflüsse bei der FE-Analyse con CTS-Proben endlicher Dicke. Proceed. Deutscher Verband für Materialforschung und –prüfung, 371-381.

Shivakumar, K. N. Tan, P. W., Newman, J. C. 1988. A virtual crack-closure technique for calculating stress intensity factors for cracked three dimensional bodies. Int. Journal of Fracture 36, R43-R50.

Experimental Mechanics, Allison (ed.)© 1998 Balkema, Rotterdam, ISBN 90 5809 014 0

Mixed mode stress intensity factors under contact stress

Teruko Aoki & Mizuho Ishida
Oita National College of Technology, Japan

Yanping Tong
Oita University, Japan

Susumu Takahashi
Kanto Gakuin University, Yokohama, Japan

ABSTRACT:Mixed mode stress intensity factors were studied in two discs combined by shrink fit, with the larger one having an artificial crack in its inner surface. The caustic experiments were performed by varying the ratio of outer to inner diameters of the larger disc and the inclination angle of the crack, aiming to clarify their effects on mixed mode stress intensity factors K_I and K_{II}. The inclination angle was found to cause an increase in the ratio of K_{II} to K_I, and the larger the angle, the greater the ratio became.

1. INTRODUCTION

Combination of mechanical parts has been widely applied in many structures, such as crankshafts, turbines and flywheels. Around the contact area between two combined parts which can be both elastic, or one elastic(Bhonsle 1970, Hill 1968, Spillers 1964) and the other hard, various serious problems have occurred. At present, they have been treated as boundary problems by many researchers(Oote 1972, Parsons 1970, Takahashi 1976). However, little has been reported on experimental or theoretical analysis of contact problems by considering them as fracture ones, though such analysis is indispensable in safety design of structures.

As we know, shrink fit is often employed in combining mechanical parts by yielding shrink fit pressure between them. The pressure can concentrate the contact stress in the vicinity of potential cracks in those parts and thus lead to fracture and damage. Usually, the stress around cracks can be effectively evaluated by stress intensity factor. However, with parts under contact pressure, it is difficult to obtain the stress intensity factor by thoretical or numerical analysis. Therefore, in this study, experimental analysis was employed. A combination of a cracked hollow disc(bore) with a smaller disc(shaft), was prepared as a model by shrink fit. Caustic experiments were conducted and mixed-mode stress intensity factors were obtained.

2. EXPERIMENTAL

The specimen used in the experiment is illustrated in

Fig.1. The bore and the shaft were made of commercially available expoxy resin plate. Crack was artificially made by a laser cutter on the inner surface of the bore. Generally speaking, resins under heat change optically as well as physically. Therefore, cooling shrink fit was employed to fulfill the combination, by inserting the shaft dipped in liquid nitrogen into the bore preheated to 125°C (taking the transient range into account) with an electric oven, and leaving them alone in the air for them to attain a stable state. It is noteworthy that even a slight external contact may result in fracture during stabilization. Such discs were subject to contact pressure and could be considered the model of a combined mechanical structure with potential or actual cracks. The mechanical parts are generally three-dimensional, however, for the sake of simplification, discs of about 6mm thickness were employed here and thus could be considered two-dimensional.

Table-1 lists the specification of specimens. In Table-1(a), shrinkage allowance and depth of crack were varied, while in Table-1(b), inclination angle of the crack was varied to clarify their effect on stress intensity factor. In order to obtain the stress intensity factor around the crack tip, the caustics image was measured by a measuring program after its picture was scanned into a computer.

3. DETERMINATION OF STRESS INTENSITY FACTOR

The transmitted apparatus was employed to form caustics, as shown in Fig.2. The radiating light was incidented on the combined discs and then the pictures of caustics images on the screen were taken. Here, the He-Ne gas laser(of 5mW output) was utilized as the light source. Figure 3 shows a typical eddy-shaped caustics image in mixed-mode, while Fig.4 demonstrates the theoretical caustics image in mixed-mode as a result of variating $\mu(= K_{II}/K_I)$. If only K_I exists, then the caustics image should be close to a strict circle.

Because the epoxy resin used here is photoelastically sensitive, K_I and K_{II} in mixed-mode can be calculated with Eqs. (1) (Gdoutos 1984, Shimada 1979, Theocaris 1972) and (2) (Bowie 1972):

$$K_I = \frac{1.671}{Z_0 t |C_0|} \frac{1}{\lambda^{3/2}} \left(\frac{D_{min}}{\delta}\right)^{5/2} \frac{1}{\sqrt{1+\mu^2}} \quad (1)$$

$$K_{II} = \frac{1.671}{Z_0 t |C_0|} \frac{1}{\lambda^{3/2}} \left(\frac{D_{min}}{\delta}\right)^{5/2} \frac{\mu}{\sqrt{1+\mu^2}} \quad (2)$$

where
Z_o is the distance between the specimen and the screen; Z_i is the distance between the specimen and the light source; t is the thickness of the specimen; λ is the magnification of the image on the screen, expressed as $(Z_o + Z_i)/Z_i$; δ is the material constant; and C_0 is an optical constant, depending on the vertical elasticity coefficient, Poisson ratio and optical properties of the specimen. In our study, C_0 is $1.16 \times 10^{-10} m^2/N$.

Prior to calculation of K_I and K_{II}, D_{max}, D_{min} and D of caustics image were measured from Fig.3. The μ and δ were then theoretically determined by the value of $(D_{max} - D_{min})/D_{max}$(Lagarde 1987).

In addition, the dimensionless stress intensity factors F_I, F_{II} and modified dimensionless stress intensity factor F_I', F_{II}' were given by the following equations(Bowie 1972):

$$F_I = \frac{K_I}{p\sqrt{\pi a}} \qquad F_I' = \frac{\beta^2 - 1}{2.24\beta^2} F_I \quad (3)$$

$$F_{II} = \frac{K_{II}}{p\sqrt{\pi a}} \qquad F_{II}' = \frac{\beta^2 - 1}{2.24\beta^2} F_{II} \quad (4)$$

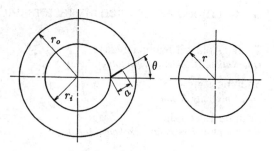

Hollow disk   Inner disk

Fig. 1 Illustration of specimen

Table 1 Specification of specimens

(a:Variation of shrinkage allowance and depth of crack)

r_o [mm]	r_i [mm]	θ [deg]	$d = 2(r - r_i)$ [mm]	a [mm]
35	17.5	0	0.10	2, 3, 4, 5
			0.20	
			0.25	
			0.30	

(b:Variation of inclanation angle of crack)

r_o [mm]	r_i [mm]	d [mm]	a [mm]	θ [deg]
35	17.5	0.2	5	0,30,45,60
35	20	0.2	5	0,30,45,60

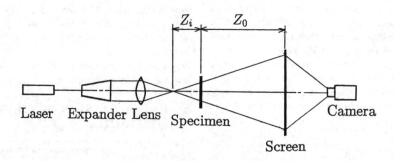

Fig. 2 Experimental setup of transmitted caustics

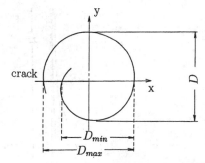

Fig. 3 Caustics image in mixed mode

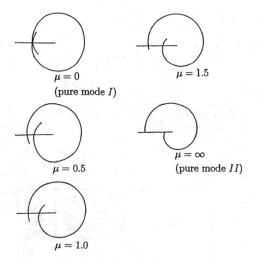

$\mu = 0$
(pure mode I)

$\mu = 1.5$

$\mu = 0.5$

$\mu = \infty$
(pure mode II)

$\mu = 1.0$

Fig. 4 Theoretical caustics images

where

p is the contact pressure between the two discs, calculated from shrinkage allowance; a is the depth of the crack and β is the ratio of outer to inner diameters of the hollow disc.

4. RESULTS AND DISCUSSION

Pictures of caustics images on the screen were taken of all combined discs with a crack in the inner surface of the hollow disc. Figure 5 demonstrates the relationship between modified dimemsionless stress intensity factor and shrinkage allowance at $\beta = 2.0$. It can be seen that at a given a/r, F_I is independent of the shrinkage allowance. Since the caustics image is not so clear at small shrinkage allowance, the shrinkage allowance is fixed at 2.0mm in further mixed-mode experiments.

Figure 6 illustrates the caustics images for different inclination angles of crack at $\beta = 1.75$ and $a/r = 0.57$.

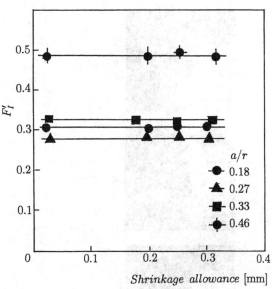

Fig. 5 Relationship F_I' and shrinkage allowance at various depth of crack $(r_o/r_i = 2.0)$

The angle is relative to the radial direction of the crack tip on the inner surface of the hollow disc. At $\theta = 0°$, the caustics image appears to be a strict circle, which indicates that only K_I exists. The caustics image becomes more eddy-shaped with increasing inclination angle, suggesting that both K_I and K_{II} exist. Note that the image at $\theta = 60°$ is very similar to the theoretical image in pure mode II shown in Fig.4, i.e., only K_{II} exists in this case. Similar pattern was followed at $\beta = 2.0$.

Figure 7 presents the relationship between μ and inclination angle of the crack. It is obvious that μ increases with θ quadratically. At $\beta = 2.0$, μ rises to about 1 at $\theta = 30°$, increases rapidly thereafter and approaches ∞ at $\theta = 60°$. Therefore, it can be concluded that with the inclination angle of crack increased, K_{II} increases while K_I decreases, as can also be seen in Fig.6.

Figure 8 shows the relationship between the modified dimensionless stress intensity factors F_I', F_{II}' and the inclination angle of the crack. It can be seen that at either β, F_I' is greater than F_{II}' before a certain value of θ is reached, becomes equal to F_{II}' at the certain value and after that becomes less than F_{II}'. The certain value of the inclination angle at $\beta = 1.75$ seems slightly larger than that at $\beta = 2.0$. This finding indicates that the contribution of mode I stress intensity factor decreases with increasing inclination angle of the crack, whereas that of mode II increases

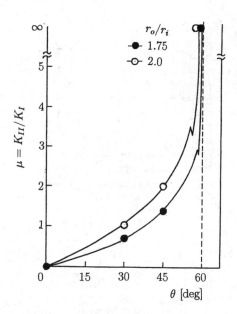

Fig. 7 Relationship between ratio of K_{II} to K_I and inclination angle of crack

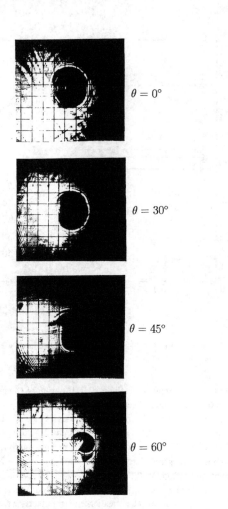

Fig. 6 Caustics images at different inclination angles of crack $(r_o/r_i = 1.75, a/r = 0.57)$

with increasing inclination angle. In addition, F'_I and F'_{II} are greater at smaller β. By now, it has become clear that whether mode I or mode II stress intensity factor takes up a larger proportion depends on the inclination angle of crack. Here more attention should be paid to the case where mode I and mode II stress intensity factor are equally large.

5. CONCLUSIONS

A combination of two discs were prepared by fitting a disc with shrinkage allowance into a hollow disc with inclined crack. The two discs were thus subject to contact pressure. The stress intensity factor of the combined discs was experimentally determined by caustics method and the following points were clarified:

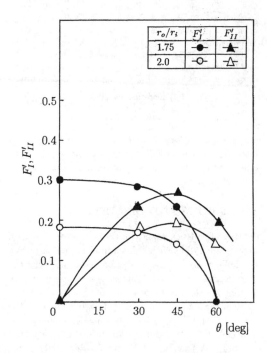

Fig. 8 Relationship between modified dimensionless stress intensity factors and inclination angle of crack

(1) Stress intensity factor of discs combined by cooling shrink fit can be obtained by caustics method.

(2) Stress intensity factor is independent of the shrinkage allowance at a give depth of crack.

(3) In the case of combined discs in mixed mode, the contribution of mode I to stress intensity factor decreases with increasing inslination angle of the crack, while that of mode II increases.

(4) There exists a certain value of inclination angle where F_I' is equal to F_{II}'. In this case, more attention should be paid to avoid fracture as a result of large stress.

(5) Both F_I' and F_{II}' are greater at larger ratio of outer to inner diameters the hollow disc.

REFERENCES

Bhonsle,S.R. 1970, Experimental Mechanics, March,19N

Bowie,O.L. and Freese,C.E. 1972, Engineering Fracture Mechanics, Vol.4,315

Gdoutos.E. 1984, Problems of Mixed Mode Crack Propagations,41

Hill, L.R. 1968, Trans. ASME Journal of Applied Mechanics, December,729

Lagarde,A., Static and Dynamic Photoelasticity and Caustics(1987),407

Oote,A, 1972Journal of JSME, Mech,38-3138,2210

Parsons,B. 1970, Trans. ASME Journal of Engineering for Industry, February,208

Shimade,H. & Shimizu,K. 1979, Journal of JSP, 1,29

Spillers,W.R. 1964, Journal of Math. Physics, 43, March,65

Takahashi,S., Shimamoto,A. & Nogata,F. 1976, Journal of JSWDI, 25(8),478

Theocaris.P.S. and Gdoutos.E. 1972, Trans. ASEM, March,91

Experimental Mechanics, Allison (ed.)© 1998 Balkema, Rotterdam, ISBN 90 5809 014 0

Real-time observation of crack extension at the interface of diamond film and transparent substrate

Shoji Kamiya, Shinji Asada & Masumi Saka
Department of Mechanical Engineering, Tohoku University, Japan

Hiroyuki Abé
Tohoku University, Japan

ABSTRACT : In-situ observation of the crack extension behavior is made from the backside of the quartz glass substrate during the indentation process of Vickers indenter onto the diamond film produced by chemical vapor deposition, using an optical microscope with a lens which is able to focus to the interface through the substrate.   Then, discussion is focused on the dissipated energy obtained form the hysteresis of load-displacement curve.   By subtracting the amount of energy dissipated by the plastic deformation of the substrate, the energy released for the crack extension along the interface, i.e., the interface fracture toughness, is estimated on the basis of the observed extension behavior of the interface crack.

1 INTRODUCTION

Hard coatings, such as synthetic polycrystalline diamond films, on the substrates are now being widely used for wear protection and many other applications.   For the practical use of thin films, their strength is one of the important engineering parameter to consider in order to ensure their mechanical integrity.   However, the strength of thin films is also one of the difficult parameters to evaluate, especially for the case of the adhesive strength.   A number of conventional methods are known for the measurement of adhesive strength of thin films, e.g., scratch tests, whose data have unfortunately little physical meaning and cannot be used as an absolute quantity for the comparison of various films on various substrates.

Recently, many researchers (e.g., Bushan 1995) have tried to evaluate mechanical properties of thin films by using small scale indentation onto the film surface.   Unfortunately, the fracture process of films and substrates due to indentation is rather complex and has not been sufficiently clarified yet. In spite of these difficulties, a few reports are found in which they tried to apply indentation tests for the estimation of adhesive strength.   In these reports, adhesive strength could be evaluated in terms of toughness, i.e., the critical energy release rate with respect to the extension of interface crack, which has a clear meaning of the macroscopic bonding energy between films and substrates.   Marshall & Evans (1984) and Rossington et al. (1984) examined a concept of interface toughness measurement for ductile films, where the plastic deformation of films were expected to play a dominant role to drive the interface crack.   On the other hand, Drory(1994) and Weppelmann et al. (1995) tried to estimate the interface toughness for the case of hard brittle films on soft substrates.   However, successful methods to evaluate the interface toughness has not yet been established.

In this report, we present a new trial to establish an experimental method based on the indentation test in order to estimate the interface toughness of hard brittle thin films on soft substrates.   Diamond films produced by chemical vapor deposition (CVD) on quartz glass substrates are picked up as a representative example.   The methods mentioned in the previous paragraph are more or less based on the simplified analytical prediction of local deformation around the indentation point, where elastic plastic properties of films and substrates are required to be known.   On the contrary, our method presented here is based on the load-displacement curve obtained experimentally during indentation tests, which does have a lot of information about the process of deformation including the amount of dissipated energy.

At first, detailed observation of the fracture process is carried out to understand how cracks extend during the indentation test.   Taking account of the observation results, the amount of energy dissipated by the plastic deformation of the substrate is separately estimated.   By subtracting energy for the plastic deformation, energy released for the crack extension is evaluated as the toughness of interface between the film and substrate.

2 EXPERIMENTAL PROCEDURE

2.1 *Specimen preparation*

The substrates were made of quartz glass with the dimension of 12.5 × 12.5 × 2 mm.   Prior to deposition, they were lightly scratched by 2 μm diamond powder in order to enhance the diamond nucleation, and rinsed in water.   Diamond growth was achieved in a microwave plasma reactor at the excitation frequency of 2.45 GHz in the gas mixture of 99% hydrogen and 1% methane.   The total gas flow rate was 10 sccm.   Substrate temperature was controlled to be 1120K.

After the deposition for an hour, we obtained specimens with the diamond film whose thickness is 0.4 μm.   Film thickness was measured by a laser scanning microscope as the difference in level where a part of the film was scratched off.

2.2 *Indentation test*

Figure 1 shows the schematic illustration of indentation test and our experimental set up.   A Vickers diamond indenter was driven upward onto the specimen which was placed with the filmside

down.   The interface of film and substrate could be seen with a microscope situated over the specimen, whose objective lenses were able to focus on the interface through the glass substrate up to the magnification of 600x.   Detailed observation was carried out in order to understand the fracture process.   Indentation load was plotted against displacement of the indenter throughout the experiment by using an x-y recorder.   The area of this load-displacement diagram, i.e., the area of hysteresis loop surrounded by the loading and unloading curves, was measured, which gave the energy dissipated during the indentation test. Photographs were also taken in order to measure the area of interface crack.

A large number of indentation tests were carried out with various amount of maximum displacement from 400 nm up to about 2000 nm which corresponded to the load from 40 mN to 500 mN. The amount of dissipated energy and the resulting area of interface crack were obtained as a function of maximum displacement.   The same indentation tests were also carried out onto the bare substrate surface without the film for the estimation of energy dissipated by the plastic deformation of substrate.

3 OBSERVATION OF FRACTURE BEHAVIOR

The representative load-displacement curve obtained during an indentation test on the specimens as explained in Section 2.1 is shown by the solid line in Figure 2, where the indentation load is plotted on the vertical axis and the displacement of indenter is on the horizontal axis.   The area surrounded by the load-displacement curve, i.e., the hysteresis of the load-displacement curve, indicates the amount of energy dissipated during the indentation test.

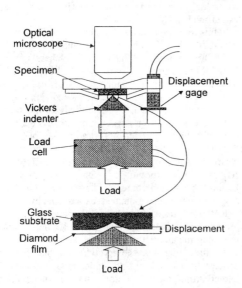

Figure 1. Schematic illustration of indentation test and its experimental set up

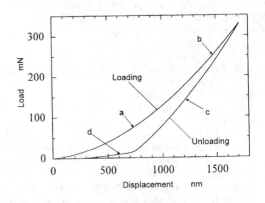

Figure 2.   Load-displacement curve

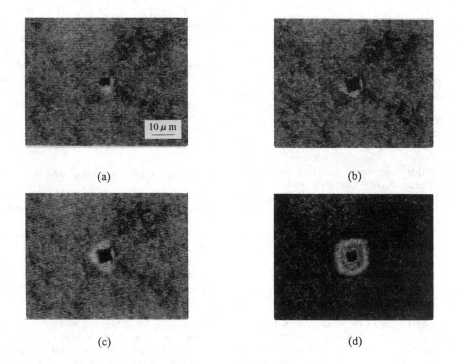

(a)

(b)

(c)

(d)

Figure 3. In-situ microscopic observations of fracture behavior

Typical observations of the fracture process are presented in Figure 3. These photographs were taken at the points a, b, c and d, respectively, as indicated in Figure 2. At an early stage of loading around the point a, a small black square appeared at the indentation point where the film was supposed to be broken by the tip of the indenter as shown in Figure 3(a). Complicated distribution of microscopic film cracks might be expected in this region. As the load increased above this load level, the interface crack extension could be observed while film cracks along the edges of the indenter also extended, which were unfortunately difficult to find clearly in the photograph Figure 3(b). The area of interface crack was far larger than the area of contact to the indenter and in approximately square shape. Abrupt increment of interface crack was frequently observed, while seldom discontinuity was found in the load-displacement curve. In the unloading process, the interface crack appeared to be clearly seen as shown in Figure 3(d) which was taken at point d in Figure 2, maybe because the crack opening displacement was becoming larger. After the unloading was completed, a permanent impression mark could be observed on the substrate. In addition, no crack was seen inside the glass substrate within the load level examined in our study.

Putting together all the observations above, the appearance of fracture may be schematically illustrated as presented in Figure 4. In the lower side of Figure 4 is presented the cross sectional view of the indentation point. The small square C as lightly shaded in the figure represents the region of black square mentioned before where complicated film fracture was expected to take place when the indenter tip broke the film at the initial stage of loading. Although the area of interface crack increases with the maximum indentation displacement, the size of this region shows little difference among all the indentation tests in this study. An additional observation using a scanning electron microscope revealed that the microscopic appearance of film surface at the indentation point never changed after the indentation test, i.e., the apices of crystals were not crushed by the pressure given by the indenter. This fact means that the deformation of diamond film was almost perfectly elastic and resulted in brittle fracture behavior.

4 EVALUATION OF DISSIPATED ENERGY

In this section, our attention is focused on the

hysteresis of the load-displacement curve obtained. The area of loading-unloading hysteresis loop indicates the amount of energy dissipated during the indentation tests. Solid circles in Figure 5 show the amount of dissipated energy, U, for the case of indentation tests of the specimen with the diamond film. In Figure 5, the vertical axis means the amount of energy, while the maximum indentation displacement is plotted on the horizontal axis.

Let us here consider the components of the dissipated energy. The model presented in Figure 4 suggests that two major kinds of irreversible process take place during indentation tests. One is the crack extension inside the film and along the interface. Another is the plastic deformation of the substrate as evident form the permanent impression just under the indentation point. The composition of these kinds of dissipated energy may be expressed in the following equation.

$$U = U_s + G_i A + 2G_f h\sqrt{A} + U_c \qquad (1)$$

where G_i and G_f mean the toughness of interface and film, respectively. The area of interface crack and the film thickness are indicated by A and h, respectively. Note that the length of film cracks is considered to be equal to the diagonal of square shaped interface crack in Equation (1). The last term U_c means the energy consumed to create the film fracture inside the region C as indicated in Figure 4. The amount of energy dissipated for the plastic deformation of the substrate is here represented by the first term U_s. Since the size of

the region C showed little change with respect to the area A as discussed in Chapter 3, U_c is expected to be a constant in Equation (1), which could be estimated as the limiting value of dissipated energy at $A = 0$. However, U_s is indeed an unknown function of A, which must be specified in order to estimate the toughness on the basis of total dissipated energy U obtained by the experiment. In the next section, a brief discussion will be presented about the plastic deformation of substrate.

4.1 Plastic deformation of the substrate

Indentation load is plotted in Figure 6 against the square of displacement, where the solid and dashed lines represent indentation tests onto the specimen with diamond films and the bare substrate surface, respectively. It is well known (Bushan 1995) that indentation load is proportional to the square of displacement for the case of pyramidal indenters impressed onto the surface of homogeneous materials as the dashed line seems to be in the figure, since the load should be proportional to the contact area of the indenter and the material surface. In the case of the solid line in Figure 6, the load is at first proportional to the square of displacement when the displacement is quite small compared to the film thickness, where its gradient would be equal to that for the case of indentation onto the bulk diamond. As the displacement increases, however, the influence of the substrate is getting more and more dominant, which makes the solid line run parallel to the dashed line (Yanai et al. 1994) when the displacement becomes sufficiently large compared to the film thickness.

It is expected from Figure 6 that the increment of plastic deformation of substrate would be almost identical in these two cases under sufficiently large

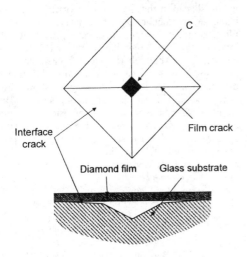

Figure 4. Schematic model of fracture in the specimen

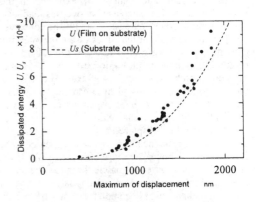

Figure 5. Dissipated energy during indentation test

displacement. After the film is debonded and broken by edges of the pyramidal indenter, it may behave just as a thin insertion between the indenter and substrate surface, which will lead to the same extent of the plastic deformation in the substrate. As a fact, the size of permanent impressions after unloading appeared to be approximately the same in these two cases with and without the film on the substrate, at the same maximum indentation displacement.

Standing on the discussion above, the amount of energy dissipated for the plastic deformation of the substrate during the indentation test of the specimens with the film, U_s , may be evaluated to be the same as the dissipated energy for the case of indentation onto the substrate without the film. The dashed curve in Figure 5 shows the dissipated energy obtained by the indentation test of the bare substrate, which is drawn by fitting the experimental results with various maximum displacement to a cubic function since the volume of plastically deformed region should be proportional to the cube of indentation displacement in the case of a pyramidal indenter. This curve is used as the reference to estimate the energy for the plastic deformation of substrate.

4.2 Estimation of interface fracture toughness

By subtracting the energy for the plastic deformation from the total dissipated energy in Figure 5, the amount of energy released for the crack extension can be obtained. This corresponds to the sum of the second to the last terms in the right hand side of Equation (1), and is plotted against the area of interface crack in Figure 7. Although the scatter is quite severe, the data in Figure 7 has a trend that the energy gradually increases with the area of interface crack, except for those extremely

away from the major portion.

Kamiya et al. (1997) recently reported that the toughness of CVD diamond film with thickness around 0.4 μm is about 15 J/m². According to this value, the third term in Equation (1) is calculated to be 1.2×10^{-10} J when the area of interface crack is 1×10^{-10} m² , which is negligibly small. Hence Equation (1) suggests that a linear relation may be expected between the energy released for the crack extension and the area of interface crack. By setting a linear line as shown in Figure 7, the toughness of interface seems to be estimated around 10 J/m². This value may be one of seldom data ever obtained about the interface fracture toughness of CVD diamond film on quartz glass substrate.

5 CONCLUSIONS

Detailed in-situ observations of fracture behavior in the process of Vickers indentation onto the CVD diamond film on the quartz glass substrate was carried out. A trial of experimental estimation for the interface fracture toughness, which is the critical energy release rate with respect to the crack extension along the interface between the film and substrate, was presented on the basis of the observation results.

Total amount of energy dissipated during the indentation tests could be experimentally obtained as the area of loading and unloading hysteresis loop in the load-displacement curve. The energy dissipated for the plastic deformation of substrate could be separately estimated by carrying out the same indentation tests onto the substrate without the film. The difference between the total dissipated energy and the energy for the plastic deformation could be supposed to drive the interface crack, and

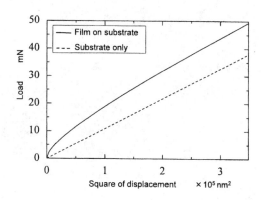

Figure 6. Relation between the load and square of displacement

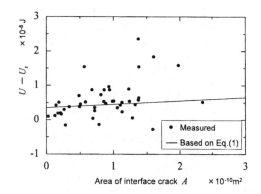

Figure 7. Energy for the crack extension

the interface fracture toughness might appear to be around 10 J/m².

Unfortunately, the severe scatter of data in Figure 7 told that the precision of experimentation in this study seemed to be insufficient for further discussion. Improved experiment should be necessary in the future for the practical application of the method proposed here. However, although transparent substrates were used in this study for the observation of fracture process, we believe that the basic concept of this pure experimental method of fracture toughness estimation will be attractive for all kinds of interfaces between thin hard films and soft substrates.

REFERENCES

Bushan, B. 1995. Nanomechanical properties of solid surfaces and thin films. In B. Bushan (ed), *Handbook of micro/nano tribology*: 321-396.

Drory, M. D. & Hutchinson, J. W. 1994. Diamond coating of titanium alloys. *Science* 263:1753-1755.

Kamiya, S., Sato, M., Saka, M. & Abé, H. (1997). Fracture toughness estimation of thin chemical vapor deposition diamond films based on the spontaneous fracture behavior on quartz glass substrates. *J. Appl. Phys.* 82:6056-6061.

Marshall, D. B. & Evans, A. G. 1984. Measurement of adherence of residually stressed thin films by indentation. I. Mechanics of interface delamination. *J. Appl. Phys.* 56:2632-2638.

Rossington, C., Evans, A. G., Marshall, D. B. & KhuriYakub, B. T. 1984. Measurement of adherence of residually stressed thin films by indentation. II. Experiments with ZnO/Si. *J. Appl. Phys.* 56:2639-2644.

Weppelmann, E. R., Hu, X.-Z. & Swain. M. V. 1995. Observations and simple fracture mechanics analysis of indentation fracture delamination of TiN films on silicon. In K. L. Mittal (ed), *Adhesion measurement of Films and Coatings*: 217-230. VSP: Utrecht.

Yanai, H., Itokazu, M., Murakami, Y. & Kishine, N. Elastic analysis of a triangular pyramidal indentation of a two-layer thin film. *Trans. JSME (A)* 60:2731-2736. (in Japanese)

Experimental Mechanics, Allison (ed.)© 1998 Balkema, Rotterdam, ISBN 90 5809 014 0

Comparison of experimental methods for determining stress intensity factors

S.J.Zhang & E.A.Patterson
Department of Mechanical Engineering, The University of Sheffield, UK

ABSTRACT: This paper presents a comparison of the reflection photoelasticity and strain gauge methods for determining the stress intensity factor (mode I) of aluminium-alloy plates with double cracks from the central hole. The three photoelastic methods used are point-by-point reading from a conventional reflection polariscope, extracting data from a colour photograph obtained using a conventional polariscope, and an automated polariscope based on the principles of phase-stepping. A two-element electrical-resistance rosette strain gauge and a single-element flat gauge were used in the experiments. The results from these methods were compared with numerical results by Newman. The comparison indicated that the automated polariscope has the smallest errors and is perhaps the easiest one to operate.

1 INTRODUCTION

Stress intensity factor (SIF) is a parameter which characterises the redistribution of the stress in a body due to the introduction of a crack. In fracture mechanics, the stress intensity factor K_I can be compared with the toughness of the material K_{IC} to predict whether the crack will be stable under an applied load or will become unstable and cause abrupt failure.

Many methods have been developed to determine stress intensity factors by experiment. However, the accuracy of individual methods is quite different when they are applied directly to a structure rather than a model. Previous investigations have revealed that the accuracy depends on a number of factors, such as the selection of the experimental methods, the instrumentation, specimen preparation and algorithms employed. Photoelasticity is one of the available techniques and has been applied in the field for more than forty years. However, the analysis of photoelastic patterns can be time-consuming and tedious, which restricts the application of photoelasticity. In recent years many efforts have been made in automated photoelastic analysis. Point-by-point automatic polariscopes (Allison & Nurse 1971, 1972; Redner 1974) and full field automatic polariscopes (Mueller & Saackel 1979;

Seguchi et al. 1979; Umezaki et al. 1984; Patterson & Wang 1991) have been proposed. The development of the technique from manual to automatic analysis makes it possible to do more complex analysis more quickly. In the present paper, results of the SIF from three methods using reflection photoelasticity are compared with those from strain gauges and from the Newman's numerical solutions for the stress intensity factor for a tensile plate with double cracks from the edge of a central hole (Murakami 1990). The three photoelastic procedures involve point-by-point readings from a conventional reflection polariscope; extracting data from a photograph obtained using a conventional polariscope (Nurse & Patterson 1990, 1992 and 1993(a)), and an automated polariscope based on the principles of phase-stepping (Patterson & Wang 1997). The different instruments and the algorithm used are described in the following subsections. The comparisons of the results will show that the automated photoelastic method has the smallest errors. In addition, the automated polariscope may be the easiest method to use.

2 EXPERIMENTAL METHODS

2.1 *Specimen*

Two aluminium-alloy plates each with cracks from a central hole were used in the experiment. Artificial cracks were machined from the edges of the hole and were perpendicular to the loading direction, as shown in Figure 1. The aluminium plates were 120 mm wide and 6.6 mm thick. The elastic modulus of the plate was 73000 MPa and Poisson's ratio 0.32. The radius of the central hole is 30 mm. The length of the crack in specimen A is 15 mm and in specimen B is 9 mm. The crack was machined using a saw which resulted in a crack thickness of 0.2 mm, with a semi-circular profile at its tip. Photoelastic coating (PS-1B) was cut to fit the geometry of the plate and cracks and were bonded on the plates but not over the cracks using PC-1 adhesive. (Both PS-1B and PC-1 were supplied by Measurement Group Inc. Raleigh, N. C. USA). The thickness of the photoelastic coating was 2 mm , strain optical coefficient 0.15, elastic modulus 2500 MPa, Poisson's ratio 0.36 and material fringe constant 200 N/fringe/mm. Strain gauges are bonded on the back of the specimen A and B in the region near the crack tip.

The specimens were subjected to tensile load of 1, 1.5, 2, 2.5 and 3 tonnes respectively on a large dead-weight loading rig. Several different methods were employed to obtain data for the subsequent calculation of the SIF of the specimens as described below.

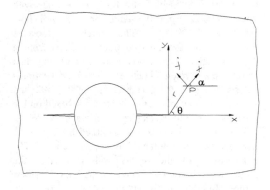

Figure 1. Orientation of single element strain gauge on specimen B (not to scale), with loading applied in the Y-direction.

2.2 *Strain gauge method*

A linear, two-element stacked rosette, electrical-resistance strain gauge (gauge factor f = 2.12, gauge length l = 5 mm) was bonded on to specimen A. The distance from geometric centre of the gauge to the crack tip was 7.6 mm on a line through the crack tip and perpendicular to the crack plane. The strain gauge was connected to a Switch & Balance Unit and a strain indicator of Model P-3500 (Vishay Instrument Measurement Group, U.S.A). The stress intensity factor was obtained by calculating data from strain gauge reading using the following expression (Dally & Sanford 1987)

$$K_I = \frac{2E}{1+\upsilon}\left(\varepsilon_{y'y'} - \varepsilon_{x'x'}\right)\sqrt{\pi r} \qquad (1)$$

where r is the distance from the gauge to the crack tip. E and υ are elastic modulus and Poisson's ratio of the material.

A linear, one element, flat strain gauge (gauge factor f = 2.09, gauge length l = 2 mm) was bonded on to specimen B. The position and orientation of the gauge are shown in Figure 1. The distance from the geometrical centre of the gauge to the crack tip was 3.75 mm. The instrumentation in this experiment was same as for specimen A, but the results were obtained from expression (Dally & Riley 1991)

$$K_I = (\frac{E}{1+\upsilon}\varepsilon_{x'x'}\sqrt{2\pi r_c}) \div$$
$$(\frac{1-\upsilon}{1+\upsilon}\cos\frac{\theta}{2} - \frac{1}{2}\sin\theta\sin\frac{3\theta}{2}\cos2\alpha$$
$$+\frac{1}{2}\sin\theta\cos\frac{3\theta}{2}\sin2\alpha) \qquad (2)$$

where α and θ are angles shown in Figure 1, and equal to 60.5° and 62° respectively.

2.3 *Conventional reflection photoelasticity method*

A reflection polariscope of Model 031 (Vishay Instruments Measurement Group, U.S.A) with a white light source (λ=575 nm) and a Null-Balance Compensator of Model 232 (Vishay Instruments Measurement Group, U.S.A) and a colour CCTV camera of Model WV-CP210/B (Panasonic, Japan) were used in the experiment. According to Smith & Kobayashi (1987), the stress field around a crack subjected to Mode I loading can be described by the following truncated series

$$\sigma_x = \frac{K_I}{\sqrt{2\pi r}}\cos\frac{\theta}{2}\left(1 - \sin\frac{\theta}{2}\sin\frac{3\theta}{2}\right) + \sigma_{ox} \qquad (3)$$

$$\sigma_y = \frac{K_I}{\sqrt{2\pi r}}\cos\frac{\theta}{2}\left(1 + \sin\frac{\theta}{2}\sin\frac{3\theta}{2}\right) \qquad (4)$$

$$\sigma_{xy} = \frac{K_I}{\sqrt{2\pi r}} \sin\frac{\theta}{2}\cos\frac{\theta}{2}\cos\frac{3\theta}{2} \qquad (5)$$

where σ_{ox} is a second order non-singular term appearing only in the σ_x component to take account of far-field effects.

From a Mohr's circle of stress, the following expression can be obtained

$$\tau_{max} = \frac{K_I}{\sqrt{8\pi r}} + f(\sigma_{ij}^o) \qquad (6)$$

where $f(\sigma_{ij}^o)$ is a set of constants which account for the contribution of the non-singular stress field. Hence, using expression (6), it can be shown that

$$\frac{K_{AP}}{\sigma\sqrt{\pi a}} = \frac{K_I}{\sigma\sqrt{\pi a}} + \frac{f(\sigma_{ij}^o)}{\sigma}\left(\frac{r}{a}\right)^{\frac{1}{2}} \qquad (7)$$

where σ is the nominal applied stress and

$$K_{AP} = \tau_{max}(8\pi r)^{\frac{1}{2}} \qquad (8)$$

Linear elastic fracture mechanics is considered to be valid when measurement is carried out in a annular region of inner radius of 10ρ and outer radius of $0.4a$ where ρ is crack tip radius and a is crack length. Hence, in this experiment the data points should be taken in an annular ring ($2mm \le r \le 12mm$). For reflection photoelasticity, data points must be taken $4t$ (coating thickness $t = 2mm$) away from edge of the coating in order to take account of the influence of the Poisson's ratio mismatch. According to the factors described above, the experimental data in the present case should be taken in an annular ring of inner radius of 8 mm and outer radius of 12 mm. The normalised apparent K varies linearly with the square root of the normalised distance of the measured data from the crack tip. The intercept on the normalised K_{AP} axis provides the value of the geometric factor, Y. The stress intensity factor is readily obtained from the relationship

$$Y = \frac{K_I}{K_O} = \frac{K_I}{\sigma_o\sqrt{\pi a}} \qquad (9)$$

2.4 Photographic method (Nurse & Patterson, 1991)

Colour photographs were taken in the dark field, circular setting of the reflection polariscope, using a SLR camera (Pentax K1000, Japan) and a colour film (Fuji, Japan). The colour photographs were produced as 5×3 inch prints and then enlarged using a colour photocopier to three times their original size. A colour calibration chart supplied with the reflection polariscope was used to assign the isochromatic fringe order on the colour photocopies. About 70 points were identified along the fringe loops in an annular region as defined in section 2.3. In order to digitise this information, each photocopy was placed on a digitising tablet connected to a personal computer. In conjunction with the AUTOCAD software, the puck on the tablet was used to define an axis system centred on the crack tip and to identify the location of the data points on the fringe loops. This information, together with the fringe order associated with each loop was stored in a data file. The file was subsequently used as input to a purpose-written program known as CHOPIN, which used the data to determine the stress intensity factors. The program CHOPIN employs the modified Multi-Point Over-Deterministic Method (MPODM) in which the Westergaard stress field equations are replaced by those of Muskhelishvili to allow a non-uniform passing stress field through the use of Fourier series to describe the stress field. The resulting stress field equations are fitted to the experimental data using an iterative, least square approach (Nurse & Patterson 1993(b)).

2.5 Automated method

In this method, a new automated reflection polariscope was used. This system uses phase stepping method to establish fractional isochromatic fringe order and isoclinic parameter at all individual points in the field without reference to their neighbouring points (Patterson and Wang 1991).

The principles used in this newly designed automated polariscope have been described by Patterson and Wang (1997). However, in brief, four phase-stepped images are observed simultaneously using a beam-splitting arrangement. The system consists of a light source that produces a beam of circularly polarised light which is directed onto the

Table 1. Arrangement of output quarter wave plates and analysers in the automated polariscope PSIOS II .

Images	Orientation of output quarter wave plate β	Orientation of analyser ϕ
1	$\dfrac{\pi}{4}$	$\dfrac{\pi}{4}$
2	0	0
3	$\dfrac{\pi}{2}$	$\dfrac{\pi}{2}$
4	0	$\dfrac{\pi}{4}$

specimen. The beam passed through the coating and is reflected back to the instrument where is collected by a focal lens. Inside the instrument the beam is collimated, then twice divided in two using cube beam splitters to produce four beams of nominally identical intensity. Each of the four beams is focused on the chip of a CCD camera by a standard camera lens. An appropriately orientated quarter-wave plate and polariser are placed in the path of each beam so that four phase-stepped image are generated. The output from the cameras are combined in a multiplex device and supplied to a digitiser as a single signal. For a circular polariscope, the light intensities observed when orientation of output quarter wave plate and analyser are fixed as in Table 1, would be

$$i_1 = \frac{a}{2}(1 - \cos 2\theta \sin \delta) \tag{10}$$

$$i_2 = \frac{a}{2}(1 + \sin 2\theta \sin \delta) \tag{11}$$

$$i_3 = \frac{a}{2}(1 + \sin 2\theta \sin \delta) \tag{12}$$

$$i_1 = \frac{a}{2}(1 - \cos \delta) \tag{13}$$

Combining the expressions of (9) - (12), isoclinic angle, θ and the retardation, δ can be found using the following expressions

$$\theta = \frac{1}{2}\tan^{-1}\left(\frac{2(i_2 - i_3)}{i_2 + i_3 - 2i_1}\right) \tag{14}$$

and

$$\delta = \tan^{-1}\left(\frac{i_2 - i_3}{\sin 2\theta(i_2 + i_3 - 2i_4)}\right) \tag{15}$$

To account for the effect of the beamsplitters on the polarised light, the beamsplitters was considered as partial linear polarisers with transmission coefficients, k_x and k_y along their principal axes. Now the intensities observed on each arm of the instrument are

$$i_1 = \frac{a}{4}\{k_{x1} + k_{y1} + (k_{x1} - k_{y1})\sin 2\theta \sin \delta$$

$$-2\sqrt{k_{x1}k_{y1}}\cos 2\theta \sin \delta\} \tag{16}$$

$$i_2 = \frac{a}{4}k_{x2}(1 + \sin 2\theta \sin \delta) \tag{17}$$

$$i_3 = \frac{a}{4}k_{x3}(1 - \sin 2\theta \sin \delta) \tag{18}$$

$$i_4 = \frac{a}{4}\{k_{x4} + k_{y4} + (k_{x4} - k_{y4})\sin 2\theta \sin \delta$$

$$+2\sqrt{k_{x4}k_{y4}}\cos \delta\} \tag{19}$$

Cubes with transmission:reflection ratio of 53:46 were selected for the system, thus the isoclinic angle and fractional isochromatic fringe order can be obtained using the following expressions

$$\theta = \frac{1}{2}\tan^{-1}\left(\frac{\frac{i_2}{2Rk_{x2}} - 1}{1 - \frac{i_1}{2Rk_{x1}}}\right) \tag{20}$$

and

$$\delta = \tan^{-1}\left(\frac{2Rk_{x1} - i_1}{\cos 2\theta(2Rk_{x1} - i_4)}\right) \tag{21}$$

where

$$R = \frac{i_2 + i_3}{4k_{x3}} \tag{22}$$

This analysis is implemented in a specially written computer program which allows a set of images each 256×256 pixels to be processed in a matter of minutes. The periodic distributions generated by expression (20) and (21) are unwrapped using the algorithms discussed by Wang and Patterson (1995). The operator has to seed a calibration of the full-field map of isochromatic parameter. In order to perform analyses of the fringe pattern around a crack tip, the location of the tip has to be identified in the image. The program creates an array of points within the annular region defined in section 2.3 and creates a data file in a similar way to that described in the section 2.4. This program and CHOPIN are linked together, so that the solution procedure is identical to that used in section 2.4.

3 RESULTS AND DISCUSSION

The experimental results for the SIF and the numerical solutions are presented in Figure 2. The numerical values of the SIF were calculated according to equation (9) where Y=1.2853 for $2R/_w = 0.25$, $2a/_w = 0.5$ and Y=1.2156 for $2R/_w = 0.25$, $2a/_w = 0.4$ for cracks from central hole in a plate (Murakami 1990, Newman 1971).

According to Figure 2 (a), (b), the results using the automated reflection polariscope are closer to the numerical results than those from the other methods used in this experiment. The relative errors for the automated method compared with the numerical results are in a range of 2.25 percent to 3.38 percent for specimen A and of 0.54 percent to 6.28 percent for specimen B. These errors arise from both systematic and random causes. The latter includes defects in the photoelastic coating which affected the quality of images and so caused some errors in the image processing, in the automated method. Whereas these errors were avoided in the manual and photographic method, almost sub-consciously through the selection of data points by the operator; in the automated method the array of selected points is chosen based on geometry relative to the crack tip and length and without regard to imperfections in the data field. These imperfections cause errors in the results, and increase the sensitivity of the MPODM to the initial estimate, tolerance of the fit, and the number of terms in the series describing the stress field. Consequently, the solution must be repeated with a number of increments of these parameters to ensure that a stable solution has been achieved.

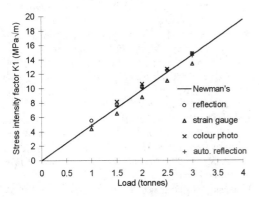

Figure 2 (a) Stress intensity factors found by five methods for various loads on specimen A

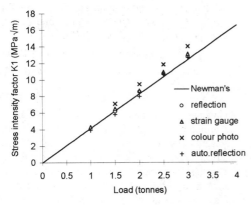

Figure 2 (b) Stress intensity factors found by five methods for various loads on specimen B

The results using conventional reflection polariscope with Null-balance compensator are close to the numerical results although the difference becomes bigger at smaller loads. The relative errors are in a range of 1.16 percent to 13.5 percent for specimen A and of 1.16 percent to 8.92 percent for specimen B (there is no result for 1 tonne loading because the fringe (colour) on the specimen B is too faint to recognise). The method is considered to be suitable for the clearly defined fringes, which means the larger loads in this experiment. In the small loading condition the fringes were less than 0.5 order within the valid annulus so that the recognition of fringes was difficult.

The photographic method yields results with relative errors in a range of 1.23 percent to 11.19 percent for specimen A and of 12.97 percent to 14.51 percent for specimen B (there are no results for 1 tonne loading because the fringe on the photocopies were too indistinct) compared with the numerical results. The errors are considered to be caused by poor data collection. The large error occurred at smaller loads because there were no integer fringes on the picture, and all points were selected by comparing the colour with the calibration chart, therefore the accuracy of the fringe order assigned was small. The truncation of the series terms and the initial estimated value of SIF in the MPODM are also potential sources of error in this method.

The results using the rosette strain gauge on specimen A are consistently lower than the numerical results (i.e. 8.25 percent to 10.91 percent). The results from single strain gauge on specimen B are higher than the numerical results with relative error from 3.38 percent to 5.5 percent. The errors may be caused by geometry (width, tip radius) and strain gradient effect as reported by Dally and Sanford (1987). The comparison of results from two gauges showed that the error increased with length of gauge. The results are consistent with the theory on strain gauges that the absolute error is directly proportional to square of the length of the strain gauge (Dally & Riley 1991). Although the strain gauge yielded good results for specimen B in this experiment, the use of strain gauges is constrained by many factors. For example, a tiny deviation of the position of the gauge from the correct place will cause large errors.

4 CONCLUSIONS

Among the experimental methods described above, the strain gauge method and photographic method are inexpensive, but the use of the two methods are limited by ability to achieve the desired accuracy. Since the evaluation of the photoelastic patterns can be time-consuming and tedious, and particularly for those specimens with low fringe order, the conventional reflection polariscope is not considered practicable. However, for high stresses, it yielded accurate results as in this experiment. By comparison, the automatic reflection polariscope has yielded the value of the SIFs with a smaller relative error and was less time-consuming, and may be the easiest one to use.

ACKNOWLEDGEMENTS

The authors would like to thank the University of Sheffield for financial support to S. J. Zhang.

REFERENCES

Allison, I. M. & Nurse, P. 1971. Optimal data acquisition for an automatic polariscope. *7th All-Union Conference on Photoelasticity, Tallinn, Nov. 1971, Proc. pt. 1, pp.93-105.*

Allison, I. M. & Nurse, P. 1972. Automatic acquisition of photoelastic data. *Proc. JBCSA Conference on the Recording and Interpretation of Engineering Measurements, Inst. Mar. Engrs., London, pp. 203-207.*

Dally, J. W. & Sanford, R. J. 1987. Strain-gage methods for measuring the opening mode stress intensity factor K_I, *Experimental Mechanics, Vol.27, No.4, pp. 381-389.*

Dally, J. W. & Riley, W. F. 1991. *Experimental stress analysis. (3rd edition), p. 323 & p. 131 McGraw-Hill, New York.*

Mueller, R. K. & Saackel, L. R. 1979. Complete automatic analysis of photoelastic fringes. *Expt. Mech., Vol. 19, pp. 245-252.*

Murakami, Y. 1990. *Stress intensity factor handbook 1, Pergamon Press, Oxford.*

Newman, J. C. 1971. An improved method of collection for the stress analysis of cracked plates with various shaped boundaries, *NASA Report TN D-6376.*

Nurse, A. D. & Patterson. E. A. 1990. Photoelastic determination of stress intensity factors for edge cracks under mixed-mode loading, *Proc. Conf. 9th Int. Conf. Experimental Mechanics, Copenhagen, 20-24 Aug., pp. 948-957.*

Nurse, A. D. & Patterson. E. A. 1992. Photoelastic coatings applied to cracked plates, *Proc. Conf.*

Technology Transfer between High Tech Engineering and Biomechanics, Ed. E.G. Little, Limerick, 4-5 Sept., pp. 477-490.

Nurse, A. D. & Patterson. E. A. 1993 (a). Photoelastic determination of fatigue crack stress intensity factor, *Proceedings of SPIE, Annual Symposium on Opt. Engineering, San Diego, 1993.*

Nurse, A. D. & Patterson. E. A. 1993 (b). Determination of predominantly mode II stress intensity factors from isochromatic data. *Fatigue fract. engng. mater. struct., Vol. 16, pp. 1339-1354.*

Patterson, E. A. & Wang, Z. F. 1991. Towards full field automated photoelastic analysis of complex components. *Strain., Vol. 27, pp. 49-53.*

Patterson, E. A. & Wang, Z. F. 1997. Simultaneous observation of phase-stepped images for automated photoelasticity, *in press, Journal of Strain Analysis.*

Redner, S. 1974. A new automatic polariscope system. *Expt. Mech., Vol. 14, pp. 486-491.*

Seguchi, Y., Tomita, Y. & Watanbe, M. 1979. Computer aided fringe pattern analysis - a case of photoelastic fringe. *Expt. Mech. Vol. 19,No.10, pp. 362-370.*

Smith, C. W and Kobayashi, A. S. (1987) Fracture Mechanics, *Chapter 20 in Hand book on experimental mechanics (edited by Kobayashi, A. S.), p. 893, Prentice-Hall, Englewwod Cliffs, New Jersey.*

Umezaki, B., Tamaki, T. & Takahashi, S. 1984. Automatic stress analysis from photoelastic fringe processing using a personal computer. *Proc. Soc. Photo. Vol.504, pp. 127-134.*

Wang, Z. F. & Patterson, E. A. 1995. Use of phase stepping with demodulation and fuzzy sets for birefringence measurement. *Optics and Lasers in Engineering, Vol. 22,pp. 91-104.*

Experimental Mechanics, Allison (ed.)© 1998 Balkema, Rotterdam, ISBN 90 5809 014 0

Micromechanical aspects of brittle fracture initiation

I. Dlouhý, V. Kozák & M. Holzmann
Institute of Physics of Materials, Academy of Sciences of the Czech Republic, Brno, Czech Republic

ABSTRACT: Based on the deterministic approach to the analysis of cleavage failure, and comparing this concept with the statistical approach of Beremin, the local fracture criteria have been evaluated using fracture micromechanism assessment combined with FEM calculations. Two model ferritic microstructures that were selected for presentation differ by carbide thickness and fracture toughness scatter. For these microstructures, the Weibull stresses were related to the cleavage fracture stresses. The data from static three-point bending and low blow impact testing of the Charpy V-notch specimen were used for these purposes. Relationships between the microstructural and/or fractographic features and the local fracture criteria were analysed.

1 INTRODUCTION

The origin of brittle fracture initiation in low alloyed structural steels is frequently associated with cleavage of ferritic (sub) grains. The nucleating event of macroscopic cleavage initiation results from the successive occurrence of several simple phenomena (Knott J.F. 1966, Lin T. et al 1987, Martín-Meizoso A. 1994): (i) slip induced cleavage separation of ferrite grain, (ii) carbide (brittle particle) cracking inside or at the grain boundary, and (iii) transmission of microcrack on the cleavage plane of the neighbouring grain across the (sub) grain boundary. Cleavage fracture stress (CFS) was introduced which is a local parameter characterising the microscopic resistance of the given microstructure against cleavage failure (Ritchie R.O. et al 1973). Based on this *deterministic* concept, brittle fracture initiation occurs when the local maximum principal stress, σ_1^{max}, exceeds the CFS. Macroscopically, the CFS was proven to be independent of temperature, hydrostatic stress and strain rate (Knott J.F. 1966, Holzmann M. et al 1996 etc.). Microscopically, the value of the CFS is strongly affected by microstructure (Smith E. 1966, Petch N.J. 1986) and the range of quantitative relations was introduced.

Almost all of the microstructural features and events preceding the fracture initiation have a *statistical* nature. Firstly, the causes can be seen in stochastic nature of cleavage separation itself (Lin T. 1987). Secondly, in the case of carbide controlled cleavage, the parameters of the particle distribution

directly govern the fracture micromechanism (Curry D. & Knott J.F. 1979, Strnadel 1995). Thirdly, some uncertainty in the exact description of real microstructures still exists, normally associated with random non-homogeneities in microstructural features (Chen J. H. et al. 1991). While in the deterministic concept, the initiation location is pre-determined by the site of local maximum stress, in the latter, the cleavage microcracks should form in a small volume located anywhere in the plastic zone ahead of crack tip. *Physically based* models may be developed enabling the fracture toughness to relate to the carbide thickness, other microstructural parameters distribution and/or different sized crack nuclei (Bowen P. 1986, Strnadel B. 1995) in changing stress gradients at the crack tip. *Statistical models* (e.g. Beremin F.M. 1983, Fontaine A. & Maas E. 1987, Wallin L. K. 1995) are usually based on Weibull stress, σ_w, that was accepted as a local fracture parameter for cleavage postulated to be independent of specimen geometry. The specimen or components loaded to the same value of σ_w have the same probability of cleavage fracture. The concept resulted in ability to predict the fracture toughness-temperature diagram. The continuum view in these approaches still cannot provide deeper physical insight into the fracture process, however.

The aim of this contribution can be seen in experimental verification of the physical nature of cleavage fracture stress and Weibull stress for two different microstructures. The susceptibility of local fracture criteria to steel microstructure should be also discussed.

2 EXPERIMENTAL PROCEDURES

2.1 *Material characteristics*

A CrMoV rotor steel has been utilized for experiments having chemical composition (in wt. %): 0.23 C, 0.64 Mn, 0.28 Si, 0.022 P, 0.028 S, 1.23 Cr, 0.55 Mo, 0.16 V. Two model microstructures have been selected for this investigation: (i) ferrite with fine carbides (FC, bainite tempered for 2 h at 680 °C) and (ii) ferrite with coarser carbides (CC, bainite tempered for 10 h at 720 °C). Some materials characteristics are shown in Tables 1 and 2. For a more detailed description see (Dlouhý I. 1996). In the table 1, the value of d_{AG} represents the prior austenitic grain size, d_p is the packet size, d_c is the mean value of the carbide size, while $d_c(c)$ falls into the 95 % percentile of carbide diameter distribution that was measured.

Table 1: Microstructural parameters of studied materials

State	Tempering	d_{AG} [μm]	d_p [μm]	d_c [μm]	$d_c(c)$ [μm]
FC	680°C/2h/AC	29	15	0.09	0.33
CC	720°C/10h/AC	29	12	0.17	0.47

2.2 *Mechanical testing*

Tensile properties and stress-strain curves have been determined using cylindrical specimens with diameter of 6 mm being loaded over temperature range of -196 to +200 °C at a cross-head speed of 2 mm.min^{-1}. The yield stress (YS) was taken to be the lower yield stress value (for FC) or the 0.2 % proof stress (CC).

Temperature dependencies of fracture toughness have been measured using precracked 25mm thick specimen loaded in three point bending (loading span 120 mm) at a cross head speed of 1 mm.min^{-1}.

Charpy type specimens were tested under the two loading rates: (i) dynamically, using an instrumented impact tester by low blow method (with a hammer drop rate of about 1.5 m.s^{-1}) and (ii) statically, by the three point bend method. For one selected temperature in the lower shelf region (below t_{GY} - see later) a range of bend tests were performed to obtain data for statistical treatment. The maximum force, F_M, fracture force, F_{FR}, and general yield force, F_{GY}, respectively, have been obtained from load - deflection traces.

Axisymmetrically notched tensile specimens with a diameter of 16 mm were also tested to generate data for local parameter calculation. The notch of Charpy type geometry was machined circumferencially, where the specimen diameter in notch root section was 8 mm.

In the table 2, the HV10 represent Vickers hardness. The tensile properties are represented by the yield stress, YS, and the ultimate tensile stress, UTS. The room temperature impact energy and the fracture appearance transition temperature, FATT, both measured on CVN specimen are here also introduced.

Table 2: Mechanical properties of investigated materials

State	HV10	YS [MPa]	UTS [MPa]	Impact energy [J]	FATT [°C]
FC	278	771	886	18	110
CC	200	511	644	51	50

2.3 *Fractography and supporting procedures*

For CVN specimens tested around and below the brittleness transition temperature, t_{GY}, the distances between the fracture initiation sites and the notch root were evaluated using SEM micrographs. To identify the microstructural causes of cleavage nucleation the triggering origins of facets in failure initiation sites were assessed. Microstructural parameters were determined by SEM, thin foils and carbide extraction replicas were examined by TEM analysis.

3 CALCULATION PROCEDURES

For Charpy type specimens and test temperature below t_{GY} the principal stress distributions below the notch were calculated. A 2D model under plane strain conditions was used for the elastic-plastic analysis using the symmetry of the specimen three point bending (Kozák V. & Malý J. 1993). The FEM package ANSYS 5.1 was used for calculations.

For the Charpy type specimen and for the notched tensile specimen, the Weibull stress has been evaluated using a procedure suggested by (Sainte-Catherine C. 1995). The flow behaviour was computed using incremental plasticity (with Von Misses criterion). The true stress - strain curves have been implemented. Weibull stresses, σ_w, and shape parameters, m, were determined by integration in the plastic zone ahead of the crack tip for the estimated values of characteristic cell size V_0. The scale parameters, σ_u, represent the value of σ_w corresponding to the 63.2 percentile on the cumulative distribution function of σ_w.

4 RESULTS DESCRIPTION

4.1 *Fracture behaviour of CVN and precracked bend specimen*

To characterise the material fracture behaviour the temperature dependencies of fracture toughness are introduced in Figure 1a, b. A specimen set was tested at one temperature in order to determine the scatter band of fracture toughness values. Data are also given in figures, commonly with the value of scatter parameter m'. Dashed lines represent the scatter band with a 5 and 95 % probability of all values in transition region. The comparison of both plots demonstrates that except for a shift of transition region toward lower temperatures, the scatter in fracture toughness decreases with an increase in average carbide diameter.

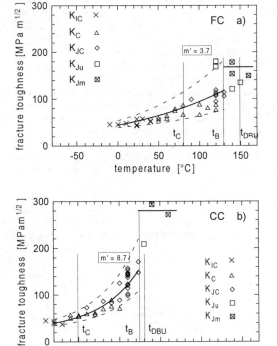

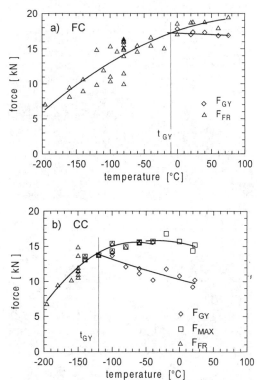

Figure 1. Fracture toughness - temperature diagrams for microstructure with fine carbides (FC - a)) and for ferrite with coarse carbides (CC – b))

In both cases, the center of the river pattern was assumed to be the cleavage initiation origin.

Figure 2. Temperature dependencies of characteristic forces obtained from load-displacement traces of CVN specimen loaded in static bending.

Temperature dependencies of the general yield force, F_{GY}, the fracture force, F_{FR}, and the maximum force F_{MAX}, obtained from statically loaded CVN specimens, are shown in Figures 2a, b. Similar dependencies were generated for specimens tested using an instrumented impact machine (Holzmann M. 1996). Two significant results can be drawn from a comparison of both plots: the lower brittleness transition temperature, t_{GY}, and lower general yield force, F_{GY}, at this temperature are observed for the microstructure with coarse carbides. The above mentioned findings are well supported by fractographic observations:

At lower magnification (Figures 3 and 4), the fracture surfaces of CVN bend specimen exhibited macroscopic river patterns. For the microstructure with fine carbides (FC) this pattern originates at one small spot in the fracture surface (see Figure 3) which contains one or a few cleavage facets. For microstructures with coarse carbides (CC) the river patterns are traced to the initiation region (Figure 4).

At a higher magnification (SEM), four types of initiation origins could be identified:
(i) One cleavage facet without a distinct triggering point, typical for FC microstructure. The average facet size was about 20 to 30 μm. Dislocation micromechanisms of cleavage were assumed.
(ii) Two or three facets arranged in initiation site, all facets without clear triggering point. Cleavage starting from grain boundaries or triple points was supposed.
(iii) Facets with one large or with a few smaller inclusions located on the middle of the facet.
(iv) Cleavage facet with carbide particles. Carbide mechanism of cleavage triggering was assumed in these cases that were essentially observed in heavy tempered microstructures (CC).

With an increase of carbide diameter, the substantial changes in cleavage micromechanism and macroscopic appearance of fracture surface were thus observed. The fracture appearance changes from morphology with single initiation site (for FC) to one with more multiple initiation sites (CC). Dislocation

1193

micromechanisms of cleavage triggering were assumed for microstructure with fine carbides, while almost carbide origins of cleavage were observed for microstructures with coarse carbides.

Figure 3. Fracture appearance of CVN specimens tested at - 80°C (a microstructure with fine carbides - FC)

Figure 4: Fracture appearance of CVN specimens tested at -150°C (a microstructure with coarse carbides - CC)

The fracture behaviour of FC microstructure corresponds to weakest link sites concept. Activating one weak link causes unstable fracture. Governing fracture mechanisms in CC microstructure fit well to concept of critical damage sites. Activating one critical damage site does not cause unstable cleavage failure; a critical number of these sites must be activated in order to obtain cleavage.

4.2 Cleavage fracture stress

To determine the cleavage fracture stress using experimental data two procedures were used:

(i) For CVN specimens loaded statically and by the low blow technique, the cleavage fracture stress was evaluated as a local maximum tensile stress ahead of the notch at fracture. At temperatures where fracture and general yield force were coincident (i.e. at brittleness transition temperatures t_{GY} – see figure 2)

the same procedure of local maximum tensile stress calculation for CVN specimen as used by (Brozzo P. 1977) was applied. As plastic stress concentration factor $k_{\sigma p}$ the value of 2,24 was taken. The relationship between yield strength and general yield force (static and/or dynamic) was measured. Arising from this - linear relationship between yield strength and general yield force the simplified approach could be used for direct calculation of σ_{CF} from general yield force at t_{GY}, thus, $\sigma_{CF} = 106.F_{GY}$.

(ii) For fracture forces of a range of CVN specimens tested statically in bending at one selected temperature the tensile stress distribution below the V notch has been calculated using FEM. The distances between the initiation origin and the notch root were measured from SEM micrographs. These distances have been applied to obtain the local tensile fracture stress from the maximum principal stress distribution corresponding to fracture. The results are shown in Figure 5. The mean values of cleavage fracture stress, determined according to procedure (i), are depicted as dashed lines in these figures.

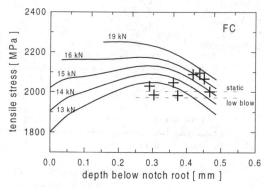

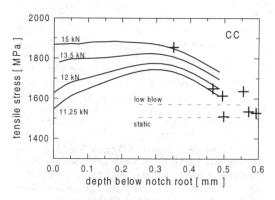

Figure 5. Principal tensile stress distribution below notch and cleavage fracture stresses obtained for CVN specimen by static and dynamic method

1194

For FC microstructure, a strong data correlation was obtained with both procedures used, and an independence of CFS on temperature and strain rate was proved. An estimation of local cleavage stress for the CC microstructure is associated with small uncertainty of localisation of initiation origin, although the values obtained for deepest distance of cleavage origin correspond well to data calculated from general yield force.

Good susceptibility of CFS to microstructural differences is the most important result from the previously described observation. From the current theories regarding cleavage fracture, only the equation derived by (Petch N.J. 1986) was found to correlate with our data for the CC microstructure.

4.3 Local fracture parameters determined using the statistic approach

Based on the experimental results generated on nine Charpy V-notch specimens for both the FC (tested at -80 °C) and the CC microstructures (tested at -150 °C) the local parameters, σ_u and m, were calculated. The results, including dependencies of cumulative probability on Weibull stress σ_w, have been analysed elsewhere (Kozák 1996). For purposes of this paper a summary illustrating the influence of V_0 on scale parameter σ_u and, at the same time, the comparison of values obtained from both tensile and bend specimen a diagram is presented in Figure 6.

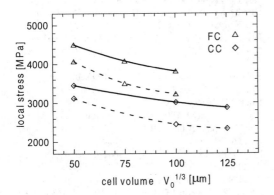

Figure 6. The dependence of Weibull stress on cell size for the notched tensile (dashed line) and CVN bend specimen (full)

Allowing for the variation of Weibull stress, σ_u, with cell size, V_0, and the shape of cumulative probability curve, the following findings could be summarised:

FC microstructure exhibits higher values of σ_u compared to the CC one, this fact is in consistency with the values of cleavage fracture stresses σ_{CF}. The values of Weibull stress, σ_u, seem to be more susceptible to the fracture micromechanism than global fracture toughness characteristics.

The scale parameter, stress σ_u, is a strong function of selected cell size V_0. The decrease of σ_u with increasing V_0 is approximately the same for both microstructures. In order to accurately assess microstructures containing defects the value of V_0 must be carefully chosen, with respect to the mechanisms controlling the cleavage nucleation and the microcrack transmission within the elementary cell. For the FC microstructure, the dimension of cell $(V_0)^{1/3} \approx 50$ µm corresponding to the size of two to three cleavage facets is assumed to involve just this controlling fracture initiation micromechanism. For the CC microstructure, according to fractographic observation, the dimension of cell size $(V_0)^{1/3} \approx 100$ µm covering up to ten initiating cleavage sites seems to be more appropriate.

Comparing the shape parameters, m, the ferrite with coarse carbides exhibits higher values of these parameters corresponding to the smaller scatter. At lower Weibull stress this means that the probability of cleavage failure initiation in the elementary cells is higher than in the cells of microstructure with fine carbides. This probability enhancement should be associated with high carbide thickness and consequently the carbide-induced cleavage.

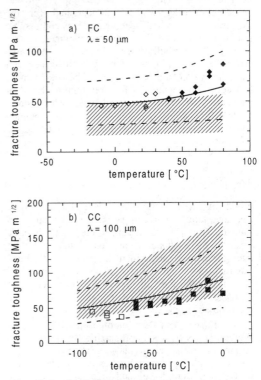

Figure 7. Predictions of fracture toughness - temperature dependencies calculated for right values of cell size (designated by full and dashed curves) and for evidently wrong values of $(V_0)^{1/3}$ (designated by shadowed area)

In order to test the reliability of determined local parameters, the comparison of predicted fracture toughness - temperature diagrams (including scatter bands) with those obtained experimentally have been carried out. The procedure used for calculations has been given elsewhere (Dlouhý I. et al 1996). The final results are summarized in Fig 7. For microstructure with fine carbides (FC) the prediction calculated for cell size $(V_0)^{1/3} \approx 50$ μm seems to be the quite reasonable. As mentioned above, this is the value which corresponds well to average dimension of cleavage facet(s) initiating the fracture. The scatter band (dashed lines) for 5 - 95 % probability is slightly wider than that designated in Fig. 1a). On the contrary, the prediction made for $(V_0)^{1/3} \approx 100$ μm has provided unacceptable data (represented by shadowed area in Fig, 7a). For microstructures with coarse carbides (CC) similar findings could be concluded. In this case, the prediction calculated for cell size $(V_0)^{1/3} \approx 100$ μm corresponds better with experimental values than the calculation for $(V_0)^{1/3} \approx 50$ μm. Also in this case a more accurate prediction can be obtained for cell size corresponding well to real fracture initiation mechanism.

The validity of local fracture parameters determined by using the CVN bend specimen was tested comparing with data generated for the axisymmetric notched specimens. For comparable test conditions and calculation parameters, the values of Weibull stress have shown some differences, the decrease of σ_u with increasing cell size is fully comparable however. The differences between both geometries for determining the local parameters could be associated with a sharp stress gradient below the V notch of the bend specimen when compared with more homogenous stress distribution of tensile specimen.

5 CONCLUSION

Based on deterministic approach to the analysis of cleavage failure and comparing this concept with statistic approach of Beremin the local fracture criteria have been evaluated using fracture micromechanism assessment combined with FEM calculations. For the two microstructures investigated the Weibull stress values were related to cleavage fracture stresses determined deterministically.

The ferrite with fine carbides exhibits higher Weibull stress σ_u when compared to the ferrite with coarse carbides, this is in consistency with the relation of cleavage fracture stresses σ_{CF}. This stress is a strong function of cell size V_0 chosen. The decrease in σ_u with increasing V_0 is approximately the same for both microstructures. Taking into account the values of σ_u, and comparing the Weibull

shape parameter the microstructure with coarse carbides exhibits higher values of this parameter corresponding to the smaller scatter and to the higher probability of cleavage failure initiation in the elementary cells than the microstructure with fine carbides. This increase of cleavage probability in cell volume strongly corresponds with carbide induced cleavage due to higher carbide thickness. The fracture toughness prediction calculated for cell sizes close to the measured size of observed initiation sites has provided that are fully comparable with experimental values. From the comparison of local parameters determined using the notched tensile specimen with those obtained using CVN bend specimen, certain differences occured, although basic quantitative relation between model microstructures remained the same.

Acknowledgments

Support to this research by Grant Agency of the Czech Republic under grant number 106/96/0821 and 101/96/K264 is gratefully acknowledged.

References

Beremin F. M. 1983. *Metall. Trans.* 14A: 2277-2287.
Bowen P. 1986, *Acta Metal.* 34: 677-684.
Brozzo P. 1977. *Metal Science.* 11: 123-128.
Chen J.H. Wang G.Z., Wang Z., Zhu L. & Gao Y.Y. 1991. *Met. Trans.* 22A: 2287-2296.
Curry D., Knott J.F. 1979. *Met. Science.* 13: 341-345.
Dlouhý I., Kozák V & M. Holzmann 1996. *Journal de Physique III.* 6: C6-205-214.
Fontaine A., Maas E. 1987. In *The Fracture Mechanics of Welds*, EGF Pub. 2: 43-58., J.G. Blauel & K.H. Schwalbe.
Holzmann M., Dlouhý I., Válka L. 1996. In *Proc. of the ECF 11: 675-680.*
Knott J.F. 1966. *J. Iron & Steel Inst.* 204: 104-111.
Kozák V., Malý J. 1993. *Engng. Mechanics* 3: 46 -52.
Lin T., Evans A.G. & Ritchie R.O. 1987. *Met. Trans.* 18A: 641-651.
Martín-Meizoso A. 1994. *Acta Metall. Mater.* 42: 2057-2068.
Petch N.J. 1986 *Acta Metal.* 34: 1387-1394.
Ritchie R.O., Knott J.F. & Rice J.R. 1973. *J. Mech. Phys. Solids* 21: 395-410.
Sainte-Catherine C. 1995: Draft procedure to measure and calculate local fracture criteria on notched tensile specimens, *Meeting ESIS TC1*, Appendix 1.
Smith E. 1966. In *Proc. of Conf. on Physical Basis of Yield and Fracture*: 36-45. Inst. of Physics, Oxford.
Strnadel B. 1995. Modelling of Fracture Events in Structural Steels. *Thesis for Doctor of Science Degree*. Technical University of Ostrava.
Válka L. Holzmann M & Dlouhý I. 1997 *Materials Science & Engineering*. A234-236: 723-726.
Wallin L K. 1995. Optimized analysis of brittle fracture toughness data, *Meeting ESIS TC1*, P10.

Mixed mode fracture

Experimental Mechanics, Allison (ed.) © *1998 Balkema, Rotterdam, ISBN 90 5809 014 0*

Griffith criterion for mode II fracture of ceramics

Hideo Awaji & Toshiya Kato
Nagoya Institute of Technology, Japan

ABSTRACT: The Griffith energy criterion for brittle fracture is extended to mode II fracture on the postulation that crack extension occurs when the maximum energy release rate in non-coplanar crack extension is equal to the fracture energy rate required for mode II crack extension. This fracture energy rate is estimated from the magnitude of the area of the frontal process zone at the crack tip. It is seen that the estimated K_{IIC}/K_{IC} ratio is 1.21 for quite brittle materials. New experimental measurements on soda-lime float glass have been performed to measure values of mode I and II fracture toughness using the circular disk technique. The results obtained for the K_{IIC}/K_{IC} ratio for float glass is 1.27 which is approximately equal to the theoretically calculated value.

1 INTRODUCTION

While a fluid withstands only normal stress, a solid body has resistance against both, normal stress and shear stress. A crack in a solid under multiple stress state then yields to a mode I stress intensity factor which is caused by tensile stress, and mode II and mode III stress intensity factors caused by shear stress. Therefore, three critical values of these stress intensity factors are defined to express the crack propagation resistance of solid materials, i.e. mode I, mode II and mode III fracture toughness correspond-ing to each mode, and all of these values might be mechanical properties of materials as far as the brittle crack propagation is concerned.

Erdogan and Sih (1963) in their pioneering work of combined mode fracture have pointed out the following three questions:

(a) Do or can mode II and mode III crack extension take place in actual structures ?

(b) Are K_{IIC} and K_{IIIC} also mechanical properties of the material ?

(c) What is the criterion for fracture if the material is subjected to a combination of various simple loading ?

In spite of much research on combined mode fracture mechanics, no conclusive answers have been found for the questions (b) and (c).

A number of testing techniques have been proposed and used on a wide variety of materials to estimate mode I and mode II fracture toughness. The previous work by the present authors (Awaji et al. 1978, 1979) found that the ratio of K_{IIC}/K_{IC}, using the

disk test, was *1.1* to *1.3*, almost independent of the materials tested. However, the mode II fracture toughness for ceramics, K_{IIC}, has been found to vary between *0.5* K_{IC} and *2* K_{IC} with various testing configurations. For example, Li and Sakai(1996) reported a value for K_{IIC}/K_{IC} ratio of *0.53* for soda-lime glass, while Tikare's result (1997) was quite different *1.05* for ceria-doped tetragonal zirconia polycrystalline, even though they used the same technique of asymmetric four-point bending. Shetty and co-workers (Shetty et al. 1985, Singh & Shetty 1989) used the disk test for several ceramic materials including their new techniques such as chevron notch or Knoop indentation flaw in the disk specimen, but they suspected that the critical value of K_{II} for pure mode II loading is not a unique material property (Shetty 1987).

In this paper, the discussion will be limited to inorganic materials which are able to emit only very few dislocations at a crack tip. The Griffith energy criterion for brittle fracture which was originally formulated for mode I cracking, is extended for mode II cracking on the supposition that crack extension occurs when the maximum energy release rate in non-coplanar crack extension is equal to the fracture energy rate required for mode II crack propagation. The fracture energy rates of mode I and mode II crack extension are estimated from the magnitude of the areas of the process zone ahead of a crack tip. The process zone areas are calculated as a primary approximation from the linear elastic stress fields in front of the crack tip. The ratio of K_{IIC}/K_{IC} is then obtained from the fracture energy rates of a crack subjected to mode I and mode II loading.

The values of the mode I and mode II fracture toughness are also obtained experimentally for a soda-lime float glass using the disk test technique, and the ratio of K_{IIC}/K_{IC} is compared to the theoretically expected value. The soda-lime float glass used here is quite brittle and has an isotropic homogeneous microstructure, and is expected to have an extremely small process zone at the crack tip. Also negligibly small effects of crack-face bridging on the mode II fracture toughness measurements are expected. Linear elastic fracture approximation, therefore, is successfully applicable for this material.

2 THEORY

2.1 Mode I/II fracture criteria

The maximum hoop stress theory (Erdogan & Sih 1963) and the maximum energy release rate theory (Hussain 1975) are widely used as relevant criteria for combined mode fracture. The maximum hoop stress theory states that a crack will propagate when the maximum hoop stress at the crack tip reaches a critical strength of the material considered. The hoop stress at the crack tip under combined mode I/II loading is expressed as

$$\sigma_\theta = \frac{1}{\sqrt{2\pi r}} \cos\frac{\theta}{2}(K_I \cos^2\frac{\theta}{2} - \frac{3}{2}K_{II}\sin\theta), \quad (1)$$

and the maximum hoop stress exists in the following direction

$$\theta_m = \mp\cos^{-1}(\frac{3K_{II}^2 + K_I\sqrt{8K_{II}^2 + K_I^2}}{9K_{II}^2 + K_I^2}), \quad (2)$$

where, r, θ: coordinates at the crack tip shown in Fig.1, K_I, K_{II}: stress intensity factors of mode I and mode II, respectively, θ_m: fracture angle at the onset of crack extension. From eqs.(1) and (2), the ratio of mode II and mode I fracture toughness estimated by the maximum hoop stress theory is derived as

$$\frac{K_{IIC}}{K_{IC}} = 0.87. \quad (3)$$

The maximum energy release rate in non-coplanar crack extension under combined mode loading was investigated analytically by many researchers (Hussain et al. 1975, Nuismer 1975, Wang 1977, Ichikawa & Tanaka 1982, Kageyama & Okamura 1982, Wang 1985). After much debate, the following Ichikawa's equation (1987) has been proposed to estimate the maximum energy release rate for a

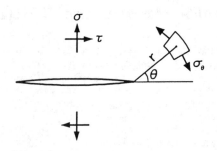

Fig.1 Hoop stress around a crack tip under combined mode.

kinked crack

$$G_{max} = \frac{1}{2E'}(K_I^2 + 3K_{II}^2 + K_I\sqrt{K_I^2 + 6K_{II}^2}). \quad (4)$$

$$E' = \begin{cases} E & \text{for plane stress} \\ E/(1-v^2) & \text{for plane strain} \end{cases}$$

Under pure mode I loading, crack extension is coplanar (along $\theta_m=0$), and the maximum energy release rate is expressed by the following well-known equation

$$G_{max} = \frac{K_I^2}{E'}. \quad (5)$$

The maximum energy release rate for pure mode II corresponds to a non-coplanar crack extension with the kinked angle $\theta_m=-76°$, and is expressed as

$$G_{max} = \frac{3K_{II}^2}{2E'}. \quad (6)$$

If these values of G_{max} are equal at the onset of crack extension, the following ratio of mode II and mode I fracture toughness is obtained

$$\frac{K_{IIC}}{K_{IC}} = 0.82. \quad (7)$$

Equations (3) and (7) give almost similar predictions of the ratio of K_{IIC}/K_{IC}, namely, the value of mode II fracture toughness, K_{IIC}, is 0.87 or 0.82 times the mode I fracture toughness, K_{IC}. The crucial oversights in these theories are, however, that a only linear elastic stress fields are postulated even in the vicinity of the crack tip, and that the energy balance between the strain energy release and the fracture energy are not considered. The idea of energy balance for the crack propagation problem was proposed in the early work of Griffith (1921) for mode I crack extension.

2.2 Griffith criterion for mode II fracture

The Griffith energy criterion for brittle fracture states that crack growth can occur if the energy required to form an additional crack of a size da can be delivered by the system, (Broek 1982). Accordingly, when the energy release rate, $G = dU/da$ (U: elastic strain energy), is equal to the fracture energy rate required for the crack growth, $R = dW/da$ (W: fracture energy), crack growth can occur. For mode I loading, the fracture energy rate is expressed as

$$R = 2\gamma_I, \tag{8}$$

where γ_I is the fracture energy per unit area when the crack extends on its own direction. This fracture energy includes the surface energy of the material and also the energy to create a frontal process zone ahead of the crack tip. The Griffith energy criterion for mode I crack extension then yields to the following relation derived from eqs.(5) and (8)

$$\frac{K_{IC}^2}{E'} = 2\gamma_I, \tag{9}$$

To extend the Griffith criterion to mode II crack growth, it is necessary to know the fracture energy per unit area for the kinked crack growth under mode II loading, γ_{II}, where the fracture energy rate for mode II fracture is expressed as follows

$$R = 2\gamma_{II}. \tag{10}$$

If we know the fracture energy per unit area, the energy balance for mode II fracture yields to the following relation derived from eqs.(6) and (10)

$$\frac{3K_{IIC}^2}{2E'} = 2\gamma_{II}. \tag{11}$$

Hence, by considering eqs.(9) and (11), the ratio of mode II and mode I fracture toughness is expressed as

$$\frac{K_{IIC}}{K_{IC}} = \sqrt{\frac{2\gamma_{II}}{3\gamma_I}}. \tag{12}$$

Ceramic materials generally have an icnic and/or a covalent bonding and only a few dislocations are emitted from the crack tip. The frontal process zone ahead of a crack tip in ceramics is therefore considered to be formed by many microcracks (Evans et al.1985). If the fracture energy is equal to the energy required to form these microcracks, the fracture energy may be proportional to the size of the process zone area formed in front of the crack tip.

2.3 Process zone area

The frontal process zone area at the crack tip is simply estimated here from the area enclosed in the iso-stress contours of both the maximum principal stress and the maximum shear stress. The maximum principal stress, σ_1, and the maximum shear stress, τ_{max}, at the crack tip are expressed as

for mode I loading

$$\sigma_1 = \frac{K_I}{\sqrt{2\pi r}}(\cos\frac{\theta}{2} + |\sin\frac{\theta}{2}\cos\frac{\theta}{2}|), \tag{13}$$

$$\tau_{max} = \frac{K_I}{\sqrt{2\pi r}}|\sin\frac{\theta}{2}\cos\frac{\theta}{2}|, \tag{14}$$

and for mode II loading

$$\sigma_1 = \frac{K_{II}}{\sqrt{2\pi r}}(-\sin\frac{\theta}{2} + \frac{1}{2}\sqrt{1+3\cos^2\theta}), \tag{15}$$

$$\tau_{max} = \frac{K_{II}}{2\sqrt{2\pi r}}\sqrt{1+3\cos^2\theta}. \tag{16}$$

The radii of the iso-stress contours are derived from eqs.(13) to (16)

for mode I loading

$$r_1 = \frac{K_I^2}{2\pi\sigma_1^2}\cos^2\frac{\theta}{2}(1+|\sin\frac{\theta}{2}|)^2, \tag{17}$$

$$r_2 = \frac{1}{2\pi}\frac{K_I^2}{\tau_{max}^2}\sin^2\frac{\theta}{2}\cos^2\frac{\theta}{2}, \tag{18}$$

and for mode II loading

$$r_1 = \frac{K_{II}^2}{2\pi\sigma_1^2}(-\sin\frac{\theta}{2} + \frac{1}{2}\sqrt{1+3\cos^2\theta})^2, \tag{19}$$

$$r_2 = \frac{1}{2\pi}\frac{K_{II}^2}{4\tau_{max}^2}|1+3\cos^2\theta|. \tag{20}$$

Where r_1, r_2 are the radii of the iso-stress contours for the maximum principal stress and the maximum shear stress, respectively. Figure 2 shows these contours of the maximum principal stress and the maximum shear stress under mode I loading, and Fig. 3 those under mode II loading. Comparing Figs. 2 and 3, the extent of the contours of maximum shear stress are smaller than that of maximum principal stress. Also the contours under mode I loading are smaller than that under mode II loading.

Calculating the area enclosed with the contours numerically, the ratio of fracture energy, γ_{II}/γ_I, was found to be estimated as

$$\frac{\gamma_{II}}{\gamma_I} = 2.18. \tag{21}$$

Hence, from eqs.(12) and (21), the ratio of mode II and mode I fracture toughness is obtained as

$$\frac{K_{IIC}}{K_{IC}} = 1.21. \tag{22}$$

This value of 1.21 is very close to our experimental results of 1.05 to 1.3 (Awaji et al. 1978, Awaji et al. 1979) for several inorganic materials presented previously, and new experimental results for float glass which will be explained in the next chapter.

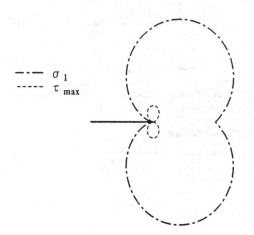

Fig.2 Iso-stress contours of maximum principal stress, σ_1, and maximum shear stress, τ_{max}, under mode I loading.

3 EXPERIMENT

3.1 *Specimen*

To confirm the theoretical prediction of the ratio of K_{IIC}/K_{IC} value mentioned above, we examined the values of the mode I and the mode II fracture toughness of soda-lime float glass.

A commercially used float glass plate with a thickness of 2 mm was cut into circular disk shaped specimens with 40 mm in diameter. The specimen configuration is shown in Fig.4. The central artificial crack was made by machining a slit with chevron notches at the tips using a thin round saw of 13 mm diameter of 0.2 mm thickness. The cracks were then extended by thermal shock using the soldering copper technique so that a crack length ratio, c/R, of around 0.4 was reached, where c: half crack length and R: radius of the disk specimens. Red ink with aceton was allowed to permeate through the crack during the heat treatment in order to watch the crack tip.

3.2 *Mode I/II fracture toughness*

Awaji et al. (1971, 1978) proposed a diametral compression method (disk test) for determining the mode I and mode II fracture toughness and the combined mode I/II fracture resistance of several kinds of inorganic materials such as graphite, rock and others. The disk specimens shown in Fig.4 have an inclined through-thickness central notch and are loaded in diametral compression. This test geometry has the advantage that all stress states ranging from pure mode I to pure mode II can be obtained in these disk shaped specimens with a simple alignment of the crack, therefore, the examination can be conducted under the same condition. The stress intensity factors for a central crack in disk shaped specimens are reported before (Awaji et al. 1978).

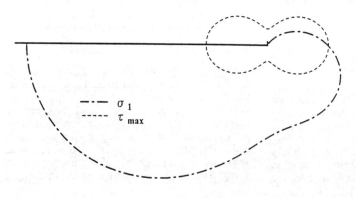

Fig.3 Iso-stress contours of maximum principal stress, σ_1, and maximum shear stress, τ_{max}, under mode II loading.

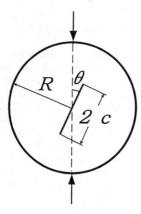

Fig.4 The disk test for combined mode fracture toughness.

4 RESULTS AND DISCUSSIONS

4.1 *Results*

The values for mode I, the mode II fracture tough-ness and the combined mode fracture resistance for the soda-lime float glass are shown in Fig.5, where the vertical axis is the mode I fracture toughness, and horizontal axis the mode II fracture toughness. Both axes are normalized by K_{IC}. The mean values and the standard deviations of mode I and mode II fracture toughness are $0.688 \pm 0.012\ MPam^{1/2}$, and $0.875 \pm 0.044\ MPam^{1/2}$, respectively, and the ratio of K_{IIC}/K_{IC} calculated from these values is 1.27 which is very close to the theoretically predicted value shown in eq.(22).

4.2 *Discussion*

The experimental results are better described by the new Griffith criterion for mode II cracking rather than by the maximum hoop stress theory or the maximum energy release rate theory. There is another theory called the minimum strain energy density criterion proposed by Sih (1974) who considered that crack propagation would occur in the direction of minimum strain energy density, and the criterion leads to the following relation for mode II and mode I fracture toughness,

$$\frac{K_{IIC}}{K_{IC}} = \sqrt{\frac{3(1-2v)}{2-2v-v^2}}, \qquad (23)$$

where v: Poisson ratio. Equation (23) indicates that the value of K_{IIC}/K_{IC} varies with the Poisson ratio. Tikare et al. (1997) reported that the minimum strain energy density criterion best describes the combined mode fracture behavior of ceramics. However, our experimental results presented previously (Awaji et al.1979) for materials with different Poisson ratios in the range of 0.07 to 0.33 showed significant differences from the values calculated by using eq.(23).

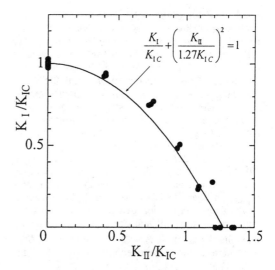

$$\frac{K_I}{K_{IC}} + \left(\frac{K_{II}}{1.27K_{IC}}\right)^2 = 1$$

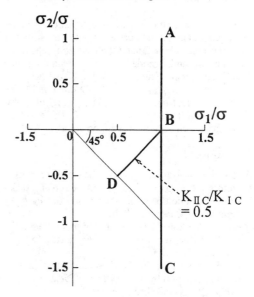

Fig.5 Normalized mode I and mode II fracture toughness envelope for soda-lime float glass.

Fig.6 Macroscopic fracture criterion under biaxial stresses in the region where the tensile stress rules predominate.

The experimentally observed values of K_{IIC}/K_{IC} were ranging from 1.11(for $v =0. 07$) to 1.13(for $v =0. 33$) instead of 1.18 to 0.91 which are predicted by eq.(23). The crucial oversights in this theory are the same as those in the maximum hoop stress theory and in the maximum energy release rate theory which were pointed out in the earlier part.

The next concern is the magnitude of the ratio of K_{IIC}/K_{IC}. The important question is whether the ratio of K_{IIC}/K_{IC} should be less than 1 or not. A criterion of macroscopic brittle fracture may suggest the answer.

Such a fracture criterion for loading under bi-axial stress state is shown in Fig.6 (Awaji 1980). It is well known that the maximum principal stress theory shown by the line A-B-C in the figure, is applicable to the stress region where tensile stress rules predominate, where σ_1: maximum principal stress, σ_2: minimum principal stress and σ: uniaxial tensile strength. The third principal stress is assumed to be ignorable value in each quadrant. The stress state at point A is ($\sigma_1/\sigma = 1$, $\sigma_2/\sigma = 1$), at point B is ($\sigma_1/\sigma = 1$, $\sigma_2 = 0$), and at point C is ($\sigma_1/\sigma = 1$, $\sigma_2/\sigma = -1$). Much brittle materials follow line A-B-C(Stout et al. 1984). If a material with $K_{IIC}/K_{IC} = 0.5$, for example, is assumed to exist and a crack in the material propagates in catastrophic manner, the fracture criterion in a tension-compression quadrant must follow the line B-D instead of line B-C, because that the maximum shear stress is $\tau_{max} = 0.5$ on the line B-D. But there is no experimental evidence for this behavior.

5 CONCLUSIONS

The Griffith criterion is extended to mode II crack propagation, and the fracture energy rate is estimated from the area of the frontal process zone ahead of the crack tip under mode II loading. This consideration leads to a predicted ratio of K_{IIC}/K_{IC} of 1.21.

The fracture toughness values of mode I and II for a soda-lime float glass are estimated experimentally using a disk test. The results of K_{IIC}/K_{IC} value is 1.27 which is very close to the predicted value.

REFERENCES

Awaji, H. & A. Kamei 1971, On the stress intensity factors at the tip of the crack in a circular disk subjected to compression, *J. Jpn. Soc. Strength & Fracture of Mater.*, 6:100-108.

Awaji, H. & S.Sato 1978, Combined mode fracture toughness measurement by the disk test, *J.Engng. Mater. & Tech.*, 100:175-182.

Awaji, H. & S.Sato 1979, Study of combined mode fracture criterion by the disk test, *J.Soc.Mater.Sic. Japan*, 28: 244-50,(in Japanese).

Awaji, H. 1980, Macroscopic brittle fracture criterion under multiaxial stresses, *J. Engng. Mater. & Tech.*, 102: 257-263.

Broek, D. 1982, *Elementary engineering fracture mechanics*, Third revised ed., Martinus Nijhoff Publishers.

Erdogan, F. & G.C.Sih 1963, On the crack extension in plates under plane loading and transverse shear, *J.Basic Engng.*,85: 519-527.

Evans, A.G. & Y. Fu 1985, Some effects of micro-cracks on the mechanical properties of brittle solids, II. Microcrac toughening, *Acta metall.*, 33:1525-31.

Griffith, A.A. 1921, The phenomena of rupture and flow in solids, *Phil.Trans.Roy.Soc. of London*, A221: 163-97.

Hussain, M.A., S.L.Pu, & J.Underwood 1975, Strain energy release rate for a crack under combined mode I and mode II, *Int.J. Fracture*, 11: 245-50.

Kageyama, K. & H. Okamura 1982, *Trans. Japan Soc. Mech. Engng.*, 48:783-91,(in Japanese).

Ichikawa, M. 1987, A note on mixed mode energy release rate, *Engng. Fracture Mech.*, 26: 311-12.

Ichikawa, M. & S.Tanaka 1982, A critical analysis of the relationship between the energy release rate and the stress intensity factors for non-coplanar crack extension under combined mode loading, *Int.J.Fracture*, 18[1]:19-28.

Li, M. & M. Sakai 1996, Mixed-mode fracture of ceramics in asymmetric four-point bending: Effect of crack-face grain interlocking/bridging, *J.Am. Ceram.Soc.*, 79: 2718-2726.

Nuismer, R.J. 1975, An energy release rate criteria for mixed mode fracture, *Int.J.Fracture*,11:245-250.

Shetty, D.K., A.R. Rosenfield & W.H. Duckworth, 1985 , Fracture toughness of ceramics measured by a chevron-notch diametral-compression test, *J.Am.Ceram.Soc.* 68[12]: C325-27.

Shetty, D.K. 1987, Mixed-mode fracture criteria for reliability analysis and design with structural ceramics, *J. Engng. Gas Turbines & Power*, 109: 282-289.

Sih, G.C. 1974, Strain-energy-density factor applied to mixed mode crack problems, *Int J. Fract.* 10: 305-321.

Singh, D. & D.K.Shetty 1989, Fracture toughness of polycrystalline ceramics in combined mode I and mode II loading, *J.Am.Ceram.Soc.*, 72:78-84.

Stout, M.G. & J.J. Petrovic 1984, multiaxial loading fracture of Al_2O_3 tubes: I, Experiments, *J. Am. Ceram. Soc.* 67:14-23.

Suresh, S., C.F.Shih, A.Morrone and N.P.O'Dowd, 1990, Mixed-mode fracture toughness of ceramic materials, *J.Am.Ceram.Soc.*,73[5]:1257-67.

Tikare, V. & S.R.Choi, Combined mode I-mode II fracture of 12-mol%-ceria-doped tetragonal zirconia polycrystalline ceramics, *J. Am. Ceram. Soc.*, 80:1624-1626.

Wang, T. C. 1977, Fracture criteria for combined mode cracks, *Fracture 1977*, Vol.4:135-54, University of Waterloo, Canada.

Wang, M. H. 1985, A theory for the mixed energy release rate, *Engng. Fracture Mech.*, 22:661-671.

Experimental Mechanics, Allison (ed.)© 1998 Balkema, Rotterdam, ISBN 90 5809 014 0

Fracture of a brittle material from cracks and sharp notches under mixed-mode conditions

T.H.Hyde & A.Yaghi
Department of Mechanical Engineering, University of Nottingham, UK

ABSTRACT: Three point bend (3PB) and compact-mixed-mode (CMM) specimens with real cracks and very sharp notches have been manufactured from Araldite CT200 with HT907 hardener. The plane strain fracture toughness, K_{IC}, and apparent plane strain fracture toughness, K_{IC}^{App}, have been obtained from the cracked and notched specimens, respectively. The combinations of mode-I and mode-II stress intensity factors, K_I and K_{II} which cause fracture in the cracked specimens and the combination of K_I^{App} and K_{II}^{App} which cause fracture in the notched specimens have been obtained from the CMM specimen tests.

The mixed-mode fracture properties (magnitude of K_{If}^{App} and K_{IIf}^{App} together with the crack propagation angle) are consistent with published data for a similar material. A method of predicting the mixed-mode fracture behaviour on the basis of the K_{IC}^{App} value, for a component with a sharp notch, is proposed. Comparison between experimental results and predictions indicate that the method is reasonably accurate.

1   INTRODUCTION

During inspection of engineering components and structures, cracks and/or crack-like flaws may be detected. If the material from which the component or structure is made is brittle, a linear elastic fracture mechanics approach would be appropriate for a safety assessment. The work reported in this paper is related to two aspects of such an assessment. Firstly, the effect of mixed-mode conditions on fracture behaviour is assessed. Secondly, the fracture behaviour of sharp notches (crack-like flaws) under mixed-mode conditions is compared with that of real cracks.

2.   SPECIMENS AND TEST CONDITIONS

An epoxy resin, Araldite CT-200 with HT-907 hardener, was chosen for the investigation because of its brittle nature. The material is often used for the manufacture of photoelastic models (eg. Fessler 1989, Hyde 1993) and hence the procedures for casting, curing etc. to minimise inhomogeneity, cast-to-cast variations and residual stresses have been established. The pure mode-I fracture behaviour was investigated using a standard (BS5447) three-point-bend (3PB) specimen, Fig. 1(a), and for the mixed-mode (I and II) fracture behaviour a compact-mixed-mode (CMM) specimen, Fig. 1(b), was used. The relationship between the mode-I and mode-II stress-intensity factors and the CMM specimen dimensions, the load and the loading angle, α, were obtained using the

results of finite element analyses (Hyde and Chambers 1988).

Sharp notches, with four different widths, were created in the specimens (3PB and CMM) by incorporating 0.05 mm, 0.1 mm and 0.2 mm thick shims and a sharpened 0.1 mm thick shim in the moulds used to cast the specimens. For the specimens with real cracks, chevron notches (see Fig. 1) were incorporated in the moulds used to cast the specimens; details of the manufacturing method have been published (Hyde and Yaghi 1995). These chevron notched specimens were subjected to cyclic loading to create fatigue cracks from the chevron notches; the loads used to produce the cracks were significantly lower than those used in the subsequent fracture tests. The loading conditions were such that the fatigue cracks were produced under mode-I conditions. Crack front curvature was checked after each test and the results for those tests for which the crack front curvature did not comply with BS5447 requirements were rejected.

The shapes of the notch tips of the crack-like flaws were viewed in a microscope and found to vary significantly; photographs of typical notch tip shapes have been published (Hyde and Yaghi 1995). Typically, the notches produced by the sharpened 0.1 mm shims had an effective width of 0.02 mm and they either had semi-circular notch tips or they were square-ended with corner radii ranging between 0.01 and 0.004 mm approximately. The 0.05 mm, 0.10 mm and 0.20 mm wide notches were square-ended with corner radii in the range 0.02 to 0.004 mm; see Fig. 2.

3. RESULTS

3.1 Mixed-mode fracture behaviour from cracks

In a previous paper (Hyde and Yaghi 1995) , the mode-I fracture toughness, K_{IC}, for the material was reported. From 63 three-point-bend tests, the mean K_{IC} value was found to be 24.5 $Nmm^{-3/2}$, with standard deviation and 95% confidence limits of 1.59 $Nmm^{-3/2}$ and 0.41 $Nmm^{-3/2}$, respectively; the results ranged from 21.5 to 27.7 $Nmm^{-3/2}$. Nine further tests, on CMM specimens, with fatigue induced cracks, were performed with a range of K_I/K_{II} ratios. It should be noted that the CMM specimen is not appropriate for mode-I, or near mode-I testing, due to the difficulty of controlling the loading arrangement to produce pure or near pure mode-I stress field without the involvement of undesirable mode-II loading.

The combinations of K_I and K_{II} at fracture, ie. K_{If} and K_{IIf}, have been normalised with respect to the K_{IC} value ($= 24.5$ $Nmm^{-3/2}$) and are included in Fig. 3. Also shown in Fig. 3 are the results of the 63 three-point-bend tests for pure mode-I. The curve drawn through the data points (obtained from the present investigation) in Fig. 3 is a quadratic fit. It can be seen, from Fig. 3, that the mixed-mode data fall within a scatter band similar to that obtained with the three-point-bend tests, ie. $\pm$ 3 $Nmm^{-3/2}$. The normalised data, ie. K_{If}/K_{IC} versus K_{IIf}/K_{IC}, obtained for a similar material (Maccagno and Knott 1989) are compared with the present results in Fig. 3. This comparison shows good agreement between the present data and other published data. The results indicate that K_{IIC} is smaller than K_{IC} (ie. $K_{IIC} \approx 0.9 K_{IC}$). Also, for $K_{II}/K_I < 0.4$ the K_{If} values are close to K_{IC}.

The direction of crack propagation, θ_p, is plotted against $K_I/(K_I + K_{II})$ in Fig. 4(a). It can be seen that the crack propagation angle varies in a systematic way. An alternative method of presenting the crack propagation direction data, ie. θ_p versus tan^{-1} (K_I/K_{II}), is shown in Fig. 4(b), where tan^{-1} (K_I/K_{II}) is termed the equivalent crack angle. Also shown in Fig. 4(b) are the results obtained by Maccagno and Knott (1989) and predictions of crack propagation direction based on the maximum tangential tensile stress (TTS), $\hat{\sigma}_{\theta\theta}$, the maximum energy release rate, G, and the minimum strain energy density, Š, criteria. It can be seen that the present results correlate closely with previous results. Also, the $\hat{\sigma}_{\theta\theta}$ and G criteria both fit the data reasonably well but the prediction based on the Š criterion is relatively poor, particularly near pure mode-II conditions. Maccagno and Knott (1989) conclude that the mixed-mode behaviour is described best by a maximum tangential stress, $\hat{\sigma}_{\theta\theta}$, criterion based on the linear elastic stress field and the present results reinforce this conclusion.

3.2 Mixed-mode fracture behaviour from sharp (crack-like) notches

Four different types of sharp notches were created in 3PB and CMM specimens by incorporating shims in the moulds used to cast the specimens. The typical shapes of these notches, established from photographs taken through a microscope, are described by the ratio s/ρ, where s and ρ are defined in Fig. 2. The shapes varied significantly and, in particular, the radii, ρ, shown in Fig. 2 were found to vary between 0.02 and 0.004 mm. For each of the crack-like flaw geometries an extensive series of 3PB tests were performed (Hyde and Yaghi 1995) under pure mode-I conditions; 147, 46, 58 and 47 three-point bend tests were carried out for the specimen notch types with the equivalent notch width, 2s, equal to 0.02, 0.05, 0.10 and 0.20 mm respectively. The apparent K_{IC} (and standard deviations) for the notch types were found to be 36.8 ($\pm$5.96) $Nmm^{-3/2}$, 50.5 ($\pm$ 6.02) $Nmm^{-3/2}$, 58.5 ($\pm$7.04) $Nmm^{-3/2}$ and 64.4 ($\pm$ 8.56) $Nmm^{-3/2}$, respectively, where the figure in brackets is the standard deviation.

The combinations of K_I^{App} and K_{II}^{App} at fracture, ie. K_{If}^{App} and K_{IIf}^{App}, obtained from the CMM tests, for each of the four notch types and for the real cracks, are shown in Figs. 5(a)-(e).

It has been shown (Hyde and Yaghi 1992, 1993) that the peak stresses at narrow notch tips can be related to the notch geometry and the K_I and K_{II} values for the equivalent cracks. By assuming that fracture from the narrow notches occurs when the peak stress in the notch reaches a critical value, the combinations of K_{If}^{App} and K_{IIf}^{App} can be obtained in terms of the K_{IC}^{App}. The predicted combinations, which depend on the s/ρ value, are given in Figs. 5 for comparison with the experimental data. By presenting the predictions in this way, the exact s/ρ value is not required, and since ρ varies significantly a precise s/ρ value could not be obtained. However, the predictions (Figs. 5(a)-(e)) for s/ρ > 2 are seen to be practically independent of s/ρ, provided they are normalised with respect to the K_{IC}^{App} for the same notch. The theoretical basis of these predictions is described in Appendix 1. The average experimental results for the rectangular notches (s = 0.2, 0.1 and 0.05 mm wide) fall reasonably close (within about + or - one standard deviation) to the predictions for s/ρ > 2; see Figs. 5(a)-(c). The average experimental results for the effectively 0.02 mm wide notch (produced from a sharpened 0.1 mm wide shim) fall reasonably close (within about + or - one standard deviation) to the predictions for s/ρ = 1, shown in Fig. 5(d). The experimental data for the real cracks, Fig. 5(e), do not fit the predictions for s/ρ > 2 or s/ρ = 1 particularly well, as would be expected for the behaviour of real cracks.

The predictions, Figs. 5(a)-(e), do not exhibit a linear relationship between K_{If}^{App} and K_{IIf}^{App} because the peak stress positions at the notch tip for pure mode-I and pure mode-II are different (Hyde and Yaghi 1993). Hence the peak stress position varies as the combination of mode-I and mode-II varies and so the simple superposition of two peak stresses (one for mode-I and one-for mode-II) cannot be used. Instead, it is necessary to combine the stress distributions for each combination of mode-I and mode-II to determine the position as well as the magnitude of the peak stress. Approximately linear relationship between K_{If}^{App} and K_{IIf}^{App} exists for dominantly mode-I and dominantly mode-II conditions because the position of the peak stresses does not change significantly under these conditions.

The directions of crack propagation, θ_p, obtained from all of the sharp notch specimens, and for the real crack specimens, are included in Fig. 6. It can be seen that there is good correlation of the data and that the propagation angle is relatively independent of the notch shape. Also, the crack propagation angles for the notches and the real cracks are the same for the same mixed-mode conditions.

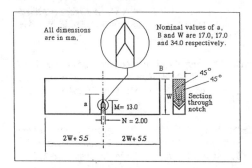

Fig. 1(a)   The three-point bend (3PB) specimen with a chevron notch and a real crack

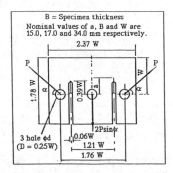

Fig. 1(b)   The compact-mixed-mode (CMM) specimen with chevron notches and a real crack

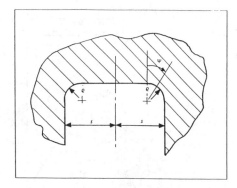

Fig. 2   The notch geometry

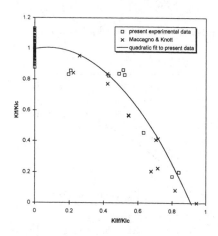

Fig. 3 Normalised mixed-mode fracture data for cracks;
Kif/KIC versus K_{IIf}/K_{IC}
($\square$ results of present investigation
x results of previous investigation
_____ quadratic fit to the results of the present investigation).

4.   CONCLUSIONS

(i)   The mixed-mode fracture behaviour of Araldite CT200/HT907 hardener has been determined. The combinations of K_I and K_{II} which cause fracture and the direction of crack propagation correspond to those obtained for a similar material (Maccagno and Knott 1989).

(ii)   The directions of crack propagation under mixed-mode conditions are the same for narrow notches and real cracks (Fig. 6).

(iii)   The relationship between the K_{If}^{App} and K_{IIf}^{App} values for narrow notches depends on the exact notch tip geometry. However, if the K_{If}^{App} and K_{IIf}^{App} values are normalised with respect to the

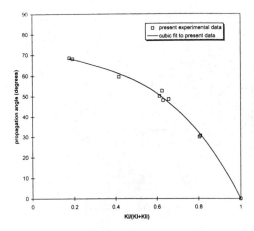

(a) propagation angle versus $K_I/(K_I+K_{II})$.

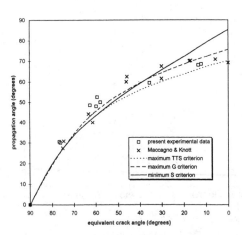

(b) propagation angle versus equivalent crack angle
$(\tan^{-1} (K_I/K_{II}))$

Fig. 4 Propagation angles for cracks

(□ results of present investigation
x results of previous investigation
____ prediction based on Š criterion,
------ prediction based on Ĝ criterion,
...... prediction based on $\hat{\sigma}_{\theta\theta}$ criterion)

K_{IC}^{App} value for the same notch, the relationship between $K_{If}^{App}/K_{IC}^{App}$ and $K_{IIf}^{App}/K_{IC}^{App}$ is insensitive to the exact notch geometry for $s/\rho > 2$.

(iv) The brittle fracture behaviour from narrow notches under any combination of opening (mode-I) and shear (mode-II) conditions, can be reasonably accurately predicted from the results for fracture tests under mode-I conditions. This

requires an estimate for s/ρ, but if $s/\rho > 2$ then the prediction is independent of the s/ρ value, as indicated in Figs. 5. This is based on the assumption that fracture occurs when the peak stress in the notch reaches a critical value.

REFERENCES

BS5447: British Standard Methods of Test for Plane Strain Fracture Toughness (K_{IC}) of Metallic Materials, 1977, British Standards Institution.

Fessler, H. and Hyde, T.H., *Stress Analysis of a Complex, Cast Node for Offshore Structures*, Applied Solid Mechanics - 3 Conference, (Eds. I.M. Allison and C. Ruiz), 427- 443, Guildford, UK, 1989, Elsevier Applied Science, 1989.

Hyde, T.H., *Photoelastic Techniques*, The Inst. Physics Stress Analysis Group Conference on "Modern Practice in Stress and Vibration Analysis, University of Sheffield, 20-22 April 1993, (Ed. J.L. Wearing),   Published by Sheffield Academic Press Ltd, 15-33.

Hyde, T.H. and Chambers, A.C., *A Compact Mixed-mode (CMM) Fracture Specimen*, J. Strain Analysis, 1988, Vol. 23, 61-66.

Hyde, T.H. and Yaghi, A., *An Assessment of a Model Method for Predicting the Fracture of Cracked Components*, J. Strain Analysis, 1995, Vol. 30, 9-14.

Hyde, T.H. and Yaghi, A. *Stresses Near Narrow Rectangular Notches, with Rounded Corners, in Beams in Bending*, J. Strain Analysis, 1992, Vol. 19, 227-234.

Hyde, T.H. and Yaghi, A., *Peak Stresses Near Narrow Rectangular Notches, with Rounded Corners, Subjected to Tensile and Shear Loading*, J. Strain Analysis, 1993, Vol. 28, 5-11.

Maccagno, T.M. and Knott, J.F. *The Fracture Behaviour of PMMA in Mixed-modes I and II*, Engineering Fracture Mechanics, 1989, Vol. 34, 65-86.

APPENDIX 1

Basis of the predictions on Figs. 5

The failure of loaded components with narrow notches has been analytically predicted in Figs. 5. The notches, at the tip, are either semi-circular or rectangular with varying corner radii. The applied loading is any combination of mode-I and mode-II. The prediction is based on the assumption that fracture occurs when the

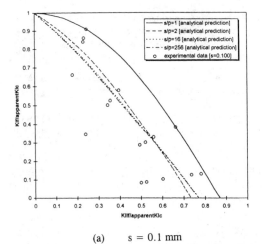

(a)   s = 0.1 mm

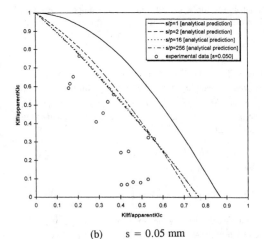

(b)   s = 0.05 mm

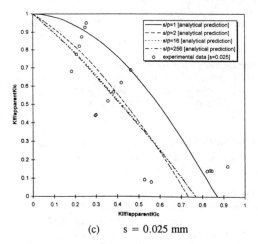

(c)   s = 0.025 mm

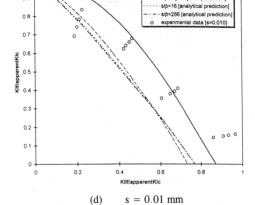

(d)   s = 0.01 mm

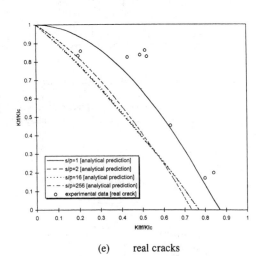

(e)   real cracks

Fig. 5 Normalised mixed-mode fracture data and predictions for the sharp notches and cracks
$(s/\rho = 1$ (____),
$(s/\rho = 2$ (____), $s/\rho = 16$ (.....) and
$s/\rho = 256$ (_._._).

peak stress, $\hat{\sigma}$, in the notch reaches a critical value.

In order to predict fracture analytically, $K_{If}^{App}/K_{IC}^{App}$ and $K_{IIf}^{App}/K_{IC}^{App}$ have been calculated, in a number of steps, from the finite element analyses and the solutions presented in a previous publication (Hyde and Yaghi 1993). Firstly, the peak stress, $\hat{\sigma}$, for a semi-circular notch, was discretely expressed in terms of K_I and K_{II}. Secondly, $\hat{\sigma}$ for the semi-circular notch was modified to describe the peak stress for three rectangular notches with s/ρ of 2, 16, and 256. Thirdly, $\hat{\sigma}$ was normalised with respect to the mode-I peak stress.

(s = 0.1 mm (*), s = 0.05 mm (x), s = 0.025 mm (Δ),
s = 0.01 mm (□) and real cracks (□);
____ cubic fit to the data).

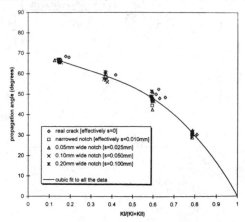

Fig. 6 Propagation angles for sharp notches and cracks; propagation angle versus $K_I/(K_I+K_{II})$

1. Solving the semi-circular mixed-mode problem

The maximum stress at a point, σ_{max}, is given (Hyde and Yaghi 1993) for a semi-circular notch in pure mode-I and pure mode-II loading by

$$\sigma_{max} \sqrt{(\pi \rho)}/K_I = 2.107 + 5.8 \times 10^4 \ \psi - 3.4 \times 10^4 \ \psi^2 + 1.47 \times 10^6 \ \psi^3 \tag{1}$$

$$\text{and} \quad \sigma_{max} \sqrt{(\pi \rho)}/K_{II} = 6.34 \times 10^2 \ \psi - 1.49 \times 10^4 \ \psi^2 - 3.94 \times 10^6 \ \psi^3 \tag{2}$$

respectively.

The pure mode-I and pure mode-II components of σ_{max} are added together to give the total σ_{max} which is then differentiated to obtain the value of ψ causing a peak value of σ_{max}, $\hat{\sigma}$. This value of ψ is replaced in equations (1) and (2) to give an expression for $\hat{\sigma}$ for a semi-circular notch in the form

$$\left(\hat{\sigma} \sqrt{(\pi \rho)}\right)_{semi} = \left(\gamma_I \ K_I + \gamma_{II} \ K_{II}\right) \tag{3}$$

$$\text{where} \quad \gamma_I = 2.107 + 5.8 \times 10^4 \ \psi - 3.4 \times 10^4 \ \psi^2 + 1.47 \times 10^6 \ \psi^3 \tag{4}$$

$$\text{and} \quad \gamma_{II} = 6.34 \times 10^2 \ \psi - 1.49 \times 10^4 \ \psi^2 - 3.94 \times 10^6 \ \psi^3 \tag{5}$$

2. Modifying the semi-circular solution to solve the rectangular mixed-mode problem

The ratio of $\hat{\sigma}\sqrt{\rho}$ for a rectangular notch divided by that for a semi-circular notch is termed η and is given (Hyde and Yaghi 1993) for s/ρ of 2, 16 and 256 for any value of $K_I/(K_I + K_{II})$. Therefore, an expression for $\hat{\sigma}$ for a rectangular notch at failure can be written in the form

$$\left(\hat{\sigma} \sqrt{(\pi \rho)}\right)_{rect} = \eta \ (\gamma_I \ K_{If} + \gamma_{II} \ K_{IIf}) \tag{6}$$

3. Normalising the results with respect to the pure mode-I results

Under pure mode-I conditions, ψ_{peak} is equal to zero (Hyde and Yaghi 1993) and therefore

$$\left(\hat{\sigma} \sqrt{(\pi \rho)}\right)_{semi} = \gamma_I^o \ K_{IC}^{App} \tag{7}$$

where γ_I^o is equal to 2.107 (from equation (4)). In order to make the results applicable to a rectangular notch, a factor η_I is included in the equation giving the expression

$$\left(\hat{\sigma} \sqrt{(\pi \rho)}\right)_{rect} = \eta_I \ \gamma_I \ K_{IC}^{App} \tag{8}$$

where η_I is equal to η when $K_I (K_I + K_{II})$ is unity.

For a particular notch, ie. for a specific value of ρ, if $\hat{\sigma}$ is assumed to be the same at failure irrespective of the failure mode, then from equations (6) and (8)

$$\eta_I \ \gamma_I^o \ K_{IC}^{App} = \eta \ (\gamma_I \ K_{If} + \gamma_{II} \ K_{IIf}) \tag{9}$$

Manipulating equation (9) gives

$$\frac{K_{If}}{K_{IC}^{App}} = \frac{\eta_I \ \gamma_I^o}{\eta \ \gamma_I} - \frac{\gamma_{II}}{\gamma_I} \left(\frac{K_{IIf}}{K_{IC}^{App}}\right) \tag{10}$$

Equation (10) has been used to provide the analytical predictions depicted in Figs. 5 for s/ρ of 1, 2, 16 and 256.

Experimental Mechanics, Allison (ed.) © 1998 Balkema, Rotterdam, ISBN 90 5809 014 0

Improved evaluation of stress intensity factors from SPATE data

M.C. Fulton & J.M. Dulieu-Barton
The University of Liverpool, Mechanical Engineering, UK

P. Stanley
The University of Manchester, School of Engineering, UK

ABSTRACT: Experimental procedures for the determination of crack-tip stress intensity factors have been developed and incorporated into a computer program. The program is used in a quantitative study of mode 1 centre cracks and mixed mode simulated cracks in thin aluminium plates. The work is based on the thermoelastic stress analysis technique and involves the use of the SPATE equipment.

1. INTRODUCTION

The thermoelastic stress analysis (TSA) technique centres on the use of the SPATE (Stress Pattern Analysis by Thermal Emissions) equipment to provide full-field stress data from components subjected to cyclic loads (Harwood & Cummings 1991). Under such conditions materials experience a small cyclic temperature change which is measured by the SPATE equipment by means of a highly sensitive infra-red detector. The thermoelastic response of the material and thus the received SPATE signal, S, are directly related to the sum of the principal surface stresses (σ_1 and σ_2), i.e.

$$\sigma_1 + \sigma_2 = AS \qquad (1)$$

where A is a calibration factor. By scanning over a selected area of the specimen surface the SPATE signal is displayed as a contour plot of ($\sigma_1 + \sigma_2$) on a computer monitor.

The high resolution and non-contact nature of TSA makes it particularly suitable for crack-tip studies. It was shown (Stanley & Chan 1986a) that the thermoelastic technique could be used to determine the stress intensity factor, K_1, for a central crack in a flat plate under mode 1 loading. An expression for the first stress invariant in the region of the crack-tip was derived from the Westergaard equations (Sih 1966). A method for determining the Paris law parameters (Paris & Erdogan 1963) for a propagating crack was also proposed. This work was developed for mode 2 loading (Stanley & Chan 1986b) using similar techniques to obtain values for K_2. A similar treatment has been applied to the case of a toe crack in a fillet weld under mixed mode

conditions (Chan & Tubby 1988), which produced results that agreed with finite element analysis. Stanley and Dulieu-Smith (1993) described a procedure for obtaining mixed mode stress intensity factors based on the geometry of the isopachic contours around the crack-tip. This work showed that in general the isopachic contour in the crack-tip region took the form of a cardioid curve (Lawrence 1972). By determining the area and orientation of a typical cardioid, values for K_1 and K_2 were found. The Muskhelishvili stress function has also been used (Tomlinson et al 1996) as the basis of a thermoelastic approach to mode 1 and mixed mode cracks. This technique used an array of approximately 100 discreet points around the tips of simulated cracks in aluminium plates. As with Stanley and Dulieu-Smith (1993) the results were reliable for mode 1 cracks, but only moderately so for mixed mode cases.

In this paper the procedures described by Stanley and Dulieu-Smith (1993) have been enhanced to include all the data around the crack-tip. The data are analysed using a computer program to calculate the area and orientation of a number of cardioid contours of constant signal using grown centre cracks in wide aluminium plates.

1.1 *Notation*

a	Semi-crack length (mm)
A	Calibration factor ($MPaU^{-1}$)
k_1, k_2	Non-dimensional stress intensity factors.
K_1, K_2	Stress intensity factors ($MPa\,m^{1/2}$)
r, θ	Polar coordinates (mm, degrees)
S	Thermoelastic signal (U)

U   Uncalibrated signal unit
w   Semi-plate width (mm)
β   Crack inclination w.r.t. x-axis (degrees)
σ_1, σ_2 Principal stresses (MPa)
σ_{app} Applied stress range (MPa)
Σ   Area within cardioid (mm²)

2. THEORY

Substituting the stresses obtained from the first-order form of the Westergaard equations into equation (1) gives the thermoelastic equation for the stress field in the region of a crack-tip, i.e.

$$AS = \frac{2}{\sqrt{2\pi r}}\left[K_1\cos(\theta/2) - K_2\sin(\theta/2)\right] \qquad (2)$$

where K_1 and K_2 are the mode 1 and mode 2 stress intensity factors, and r and θ are polar coordinates relative to the crack-tip (see Figure 1). For a constant signal, S, equation (2) represents a closed curve in r, θ coordinates, known as a cardioid, centered on the crack-tip. A typical cardioid is shown in Figure 1. Stanley and Dulieu-Smith (1993) derived two important relationships from the form of the curve to give the following two independent equations in K_1 and K_2

$$\frac{K_1}{K_2} = \tan\left[\frac{3\theta_{max}}{2} - \beta\right] \qquad (3)$$

where θ_{max} is the angular coordinate of the contact point of the horizontal tangents to the cardioid (θ_{max1} or θ_{max3} in Figure 1), and β is the angle between the crack and the x-axis, and

$$K_1^2 + K_2^2 = A^2S^2(2\pi\Sigma/3)^{1/2} \qquad (4)$$

where S is the signal value which defines the cardioid curve and Σ is the area enclosed within curve. Equations (3) and (4) allow K_1 and K_2 to be derived directly from the SPATE data.

3. DATA ANALYSIS

Stanley and Dulieu-Smith (1993) used the data from a single cardioid in the crack-tip field; the data analysis was essentially manual. The data processing technique has been considerably improved in this paper by automating the procedure with the use of a specially developed computer program (FACTUS – Fracture Analysis of Crack-Tips Using SPATE, which operates in the Windows 95[TM] environment) so all the data from the area of the crack-tip can be used. Thus more accurate area data for each

cardioid can be calculated and θ_{max1} and θ_{max3} can be averaged over a number of maxima.

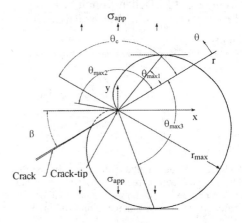

Figure 1. The cardioid curve

Firstly, it is necessary to export the SPATE 4000 data to a text file which is then imported directly into the FACTUS program. The program gives the option of smoothing the raw data to minimise the effects of signal noise. The smoothing regime uses a two-dimensional Gaussian distribution over a nine-point array. The data are displayed graphically so that any incomplete cardioid contours can be discounted from the analysis. The analysis examines nine cardioids defined by signal values which cover the entire signal range in the crack-tip field. Having established the individual cardioid signal levels the area enclosed by a given cardioid (Σ - see equation (4)) is obtained by simply counting the number of pixels in the data field that are greater than or equal to the corresponding signal value; allowances are made for edges. (The absolute evaluation of the area is crucially dependent on an accurate knowledge of the pixel size which follows from the working distance of the detector and the preset resolution of the scanning spot. This calculation is a built-in feature of the SPATE 4000 system. Errors in the pixel size may result in possibly serious systematic errors in the derived area value and hence the K values.)

The program calculates θ_{max1} and θ_{max3} by first finding the position of the maximum signal in the data field. The maximum signal on each horizontal line of data is then found and indicated graphically in the signal display. This results in two clearly defined lines of maxima, one above and one below the position of the maximum signal, and each containing up to 50 points. These are loci of the positions of the horizontal tangents to the cardioids. The gradients of the best-fit lines through the two sets of points are then calculated, thus providing

1212

$\theta_{\text{max}1}$ and $\theta_{\text{max}3}$. The number of points used to calculate the gradient can be adjusted according to the size of the scan area, and points very close to the crack-tip can be excluded as these may fall within the plastic zone. Also only points within the range of validity of the Westergaard equations (i.e. for points where r is less than the semi-crack length) are used.

4. EXPERIMENTAL PROGRAMME

All of the test specimens were large aluminium alloy plates, 1.5 mm thick, with a central crack as shown in Figure 2. Two sets of specimens were tested, the first containing real cracks, and the second containing spark eroded slots. The real cracks were grown to different lengths (shown in Table 1) from a machined 20 mm starter slot, and tested under mode 1 loading (i.e. $\beta = 0°$). The spark eroded slots were 0.33 mm wide with a semi-circular tip, both mode 1 and mixed mode loading were examined by orientating the slots at 90°, 45° and 60° to the applied load (i.e. $\beta = 0°$, 45° and 30°). The test rig, as shown in Figure 2, was designed to spread the applied load evenly across the width of the plates with the use of slots in the steel end plates. Bending in the plates was prevented with the use of 'knuckle' joints at each end.

The data were collected for plates 1 to 5 using a PC based SPATE 4000 system. The area in the region of the crack-tip was scanned, each scan being approximately 25 x 25 mm and containing over ten thousand data points. Plates 6 and 7 were tested using a SPATE 9000 system with a scan area of 60 x 60 mm. For all specimens the distance between the SPATE detector and the specimens was around 500 mm, giving a spot diameter of approximately 0.7 mm; the spatial resolution was set to 0.2 mm. The calibration factor (see equation(4)) was calculated using a standard procedure (Dulieu-Smith 1995). Two "dog-bone" specimens were manufactured from the specimen material and tested in tension; the average A value obtained was 0.052 MPaU^{-1}.

Table 1. Test Specimen Details

Plate	a (mm)	β (mm)	w (mm)
1 (c)	20	0	270
2 (c)	30	0	270
3 (c)	40	0	270
4 (c)	50	0	270
5 (s)	27	0	270
6 (s)	25	30	150
7 (s)	25	45	150

c = crack   s = slot

Plates 1 to 4 were tested with an applied stress of 7.4 ± 3.7 MPa with a minimum to maximum stress ratio, R, of 0.33 so as to prevent crack closure. The effects of crack closure were examined by testing plate 1 at a number of stress ranges (0.11 < R < 0.56). Plate 5 was tested with an applied stress of 10.67 ± 5.3 MPa. Plates 6 and 7 were tested with an applied stress of 12.5 ± 6.25 MPa. All the specimens were loaded cyclically at a frequency of 10 Hz.

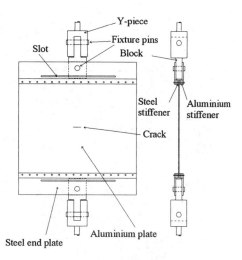

Figure 2. Test Specimen Arrangement

5. RESULTS

Non-dimensional values of stress intensity factors were derived by dividing K_1 and K_2 by the quantity $\sigma_{\text{app}}(\pi a)^{1/2}$ to give k_1 and k_2 respectively. Table 2 gives a comparison of theoretical and experimental results for the configurations shown in Table 1. The theoretical values for mode 1 loading were obtained using elementary solutions for a centre-cracked plate. For mixed mode loading, the theoretical values were taken from Kitigawa & Yuuki (1977). The theoretical values for θ_{max} were calculated from equation (3) using the theoretical k_1/k_2 values shown in Table 2.

The results for the increasing stress range test are shown in Figure 3 where values of the ratio k_{exp} to $k_{\text{theoretical}}$ are plotted. There is an obvious downward trend in the values of $k_{\text{exp}}/k_{\text{theory}}$ with increasing stress range. This is possibly due to the effect of crack closure. This observation is consistent with the comments of Ewalds & Wanhill (1991).

Table 2. Experimental and Theoretical Results

Plate	σ_{min} (MPa)	σ_{max} (MPa)	R	k_1 T	k_1 E	k_2 T	k_2 E	k_1/k_2 T	k_1/k_2 E	θ_{max}^{*} (deg) T	θ_{max}^{*} (deg) E
1	3.7	11.1	0.33	1.00	1.02	0	0.04	n/a	n/a	60	62
										-60	-45
2	3.7	11.1	0.33	1.01	0.91	0	0.09	n/a	n/a	60	64
										-60	-67
3	3.7	11.1	0.33	1.01	1.09	0	0	n/a	n/a	60	60
										-60	-57
4	3.7	11.1	0.33	1.02	0.91	0	0.06	n/a	n/a	60	63
										-60	-59
5	12.3	22.2	0.56	1.00	1.03	0	0.07	n/a	n/a	60	61
										-60	-52
6	2.1	14.6	0.14	0.77	0.67	0.44	0.41	1.76	1.63	20	19
										-100	-72
7	2.1	14.6	0.14	0.51	0.52	0.51	0.53	1.02	0.98	0	0
										-120	-81

T=theory   E=experimental $^{*}\theta_{max1}$ is shown first, then θ_{max3} for each plate

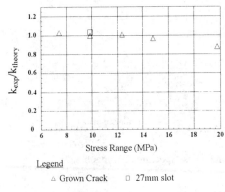

Legend

△ Grown Crack   ▢ 27mm slot

Figure 3. k_{exp}/k_{theory} for mode 1 loading over a number of load ranges

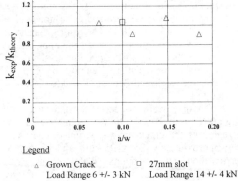

Legend

△ Grown Crack   ▢ 27mm slot
Load Range 6 +/- 3 kN   Load Range 14 +/- 4 kN

Figure 4. Normalised stress intensity factors for increasing crack length under mode 1 conditions

The ratio k_{exp}/k_{theory} against different values of the ratio a/w is plotted in Figure 4. The ratio is close to unity in most cases. There is a small amount of scatter, with a maximum difference of 11% for a/w = 0.185, and a minimum difference of 2% for a/w = 0.07. It is considered that these variations could be due to the geometric irregularities of real cracks. Experimental values of θ_{max1} and θ_{max3} are shown in Table 2. The former agreed well in general with theoretical values for the plates with real cracks subjected to mode 1 loading (plates 1 to 4); the maximum departure was less than 4° and the average departure was less than 0.1°. With the exception of plate 1 there was also good agreement between the theoretical and experimental θ_{max3} values for these plates.

The k values for the slotted plates under mode 1 loading (plate 5) showed good agreement with theory; the level of agreement of the θ_{max} values was much the same as for plates 1 to 4.

For the slotted plate loaded at 30° (plate 6) the experimental k_1 and k_2 values are somewhat smaller than the theoretical values (k_1 by 13%, k_2 by 7% and k_1/k_2 by 7%). However, for the 45° mixed mode loading (plate 7) the results are very good with only slight discrepancies in the k values. The θ_{max1} values for the two cases of mixed mode loading (plates 6 and 7) showed excellent agreement with theory, but the considerable discrepancies in the θ_{max3} values remain an unresolved anomaly in this data set. (It should be noted that the k_1/k_2 values shown in Table 2 are calculated from θ_{max1}.)

Table 3. Comparison with theory and previous work

Plate	k_1					k_2			
	T	E	1	2	3	T	E	2	3
2 (c)	1.01	0.91	1.06	n/a	0.98	0	0.09	n/a	n/a
3 (c)	1.01	1.09	1.10	n/a	1.00	0	0	n/a	n/a
4 (c)	1.02	0.91	1.10	n/a	1.03	0	0.06	n/a	n/a
5 (s)	1.00	1.03	n/a	n/a	0.98	0	0.07	n/a	n/a
6 (s)	0.77	0.67	n/a	0.75	n/a	0.44	0.41	0.36	n/a
7 (s)	0.51	0.52	n/a	0.59	0.52	0.52	0.53	0.56	0.54

T	Theoretical
E	Experimental
1.	Stanley and Chan (1986a)
2.	Stanley and Dulieu-Smith (1993)
3.	Tomlinson et al (1997)

6. COMPARISONS WITH OTHER DATA

Relevant k_1 and k_2 data from other sources are listed in Table 3 alongside values from the present study and from theory. Equal or closely matching a/w values were used in selecting the data from the literature. For the cracked mode 1 plates the results from the present study are, in general, somewhat smaller than those given previous experimental work. The results of Tomlinson et al (1997) are in closest agreement with the theoretical values in spite of the use of a machined slot in their specimens. It is also noted that the present result for the slotted mode 1 plate is in very good agreement with theory.

The values of k_1 and k_2 for the mixed mode slots are generally an improvement on those given by Stanley and Dulieu-Smith (1993), with the exception of k_1 for the 30° slot (plate 6). This improvement is attributable to the greater number of data points used by the computer program. (It should be noted that identical data has been used in both the present study and by Stanley and Dulieu-Smith (1993).) The k_1 value for the 45° slot (plate 7) derived in the present work concurs exactly with that given by Tomlinson. The k_2 value given by the present work is in slightly better agreement with the theory than previous work.

The general level of agreement present in the Table 3 results is encouraging. Of particular importance are the improvements in the quality of the data and the efficiency in obtaining k_1 and k_2 by using the FACTUS program.

7. FURTHER WORK

The next step in pursuing this work is to apply the FACTUS program to the analysis of data from real cracks under mixed mode loading. Preliminary tests have already been carried out. The test specimens have been produced by growing mode 1 cracks in large plates and then cutting sections out of the plates so that the cracks were at the relevant angle to provide the required mixity. Early indications have shown that frictional effects between the two crack surfaces may be important.

CONCLUSIONS

1. Previous procedures for the determination of k_1 and k_2 have been improved by the implementation of the FACTUS program which uses all the thermoelastic data within the region of the crack-tip.

2. Comparisons show that the use of the Muskhelishvili approach in place of the Westergaard approach does not affect the general quality of the results for central cracks, even though only the first order terms are retained in the Westergaard treatment and σ_{ox} has been omitted.

3. A decreasing trend in k_1 has been observed that is consistent with crack closure.

ACKNOWLEDGEMENTS

This work was funded by an Engineering Physical Sciences Research Council (EPSRC) grant.

REFERENCES

Chan, S.W.K. & Tubby, P.J. 1988. Thermoelastic determination of stress intensity factors for toe cracks in fillet welded joints and comparison with finite element results. *Welding Institute Report. No. 369.*

Dulieu-Smith, J.M. 1995. Alternative calibration techniques for quantitative thermelastic stress analysis. *Strain.* 31: 9-16.

Ewalds, H.L. & Wanhill, R.J.H. 1991. Fracture Mechanics. Edward Arnold. London.

Harwood, N. & Cummings, W.M. 1991. Thermoelastic Stress Analysis. Adam Hilger (copyright IOP Publishing Ltd.)

Kitagawa, H. and Yuuki, R. 1977. Analysis of an arbitrarily shaped crack in a finite plate using conformal mapping. 1st report – Construction of Analysis Procedure and its Applicability. *Tran. Japan Soc. Mech. Engrs.* 43: 4354-4362.

Lawrence, J.D. 1972. A catalog of special plane curves. *Dover Publications Ltd.* New York

Paris, P.C. & Erdogan, F. 1963. A critical analysis of the crack propagation laws. *Trans. ASME, J. Basic Eng.* 85:528-534

Sih, G.C. 1966. On the Westergaard method of crack analysis. *Int. J. Fracture.* 2:628-631.

Stanley, P. and Chan, W.K. 1986. The determination of stress intensity factors and crack-tip velocities from thermoelastic infra-red emissions. *Proc. Int. Conf. On Fatigue of Engineering Materials and Structures*: 105-114. Sheffield.

Stanley, P. and Chan, W.K. 1986. Mode II crack studies using the SPATE technique. *Proc. SEM Spring Conf. On Exp. Mech.*: 916-923. New Orleans.

Stanley, P. & Dulieu-Smith, J.M. 1993. Progress in the thermoelastic evaluation of mixed-mode stress intensity factors. *Proc. SEM Spring Conf.:* 617-626. Dearborn.

Tomlinson, R.A., Nurse, A.D. & Patterson, E.A. 1997. On determining stress intensity factors for mixed mode cracks from thermoelastic data. *Fatigue and Fract. Eng. Mat. Struct.* 20: 217-226.

Experimental Mechanics, Allison (ed.) © 1998 Balkema, Rotterdam, ISBN 90 5809 014 0

The application to mixed-mode cracks of a general methodology for automated analysis of caustics

J.D.Carazo-Alvarez
Escuela Universitaria Politécnica, University of Jaén, Spain

E.A.Patterson
Department of Mechanical Engineering, University of Sheffield, UK

ABSTRACT: The method of caustics has been widely applied to determine stress intensity factors by manually measuring some characteristic dimensions of the shadow patterns obtained at the crack tip. However, diffraction at the crack faces, troubles enormously this analysis, since the part of the caustic adjacent to the crack is blurred and obscured. This paper applies a general methodology based on image processing and non-linear regression to measure K_I and K_{II} at the crack tip using the whole caustic curve as the source of information, and therefore discarding the blurred and obscured parts. To perform this automated analysis, the system only requires data from the optics and the specimen employed, hence, finding the position of the crack tip and the value of both stress intensity factors. Experimental results demonstrate that the general methodology for the automated system provides an effective means of measuring mixed mode stress intensity factors in planar components.

1 INTRODUCTION

The shadow optical method of caustics is sensitive to stress gradients, and therefore was developed to investigate stress concentration problems. The method was originally introduced by Manogg (1964) for investigating crack tip stress intensities, and extended later by other authors, amongst whom it is appropriate to mention Kalthoff (1986) and Theocaris (1973)

Shadow optical images from test specimens under loading are characterised by geometric patterns that can be used to evaluate parameters relating to the loads or stresses. These patterns become rather complex with the difficulty of the problem, and to evaluate the loads or stresses usually involves a tedious mathematical apparatus to be developed for each particular case. Despite the complexity that may be inherent in the problems to be investigated, the clarity of the recorded images is essential to obtain reliable and informative data.

Mixed mode cracks produce a well-known caustic pattern that has been analysed using many different approaches. Manual analysis is far more developed than any other method. Basically, manual analysis is carried out by measuring some characteristic dimensions of the caustic curve, for which only two or three points are required, effectively wasting large amounts of information. Moreover, for mixed mode cracks, two of the three points required to perform the analysis are adjacent to the faces of the crack, and due to diffraction at the crack faces, this part of the caustic curve is often blurred and obscured.

Other methods of analysis have used digital image processing and least squares techniques to analyse the caustic at a crack tip. These methods require different levels of operator participation, *i.e.* some (Younis & Njock Libii 1992) require the operator to input a good initial estimate of the parameters in the caustic curve, some others assumed that the crack tip position was known (Herrmann & Noe 1992), or utilise only a limited number of points (Younis & Zachary 1987), or even require the operator to manually select those points from the caustic curve to be included in the analysis (Rossmanith & Knasmillner 1991). Summarising all these methods, the common characteristic is their exclusive development to solve the particular case of mixed mode cracks, and therefore they can not be used to solve other caustic patterns.

This paper applies a general methodology that has been previously and successfully used in studying contact problems (Carazo-Alvarez & Patterson 1995) and which is capable of dealing with different problems, in particular with mixed mode cracks.

This methodology will be explained in different parts, starting with the experimental procedure for obtaining caustic images, carrying on with the image treatment and finishing with the numerical approximation given by a non-linear least squares method. Finally experimental results obtained from mode I and mixed mode cracks are obtained using this methodology and discussed.

2 THE METHOD OF CAUSTICS

Stresses alter the optical properties of a solid. Tensile stresses decrease the thickness of the body due to Poisson's effects and they also decrease the refractive index of the material. When a plane specimen is illuminated by a parallel incident light beam, the stress-free parts of the plate are traversed by the light rays without deflection, however, the stressed areas of the model produce deflections of the light beams traversing such areas. Consequently, the light distribution in an arbitrary plane, called the reference plane or image plane, behind the specimen, is no longer uniform. Certain areas are not hit by light rays and appear dark, whereas other areas appear as areas of increased brightness. The shadow optical image exhibits a sharp boundary line between the areas of darkness and the surrounding area of light concentration. This boundary line between the two areas is call the caustic curve.

The accuracy of the method of caustics relies strongly on the correct set-up of the experimental arrangement in order to minimise systematic errors in the caustic measurement. Caustic images were acquired in transmission using the system shown on Figure 1. The incident beam is spatially and temporally coherent (Carazo-Alvarez & Patterson 1995), due to the use of a spatial filter to obtain a beam generated from a point source, and the presence of an interference filter, that transforms the white light generated from the source into monochromatic. A single lens forms the image directly on the microchip of a CCD camera, allowing accurate determination of the reference plane and magnification, and avoiding the use of a screen, which degrades the final image and serves no useful purpose. The image is finally transferred to a personal computer with a digitiser card where it is processed. This system generates well-defined thin caustic rings and high contrast images.

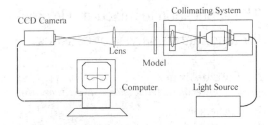

Figure 1. Experimental arrangement.

The use of white light with an interference filter, instead of using lasers, which are very common in this technique, avoids the typical laser speckle in the final image, facilitating the processing of the caustic by the computer. A caustic image from a mixed mode crack using the arrangement related above can be seen in Figure 2.

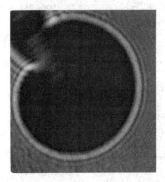

Figure 2. Caustic image from a mixed mode crack.

3 CAUSTIC ANALYSIS

Once the image has been recorded, the following step in any methodology is to analyse that image and obtain the parameters required in each case to determine the load or stresses of the problem. In the particular case of mixed mode cracks, these parameters are the stress intensity factors K_I and K_{II}.

The method employed in this paper, only requires the parametric equations describing any caustic curve to solve that problem, and has been implemented in a computer program.

The caustic ring is clearly defined as the point with higher light intensity, and it is detected by applying an intensity threshold value. Then the central line of the caustic ring can be found in order to reduce the number of data-points and obtain a single line to describe the experimental caustic.

The following stage is to perform the analysis of the caustic, by finding the theoretical curve that best fits the experimental data given by the central line. This is carried out by a non-linear least squares regression. For this analysis, some optical parameters are required in order to determine the position of the plane where the caustic has been recorded, and the magnification of the image. Also the parametric equations of the particular case to be solved, with some material and geometry properties are required at this stage of the process. Values of both stress intensity factors are finally obtained by simple mathematics from the parameters of the fitted curve.

4 NUMERICAL ANALYSIS

The non-linear regression analysis is carried out using the numerical Gauss-Newton algorithm for non-linear least squares. This method finds the value of the parameters in a particular regression model. The parametric equations of the caustic curve for a mixed mode crack are (Kalthoff 1986):

$$x = r\cos\phi + \frac{K_{\mathrm{I}}}{\sqrt{2\pi}} z_0 ct \cdot r^{-3/2} \cos\left(\frac{3}{2}\phi\right)$$
$$- \frac{K_{\mathrm{II}}}{\sqrt{2\pi}} z_0 ct \cdot r^{-3/2} \sin\left(\frac{3}{2}\phi\right) \tag{1}$$

$$y = r\sin\phi + \frac{K_{\mathrm{I}}}{\sqrt{2\pi}} z_0 ct \cdot r^{-3/2} \sin\left(\frac{3}{2}\phi\right)$$
$$+ \frac{K_{\mathrm{II}}}{\sqrt{2\pi}} z_0 ct \cdot r^{-3/2} \cos\left(\frac{3}{2}\phi\right)$$

where r, ϕ are the polar co-ordinate, z_0 is the image plane, c is the shadow optical constant of the material and t is the specimen thickness. Imposing the condition given in the general mapping procedure of caustic curves (Kalthoff 1986), the equation of the initial curve gives:

$$r = \left[\frac{3}{2}\frac{\sqrt{K_{\mathrm{I}}^2 + K_{\mathrm{II}}^2}}{\sqrt{2\pi}}|z_0\|c|t\right]^{2/5} \equiv r_0 \tag{2}$$

Therefore, the initial curve is a circle around the crack tip where radius r_0 is dependent on the stress intensity factors.

When working with experimental caustic data, the centre of the stress concentration, *i.e.* the crack tip, appears as additional unknown parameters to be found by the least squares fitting. These parameters come into the parametric equations as:

$$x' = x(\phi, K_{\mathrm{I}}, K_{\mathrm{II}}) + a$$
$$y' = y(\phi, K_{\mathrm{I}}, K_{\mathrm{II}}) + b \tag{3}$$

The least squares method of Gauss-Newton can be applied by determining the following derivatives for each data point:

$$\frac{dy'}{dK_{\mathrm{I}}}, \frac{dy'}{dK_{\mathrm{II}}}, \frac{dy'}{da}, \frac{dy'}{db} \tag{4}$$

However, to obtain equation 4 is not straight, since the polar variable ϕ should not be present in these derivatives (Carazo-Alvarez & Patterson 1995). Evaluating the derivatives in equation 4 in the following way solves this problem:

$$\frac{dy'}{dK_{\mathrm{I}}} = \frac{\partial y'}{\partial K_{\mathrm{I}}} - \frac{\partial y'}{\partial \phi} \cdot \frac{1}{\frac{\partial x'}{\partial \phi}} \cdot \frac{\partial x'}{\partial K_{\mathrm{I}}} \tag{5}$$

$$\frac{dy'}{dK_{\mathrm{II}}} = \frac{\partial y'}{\partial K_{\mathrm{II}}} - \frac{\partial y'}{\partial \phi} \cdot \frac{1}{\frac{\partial x'}{\partial \phi}} \cdot \frac{\partial x'}{\partial K_{\mathrm{II}}} \tag{6}$$

$$\frac{dy'}{da} = -\frac{\partial y'}{\partial \phi} \cdot \frac{1}{\frac{\partial x'}{\partial \phi}} \cdot \frac{\partial x'}{\partial a} = -\frac{\partial y'}{\partial \phi} \cdot \frac{1}{\frac{\partial x'}{\partial \phi}} \tag{7}$$

$$\frac{dy'}{db} = 1 \tag{8}$$

Equations 5-8 allow the method of Gauss-Newton to be applied to find the theoretical best fit to experimental data. Operating with exactly the same methodology any other problem can be solved using Gauss-Newton from the parametric equations describing the caustic curve.

The numerical method of Gauss-Newton also requires an initial estimation of the parameters defining the caustic curve. This initial estimation of parameters is carried out in the following way:

1. The position of the crack tip is estimated as the mean of all points in the centre line of the caustic curve, from the expression:

$$a = x = \frac{\sum_{i=0}^{n} x_i}{n}$$

(9)

$$b = y = \frac{\sum_{i=0}^{n} y_i}{n}$$

2. The mode I stress intensity factor is estimated using the following expression (Kalthoff 1986):

$$K_1 = \frac{2\sqrt{2\pi}}{3(3.17)^{5/2} z_0 ct} D^{5/2}$$

(10)

where D, diameter of the caustic is estimated as the maximum difference of points in the caustic curve in the direction of the crack.

3. Finally the mode II stress intensity factor is estimated from the stress intensity factor ratio η, that is initially set to a constant value.

Figure 3 shows an example of the analysis of a caustic image for a mixed mode crack. The fitting of the theoretical caustic, represented by the thin grey line, was obtained using the method proposed above.

5 MANUAL ANALYSIS

The manual analysis of mixed mode cracks is not obvious, and the stress intensity factors K_1 and K_{11} can be obtained from the caustic image by measuring the two characteristic length parameters

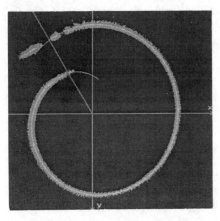

Figure 3. Experimental solution of a mixed mode crack using the proposed methodology.

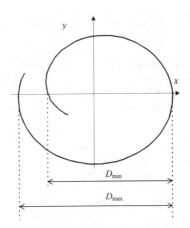

Figure 4. Definition of magnitudes D_{max} and D_{min} of the real caustic curve in transmission for a mixed mode crack.

D_{max} and D_{min} as defined in Figure 4. From the parametric equations of the caustic curve, given by equations 1-2, the distances D_{max} and D_{min} can be related to the ratio of stress intensity factors η (Kalthoff 1996):

$$\frac{(D_{max} - D_{min})}{D_{max}} = F\left(\eta = \frac{K_{11}}{K_1}\right)$$

(11)

Once η has been obtained, the numerical factor δ, defined as the ratio of the distance D_{max} over the initial radius r_0 can be determined:

$$\eta = G\left(\delta = \frac{D_{max}}{r_0}\right)$$

(12)

Relations expressed by equations 11-12 are obtained from the discrete numerical solution of the parametric equations 1 when $y = 0$, $i.e.$ those points where the caustic curve cuts the x-axis, and which define the characteristic lengths D_{max} and D_{min}. Thus the absolute value of the mode I stress intensity factor K_1 can be determined from the measured distance D_{max} as (Kalthoff 1996):

$$K_1 = \frac{2\sqrt{2\pi}}{3\delta^{5/2} z_0 ct} \frac{1}{\sqrt{1+\eta^2}} D_{max}^{5/2}$$

(13)

6 EXPERIMENTAL RESULTS

Two sets of experiments were performed: the first one with a mode I crack, and the second one with a mixed mode crack.

The first model, a horizontal edge crack on a plate under unixial tension, will produce mode I caustic images. The theoretical stress intensity factor K_I can be determined from the tensile load applied, by equation (Rooke & Cartwright 1976):

$$K_I = 1.5\sigma_I \sqrt{\pi a} = 1.5 \frac{P}{W \cdot t} \sqrt{\pi a} \qquad (14)$$

where a is the crack length, W is the width of the plate, and t is the thickness. The specimen was loaded in traction, ranging the applied load from 300 to 1500 N. Thirteen different caustic images were recorded in transmission by forming the real image at $z_0 = -566.71$ mm from the specimen plane. Experimental results were obtained through performing the manual analysis by measuring the caustic diameter, and the least squares analysis, by analysing the image as a mixed mode crack. Figure 5 shows the analysis of one of these images, and Figure 6 shows the results obtained.

The second specimen, an oblique edge crack in a plate under tension, will produce a mixed mode caustic image. A 20 mm crack forming an angle of 30° with the direction of the uniaxial stress was machined at the centre of the edge of a 80 mm width plate. This configuration has been previously analysed (Rooke & Cartwright 1976), and the shape factors for the specimen geometry are:

$$K_I = 0.51\sigma_I \ \pi a$$
$$K_{II} = 0.64\sigma_I \ \pi a \qquad (15)$$

The specimen was loaded in increments between 1300 and 2100 N. Nine different caustic images were recorded in transmission at a distance $z_0 = -599$ mm from the specimen plane. Experimental results were obtained by performing manual analysis following the procedure described above, and automated analysis. Figure 7 shows a comparison of the results obtained with their theoretical solution. The solution given by the least squares fitting for one of this experiment is also shown in Figure 3.

7 DISCUSSION

The typical error of the method was established (Kalthoff 1986) to be close to 5%, this error has been confirmed by the experimental work of other authors (Theocaris 1973). The value of 5% is therefore the reference value when evaluating experimental errors obtained with the shadow optical method of caustics.

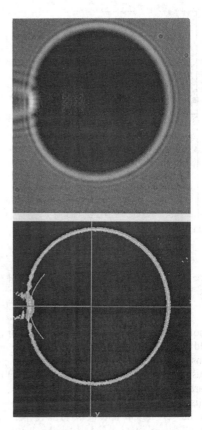

Figure 5. Experimental image, and its corresponding solution for a mode I crack.

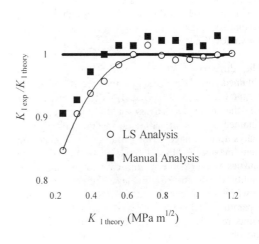

Figure 6. Experimental results from a mode I crack.

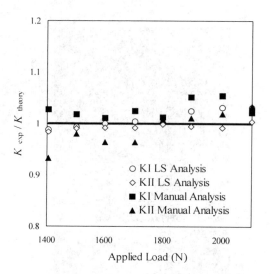

Figure 7. Experimental results from a mixed mode crack.

Kalthoff (1986) also explained that the initial curve should avoid the area of influence of three-dimensional stresses, since the mapping procedure of caustic curves is based on plane stress conditions. The area affected by three-dimensional stresses was established (Kalthoff 1986) to be smaller than 0.3 times the specimen thickness. The experimental results obtained for the mode I cracks presented in this paper, show errors below 5% for values of r_0/t above 0.25. This can be seen clearly in Figure 6, where the trend clearly indicates the convergence to the theoretical solution. Therefore, the size of the area affected by three-dimensional stresses is not clearly quantified and 0.25 times the specimen thickness is used in this paper as the reference. The second set of results shown in Figure 7 was obtained for values of r_0/t above 0.25.

The difference between theory and experiment obtained with the mode I specimen are below 1.64% for the least squares (LS) analysis and below 2.96% for the manual analysis. The maximum error obtained with the mixed mode specimen was 3.15% analysing the images with LS and 6.69% analysing them manually. Manual analysis of mixed mode caustics was possible in this case due to the good quality of the caustic images obtained (Fig. 3). However, other experiments recorded with the same apparatus could not be analysed manually, since the points required in this type of analysis are blurred and obscured by the diffraction at the crack faces (Fig. 2).

8 CONCLUSIONS

A general methodology for automated analysis of caustic images using image processing and non-linear least squares has been used in the determination of both stress intensity factors from mixed mode cracks. The method has the advantage of requiring only the parametric equations of the caustic curve to solve the problem, and uses the whole caustic ring as the source of information.

Experimental errors obtained with this automated methodology are below 4%, and therefore smaller than those obtained by manual analysis. Also, this method allows the analysis of mixed mode cracks even when the area close to the crack faces is blurred and obscured.

REFERENCES

Carazo-Alvarez, J.D. & E.A. Patterson 1995. Study of contact loads by using image processing and non-linear regression with caustics. *Proceedings of the 1995 SEM spring conference on experimental mechanics*. 1: 812-819. Grand Rapids.

Herrmann, K.P. & A. Noe 1992. Analysis of quasi-static and dynamic interface crack extensions by the method of caustics. *Engineering fracture mechanics*. 42(4): 574-588.

Kalthoff, J.F. 1986. Shadow optical method of caustics. *Handbook on experimental mechanics*: 430-500. Englewood Cliffs: Prentice Hall.

Manogg, P. 1964. *Anwendung der schattenoptik zur untersuchung des zerreissvorgangs von platten*. Freiburg

Rooke, D.P: & D.J. Cartwright 1976. *Compendium of stress intensity factors*. London: Procurement Executive, Ministry of Defence, H.M.S.O.

Rossmanith, H.P. & R.E. Knasmillner 1991. Recent advances in data processing from caustics. *Dynamic failure of materials, theory, experiments and numerics*. 27: 260-272.

Theocaris, P.S. 1973. Stress singularities at concentrated loads. *Experimental mechanics*. 13: 511-518.

Younis, N.T. & J. Njock Libii 1992. Determination of KI and KII by using a least-squares method with caustics. *Engineering fracture mechanics*. 42(4): 629-641.

Younis, N.T. & L.W. Zachary 1987. A non-linear least-squares method for the determination of stress intensity factors using the experimental method of caustics. *Proceedings of SEM Conference*: 210-214. Houston.

Interface fracture

Experimental Mechanics, Allison (ed.) © 1998 Balkema, Rotterdam, ISBN 90 5809 014 0

Analysis of bond line effects on cracks within and parallel to bond lines by stress freezing methods

C.W.Smith & E.F.Finlayson
Virginia Polytechnic Institute and State University, Blacksburg, Va., USA

C.T.Liu
Phillips Laboratory, Edwards AFB, Calif., USA

ABSTRACT: Frozen stress photoelastic tests were conducted using a three specimen test procedure to try to separate residual bond stress from other effects for two different bimaterial combinations. The first bimaterial consisted of araldite and aluminum filled araldite (25% by wt. of aluminum) which we call aral-alar provided by Survey Technologies, and the second consisted of two commercial photoelastic materials PSM9 and PLM4B manufactured by Micromeasurements Inc. Both bimaterial specimens contained nominal property mismatches in addition to modulus mismatch. Results suggest the aral-alar specimen yields very good results but that the PSM9-PLM4B specimen is inaccurate for cracks near the bond line but useful for cracks within the bond line or away from it.

1. INTRODUCTION

Traditionally, when frozen stress photoelastic experiments are conducted on complex shapes requiring glued joints for assembly, joint bond lines are kept very thin to localize self-equilibrating stresses and are generally avoided in the stress analysis. In recent years, in the area of fracture mechanics, considerable interest has developed in cracks which lie close to or within the bond line or "interface" between two materials, and analysis for a zero thickness bond line, or interface has shown that the Mode I (K_1) and Mode II (K_2) stress intensity factors (SIF) cannot be decoupled. However for a state of plane strain in incompressible materials, the interface fracture equations reduce to their classical homogeneous form. (Hutchinson & Suo 1992).

Since little is known of the SIF distribution along crack borders near to and within bond lines, the authors have been studying the feasibility of applying the frozen stress method to such problems.

After briefly describing the methods of testing and analysis used, results will be summarized and conclusions drawn.

2. METHODS OF ANALYSIS AND TEST PROCEDURES

The test specimen selected for all tests was an edge cracked specimen (Fig. 1). Three specimens

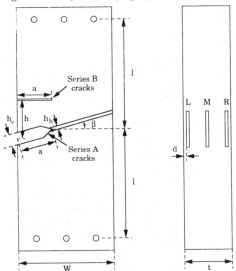

Series A: β = 0, 15, 30, & 45 degrees, h=0

Units	a	W	l	t	h_b	h_c	d
m.m.	9.53	38.1	88.9	12.7	0.356	1.59	1.02
in.	0.375	1.50	3.50	0.500	0.014	0.063	0.040

Series B: h = 0, 3.175 (0.125), 6.350 (0.250),
& 12.70 (0.500) [mm. (in.)]
$h_b = h_c = 0.305$ (0.012) [mm. (in.)]

Fig. 1. Test Specimens Dimensions

were tested for each condition; i) A homogeneous edge cracked control, ii) A homogeneous bonded edge cracked specimen, and iii) A cracked bimaterial specimen. In Test Series A, machined cracks were located in the bond line between aral and alar, (both made by Survey Technologies) which was inclined at various angles β to a normal to the load direction. In Test Series B, $\beta = 0°$ for all bond lines and all cracks were above and parallel to the bond lines at three different distances (h) from the bond line except for one in the bond line. The bimaterial specimens were made from PSM9 on the cracked side and PLM4B on the opposite side. Both materials were made by Micromeasurements Inc. Material Properties for both Test Series A and B are given in Table 1. The bond line adhesive was liquid PSM9.

Table 1
Series A - Material Properties

Material	T_{critical}	E_{Hot}	f
Araldite	240°F	(18.60 MPa)	(286.9 Pa-m)
Aral-Alar	240°F	(36.88 MPa)	-

Matched thermal coefficients at $68°F$ $\alpha = 15.3 \times 10^{-6}$ per $°F$. At critical temperature (T_c) thermal coefficients were $119 \times 10^{-6} \pm 20 \times 10^{-6}$ per $°F$.

Series B - Material Properties

Material	T_{critical}	E_{Hot}	f
PLM-4B	180°F	(12.96 MPa)	(420.3 Pa-m)
PSM-9	205°F	(47.67 MPa)	(520.1 Pa-m)

These materials have the same thermal coefficients at room temperature (T_R) and critical temperature (T_c). They were $\alpha_{RT} = 39 \times 10^{-6}/°F$ and $\alpha_{CT} = 90 \times 10^{-6}/°F$ respectively.

The test procedure involved a no load thermal cycle in order to cure the bond line glue followed by the stress freezing cycle. After stress freezing, thin slices L,M,R (Fig. 1) were removed and analyzed photoelastically in a refined polariscope using the Post and Tardy Methods in tandem for increased fringe sensitivity. SIF values were obtained from the algorithm in the Appendix.

3. RESULTS

Figure 2 shows fringe patterns for the homogeneous bonded specimens for $\beta = 30°$ in Series A.

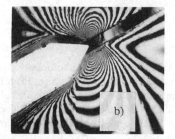
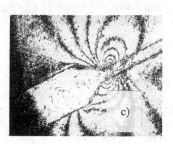

Fig. 2 Near Tip Fringe Patterns for Homogeneous Bonded Specimens Series A, $\beta = 30°$ a) No load b) Loaded, through thickness photo c) Slice (Bright Field)

Typical results from Test Series A are shown in Figure 3 where all SIF's have been normalized with respect to the Mode I values from the control specimens. The expected convex shape exhibited by the SIF distribution through the thickness is shown. Moreover, for the cases where SIF levels increased in the bimaterial specimens, it is seen that comparable increases occurred in the homogeneous bonded specimens, indicating that the increase was due to bond line residual stress and not modulus mis-match.

It is interesting to note that, in the cracks inclined 30° or more to the normal to the load no increase in the SIF resulted. The reason for this

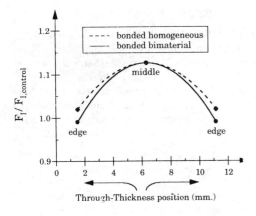

Fig. 3. F_1/F_{IC} Distributions Through the Thickness for Series A, $\beta = 30°$ where $F_1 = K_1/\bar{\sigma}(\pi a)^{1/2}$, and $F_{1C} = K_{1C}/\bar{\sigma}(\pi a)^{1/2}$, C = control, $\bar{\sigma}$ = average stress.

is pictured in Figure 4. For the smaller angles of

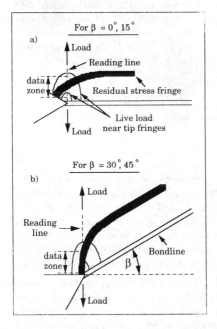

Fig. 4. Artist's Version of Near Tip Bond Line Residual Stress Fringes

inclination $(\beta = 0°, 15°)$ the residual stress fringe crossed the reading line (Fig. 4a) causing an increase in the fringe gradient and thus increasing the SIF. However, for larger angles $(\beta = 30°, 45°)$ the residual fringe became coincident with the reading line producing no gradient to affect the SIF.

The largest increase in the SIF in the bimaterial specimens above the homogeneous control specimen value occurred for $\beta = 0°$ and amounted to 19%. However, the corresponding homogeneous bonded specimen registered an increase of 22%, suggesting that the former increase is due to residual stress and not modulus mismatch. A complete set of normalized SIF values from Test Series A is found in (Smith, Finlayson & Liu 1997a).

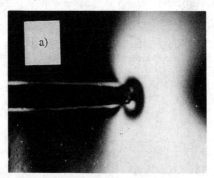

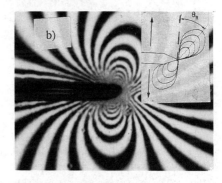

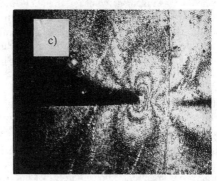

Fig. 5. Mixed Mode Fringe Pattern for Crack Parallel Bond Line photo Series B $\beta = 0°$, h = 6.35mm a) No load, b) Loaded, through thickness photo, c)Slice (Dark Field) Θ_R = reading angle

For Test Series B, involving PSM9 and PLM4B materials, the largest no load residual stress fringe order on the side of the crack was one, and no significant increases in K_1 occurred for any of the cracks in the homogeneous bonded specimens but in the bimaterial specimen a 29% increase in the SIF over the control value was observed for the crack 3.18mm from the bond line. It was also found that, in all of the bimaterial tests, some K_2 was present (7 to 16%), rotating reading angle (Θ_R) 8° for h = 12.7mm, 13° for h = 6.35mm (Fig. 5b) and 17° for h=3.18mm. However, as shown in Fig. 6, we again find a convex distribution in the normalized K_1 through the thickness as before. A complete set of test data for Test Series B is found in (Smith, Finlayson and Liu 1997b).

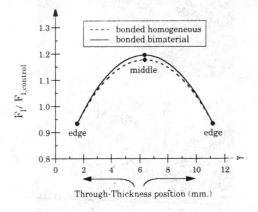

Fig. 6. F_1/F_{IC} Distributions Through the Thickness for a Crack Parallel to the Bond Line Series B $\beta = 0°$, h=6.35mm

Since it was not clear whether the shear mode and elevated SIF were due to T_c mismatch or modulus mismatch, an additional test specimen was made from aral- alar with a crack 3.18mm from the bond line and tested. The results showed neither a shear mode nor an increase in the SIF, suggesting that the shear mode and SIF increases in the series B models were due to T_c mismatch.

Two additional tests were run in Series B with $(\beta°, h = 0)$. A homogeneous bonded test using PLM4B showed an increase in F_1/F_c of 11% and a bimaterial specimen showed only 6% increase which is the order of experimental scatter. As in Series A with $\beta° = 0$ no shear mode was observed. Moreover, the above tests further support the absence of a bimaterial effect.

4. DISCUSSION OF RESULTS

For Test Series A, residual stresses created as many as two fringes prior to application of live load. It may be conjectured that part of this stress field may have been due to the ±17% variation in the thermal coefficients above T_c but it is difficult to conclude anything definite from these tests since bond line residual stress was also involved.

In Test Series B, there is also a source of disturbance due to the T_c mismatch. By placing the cracks in the higher T_c material (PSM9), it was thought that this effect might be minimized since when PSM9 turns glassy, PLM4B should offer little restraint at it's still hot modulus (i.e. $E_{RUBBERY4}/E_{GLASSY9} \leq 0.5\%$). Then when the PLM4B goes glassy, it should suffer little restraint from the PSM9 which is already glassy. However, some live stress will be induced in the PSM9. When slices are extracted, the live stresses in the PSM9 are released and this has been documented. However, as noted in the results, the T_c mismatch appears to be responsible for Mode II and a large increase in the SIF for the crack near the bond line.

5. CONCLUSIONS

As a result of the tests described in the foregoing for evaluating bond line stress effects, it appears that:

1) The aral-alar specimens gave good results and showed that, for cracks in bond lines, a) SIF elevations for cracks in bond lines over control values are due solely to residual stresses and are not affected by modulus mismatch. b) Residual stress effects are altered when the bond line orientations relative to load are altered. c) Reasonable SIF distributions through the thickness of the specimens can be determined.

2) The PLM4B-PSM9 specimens revealed a shear mode and an elevation of the SIF for a crack near the bond line likely due to T_c mismatch. However, these effects are small outside a distance of two thirds of the crack length from the bond line. It was also shown that these effects do not occur when the crack is in the bond line normal to the load.

3) Residual stress effects on SIF distributions can be accounted for by the frozen stress method. They may be minimized by no load thermal cycling and careful post curing.

ACKNOWLEDGEMENTS

The authors wish to acknowledge the support of Hughes STX Corp. under subcontract No. 96-7055-LI467, laboratory facilities of the Department of Engineering Science and Mechanics at VPI & SU and advice of Dr. I. Allison.

REFERENCES:

Hutchinson, J.W. & Suo, Z., 1992 "Mixed Mode Cracking in Layered Materials" *Advances in Applied Mechanics* Vol. 29.

Smith, C.W., Finlayson, E.F., & Liu, C.T. a 1997, "A Method for Evaluating Stress Intensity Distribution for Cracks in Rocket Motor Bond Lines" *Engineering Fracture Mechanics*, Vol. 58, No. 1/2 p. 97-105.

Smith, C.W., Finlayson, E.F., & Liu, C.T. b 1997 "Bond Line Residual Stress Effects on Cracks Parallel to the Bond Lines in Rubberlike Materials" *Symposium on Recent Advances in Mechanics of Solids and Structures (Proceedings)* Y.W. Kwon, D.C. Davis, H.H. Chung, L. Librescu Eds. ASME-PVP V369, p. 1-10.

Smith, C.W. & Kobayashi, A.S. 1993 "Experimental Fracture Mechanics" Ch. 20 of *Handbook on Experimental Mechanics* 2nd Ed. A. S. Kobayashi, Ed. VCH, Publishers, p. 905-968.

APPENDIX (Mode I Algorithm)

Beginning with the Griffith-Irwin Equations, we may write, for Mode I, for the homogeneous case

$$\sigma_{ij} = \frac{K_1}{(2\pi r)^{\frac{1}{2}}} f_{ij}(\theta) + \sigma_{ij}^{\circ} \qquad (i.j. = n, z) \quad (1)$$

where: σ_{ij} are components of stress, K_1 is SIF, r, θ are measured from crack tip (Fig. I-1), σ_{ij}° are non-singular stress components.

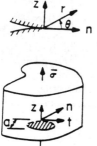

Fig. I-1 Mode I Near Tip Notation

Then, along $\theta = \pi/2$, after truncating σ_{ij}

$$(\tau_{nz})_{\max} = \frac{K_1}{(8\pi r)^{\frac{1}{2}}} + \tau^{\circ} = \frac{K_{AP}}{(8\pi r)^{\frac{1}{2}}} \quad (2)$$

where $\tau^{\circ} = f(\sigma_{ij}^{\circ})$ and is constant over the data range, K_{AP} = apparent SIF, τ_{nz} = maximum shear stress in nz plane

$$\therefore \frac{K_{AP}}{\bar{\sigma}(\pi a)^{\frac{1}{2}}} = \frac{K_1}{\bar{\sigma}(\pi a)^{\frac{1}{2}}} + \frac{\sqrt{8}\tau^{\circ}}{\bar{\sigma}} \left(\frac{r}{a}\right)^{\frac{1}{2}} \quad (3)$$

where (Fig. I-1) a = crack length, and $\bar{\sigma}$ = remote normal stress

i.e. $\dfrac{K_{AP}}{\bar{\sigma}(\pi a)^{\frac{1}{2}}}$ vs. $\sqrt{\dfrac{r}{a}}$ is linear.

Since from the Stress-Optic Law $(\tau_{nz})_{max} = nf/2t$ where, n = stress fringe order, f = material fringe value, t = specimen (or slice) thickness, then from Eq. 2,

$$K_{Ap} = (8\pi r)^{\frac{1}{2}} (\tau_{nz})_{max} = (8\pi r)^{\frac{1}{2}} nf/2t$$

A typical plot of normalized K_{AP} vs. $\sqrt{r/a}$ for a homogeneous specimen is shown in Fig. I-2.

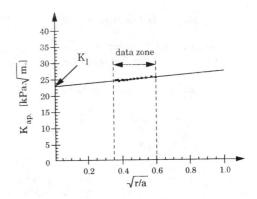

Fig. I-2 Middle Slice Data

Mixed Mode Algorithm

The mixed mode algorithm was developed (see Fig. I-3) by requiring that:

$$\lim_{\substack{r_m \to 0 \\ \Theta_m \to \Theta_m^{\circ}}} \left\{ (8\pi r_m)^{1/2} \frac{\delta(\tau)_{nz}^{max}}{\delta\Theta} (K_1, K_2, r_m, \Theta_m, \tau_{ij}) \right\} = 0$$

$$(4)$$

which leads to:

$$\left(\frac{K_2}{K_1}\right)^2 - \frac{4}{3}\left(\frac{K_2}{K_1}\right) cot2\Theta_m^{\circ} - \frac{1}{3} = 0 - - - \quad (5)$$

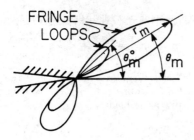

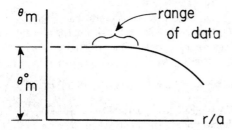

Fig. I-3 Determination of θ_m°

By measuring Θ_m° which is approximately in the direction of the applied load, K_2/K_1 can be determined.

Then writing the stress optic law as:

$$\tau_{nz}^{max} = \frac{fn}{2t} = \frac{K_{AP}^*}{(8\pi r)^{\frac{1}{2}}}$$

one may plot $\dfrac{K_{AP}^*}{\bar{\sigma}(\pi a)^{\frac{1}{2}}}$ vs. $\sqrt{r/a}$ as before, locate a linear zone and extrapolate to $r = 0$ to obtain K^*. Knowing, K^*, K_2/K_1 and Θ_m°, values of K_1 and K_2 may be determined since

$$K^* = [(K_1 sin\Theta_m^\circ + 2K_2 cos\Theta_m^0)^2 + (K_2 sin\Theta_m^\circ)^2]^{\frac{1}{2}} ---$$

(6)

A plot of $K_{AP}^*/\sigma\sqrt{\pi a}$ vs. $\sqrt{r/a}$ will yield a linear zone from which K^* can be extracted. Knowing K^* and θ_m°, K_1 & K_2 can be determined from Eqs. 5 and 6. Details are found in (Smith and Kobayashi, 1993).

Although the Series A cracks were vee notches and the Series B cracks were made with a Buehler saw with rounded tip, the linear zone in the graphs such as Fig. I-2 were both between $\sqrt{r/a} = 0.30$ and 0.60.

Experimental Mechanics, Allison (ed.) © 1998 Balkema, Rotterdam, ISBN 90 5809 014 0

Fracture strength of explosion clad plate with parallel cracks close to bonded interface

I. Oda, Y. Tanaka, T. Nakajo & M. Yamamoto
Kumamoto University, Japan

ABSTRACT: Tensile tests are carried out at various temperatures by using rectangular plate specimens extracted from the clad plate composed of copper and mild steel. Each specimen has two parallel artificial through-the-thickness edge cracks close and perpendicular to the explosive interface. As the plastic zone grows from the crack tip at a higher stress level, the critical crack opening displacement of the inhomogeneous specimen is influenced by not only crack-tip material but also the material inhomogeneity, the residual stress and the hardened zone caused by explosion bonding, and approaches to that of the material on the opposite side with respect to the explosion interface. The fracture strength of the inhomogeneous specimen whose crack-tip material is copper is much higher than that of the homogeneous copper specimen throughout all the test temperature.

1 INTRODUCTION

The bonded dissimilar plates are often used for chemical apparatus and pressure vessels for the purpose of corrosion resistance and heat resistance. The failures of those apparatus and vessels can lead to serious accidents. Therefore, it is highly important to examine the deformation and the fracture of the dissimilar plate. A bonded dissimilar plate has material inhomogeneity, residual stresses and changes in the material characteristics. Relating to the deformation and the fracture of the dissimilar plate with a crack, some papers have been published which examine analytically (Sato et al. 1983, Vijayakumar & Cormack 1983) and experimentally (Nishioka et al. 1993) the effect of the material inhomogeneity. A paper on J integral analysis in the dissimilar plate also has been reported (Kikuchi & Kobayashi 1994). The fracture toughness (Oda et al. 1982) and the fracture strength (Oda et al. 1985) of stainless-clad steel have been studied. The deformation behavior of stainless-clad steel with a crack has been also reported (Oda & Shiraishi 1996). The study which examines totally the effects of the material inhomogeneity, the residual stress and the change of material characteristics on the deformation and the fracture, however, has not been published. Furthermore, there have been few papers on the deformation and the fracture of a dissimilar plate with plural cracks although plural cracks are more important than a single crack from a practical

point of view.

In the present study, an explosion clad plate is used as a typical example of the bonded dissimilar plate. The clad plate is composed of C1100 copper and SS400 mild steel which have considerably different mechanical properties. The two-dimensional deformation behavior in the vicinity of two parallel cracks close to the bonded interface and fracture strength are examined by experiment as well as elasto-plastic finite element analysis. The effects of the material inhomogeneity, the residual stress, the hardened zone and the interaction between cracks on the deformation and the fracture are revealed.

2 EXPERIMENT

Figure 1 shows the result of the microhardness test on a vertical section of the clad plate. A remarkably hardened zone develops in the vicinity of the bonded interface. Tensile tests were carried out by using rectangular plate specimens extracted from the clad plate. Figure 2 shows the shape and dimensions of the specimen. The surface plane of the specimen is perpendicular to the plane of the bonded interface. Figure 3 shows the shape and dimensions of the central portion of the specimen. The thickness of the central portion is 3 mm. Each specimen has two parallel through-the-thickness edge notches perpendicular to the bonded interface. The notch tip

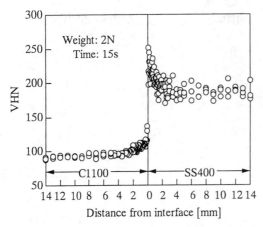

Figure 1. Microhardness profile on section of clad plate.

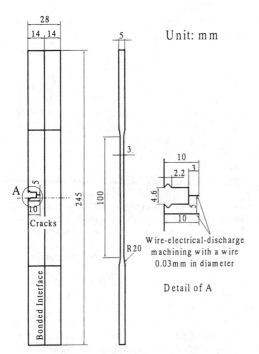

Figure 2. Shape and dimensions of test specimen.

portions of 3 mm long were prepared by the wire-electrical-discharge machining with a wire of 0.03 mm in diameter. The specimens which have a single notch are also used for comparison. In the present study, these notches are called hereafter the cracks for convenience. Eight types of specimen were used. A homogeneous specimen, Type CH and an inhomogeneous specimen, Type CI have their crack

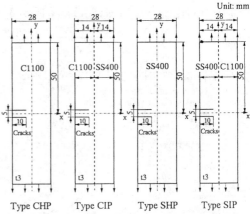

Figure 3. Central portion of test specimen.

tips in C1100 copper. A homogeneous specimen, Type SH and an inhomogeneous specimen, Type SI have their crack tips in SS400 mild steel. The distribution of residual strain, ε_{ry} parallel to the plane of the bonded interface in a 3mm thick plate extracted from the clad plate was measured by the stress-relaxation method using 1mm gauge-length strain gauges. The dotted line in Figure 4 shows the distribution of ε_{ry} in Type CI. The distribution of residual stress, σ_{ry} parallel to the bonded interface which was redistributed after cracking as the cracked specimen shown in Figure 3 was evaluated by FEM analysis. The solid line in Figure 4 shows the distribution of residual stress σ_{ry} on the crack growth in Type CIP. The residual stress σ_{ry} is compressive near the crack tip, tensile in the bonded region and compressive in the plate edge region opposite to the crack.

A uniformly tensile load perpendicular to the crack planes was applied to the specimens at various temperature ranging from −180℃ to 20℃. The load and the crack mouth opening displacement were

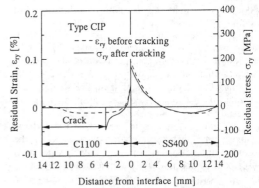

Figure 4. Distribution of residual stress in Type CIP.

1232

Table 1. Mechanical properties of materials at room temperature.

Materials	σ_{ps} [MPa]	T.S. [MPa]	El. [%]	E [GPa]	ν
C1100	181	225	46	106	0.3
SS400	344	481	32	200	0.3
Hardened zone	281 - 957	-	-	200	0.3

σ_{ps} : Proof stress (0.2%)
T.S. : Tensile strength
El. : Elongation, Gauge-length = 20 mm
E : Young's modulus
ν : Poisson's ratio

Materials	C [MPa]	α	n
C1100	405	1×10^{-5}	0.15
SS400	865	4×10^{-3}	0.2
Hardened zone	600 - 1400	0.5×10^{-3} - 0.2×10^{-2}	0.05 - 0.13

measured. A clip gauge was used for the measurement of crack mouth opening displacement. The temperature was measured with Copper-Constantan thermocouple welded to the specimen.

3 ANALYSIS

The deformation at room temperature near the tip of a crack in a plane stress field, which is uniaxial tension in the y direction, was analyzed. An elasto-plastic finite element analysis with an incremental theory of plasticity was used. The material was assumed to harden according to the following power law relation between stress and strain suggested by Swift (Swift 1952).

$$\bar{\sigma} = C(\alpha + \bar{\varepsilon})^n \tag{1}$$

where $\bar{\sigma}$ (MPa) is the equivalent stress, $\bar{\varepsilon}$ is the equivalent plastic strain and C, α, n are constants. Table 1 shows mechanical properties and material constants used in the analysis. Material constants for base materials were obtained by the tensile tests carried out using small specimens of C1100 and SS400 parent materials at room temperature. C1100 copper has much lower strength and somewhat higher elongation percentage than those of SS400 steel. The material constants for the region hardened by bonding were obtained based on the result of the hardness test and the relation between strength and hardness. For the inhomogeneous specimens, the residual stresses were considered in the analysis.

4 RESULTS AND DISCUSSION

Figure 5 shows the relationship between the crack mouth opening displacement, V_g measured by a clip gauge and the applied net stress for parallel cracks at room temperature. The analytical results are also shown in the figure. Both results show the same

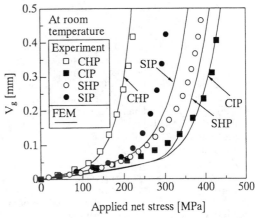

Figure 5. Relationship between crack mouth opening displacement and applied net stress at room temperature.

tendency. The crack mouth opening displacement, in the dissimilar plate, can not necessarily be used as a determined parameter for the fracture mechanics. In the present paper, however, the crack mouth opening displacement is used as a measure concerned with the deformation around the crack tip. The V_g value of the inhomogeneous specimen, Type CI is lower than that of the homogeneous specimen, Type CH, even if the crack tip material of these two specimens is exactly the same. This phenomenon is caused by the hardened zone produced by bonding, SS400 steel which has higher strength than the crack-tip material and the residual stress ahead of the crack tip in the inhomogeneous specimen. The V_g value of the inhomogeneous specimen, Type SI is higher than that of the homogeneous specimen, Type SH, even though the crack tip material of these two specimens is exactly the same. This phenomenon is mainly caused by C1100 copper, which has lower strength than the crack-tip material, ahead of the bonding

1233

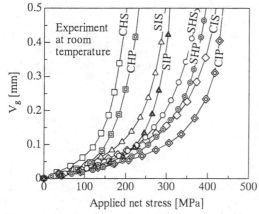

Figure 6. The interaction effect of parallel cracks on crack opening displacement at room temperature.

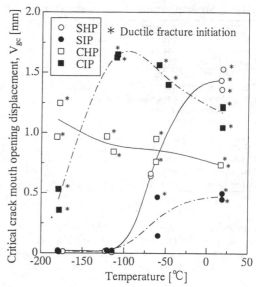

Figure 7. Relationship between critical crack mouth opening displacement and temperature.

interface in the inhomogeneous specimen. The V_g values of the single crack measured by the clip gauge are shown in Figure 6 for comparison with those of the parallel cracks. The V_g value of the parallel crack is lower than that of the single crack in both homogeneous and inhomogeneous specimens. This phenomenon is caused by the interaction between plural cracks.

In the inhomogeneous specimen, it is expected that the so-called COD criterion (Wells 1961) using the crack mouth opening displacement can not be simply applied to the fracture behavior. In the present paper, however, the critical crack mouth opening displacement is examined as a measure concerned with the fracture toughness.

Figure 7 shows the relationship between the critical crack mouth opening displacement, V_{gc} and temperature. The value of V_{gc} is determined as V_g at the onset of macroscopic unstable fracture. The symbol (*) indicates that the ductile fracture occurs at the crack tip. The V_{gc} value of Type SH at very low temperature is extremely low because of the low stress brittle fracture. On the other hand, the V_{gc} value of Type CH, the homogeneous C1100 specimen, is almost the same throughout all the test temperature range because the brittle fracture does not take place in Type CH specimen. The V_{gc} value of the inhomogeneous specimen Type CI is, at very low temperature, intermediate between those of two homogeneous Types. The V_{gc} value of Type CI increases with increasing temperature and then, at room temperature, approaches that of Type SH which is identical with the material in the opposite side with respect to the explosion interface of Type CI. The V_{gc} value of the inhomogeneous specimen Type SI is almost the same at very low temperature as that of Type SH specimen which is identical with crack tip material in Type SI. The V_{gc} value of Type

SI approaches, with increasing temperature, that of Type CH which is identical with the material ahead of the explosion interface of Type SI.

It can be expected that the critical crack mouth opening displacement, V_{gc} of the inhomogeneous specimen is governed by the crack tip material as long as the crack tip material behaves elastically at a low applied stress level. But, as the plastic zone grows from the crack tip at a higher stress level, the V_{gc} value of the inhomogeneous specimen is influenced by not only crack tip material but also other factors such as the residual stress, the hardened zone and the different material ahead of the explosion interface.

The V_{gc} values of the single crack are shown in Figure 8 for comparison with those of the parallel cracks. In homogeneous specimens, Type CH and Type SH, the V_{gc} values of parallel cracks are almost the same or somewhat higher when compared with those of the single crack. In the inhomogeneous specimens, Type CI and Type SI, the V_{gc} values of parallel cracks are almost the same at low temperature and somewhat lower at high temperature when compared with those of the single crack.

Figure 9 shows the relationship between fracture strength σ_{nf} and temperature. The value of σ_{nf} is the critical value of net stress at the moment of the macroscopic unstable fracture initiation. At very low temperature, the low stress brittle fracture takes place in Type SH specimen although it does not occur in Type CH specimen. The σ_{nf} value of the

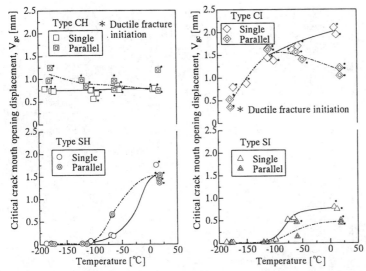

Figure 8. Comparison of critical mouth opening displacement of parallel cracks with that of single crack.

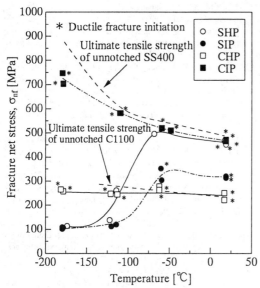

Figure 9. Relationship between fracture net stress and temperature.

inhomogeneous specimen Type CI is much higher than that of the homogeneous specimen Type CH over all the test temperature. Furthermore, the σ_{nf} value of Type CI is the highest among those of all the specimen types throughout all the test temperatures. Consequently, if an inhomogeneous plate like Type CI is used, the clad plate can contribute to the increase of the fracture strength.

The fracture strength σ_{nf} of the inhomogeneous specimen Type SI is roughly equal at low temperature and lower at high temperature when compared with that of the homogeneous specimen Type SH.

These phenomena of the fracture strength of the inhomogeneous specimen can be explained from the combination of the relationship between crack mouth opening displacement V_g and applied stress and the critical value of V_g.

The fracture strength σ_{nf} of the specimen with single crack is shown in Figure 10 for comparison with that of the specimen with parallel cracks. The σ_{nf} values for parallel cracks are almost the same or somewhat higher when compared with those for single crack in both homogeneous and inhomogeneous specimens.

5 CONCLUSIONS

The following conclusions may be drawn.

1. The deformation near cracks and the crack opening displacement in the explosion clad plate are basically influenced by the material inhomogeneity, the residual stress, the hardened zone ahead of crack tip and the interaction between cracks.

2. As long as the crack–tip material behaves elastically, the critical crack mouth opening displacement of the inhomogeneous specimen is governed by the crack-tip material and roughly equal to that of the homogeneous crack-tip material. As the plastic zone grows from the crack tip at a higher stress level, the critical crack mouth opening

1235

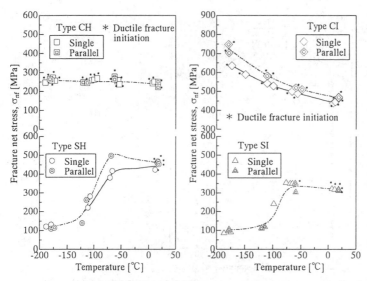

Figure 10. Comparison of fracture strength of parallel-cracked specimen with that of single-cracked specimen.

displacement of the inhomogeneous specimen is influenced by not only crack-tip material but also the material inhomogeneity, the residual stress and the hardened zone caused by explosion bonding, and approaches to that of the material in the opposite side with respect to the explosion interface.

3. The critical crack mouth opening displacement of parallel cracks in the homogeneous specimen is almost the same or somewhat higher when compared with that of the single crack. In the inhomogeneous specimen, the critical crack mouth opening displacement of parallel cracks is almost the same at low temperature and somewhat lower at high temperature when compared with that of the single crack.

4. The fracture strength of the inhomogeneous specimen whose crack-tip material is copper is much higher than that of the homogeneous copper specimen throughout all the test temperature. The fracture strength of the inhomogeneous specimen whose crack-tip material is mild steel is lower at relatively high temperature although it is almost the same at a low temperature when compared with that of the homogeneous mild steel specimen.

5. In both homogeneous and inhomogeneous specimens, the fracture strength of the specimen with two parallel cracks is almost the same or somewhat higher when compared with that of the specimen with a single crack.

REFERENCES

Kikuchi, M. & M.Kobayashi 1994. Two-Dimensional Crack Growth Simulation across the Fusion Line. *Transaction of JSME* 60:25-30 (in Japanese).

Nishioka, T. et al. 1993. Measurements of the Near-Tip Deformation in Inhomogeneous Elastic-Plastic Fracture Specimens Using the Moire Interferometry. *Transaction of JSME* 59:558-565 (in Japanese).

Oda, I. et al. 1982. Fracture Toughness of Explosive Stainless-Clad HT80 High Strength Steel. *Transaction of Japan Welding Society* 13:45-50.

Oda, I. et al. 1985. Fracture Strength of Explosive Clad Steel with Suface Notch. *Transaction of Japan Welding Society* 16:55-60.

Oda, I. & K.Shiraishi 1996. Deformation of Explosion Clad Plate with a Crack. *Abstract Proceedings VIII Int. Cong. on Expecimental Mechanics*:431-432. USA: Nashville.

Sato, K. et al. 1983. Crack Tip Plastic Deformation of Notched Plates with Mechanical Heterogeneity. *Journal of Japan Welding Society* 52:154-161 (in Japanese).

Swift, H.W. 1952. Plastic Instability under Plane Stress. *J. Mech. Phys. Solids.* 1:1.

Vijayakumar, S. and D.E.Cormack 1983. Stress Behavior in the Vicinity of a Crack Approaching a Bimaterial Interface. *Engineering Fracture Mechanics* 17:313-321.

Wells, A.A. 1961. Unstable Crack Propagation in Metals-Cleavage and Fast Fracture. *Proceedings, Crack Propagation Symposium* 1:210. Cranfield College of Aeronautics.

Experimental Mechanics, Allison (ed.)© 1998 Balkema, Rotterdam, ISBN 90 5809 014 0

Caustics at bimaterial interface crack tips – A symbiosis of finite element simulation and digital image analysis

F. Ferber & K. P. Herrmann
Laboratorium für Technische Mechanik, University of Paderborn, Germany

ABSTRACT: In this research contribution, an overview about the experimental and numerical modelling of crack systems arising in thermomechanically loaded models of two-phase composite structures will be given. In the scope of an experimental failure analysis of brittle composites the method of caustics is applied to determine stress intensity factors or related quantities at crack tips situated in homogeneous components or at the interfaces of composites. By applying a generated finite element mesh numerically simulated caustics at the tips of straight matrix cracks and interface cracks, respectively, were obtained. By using the components of the stress field or the displacement vector field of a specimen under plane stress conditions obtained by means of a finite element or an analytical calculation the image points for the generation of the associated shadow spots can be determined.

1 INTRODUCTION

In the past a great deal of effort was focused on the problem of interface cracks. Owing to the importance and complexity of this mixed-mode fracture phenomenon a very large volume of literature was accumulated within the past three decades [Ferber 1986, Ferber & Herrmann 1991, Dong 1993, Hinz 1993, Noe 1994, Ferber & Herrmann 1994, Ferber et al. 1996].

In this research contribution, an overview about the numerical modelling of crack systems arising in a plane disk-like model of a two-phase composite structure will be given.

2 THE SHADOW OPTICAL METHOD

In the scope of an experimental failure analysis of brittle composites the method of caustics is applied to determine stress intensity factors or related quantities at crack tips situated in homogeneous components or at the interfaces of composites. The shadow optical method represents an important tool for the experimental determination of stress intensity factors at the tips of quasistatically extending and fast running cracks, respectively. The physical principle underlying the method of shadow patterns is illustrated in Figure 1. A specimen containing a crack is illuminated with light generated by a point light source.

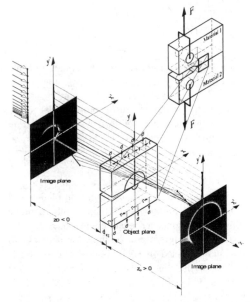

Figure 1. Experimental set-up for the shadow optical method of caustics.

In this case, a specimen of a transparent composite material is considered. The stress intensification in the region surrounding the crack tip leads to a reduction of both the thickness of the specimen and the refractive index of the material. As a consequence, in the transmission case, the light passing through the specimen is deflected outwards. On an image plane at any distance z_o behind the specimen, therefore, a dark shadow spot is formed. The spots are bounded by bright light concentrations, the caustics. Figure 2 shows the light deviation induced by the stress field singularity which generates shadow spots in the reference plane. The bright limit curve of the shadow spot is called caustic. A treatment of cracked opaque model materials can also be performed because the method of caustics can be used in transmission as well as in reflection.

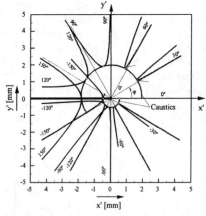

Figure 2. Light concentrations and geometry of the caustics.

The geometry of the caustic is proportional to the stress field gradient and therefore the caustic contour can be taken as a quantity for experimental measurements (Manogg 1964, Kalthoff 1985, Theocaris 1976, Rossmanith 1982, 1979).

Computer vision and computer animation deal with the manipulation and interpretation of pictorial information by computers. Pictures result in many application areas in science and engineering.

This paper presents a method for a crack tip stress analysis based on a combination of two techniques, namely the classical optical methods of stress analysis and modern digital image analysis, respectively. The experimentally obtained shadow spots are imaged by a CCD video camera and this video signal is digitized into an 8-bit (256-level) light-intensity distribution by the digital video frame store. The digital image is stored as a 512x512 pixel array in the frame-memory. A digital image processing computer program *KARL* has been developed in order to implement the different measurement algorithms and simulation techniques.

3 CRACK IN A MODIFIED CT-SPECIMEN

The numerical simulation of shadow spots is based on the fundamental equations of optics. By using the components of the stress field or the displacement vector field obtained by means of a finite element (Figure 3) or an analytical calculation the image points for the generation of the associated shadow spots can be determined [Rossmanith 1982, 1979, Ferber et al. 1996]. For the sake of a comparison of numerically and experimentally gained fracture mechanical parameters concerning a loaded crack associated caustics are simulated based on the experimental set-up for the shadow optical method by applying the reflection principle (Figure 1). Due to the distortion of the model surface the incoming light beams are reflected according to the surface normal vector of a certain point. Considering the distribution of the light intensity in a reference plane a shadow spot surrounded by a caustic arises.

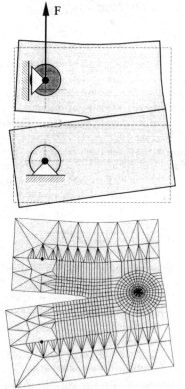

Figure 3. Geometry and finite element mesh for a modified CT-specimen.

By applying suitable near tip solutions for the stress and strain fields fracture mechanical parameters like energy release rates or stress intensity factors can be gained by evaluating the caustic contour. The measuring algorithm has been tested for interface cracks as well as cracks in a homogeneous material.

Figure 3 shows the geometry and finite element mesh for a modified Compact Tension specimen containing a straight interface crack. Thereby the given total finite element mesh consists of several substructures which can be either generated by means of special programs or implemented by hand through the user (Figure 4). By using the components of the displacement vector field of a specimen under plane stress conditions obtained by means of a finite element calculation the image points for the generation of the associated shadow spots can be determined.

Table 1. Material properties

Material	E [N/mm^2]	ν [1]	$\alpha \cdot 10^{-6}$ [K^{-1}]
Aluminium	72200	0.350	23.9
Araldite B	3500	0.370	53.0
Araldite F	2600	0.390	53.0
PMMA	3300	0.450	70.0
Glass	74850	0.170	5.0
Steel	210000	0.300	12.0
SF11 (Glass)	66000	0.237	6.1
SSKN5 (Glass)	88000	0.278	6.8

Table 2. Material combination

Combination	Bimaterial constant β	E_2/E_1
Araldite B/ Aluminium	0.09344	20.63
Araldite B/ Araldite B	0.0	1.0
Araldite F/ Glass	0.09314	28.79
PMMA/ Aluminium	0.08088	21.88
Araldite B/ Steel	0.09996	60.0
SF11/ SSKN5 (Glass)	0.02017	1.33

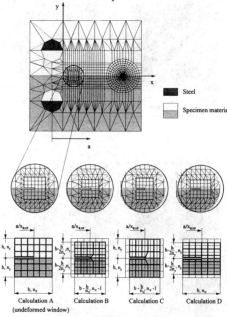

Figure 4. Finite element mesh and substructures.

The Figures 5-7 show the results of some examples, which contain energy release rates, mixed-mode stress intensity factors and numerically simulated caustics, for straight interface cracks situated in the discontinuity area of the two-phase compound. Thereby the results were obtained for different material combinations and loading conditions (Table 1 and Table 2).

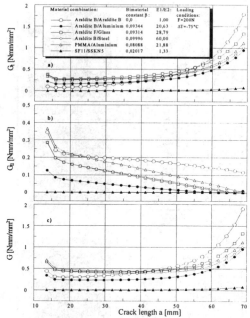

Figure 5. Energy release rates.

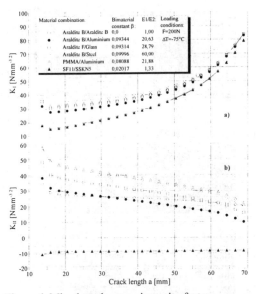

Figure 6. Mixed-mode stress intensity factors.

Figure 5 demonstrates the influence of the material inhomogeneity at the energy release rates around the tip of an interface crack. The mixed-mode stress intensity factors are given in Figure 6.

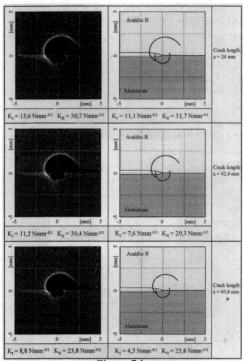

Figure 7-b.

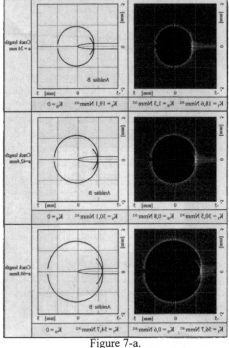

Figure 7-a.

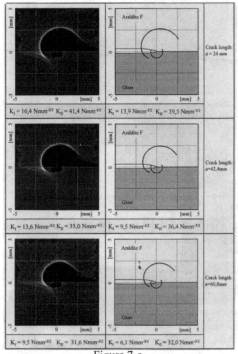

Figure 7-c.

1240

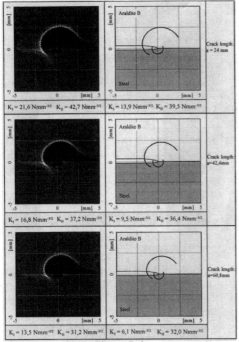

Figure 7-d.

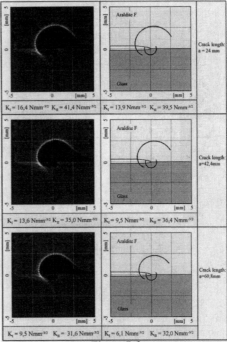

Figure 7-f.

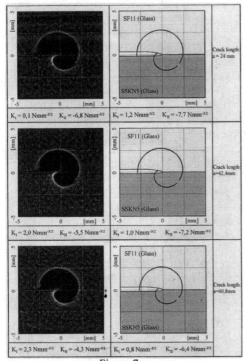

Figure 7-e.

Figure 7-a-f: Determination of mixed-mode stress intensity factors from shadow spots around a straight interface crack tip by using the multiple-point method of caustics.

(Modified CT-specimen; external load F = 200 N, thermal load ΔT=-75°C).

Figure 7 gives the mixed-mode caustics around the tip of an interface crack at both sides of a material interface of a two-phase composite structure in case of optical isotropy of the material. The left picture is a simulated shadow spot from a finite element calculation and the right picture is the caustic curve form the energy release rates.

The stress intensity factors which have been calculated from a set of measuring points recorded from the caustics by using a digital image system have been compared with the stress intensity factors assumed for the associated simulation (Figure 8).

Further, it could be shown that there exists a good agreement between a numerically simulated caustic for a modified CT-specimen with a straight interface crack and the associated experimental caustic gained by the shadow optical method (Figure 9).

1241

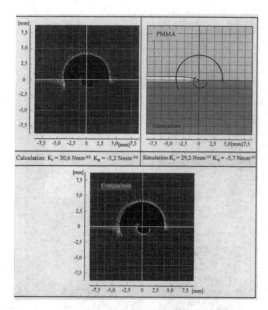

Figure 8: Determination of mixed-mode stress intensity factors from a numerically simulated shadow spot around a straight interface crack tip by using the multiple-point method of caustics.

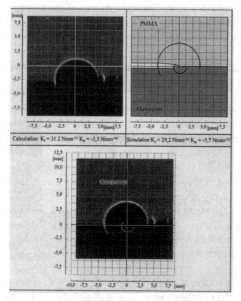

Figure 9: Determination of mixed-mode stress intensity factors from a exerimentally gained shadow spot around a straight interface crack tip by using the multiple-point method of caustics.

CONCLUSIONS

An overview about the experimental and numerical modelling of cracks arising in plane disk-like models of two-phase composite structures is given. Shadow optical data were collected from digitally sharpened caustics by using a digital image analysis system.

Further, the caustics simulation developed in the vicinity of a crack tip lying in the interface of two bonded dissimilar materials is studied. Based on the derived caustics equations a measuring algorithm can be formulated in order to determine stress intensity factors from numerically simulated caustics.

REFERENCES

Ferber, F. 1986. Dissertation, Paderborn University.

Ferber, F. & K.P. Herrmann 1991. Brittle Matrix Composites 3, eds. A.M. Brandt and I.H. Marshall, Elsevier Applied Science Publishers, London/New York:403-412.

Dong, M. 1993. Fortschritt Berichte, VDI-Reihe 18, No 138.

Noe, A. 1994. Dissertation, Paderborn University.

Ferber, F. & K.P. Herrmann 1994. Brittle Matrix Composites 4, eds. A.M. Brandt, V.C. Li and I.H. Marshall, IKE, Warschau/Woodhead Publ. Ltd., Cambridge, UK.:190-199.

Manogg, P. 1964. Dissertation, Freiburg University.

Kalthoff, J.F. 1985. The Shadow Optical Method of Caustics. In A.S. Kobayashi (ed.), Handbook of Exp. Mech.:430-500. Englewood Cliffs, Prentice Hall.

Theocaris, P.S. 1976. Partly unbonded interfaces between dissimilar materials under normal und shear loading. Acta Mechanica, Vol. 24:99-115.

Rossmanith, H.P. 1982. The method of caustics for static plane elasticity problems. Journal of Elasticity, Vol. 12:193-200.

Rossmanith, H.P. 1979. Determination of stress intensity factors by the dynamic method of caustics for optically isotropic materials. Ingenieur-Archiv, Vol. 48:363-381.

Hinz, O. 1993. Dissertation, Paderborn University.

Ferber, F., K.P. Herrmann & K. Linnenbrock 1996. Elementary failure analysis of composite models by using optical methods of stress analysis and modern digital image systems. In: ESDA 1996, (Eds. A. Large and M. Raous) ASME, New York, Vol. 4:55-68.

Experimental Mechanics, Allison (ed.) © 1998 Balkema, Rotterdam, ISBN 90 5809 014 0

Anisotropic intralaminar fracture properties of unidirectional composites

Ulrich Hansen
Department of Engineering, University of Reading, UK

John W.Gillespie, Jr
Center for Composite Materials, University of Delaware, Newark, Del., USA

ABSTRACT: Mode I intralaminar fracture toughness, G_{IC}, of graphite fibre reinforced bismaleimide (X5260/ G40-800) has been studied experimentally using notched three-point-bend specimens. The intralaminar fracture toughness was found to depend significantly on the direction of fracture propagation. Examination of the fracture surface showed the fracture surface to be relatively smooth or ragged depending on the direction of the fracture propagation, thus substantiating the measured differences in G_{IC}. Furthermore, the material exhibited a substantial R-curve behaviour, which appeared related to fibre pull-out.

1 INTRODUCTION

A great number of studies have been done to determine the fracture toughness of composite materials. Because the interface between plies usually is not reinforced with fibres, it is traditionally recognised as the weakest part of the composite. Therefore, the vast majority of earlier works have been aimed at determining the interlaminar fracture toughness. A comprehensive review of the field is presented by Sela & Ishai (1989).

However, the intralaminar fracture mode also represents a matrix failure mode with correspondingly low fracture toughness. Contrary to the interlaminar mode the intralaminar fracture toughness has only received limited attention. The interlaminar fracture is related to delamination which for most applications implies failure of the component, whereas the intralaminar fracture is associated with initial damage of the material indicating the onset of microcracking. In order to understand the damage evolution eventually leading to final failure, it is accordingly necessary to understand the intralaminar fracture mode. Furthermore, the microcracks, leading to reduced stiffness (Talreja 1985) and pathways for diffusion, often constitute a failure criteria themselves.

Wang (1984) pointed out that macroscopically the intralaminar fracture mode could depend significantly on the direction of fracture propagation. The two extreme fracture paths are illustrated in Figure 1. In Figure 1a, the crack grows orthogonally to the orientation of the fibres, whereas in Figure 1b the crack grows parallel to the fibres. Consequently, it is to be expected that depending on the direction of the crack growth, the material will exhibit a variation in the fracture toughness.

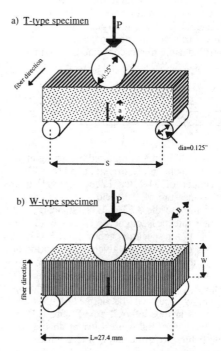

Figure 1. Notched specimen for 3-pt.-bend testing. S=4W, B=W, a+Δa=[0.4W; 0.7W], W=6.7mm.

Cullen (1981) and Williams (1981) investigated the interlaminar fracture toughness using a compact tension specimen and a splitting cantilever beam, and reported marked differences in the measured G_{IC} depending on the direction of the crack propagation. However, since the works by Cullen and Williams investigated the interlaminar fracture mode their result are only indicative of a similar directional effect in the interlaminar case.

The notched three-point bend ASTM E399 standard for G_{IC} testing was used in the present work since this test configuration allows us to address the directional effects of the intralaminar fracture mode. This method was also used by Parvizi et al. (1978), although, the directional characteristic of G_{IC} was not addressed by Parvizi.

2 MATERIAL AND SPECIMEN DESCRIPTION

BASF's graphite fibre reinforced bismaleimide prepreg material (X5260/G40-800) was used for all specimens. The elastic properties of this material is presented in Table 1. A unidirectional panel, 50 plies thick, was laid up. This panel was then processed in an autoclave according to the manufacturer's specifications. Within the relatively thick panel a thermocouple had been embedded to ensure that the material followed the specified cure cycle. One inch was trimmed off both panels. Panel quality was verified using microscopy and C-scanning.

Table 1. Material properties for X5260/G40-800.

E_L	E_T	G_{LT}	G_{TT}	v_{LT}	v_{TT}
160 GPa	9.3 GPa	6.0 GPa	3.5 GPa	0.3	0.49

Using a G_{IC} of 380 J/m^2 as reported by Huang et al. (1993), and a tensile failure strength of a 90° X5260/G40-800 coupon of 54 MPa (Pederson 1991) as a conservative measure of the yield stress, we determined, according to the ASTM E399 standard, that the specimen thickness ought to be larger than 6 mm, resulting in the specimen geometry of Figure 1. The specimen dimensions are given in the caption of Figure 1. A total of 14 specimens were machined. Of these 14 specimens 7 specimens were of the T-type and the rest were of the W-type. This terminology (T-type, W-type) will be defined in the next section but may also be inferred from Figure 1. Prior to testing, the specimens were dried in an oven at 100 °C for 24 hours to eliminate moisture effects.

The machined notch lengths varied between 2.5 and 3.4 mm. To avoid crack tip geometry effects resulting in artificially high G_{IC} values, the crack tip radius should, ideally, be zero. The specimens were notched with a wire blade saw resulting in a crack tip radius of only 0.05 mm. Even this very small crack tip radius, depending on the notch sensitivity of the material, will result in artificially high G_{IC} values (Hansen 1995). However, in the displacement controlled version of the Three-Point-Bend test, the crack growth is stable. Therefore, as we will show, this test configuration enables us to calculate G_{IC} based on a specimen with a natural crack tip geometry.

The three-point-bend configuration also allows us to investigate the effect of fibre orientation on G_{IC}. Two specimen configurations were used, labelled "W" and "T" in Figure 1. In the W-type specimens, the crack front will progress in the fibre direction, whereas in the T-type specimens the crack will grow orthogonally to the fibre direction. Notice, that in both cases the fracture surface will be parallel to the plane of the fibres, and in a post-failure inspection of the two specimen types, the fracture surface will, at least from a macroscopic viewpoint, appear identical. We can, however, expect the two types of specimens to exhibit different fracture toughness values.

3 DATA REDUCTION SCHEME

The three-point-bend specimen configuration of this study is relatively thick (see caption of Fig. 1) and is designed according to ASTM E399 in order to obtain a state of plane strain.

Incorporating the material anisotropy of the W-type specimen and of the T-type specimen in a plane strain finite-element-model, ABAQUS (Hibbitt et al.), we simulated the three-point-bend test using the virtual-crack-closure technique for calculating G_I. The result of this analysis (unit load) is shown in Figure 2 as small hollow circles. It is observed that the anisotropy of the T-type specimen led to G values significantly higher than the G values of the W-type specimen at the same load level.

A mesh convergence study of the finite-element-model is also included in Figure 2. The result of using a refined mesh is shown by diamond symbols. The fine mesh includes twice as many elements in the path of the crack as compared to the relatively coarse mesh (only G values at corresponding values of a/w of the two analysis are shown in the figure). From Figure 2 we observe that it is almost impossible to distinguish the results of the two FE-analysis from each other, in effect indicating mesh convergence.

In order to gain confidence in the implementation of the finite-element-model a three-point-bend specimen with isotropic material properties was also simulated (using the same coarse mesh as in the anisotropic analysis just mentioned). The accuracy of this finite-element solution can be evaluated from a comparison with the following standard closed

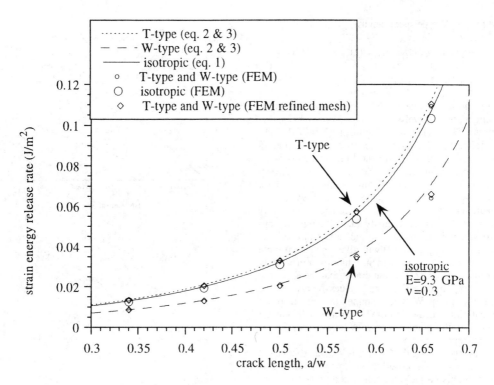

Figure 2. Anisotropic energy release rate including a mesh convergence study. Applied force = 1 N.

form expressions, eq. (1) and eq. (2), which are valid for isotropic materials

$$G_I = (1 - v^2)K_I^2 / E \qquad (1)$$

where K_I is approximated according to eq. (2) and P, S, B, W and 'a' are defined in Figure 1:

$$K_I = \frac{3PS}{2B\sqrt{W}} \sqrt{\frac{a}{W}}$$

$$\times \frac{\left[1.99 - \frac{a}{W}\left(1 - \frac{a}{W}\right)\left(2.15 - 3.93\frac{a}{W} + 2.7\frac{a^2}{W^2}\right)\right]}{\left(1 + 2\frac{a}{W}\right)\left(1 - \frac{a}{W}\right)^{3/2}} \qquad (2)$$

The results of these two isotropic analysis are also shown in Figure 2. The result of using eq. (2) in combination with eq. (1) is shown with a solid line and the finite-element result is shown as large hollow circles. We find the correlation to be convincing. Figure 2 also shows that the response of the T-type specimen and of the isotropic specimen is very similar. This is to be expected considering that the T-type specimen is transversely isotropic with respect to the orientation of the crack.

Suo (1990) mentions that the stress intensity factors for the standard test geometries used in composite fracture testing are almost independent of any anisotropic factors. Thus, eq. (2) should supply us with a reasonable estimate of the stress intensity factor for both the T-type and the W-type specimen. Having an expression for K_I we can get an estimate of the anisotropic G_I from eq. (3) where α is defined in eq. (4).

$$G_I = \alpha K_I^2 \qquad (3)$$

$$\alpha = \sqrt{a'_{11}a'_{22}/2}\left(\sqrt{a'_{22}/a'_{11}} + \left(2a'_{12} + a'_{66}\right)/2a'_{11}\right)^{1/2} \qquad (4)$$

In eq. (3) and eq. (4) a'_{ij} is the compliance matrix for plane strain. a'_{ij} is defined as

$$a'_{ij} = a_{ij} - a_{i3}a_{j3}/a_{33} \qquad (5)$$

For the T-type specimen a_{ij} is defined as

$$a_{11} = 1/E_T, \ a_{22} = 1/E_T, \ a_{33} = 1/E_L, \ a_{66} = 1/G_{TT} \qquad (6)$$

$$a_{12} = -v_{TT}/E_T, \ a_{13} = -v_{LT}/E_L, \ a_{23} = -v_{LT}/E_L$$

and for the W-type specimen as

$$a_{11} = 1/E_L \ , \ a_{22} = 1/E_T \ , \ a_{33} = 1/E_T \ , \ a_{66} = 1/G_{LT}$$

(7)

$$a_{12} = -\upsilon_{LT}/E_L \ , \ a_{13} = -\upsilon_{LT}/E_L \ , \ a_{23} = -\upsilon_{TT}/E_T$$

The result of simulating the T-type and the W-type specimen, respectively, using eq. (2) in combination with eq. (3) is shown in Figure 2 as dashed lines. We notice that for the T-type specimen the predictions of eq. (2 & 3) and of the finite-element-model are almost identical, whereas, the correlation between eq. (2 & 3) and the finite-element-model for the W-type specimen for large values of a/w is reasonable but not as good. This difference is probably caused by the stress intensity factor being only approximately independent of the anisotropy of the material. Isotropic eq. (2) will give a better estimate of K_I of the T-type specimen, which is transversely isotropic with respect to the crack, than of K_I of the W-type specimen. For large values of a/w in the simulation of the W-type specimen the anisotropy of the material and the finite-width effect will interact resulting in the small discrepancy between the finite-element-model and eq. (2 & 3).

It is worth noting that, whereas we find that eq. (2 & 3) does not give a perfect estimate for large a/w values in anisotropic materials, standard test geometries for fracture testing usually specifies a specific and moderate value of a/w.

4 TEST PROCEDURE

The experimental set-up was indicated in Figure 1 and the testing was performed in an INSTRON 1125 testframe. In order to maximise our ability to control the crack growth, a very low crosshead displacement speed of 0.05 mm/min. was used in all tests. An optical microscope with a video hook-up enabled us to monitor and record the crack growth versus time.

Correlating the crack growth versus time with the load versus time, we obtained critical loads at increasing crack lengths. These corresponding loads and crack lengths can then be related to G_{IC} either through eq. (3) or via the orthotropic FEA results of Figure 2.

5 RESULTS AND DISCUSSION

Figure 3 shows typical load versus displacement curves with corresponding optically observed crack lengths, for each of the two specimen types. The accuracy of the displacement measure is not critical to our calculations and the displacement in Figure 3 is simply the cross head displacement. The crack lengths were measured directly from the recorded video.

(a)
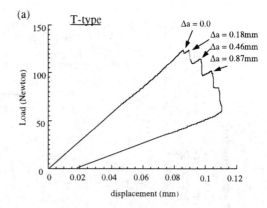

(b)
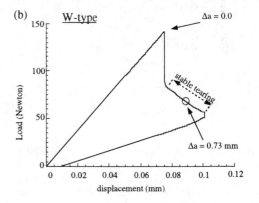

Figure 3. Typical load versus displacement curves.

We observe a very characteristic difference in the behaviour of the two types of specimens. The initial crack extension and load drop in the W-type specimens were significantly larger than in the T-type specimens. Additionally, subsequent crack extension was more stable in the T-type specimen. This implies that the material tested in the direction parallel to the fibres, i.e. the W-type specimen configuration, is more brittle and notch sensitive than when tested in the direction orthogonal to the fibres.

Using test results like those in Figure 3 and the orthotropic FEA results in Figure 2, we obtained the relationship between the critical energy release rate, G_{IC}, and the crack extension, Δa, shown in Figure 4. Results are shown for 14 specimens, 7 T-type specimens and 7 W-type specimens. In both the W-type and T-type tests, the material exhibited an R-curve behaviour. From in-situ observations using the optical microscope, it appears that the R-curve can be explained as a fibre bridging effect.

For small values of Δa, the R-curve effect is only one of two competing effects, the other being the

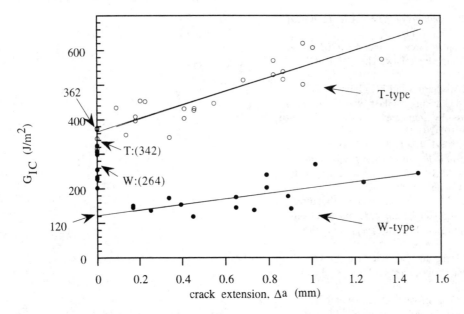

Figure 4. G_{IC}'s variation with crack orientation, as well as with crack growth.

previously mentioned geometry effect of the machined notch, which tends to over estimate G_{IC}. The values in brackets are based on this initial machined notch length (each value being the average of 7 specimens). The significant geometry effect on the G_{IC} values of the W-type specimens supports the earlier speculation that the material tested in this direction is relatively notch sensitive.

The general assumption relevant for microcracking is that G_{IC} is dependent only on the material; i.e., all geometry effects should be excluded. Furthermore, we consider the G_{IC} value at zero crack extension, i.e. no R-curve effect, to be the most appropriate choice. While the results shown in Figure 4 do not offer this value explicitly, we can, however, obtain an approximation by extrapolating the result from $\Delta a > 0.2$ mm, i.e., the results with no geometry effects, back to $\Delta a = 0$, as shown by the solid lines in Figure 4. Following this approach, we obtained a G_{IC} of 120 J/m² for the W-type specimens and a G_{IC} of 362 J/m² for the T-type specimens.

The significant anisotropy of the intralaminar G_{IC} seems qualitatively justified from an examination of the fracture surface of the two types of specimens. From simple visual, as well as SEM inspection of the W-type specimen it was obvious that the fracture surface was relatively smooth while pulled-out fibre bundles were in abundance on the fracture surface of the T-type specimen. We propose that the crack front in the W-type specimen is relatively constrained by the fibres, and the potential for crack branching is less due to the fibre orientation relative to the crack growth direction.

6 SUMMARY

In summary, the directional effects were significant and the material exhibited a significant R-curve effect. Using the procedure suggested in this work we determined a G_{IC} equal to 362 J/m² for a crack propagating orthogonally to the fibre direction. For this kind of crack growth, the material appeared to be relatively notch insensitive but exhibited a significant R-curve effect. For a crack propagating along the fibre direction a G_{IC} of 120 J/m² was determined. For this fracture mode in addition to an R-curve effect, the material demonstrates a significant notch sensitivity. The measured difference in G_{IC} depending on the direction of fracture propagation is supported by a marked difference in the appearance of the fracture surfaces in the two cases.

REFERENCES

Cullen, J.S. 1981. *Mode-I Delamination of Unidirectional Graphite Epoxy Composite Under Complex Load Histories*. M.S. thesis, Texas A&M University, College Station, TX.

Hansen, U. 1995. *Transverse Cracking of Laminated Composite Materials with Interleaves*. Ph.D. dissertation, University of Delaware, Dept. of Mechanical Engineering.

Huang, X.G., J.W. Gillespie and R. Eduljee 1993. Fracture Test and Prediction of High Speed Civil Transport Composite Materials in Service Temperature. *Proceedings of the 8th ASC Technical Conference*, Cleveland.

Hibbitt, Karlsson & Sorensen, Inc. *ABAQUS manual*. 1080 Main Street, Pawtucket, RI 02860-4847, USA.

Parvizi, A., K.W. Garrett and J.E. Bailey 1978. Constrained Cracking in Glass Fiber-Reinforced Epoxy Cross-Ply Laminates. *Journal of Materials Science 13*: 195-201.

Pederson, C.L. 1991. *The Effect of Temperature on the Long term Behavior of High-Performance Composites*. M.S. thesis. Center for Composite Materials, University of Delaware.

Sela, N. and O. Ishai 1989. Interlaminar Fracture Toughness and Toughening of Laminated Composite Materials: a Review. *Composites* 20(5): 423-435.

Suo, Z. 1990. Delamination Specimens for Orthotropic Materials. *Journal of Applied Mechanics* 57:627.

Talreja, R. 1985. Transverse Cracking and Stiffness Reduction in Composite Laminates. *Journal of Composite Materials* 19: 355-375.

Wang, A.S.D. 1984. Fracture Mechanics of Sublaminate Cracks in Composite Materials. *Composites Technology Review* 6(2):45-62.

Williams, D. 1981. *Mode-I Transverse Cracking in a Epoxy and a Graphite Fiber Reinforced Epoxy*. M.S. thesis. Texás A&M University, College Station, TX.

Experimental Mechanics, Allison (ed.) © 1998 Balkema, Rotterdam, ISBN 90 5809 014 0

A fractographic study of an edge crack growth near a ceramic-metal interface

Y. Arai, E. Tsuchida & J. Miyagaki
Saitama University, Urawa, Japan

ABSTRACT: Four point bending tests for Si_3N_4 - SUS304 joints after thermal cycling were conducted and their fracture surfaces were examined by a fractographic method. Based on these results, the fracture path and the criterion for unstable fracture is discussed by a interface fracture mechanics approach. The crack initiates in the ceramic very close to the ceramics - brazing filler interface. The unstable fracture occurs when K_1 of the subinterface crack reaches K_{IC} for the ceramic material. Consequently, strength degradation during thermal cycling is caused by micro-cracking. A Crack propagation model based on an interface fracture mechanics is proposed. Using the model, effects of interface crack length on the strength of ceramic - metal joints are studied. From comparison of the predicted results with experimental results the validity of the fracture mechanics model are demonstrated.

1 INTRODUCTION

An introduction of ceramic-metal joining is a promising technology which enables us to use ceramic materials as structural components. A fracture strength of the ceramic-metal joints is, however, influenced by a thermal loading history and defects around the joint part. The defects are induced by residual stresses and a stress concentration(a singular stress field) due to the difference of thermal expansion coefficients and elastic moduli [1] [2] [3] [4].

In this study, four point bending tests for Si_3N_4 - SUS304 joints after thermal cycling were conducted and their fracture surfaces were examined by a fractographic method. Based on these results, the fracture path and the criterion for unstable fracture is discussed by an interface fracture mechanics approach. A Crack propagation model based on an interface fracture mechanics is proposed. Using the model, effects of interface crack length on the strength of ceramic - metal joints

are studied. From comparison of the predicted results with experimental results the validity of the fracture mechanics model are demonstrated.

2 EXPERIMENTAL PROCEDURE

2.1 *Fracture test and SEM analusis*

Si_3N_4 and Japan Industrial Standards(JIS) SUS304 (stainless steel) were joined by the activation metal, vacuum brazing method. A copper(Cu) sheet was used as the interlayer and a Ti-Ag-Cu alloy was used as the brazing filler metal. Material properties and conditions of joining are listed in Tables 1 and 2, respectively.

Table 1 Mechanical properties.

	Si_3N_4	Cu	SUS304
E(GPa)	304	108	196
ν	0.27	0.33	0.30
$\alpha(\times 10^{-6})$	3.0	17.7	15.0

Table 2 Typical joining conditions.

Brazing filler	:	Ti-Ag-Cu
Temperature	:	$1073 \sim 1123K$
Atmosphere	:	Vacuum. 1×10^{-5} torr
Interlayer	:	Cu (thickness 0.2mm)

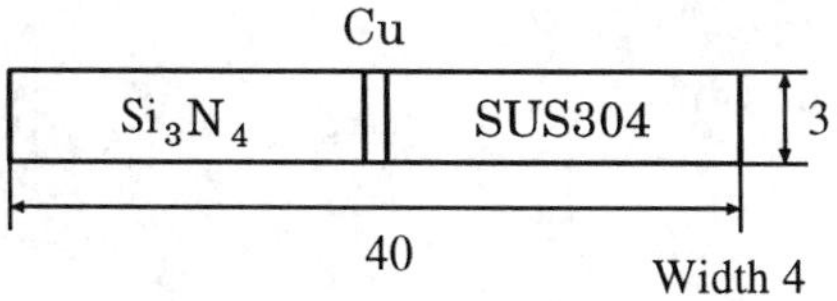

Fig. 1 Specimen configuration(unit:mm).

Fig. 1 shows the configuration and dimensions of the specimen. Four point bending test procedures in JIS were adhered to [5]. Fig. 2 shows the bending test system used. The strains near the interface were measured by a strain gauge with gauge length 2 mm. The specimens were subjected to a thermal cycle of which the maximum temperatures were between 573 and 1073 K in air and vacuum environments before testing.

Observations on the fracture surfaces for measurement of interface crack length at fracture origin were conducted by scanning electron microscopy (SEM). The crack length and inclined angle of the crack face at the fracture origin are examined precisely. Compositional analyses by X-ray method were also conducted on the fracture surface and the side surface of unfractured specimen. On the normal SEM photograph, crack initiations were identified with the hackle pattern in Si_3N_4 as schematically illustrated in Fig. 3.

According to the SEM observation, the fracture origins was on the ceramic-interlayer interface. The heights of points on the fracture surface were calculated by ordinary stereo photograph technique. Fig. 4 shows the definition of height, 'h_f', from brazing filler-ceramic interface to fracture surface in a cross section of fracture surface. Fig. 5 shows the configuration of height measurement using SEM images. Using a set of SEM images, a normal and an inclined image, the crack deflection angle from the ceramic-brazing filler interface, θ, is calculated by following equation.

$$\theta = \tan^{-1}\left\{ \frac{\frac{a}{b}\sin\theta_2 - \sin\theta_1}{\cos\theta_1 - \frac{a}{b}\sin\theta_2} \right\}, \qquad (1)$$

where $\theta_1 = 10°$, $\theta_2 = 20°$ were used in this study.

2.2 Estimation of stress intensity factor for interface crack

The joining residual stresses were analysed by the elastic-plastic finite element method [6]. From the FEM calculation, joining residual stress distributions are expressed as follows(Fig. 6).

$$\sigma_x^R = I_1^R y^{-\lambda} - c, \qquad 0 \leq y \leq B/2, \qquad (2)$$

$$\sigma_x^R = I_1^R (B - y)^{-\lambda} - c, \qquad B/2 \leq y \leq B, \qquad (3)$$

$$\tau_{xy}^R = I_2^R y^{-\lambda} - d, \qquad 0 \leq y \leq B/2, \qquad (4)$$

$$\tau_{xy}^R = -I_2^R (B - y)^{-\lambda} - d, \qquad B/2 \leq y \leq B, \qquad (5)$$

where c is determined by the equilibrium condition of force to x direction, d is determined by the symmetric property of the distribution. B is specimen height.

$$c = \frac{I_1^R}{1 - \lambda}\left(\frac{B}{2}\right)^{-\lambda}. \qquad (6)$$

$$d = I_2^R\left(\frac{B}{2}\right)^{-\lambda}. \qquad (7)$$

The stress intensity factor, K, for the interface crack is given by Hutchinson [7].

$$(\sigma_x - i\tau_{xy}) = \frac{K(y - a)^{i\varepsilon}}{\sqrt{2\pi(y - a)}}, \qquad x = 0, \qquad (8)$$

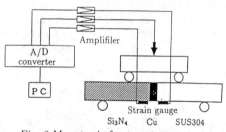

Fig. 2 Monotonic fracture test system.

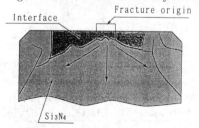

Fig. 3 Determination of fracture origin.

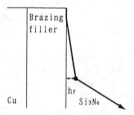

Fig. 4 Definition of height, "h_f".

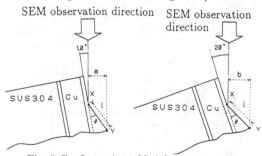

Fig. 5 Configuration of height measurement.

$$K = K_1 + iK_2, \qquad (9)$$

where a is the edge crack length in Fig. 6, $i = \sqrt{-1}$.

$$\epsilon = \frac{1}{2\pi} \ln[(\frac{\kappa_1}{\mu_1} + \frac{1}{\mu_2})/(\frac{\kappa_2}{\mu_2} + \frac{1}{\mu_1})], \qquad (10)$$

where μ_j is shear modulus, κ_j is $3 - 4\nu_j$ for plane strain, $(3 - \nu_j)/(1 + \nu_j)$ for plane stress, ν_j is pois-son's ratio, j=1 for material 1 and j=2 for material 2.

We estimated the stress intensity for interface crack at the final fracture using a principle of superposi-tion.

$$K = K^R + K^L, \qquad (11)$$

where, K^R is a residual stress intensity factor in-duced by the joining residual stresses, K^L is an applied stress intensity factor with correction fac-tor. The stress intensity factor for an edge inter-face crack subjected to distributed traction on the crack surface is given as following integrals[8]. The magnitudes of the distributed tractions are equal to the initial residual stresses, σ_x^R and τ_{xy}^R and its sign is opposite to the residual stress.

$$K_1^R = \frac{m_1}{m_2} \frac{\cosh(\pi\epsilon)}{\sqrt{\pi}} \int_0^a \sqrt{\frac{2}{\xi}} [\sigma_x^R(\xi)\cos(\epsilon\ln\xi) \quad (12)$$
$$- \tau_{xy}^R(\xi)\sin(\epsilon\ln\xi)]d\xi,$$

$$K_2^R = \frac{m_1}{m_2} \frac{\cosh(\pi\epsilon)}{\sqrt{\pi}} \int_0^a \sqrt{\frac{2}{\xi}} [-\tau_{xy}^R(\xi)\cos(\epsilon\ln\xi) \quad (13)$$
$$- \sigma_x^R(\xi)\sin(\epsilon\ln\xi)]d\xi,$$

where $\xi = a - y$ is the axis along the crack surface and its origin is at the crack tip, m_1 is the correc-tion factor for edge crack and m_2 is the correction factor for semi-infinite crack.

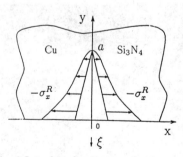

Fig. 6 Interface fracture mechanics model.

The stress intensity factor for edge interface crack subjected to pure bending stress, σ, is as follows[8] [9].

$$K_1^L = m_1\sigma\sqrt{\pi a}\{\cos(\epsilon \ln 2a) + 2\varepsilon \sin(\epsilon \ln 2a)\}, \qquad (14)$$

$$K_2^L = m_1\sigma\sqrt{\pi a}\{-\sin(\epsilon \ln 2a) + 2\varepsilon \cos(\epsilon \ln 2a)\}, \qquad (15)$$

In this study we considered th elastic-plastic ef-fect on the residual stresses using FEM. However we neglected this elastic-plastic effect on the stress intensity factor for the difficulty of calculation due to the residual stress singularity. The real part of total stress intensity factor, K_1, or that of exter-nal stress intensity factor, K_1^L, are examined their validness as mechanical parameter to evaluate the interface fracture.

$$K_1 = K_{Ic}. \qquad (16)$$

$$K_1^L = K_{Ic}. \qquad (17)$$

3 EXPERIMENTAL RESULTS

Table 3 shows a bending strength, an interface crack length, a height from brazing filler - ceramic interface and a crack deflection angle at the un-stable fracture from the parallel direction with the interface into ceramic.

Table 3 Summary of strength tests and SEM ob-servations.

TP No.	Bending strength [MPa]	Crack length [μm]	Deflect. angle [deg.]	h_f [μm]	T_{max} [K]
1	362.4	92.3	13.5	2.4	A. J.
2	375.8	14.4	8.6	2.5	A. J.
3	425.7	31.0	11.0	1.0	A. J.
4	432.1	66.7	8.6	1.5	A. J.
5	422.4	73.3	9.2	3.0	A. J.
6	349.2	56.4	15.0	2.5	573
7	395.8	68.8	18.9	2.4	573
8	382.5	653.1	6.9	3.0	573
9	329.9	123.5	7.3	1.2	773 (vacuum)
10	261.7	117.5	10.2	1.5	773
11	356.1	256.4	8.6	2.0	773
12	312.6	47.6	11.4	2.5	773
13	266.1	24.2	11.4	3.0	773
14	245.7	692.3	11.7	2.5	1073

A. J. : As joined.

Relation between bending strength and crack length is shown in Fig. 7. The "crack length" means the length of an interface crack(including microscopi-cally sub-interface crack) around the fracture ori-gin. The bars in Fig. 7 show the variation of the

1251

crack length along the crack front of a specimen. The plot symbols show the estimated crack length at the fracture origin. The bending strength of ceramic-metal joints decreased with increasing the maximum temperature of thermal cycle.

The solid line shows a predicted relation based on equation (16), the doted line shows a predicted relation based on equation (17). The agreement between the predicted values based on the total stress intensity factor and the measured values is fairly good. Degradation of the strength of ceramic-metal joints can be estimated from the increase of interface crack growth.

Fig. 8 is a typical scanning electron macro - fractograph showing ceramic - interlayer interface fracture in specimen 5 which was subjected no thermal cycling. This specimen exhibited a through-thickness interface crack growth. Fig. 9 is the medium magnification fractograph showing the detail image of "Fracture origin " designated in Fig. 8. Arrow A and arrow B show portions of fracture origin. A step exists between the portion A and B. Fig. 10 is the inclined image of the same portion of Fig. 9. The inclined image indicated that the portion A was the main crack growth and the portion B was included by the main crack growth at about 200 μm far from the edge. Fig. 11 is the inclined medium magnification fractograph showing the same fracture origin. No micro-crack was observed around brazing filler-ceramic interface in the specimen subjected no thermal cycling.

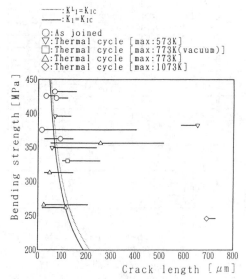

Fig. 7 Relation between bending strength and crack length.

Fig. 12 is normal high magnification fractograph showing "interface fracture surface". In the fracture origin area, the fracture is composed of facets and dark area which most likely indicate a brittle fracture of β-Si$_3$N$_4$ grains. No ductile fracture was observed.

Fig. 13 is a typical scanning electron macro-fractograph showing ceramic-interlayer interface fracture subjected a thermal cycling which maximum temperature is 773K. Though this specimen also exhibited a through-thickness interface crack growth, the crack front shape is irregular.

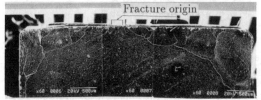

Fig. 8 TP5 SEM macro-fractograph.

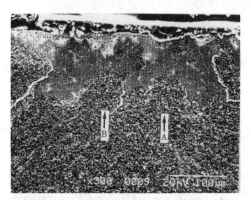

Fig. 9 Detail of fracture origin(TP5, 0°).

Fig. 10 Inclined image of fracture origin(TP5, 20°).

Fig. 14 is the inclined high magnification fractograph of "C" in Fig. 13. The inclined image indicated that the interface crack dose not continue smoothly into ceramic region. Fig. 15 is the inclined high magnification fractograph showing the detail image of "Fracture origin " designated in Fig. 13. Arrow A and arrow B show portions of fracture origin. There is an interface crack region in the portion A while there is no interface crack region in the portion B. Decohesions at ceramic-brazing filler interface is clearly visible in Fig. 16. The thermal cycle developed the microdefects on the ceramic-brazing filler interface. Fig. 17 shows the result of compositional analysis on the fracture surface for the specimen without thermal cycling. The composition of Si was detected even at the region called "interface crack" that the region looks like the interface between brazing filler and ceramic by SEM observation.

Above SEM observation and X-ray compositional analysis indicate that the location of fracture origin is in the ceramic closely along the brazing filler - ceramic interface. Fig. 18 illustrates the crack path initiated at ceramic-brazing filler interface in the cross section. Distances from "Brazing filler" to thick line show the results of height measurement by stereo analysis.

The crack initiated at the brazing filler - ceramic interface and grew as sub-interface crack apart from the interface in several micro meters. The sub-interface crack deflects into ceramic side in an angle of about 10°.

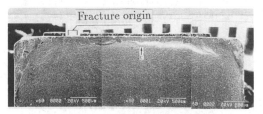

Fig. 13 TP12 SEM macro-fractograph.

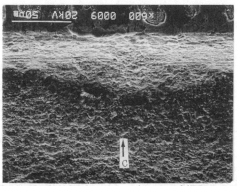

Fig. 14 Inclined image around region C(TP12, 20°).

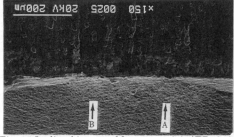

Fig. 15 Inclined image of fracture origin(TP12, 20°).

Fig. 11 Inclined image of fracture origin(TP5, 20°, medium magnification).

Fig. 12 Micro-fractograph of fracture origin(TP5, 0°).

Fig. 16 Microcracks along interface(TP12, 20°).

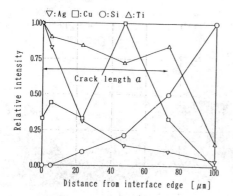

Fig. 17 Result of compositional analysis.

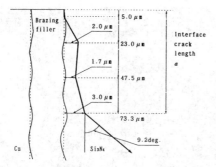

Fig. 18 Crack path in cross section.

CONCLUSIONS

Four point bending tests for Si_3N_4/SUS304 joints after thermal cycling were conducted and their fracture surfaces were examined by a fractographic method. Based on these results, the fracture path and the criterion for unstable fracture is discussed by an interface fracture mechanics approach. A Crack propagation model based on an interface fracture mechanics is proposed. Using the model, effects of interface crack length on the strength of ceramic/metal joints are studied. From comparison of the predicted results with experimental results the validity of the fracture mechanics model are demonstrated. The results obtained are as follows:

(1) The bending strength of ceramic-metal joints decreased with increasing the maximum temperature of thermal cycle. The agreement between the predicted strength based on the total stress intensity factor and the measured values is fairly good. Degradation of the strength of ceramic-metal joints can be estimated from the increase of interface crack growth.

(2) SEM observation and X-ray compositional analysis indicate that the location of fracture origin is in the ceramic closely along the brazing filler - ceramic interface.

(3) Results of measured height from "Brazing filler" to fracture surface by stereo analysis show that the crack initiated at the brazing filler - ceramic interface and grew as sub-interface crack apart from the interface in several micro meters. The sub-interface crack deflects into ceramic side in an angle of about 10°.

References

[1] Dalgleish, B. J., Lu, M. C. and Evans, A. G., Acta Metall., 36-8 (1988), 2029.

[2] Cao, H. C., Thouless M. D. and Evans, A. G., Acta Metall., 36-8 (1988), 2037.

[3] Evans, A. G., Lu M. C., Schmauder S. and Ruhle, M., Acta Metall., 34-8 (1986), 1643.

[4] Kobayashi, H., Arai, Y., Nakamura, H. and Nakamura, M., Proc. KSME/JSME Joint Conf. Fracture and Strength '90, (1990), 236.

[5] Jpn. Ind. Stand. JIS-R1601-1981.

[6] Kobayashi, H., Arai, Y., Nakamura, H. and Sato, T., Mater. Sci. and Engng, A143, (1992), 91.

[7] Hutchinson, J. W., Mear, M. E. and Rice, J. R., Trans. ASME, J. Appl. Mech., 54 (1987), 828.

[8] Rice, J. R. and Sih, G. C., Trans. ASME, J. Appl. Mech., 32, 418(1965).

[9] Rice, J. R., Trans. ASME, J. Appl. Mech., 55 (1988), 98.

[10] Bogy, D. B., Trans. ASME, J. Appl. Mech., 35 (1968), 460.

[11] Blanchard, J. P. and Ghoniem, N. M., Trans. ASME, J. Appl. Mech., 56 (1989), 756.

Elastic plastic fracture

Experimental Mechanics, Allison (ed.)© 1998 Balkema, Rotterdam, ISBN 90 5809 014 0

The effect of the release-film thickness on the behaviour of artificial delaminations used in composite materials testing

J. McKelvie & D. N. Balmforth
Department of Mechanical Engineering, University of Strathclyde, Glasgow, UK

ABSTRACT: The compressive load at which an artificial delamination will propagate, or "spread", is examined with respect to its dependence upon the thickness of the release film used in its creation.   It is found that, counter-intuitively, the spreading load increases with increasing film thickness.   An explanation is given in terms of delamination toughness, the indication being that the "real" delamination will tend to propagate at lower loads than an artificial one created by the usual release-film method. This represents a non-conservative situation.

1.   INTRODUCTION

Efforts to produce predictive systems for residual strength of damaged composites have been widespread and intensive.   In particular, attention has been focussed in many instances on the compression-after-impact (CAI) strength, and – because it is deemed to play a crucial role in strength reduction – on the growth of delamination damage under compressive load.   Examples of the types of investigations are reported in Pavier & Chester 1990, Ilcewicz et al. 1991, Dost et al. 1993, Greenhalgh 1993.

Central to a system for predicting growth in multiply-delaminated material is a properly validated predictor for the growth of a single delamination, and therefore there has been experimental work in numerous centres in order to provide practical data for comparison with the predictive tools proposed.   There has been no report of any method for the production of natural single delaminations of particular geometry, and the normal procedure is to create artificial delaminations by the insertion of a non-stick material such as PTFE at the manufacturing stage.   This is particularly convenient since the shape of the "delamination" and its positioning are readily controlled.   The first author has reported the results of a number of such tests, - McKelvie et al. 1995.

It is, however, a matter for concern as to the extent to which such artificial devices behave in the same manner as the real delamination of the same geometry, and this present work reports an investigation related to that particular concern.

2.   THE INTUITIVE EXPECTATION

The introduction of the non-stick film leaves the cured material with a local bulge, which may vary in degree depending upon the thickness of the film and the curing conditions (notably the pressure). Intuitively, the more pronounced the bulge the more readily will a compressive load induce further bulging of the minor sublaminate, leading to the initiation of "real" delamination, spreading, and collapse.

3.   EXPERIMENTAL OBJECTIVE AND PROCEDURES

The approach was to manufacture specimens with different thicknesses of artificial delamination material and test them in compression, to see what difference, if any, they made to the behaviour.

The testing procedure followed that described in McKelvie et al. 1995, using the same anti-buckling rig, and specimen material and geometry as follows:-

Material: Ciba Carbon/Epoxy T300/914C
Length: 170mm + ends with GRP tabs
Width: 45mm
Thickness: 2mm nominal
Stacking sequence: $[O_2^\circ/45^\circ/O_2^\circ/-45^\circ/O^\circ/90^\circ]_s$

A first plaque was manufactured by standard bagging, evacuating, and autoclaving, and from it 5 specimens were cut. These had artificial delaminations positioned centrally length-wise and width-wise, between layers 3 and 4. The delamination release films were of PTFE either 120 microns or 20 microns.

3.1 Shadow Moiré measurements

In order to evaluate the degree of "bulge" in each case, shadow moiré was employed. Fig 1 shows a typical fringe pattern, the contour interval being 100 microns. By slightly inclining the grating, a uniform field was superimposed, as can be seen. This allowed the displacement field to be plotted and the maximum height of the bulge determined by interpolation and subtraction of the superimposed field. The results are shown in the second and third columns for Specimens 1 to 5 in Table 1 (which also gives the findings on two specimens previously manufactured at NEL in a hot press under very high pressure). These results indicated that the bulge height was in fact not related at all to the release film thickness and showed that the bulge may be much larger than the film thickness (and, in fact, as in the case of the NEL specimens, may be *smaller* than the film thickness, due presumably to the component materials deforming under the high pressure).

3.2 Micro sections

A second plaque was manufactured with the same processing, and the bulge areas examined by microsectioning. It was evident that despite the application of vacuum, there was still air trapped in the region, as can be seen in Fig 2.

A further plaque was then produced, this time the material being evacuated after only one ply had been laid over the release film. Subsequent microsectioning showed that this modification produced the desired result, -fully consolidated material with no air entrapment.

Four specimens were made by this process, each having a 10 micron thick release film. Shadow moiré revealed no evidence of any bulging – specimens 6 to 9, Table 1.

3.3 The results

The five specimens originally manufactured, and the second four, were loaded slowly in compression, and notes were made of the loads as follows.

1. P_B - When "blistering" occurred. (Blistering being when a first bulge, or increased bulge, occurred within the dimensions, approximately, of the release film).

2. P_S - When the delamination first "spread" – that is, beyond the area of the release film.
(Both blistering and spreading are accompanied by typical noise, as had been previously confirmed by US C-scan).

3. P_F - When the specimen collapsed completely.

These results too are shown in Table 2.

Specimen 1 did not exhibit any blistering or spreading prior to collapse – as was expected on the basis of previous experience with small delaminations. Specimens 2 to 5 did not exhibit any blistering, - as could be expected, since they already had a "blister" – but they did exhibit obvious spreading. Specimen 7 went into a "strong" buckling mode, with the whole laminate bending convexedly as viewed from the thin sublaminate side and with no delamination spread.

4. THE STATISTICAL EVALUATION

The important load, from a modelling point of view is the spreading load, as this is the point at which irreversible damage commences; (blistering tends to be elastic). There is a suggestion in the results that, contrary to expectation, the "no bulge" cases had a lower spreading load than the larger bulge ones, but for the purpose of drawing obvious conclusions the numbers of specimens were rather low. It was therefore decided to carry out a standard statistical analysis, using the specimens 2 to 5 on the one hand, and 6,8,9 on the other. To reinforce these statistics, pre-existing data for two other NEL specimens (identical to those in Table 1) were added to the results for specimens 2 to 5. The data used were therefore as follows.

Sample 1. Thick delamination spreading loads (kN)	Sample 2. Thin delamination spreading loads (kN)
35.0	33.6
36.8	35.0
37.9	32.5
39.1	
35.0	
36.4	
Mean 36.7	33.7

A t-test on the means gave the value of t = 2.79, which, with the seven degrees of freedom, indicated that the difference is significant at the 97.5% confidence level. (It is to be noted that the results from the earlier work actually tend to reduce the significance, rather than increase it).

5. AN EXPLANATION

As suggested already, the finding is quite counter-

Table 1: Defect sizes and failure loads.

Specimen number	Release film: Size (mm)	thickness (μm)	Moiré "bulge" result (μm)	P_B (kN)	P_S (kN)	P_F (kN)
1	5x5	125	95	-	-	44.4
2	10x10	20	130	-	37.9	47.0
3	10x10	20	130	-	39.1	48.0
4	10x10	125	90	-	35.0	47.3
5	10x10	125	120	-	36.4	44.5
NEL No1	10x10	125	50	n/a	n/a	n/a
NEL No2	10x10	125	50	n/a	n/a	n/a
6	10x10	10	0 (not detectable)	33.6	33.6	45.3
7	10x10	10	"	-	-	52.1
8	10x10	10	"	31.0	35.0	42.9
9	10x10	10	"	31.5	32.5	45.0

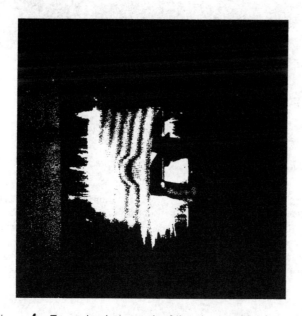

Figure 1. Example photograph of the shadow Moiré results.

intuitive, and an explanation from the mechanics is proposed:-

It is known in delamination toughness testing, (see for example Davies et al. 1989 and 1990), that the value for G at initiation from a thick "starter" is higher than the value from a thinner one and than that determined once the delamination has propagated. This is ascribed to the existence of a region of resin-rich material ahead of the release film – as indeed can be seen in Fig.3, which shows a properly consolidated artificial delamination. This means that the "crack" is effectively blunt, and therefore returns a high toughness, in comparison to the much sharper crack-tip of a delamination

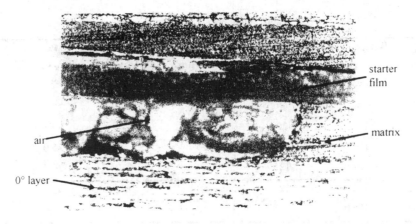

Figure 2.   20 μm release film centre (750x).

Figure 3.   - 125 μm release film edge (200x).

extending into normal interlaminar conditions.

The analogy is fairly clear in the current context:- a larger bulge will almost inevitably imply a thick resin-rich region, and therefore will endure a higher G-value (which requires a higher load) prior to propagation, than a smaller bulge.

6.   CONCLUSION

Contrary to expectation, it has been found that thicker artificial delaminations are more difficult to cause to spread in compression than thinner ones. This has implications for experimental modelling of the residual strength of damaged composites, since experimental values for the spreading load will tend to be higher with artificial delaminations created with typical release film material than for real delaminations, in which the critical energy release rate for propagation will be lower.

This means that such experimental values will be non-conservative, which is a particularly undesirable situation.

7.   REFERENCES

Davies,P.,   Cantwell,W.,   &   Kausch,H.H. Measurement of Initiation Values of G_{IC} in IM6/PEEK Composites. 1989.   Compos, Sci, Technol.. 35: 301-313.

Davies,P. , Moulin,C., Kausch,H.H., & Fischer,M. 1990. Measurement of G_{IC} and G_{IIC} in Carbon Epoxy Composites. Compos. Sci. Technol. 38: 211-227.

Dost,E.F., Ilcewicz,L.B., Avery,W.B. and Coxon,B.R. 1996. Effects of stacking sequence on impact damage resistance and residual strength of quasi-isotropic laminates. In *Composite Materials: Fatigue and Fracture third ASTM 1110*, T.K.O'Brien (Ed.): 476-500.

Greenhalgh,E.S. 1993. Delamination growth in carbon-fibre composite structures. *Composite Structures*, 23: 165-175.

Ilcewicz,L.B., Dost,E.F., and Coggeshall,R.L. 1996. A model for compression after impact strength evaluation. *Proc. 21st Int.SAMPE Conf., Altantic City*, Chap 97: 130-140

McKelvie,J., Jam,M. and Fort,D. 1995. Compression Testing of Damaged Graphite/Epoxy Composite. *Proc. 3rd Int Conf on Deformation and Fracture of Composite, Guildford,:* 180-189.

Pavier,M.J. and Chester,W.T. 1990. Compression failure of carbon fibre-reinforces coupons containing central delaminations. *Composites*, 21: 23-31.

Experimental Mechanics, Allison (ed.) © 1998 Balkema, Rotterdam, ISBN 90 5809 014 0

Extension resistance of a semi-infinite crack in a viscoelastic strip

Akihiro Misawa
Kanagawa Institute of Technology, Atsugi, Japan

Takahisa Nakamine
TOK Bearing Company Limited, Itabashi-ku, Tokyo, Japan

Masahisa Takashi
Aoyama Gakuin University, Setagaka-ku, Tokyo, Japan

ABSTRACT: Crack tip stress field in a viscoelastic strip is complicated dependent not only on crack tip configuration, length and loading condition but also on the histories of crack extension and loading. Extension resistance of a semi-infinite crack traveling in a viscoelastic strip is investigated standing on precise evidences observed in experiments, in which crack extends with some constant velocities under several constant temperatures. Time variation of J'-integral in the process of crack propagation was evaluated. The results obtained are summarized that during crack propagation with a constant velocity J'-integral keeps a constant value, which is considered as one of the material characteristics dependent on temperature and crack velocity.

1 INTRODUCTION

Studies on the critical condition of crack growth and subsequent behaviors in viscoelastic materials are very difficult to compare with those in elastic materials because these materials show the dependence on not only temperature but also loading history. It is obvious that strain ε at time t at which stress comes to a value σ_0 couldn't be unique even in uni-axial tension test depending on the history of loading. In the case of crack growth behaviors, stress and strain are dependent on not only crack size but also crack extension rate. In addition, stress and strain analysis on crack growth problems belonging to the mixed boundary value problems is very difficult due to the inapplicability of the well-known correspondence principle (Lee 1955) in linear viscoelasticity.

The authors have already proposed a method for introducing an artificial crack which can simulate a mathematical model in a specimen (Misawa et al. 1982), the critical condition of crack growth obtained by explicit experiments using this technique (Misawa et al. 1982) and an equation predicting the crack growth rate (Misawa et al. 1984).

In this paper, the extension resistance of a semi-infinite crack propagating with a constant velocity in a viscoelastic strip is discussed standing on experimental evidences.

2 CRACK EXTENSION RESISTANCE IN A VISCOELASTIC MATERIAL

Although stress and strain in an elastic strip having a crack are dependent on the shape of crack tip, crack size and loading conditions, those in a viscoelastic strip are influenced by the histories of crack size variation and loading conditions, too. As mentioned before, it is very difficult to analyze stress and strain around propagating crack tip in a viscoelastic material. Therefore, it is next to impossible to discuss strictly the crack extension resistance under the general conditions. Consequently, crack extension resistance has to be started from the discussion on more simplified problem standing on experimental results and evidences.

Stress and strain in an infinite viscoelastic strip with a non-propagating semi-infinite crack are not dependent on crack size but only on loading history. If the crack has once extended then those depend not only on the loading history but also on the history of crack velocity after that instant. In the case if a crack extends with a certain constant velocity, stress and strain around crack tip are only affected by the loading history in order to preserve the constant crack velocity. In addition, on the other hand, if a semi-infinite crack extends in a relaxed stress state, stress and strain in the vicinity of the crack tip do not vary with time. The crack extension resistance in this case will be constant having crack velocity as a parameter. In other words, in a problem where a semi-infinite crack propagates with a constant velocity in a stress-relaxed infinite viscoelastic strip, crack extension resistance could be measured and evaluated much easier.

In this paper, the extension resistance of a semi-infinite crack with a certain constant velocity in a stress-relaxed infinite viscoelastic strip is studied under the condition of several constant temperatures.

2.1 J'-integral

The well-known correspondence principle is inapplicable to theoretical analysis of crack growth behavior in a viscoelastic body belonging to the mixed boundary value problems because the type of boundary condition at a point on the expected crack surface changes before and after the passage of the crack. Only the values of normal stress and normal displacement on the crack surface are known even if the extended correspondence principle (Graham & Sabin 1973) could be applied. Therefore the crack extension resistance which is defined as the energy required to make a unit area of crack surface couldn't be measured because stress and strain in the vicinity of propagating crack tip would not be known. The authors have paid attention to the so-called J-integral that could be applicable to the problem even in the case that the values of stress and strain in the vicinity of crack tip are still unknown. This concept, however, can not be applicable to the case of viscoelastic body in any extent. The authors have already proposed a new concept of J'-integral in one of previous papers (Misawa et al. 1990), in which the strain energy density function in J-integral was replaced by the free energy per unit volume. Let $\sigma_{ij}, \varepsilon_{ij}$ and u_i denote the Cartesian components of stress, strain and displacement. Γ is a curve

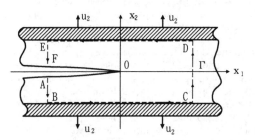

Figure 1 Path of integral.

surrounding the crack tip. ds is the infinitesimal arc length along Γ. n_j is a component of a unit outer normal vector n on Γ. Then, J'-integral is defined by

$$J' = \int_{\Gamma} \psi dx_2 - \sigma_{ij} n_j \frac{\partial u_i}{\partial x_1} ds,$$

where ψ is the free energy per unit volume. Integral is evaluated in a counterclockwise along Γ from the lower crack surface to the upper crack surface. ψ is also defined by

$$\psi = \frac{1}{2} \int_{-\infty}^{t} \int_{-\infty}^{t} G_1(2t-\tau-\eta) \frac{\partial \varepsilon_{kl}(\tau)}{\partial \tau} \frac{\partial \varepsilon_{kl}(\eta)}{\partial \eta} d\tau d\eta$$

$$+ \frac{1}{6} \int_{-\infty}^{t} \int_{-\infty}^{t} \{G_2(2t-\tau-\eta) - G_1(2t-\tau-\eta)\}$$

$$\times \frac{\partial \varepsilon_{kk}(\tau)}{\partial \tau} \frac{\partial \varepsilon_{ll}(\eta)}{\partial \eta} d\tau d\eta,$$

where G_1 and G_2 are the relaxation functions in shear and in isotropic compression respectively. If J'-integral is applied to an infinite viscoelastic strip with a semi-infinite crack and Γ is adopted as shown Figure 1, the J'-integral is expressed as,

$$J' = \int_{CD} \psi dx_2.$$

In this study crack extension resistance is evaluated by the value of J'-integral because not only theoretical analysis of stress and strain in a viscoelastic body with propagating crack is very difficult, but also other reasonable measure couldn't be found.

2.2 Procedure of Experiment

Figure 2 shows the geometry of specimen. Specimens having an initial edge crack a_0=30mm is made of polyurethane rubber sheet with 1mm in thickness of which the glass temperature T_g is 220.5K. The ratio of initial crack length to the other dimensions of specimen is chosen to satisfy approximately the conditions of a semi-infinite

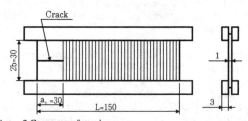

Figure 2 Geometry of specimen.

crack. The fine grid patterns with the space of 1mm are printed on the specimen in order to measure the crack length variation with time. Figure 3 shows the relaxation function in tension $E_r(t)$ at T_0=267.2K. This material is thermo-rheologically simple and the time-temperature shift factor of this material can be estimated by W.L.F. equation (Williams et al. 1955).

Experiments were carried out at four different constant temperatures, 248.2, 243.2, 233.2 and 230.2K. These temperatures are chosen by consideration on the relaxation time, crack velocity

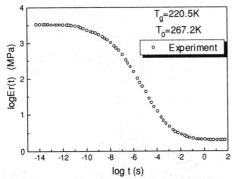

Figure 3 The relaxation function in tension at 267.2K.

and length of specimen L taking evidences and results in the pretest into account. If it could be regarded that relaxation process is nearly completed within time $t=10^{1.5}$s (i.e. about 3.16×10^{2}s) judging from Figure 3, the relaxation times at 248.2, 243.2, 233.2 and 230.2K are read about 5, 30, 2200 and 10,700s respectively.

The moment constrained type grips were used in order to extend the crack perpendicularly to loading. The displacement loading was applied to the lower edge of specimen and controlled by on-off to keep a constant crack velocity.

3 RESULTS AND DISCUSSIONS

Some examples of crack growth curves obtained at 243.2 and 230.2K are shown in Figures 4 and 5.

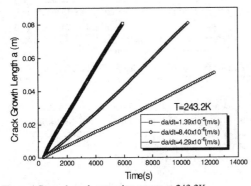

Figure 4 Several crack extension curves at 243.2K.

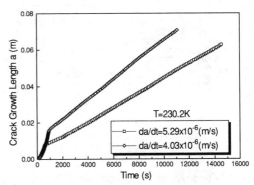

Figure 5 Several crack extension curves at 230.2K.

These figures show that crack has extended with a nearly constant velocity from early stage of growth behavior. The similar crack growth behavior has been obtained in other cases at different temperatures. Three kinds of constant crack velocity were selected at each temperature, 248.2, 243.2, 233.2K, but only the first digit of velocity could be controlled as same at best. Slower crack velocity is chosen at 230.2K because relaxation time is very long. The time variation of the displacement loading and the calculated values of J'-integral were shown in Figures6a,b to Figures 9a,b. Here, it was assumed that Poisson's ratio v is constant. Also, relaxation function in tension shown in Figure 3 is approximated by Prony series to be applied to the computation of J'-integral. Each plot in Figure 6a to Figure 9a shows the starting time or ending time of

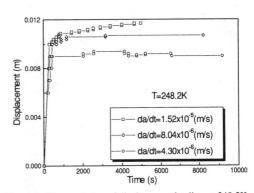

Figure 6a Time variation of displacement loading at 248.2K.

loading control and displacement at that instant. These figures show that in order to keep a constant crack velocity, relatively large adjustment by displacement loading control has to be added frequently in the early stage of crack growth at higher temperatures, 248.2, 243.2 and 233.2K. After

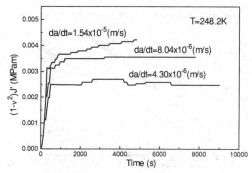

Figure 6b Time variation of J'-integral value at 248.2K.

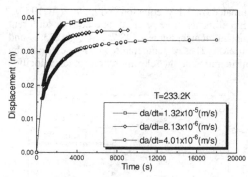

Figure 8a Time variation of displacement loading at 233.2K.

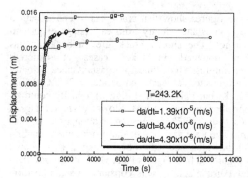

Figure 7a Time variation of displacement loading at 243.2K.

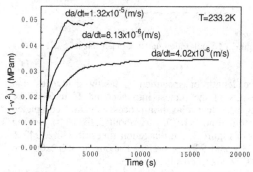

Figure 8b Time variation of J'-integral value at 233.2K.

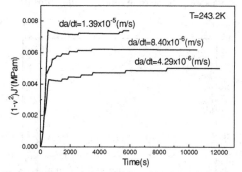

Figure 7b Time variation of J'-integral value at 243.2K.

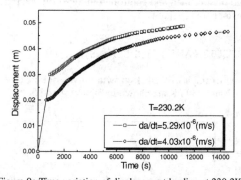

Figure 9a Time variation of displacement loading at 230.2K.

the crack has extended with a constant velocity for a relatively long time, only small adjustment of displacement loading control is occasionally required. It would be pointed out that crack in this circumstance is propagating with a constant velocity in a relaxed stress state. To confirm this prospect, additional experiments were carried out at 233.2K, in which different histories of displacement loadings

in the early stage of crack growth were applied. The time variation of the displacement loadings and crack growth curves are shown in Figure 10 and 11. Each crack shows almost the same crack velocity except in very early stage of crack growth. On the other hand, the history of displacement loading is very different each other in the early stage although each displacement becomes almost the same value

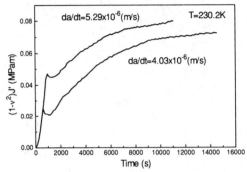

Figure 9b Time variation of J'-integral value at 230.2K.

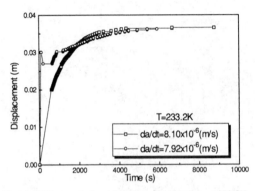

Figure 10 Time variation of displacement loading at 233.2K.

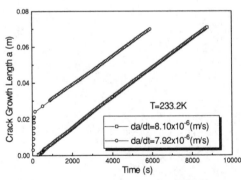

Figure 11 Several crack extension curves at 233.2K.

after crack propagates with a constant velocity for long period. Therefore the difference of loading history could be ignored in this circumstance. The control of displacement loading is adjusted quite often up to the final stage of crack propagation at the lowest 230.2K as shown in Figure 9a. This fact

shows that crack extension in relaxed stress state couldn't be carried out at 230.2K because relaxation time is too long.

When the displacement loading for almost constant crack velocity is successfully controlled, the value of J'-integral seems to keep a constant value corresponding to each constant crack velocity as shown in Figure 6b to Figure 8b. Therefore, when crack extends with a constant velocity in a relaxed stress state, the experimental evidences mentioned above could lead to the following equation for the case of constant crack velocity under constant temperature,

$$J' = J'_R.$$

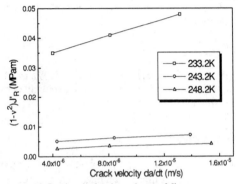

Figure 12 Crack velocity dependence of J'_R.

Figure 12 shows the relationship between these constant values of J'-integral, J'_R and crack velocities. From this figure it is pointed out that J'_R is linearly dependent on crack velocity at each temperature and this dependence increases with temperature decrease. The dependence of J'_R with the same crack velocity on temperature is shown in Figure 13. J'_R increases with temperature in each crack velocity.

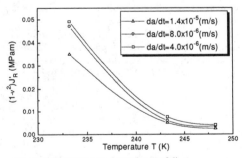

Figure 13 Temperature dependence of J'_R.

It is already pointed out that crack growth behavior at 230.2K can not be regarded as same as that in relaxed stress state. But Figure 9b suggests that the value of J'-integral will become constant if crack extends with a constant velocity in a relaxed stress state and when a smooth extrapolation could be allowed.

4 CONCLUSIONS

1. Crack velocity is possibly controlled constant by adjusting of displacement loading. But the designated velocity with two significant figures could not be realized in a strict sense by this method

2. Extension resistance of a semi-infinite crack propagating with a constant velocity in a stress-relaxed infinite viscoelastic strip is described by J'-integral as follows;

$$J' = J'_R ,$$

where J'_R is considered as one of the material characteristics dependent only on crack velocity and temperature.

REFERENCES

Graham, G.A.C. & G.C.W. Sabin 1973. The correspondence principle of linear viscoelasticity for problems that involve time-dependent regions. *J. Engng. Sci.* 11:123-140.

Lee, E.H. 1955. Stress analysis in viscoelastic bodies. *Quart. Appl. Math.* 13:183-190.

Misawa, A., M.Takashi & T.Kunio 1982. Threshold condition of crack growth in a viscoelastic plate. *Proc. 1982 Joint Conf. On Exper. Mech., Part II*, 798-813.

Misawa, A., M.Takashi & T.Kunio 1984. New criteria for onset and subsequent crack growth in a viscoelastic strip. *Proc. the 6th ICF, 2571-2578.*

Misawa, A., M.Takashi & T.Kunio 1990. Application of conservation law to the threshold of crack growth in viscoelastic strip. *Proc. the 9th Int. Conf. Exper. Mech.* 5:1786-1794.

Williams, M.L., R.F. Landel & J.D. Ferry 1955. The temperature dependence of relaxation mechanisms in amorphous polymers and other glass-forming liquids. *J. Amer. Chem. Soc.* 77:3701-3707.

Experimental Mechanics, Allison (ed.) © 1998 Balkema, Rotterdam, ISBN 90 5809 014 0

Factors affecting J_{IC}-value of ABS resin: Crosshead speed and retention time of displacement

H. Fujino
Department of Mechanical Engineering, Kitakyushu College of Technology, Japan

S. Kawano
Department of Mechanical Engineering, Yamaguchi University, Japan

M. Fujii
Ube Industries Limited, Japan

ABSTRACT: This paper deals with the factors having influence on the fracture toughness of ABS resin which is one of the rubber-toughened solid polymer. For this purpose, three-point bending moment test was performed and the dependence of J integral (J_{IC}-value) upon the loading rate at loading point (that is, crosshead speed symbolized by HS) and the duration time for holding the given displacement constant (represented as Tr) were studied. In addition, the observation of the surface of specimens broken forcibly after testing was made with the scanning electron microscope (SEM) to determine the propagating crack .The mechanical model was proposed to explain the dependence of J_{IC}-value on HS and Tr qualitatively.

1 INTRODUCTION

Fracture mechanics has evolved for understanding and preventing the brittle fracture of metallic materials containing crack-like defects based on continuum mechanics and it has been a very useful approach for evaluating the crack-strength of metallic materials. But in general the toughness of solid polymers with a sharp notch or a keen crack is evaluated by Izod or Charpy impact test. The method of fracture toughness testing for brittle polymers has been standardized as ASTM-D-5045-91 in 1991 (William & Cawood.1990), but for ductile polymers has not yet established (Narisawa. 1993). This method estimates by stress intensity factor, termed K_c. On the other hand, ductile polymers that produce significant crack tip plasticity should be assessed by J-integral which means energy release rate during a crack extension. On the remarkable increase in use of polymers as structural materials, it is expected to enact the standards for ductile polymers as soon as possible. Thus, it should be made haste now to collect the data on factors affecting the fracture toughness, such as loading speed (Hobbs & Bopp.1980), ambient temperature, duration for holding the given displacement constant (retention-duration) and etc.

ABS resin used in this study is one of the typical rubber-toughened amorphous polymer alloys and widely applied to exterior parts of electric products and so on. Numerous papers have ever been presented concerning the Izod impact test in a wide range of temperature from structural viewpoint and the dependence of orientation and anisotropy (Yokota & Asano.1991). But the papers on the fracture toughness test for ductile polymers do not seem to be sufficient.

This paper makes report on the results of 3-point bending test performed to look over the effect-factors in the fracture toughness of ABS resin. The matters investigated in this study are as follows: (1) Observation of the growth of craze and the propagation of crack during deformation (2)SEM observation of the section fractured forcibly after testing to discuss the propagating crack (3)The dependence of fracture toughness J_{IC}-value upon cross head speed HS and retention-duration Tr. (4) The mechanical model is proposed to explain the obtained results qualitatively.

2 PREPARATORY EXPERIMENT

A series of tensile and compressive tests was done at a variety of strain rates and ambient temperatures. Dumbell-shaped specimens were used in tensile test, and the temperature-environment test was performed in environment chamber adjustable to -70 to 250°C.

2.1 *Upper yield stress in tension*

Figure 1 describes the relationship between upper yield stress σ_y and strain rate $\dot{\varepsilon}$ with regard to different ambient temperatures t. This indicates that the upper yield stress is proportional to the logarithm of strain rates and in inverse proportion to ambient temperature. This relationship is written by a linear expression.

$$\sigma_y = A \log \dot{\varepsilon} + B t + C \qquad (1)$$

, where on this material, A=5.0 [MPa], B=−0.37 [MPa/°C] and C=73 [MPa] given from Figure 1. It is well known that the coefficient of viscosity grows

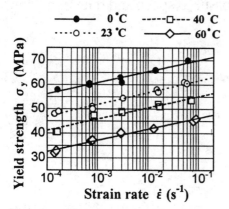

Figure 1. $\sigma - \dot{\varepsilon}$ curve at different temperatures

2.3 Stress-whitening and behavior of crazing

Figure 3 shows the photographs of whitening taking place by two varieties of strain rates, that is, 3.0×10^{-5} /s (a), 7.0×10^{-2} /s (b), respectively. Low stress rate causes many clear stripes-like whitening which scatter on the specimen, whereas the whitening by high strain rate spreads uniformly throughout the specimen and appears like surface mist as indicated in the Figure (b).

Figure 4 depicts the SEM micrographs of crazes, in which Figure (a) and (b) show the crazes caused by strain rate of 1.5×10^{-4} /s and 10^3 /s, respectively. The crazes are different in size and in intensity by the strain rate, that is, the crazes are few in number and big as the strain rate is decreased. This shows that formation-mechanics of crazes depends upon the strain rate.

larger with strain rate and decreases with temperature and such behavior is considered to result from viscosity. In general, the non-linear relationship exists among σ_y, $\dot{\varepsilon}$ and t. However, the results above does not consist with the general relationships.

2.2 Internal temperature-variation in compression

The compressive test was made in order to investigate how internal temperature varied during deformation. The temperature was measured in means of ϕ 0.5 mm thermocouple (Chromel-alumel) which was inserted in a tiny hole on the side of specimen. The variation of the increment of internal temperature (from 23 ℃) with strain rates for different strains is depicted in Figure 2. It shows that the increment grows larger with strain rate and strain. The maximum increment is approximately 18 ℃ at strain rate of 2.5×10^{-1} /s and strain 0.5 mm/mm. For ambient temperature of 60 ℃. It corresponds to about 78℃.

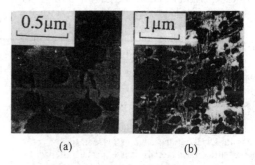

(a)   (b)

Figure 3. Whitening for different strain rates

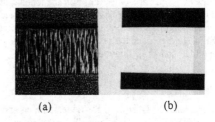

(a)   (b)

Figure 4. SEM micrograph of micro-craze
for different strain rates

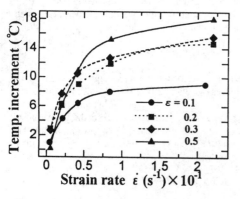

Figure 2. Temperature – strain rate measurements
as a function of true strains

3 EXPERIMENTAL

3.1 Specimen

ABS polymer is composed of a grafted polybutadiene rubber and a styrene/acrylonitrile copolymer (AS). The material used in this investigation is a commercial ABS of fire-retardant type. The tensile strength is 54 MPa (ambient temperature 23℃,loading speed20 mm/min).

The specimen is a rectangular bar with a V-shaped single-edged crack in the middle along the length whose dimensions are of 63 in length , 13 in depth and 6.25 in width (mm), as shown in Figure 5.

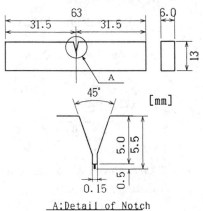

[mm]

A:Detail of Notch

Figure 5. Shape and Dimension of Specimen

3.2 *Jig for cutting a pre-crack*

The simple jig devised in author's laboratory for making a pre-crack is described in Figure 6. It consists of two elements of the symmetrical shape, which are joined together with two bolts as illustrated in the Figure. On the inside of elements, a shallow groove adjustable to insert the edge is cut. This groove is slant by 30 degrees to the upper plane of the jig with a guide of inverse V-shape which is just fit for the notch of the specimen. The depth of pre-crack is able to be adjusted easily and precisely by sliding the edge along the slant groove. The razor is used for the edge.

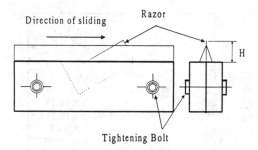

Direction of sliding   Razor

Tightening Bolt

Figure 6. Jig for cutting pre-crack

The pre-crack is made in the way that the specimen was slid carefully along the guide of the jig which was fixed strongly with a vice, pressing down with hands after milling work for cutting V-notch.

3.3 *Experiment Procedure*

The 3-point bending test was carried out for cross head speed-dependence test and retention duration-dependence test. The experiment was made in the room of 23°C and 50 %. The testing machine used

here was a ball screw tester. The load was measured through 1kN load transducer. The real displacement is estimated from HS and the time taken to the given displacement after testing.

The given displacements are 0.7, 1.0, 1.2 and 1.4 mm. The cross head speeds HS are 0.5, 10, 50, and 200 mm/min. Three kinds of 0, 2.0 and 5.0 seconds were set up as the retention-duration Tr at HS of 50 and 200 mm per minute.

J-value is calculated by the equation below.

$$J = 2 \times S/A \qquad (2)$$

, where S is the area of load-displacement curve and the capital, A, is the area of ligament (Rice et al.1973).

The R-curve was drawn from the measured points by least-squares and the fracture toughness J_{IC}-value was estimated as J-integral at crack length of zero on R-curve, not taking account of yielding of pre-crack tip. And all the J-values in duration-time test are the values at Tr =0.

To measure the growth of crack, the specimens were broken at ligament after soaked in liquid nitrogen for twenty minutes. The measurement of crack-growth was done by monitoring CCD camera. The behavior of whitening and cracking was monitored at real time through CCD camera and display, recording in the video-recorder. Combining the CCD camera and display gives the magnification of about 150 times.

4 RESULTS

4.1 *Observation of fractured surface and Behavior of whitening and cracking*

(1) *Observation of fractured surface*: Figure 7 is the photo (a) and the schematic illustration of fractured surface (b). It shows that the fractured surface consists of three parts of A, B and C as symbolized in Figure (b), which may be distinguished one from the other easily because of their characteristic appearances. Each portion in the photo and the illustration corresponds mutually. Portion of A is a notch cut with a razor. Portion B has a very velvety face of snow-white. C part is as gray as the virgin material and rough in undulation.

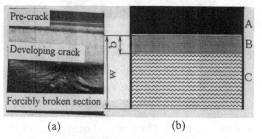

(a)   (b)

Figure 7. Ruptured section after freezing and the schematic illustration

Figure 8 depicts the SEM micrographs of Figure 7 (a). This reveals that the portion equal to B in Figure 7 presents the wavelets-pattern, whose most undulations stand regularly in the direction parallel to propagation of crack. The undulations of portion C in Figure 8 turns toward he direction normal to the growth of crack and appears to be multi-laminar in comparison .

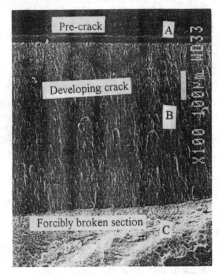

Figure 8. SEM micrograph of fractured surface

(2)*Behavior of whitening and tip of crack*: A sample of whitening is shown in Figure 9. Whitening initiates at the tip of pre-crack at about 40 % of maximum displacement 1.4 mm, extending the length and width toward the direction of crack-growth with displacement. However unloading decreases both the length and width by nearly 20 % ,that is, "*sparing* "happens. Generally, several whitening-stripes take place.

Figure 9. VTR photograph whitening at the tip
of pre-crack under loading

Crack-growth start at the tip after whitening grows to a pretty large extent. The number of occurring cracks is single or plural, depending upon the quantity of displacement. But all cracks do not develop to fracture and some stop the growth on the way.

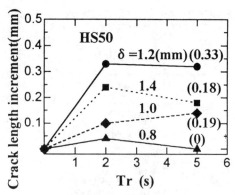

Figure 10. Crack growth during holding the
displacement constant

The observation with CCD camera shows that the cracks never develop up to more than 10 % of whitening length, which is approximately 50% of the length, b, of B portion in schematic illustration of Figure 7. Accordingly, it is considered that the remaining crack are behind B. Furthermore, since the length, b, is about 20 % of whitening length, w in Figure 7, 80% of whitening supposed to be inside of C in Figure 7.

(3) *Time-delay phenomenon of cracking*: Crack develops even when the displacement is held constant because of time-relay of strain as depicted in Figure 10. This Figure shows that crack grows with Tr except the given displacement 0.8 mm and the crack-growth stops after two seconds under every displacement. However, the final increment does not depend upon the quantity

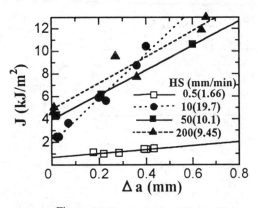

Figure 11. R−curve for different HSs

of given displacement. The case of HS 200 mm/min. indicates the tendency similar to that of HS 50 mm/min.

4.2 *Fracture toughness test*

Figure 11 reveals the R-curve at different crosshead speeds HS (mm/min). The longitudinal and horizontal

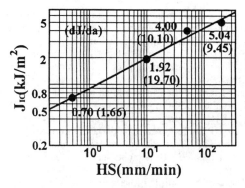

Figure 12. Variations of J_{IC}-value with various HSs

axes are J-value (kJ/m²) and crack growth Δ a (mm) respectively. The figures in parenthesis indicate the crack-growth sensitivity dJ/da at every HS. The dJ/da differs from HS to HS and there exists no fixed relationship between dJ/da-value and HS. Figure 11 tells that dJ/da-value at HS 0.5 mm/min is the least, 1.66, while dJ/da at HS 10 is the maximum, 19.7, which is approximately 11.8 times of the least and it is extremely easy for the crack at HS 10 to shift to unstable crack.

Figure 12 shows the relationship between J_{IC} and HS drawn on log-log diagram, in´which the measured points are approximated with straight lines by least-square method. The figures in diagram and in parenthesis designate J_{IC}-values and dJ/da respectively at every HS. The J_{IC}-HS diagram indicates that the maximum value of J_{IC} is 5.0 at HS 200mm/min and the minimum, 0.7 at HS 0.5 mm/min, the ratio of HS 200 to HS 0.5 being about 7.1.

The straight line give an empirical equation.

$$\log (J_{IC}) = 0.34 \log (HS) - 0.10 , $$
$$(0.5 \leqq HS \leqq 200 \text{ mm/min}) \qquad (3)$$

This reveals that J_{IC}-and dJ/da-value are both much dependent upon HS. The relationship between J_{IC}-value and retention-duration, Tr, at HS 50 and 200 (mm/min) is shown in Figure 13. J_{IC}-value for Tr=0

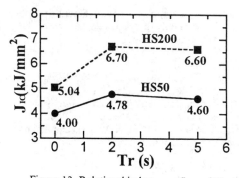

Figure 13. Relationship between J_{IC} and Tr

is the smallest and J_{IC} for Tr=2 is nearly equal to for Tr=5 at both HSs. This tells the change of J_{IC}-values by retention-duration Tr will stop by period of 2 seconds and Tr will raise J_{IC}-value by a degree. Comparing the maximum J_{IC} with the values at Tr=0 for HS 50and HS 200, the ratios are 1.2 and 1.3 respectively. These values show that the dependence of J_{IC} upon retention-duration is not so great as upon HS.

5 DISSCUSIONS

5.1 Determination of developing crack

The argument for determining the developing crack of ductile polymers is often made because of the obscurity of initiation-point or the difficulty of distinguishing the developing crack from the crack forcibly broken after testing. In (1) and (2) of Section 1 of chapter 4 it was clarified that the fractured surface was composed of three different portions. It is considered to be caused by the differences in the mechanism of fracture as follows. The fibrils initiate to orient at the tip of pre-crack and the craze propagates inwards with displacement, generating voids to turn into crack by degrees. That is to say, the crack is formed by slipping out of fibrils. The snow-white surface on B portion in Figure 7, as covered with minute particles, is presumed to be constructed of slipping-out fibrils. On the other hand, although a part of the fibriles orients, the voids do not take place yet on C portion which is brittle-destroyed forcibly. As shown in Figure 7 and Figure 8, there is an apparent boundary between B and C. Considering the crack exists within C as observed by monitoring, it is certain that B portion is the surface fractured by propagating crack and C is the surface broken forcibly after testing. That is the reason why in this study the width of B portion is regarded as the growth-quantity of crack, Δ a.

5.2 MECHANICAL MODEL

(1) Mechanical model: The behaviors indicated in chapter 4.2 can be regarded to be resulted from the viscous behavior of stress-strains in the vicinity of the tip of crack. In order to explaining the behaviors qualitatively, a mechanical model consisted of four elements is proposed as described in Figure 14(a). Here, load and displacement are equivalent to stress and strain respectively. The symbols used in Figure 13 are as follows: S_1 =soft spring; S_2 =linear spring ; Da=dashpot ; M=slider.

The construction-equation is given by Eq.(4).

$$P = \begin{cases} \kappa_1 \delta - \beta \delta^3, & (\delta \leq \delta_{in}) \\ \kappa_2(\delta - \delta_{in}) + R + \eta \, d\delta/dt, & (\delta \geq \delta_{in}) \end{cases} \qquad (4)$$

where, δ means displacement. P_{in} and δ_{in} are the load and displacement when M starts sliding respectively. R is sliding friction force, η viscosity of dashpot and $d\,\delta/dt$ displacement rate. k_1, k_2, β are fixed numbers and $k_1 > k_2$ and $\beta < 1$. But δ and η are the function of T (internal temperature), δ and $d\,\delta/dt$. And T, δ and $d\,\delta/dt$ are mutually affecting factors (refer to Figure 2).

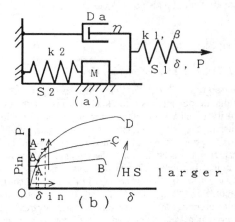

(a)

(b)

Figure 14. Mechanical model

The schematic $\delta - P$ diagram characteristic of mechanical model proposed above is illustrated in Figure 14 (b). The three arrows indicate the shifting directions of δ in , Pin (A, A', A" point) and the maximum load with displacement rate (which is equivalent to HS). In case of extremely low HS which the viscosity is able to be neglected, $\delta - P$ curve draws the course of $O \rightarrow A \rightarrow B$. As the viscosity of dashpot, Da, increases with HS, the curve will move upward along the route as illustrated as $O \rightarrow A' \rightarrow C$, furthermore, $O \rightarrow A" \rightarrow D$. The increase in displacement will rise in T to lessen η and to transform $A \rightarrow B$ from straight to curved line such as $A' \rightarrow C$ or $A" \rightarrow C$ with HS.

Let's assume that the slider, M, is equivalent to craze or crack growth. HS becoming faster and the effect of viscosity getting larger, the start in sliding of M will delay, that is, the displacement in which the formation of craze occurs will shift toward the right in Figure (b). The internal temperature, T, will rise with displacement as indicated in Figure 2 and the viscosity will be reduced to make crack-growth easy. But the internal temperature-rise is saturated with a specific displacement and HS to lessen the differences in viscosity that depends upon HS (refer to Figure 2). As a result the behavior of crack will approach a settled inclination. And the hold-duration Tr will move M toward the right because of the assumption of $k_1 > k_2$ until S_1 and S_2 balances, which means the craze or crack growth. This corresponds to stress relaxation.

(2) *Dependence of J_{IC} upon HS and Tr*: The intricate variations of J_{IC} and dJ/dt as shown in Figure 11 and 12 are considered due to the viscosity changing

with HS and displacement. Let's attempt the **qualitative** explanation of the variations of J_{IC} and dJ/dt by the application of the mechanical model .

For Tr=0, the duration of crack growth (termed "Ta") is limited within arriving at the given displacement. Ta becomes short with the increase of HS, so that M turns difficult to slide as indicated in Figure 14, i.e. the crack-growth gets small. Comparing with HS 0.5, HS 10 has greater values on both dJ/dt and J_{IC} because P_{in} shifts upwards and Da cannot move easier with HS. In comparison of HS 50 and HS 200, dJ/dt value of HS 10 is higher and J_{IC} value is lower. The reason is considered that although both P_{in} and J_{IC} grow larger with HS, the increase in displacement causes the internal temperature-rise to lessen the viscosity η, and the reduction of η makes the crack-growth easier. The dJ/dt of HS 50 is closely to that of HS 200. This is because the internal temperature is saturated and the viscosity comes uniform above a specific HS as written in section 5.2 (1).

For Tr$\neq$0, the crazing or crack growth takes place by stress relaxation with Tr, so that the gradient of R curve decreases and J_{IC}-value is raised. This tells that it is necessary to remove load as immediately as possible on arriving at the set-up displacement.

6 CONCLUSIONS

(1) The snow-white portion of fractured section after testing is formed by a propagating crack.

(2) The fracture toughness J_{IC} of ABS resin depends much upon deformation rate (crosshead speed HS in this study). The relationship is approximated by the following empirical equation.

$$\log J_{IC} = 0.34 \log HS - 0.10 \ (\ 0.5 \leqq HS \leqq 200 \text{ mm/min})$$

(3) The J_{IC}-value depends on retention-duration, which overestimates J_{IC} by a degree.

(4) Considering the variation of viscosity, the dependence of J_{IC} and dJ/dt can be explaine qualitatively by application of the proposed mechanical model.

REFERENCES

Hobbs, S.Y. & Bopp, R. C. 1980. Fracture
    Toughness of poly (butylene telephthalate).
    Polymer. 21: 559−563I
Narisawa 1993 *The fracture toughness.* 96. Tokyo.
    Sigma Publisher
William, J. G. & Cawood, M. J. 1990. European
    Group on Fracture Toughness: Kc and Gc
    Methods for Polymers. *Polymer Testing.* 9 :15-20.
Rice, J. R., Paris, P. C. & Merkel, J. G. 1973.
    Further result on J Integral Analysis and Estimate.
    ASTM · STP. 576: 231−245
Yokota, H. & Asano, H. Nov. 1991. Anisotropy of
    Strength in ABS Resin Molding Parts.*J. JSME.*57
    (543):168−173

Experimental Mechanics, Allison (ed.)© 1998 Balkema, Rotterdam, ISBN 90 5809 014 0

Evaluation of plastic deformation near crack by infrared thermography

I.Oda, S.Mitsuya & T.Doi
Kumamoto University, Japan

ABSTRACT: A nondestructive and noncontact technique for evaluating the plastic region near a crack by using infrared thermography is proposed. The welded sheet specimen with a through-the-thickness center crack is used. A uniform tensile load perpendicular to the crack is applied to the specimen. The temperature rise of specimen surface is measured by the infrared thermography. The effects of the welding residual stress and the hardened zone on the shape and dimensions of the heated region are examined. The heated region and the temperature rise are compared with the plastic deformation obtained by the elasto-plastic finite element analysis.

1 INTRODUCTION

Plastic deformation near a crack takes an important role relating to failure or fracture of the structures. Plastic deformation governs failure mechanism. The applicability of Linear Elastic Fracture Mechanics is limited by the plastic deformation. The crack opening displacement is considerably influenced by plastic deformation near a crack. In the present paper, a nondestructive and noncontact technique for evaluating the plastic and damaged region near a crack by using infrared thermography is proposed. Most of energy expended in deforming a metal plastically is converted into heat. It is assumed that the region heated by the work of plastic deformation corresponds to the plastic zone near a crack.

Papers which experimentally (Kihara et al. 1959) and analytically (Tada et al. 1983) examined the effect of residual stress on the fracture of weldments have been published. There have been, however, few works concerned with stress and strain near a crack in the welding residual stress field. Although the plastic deformation near a crack in weldments has been analyzed in a few papers (Oda et al. 1992, Oda et al. 1992), it has not been sufficiently verified by the experiment. In the present study, the welded sheet plates with a crack parallel to welding bead are dealt with. A uniform tensile load perpendicular to the crack is applied to the specimen. The temperature rise near a crack in the specimen is measured by the infrared thermography. The heated regions measured by the thermography are compared with the plastic zones obtained by an elasto-plastic finite element analysis. The usefulness and the limitations of the application of infrared thermography to the

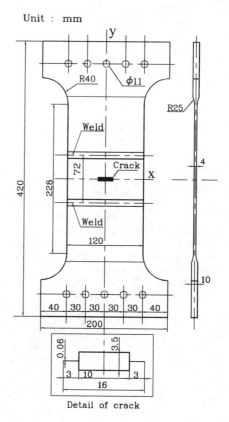

Figure 1. Specimen used in experiment (Type D).

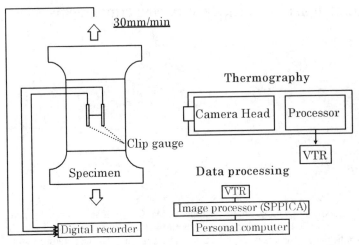

Figure 2. Schematic diagram of experiment.

Table 1. Specifications of thermographic system.

Range of measurable temperature	-40〜300℃
Resolution of temperature	0.025℃
Resolution (M×N)	76800 pixels
M : Horizontal pixels	(320×240)
N : Vertical pixels	
Number of frames	60 frames / second
Detector	In-Sb
Cooling of detector	Stirling cooling system

evaluation of the plastic region are discussed.

2 EXPERIMENT

The material supplied is austenitic stainless steel. Three types of sheet specimen were used. They are Type N, Type S and Type D. Each type of specimen has a through-the-thickness center crack of the same length. Figure 1 shows the shape and dimensions of a test specimen, Type D used in the experiment. For Type D, two V-grooves without root gap were prepared by cutting on both surfaces, in parallel to the minor axis (the x axis) of the specimen. The arc welding beads were laid on the grooves under the conventional welding condition. And then, a pre-crack parallel to the welding beads was prepared after machining down the thickness of parallel part of specimen to 4mm. Both tips of pre-crack were electrospark-machined with a wire of 0.03 mm diameter. The pre-crack was located between two welding beads. The pre-crack was desired to be in the compressive residual stress field parallel to the weld direction for Type D. For Type S, A similar V-groove was prepared on the x axis. And then, the arc welding

and the pre-cracking were made similarly to Type D. The pre-crack was located on the welding center line and was desired to be in the tensile residual stress field parallel to the weld direction for Type S. Type N is not welded. The surfaces of the specimens were hand polished on a series of polishing papers up to No. 100. A uniform tensile load was applied to each specimen parallel to the y axis. The temperature rise of specimen surface was measured by the infrared thermography. Table 1 shows the specifications of the thermographic system. The crack opening displacement was also measured by clip gauge. The schematic diagram of the experiment is shown in Figure 2.

3 ANALYSIS

The deformation near the tip of a crack in a plane stress field, which is uniaxial tension in the y direction, was analyzed. An elasto-plastic finite element analysis with an incremental theory of plasticity was used. Three types of specimen model shown in Figure 3 which correspond to those in the experiment were used in the analysis. The material was assumed to harden according to the following power law relation between stress and strain suggested by Swift (Swift 1952).

$$\sigma = c(\alpha + \varepsilon)^n \tag{1}$$

where σ (MPa) is the equivalent stress, ε is the equivalent plastic strain and c, α, n are material constants. Table 2 shows mechanical properties of materials used in the analysis. These properties were obtained from tensile tests using small uniaxial specimens extracted from a welded plate.

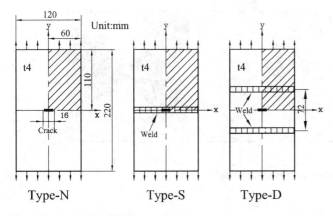

Figure 3. Models for FEM analysis.

Table 2. Mechanical properties of materials used.

	E (GPa)	v	σ_Y (MPa)
Base metal	155	0.28	305
Weld metal + HAZ	155	0.28	320~399

	c (MPa)	α	n
Base metal	607	0.5×10^{-5}	0.104
		0.5×10^{-5}	0.079 ~
Weld metal + HAZ	603~640	~	0.113
		0.4×10^{-3}	

E : Young's modulus v : Poisson's ratio
σ_Y : Yield strength

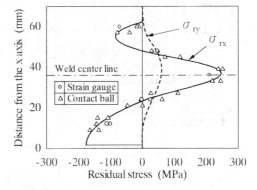

Figure 4. Distributions of residual stresses along the y axis in Type D.

For Type S and Type D, the residual stresses redistributed after pre-cracking were considered in the analysis. The distributions of the residual stress in the weld direction, σ_{rx}, and the transverse residual stress, σ_{ry}, were measured by the stress relaxation method with biaxial strain gauges and contact balls for similar welded and pre-cracked specimens. Figure 4 shows the distributions of σ_{rx} and σ_{ry} along the y axis in Type D. A relatively high compressive σ_{rx} in the vicinity of the pre-crack tip and high tensile σ_{rx} near the weld center line exist. The low tensile residual stress σ_{ry} exists near the weld center line. In Type S, both the stress σ_{rx} and σ_{ry} were highly tensile near the crack tip.

4 RESULTS AND DISCUSSION

Figure 5 shows the temperature rise in specimen surface of Type N. A moving speed of the crosshead in the present tensile test was kept constant, that is, 30 mm per minute while the strain rate of the specimen was not able to be measured. In the figure, the value of σ_n / σ_Y shows the ratio of applied net

stress to yield strength of base metal. The distribution of temperature changes as the applied load increases. The heated region near the pre-crack tip spreads at an angle of about 45 degrees to the x axis. The temperature distribution in the heated region can be also confirmed.

Figure 6 shows the temperature rise in specimen surface of Type S. For Type S, the temperature rise on the extension of the crack line is restrained and the breadth of heated region is narrower than that for Type N although the heated region spreads from the crack tip at an angle of about 45 degrees to the x axis. In addition, the size of highly heated region is smaller over all the applied stress level when compared with Type N. In Type S, the crack line is on the weld center line. The high tensile residual stress σ_{rx} and the hardened zone such as weld metal and heat-affected-zone exist near weld center line. Furthermore, near the weld center line, the compressive residual stress σ_{ry} exists away from

1277

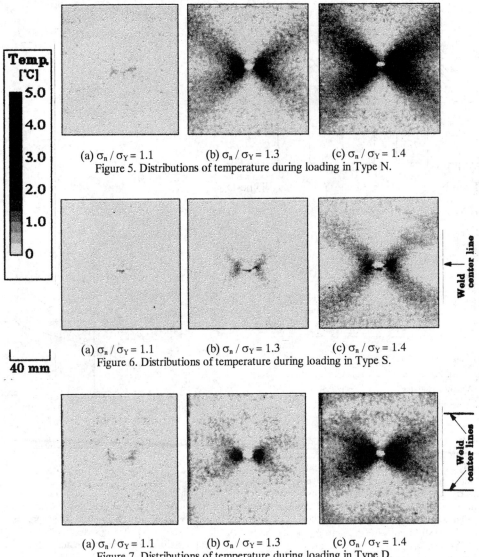

Temp.
[°C]

5.0
4.0
3.0
2.0
1.0
0

40 mm

(a) $\sigma_n / \sigma_Y = 1.1$ (b) $\sigma_n / \sigma_Y = 1.3$ (c) $\sigma_n / \sigma_Y = 1.4$
Figure 5. Distributions of temperature during loading in Type N.

(a) $\sigma_n / \sigma_Y = 1.1$ (b) $\sigma_n / \sigma_Y = 1.3$ (c) $\sigma_n / \sigma_Y = 1.4$
Figure 6. Distributions of temperature during loading in Type S.

Weld center line

(a) $\sigma_n / \sigma_Y = 1.1$ (b) $\sigma_n / \sigma_Y = 1.3$ (c) $\sigma_n / \sigma_Y = 1.4$
Figure 7. Distributions of temperature during loading in Type D.

Weld center lines

crack tip. These residual stresses and the hardened zone restrain the plastic deformation near the crack line in Type S.

The temperature rise in Type D is shown in Figure 7. At a low applied-stress level ($\sigma_n / \sigma_Y = 1.1$), the size of heated region is slightly larger when compared with that for Type N. This phenomenon in Type D occurs because the plastic deformation progresses under the influence of compressive residual stress,

σ_{rx} in the vicinity of pre-crack. At a high applied-stress level ($\sigma_n / \sigma_Y = 1.3$ or 1.4), however, the spread angle of the heated region in Type D becomes narrower in the zone away from pre-crack than that, about 45 degrees, in Type N. In addition, the size of highly heated region in Type D becomes smaller than that in Type N. These phenomena for Type D occur because the high tensile residual stress, σ_{rx}, as well as the hardened zone exists near weld center lines

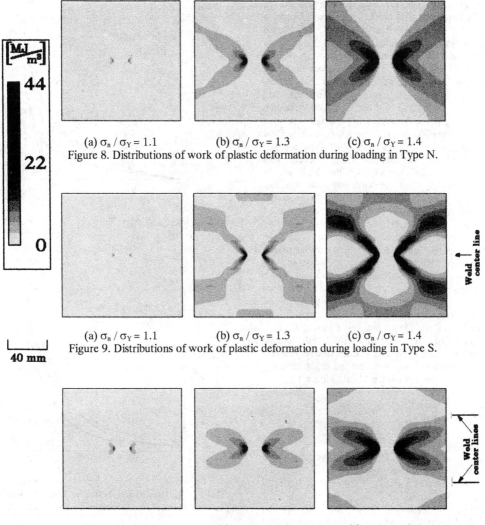

(a) $\sigma_n / \sigma_Y = 1.1$ (b) $\sigma_n / \sigma_Y = 1.3$ (c) $\sigma_n / \sigma_Y = 1.4$

Figure 8. Distributions of work of plastic deformation during loading in Type N.

(a) $\sigma_n / \sigma_Y = 1.1$ (b) $\sigma_n / \sigma_Y = 1.3$ (c) $\sigma_n / \sigma_Y = 1.4$

Figure 9. Distributions of work of plastic deformation during loading in Type S.

(a) $\sigma_n / \sigma_Y = 1.1$ (b) $\sigma_n / \sigma_Y = 1.3$ (c) $\sigma_n / \sigma_Y = 1.4$

Figure 10. Distributions of work of plastic deformation during loading in Type D.

which are 36 mm away from the x axis. The high tensile residual stress and the hardened zone restrain the plastic deformation in the zone away from pre-crack. At a high applied stress level ($\sigma_n / \sigma_y = 1.3$ or 1.4), the size of highly heated region near the crack tip becomes greater in the order, Type S, Type D, Type N.

Figure 8, Figure 9 and Figure 10 show the distributions of work of plastic deformation per unit volume during loading analyzed for Type N, Type S and Type D respectively. The behavior of plastic deformation in each specimen type qualitatively agrees with that of temperature rise. This implies that the effects of residual stress and hardened zone on plastic deformation can be inferred from the temperature rise obtained by infrared thermography.

Figure 11 and Figure 12 show the relationships between the crack opening displacement V_g at 4 mm from the crack tip and applied net stress σ_n. The experimental results and the analytical ones show the same tendency. At a low applied stress level, as shown in Figure 11, the crack opening displacement of Type D is higher than that of Type N. It is caused by the compressive residual stress, σ_{rx}, near the crack tips in Type D. At a high applied stress level above about 1.2 of σ_n/σ_Y, the crack opening displacement of Type D becomes lower than that of Type N as shown in Figure 12. This phenomenon occurs under the influence of the high tensile residual stress, σ_{rx}, and the hardened zone which are near weld center lines 36 mm away from the x axis in Type D. The crack opening displacement of Type S is the lowest over all the applied stress level. This phenomenon occurs because of the high tensile residual stress σ_{rx} and the hardened zone near the crack line in Type S. The crack opening displacement is considerably influenced by the plastic deformation near a crack. When σ_n/σ_Y is 1.1, the size of heated region for Type D is larger than that for Type N as shown in Figure 5 and Figure 7. The size of plastic zone for Type D is also larger when compared with that for Type N at σ_n/σ_Y of 1.1 as shown in Figure 8 and Figure 10. At the higher applied stress level ($\sigma_n/\sigma_Y = 1.3$ or 1.4), the size of heated region and the plastic-zone size become smaller for Type D than for Type N as shown in Figure 5 to Figure 10. These phenomena of temperature rise and plastic deformation can explain the effects of the residual stress and the hardened zone on the crack opening displacement shown in Figure 11 and Figure 12.

5 CONCLUSIONS

The following conclusions may be drawn.

1. The effects of residual stress and hardened zone on plastic deformation near a crack can be qualitatively inferred from temperature rise obtained by infrared thermography.

2. When a uniform tensile load is applied perpendicularly to the crack which is located on the weld center line, the plastic deformation is restrained mainly by the hardened zone near the weld center line.

3. When the crack is located between and parallel to two welding beads, the plastic-zone size is facilitated to grow at the low applied-stress level because of the compressive residual stress parallel to the crack. At the high applied-stress level, however, the growth of plastic deformation is restrained by the tensile residual stress and hardened zone which are parallel to the crack and located away from the crack.

4. The effects of residual stress and hardened zone on the crack opening displacement can be explained by the temperature rise or the plastic deformation near a crack tip.

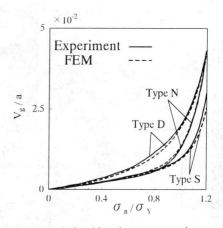

Figure 11. Relationships between crack opening displacement and applied stress at low stress level.

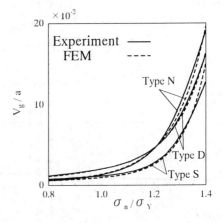

Figure 12. Relationships between crack opening displacement and applied stress at high stress level.

REFERENCES

Kihara, H. & K.Masubuchi 1959. Effect of residual stress on brittle fracture. *Welding Journal* 38(4):159s-168s.
Oda, I. & H.Sakamoto 1992. Deformation near a crack in residual weld stress field. In H.Fujiwara, T.Abe & K.Tanaka(eds), *Residual Stress-III*:482-487. London and New York: Elsvier.
Oda, I., H.Sakamoto & M.Yamamoto 1992. Effect of mechanical stress-relieving on deformation near a crack in weldment. *Proc. 7th ICEM*:51-56. SEM, INC.:USA.
Swift, H.W. 1952. Plastic instability under plane stress. *J. Mech. Phys. Solids*. 1:1.
Tada, H. & P.C.Paris 1983. The stress intensity factor for a crack perpendicular to the welding bead. *Int. J. Fract.* 21(4):279-284.

Experimental Mechanics, Allison (ed.) © 1998 Balkema, Rotterdam, ISBN 90 5809 014 0

Bixial yield locus determination with a single cruciform specimen

E. Terblanche
University of Stellenbosch, South Afrika

ABSTRACT: A standard cruciform specimen originally developed for biaxial fatique studies were used in a computer controlled biaxial test frame to obtain yield data. Strain gauge rosettes on both sides of the specimen provided feedback for control as well as actual strain. Yielding was detected by employing acoustic emission techniques. Loading was achieved by applying a certain F_x/F_y loading ratio until yielding was just detected, wherupon the strain gauge values were stored. This was then followed by loading with a different F_x/F_y combination. This process was repeated between 16 to 20 times, every time with a different F_x/F_y ratio until the complete biaxial yield locus was covered. With the 16 to 20 points it was possible to determine the complete yield locus to a fair degree.

1 INTRODUCTION

de Villiers (1966) was amongst the first to use cruciform specimens for biaxial fatigue studies. A fairly detailed photoelastic analysis was undertaken by him to prove that it was indeed a valid biaxial test specimen. Various other researchers, amongst them Wilson and White (1969), Fourie (1979), Makinde,

Thibodeau and Neale (1992) as well as the author used more detailed finite element analyses (see figure 1) to prove that it is in fact a valid test specimen for various types of biaxial studies. Figure 2 shows typically the Von Mises stress values in two different radial directions at the surface and at the centre of the specimen obtained from a finite element analysis. From results such as these it was possible to show that this is an excellent specimen.

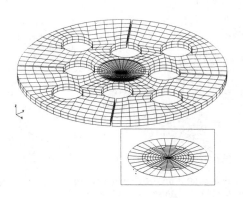

Figure 1 Finite element model of cruciform specimen, with inset showing the fine mesh in inside 5 mm diameter test section, with 8 elements thick.

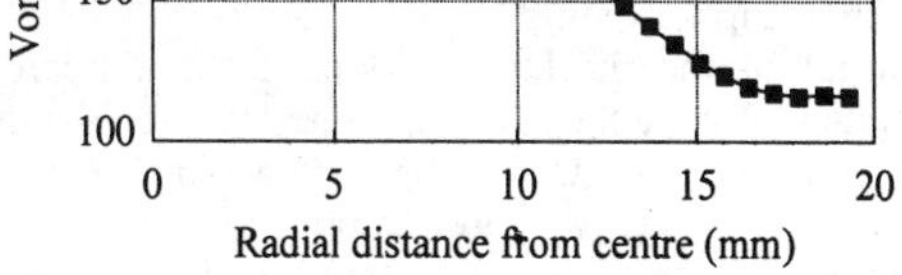

Figure 2 Radial Von Mises stress variation from centre to edge of hollowed out section for the loading case $F_x/F_y = 1$.

2 TEST EQUIPMENT

The test machine (See figure 3 for a schematic of the frame) is basically an electro-hydraulic servo-controlled unit. A test specimen can be clamped without any appreciable bending in the jaws of the test machine. Each of the four jaws are mechanically coupled to hydraulic cylinders via loadcells.

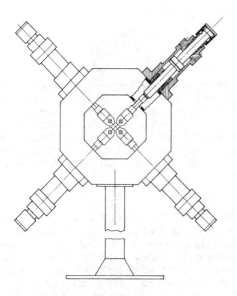

Figure 3 Biaxial test frame.

The faces of the jaws are made of phosphor bronze. The whole loading path from the test specimen to the hydraulic cylinders is in fact made from a series of coupled dissimilar materials. The reason for this arrangement is the elimination of ultrasonic acoustic noise that may originate from the hydraulics at the test specimen. This is the desired situation when use is made of an acoustic emission sensor at the specimen to detect onset of yielding and other acoustic emission phenomena that may occur at the specimen during testing.

Control of the machine is done from a micro-computer. A loading curve is generated for the X and Y directions on the test specimen. The types of signals generated for the input to the servo amplifiers can be divided into two broad categories, namely; those that can be thought of as radially radiating from the origin as shown in figure 4. Upon reaching the yield locus the process is reversed and the signal is generated so as to backtrack on itself until the yield locus on the opposite side (180° removed from the first) is reached. From here it will be generated so as to go back to the origin before it

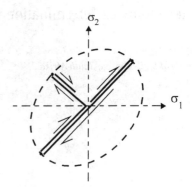

Figure 4 Potential radial load path generation.

is generated to branch out at say 90° to the first and second curves. The whole process is now repeated as before.

The second type of generated signal starts off from the origin and moves radially out from there until the yield locus is reached. From this position it will retract slightly as shown in figure 5 and will then continue to seek the yield locus only a little way away from the first encounter. If this process is kept up it is possible to move around the yield locus until the first yield encounter is reached again. In doing so the complete yield locus would have been determined.

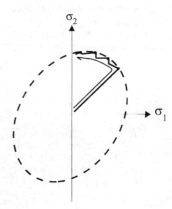

Figure 5 Potential locus hugging loading path.

This last loading path can generally be employed to determine the sensitivity of a particular material to the loading path sequence, whereas the first was the preferred path generation for the studies described in this document.

The computer generated (See figure 6 for a schematic of the complete measurement and control system.) loading paths are outputted through the digital to analogue converter via a set of optical

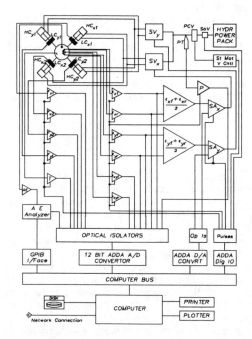

Figure 6 Schematic of biaxial test frame measurement and control system.

isolators to the servo amplifiers. There is a need for the optical isolators to prevent the formation of earth loops between the computer and the rest of the electronic instrumentation. The error outputs from the servo amplifiers are fed to the hydraulic servo valves. These valves open or close in response and in relation to the output current from the servo amplifiers. Through this action the hydraulic oil flow to the hydraulic cylinders is controlled, which in turn moves the cylinders. The cylinders are of the rod through type (equal areas on opposite sides of the piston) to have an equal amount of oil enter on the one side of the cylinder as that which is flowing out the other end. The cylinders are connected in such a way that the two opposite cylinders will always move in opposite directions so as to cause either a tensile or compressive action in that particular direction in the specimen.

Apart from a possible acoustic emission sensor on the test specimen, at least two rectangular strain gauge rosettes (the other rosette is bonded to the other side of the specimen) are also bonded to the test specimen. These rosettes are of the utmost importance for the strain control feedback to the servo amplifiers. Apart from this duty, the rosettes may of course also be used as the primary measuring sensors to determine the strain in the specimen. It is

possible to use the strain gauges for both the control and the measurement, because the signal that is either measured or used for control is both fed into very high impedance devices (In the case of both the measurement signal and the control signal the input impedance of the analogue to digital converter and the servo amplifier are in the region of 10^{10} ohm.) which makes the current drawn and then of course the subsequent voltage drop negligible. Use are made of two rosettes to determine if any bending effects are present during the original mounting of the test specimen onto the jaws of the test machine.

If any bending is present two gauges mounted in the same direction but on opposite sides of the specimen will give readings of opposite sign. A limit had been set on the magnitude of the absolute maximum difference between any two gauges. If this limit is exceeded a warning is displayed on the computer screen and the specimen will have to be re-mounted.

When a load is applied to the test specimen the strain in the specimen is picked up by the strain gauge rosettes. If say, the X direction gauges are taken as an example. The instrumentation amplifiers will amplify the signals from these two strain gauges. Being in the same direction the outputs from the amplifiers can be summed in the summation amplifiers and the mean of the summed values determined. These mean values are used as the feedback signals to the servo amplifiers. Here only the X direction gauges were under consideration. The Y direction gauges are treated in exactly the same way to eventually act as feedback to the Y direction servo amplifier.

The outputs from all six strain gauges (three from each rosette) are fed via an optical isolator to the 12 bit analogue to digital converter on the computer bus. From the computer the data is stored on disk for later manipulation.

The acoustic emission system data is handled in a different manner. From the acoustic emission sensor the signal travels to an amplifier system where filter discrimination can be applied to the signal so as to get rid of some of the unwanted noise that may be present in the signal. The amplifier feeds the signal to the acoustic emission analyzer where certain actions are taken. The analyzer is coupled to the analogue to digital converter via the optical isolators.

Other signals that are handled are the four loadcells via the instrumentation amplifiers and eventually via the optical isolators to the analogue to digital converter. The temperature of the test specimen is also measured by a bonded nickel sensor, the signal is handled in the standard manner via the amplifier.

One other signal that is used is the hydraulic oil pressure measured by pressure transducer and amplifier. This particular signal is only used during the initial startup of the system when the operating pressure is set from the computer.

Other electronics used are the stepper motor controls for opening or closing the primary pressure control valve.

3 RESULTS

The reasoning behind the approach followed can best be understood with reference to figure 7. This figure is the force vs. time curve for a simple tensile test, but also with an acoustic emission sensor on the specimen. It can clearly be seen that the acoustic emission activity reaches an abrupt peak when the

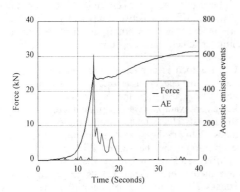

Figure 7   Force and acoustic emission events vs. time in a simple tensile test on a low carbon steel.

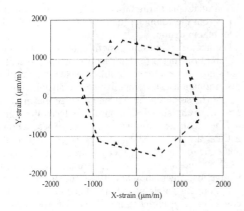

Figure 8 Best fit Tresca strain hexagon superimposed on actual data points obtained from acoustic emission yield detection.

steel yields at the upper yield point. Some acoustic activity can still be detected during the plastic phase,

but this quickly reaches a zero value when work hardening starts.

If the yield point of a material such as this can be detected with the aid of the acoustic emission activity present during yielding the same approach can be used during biaxial yield studies. Figure 8 depicts the data obtained by probing the biaxial field until yielding had been detected. The hexagon superimposed on the data points is the best fit Tresca hexagon in the strain domain.

4 DISCUSSION

The single biggest problem regarding the complete system is the specimen or more particular the yield volume of the specimen. In the specimen used this volume is a cylinder of diameter 5 mm with a thickness of 1.2 mm. Acoustic emission is really a volume based sensing system. The greater the volume of the material that yields the easier it will be for the sensor to pick up the acoustic emission events. The data that figure 7 was generated from was a tensile specimen with a total volume of 6000 mm^3 that could theoretically yield, as compared to the 23.6 mm^3 for the biaxial specimen. This made the sensitivity of the sensing fairly low as compared to the case for the mentioned tensile specimen. The acoustic emission equipment was also fairly primitive as compared to more modern instruments where it is possible to set the analyzer portion of the equipment to discriminate between, say, slip and crack propagation. The system that was used basically just counted the events, regardless if they originated from crack formation, dislocation movements etc.

Some of the problems arose from the fact that the particular specimen that was used (See figure 9 for a detailed drawing.) was originally developed to study other biaxial phenomena where strain gauges were the only sensing elements employed. In fact the total test frame (See figure 3.) was to a large degree developed around the test specimen instead of the other way around. It is possible to use larger specimens, but then shorter loadcells will have to be used in place of the current column type loadcells.

The control of the system was problematic at first, especially after the hydraulic pressure had been applied because analogue integrators were then switched in to get rid of small fluctuations. This problem was overcome by employing digital control instead of the analogue control that was used originally.

Acoustic emission detection holds promise especially for materials with no definite yield point like pure aluminium for instance. If this has to be used the yield point for these materials will have to be re-defined. Curves like figure 7 obtained for aluminium shows the same tendencies, but the initial increase in acoustic emission activity is not as sudden as for the low carbon steels. Preliminary curves obtained for aluminium shows the acoustic emission activity peaking at about a 0.15% to 0.18% permanent set, which is not too far off the mark of the 0.20% set which are the value normally used in practice.

Although the acoustic emission detection of yielding holds promise if some of the changes can be made that had been discussed above it is still fairly difficult to obtain nice clean acoustic emission signals. All the precautions of blanking the noise sources off from the specimen had been taken, but even so some noise was always present. The equipment that was used did unfortunately not have the function for using screening sensors to get rid of some of the noise. This approach would have been very difficult with the fairly small specimen and hence the fairly rapid wave propagation between the screening sensors and the measurement sensor. With a larger specimen this problem may be overcome.

The Tresca shape of the yield locus obtained hint at the yielding of the low carbon steel at the lower yield point. Fourie (1979) found that 0.3% carbon steels tend to follow a Tresca yield locus for the lower yield point and a Von Mises ellipse for the higher yield point

CONCLUSION.

The biaxial test frame and the associated hard and software that had been developed had proved to function within the original specifications. The system can become a link in the chain of standard biaxial testing, both for yield locus determination as well as for biaxial fatigue studies.

The system with the acoustic emission sensing ought to be used for biaxial failure studies for composite materials. The equipment will then have to be changed as discussed above.

Other methods like the post yield probing technique used by Fourie (1979) gives very good results, but the method is fairly cumbersome, especially for testing under other than under laboratory conditions. A method that can eventually be totally automated is currently under development. This sytem unfortunately does not rely on acoustic emission for sensing the onset of yielding.

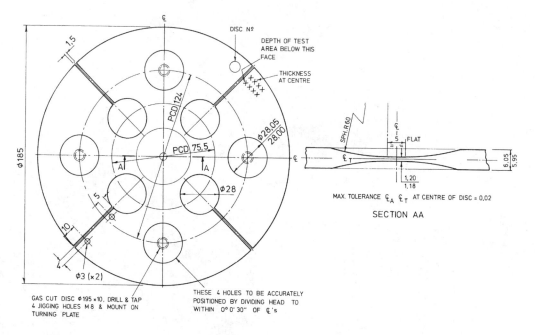

Figure 9 Biaxial test specimen that were used for all the tests described in this document.

REFERENCES

de Villiers, J.W.R. 1966. Low cycle fatigue failure of steels under biaxial straining. *Ph.D Thesis , Cambridge University.*

Fourie, E. 1979. Biaksiale swigting met behulp van die kruistoetsstuk. *Ph.D Thesis, Stellenbosch University.*

Makinde, A., Thibodeau, L., Neale, K.W. 1992. Development of an apparatus for biaxial testing using cruciform specimens. *Experimental Mechanics, Vol 32 No 2, pp 138-144.*

Wilson, I.H., White, D.J. 1969. Cruciform Specimens for Biaxial fatigue tests. An investigation using finite element analysis and photoelastic coating techniques. *English Electric Mechanical Engineering Laboratory Whetstone.*

Residual stress

Experimental Mechanics, Allison (ed.)© 1998 Balkema, Rotterdam, ISBN 90 5809 014 0

Reduction of residual stress using vibrational load during welding

S. Aoki & T. Nishimura
Tokyo Metropolitan College of Technology, Japan

ABSTRACT: A new reduction method of residual stress in welding joint is proposed where welded metals are shaken during welding. By an experiment using a small shaker, it can be shown that tensile residual stress near the bead is significantly reduced. The proposed method is examined by analysis. Since reduction of residual stress is considered to be caused by local plastic deformation, model with mass and preloaded springs with elasto-plastic characteristic is used. From analysis, it can be shown that residual stress is reduced when vibrational load is used during welding operation.

1 INTRODUCTION

Welding is widely used for construction of many structures. There exists residual stress near the bead because of locally given heat. Tensile residual stress on the surface degrades fatigue strength. Some reduction methods of residual stress, using heat treatment, shot peening so on, are used (AST 1983, AST 1993). However, much energy and time are consumed in these method. A method using vibrational load after welding is proposed. However, this method is not always effective (Gnirss 1988). It is important to develop a practical method for reduction of residual stress.

In this paper, a new method is proposed in which vibrational load is used during welding operation. Advantage of this method is that vibrational load is generated by a small shaker. Hence, it is easy to move the shaker. This method is expected to be practical.

First, reduction of residual stress is examined by an experiment in which vibrational load is applied to the specimens during welding. It is found that tensile residual stress near the bead can be reduced when this method is used. Then, an analytical method for reduction of residual stress by the proposed method is examined. It is obvious that reduction of residual stress is caused by local plastic deformation because initial residual stress is great and nearly equal to yield stress. Using an analytical model which consists of mass and preloaded springs with elasto-plastic characteristic,

reduction of residual stress is evaluated. The natural frequency changes as the length of the bead increases This point is also considered.

2 EXPERIMENT

A new method for reduction of residual stress in which vibrational load is used during welding is examined by an experiment.

2.1 *Experimental method*

Figure 1 shows the specimen used in this study. Material of specimen is rolled steel for general use (JIS SS400). Two plates of 280mm length, 100mm width and 6mm thickness are fixed on the supporting devices by bolts and shaken by a small shaker during welding. Specimens are butt welded using automatic CO_2 gas shielded arc welding machine. Groove is V-shaped. Diameter of the

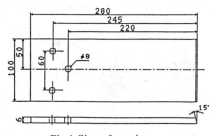

Fig.1 Size of specimen

wire is 1.2mm. Voltage is 25V and current is 200A. Welding is completed through one pass and root opening is 1mm. Groove angle is 30 degrees. Velocity of welding is 30cm/min. Thus, it takes 20 seconds to complete welding. Specimens are fixed on the supporting devices as shown in Figure 2 and shaken at the points 75mm from the weldment. In order to transmit vibrational load from the shaker effectively, arms are connected with driving part of the shaker using plates of 45mm width. Excitation frequency is chosen as 60Hz which is the fundamental natural frequency of specimens installed as shown in Figure 2. In order to examine reduction effect of residual stress at non-resonance frequency, excitation frequency is also chosen as 100Hz.

The amplitudes of excitation are determined by current indicated at the amplifyer of the shaker. The amplitudes of free edges of the specimens before welding are almost proportional to indicated current. For each current, peak to peak values of acceleration at free edge of the specimen before welding are measured by accelerometer and listed in Table 1. Peak to peak values of displacement are calculated and listed in Table 1.

Residual stress is measured by removing the quenched scale chemically and using a pararelled beam X-ray diffractometer with a scintillation counter. The conditions of the X-ray stress measurements are shown in Table 2. Measuring points are shown in Figure 3. 5 points at the center

Table 1 Relation between current and amplitude

f (Hz)	Current (A)	Amplitude Acc. (G)	Amplitude Disp. (mm)
60	0.4	1.3	0.090
	2.0	5.5	0.379
	4.0	9.7	0.669
100	2.0	2.5	0.062
	4.0	5.2	0.129
	5.8	7.4	0.184

Table 2 Conditions for X-ray stress measurements

Characteristic X-rays	Cr-K α
Diffraction plane	α-Fe (211)
Filter	Vanadium foil
Stress determination	$\sin^2 \psi$ method
Irradiated area	2x4mm^2
Tube voltage and current	30Kv, 8mA
Scan condition of 2θ	Step scanning
Divergence angle shift	1.0°
Peak determination	Half value width method

of the specimens are selected. Point C is on the bead. Points B and D are 10mm from the bead. Points A and E are 50mm from the bead. Stresses in the direction of the bead are measured.

2.2 Results of experiment

Figure 4 shows the results for residual stress of the specimens shaken by 60Hz vibrational load. Symbols ● are the results for specimen without vibrational load. For this condition, experiments are performed three times. From Figure 4, on the bead, tensile residual stress is significantly reduced when vibrational load is acting during welding. At points B and D, residual stress is also reduced. At points A and E, compressive residual stress tends to be tensile. Residual stress decreases as the amplitude of excitation increases. For the case where the excitation frequency is 100Hz, almost the same results as shown in Figure 4 are obtained. Thus, even if the excitation frequency is not equal to the natural frequency of the specimen, the proposed method is also effective. In this case, the amplitude at the free edge of the specimen is small.

Table 3 shows residual stress on the bead and ratio of residual stress with vibrational load to that without vibrational load. When vibrational load is used during welding, residual stress is reduced more than 50% for some condition.

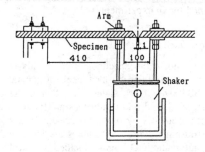

Fig.2 Experimental set-up

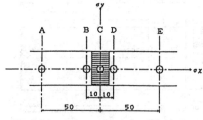

Fig.3 Measuring points of residual stress

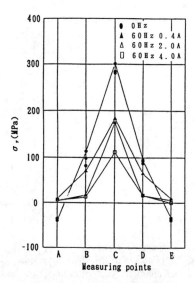

Fig.4 Residual stress （60Hz）

Table 3　Relation between current and amplitude

f (Hz)	Current (A)	Residual stress	
		σ_y (MPa)	σ_{yv}/σ_{ys}
60	0.0	303	1.00
	0.4	184	0.61
	2.0	174	0.58
	4.0	110	0.37
100	0.0	303	1.00
	2.0	231	0.76
	4.0	161	0.53
	5.8	271	0.90

Thus, it is obvious that when the specimens are shaken during welding, residual stress near the bead can be reduced.

3 ANALYTICAL METHOD

Since welding is joint method by melting base metal and weld metal, large residual stress and deformation are generated by shrinkage of metal during resolidification process. On the other hand, yield stress of metal immediately after welding is generally very low. It is considered that permanent deformation can be generated by very low external load. In this paper, reduction of tensile residual stress on the bead is dealt with.

3.1 Analytical model

As an analytical model, a single-degree-of-freedom system shown in Figure 5 is used. As shown in Figure 5 (b), the single-degree-of-freedom system is excited on the condition where springs are extended by Z_e from the equilibrium position. In this case, kZ_e is initial residual stress. It is assumed that restoring force-deformation relation of the springs is represented by the perfectly-elasto-plastic model as shown in Figure 6. Z_r is displacement in which sign of velocity changes from positive to negative vice versa. The amplitude of response is assumed to be small and the spring is yielding only when tensile force is acting. Equation of motion in the elastic range is expressed as:

$$m\ddot{x}+k\left(x-y-Z_e-Z_p^-\right)+k\left(x-y+Z_e-Z_p^+\right)=0 \quad (1)$$

where Z_p^+ and Z_p^- are permanent deformation of left spring and right spring, respectively. Equation of motion with respect to relative displacement $z\ (=x-y)$ can be written as:

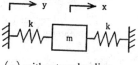

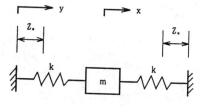

（a）without preloading

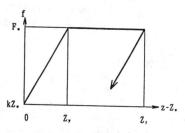

（b）with preloading
Fig.5 Single-degree-of-freedom model

Fig.6　Perfectly-elasto-plastic　restoring　force-deformation relation

$$m\ddot{z}+k\,(z-Z_c-Z_p{}^-)+k\,(z+Z_c-Z_p{}^+)=-m\ddot{y} \qquad (2)$$

When left spring is yielding, that is, $z+Z_c-Z_p{}^+$ $>Z_y$ and $\dot{z}>0$, equation of motion is given as:

$$m\ddot{z}+k\,(z-Z_c-Z_p{}^-)+F_c=-m\ddot{y} \qquad (3)$$

where F_c is yielding force. When $\dot{z}$ becomes negative, displacement at the time when $\dot{z}=0$ is defined as Z_r+Z_c and $Z_p{}^+$ is given as follows.

$$Z_p{}^+=Z_r+Z_c-Z_y \qquad (4)$$

Then, Eq. (1) is used. When right spring is yielding, that is, $z-Z_c-Z_p{}^-<-Z_y$ and $\dot{z}<0$, equation of motion is given as:

$$m\ddot{z}-F_c+k\,(z+Z_c-Z_p{}^+)=-m\ddot{y} \qquad (5)$$

When $\dot{z}$ becomes positive, displacement at the time when $\dot{z}=0$ is defined as Z_r-Z_c and $Z_p{}^-$ is given as follows.

$$Z_p{}^-=Z_r-Z_c+Z_y \qquad (6)$$

Then, Eq. (1) is used.

$kZ_p{}^+$ and $kZ_p{}^-$ are residual stresses which are released during shaking in left spring and right spring, respectively. When $kZ_p{}^+$ is not equal to $\mid kZ_p{}^-\mid$, they are statically equilibrated. Residual stress after shaking is evaluated as:

$$\sigma_{sy}=0.5\,\{k\,(Z_c-Z_p{}^+)+k\,(Z_c+Z_p{}^-)\} \qquad (7)$$

Unit of right hand side of Eq. (7) is force and that of left hand side is force per area. Model shown in Figure 5 is used in this study. Thus, when force generated by spring per unit area is considered, units of both sides of Eq. (7) are equal.

The dynamic characteristics are changes as the length of the bead increases. In order to consider this point, change of the natural frequency is obtained using a beam element model shown in Figure 7. The eigenvalue problem as follows is considered.

$$\mathbf{M}\,\ddot{z}+\mathbf{K}\,z=0 \qquad (8)$$

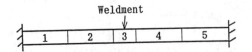

Fig.7 Beam element model of specimen

where $\mathbf{M}$ is mass matrix and $\mathbf{K}$ is stiffness matrix, respectively. $\ddot{z}$ and z are acceleration vector and displacement vector, respectively. Mass matrix and stiffness matrix of i-th element are square matrix of order 4 given as:

$$\mathbf{M}_i=\frac{\rho L}{420}\begin{bmatrix} 156 & 22L_i & 54 & -13L_i \\ 22L_i & 4L_i{}^2 & 13L_i & -3L_i{}^2 \\ 54 & 13L_i & 156 & -22L_i \\ -13L_i & -3L_i{}^2 & -22L_i & 4L_i{}^2 \end{bmatrix} \qquad (9)$$

$$\mathbf{K}_i=\frac{2EI}{L_i{}^3}\begin{bmatrix} 6 & 3L_i & -6 & 3L_i \\ 3L_i & 2L_i{}^2 & -3L_i & L_i{}^2 \\ -6 & -3L_i & 6 & -3L_i \\ 3L_i & L_i{}^2 & -3L_i & 2L_i{}^2 \end{bmatrix} \qquad (10)$$

where ρ is mass of unit length, L_i is length of element, EI is modulus of flexural rigidity. It is assumed that characteristics of material of the bead are same as those of the specimen and the length of the bead changes with time. Thus, ρ and I of the bead change. Number of element for specimen is 4 and that for weldment is 1. Length of element for weldment L_3 is 2mm. Considering the fundamental natural frequency is 60Hz, length of each element of specimen is $L_1=L_2=L_4=L_5$ $=143.5$mm.

3.2 Results of analysis

Excitation term y is given as $y=Y\sin\omega t$ in equation of motion. The natural circular frequency in elastic range is defined as $\omega_n=\sqrt{2k/m}$. The natural frequency within elastic limit $\omega_n=\sqrt{2k/m}$ changes as the length of the bead increases. Table 4 shows relation between the length of the bead L_w and the natural frequency f_n. When $L_w\geq 10$mm, f_n becomes constant value 95Hz. f_n is approximately given as:

$$f_n=\begin{cases} 60+3.5L_w\ (\text{Hz}) & : \ 0\leq L_w<10\text{mm} \\ 95\ (\text{Hz}) & : 10\leq L_w\leq 100\text{mm} \end{cases} \qquad (11)$$

Table 5 shows ratio of residual stress after shaking σ_{sy} to that before shaking σ_{yi} for some values of ratio of yield force to excitation force F_c/mY and ratio of σ_{yi} to F_c. Excitation frequencies f are 60Hz and 100Hz corresponding to experiment. From this table, reduction rate increases as F_c/mY decreases, that is, amplitude of excitation increases. This characteristic corresponds to experimental results.

Table 4　Length of bead and natural frequency of specimen

L_w (mm)	0.01	0.1	1	5
f_n (Hz)	60.2	64.7	81.3	91.0

Table 5　Ratio of residual stress before and after shaking

f_n (Hz)		60		100
		F_e/mY		
σ_{yi}/F_e	20	40	1	2
0.9	0.45	0.73	0.62	0.81
0.8	0.50	0.82	0.64	0.88
0.7	0.56	0.98	0.66	0.94

Residual stress is not reduced after about 0.3s (18 periods) passes for the case of f=60Hz and about 0.06s (6 periods) for 100Hz. Reduction rate also increases as σ_{yi}/F_e tends to be 1, that is, residual stress before shaking approaches to yield force. Since it is considered that yield force of metal immediately after welding is very low, condition where residual stress approaches to yield force is satisfied. Thus, effectiveness of the proposed method can be shown by the response analysis using the simple mechanical model as shown in Figure 5.

3.3 *Analysis using yield criterion*

As residual stress and vibrational load are acting on the specimen, another model is proposed. Since specimens are thin plates, plane stress state is considered. Hence, as an analytical model shown in Figure 8 (a) is used considering actual stresses in plane. x-axis is longitudinal direction of the specimen, direction of vibrational load, and y-axis is transverse direction, direction of the bead. As shown in Figure 8 (b), springs in transverse direction are extended by Z_e from the equilibrium position. In this case, $k_y Z_e$ is initial residual stress. It is assumed that restoring force-deformation relation of the springs is represented by the perfectly-elasto-plastic model as shown in Figure 6. Hence, when stresses in x-axis and y-axis are considered to be principal stresses, for example, springs are yielding according to Tresca yield criterion as shown in Figure 8. This model is excited in x-axis. Equation of motion in the elastic range is expressed as:

$$m\ddot{x}+2k_x(x-u-Z_{px})=0 \qquad (12)$$

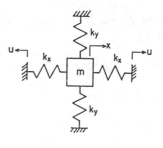

(a) without preloading

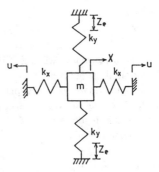

(b) with preloading

Fig.8 Two dimensional model

where Z_{px} is permanent displacement of springs in x-axis. Equation of motion for relative displacement z (=x-u) is written as:

$$m\ddot{z}+2k_x(z-Z_{px})=-m\ddot{u} \qquad (13)$$

Z_{fx} is defined as yield displacement. In the case, where springs in x-axis are yielding when springs are subjected to tensile stress, that is z-Z_{px} > Z_{fx}, $\dot{z}$ > 0, equation of motion is,

$$m\ddot{z}+2F_x=-m\ddot{u} \qquad (14)$$

where F_x is yield force in x-axis. Permanent displacement is given as:

$$Z_{px}=Z_{pr}-Z_{fx} \qquad (15)$$

where Z_{pr} is displacement when $\dot{z}$ < 0. Then Eq. (12) for elastic range is used. In the case, where springs in x-axis are yielding when springs are subjected to compressive stress, yield force changes as residual stress σ_{sy} changes. When $z-Z_{px} < Z_{ff}$, where Z_{ff} is yield displacement, and $\dot{z} < 0$, equation of motion is written as:

$$m\ddot{z}-2\,(F_x-\sigma_{ys})=-m\ddot{u} \qquad (16)$$

where Z_{ff} is given as:

$$Z_{ff}=(\sigma_{ys}-F_x)/\omega_s \qquad (17)$$

ω_s is initial natural circular frequency and initial σ_{ys} is $k_y Z_e$. Permanent displacement is given as:

$$Z_{px}=Z_{mr}+Z_{fx} \qquad (18)$$

where Z_{mr} is displacement when $\dot{z} > 0$. Then, Eq. (12) for elastic range is used.

It is assumed that plastic displacement in x-axis is equal to that in y-axis. When sum of plastic deformation is defined as Z_{py}, residual stress after shaking is evaluated as:

$$\sigma_{sy}=k_y(Z_e-Z_{py}) \qquad (19)$$

Residual stress is considered to be reduced through the pass shown in Figure 9.

Table 6 shows ratio of residual stress after shaking σ_{sy} to that befor shaking σ_{yi} for some values of F_e/mY. Excitation frequency is 60Hz. From this table, reduction rate of residual stress increases as σ_{yi}/F_e tends to be 1, that is, residual stress before shaking approaches to yield force. This result is the same as that for one

Table 6 Ratio of residual stress before and after shaking considering yield criterion

f_n (Hz)	60	
	F_e/mY	
σ_{yi}/F_e	40	45
0.9	0.02	0.01
0.8	0.21	0.20
0.7	0.45	0.42

dimensional model. However, reduction rate is almost same for different values of F_e/mY. Although two dimensional model has advantage which represents excitation force and residual stress in plane, the experimental results can not be represented. Hence, the proposed two dimensional model should be modified by considering other conditions.

4 CONCLUSION

A new reduction method of residual stress where welded metals are shaken during welding is proposed. Effectiveness of this method is examined. By the experiment, it is found that tensile residual stress near the bead is significantly reduced and reduction rate of residual stress increases as the amplitude of excitation increases. These properties are examined by an analytical method using a single-degree-of-freedom system with pre-loaded springs having elasto-plastic characteristics. From the analysis, it is obvious that reduction rate of residual stress increases as the amplitude of excitation increases. This property is also examined by using two dimensional model considering yield criterion. Thus, effectiveness of the proposed method can be demonstrated by experiment and analysis.

The authors thank Associate Professor T.Hiroi and the late Professor Y.Amano of Tokyo Metropolitan College of Technology for their help in experiment.

REFERENCES

American Society of Metals 1983. *Metals Handbook* : 856-895
American Society of Metals. 1993. *ASM Handbook*: 1094-1102
Gnirss,G. 1988. Vibration and vibratory stress relief. Historical development, theory and practical application. *Welding in the World* 26-11/12:4-8.

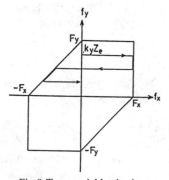

Fig.9 Tresca yield criterion

Experimental Mechanics, Allison (ed.) © 1998 Balkema, Rotterdam, ISBN 90 5809 014 0

X-ray residual stress measurement of an induction heated short tube

T. Nishimura
Department of Production Systems Engineering, Tokyo Metropolitan College of Technology, Japan

ABSTRACT: An experiment was carried out with a steel tube which had axially symmetric and longitudinally nonuniform residual stress distribution. After machining and annealing, the tube was heated by an induction coil and quenched in water. The longitudinal distribution of residual stresses was measured by the X-ray diffraction technique at each surface after removing thin concentric layers from the tube. The initial distribution of longitudinally nonuniform residual stresses including the shearing stress in the tube is calculated by using new correction formulas developed by the author.

1 INTRODUCTION

Mechanical methods of residual stress determination are developed by many investigators for tubes. An exact method of determining the longitudinal, circumferential and radial residual stresses in a solid or a hollow long cylinder is developed by Mesnager(1919) and modified by Sachs(1927). Lambert (1953) proposed a method of determining non symmetric stresses in bars. Rammerstorfer and Fischer(1992) derived a method for axisymmetric composite cylinders. All these methods have a common assumption that the stress distributions in cylinders are uniform in the longitudinal direction.

However, longitudinally nonuniform stress distributions would frequently occur in tubes, particularly in ones which have been machined, ground, welded or received other partial heat or mechanical treatment. Most residual stress distributions due to welding are highly localized.

A determination of nonuniform residual stress distribution in girth welded pipes has been given by Rybicki and Shadley (1986). Ueda and Fukuda (1989) have proposed a method for measurement of three-dimensional residual stresses using inherent strains in welded thick plate. Cheng and Finnie (1986A) have proposed a compliance method based on fracture mechanics for measurement of the hoop or longitudinal residual stresses in cylinders as functions of radius. This simple method may be applied for longitudinally nonuniform residual stresses when the gradient of stress is not too large. These methods involve

splitting or cutting steps into small pieces. The repetitive process of separating material makes the chance of introducing new residual stresses (Masubuchi, 1980).

Determination of subsurface residual stresses by the X-ray diffraction techniques must be corrected in order to obtain the true stress distribution that had existed before the layers were removed. Moore and Evans (1958) have developed a method to determine axisymmetric residual stresses in a solid and a hollow long cylinder. However, this method has a assumption that the stress distributions in cylinders are uniform in the longitudinal direction.

In previous papers by Nishimura et al (1976, 1993), new methods for determination of axially symmetric and longitudinally nonuniform residual stresses in bars and tubes was proposed to obtain the true stress distribution that existed the layers were removed, by using X-ray diffraction techniques. The layers are removed by chemical etching. Determination of axisymmetric and longitudinally nonuniform residual stresses in pipes using strain-gage and layer-removal was proposed by Nishimura (1996). Chemical etching requires skill in experiment and a long time. However, the use of chemical etching seems to be justified as a way of eliminating one further possible source of error.

This study is carried out to determine an axially symmetric and longitudinally nonuniform residual stresses in a tube which was locally induction heated and quenched in water. The original residual stresses are related to the stresses measured at a surface after removing some layers of material in this method. Since determination of the residual stress by X-ray methods is achieved by integrating the stress change, the X-ray methods is more desirable than the strain measurement methods. Especially, when a pipe has steep longitudinal gradients in residual stress distributions and has no metallurgical limitations, it is considered that the X-ray methods is very useful.

2 ANALYSIS PROCEDURE

Determination of subsurface residual stresses by the X-ray diffraction techniques must be corrected in order to obtain the initial residual stress distribution that had existed before the layers were removed. Equations for correction of longitudinally nonuniform residual stress distributions are analyzed as follows. Consider an isotropic and homogeneous hollow cylinder with an outer radius b, inner radius a and length $2h$. Let the z-axis in a $r\,\theta z$-coordinate system be the axis of the tube. Under condition of axial symmetry and the absence of body force, the following equations of equilibrium in a cylindrical coordinates hold for stresses including residual stresses.

$$\frac{\partial \sigma_r}{\partial r} + \frac{\partial \tau_{rz}}{\partial z} + \frac{\sigma_r - \sigma_\theta}{r} = 0 \qquad (1)$$

$$\frac{\partial \tau_{rz}}{\partial r} + \frac{\tau_{rz}}{r} + \frac{\partial \sigma_z}{\partial z} = 0 \qquad (2)$$

At the surface of a cylinder, the residual stress $\sigma_r = \tau_{rz} = 0$ and $\partial \tau_{rz} / \partial z = 0$, therefore the Eqs. (1) and (2) reduces to as follows:

$$\frac{\partial \sigma_r}{\partial r} = \frac{\sigma_\theta}{r} \qquad \frac{\partial \tau_{rz}}{\partial r} = -\frac{\partial \sigma_z}{\partial z}.$$

During the removing process, the radius and stresses are changing continuously and these are measured step by step after removing thin layers. Consider the mth step in which the outer radius of the specimen reduces to ρ from b. On removing the surface layer of small thickness $\Delta \rho$, the force, which has been removed from each square unit of surface of remaining cylinder, is the small outward force

$$-\Delta \sigma_r = -\Delta \rho \frac{\sigma_{\theta m}}{\rho} \qquad (3)$$

$$-\Delta \tau_{rz} = \Delta \rho \frac{\partial \sigma_{zm}}{\partial z} \qquad (4)$$

where $\sigma_{\theta m}$ and σ_{zm} are circumferential and longitudinal measured stresses at the side surface of radius ρ.

The stress distribution of hollow cylinder of finite length subjected to arbitrary axisymmetric external forces has been derived by Shibahara and Oda (1968, 1970). The small stress changes caused by the removal of the thin layers can be derived from their stress function. The total stress changes which are caused by the removal of outer-surface layers form radius b to radius ρ, can be obtained by summing the small stress changes, as functions r, z and ρ. Therefore, the initial distribution of the stresses including the shearing stress is computed from the longitudinal distributions of residual stresses $\sigma_{\theta m}$ and σ_{zm} measured by the X-ray methods at the surface after removal of successive concentric layers of material. The equations consist of a series of trigonometric, Bessel and hyperbolic functions.

3 EXPERIMENT

An experiment was carried out with a carbon steel tube (JIS S45S) which had longitudinally nonuniform residual stress distribution. A tube with an outer diameter 80 mm, inner diameter 40 mm and length 150 mm was machined from a commercial steel bar. After machining the tube was annealed in vacuum at 973 k for 1 hour. This tube was then heated to 1073 K by an induction quenched in water.

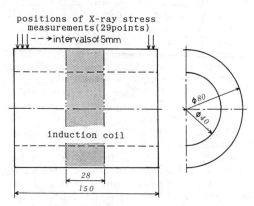

positions of X-ray stress
measurements(29points)
--→intervals of 5mm

induction coil

28

150

All dimensions in mm

Fig. 1 Shape of the specimen and locations measured by the X-rays.

The longitudinal distribution of residual stresses was measured by the X-ray diffraction technique at each surface after removing thin concentric layers from the tube. The shape of the specimen and locations measured by the X-ray diffraction method are shown in Fig. 1. The layers were removed 30 times by chemical etching. Etching solution was commercial chemical polishing liquid and the inner

Table 1 Conditions for X-ray stress measurements

Characteristic X-rays	Cr-K α
Diffraction plane	α -Fe(211)
Filter	Vanadium foil
Stress determination	$\sin^2 \phi$ Method
Irradiated area	2x4 mm^2
Tube voltage / current	30Kv / 8mA
Scan condition of 2 θ	Step scanning
Divergence angle of slit	1.0°
Peak determination	Half value width

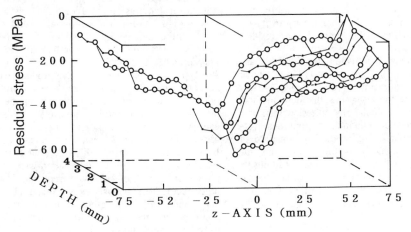

Fig. 2 Axial distributions of longitudinal residual stress at surfaces after removal of some layers of induction quenched tube. Values shown by symbols are measured using X-ray methods. Stress date at the three radii 40, 39, 38 and 37 mm are shown open circles.

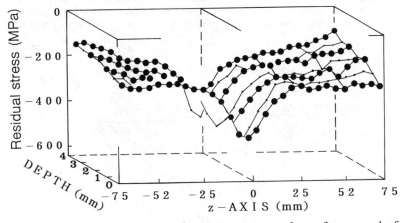

Fig. 3 Axial distributions of circumferential residual stress at surfaces after removal of some layers of induction quenched tube. Values shown by symbols are measured using X-ray methods. Stress date at the three radii 40, 39, 38 and 37 mm are shown solid circles.

and outer surfaces of the specimen were protected by vinyl tape. The longitudinal and circumferential residual stress were measured by a parallel beam X-ray diffractometer with a scintillation counter.

The conditions of the X-ray stress measurements are shown in Table 1. The distributions of the longitudinal and circumferential stresses measured are shown in Figs. 2 and 3, respectively. Each figure shows a plot of the residual stress as a function of radius and longitudinal position. The measured stresses at shown every four removals in these figures for simplicity. In these figures, the 68.3% percent confidence intervals of the measured stresses were within ±10 MPa. The difference of measured stresses at four positions, circumferentially 90 deg apart were within the range of ±10 MPa. The measured stresses at radii 40 mm, 39 mm, 38 mm and 37 mm are denoted by the solid circles to which special attention should be paid. Figures 2 and 3 show clearly that longitudinal distributions are not uniform, and the true residual stress distribution should be obtained by using the method for longitudinally nonuniform residual stress distribution.

Figures 4 shows corrected longitudinal and shearing stress distributions, and Fig.5 shows corrected circumferential and radial stress distributions obtained by using the new method, respectively. The calculations of residual stresses from the observed longitudinal distributions were made by a computer. Its input data consisted of the observed stresses $\sigma_{\theta m}$ and σ_{zm} and the corresponding radial and longitudinal positions. The integration for correction of measured stresses was computed by using Simpson's rule. Poisson's ratio $\nu = 0.3$ and the number of terms of the series are 10 in the calculations.

4 DISCUSSION

Both strain measurement methods and X-ray diffraction techniques can be used to determine axially symmetric and longitudinally non-uniform residual stresses in tubes. Comparing the two methods as a technique for measuring residual stresses, it should be noted, first, that material has to be removed in both cases. Chemical etching turned out to be a reliable method for this purpose. The strain measurement methods has no metallurgical limitations but has limitations caused by Saint-Venant's principle (Timoshenko and Goodier, 1951). Further sources of inaccuracies are the calculations of the other components. Especially the derivation of longitudinal and circumferential residual stresses is somewhat critical; it requires the differentiation of a point wise given function of measured strain. Furthermore, the solution of this inverse problem is sensitive· with respect to number of terms of polynomials. It should be also mentioned that the strain measurement methods can produce usefully accurate estimates of residual stress distribution when a tube has not steep gradients in the stress distribution. Cheng and Finnie (1986B) studied analytically the accuracy of the layer-removal method for measuring localized residual stresses. In their approach, strips, which may have been cut from a pipe or a plate, have strain-gage rosettes placed on one face and layers removed from the other face. Their analysis shows that actual stress distribution may be quite different from that predicted by the computational model normally used in the layer-removal method. They recommend that the layer-removal method can be used for measuring residual stresses for cases in which the ratio of the height of the strip to half dimension of the localized residual stress zone.

Since the measurements of $\sigma_{\theta m}$ and σ_{zm} are not free of small errors, the calculated initial residual

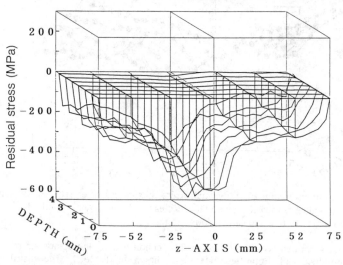

Fig. 4 Axial distributions of longitudinal and shearing residual stresses calculated from the measured date in the induction quenched tube.

1298

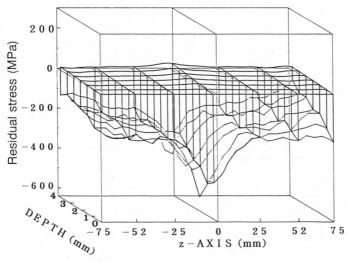

Fig. 5 Axial distributions of circumferential and radial residual stresses calculated from the measured date in the induction quenched tube.

stress distributions are not fully exact. However, determination of the residual stress by X-ray methods is carried out by integrating the stress change, the X-ray methods is more desirable than the strain measurement methods. Especially, when a pipe has steep gradients in residual stress distributions and has no metallurgical limitations, it is considered that the X-ray methods is very useful.

CONCLUSION

Longitudinally nonuniform residual stress distributions would frequently occur in tubes, particularly in ones which have been received partial heat or mechanical treatments. Determination of subsurface residual stresses by the X-ray diffraction techniques must be corrected in order to obtain the true stress distribution that had existed before the layers were removed. A new method developed by the author, is applied for a steel tube to determining axisymmetric and longitudinally nonuniform residual stress produced by induction quenching. The initial distribution of the stresses including the shearing stress is computed from the longitudinal distributions of residual stresses measured by the X-ray methods at the surface after removal of successive concentric layers of material. Since the X-ray method is not necessary to use a differential function of measured values, the methods is more desirable than the strain measurement methods.

REFERENCES

Cheng, W. and Finnie, I. , 1986A, "Measurement of Residual Hoop Stresses in Cylinders Using the Compliance Method," ASME Journal of Engineering Materials and Technology, Vol.108, pp.87-92.

Cheng, W. and Finnie, I. , 1986B, "Examination of the Computational Model for the Layer-Removal Method for Residual Stress Measurement," Experimental Mechanics, Vol.26, pp.150-153.

Lambert, J. W., 1954, "A Method of Deriving Residual Stress Equations," Proc. of Soc. for Experimental Stress Analysis, Vol.12, No.1, pp.91-96.

Masubuchi, K.,1980, Analysis of Welded Structure, Pergamon Press, 1980.

Mesnager, M., 1919, "Methods de determination des tensions existant dans un cylindre circulaire," Comptes Rendus, Vol. 169, pp. 1391-1393.

Moore, M., G. & Evans, W., P., 1958, "Mathematical Correction for Stress in Removed Layers in X-ray Diffraction Residual Stress Analysis," SAE Trans., 66, pp.340-345.

Nishimura, T., Kawada, Y. and Kodama,S., 1977, "On Axially Symmetrical Residual Stresses in Bars with Longitudinally Nonuniform Stress Distribution," Bulletin of the JSME,Vol.20, No. 150, pp.1533-1539.

Nishimura, T.,1993, " On Axisymmetric Residual Stresses in Tubes with Longitudinally Nonuniform Stress Distribution," ASME Journal of Applied Mechanics, Vol.60, pp. 300-309.

Nishimura, T.,1996, " Determination of Axisymmetric Residual Stresses in Pipes using Strain-gage and Layer-removal," Proc. of the 1996 Annual Meeting of JSME/MMD, Vol.A, pp. 461-462.

Rammerstorfer, F. G. and Fischer, F. D., 1992,"A Method for the Experimental Determination of Residual Stresses in Axisymmetric Composite Cylinders," ASME Journal of Engineering Materials and Technology, Vol.114, pp. 90-96.

Rybicki, E. F. and Shadley, J. R., 1986, " A Three-Dimensional Finite Element Evaluation of a Destructive Experimental Method for Determining Through-Thickness Residual Stresses in Girth Welded Pipes," ASME Journal of Engineering Materials and Technology, Vol.108, pp. 99-106.

Sachs, G., "Der Nachweis Innerer Spannungen in Stangen und Rohren," Zeischrift fur Metallkunde, Vol 19, 1927, pp.352-357.

Shibahara, M., and Oda, J., 1968, "Axisymmetric deformation of Hollow Cylinders of Finite Length," Trans. Japan Soc. Mech. Engrs., Vol. 34, pp388-402, (in Japanese).

Shibahara, M., and Oda, J., 1970, "Axisymmetric deformation of Hollow Cylinders of Finite Length, Part-2" Trans. Japan Soc. Mech. Engrs., Vol 36, pp168-176, (in Japanese).

Timoshenko,S.,P., Goodier, J.,N., 1951,"Theory of Elasticity," 2nd ed., International Student Edition, McGraw-Hill, Tokyo, pp.39.

Ueda, Y., Fukuda, k., Kim,Y.,C., 1986, "New Measuring Method of Axisymmetric Three-Dimensional Residual Stresses Using Inherent Strains as Parameters," ASME Journal of Engineering Materials and Technology, Vol.108, pp.328-334.

Ueda, Y., Fukuda, k., Nakacho,K. and Endo,S., 1975, "A New Measuring Method of Residual Stresses with the Aid of Finite Element Method and Reliability of Estimated Values," Trans. Welding Research Institute of Osaka Univ., Japan, Vol.4-2, pp.123-131.

Experimental Mechanics, Allison (ed.)© 1998 Balkema, Rotterdam, ISBN 90 5809 014 0

Assessment of distortions in the deep hole technique for measuring residual stresses

A.A.Garcia Granada, D.George & D.J.Smith
Department of Mechanical Engineering, University of Bristol, UK

ABSTRACT : The Deep Hole Technique for measuring residual stresses in thick section components is used to determine distributions of residual stresses, through the thickness, of a variety of welded steel components of varying geometric complexity. The technique relies on the measurement of the distortion of a reference hole drilled through the wall thickness. This paper presents recent developments in the analysis of the hole distortion. The measured distortions include diametral changes in the reference hole, together with changes in height during the trepanning of a column containing the reference hole, and are converted to residual stresses using elasticity theory. FE analyses have also been carried out to determine the influence of near surface conditions for thick walled components. These results are then incorporated into the solution procedure.

1. INTRODUCTION

In assessing the integrity of structures it is important to determine the residual stresses in the structure. Various techniques are available to measure residual stresses, although measurement through the full thickness can only be achieved by some of these techniques. One technique which has undergone substantial developments is the Deep Hole Technique. This is a method originally devised by Beaney [1] and Zhadanov and Gonchar [2], and developed further by Leggatt et al [3], and Smith and co-workers [4,5]. The technique has been used to determine through thickness residual stresses in a number of thick section welded components; principally for ferritic and stainless steel butt welded plates and pipes.

The deep hole technique relies on the measurement of the elastic distortion of a reference hole to determine the residual stresses. Earlier analysis by Leggatt et al [3], and Bonner and Smith [4,5] use two dimensional plane stress analysis to determine the distortion of the reference hole when a plate containing the hole is subjected to uniform stresses.

In this paper the analysis is extended to take account of distortions associated with conditions near to entrance and exit faces of the reference hole.

2. PROCEDURE AND THEORY

2.1. Basic Procedure

The technique is applied in four different steps shown in figure 1; (i) a small reference hole of 3.175 mm diameter is drilled through the component, (ii) then the hole diameter is measured accurately at different angles around the reference hole and at increments of about 0.2 mm through the thickness, (iii) then a column of material containing the reference hole is trepanned incrementally and the variation of height of the trepanned column is measured and finally, (iv) after relaxation of the stresses in the trepanned column, the reference hole is remeasured to obtain the change in diameter.

2.2. Theory

In order to obtain the residual stress distribution through the component, the following assumptions will be used :

i) The axis of the reference hole is a principal stress direction.

ii) Stress relaxation at the hole edge is elastic.

Cross-section
Through
Weldments

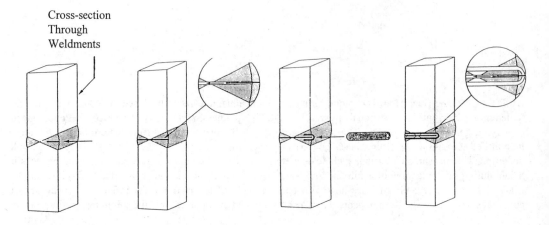

step 1 : Gun-Drill Reference Hole

step 2 : Measure Reference Hole Diameter

step 3 : Co-axially trepan Out Core by Electro-Dynamical-Machining

step 4 : Remeasure Reference Hole diameter

Figure 1 : Basic Process of the Deep Hole Residual Stress Measurement Method

For a plate (see Figure 2) containing a circular hole through the thickness 2t, the normalised radial strain or normalised radial displacement can be shown [6] to be :

$$\varepsilon_r . E = \frac{u_r . E}{d_0} \qquad (1)$$

$$\varepsilon_r . E = \sigma_{xx} . f[\theta] + \sigma_{yy} . g[\theta] + \tau_{xy} . h[\theta] - v . \sigma_{zz}$$

where u_r is the radial displacement, d_0 is the diameter of the hole, σ_{xx}, σ_{yy}, τ_{xy} and σ_{zz} are stresses acting at the edge of the plate, E is Young's Modulus and f, g and h are functions of angles around the hole. It can be shown that :

$$f[\theta] = 1 + 2\cos(2\theta)$$
$$g[\theta] = 1 - 2\cos(2\theta) \qquad (2)$$
$$h[\theta] = 4\sin(2\theta)$$

In the case of a uniform tensile stress σ_{xx} and plane stress conditions, equation (1) reduces to :

$$\varepsilon_r = \frac{\sigma}{E}\{1 + 2\cos(2\theta)\} \qquad (3)$$

Near to the entrance and exit of the circular hole, the distortion of the hole is no longer accurately described by equations (1) to (3), and we introduce initially two parameters A and B to take account of near entrance and exit faces distortions of the hole. In the case of a uniaxial stress, equation (3) is modified so that :

$$\varepsilon_r[\theta, z] = \frac{\sigma}{E} A[z].\{1 + B[z].2\cos(2\theta)\} \qquad (4)$$

1302

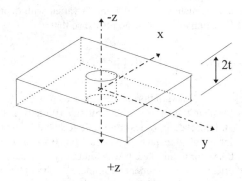

Figure 2 : Plate of through thickness 2t containing a circular hole of diameter d_0.

where :

$$A[z] = \frac{da[z] + db[z]}{d_0} \qquad (5a)$$

$$B[z] = \frac{da[z] - db[z]}{2(da[z] + db[z])} \qquad (5b)$$

A represents a measure of the uniform expansion of the hole and B the eccentricity of the hole, with a and b the major and minor diameters of the ellipse. The parameters A and B are functions of the position (± z) through the thickness of the plate.

In addition to the radial displacement of the hole, there is a change in thickness of the plate resulting from the through thickness stress σ_{zz}. The through thickness strain ε_{zz} is σ_{zz} / E.

In the Deep Hole Technique the radial displacement and change in thickness are measured when a uniaxial residual stress is released as a result of trepanning. The relaxed normalised displacement are therefore related to the residual stress using :

$$\varepsilon_r[\theta, z].E = -\sigma_{xx}.f[\theta, z] - \sigma_{yy}.g[\theta, z]$$
$$-\tau_{xy}.h[\theta, z] + \upsilon.\sigma_{zz} \qquad (6)$$

$$\varepsilon_{zz}.E = -\sigma_{zz}$$

where : $f[\theta, z] = A[z] . \{ 1 + B[z] . 2 \cos (2\theta) \}$
$g[\theta, z] = A[z] . \{ 1 - B[z] . 2 \cos (2\theta) \}$
$h[\theta, z] = 4 . A[z] . \{ B[z] . \sin (2\theta) \}$

At a given through thickness position, z, a minimum of four strains can be measured to obtain the residual stresses. For example $\varepsilon_r[\theta , z_1]$ at $\theta = 0°, 45°$ and $90°$ and also ε_{zz}.

Alternatively, as proposed by Bonner and Smith [5], if measurements are obtained at n angles a least squares fit to the diametral strains can be used to determine the stresses. For a given through thickness position z_1 equation (6) is rewritten in matrix form as :

$$\varepsilon = -[M].\sigma \qquad (7)$$

where :

$$\varepsilon = [\varepsilon_r[\theta_1 , z_1] , \varepsilon_r[\theta_2 , z_1] , .. , \varepsilon_r[\theta_n , z_1] , \varepsilon_{zz}]^T$$
$$\sigma = [\sigma_{xx} , \sigma_{yy} , \tau_{xy} , \sigma_{zz}]^T$$

and :

$$[M] = \begin{bmatrix} f[\theta_1, z_1] & g[\theta_1, z_1] & h[\theta_1, z_1] & -\upsilon \\ f[\theta_2, z_1] & g[\theta_2, z_1] & h[\theta_2, z_1] & -\upsilon \\ . & . & . & . \\ . & . & . & . \\ f[\theta_n, z_1] & g[\theta_n, z_1] & h[\theta_n, z_1] & -\upsilon \\ -\upsilon & -\upsilon & 0 & 1 \end{bmatrix}$$

By using Pseudo-Inverse or Moore-Penrose inverse matrices [7] the optimum residual stress can be expressed as :

$$\hat{\sigma} = -[M]^*.\varepsilon \qquad (8)$$

where $[M]^* = (M^T.M)^{-1}.M^T$ is the Pseudo-Inverse of the matrix [M] and $\hat{\sigma}$ is the optimum stress vector $[\sigma_{xx}, \sigma_{yy}, \tau_{xy}, \sigma_{zz}]^T$ that best fits the measured strains ε.

To determine the coefficients A and B as a function of depth z for different thickness 2t, finite element (FE) analyses were carried out.

3. FINITE ELEMENT ANALYSIS AND RESULTS

3.1. Finite Element Analysis

A 3D FE analysis using the ABAQUS code was used to model a square plate of width 50 mm containing a 3.175 mm diameter hole. The thickness of the plate was varied. 20 noded brick elements were used to construct the FE mesh. Fully elastic analyses were carried out assuming E = 200 GPa and ν = 0.33. The plate for different thicknesses was subjected to unit loads at the boundary of the plate in the plane of the hole and also normal to the hole axis.

3.2. Results

When the plate was subjected to a uniform stress in one direction e.g. $\sigma = \sigma_{xx}$, results from the FE analysis revealed that the hole distortion was a function of depth z. The diametral distortion given by equation (4) was such that A was approximately 1 for all z and B varied with depth. Equation (4) is then simplified so that :

$$\varepsilon_r[\theta, z] = \frac{\sigma}{E}\{1 + B[z].2\cos(2\theta)\} \qquad (9)$$

The variation of B[z] is shown in Figure 3 for plate thicknesses of 20, 50 and 100 mm. For a 20 mm plate B varies from 0.98 to about 0.86 and for a 100 mm plate B varies from 0.94 to about 0.85. For a thin plate it was found that near surface effects were evident completely through the thickness. For thicker plates (2t > 70 mm) B eventually becomes a constant value at a sufficient depth.

For stresses σ_{yy} and τ_{xy} the distortion of the hole can be determined from linear superposition. Consequently equation (6) applies with A[z] = 1 and B[z] as shown in Figure 3 for all in-plane stresses.

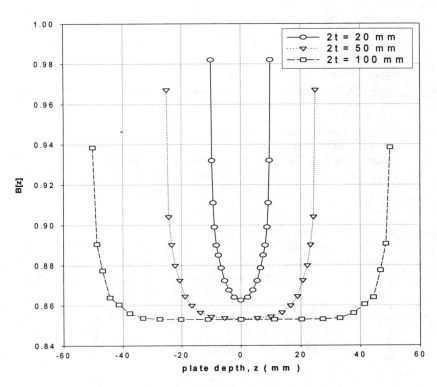

Figure 3 : Variation of parameter B[z] as a function of specimen thickness

4. DISCUSSION AND CONCLUDING REMARKS

If only four measurements are obtained, the measured distortions at a given depth z_1 will be given by :

$$\begin{bmatrix} \varepsilon_r[0°,z_1] \\ \varepsilon_r[90°,z_1] \\ \varepsilon_r[45°,z_1] \\ \varepsilon_{zz} \end{bmatrix} = -\frac{1}{E}\begin{bmatrix} 1+2B & 1-2B & 0 & -\upsilon \\ 1-2B & 1+2B & 0 & -\upsilon \\ 1 & 1 & 4B & -\upsilon \\ -\upsilon & -\upsilon & 0 & 1 \end{bmatrix}\begin{bmatrix} \sigma_{xx} \\ \sigma_{yy} \\ \tau_{xy} \\ \sigma_{zz} \end{bmatrix}$$

$$= -[M].[\sigma] \qquad (10)$$

The inversion of [M] provides a solution to the residual stresses and gives :

$$\begin{bmatrix} \sigma_{xx} \\ \sigma_{yy} \\ \tau_{xy} \\ \sigma_{zz} \end{bmatrix} = -M^*.\begin{bmatrix} \varepsilon_r[0°,z_1] \\ \varepsilon_r[90°,z_1] \\ \varepsilon_r[45°,z_1] \\ \varepsilon_{zz} \end{bmatrix} \quad ; \quad F = \frac{E}{8B(1-\upsilon^2)} \qquad (11)$$

$$M^* = F.\begin{bmatrix} 2B+(1-\upsilon^2) & 2B-(1-\upsilon^2) & 0 & 4B\upsilon \\ 2B-(1-\upsilon^2) & 2B+(1-\upsilon^2) & 0 & 4B\upsilon \\ -(1-\upsilon^2) & -(1-\upsilon^2) & 2(1-\upsilon^2) & 0 \\ 4\upsilon & 4\upsilon & 0 & 8B \end{bmatrix}$$

If plane stress conditions apply, i.e. B=1, then equation (11) reduces to a plane stress solution given by Leggatt et al. [3].

By assuming plane stress conditions there will be an error $\delta\sigma$ in the measured residual stresses. This error is given by :

$$\begin{bmatrix} \delta\sigma_{xx} \\ \delta\sigma_{yy} \\ \delta\tau_{xy} \\ \delta\sigma_{zz} \end{bmatrix} = \begin{bmatrix} \dfrac{1-B}{2}\left(1 - \dfrac{\sigma_{yy}}{\sigma_{xx}}\right) \\ \dfrac{1-B}{2}\left(1 - \dfrac{\sigma_{xx}}{\sigma_{yy}}\right) \\ 1-B \\ 0 \end{bmatrix} \qquad (12)$$

The error $\delta\sigma_{xx}$ is shown in Figure 4 as a function of depth for a plate thickness of 20, 50 and 100 mm, assuming $\sigma_{yy} = \tau_{xy} = 0$. The maximum error is for τ_{xy} and is about 15 % which occurs for a thin plate near to the entrance and exit faces of the reference hole.

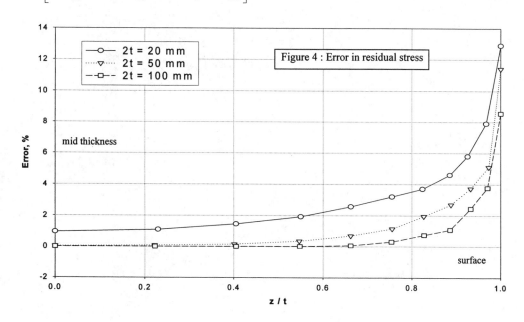

Figure 4 : Error in residual stress

The analysis presented in this paper has shown that it is possible to take account of the non-uniform distortions of the reference hole near to its entrance and exit surfaces. If it is assumed that a plane stress analysis is used to determine the residual stresses from the relaxed distortions of the reference hole the maximum error in the measured stress is about 15 %.

REFERENCES

[1] E.M. Beaney, 1976, "Accurate Measurement of Residual Stress on Any Steel Using the Centre Hole Method", Strain, vol. 2, No. 3, pp. 99-106.

[2] I.M. Zhadanov and A.K. Gonchar, 1978, "Determining the Residual Welding Stress at a Depth in Metal", Automatic Welding, vol. 31, No. 9, pp. 22-24, translation of Avt. Svarka, 1978, No.9, pp. 26-27 & 36.

[3] R.H. Leggatt, D.J. Smith, S.D. Smith and F. Faure, 1996, "Development and Experimental Validation of the Deep Hole Method for Residual Stress Measurement", Journal of Strain Analysis, vol. 31, pp. 177-186.

[4] D.J. Smith and N.W. Bonner, 1996, "Measurement of Residual Stresses Using the Deep Hole Method", ASME Pressure Vessels and Piping Conference, Florida, USA, July 1997.

[5] N.W. Bonner & D.J. Smith, 1994, "Measurement of Residual Stresses in Thick Section Steel Weld", Proc. International Conference on Engineering Integrity Assessment, Engineering Materials Advisory Services Ltd., pp. 259-274.

[6] N.W. Bonner, 1996, "Measurement of Residual Stresses in Thick Section Steel Welds", PhD thesis, University of Bristol.

[7] G.W. Stewart, 1973, "Introduction to Matrix Computation", Academic Press Inc. (London) ltd.

Experimental Mechanics, Allison (ed.) © 1998 Balkema, Rotterdam, ISBN 90 5809 014 0

Residual stress determination in plates using layer growing/removing methods

J. Kõo & J. Valgur

Chair of Structural Mechanics and Engineering, Estonian Agricultural University, Tartu, Estonia

ABSTRACT: A generalized algorithm of the layer growing/removing methods is presented for computing residual stresses in a free rectangular orthotropic inhomogeneous plate whose elastic parameters depend on its thickness coordinate. The algorithm allows calculation of residual stresses from strains or curvatures measured on the stationary surface of the plate as well as from initial stresses measured on the moving surface using X-ray diffraction. The suggested algorithm is programmed for PC and presents interest first of all in the study of residual stresses in coatings and surface layers generally. Three examples of application are presented.

1. INTRODUCTION

The layer removing method (destructive method) and the layer growing method (non-destructive method) are used for the determination of residual stresses in coated plates. The elaboration of the theory of the layer removing method started with papers by Treuting and Read (1951) and by Moore and Evans (1958) treating homogeneous plates. The theory of the layer growing method was evolved in works by Kõo (1959, 1969, 1979), by Birger and Kozlov (1974), by Doi et al. (1974, 1975) as well as by other authors.

In a thesis Kõo (1994) and in a paper by same author (1997) a common algorithm of the layer growing and layer removing methods is presented for the determination of equibiaxial residual stresses in an isotropic two-layer plate. This algorithm enables to calculate residual stresses from the strain or curvature measured on the stationary surface of the plate, as well as from initial stresses measured by the X-ray diffraction technique on the moving surface. In this study an advanced algorithm is presented that allows to calculate biaxial residual stresses in a free rectangular orthotropic inhomogeneous elastic plate whose elastic parameters depend on its thickness coordinate continuously or piecewise.

2. A GENERALIZED ALGORITHM OF LAYER GROWING/REMOVING METHODS FOR PLATES

Consider a thin layer growing on one face of a free rectangular plate (Fig. 1). Let the initial thickness of the plate be z_1, variable thickness h and final thickness z_2. Rectangular coordinates x, y and z are used, where the free stationary surface of the plate is taken as the reference surface (x, y) and coordinate z is perpendicular to the stationary surface. It is assumed that axes x and y are both orthotropic axes and principal axes of the state of residual stresses depending on the coordinate z only. It is also assumed that the edges of the plate are parallel to axes x and y.

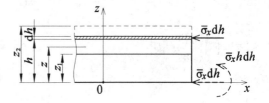

Figure 1. Layer-growing on the upper face of a rectangular plate.

The algorithm developed in this paper is based on the Kirchhoff small-deflection theory of plates (Timoshenko & Woinowski-Krieger 1959) and on the general algorithm of the layer growing/removing methods (Kõo 1994, 1997).

According to the mentioned general algorithm, residual stresses in a layer z of the coating can be calculated as a sum of initial and additional stresses:

$$\{\sigma\}=\{\bar{\sigma}\}+\{\sigma^*\} \tag{1}$$

where

$$\{\sigma\}=\begin{bmatrix} \sigma_x & \sigma_y \end{bmatrix}^T$$

$$\{\bar{\sigma}\}=\begin{bmatrix} \bar{\sigma}_x & \bar{\sigma}_y \end{bmatrix}^T$$

$$\{\sigma^*\}=\begin{bmatrix} \sigma_x^* & \sigma_y^* \end{bmatrix}^T$$

are the vectors of residual stresses, initial stresses and additional stresses, respectively.

In order to express residual stresses from the measured deformation parameters we proceed from the mechanical effect of formation of the differential superficial layer dh (Fig. 1) at variable thickness h. As is known (Kõo 1994, 1997), this effect can be expressed by applying differential edge forces to the edges of the differential superficial layer, which, when reduced to the reference surface, yield compressive edge forces $\bar{\sigma}_x dh$, $\bar{\sigma}_y dh$ and edge moments $\bar{\sigma}_x h dh$, $\bar{\sigma}_y h dh$.

Thus the problem is reduced to a problem for the free rectangular orthotropic inhomogeneous plate subjected to compressive edge forces and bending edge moments along the edges.

In order to solve the problem we use, as was already noted, the Kirchhoff approximation. Initial stresses $\bar{\sigma}_x = \bar{\sigma}_x(h)$, $\bar{\sigma}_y = \bar{\sigma}_y(h)$ in the differential surface layer dh can be expressed by strains $\varepsilon_x = \varepsilon_x(h)$, $\varepsilon_y = \varepsilon_y(h)$, and curvatures $æ_x = æ_x(h)$, $æ_y = æ_y(h)$ measured on the stationary surface as follows:

$$\{\bar{\sigma}\}=[B]\left\{\frac{d\tilde{\varepsilon}}{dh}\right\}-[C]\left\{\frac{d\tilde{æ}}{dh}\right\} \tag{2}$$

$$h\{\bar{\sigma}\}=[C]\left\{\frac{d\tilde{\varepsilon}}{dh}\right\}-[D]\left\{\frac{d\tilde{æ}}{dh}\right\} \tag{3}$$

where

$$\left\{\frac{d\tilde{\varepsilon}}{dh}\right\}=\begin{bmatrix} \dfrac{d\tilde{\varepsilon}_x}{dh} & \dfrac{d\tilde{\varepsilon}_y}{dh} \end{bmatrix}^T$$

is the vector of the derivatives of strain changes

$\tilde{\varepsilon}_x = \varepsilon_x(z_2)-\varepsilon_x(h)$, $\tilde{\varepsilon}_y = \varepsilon_y(z_2)-\varepsilon_y(h)$,

$$\left\{\frac{d\tilde{æ}}{dh}\right\}=\begin{bmatrix} \dfrac{d\tilde{æ}_x}{dh} & \dfrac{d\tilde{æ}_y}{dh} \end{bmatrix}^T$$

is the vector of the derivatives of curvature changes $\tilde{æ}_x = æ_x(z_2)-æ_x(h)$, $\tilde{æ}_y = æ_y(z_2)-æ_y(h)$, and

$$[B]=\begin{bmatrix} B_x & B_\mu \\ B_\mu & B_y \end{bmatrix} \tag{4}$$

$$[C]=\begin{bmatrix} C_x & C_\mu \\ C_\mu & C_y \end{bmatrix} \tag{5}$$

$$[D]=\begin{bmatrix} D_x & D_\mu \\ D_\mu & D_y \end{bmatrix} \tag{6}$$

are the matrices of the elastic parameters given by

$$\begin{bmatrix} B_x & B_y & B_\mu \\ C_x & C_y & C_\mu \\ D_x & D_y & D_\mu \end{bmatrix}=\int_0^h [E^0(z)]\left\{\begin{matrix} 1 \\ z \\ z^2 \end{matrix}\right\}dz \tag{7}$$

with

$$[E^0]=\begin{bmatrix} E_x^0 & E_y^0 & E_\mu^0 \end{bmatrix} \tag{8}$$

and

$$E_x^0=\frac{E_x}{1-\mu_{xy}\mu_{yx}}, \quad E_y^0=\frac{E_y}{1-\mu_{xy}\mu_{yx}}, \tag{9}$$

$$E_\mu^0=\frac{\mu_{xy}E_x}{1-\mu_{xy}\mu_{yx}}=\frac{\mu_{yx}E_y}{1-\mu_{xy}\mu_{yx}} \tag{10}$$

where $E_x = E_x(z)$, $E_y = E_y(z)$ denote the orthotropic moduli of elasticity, and $\mu_{xy}=\mu_{xy}(z)$, $\mu_{yx}=\mu_{yx}(z)$ denote orthotropic Poisson's ratios.

The expression for computing additional stresses in the coating ($z_1 \leq z \leq z_2$) is

$$\{\sigma^*\}=[E^*]\int_z^{z_2}\left[-\left\{\frac{d\tilde{\varepsilon}}{dh}\right\} \quad z\left\{\frac{d\tilde{æ}}{dh}\right\}\right]dh \tag{11}$$

where

$$[E^*]=\begin{bmatrix} E_x^0(z) & E_\mu^0(z) \\ E_\mu^0(z) & E_y^0(z) \end{bmatrix} \tag{12}$$

For computing residual stresses in the substrate ($0 \leq z \leq z_1$) the lower limit z of the integral in expression (11) should be replaced by z_1.

If measurement of strains or curvatures is not performed during coating growth then, using removing procedure, it should be assumed in the above algorithm that

$\varepsilon_x(z_2)=\varepsilon_y(z_2)=æ_x(z_2)=æ_y(z_2)=0$.

Expressions (1)-(12) form a common algorithm of the layer growing/removing methods for orthotropic inhomogeneous plates, allowing calculation of residual stresses at growing/removing on one face of the plate:

1. From strains and curvatures measured on the free stationary surface ($z = 0$) depending on thickness h. In this case initial stresses are computed by using expression (2) or (3). From equation (11) the expression for computing additional stresses is

$$\{\sigma^*\} = [E^*][\{\tilde{\varepsilon}\} - z\{\tilde{æ}\}] \tag{13}$$

2. From measured strains or curvatures only. In this case the unmeasured deformation parameter is computed from the equation

$$[[C] - h[B]]\left\{\frac{d\tilde{\varepsilon}}{dh}\right\} = [[D] - h[C]]\left\{\frac{d\tilde{æ}}{dh}\right\} \tag{14}$$

which follows from expressions (2) and (3).

3. From initial stresses measured by the X-ray diffraction technique on the moving surface depending on thickness h. In this case the derivatives of strain and curvature changes are calculated from expressions

$$\left\{\frac{d\tilde{\varepsilon}}{dh}\right\} = \frac{[[D] - h[C]]}{[[B][D] - [C]^2]}\{\bar{\sigma}\} \tag{15}$$

$$\left\{\frac{d\tilde{æ}}{dh}\right\} = \frac{[[C] - h[B]]}{[[B][D] - [C]^2]}\{\bar{\sigma}\} \tag{16}$$

which are obtained by solving equations (2) and (3).

In special cases computation of residual stresses is simplified. For example, in case of the homogeneous plate (two-layered plate with equal elastic constants $\mu_1 = \mu_2 = \mu$, $E_1 = E_2 = E$) the Treuting-Read (1951) and Moore-Evans (1958) formulas follow from the above generalized algorithm.

Initial stresses are usually assumed to be equal in the determination of residual stresses in coatings. This hypothesis allows to obtain all special algorithms published earlier (Kõo 1994 et al.) as special cases of the generalized algorithm.

3. COMPUTER PROGRAM RS-PLATE AND COMPUTATIONAL EXAMPLES

On the basis of the presented algorithm (Sect. 2) a computer program RS-PLATE is written for PC in Turbo Pascal, which enables calculation of residual stresses in orthotropic inhomogeneous plates from strains, curvatures or initial stresses measured during growing or removing process. According to the program, calculation of the derivatives of experimental data is carried out by a preliminary fitting with a polynomial.

Using the program RS-PLATE three computational examples are realized.

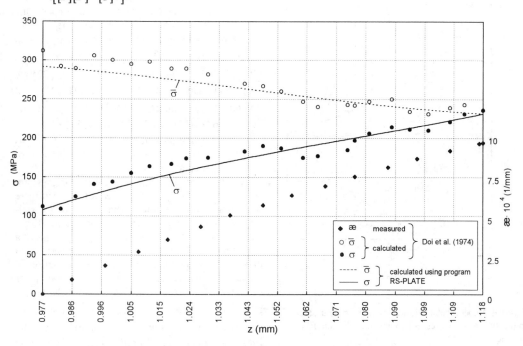

Figure 2. Experimental information and distribution of initial and residual stresses in a galvanic nickel coating.

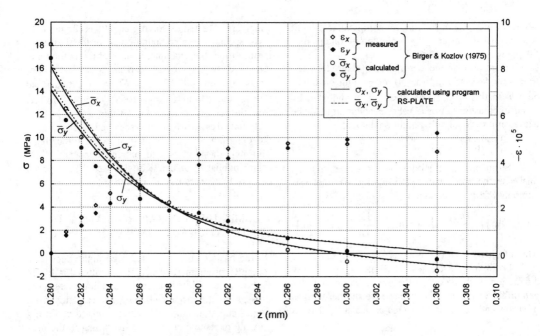

Figure 3. Experimental information and distribution of initial and residual stresses in a polycristallic tellurium layer.

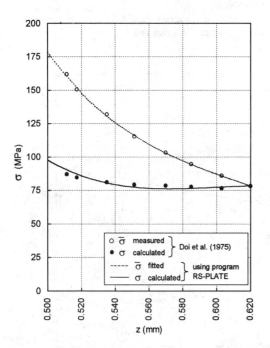

Figure 4. Experimental information and distribution of initial and residual stresses in a galvanic chromium coating.

As the first example, equibiaxial residual stresses are computed in a galvanic nickel coating from curvatures measured by Doi et al. (1974) during the growing process of the coating on a copper plate substrate. Figure 2 shows the experimental data of curvature measurements ($\mathfrak{x}_x = \mathfrak{x}_y = \mathfrak{x}$) and the distribution of initial and residual stresses ($\overline{\sigma}_x = \overline{\sigma}_y = \overline{\sigma}$, $\sigma_x = \sigma_y = \sigma$) in the coating ($z_2 = 1.118\,\text{mm}$, $\mu_2 = 0.31$, $E_2 = 207\,\text{GPa}$) deposited on a plate substrate ($z_1 = 0.977\,\text{mm}$, $\mu_1 = 0.34$, $E_1 = 122.6\,\text{GPa}$). Considerable scattering of the results of computation obtained by Doi et al. (1974) can be evidently explained by the omission of experimental data fitting.

As the second example, residual stresses are computed in the surface layer of an isotropic bimetallic (monocristallic germanium + polycristallic tellurium) plate from strains measured by Birger and Kozlov (1975) during the removing process of a tellurium layer. Figure 3 presents the experimental data of strain measurements and the distribution of initial and residual stresses in the tellurium layer ($z_2 = 0.31\,\text{mm}$, $\mu_2 = 0.188$, $E_2 = 38.95\,\text{GPa}$) on a germanium plate ($z_1 = 0.28\,\text{mm}$, $\mu_1 = 0.274$, $E_1 = 102.02\,\text{GPa}$). As one can see, initial stresses by Birger and Kozlov (1975) differ slightly from those computed using program RS-PLATE. This is due, among other things, to the averaging of the Poisson's ratios of germanium and tellurium in the calculations of Birger and Kozlov. It is also evident that initial and residual stresses differ slightly, i.e. additional stresses in thin coatings are negligible (see Kõo 1994, 1997).

As the third example, we present the results of computing equibiaxial residual stresses in a galvanic chromium coating from initial stresses measured by Doi et al. (1975) during the process of removing of the coating from a steel plate substrate using X-ray technique. Figure 4 shows the experimental data of initial stress measurements ($\overline{\sigma}_x = \overline{\sigma}_y = \overline{\sigma}$) and the distribution of initial and residual stresses ($\sigma_x = \sigma_y = \sigma$) in the coating ($z_2 = 0.62\,\text{mm}$, $\mu_2 = 0.23$, $E_2 = 181.5\,\text{GPa}$) deposited on a steel plate substrate ($z_1 = 0.5\,\text{mm}$, $\mu_1 = 0.28$, $E_1 = 192.3\,\text{GPa}$). It can be seen that residual stresses determined by Doi et al. (1975) agree satisfactorily with those computed using program RS-PLATE.

4. CONCLUSIONS

1. A generalized algorithm is elaborated for the computation of residual stresses in a free orthotropic inhomogeneous plate. The algorithm is universal and allows calculation of residual stresses at layer growing or layer removing from strains or curvatures measured on the stationary surface, or from initial stresses measured on the moving surface of the plate.

2. A computer program RS-PLATE based on the presented algorithm is introduced.

3. Using the program RS-PLATE residual stresses are computed in the galvanic nickel coating on a copper plate, in the tellurium layer on a germanium plate and in the galvanic chromium coating on a steel plate.

REFERENCES

Birger, I. A. & M. L. Kozlov 1974. Determination of residual stresses in thin coatings of orthotropic plates. (In Russian). *Zavodskaya Laboratoriya* 40 (2): 223-225.

Birger, I. A. & M. L. Kozlov 1975. Determination of residual stresses in a plate whose elastic parameters are variable through thickness. (In Russian). *Zavodskaya Laboratoriya* 41 (2): 239-241.

Doi, O., T. Ukai & A. Ohtsuki 1974. Measurements of Electrodeposition Stress in a Plate Cathode. *Bull. Jap. Mech. Eng.* 17 (105): 297-304.

Doi, O. & T. Ukai 1975. Residual Stress Measurement of Multi-Layered Plate by X-Ray Method. *Bull. Jap. Mech. Eng.* 18 (119): 493-500.

Kõo, J. 1959. Calculation of residual stresses in the electrodeposits from deformation of a thin plate cathode. (In Russian). *Trans. Est. Agricult. Acad.* 13: 63-75.

Kõo, J. P. 1969. Determination of residual stresses in coatings of bars and plates using strain measurement method. (In Russian). *Izvestiya vuzov. Mashinostroenie* 12: 47-51.

Kõo, J. P. 1979. Determination of residual stresses in coatings of plates using curvature measurement method. (In Russian). *Izvestiya vuzov. Mashinostroenie* 8: 9-12.

Kõo, J. 1994. Determination of Residual Stresses in Coatings and Coated Parts. *Dr. Eng. Thesis, Series E. Machinery and Fine Mechanics.* Tallinn Technical University, Estonia.

Kõo, J. 1997. Layer-Growing and Layer-Removing Methods for Residual Stress Analysis. Accepted for publication in *Proc. of the 4th Europ. Conf. on Residual Stresses, Cluny (France), 4-6 June 1996:* 10 p.

Moore, M. G. & W. P. Evans 1958. Mathematical Correction for Stress in Removed Layers in X-Ray Diffraction Residual Stress Analysis. *SAE Trans.* 66: 340-345.

Perakh, M. 1979. Calculation of spontaneous macrostresses in deposits from deformation of substrates and from restoring force factors. *Surf. Technol.* 8: 265-309.

Timoshenko, S. & S. Woinowsky-Krieger 1959. *Theory of Plates and Shells.* New York: McGraw-Hill.

Treuting, R. G. & W. T. Read 1951. A Mechanical Determination of Biaxial Residual Stress in Sheet Materials. *J. Appl. Phys.* 22 (2): 130-134.

Experimental Mechanics, Allison (ed.)© 1998 Balkema, Rotterdam, ISBN 90 5809 014 0

Residual stress measurement by the crack compliance technique

S. Tochilin, D. Nowell & D. A. Hills
Department of Engineering Science, University of Oxford, UK

ABSTRACT: A detailed knowledge of residual stresses in the surface layers of any safety-critical engineering component is vital to its safe operation and for a reliable estimate of its life. This paper describes a combination of experimental and theoretical procedures designed to measure near-surface residual stresses accurately, quickly and inexpensively, by a controlled relaxation technique. The relaxation is effected by using EDM to produce a narrow slit through the thickness of a specimen. Strain gauges are applied to both the top surface of the specimen, adjacent to the slit, and on the back face, directly beneath the slit. The work described builds on the pioneering contributions by Finnie and colleagues in the USA. Sample data are displayed, and compared with results obtained from the same coupons using neutron diffraction. The correlation between the results is encouraging and suggestions are made for further improvements to the experimental and analytical procedures.

1. INTRODUCTION

A knowledge of the pre-existing residual stress within the surface of any safety-critical component is essential to predict the performance of that component, both as a quality assurance check, and to enable the designer to determine its fatigue performance. The residual stresses may originate either as a largely unwanted by-product of machining processes, when they are generally tensile in character at the surface itself, or they may be deliberately introduced by a finishing process such as shot-peening, ion-implantation, or the laying down of a surface coating. Also, exceptionally, the stresses may originate from some macroscopic pre-stressing process, such as over-bending (upsetting) of springs, over-twisting of torsion bars, or autofrettage. Whatever the origin of the stresses, an efficient way of determining the magnitude of residual stresses, and their variation with depth is crucial. It is important to recognise that all of the residual stress fields mentioned are self-equilibrating, i.e. they give rise to no resultant force, that by and large they are cylindrical in nature, and that the main difference between them from the perspective of their measurement is the scale over which they exist. For ion-implantation the depth of influence may be as

little as 100 micrometres, for shot peening it may be of the order of a millimetre, whilst for the macroscopic processes it may be several millimetres. It is generally more difficult to obtain precise information about small scale (very near surface) residual stresses by mechanical methods such as the one to be described, whilst the reverse is true for X-ray diffraction-based methods.

Diffraction methods have the advantage that they are sensitive (particularly neutron diffraction), and are naturally suited to separation of stress components if the stress state is complex, but their principal disadvantage is the complexity of conducting the test, and consequent cost. X-ray methods also suffer from the drawback that they are effective only close to a free surface. The common traditional alternative has been the use of hole-drilling techniques, in which a strain rosette is glued to the surface and a hole drilled through the middle. Strain changes may then be used to infer the pre-existing residual stress state relaxed out. However, the problem here is that a circumferential stress, $\sigma_{\theta\theta}$, develops which resists changes in the hole size, and so limits the sensitivity of the method. Also, the use of mechanical conventional drilling to form the hole is known to induce residual stresses of its own, although this can be considerably reduced by careful experimental techniques, for example by the use of high speed drilling (Fathallah, et al. (1994)).

Both of these inhibiting features in the hole-drilling method may be avoided by using a different approach, in which the specimen is prepared in the form of a beam (typically of cross sectional dimensions 20mm x 10mm), and a slit is gradually made across the entire width of the specimen, by electric-discharge machining (EDM). The original work in this field was conducted by Finnie and colleagues nearly ten years ago (e.g. Cheng and Finnie (1990)), but the technique is still not widely known. The principal obstacle to its wider adoption had, perhaps been the difficulty in calculating the compliance functions necessary to establish the residual stress distribution. Application of boundary element or dislocation density procedures in recent years have considerably simplified this step in the analysis. A comprehensive review of the crack compliance method is given by Prime (1997).

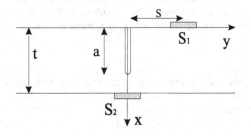

Figure 1. Geometry of a strain-gauged specimen containing a slot of depth a.

2. EXPERIMENTAL TECHNIQUE

The basic principles behind the method can be appreciated by reference to Fig. 1. A beam type specimen is prepared, ensuring that the length of the specimen is sufficiently large for there to be no end effects, and the width is chosen so that any residual stresses left by the preparation itself are likely to persist over only a small fraction of the width of the coupon. Strain gauges are applied to the beam at several surface locations, and on the back face, immediately opposite the proposed location of the slit. To obtain the best resolution very small gauges are employed, with an active measurement length of only 0.2mm, but with a wide active filament (1.4mm) to obtain high sensitivity. As will transpire from the results, the top-face gauges are most appropriate during cutting to approximately midway through the depth of the beam, and the use of two or

more staggered gauges further improves sensitivity; the highest rate of change in strain occurs when the gauge position along the top face from the slit edge is approximately equal to the current depth of the slit.

The chosen parameters of the slit-machining process are a compromise; for speed a high current is desirable, but this has the effect of (a) widening the slit, and leaving the bottom radiused, so that its idealisation as a crack in subsequent calculations becomes less realistic, (b) producing excessive wear of the electrode, so that the slit depth is imprecisely known, and (c) giving a very poor finish to the sides of the slit.

Conversely, a low current means that the measurement takes a long time to perform, although this may not be important in the case of very near surface measurements, when the greatest care is needed. Both wire and sheet form electrode have been successfully used. The former has the advantage that a cassette to provide continuous feed may be employed, so that the effects of erosion of the electrode are minimised.

As the test proceeds cutting is interrupted from time to time, the electrode is withdrawn, and strain readings are taken. The whole process is conducted under the control of a PC, including strain data logging, and a single software package has been developed which also allows post-processing of the data to obtain the residual stress profile. In future it will be possible to carry out this processing during the experiment itself so that successive estimates of the residual stress profile may be obtained on line.

3. THEORY

It is appropriate to commence this section by giving a brief introduction to the basic analysis technique employed in the crack compliance method. Let us assume that we wish to measure the variation of residual stress component with depth in a plate of thickness t. We progressively introduce a cut to a maximum depth a_K and record the strain gauge readings at a gauge position S (x_s, y_s) at $K-1$ intermediate depths of cut, as well as the final position so that we have K pairs of readings (a_k, ε_k). We then assume that the residual stress distribution on $y = 0$ can be represented as a linear combination of N suitable basis functions so that

$$\sigma_{yy}(x,0) = \sum_{i=0}^{N-1} \alpha_i p_i(x) \qquad (1)$$

where p_i are the chosen basis functions (which may

be power series polynomials, Chebyshev polynomials, piece-wise functions, or other suitable choices). The terms α_i are the weights of these basis functions which must be determined from the experimental data. We note that, for the case of plane strain, the strain change recorded at the strain gauge when the cut has a depth of a_k may be written as

$$\epsilon(S, a_k) = \frac{1 - v^2}{E} \sum_{i=0}^{N-1} \alpha_i C_i(S, a_k) \quad (2)$$

The term $C_i(S, a_k)$, which we will write in shorthand as C_{ik}, represents the change in the required stress component produced at location S when a crack is introduced to a depth a_k in a plate containing a residual stress distribution of the ith basis function, $p_i(x)$. It is this term, which is called the compliance by Finnie and others (e.g. Cheng and Finnie (1994)), which must be determined before the experimental results can be analysed. Once these compliances have been established, the values of α_i may be found by the normal procedures of least squares fitting for the case of $K>N$:

$$\frac{\partial}{\partial \alpha_j} \sum_{k=1}^{K} \left(\epsilon_k - \sum_{i=0}^{N_1} C_{ik} \right)^2 = 0 \qquad j = 0, \ldots, N \quad (3)$$

It will be apparent from the above explanation that it is possible to combine data from several gauge positions in a single least squares fit and this improves the reliability of the data. If singular value decomposition (e.g. Press et al. (1986)) is employed to solve the least squares problem, it is possible to assign different weights to data according to its likely reliability.

The most demanding step in the analysis is the determination of the compliances C_{ik}. A number of different methods have been employed (Prime (1997)) which generally consider the machined slit as equivalent to a crack. Various authors have used Castagliano's theorem, together with stress intensity factor solutions for a crack loaded by a remote force pair, boundary element methods, and finite element analysis. An alternative is to employ the displacement density method of crack analysis (Nowell and Hills (1987)), which we have found extremely effective for this type of problem. At each crack depth the crack is modelled by a continuous distribution of displacement discontinuities (or dislocations). The density of the distribution is found from the requirement that the crack faces are

free of tractions when loaded by the chosen basis function. This formulation leads to an integral equation in the unknown dislocation density, B_y, which may be solved by standard methods:

$$p_i(x) + \int_0^{a_k} B_y(c) K(x, 0, c) dc = 0 \qquad 0 < x < a_k$$

(4)

Here, the Kernel function K gives the σ_{yy} component of stress at the integration point x caused by a single dislocation located at the integration point c. Once the dislocation density has been found, it is possible to find the strain changes at any point in the specimen by integrating the effects of the individual dislocations. This means that the strain changes at a number of gauge positions may be found from a single inversion of the integral equation. Space considerations do not permit a detailed explanation of the method here, but a full description is given in Nowell et al. (1997).

The dislocation density method has a number of advantages. It is straightforward, accurate, and the Kernel function K is known for a number of basic geometries of interest. These include a half-plane (for cracks close to a single free surface), a strip (for cracks a significant proportion of the thickness of the specimen) and layers of different elastic constants (for cases where a surface coating is present).

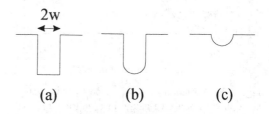

Figure 2. Possible slot geometries: (a) flat bottomed, (b) round bottomed, and (c) shallow notch.

The idealisation of modelling the slit as a crack of zero thickness is appropriate when the slit is long (i.e. its length is much greater than the distance between the strain gauges and the slit). However, when we are attempting to measure residual stress profiles close to the surface short slits will be required and the assumption of zero width is less satisfactory. This may be allowed for by using a more sophisticated formulation for the dislocation distribution. Instead of deploying the dislocations along a single straight line, and imposing the traction-free requirement there, one of the shapes shown in Fig. 2 is used. In the first of these (Fig. 2(a)), the crack is replaced by a flat-bottomed notch, and the influence on the slot width on the strain changes experienced may be found. However, changing the form of the slot root to a semi-circle, Fig. 2(b) was found to have a profound effect on the stress state at the local surface, and it is recognised that for extremely shallow cuts, as depicted in Fig.2(c), an incompletely formed slot will be present.

All of these shapes may be treated by the dislocation density method, and are being used to improve the compliance functions used in our analysis. Figure 3 shows the variation of compliance C_{ik} with gauge position for a slit in a uniform residual stress field for the sample case of $w = a/4$. It will be seen that the 'crack' and flat-bottomed slot solutions are very close, provided that the gauge position is measured from the edge of the slit. In contrast, the compliance function for a round-bottomed slot is different in form and this highlights the need to know accurately the shape of the slot root when carrying out experiments of this type.

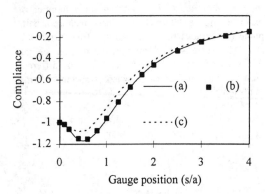

Figure 3. Variation of compliance C_{ik} with gauge position for a top face gauge in a uniform residual stress field. (a) thin slot modelled as a crack, (b) Flat-bottomed slot, $a/b=4$, (c) Round-bottomed slot, $a/b=4$.

Analysis of finite width slots has also been carried out by Cheng and Finnie (1993). Unfortunately, some of the plots of their results are mis-labelled (Finnie (1997)), but when this is taken into account our results are in good agreement with theirs. Cheng and Finnie mainly present results for strain gauges which are remote from the slot mouth and for these configurations they conclude that slot width is important. Our results, however, show that this is not the case if the gauges are close to the slot,

rather it is the geometry of the slot root which is the dominating factor. Cheng et al. (1994) suggest an empirical correction factor to take account of the notch root geometry, but their proposed correction does not give good agreement with the current method.

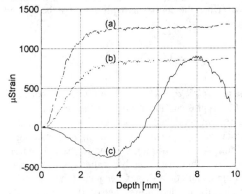

Figure 4. Strain gauge readings obtained from a Waspaloy beam with residual stresses caused by bending, $t=10$mm. (a) top face gauge, $s=1.3$mm, (b) top face gauge, $s=2.2$mm, (c) back face gauge.

4. RESULTS

Example results are shown in Fig. 4. for a Waspaloy beam bent into the elastic-plastic regime. This is one of the simplest examples on which measurements may be taken, as the residual stress profiles are macroscopic, vary relatively slowly with depth, and may be estimated from elementary beam theory. The raw strain gauge change data as a function of depth is shown in the figure. It may be noted that the gauges on the upper surface show significant changes of reading whilst the crack remains short, so that they provide reliable information, whilst the readings level off to an almost constant value as soon as the slit becomes long. On the other hand, a strain reading taken from the back face gauge initially shows only a slow variation with depth, i.e. it is relatively insensitive to the residual stress state near the surface, but starts to show a much more pronounced reading as the crack approaches the half way point. This means that, in order to obtain a complete picture of the stress variation with depth, it is prudent to place a higher reliance on the top face gauges at small depths, and higher reliance on the back face gauge at greater depths. The residual stress state inferred from the raw data is shown in Fig. 5, where it is compared with the residual stress predicted by bending theory. Also shown in Fig. 5 are neutron diffraction measurements made by our collaborators at Imperial College (Ezeilo (1997))

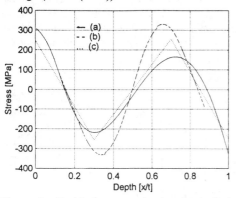

Figure 5. Residual stresses in the plastically-bent Waspaloy beam (a) measured by crack compliance (b) measured by neutron diffraction, (c) predicted by simple beam theory.

Results from a second Waspaloy specimen, which has been shot peened, are shown in Fig. 6. The predicted stress distribution close to the surface is similar to that expected. Stresses approaching the compressive yield stress of the material are found in a shallow surface layer and these are balanced by tensile stresses of a lower magnitude at greater depths.

5. CONCLUSIONS AND FURTHER WORK

The crack compliance method described here works well for macroscopic residual stress fields, and dislocation density methods provide a convenient means of evaluating the required compliance functions. More accurate results are obtained if data from a number of different strain gauges are used in the calculation of the residual stresses, weighted according to the sensitivity of the gauge at a particular crack length. For near surface residual stress fields the finite width of the slot must be considered and the geometry of the root of the notch has a significant effect on the compliance functions. Overall, the crack compliance method provides a convenient and inexpensive means of determining a range of residual stress components in simple geometric configurations. For more complex geometries it may be possible to introduce 'thumbnail' slots and we are currently investigating the feasibility of this method.

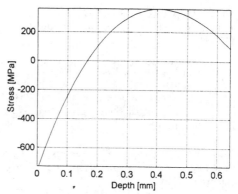

Figure 6. Example residual stress profile measured on a shot-peened Waspaloy sample.

6. REFERENCES

Cheng, W., and Finnie, I., (1990) 'Measurement of residual stresses near the surface using the crack compliance method', ASME J. of Eng Mat. and Tech., 113, 199-204.

Cheng, W., and Finnie, I. (1993) 'A comparison of the strains due to edge cracks and cuts of finite width with applications to residual stress measurement', Jnl Eng Mats and Tech., 115, 220-226.

Cheng, W., and Finnie, I., (1994) 'An overview of the crack compliance method for residual stress measurement', pp 449-458 of Proc. 4th Int. Conf. on Residual Stress, Baltimore, June 1994, Pub: Soc. Exp. Mech., Bethel, Connecticut, 1994.

Cheng, W., Finnie, I., Gremaud, M., and Prime, M.B., (1994) 'Measurement of near surface residual stresses using electric discharge wire machining', Jnl Eng Mats and Tech., 116, 1-7.

Ezeilo, A., (1997) 'Comparison of compliance and neutron diffraction measurements of residual stress', private communication, December 1997.

Fathallah, R., Cao, W., Castex, L., Webster, P.S., Ezeilo, A., Webster, G.A., and Webster, P.J., (1994) ' Comparison of residual stresses determined by X-ray diffraction, neutron diffraction and the hole drilling method in aerospace shot peened materials', Proc. 4th Int. Conf. on Residual Stresses, Baltimore, June 1994, Society for Experimental Mechanics, Bethel, CT.

Finnie, I., (1997) Private communication.

Nowell, D., and Hills, D.A., (1987) 'Open cracks at or near free edges', Jnl. Strain Analysis, 22, 3, 177-186.

Nowell, D., Tochilin, S., and Hills, D.A., (1997) 'The crack compliance method for the measurement of residual stress: dislocation density analysis', Proc. 5th Int. Conf. on Residual Stress, Linköping, Sweden, 16th-18th June 1997.

Press, W.H., Flannery, B.P., Teukolsky, S.A., and Vetterling, W.T. (1986), 'Numerical Recipes', Cambridge University Press.

Prime, M.B., (1997) 'Residual stress measurement by successive extension of a slot: a literature review', Report LA-13283-MS, Los Alamos National Laboratory.

7. ACKNOWLEDGEMENT

The work described in this paper forms part of a joint project with Imperial College of Science, Technology and Medicine, London (Prof. G.A. Webster and Dr A. Ezeilo) and is supported by the Engineering and Physical Sciences Research Council under contract GR/K75507. The Neutron diffraction results presented were obtained by Imperial College. The authors would also like to thank Rolls-Royce Plc for provision of shot-peened sample material.

Experimental Mechanics, Allison (ed.)© 1998 Balkema, Rotterdam, ISBN 90 5809 014 0

Residual stresses by Moiré Interferometry and incremental hole drilling

Zhu Wu & Jian Lu
LASMIS – Mechanical Systems and Engineering, Universite de Technologie de Troyes, France

ABSTRACT: A new method is proposed in determining residual stress in depth by single moiré interferometry and incremental hole drilling method. The relationships between the single surface displacement field of moiré interferometry and the corresponding residual stress is established by introducing a set of calibration coefficients. These coefficients are determined by 3-D finite element method, where, two specific loading cases are used. The method is accurate and convenient and is applied successfully in determining the distribution of welding residual stress in depth be the single U_x fields, where the U_x field contains more fringe orders than the corresponding U_y field. Many other applications for various residual stress problems are anticipated.

INTRODUCTION

Although X-ray diffraction method, Neutron diffraction method and the other residual stress measuring methods have found many application recently in residual stress study, the hole drilling method is still widely used in engineering since it is less sensitive to the microstructure of materials tested, less expensive and more convenient for *in-situ* measurement. Strain gage method was first adopted for residual stress measurement by Soete (1949). Numerical coefficients calibration method was developed by Schajer (1981). It was a milestone in hole-drilling method. In the finite element model, external loads could be applied on the side surface of the hole to simulate the relaxation of residual stresses. Combining the numerical coefficients calibration method and the strain gage incremental hole drilling technique, a modified method was developed (Lu, 1985; 1996). In the method, the residual stress profile in depth could be determined by drilling a hole step by step with tiny increment for each step. For a non-uniform residual stress field or an inhomogenous material, the strain gage or strain rosette hole-drilling method has inherent drawbacks. Since the 1980s, some laser interferometry techniques with very high sensitivities have been developed for the detection of small deformations, such as: holographic interferometry, laser speckle interferometry and moiré interferometry. They seem to be an improvement on the traditional strain gage method because of their capacity for whole field observation. However, most of the studies on residual stress by the optical methods were qualitative or semi-quantitative analyses, in which the surface residual strain fields or displacements fields were given but no relationships were established between the residual stresses and residual strains or residual displacements. Moiré interferometry (Nicoletto, 1991) and holographic interferometry (Antonov, 1983) were used in the determination of residual stresses when a through-hole was introduced so that a closed form solution could be found. Recently, a new development was made to treat the blind hole problems by combining single axis holographic interferometry with finite element calibration method (Makino, 1994; Nelson, 1994). However, due to the poor contrast and low signal to noise ratio of fringe pattern in holographic interferometry, the displacement data were extracted at a distance of $2r_0$ from the center of the hole. Although the theoretical basis for adopting incremental hole-drilling was given in the reference, the method was implemented only for a blind hole with a single increment because of the lack of spatial resolution and displacement measurement accuracy which were essential to detect small deformations occurred during hole drilling with a tiny increment. The newly developed technique which combined moiré interferometry and incremental hole drilling method made it practical to determine the residual stresses profile in depth (Wu, 1997). In the method, both of the in-plane displacement fields U_x and U_y are essential. For some residual stress fields, however, the U_x field in moiré interferometry may contain more fringe orders than the corresponding U_y field. It must be more convenient and accurate if the single U_x field is used in fringe order counting process. Consequently, if the single U_x field was essential in residual stress determination, The accuracy would be improved.

Welding residual stresses are created by phase transformations and temperature gradients arising when the weld and neighboring material melts or heats up and cools down. Due to the variety of factors may involved in a welding part, prediction and control of welding residual stresses has proved difficult. Apart from theoretical and numerical analyses, numerous experimental methods were developed and applied to welding residual stress research. Such as: X-ray diffraction method (Ruud, 1993), neutron diffraction method (Allen, 1985) and strain gage rosette hole drilling method (Lu, 1994) etc. The grain size and phase transformation in the weld and neighboring material as well as the non-uniform distribution of welding residual stresses made these methods inaccurate. In this study, the single field moiré interferometry combined with incremental hole drilling method will be developed to deal with the welding residual stress problem.

SURFACE DISPLACEMENT FIELD IN HOLE DRILLING METHOD

For blind hole problems, a Fourier expansion solution was proposed by Schajer (1981) to account for the effect of out of plane displacement.

As the first order approximation, n=0, 2, for a bi-axial residual stress field, the Fourier expansion solution can be expressed as:

$$u_r(r,\theta) = A(\sigma_{xx}+\sigma_{yy})+B[(\sigma_{xx}-\sigma_{yy})\cos 2\theta + 2\tau_{xy}\sin 2\theta]$$
$$u_\theta(r,\theta) = C[(\sigma_{xx}-\sigma_{yy})\sin 2\theta - 2\tau_{xy}\cos 2\theta]$$
$$u_z(r,\theta) = F(\sigma_{xx}+\sigma_{yy})+G[(\sigma_{xx}-\sigma_{yy})\cos 2\theta + 2\tau_{xy}\sin 2\theta]$$
$$(1)$$

where, σ_{xx}, σ_{yy} and τ_{xy} are the residual stress components in the Cartesian coordinate system; u_r, u_θ and u_z are the surface displacements around the hole in the cylindrical coordinate system; A, B, C, F and G are undetermined coefficients which depend on the material constants and the geometrical parameters of the blind hole.

Eq. (1) can be re-written in the Cartesian coordinate system as:

$$u_x(x,y) = (A\cos\theta + B\cos 2\theta\cos\theta - C\sin 2\theta\sin\theta)\sigma_{xx} +$$
$$(A\cos\theta - B\cos 2\theta\cos\theta + C\sin 2\theta\sin\theta)\sigma_{yy} +$$
$$(2B\sin 2\theta\cos\theta + 2C\cos 2\theta\sin\theta)\tau_{xy}$$
$$(2\text{-a})$$

$$u_y(x,y) = (A\sin\theta + B\cos 2\theta\sin\theta + C\sin 2\theta\cos\theta)\sigma_{xx} +$$
$$(A\sin\theta - B\cos 2\theta\sin\theta - C\sin 2\theta\cos\theta)\sigma_{yy} +$$
$$(2B\sin 2\theta\sin\theta - 2C\cos 2\theta\cos\theta)\tau_{xy}$$
$$(2\text{-b})$$

$$u_z(r,\theta) = F(\sigma_{xx}+\sigma_{yy})+G[(\sigma_{xx}-\sigma_{yy})\cos 2\theta + 2\tau_{xy}\sin 2\theta]$$
$$(2\text{-c})$$

For a blind hole with a single increment, the coefficients A, B, C, F and G in Eq. (1) can be determined by three dimensional finite element method when two specific loading cases were considered.

(1) $\sigma_{xx} = \sigma_{yy} = \sigma$, $\tau_{xy} = 0$, an equibiaxial residual stress field, which is equivalent to a uniform pressure p=σ acting on the hole boundary.

(2) $\sigma_{xx} = -\sigma_{yy} = \sigma$, $\tau_{xy} = 0$, a pure shearing residual stress field, which is equivalent to the harmonic distributions of the normal stress $\sigma_{rr} = -\sigma\cos 2\theta$ and the shear stress $\tau_{r\theta} = \sigma\sin 2\theta$ acting on the hole boundary.

For an incremental hole drilling problem, Eqs. (2-a), (2-b) and (2-c) were used repeatedly to determine a set of calibration coefficients when the specific loading cases were applied to the boundary of each layer, respectively. It should be noted that other load cases can also be used for coefficients calibration.

MOIRÉ INTERFEROMETRY APPLIED IN HOLE DRILLING METHOD

Moiré interferometry is an optical methods, providing real time and whole field contour maps of in-plane displacements. In moiré interferometry, the specimen grating of frequency f_s is replicated onto the surface of the specimen containing residual stress. When a blind hole is drilled, residual stresses will be released, the specimen grating deforms correspondingly. The surface displacement field due to the relaxation of residual stress should be localized in a small region around the hole. When the surrounding region beyond a few times of hole diameter is adjusted as a bright background of null field. The resulting interference patterns represent the absolute contours of constant U_X and U_y displacements. The displacements can then be determined from the fringe orders by the following relationships:

$$U_X(x,y) = \frac{1}{2f_s}N_X(x,y),$$

$$U_y(x,y) = \frac{1}{2f_s}N_y(x,y) \qquad (3)$$

where N_X and N_y are fringe orders in U_X and U_y fields, respectively. In routine practice, a frequency of 1200 lines/mm is used, providing a contour interval of 0.417 μm/fringe order.

DETERMINATION OF RESIDUAL STRESS

By combining Eqs. (2-a) and (2-b) with Eq. (3), $u_x = U_X$ and $u_y = U_y$, the relationships between the fringe orders obtained by moiré interferometry and the residual stress can be written as:

$$N_X(x_k, y_k) =$$
$$2f_s[(A\cos\theta_k + B\cos 2\theta_k \cos\theta_k - C\sin 2\theta_k \sin\theta_k)\sigma_{xx}$$
$$+ (A\cos\theta_k - B\cos 2\theta_k \cos\theta_k + C\sin 2\theta_k \sin\theta_k)\sigma_{yy}$$
$$+ (2B\sin 2\theta_k \cos\theta_k + 2C\cos 2\theta_k \sin\theta_k)\tau_{xy}]$$
$$(4\text{-}a)$$

$$N_y(x_k, y_k) =$$
$$2f_s[(A\sin\theta_k + B\cos 2\theta_k \sin\theta_k + C\sin 2\theta_k \cos\theta_k)\sigma_{xx}$$
$$+ (A\sin\theta_k - B\cos 2\theta_k \sin\theta_k - C\sin 2\theta_k \cos\theta_k)\sigma_{yy}$$
$$+ (2B\sin 2\theta_k \sin\theta_k - 2C\cos 2\theta_k \cos\theta_k)\tau_{xy}]$$
$$(4\text{-}b)$$

where, the coefficients A, B, C, F and G are calibrated at the radial coordinate r_c,
$$(x_k)^2 + (y_k)^2 = (r_c)^2 \text{ and } \theta_k = \tan^{-1}\left(\frac{y_k}{x_k}\right), \ k=1,2,3$$

Generally, only one of Eq. (4-a) and (4-b) is used. For an incremental hole drilling problem, Eq. (4-a) can be written as:

$$N_X^i(x_k, y_k) = \sum_{j=1}^{i}$$

$$\left\{ \begin{array}{l} 2f_s[(A^{ij}\cos\theta_k + B^{ij}\cos 2\theta_k \cos\theta_k - C^{ij}\sin 2\theta_k \sin\theta_k)\sigma_{xx}^j \\ + (A^{ij}\cos\theta_k - B^{ij}\cos 2\theta_k \cos\theta_k + C^{ij}\sin 2\theta_k \sin\theta_k)\sigma_{yy}^j \\ + (2B^{ij}\sin 2\theta_k \cos\theta_k + 2C^{ij}\cos 2\theta_k \sin\theta_k)\tau_{xy}^j] \end{array} \right\}$$

$$(5)$$

where, k=1, 2, 3; i=1, 2, ..., n; n is the total number of incremental steps; j=1, 2, ..., i.

Similar expression can be written from Eq. (3-b), and is omitted here.

EXPERIMENTAL PROCEDURE

The specimen of Titanium alloy plate, containing a V-groove weld, was 200.0 mm long, 40.0 mm wide and 2.0 mm thick, and was welded by a laser welding technique. The material properties of the plate were: $E = 1.1 \times 10^5$ MPa, $v = 0.3$. The grating with a frequency of 1200 lines/mm was replicated onto the surface of the specimen covering the weld and its neighboring material, which can provide very high sensitivity to detect the tiny deformation in incremental hole drilling method. The thickness of the grating was less than 20 μm, which was much smaller than the incremental depth used in the experiment, and thus the influence of the extra grating layer can be ignored. A schematic drawing of the specimen with the replicated grating is shown in Fig. 1. The typical regions R1, R2, R3 and R4 are also marked in the figure. They were the regions for residual stress analysis.

A combined system of moiré interferometry and incremental hole drilling device (Wu, 1997) was used in the experiment. After mounting the specimen to a desired position, the optical system was first adjusted to produce an initial null field condition, which was devoid of fringes. The drilling machine subassembly was mounted and a flat bottom hole with a diameter of 2.0 mm and an incremental depth of 0.1 mm was drilled carefully. The location of the hole is also shown in Fig. 1. After a hole-drilling procedure, the drilling machine subassembly was removed and the U_x displacement field was recorded by moiré interferometry. The experiment continued with rotating the specimen by 90° and the corresponding U_y displacement field was obtained. The total number of seven incremental steps were used and the above procedure was repeated at each incremental hole-drilling step.

The typical moiré fringe patterns, representing U_x and U_y displacement contours obtained from the

TA6V alloy specimen are presented in Fig. 2, where a contour interval was 0.417 μm per fringe order, the black fringes and white fringes represent the half fringe orders and integral fringe orders, respectively.

The corresponding regions of R1, R2, R3 R4 in Fig. 1 are marked in Fig. 2.

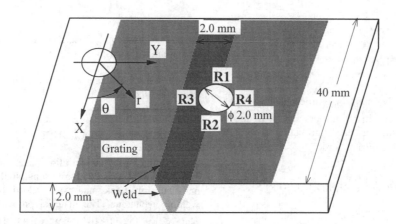

Fig. 1 A schematic drawing of the laser welded TA6V alloy specimen

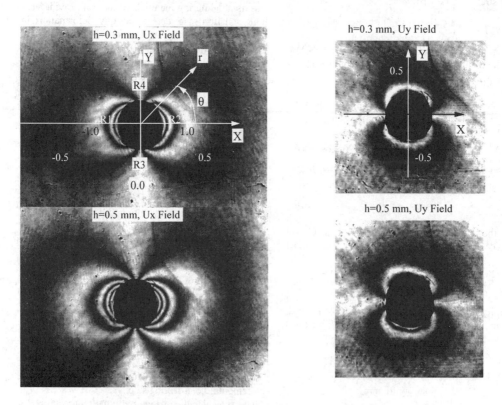

Fig. 2 Typical U_x and U_y fringe patterns obtained from TA6V alloy specimen

ANALYSIS AND RESULTS

Obviously, in Fig. 2, U_x displacement field in the longitudinal direction of the weld contained more fringe orders than the corresponding U_y displacement field in the transverse direction for each incremental depth. The U_x displacement increased progressively with hole depth. Whereas, the U_y displacement increased very slowly when hole depth was less than 0.4 mm and decreased slightly in the following steps. According to the analyses of previous sections, only the U_x displacements were used in the determination of residual stresses.

According to Eq. (2-a), if a blind hole was drilled in a uniform residual stress field, the released surface displacement field should be symmetric about X-axis and Y-axis in the Cartesian coordinate system. However, for the U_x field as shown in Fig. 2, it is obvious that the fringe orders in region R1 and R2, and the fringe gradient $\partial u/\partial x$ in region R3 and R4 were different, i. e., the axial symmetry was destroyed. It has been proved that (1) the asymmetry in the displacement field was ascribed to the non-uniform residual stress field in the hole-drilling region; (2) the displacement field in the close to hole boundary region was ascribed mainly to the effect of local non-uniform residual stress. Therefore, in this study, the residual stresses were calculated from the local displacement data in the regions R1, R2, R3 and R4, respectively.

After confirming the sign of fringe gradient by using the carrier fringe technique, the fringe orders at the desired points were determined accurately. Finally, the distribution of residual stresses in depth, was determined by Eq. (5) respectively for regions R1, R2, R3 and R4, and are plotted in Figs. 3-a~3-d.

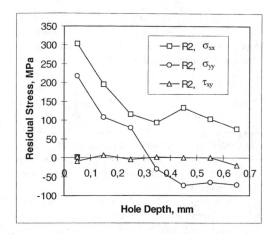

Fig. 3-b Residual stress profile in depth in region R2

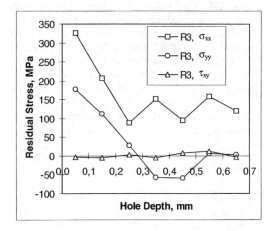

Fig. 3-c Residual stress profile in depth in region R3

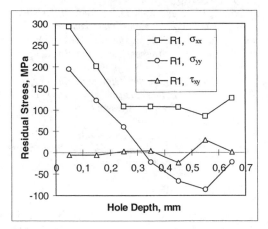

Fig. 3-a Residual stress profile in depth in region R1

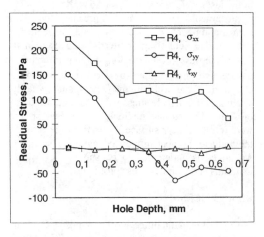

Fig. 3-d Residual stress profile in depth in region R4

1323

The following results were displayed. (1) The residual stress distribution law were similar in regions R1, R2, R3 and R4. The residual stress σ_{xx} in the longitudinal direction of the weld was always tensile stress. The maximum value occurred in the first 0.1 mm layer. It decreased very fast in the second and the third 0.1 mm layers to 1/3 the maximum value, and maintained at this level in the following incremental layers. σ_{yy} in the transverse direction of the weld was always smaller than the σ_{xx} in the corresponding layers. σ_{yy} had its maximum value in the first 0.1 mm layer and decreased more faster than σ_{xx} in the corresponding layers. It decreased to zero in the third or fourth 0.1 mm layer and then became compressive stress in the deeper layer. (2) The residual stress level in region R3—center of the weld was greater than that in the region R4—heat influenced region. The residual stress gradients were obvious between the region R3 and R4 in their corresponding incremental layers. (3) As regions R1 and R2 were at the same location of the weld, the similar residual stress level and distribution were found between regions R1 and R2. Thus, the accuracy of the method was verified. (4) The shearing stresses were very small in each incremental layers and thus the σ_{xx} and σ_{yy} can be regarded as the principal stresses in all the regions.

CONCLUSIONS

A method was developed combining single field moiré interferometry and incremental hole-drilling technique. The method was applied successfully for determination of residual stress distribution in depth of the laser welded specimen- TA6V alloy, where U_x displacements were much greater than the corresponding U_y displacements and consequently, only the U_x field fringe patterns were used in the analysis. Stress σ_{xx} in the longitudinal direction of the weld was greater than stress σ_{yy} in the transverse direction of the weld in the corresponding layers. The maximum tensile stress was found in the surface of weld. The strong gradients of residual stresses were in all the near surface layers from the center of the weld to the heat influenced region. The method developed is anticipated to be suitable for various materials and residual stress problems.

REFERENCES

Allen, A. J., Hutchings, M. T.; Windsor, C. G. and Andreani, C., *Neutron Diffraction Method for the Study of Residual Stress Fields*, Advanced Physics, 34(4), July-Aug., 1985, PP. 445-473.

Antonov, A., *Inspecting the Level of Residual Stresses in Welded Joints by Laser Interferometry*, Weld Prod., 30(9), 1983, PP. 29-31.

Lu, J., Niku-Lari, A. and Flavenot, J. F., *Mesure de la Distribution des Contraintes Residuelles en Profondeur par la Méthode du Trou Incrémentale*, Mémoire et Etudes Scientifiques, Revue de la Métallurgie, Feb. 1985, PP. 69-81.

Lu, J., Bouhelier, C., Lieurade, H. P., Miege, B. and Flavenot, J. F., *Study of Residual Welding Stress Using the Step-by-Step Hole Drilling and X-ray Diffraction Method*, Welding in the World/Le Soudage dan le Monde, Vol. 33, No. 2, 1994, PP. 118-128.

Lu, J., Editorial Board, James, M., Lu, J. and Roy, G., *Handbook of Measurement of Residual Stresses*, Published by The Fairmont Press, INC, 1996.

Makino, A. and Nelson, D., *Residual stress Determination by Single Axis Holographic Interferometry and Hole Drilling, Part I: Theory*, Exp. Mech., 3,1994, PP. 66-78.

Nelson, D, Fuchs, E., Makino, A. and Williams, D., *Residual Stress Determination by Single Axis Holographic Interferometry and Hole Drilling, Part II: Experiments*, Exp. Mech., 3,1994, PP. 79-88.

Nicoletto, G., *Moiré Interferometry Determination of Residual Stresses in the Presence of Gradients*, Exp. Mech., 31(3), 1991, PP. 252-256.

Ruud, C. O., Josef, J. A. and Snoha, D. J., *Residual Stress Characterization of Thick-Plate Weldments Using X-Ray Diffraction*, Welding Journal, Vol. 72(3), March 1993, PP. 87-91.

Schajer, G. S., *Application of Finite Element Calculations to Residual Stress Measurements*, J. Eng. Mater. Tech., 4, 1981, Vol. 103, PP. 157-163.

Soete, W., *Measurement and Relaxation of Residual Stress*, Sheet Met. Ind., 26, No. 266, 1949, PP. 1269-1281.

Wu, Z., Lu, J. and Joulaud, P., *Study of Residual Stress Distribution by Moiré Interferometry Incremental Hole Drilling Method*, The Fifth Int. Conf. on Residual Stresses, Linkoping, Sweden, June 1997.

Wu, Z. and Lu, J., *A New Optical Method for Shot Peening Residual Stress Measurement*, MAT-TEC 97, 'Analysis of Residual Stresses from Materials to Bio-Materials', Nov. 1997.

Wu, Z. and Lu, J., *Distribution of Welding Residual Stress in Depth with in-Plane Stress Gradient*, Submitted to Exp. Mechanics, 1997.

ACKNOLEDGEMENT

Special thanks to SNECMA and CETIM in France for their cooperation in the preparation of Specimen.

Experimental Mechanics, Allison (ed.)© 1998 Balkema, Rotterdam, ISBN 90 5809 014 0

Viscoelastic analysis of residual stress in thermosetting resins introduced during curing process

Minoru Shimbo, Masashi Yamabe & Yasushi Miyano
Department of Materials Science and Engineering, Advanced Materials Science R&D Center, Materials System Research Laboratory, Kanazawa Institute of Technology, Japan

ABSTRACT: In this study, the generation mechanism of residual stress generated in thermosetting resin due to the curing shrinkage during the molding process was investigated. The basic concept of the reciprocation law of time, temperature and degree of curing which can be expressed mechanical behavior of thermosetting resin during curing process is proposed. The propriety of this reciprocation law is evaluated experimentally using thermosetting resin. Comparing the theoretical results based on this reciprocation law and the linear-viscoelastic theory and experimental results by the layer removal method, the generation mechanism of residual stress generated in thermosetting resin due to the curing shrinkage as well as the thermal shrinkage is discussed.

1 INTRODUCTION

The residual stress generated by the molding process of thermosetting resins exerts serious influence on their mechanical properties. Therefore, it is very important for the reliability of thermosetting resin to clarify the generating mechanism and a prevention method of residual stress. Generally, residual stress generated in thermosetting resin during the molding process is classified widely into two process. One is produced due to the thermal shrinkage in the cooling process. The other is produced due to the curing shrinkage in the curing process. Authors have already performed the theoretical and experimental analysis of the residual stress induced in thermosetting resin during cooling process. Heretofore, the theoretical work has been developed in accordance with thermo-viscoelastic behavior. The layer removal method has been developed to obtained experimentally the residual stress in the thermosetting resin. The relationship between the residual stress and thermo-viscoelastic behavior has been verified [1-5]. However, there has been a little quantitative study on the generation mechanism of residual stress generated in thermosetting resin due to the curing shrinkage [6-8]. The reason is due to the difficulty of the formulation of the mechanical behavior during curing process. It is necessary to find rules explaining the mechanical behavior during the curing process.

The purpose of this paper is to study the generation mechanism of residual stress generated in thermoset-ting resin due to the curing shrinkage during molding process. First, the basic concept of the reciprocation law of time, temperature and degree of curing which can be expressed mechanical behavior of thermosetting resin during curing process is proposed and this concept is confirmed with the experimental data. Next, the fundamental equations are derived for the residual stress in the laminates cured and cooled during molding process based on the linear-viscoelastic theory and the above reciprocation law. The residual stresses in the laminate molded by casting epoxy resin between two epoxy plates are calculated under various molding conditions using the viscoelastic characteristics of epoxy resin in the curing and cooling processes. Finally, the experiments are carried out under the above same conditions and the residual stresses in the laminate are measured carefully using the layer removal method. Comparing the theoretical and experimental results, the generation mechanism of residual stress generated in thermosetting resin due to the curing shrinkage as well as the thermal shrinkage is discussed.

2 THE RECIPROCATION LAW OF TIME, TEMPERATURE AND DEGREE OF CURING

The reciprocation law of time and temperature for the mechanical behavior of polymers is known well as the equivalency of time and temperature of mechanical behavior. For example each relaxation modulus measured under various temperatures is shifted parallel along logarithmic time scale and aligned with the relaxation modulus at an arbitrary reference temperature. This results in a master curve on which the mechanical behavior of long term over low temperature can be estimated by one of short term over high temperature [7]. This idea will be expanded as following the reciprocation law of time, temperature and degree of curing for the mechanical behavior of thermosetting resin.

Figure 1 explains the reciprocation law of time, temperature and degree of curing proposed in this paper. First, applying the reciprocation law of time and temperature, the master curves for the relaxation modulus at various degrees of curing ξ are obtained at the reference temperature T_0 as shown on the above side in Figure 1. It can be assumed that the master curves of relaxation modulus having various ξ have the same configuration as each other. Therefore if each master curve of relaxation modulus is shifted parallel along the logarithmic time scale, these master curves can be aligned into one master curve at the reference temperature which equals the glass transition temperature $T_g(\xi)$ determined by degree of curing ξ. In this case, the shift factor of time, temperature and degree of curing can be described as shown on the below side of the same Figure. Figure (a) shows the time-temperature shift factors $a^*_{T0}(T)$ of relaxation modulus having various degrees of curing ξ at the reference temperature T_0 can be expressed by the Arrhenius equation. Since these shift factors have the same configuration, if the reference temperature T_0 is replaced by the glass transition temperature $T_g(\xi)$ determined by

degree of curing ξ, these shift factors can be rewritten as shown in Figure (b). These are the time and degree of curing shift factors $a'_{Tg(\xi)}(T_0)$. In addition, writing $1/T$ to replace $(1/T-T_g(\xi))$ these shift factors result in a curve as shown in Figure (c). This shift factor $a_{Tg(\xi)}(T)$ is defined as the time, temperature and degree of curing. These shift factors mentioned above can be expressed by the following equations.

The time and temperature shift factor $a^*_{T0}(T)$

$$\log \frac{t}{t^*} = \log a^*_{T0}(T) = \frac{\Delta H}{2.303R}\left(\frac{1}{T}-\frac{1}{T_0}\right) \qquad (1)$$

The time and degree of curing shift factor $a'_{Tg(\xi)}(T_0)$

$$\log \frac{t^*}{t'} = \log a'_{Tg(\xi)}(T_0) = \frac{\Delta H}{2.303R}\left(\frac{1}{T_0}-\frac{1}{T_g(\xi)}\right) \qquad (2)$$

The time, temperature and degree of curing shift factor $a_{Tg(\xi)}(T)$

$$\log \frac{t}{t'} = \log a_{Tg(\xi)}(T) = \frac{\Delta H}{2.303R}\left(\frac{1}{T}-\frac{1}{T_g(\xi)}\right),$$

$$t' = \frac{t}{a_{Tg(\xi)}(T)} \qquad (3)$$

where, t is time, t',t^* is reduced time, ξ is degree of curing, T is temperature K, T_0 is the reference temperature K, $T_g(\xi)$ is the glass transition temperature determined by degree of curing K, ΔH is activation energy kJ/mol, R is Gas constant kJ/mol.

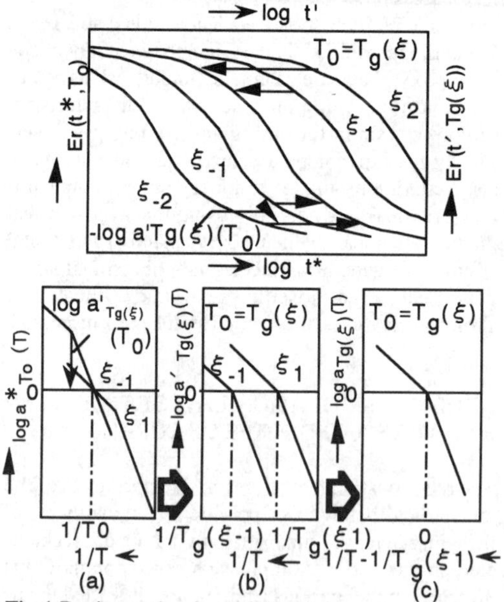

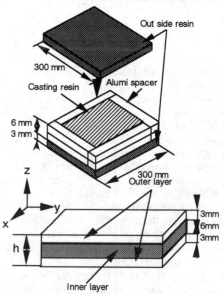

Fig.1 Reciprocation law of time-temperature-digree of curing.

Fig.2 Molding procedure of laminate molded by casting epoxy resin.

1326

3 FUNDAMENTAL EQUATIONS OF RESIDUAL STRESS

Consider the thin laminate plate of thermosetting resin molded by casting resin between epoxy plates as shown in Figure 2 which is cured at a uniform holding temperature T_h, and free of any tractions and body forces. The Cartesian coordinate axes x, y and z are taken in the direction of the longitudinal and transverse width and the depth, respectively, as shown in the same Figure It is assumed that the width is sufficiently larger than the depth.

It is assumed that the thermosetting resin always shows linear-viscoelastic behavior and that the reciprocation law of time, temperature and degree of curing is holds which was presented by previous chapter in this paper. This law is defined by assuming that the master curve of relaxation modulus E_r for all of degree of curing coincides with each other, when the glass transition temperature $T_g(\xi)$ decided by the degree of curing ξ is chosen as the reference temperature of the master curve. The stress $\sigma(z,t)$ distributed uniformly in the x-y plane is indicated by the following equations.

$$\sigma(z,t) = \frac{1}{1-v}\int_0^t Er(t'-\tau_1, Tg(\xi))\frac{\partial}{\partial\tau}\{\varepsilon(z,t) - \varepsilon_S(T(z,\tau), \xi(z,\tau))\}d\tau \tag{4}$$

where $E_r(t', T_g(\xi))$ is the relaxation modulus at the reference temperature $T_g(\xi)$ and $\varepsilon(z,t)$ is the strain in x-y plane, ε_s is the thermal and curing shrinkage and v is Poisson's ratio. The $t'(z,t)$ is herein defined as reduced time, and is given by the following equation.

$$t'(z,t) = \int_0^t \frac{du}{a_{Tg(\xi)}(T(z,u))} \tag{5}$$

where $a_{Tg(\xi)}(T)$ is defined as the time-temperature-curing shift factor.

Knowing that the molding process is symmetric with respect to the central plane of the plate and that the plate is free of traction at its edge, the equilibrium equation of forces are given as follows.

Table.1 Composition and cure schedule of epoxy resin.

Composition	Epoxy resin:EPIKOTE604	
	Hardener:MHAC-P	
	Cure acceleror:2E4MZ	
Weight ratio	Epoxy resin:Hardener:Cure acceleror	
	120 : 200 : 2	
Cure schedule	Condition I	70°C(12h)+150°C(6h)
	Condition II	70°C(12h)
	Condition III	70°C(24h)
	Condition IV	70°C(48h)
	Condition V	70°C(96h)

$$\int_{-\frac{d}{2}}^{\frac{d}{2}}\sigma(z,t) = 0 \tag{6}$$

The condition of compatibility is shown by the following equation.

$$\varepsilon(z,t) = \varepsilon(t) \tag{7}$$

It can be verified that eqs. (4)-(7) yield the following equation.

$$\int_{-\frac{d}{2}}^{\frac{d}{2}}\int_0^t Er(t'-\tau, Tg(\xi))\frac{\partial}{\partial\tau}\{\varepsilon(z,\tau) - \varepsilon_S(T(z,\tau), \xi(z,\tau))\}d\tau dz = 0 \tag{8}$$

If $\varepsilon(t)$ is obtained by solving eq. (8), the stress $\sigma(z,t)$ can be obtained by eq. (4). The stress $\sigma(z,t)$ is the resulting residual stress when the whole plate attains room temperature after molding.

4 EXPERIMENTS

4.1 Preparation of specimen

Table 1 shows the composition, weight ratio and cure schedule of epoxy resin. The epoxy resins were cured under various curing conditions as shown in this Table in order to change the degree of curing. The epoxy resin of condition I (70 °C/12h+ 150 °C /6h) in this Table was regarded as complete curing and the other epoxy resins were considered to have various different degrees of curing.

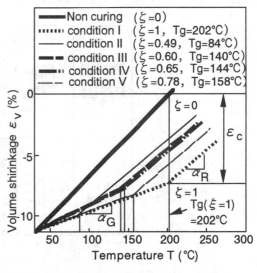

Fig.3 Relationship between volume shinkage and temperature.

1327

4.2 Configuration of specimen and molding method

The dimension of longitudinal and transverse width is 300 mm, the depth of outer epoxy plates is 3 mm and the depth of casting layer is 6 mm. After casting epoxy resin into the middle layer between two epoxy plates shown in Figure 2, the specimens were molded under the condition of 70 °C/12h.

4.3 Measurements of mechanical behavior

The thermal and mechanical properties controlling the generation of residual stress are mainly the volume expansion and stress-strain relation during molding. First, the relationship between the volume expansion and temperature of epoxy resins with various degrees of curing were measured by using dilatometer. Next, the time and temperature dependences of the epoxy resins storage moduli were measured by dynamic viscoelastic testing using a viscoelastic analyzer.

4.4 Measurements of residual stress

The residual stress distributions in the plate were measured by using the layer removal method. Two strips of width 7 mm and length 100 mm were cut from both the x and y directions from the middle of the plate after molding. The lower surface of the strip was removed successively with emery paper at room temperature. The changes of strain on the upper surface of the strip which is a function of z were measured by using strain gages. The residual stresses of the plate were obtained by substituting measured strains into the equation from the layer removal method [1,2].

5 RESULTS AND DISCUSSION

5.1 The reciprocation law of time, temperature and degree of curing

Figure 3 shows the relationship between volume shrinkage and temperature of the epoxy resins cured at various conditions. In this Figure, the dotted line indicates the measured value of epoxy resin cured under the condition of 70 °C/12h + 150 °C/6h. The glass transition temperature of this resin is 202 °C and the degree of curing ξ defined as 1 which means complete curing. On the other hand, the solid line indi-

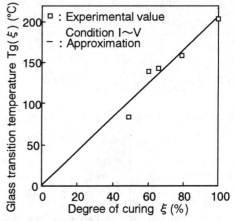

Fig.4 Relationship between glass transition temperature and degree of curing.

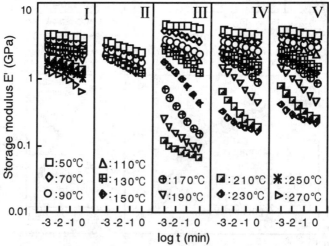

Fig.5 Time and temperature dependence of storage modulus of EPIKOTE 604 cured by various conditions.

cates that of the uncured epoxy resin and in this case the degree of curing ξ is defined as 0. The other lines indicate epoxy resin cured at various conditions and the degree of curing ξ of these resins were obtained by proportional allotment from an intersection of vertical line at 202 °C and the line of volume shrinkage of each resin. The glass transition temperatures found to be a function of degree of curing are also defined as the inflection point of the relationship between the volume shrinkage and temperature as shown in this Figure. The slope above T_g of the volume shrinkage against temperature does not change with the degree of curing, although the slope below T_g is evidently different from that above T_g. The relationship between glass transition temperature and degree of curing obtained from Figure 3 are shown in Figure 4.

Figure 5 shows the relationship between the storage modulus and time at various temperatures for the epoxy resins having various degrees of curing. The time scale is logarithmic and is equal to the reciprocal of the frequency (rad/min). It is found that the time and temperature dependence of storage modulus changes with degree of curing.

Figure 6 shows a master curve of storage modulus and time and temperature shift factor at the reference temperature $T_0=T_g(ξ=1)$ for completely cured resin which is obtained by shifting horizontally the storage modulus at condition I. It is confirmed that the reciprocation law of time and temperature holds for the storage modulus because one smoothed master curve can be obtained.

Figure 7 is obtained by shifting horizontally the master curves of storage modulus having various degree of curing ξ gotten from the same means as Figure 6 by using the amount of the time-temperature shift factor $a_{Tg(ξ)}(T)$. From this Figure, it can be confirmed that the reciprocation law of time, temperature and degree of curing holds for the storage modulus of epoxy resin during curing, because one smooth master curve can be obtained for storage modulus at various degrees of curing.

5.2 Residual stress

Figure 8 shows molding temperature and degree of curing of inner layer for three layer laminates and Fig-

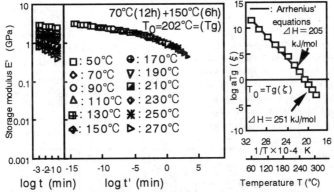

Fig.6 A master curve of epoxy resin molded condition I and time-temperature shift factor.

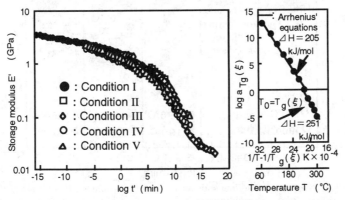

Fig.7 Master curves and time-temperature shift factors by reciprocation law of time-temperature-degree of curing.

ure 9 shows the contraction with curing time of epoxy resin. The numerical results of residual stress were calculated by equations (1) to (8) using these datum.

Figure 10 shows the residual strain and stress distributions of three layers laminate made from epoxy resin generated during molding process under the above condition. The amount of scatter in the experimental values showed by the square marks is very small, indicating that the experiments are repeatable. The residual stress is compressive in the outer layers while tensile in the inner layer, and they are balanced. The experimental results agree fairly with the theoretical ones. It is clarified that the generation mechanism of residual stress due to the curing shrinkage as well as thermal shrinkage can be quantitatively obtained by assuming the reciprocation law of time, temperature and degree of curing.

6 CONCLUSION

In this study, the generation mechanism of residual stress generated in thermosetting resin due to the curing shrinkage during the molding process was investigated. The results obtained lead to the following conclusions.

1) The reciprocation law of time, temperature and degree of curing was evaluated by experiments, and the fundamental equations were derived for the residual stress in the laminate cured and cooled during molding process based on the linear-viscoelastic theory including this law. The residual stresses in the laminate molded by casting epoxy resin between two epoxy plates were calculated under the condition of 70 °C /12h, using the viscoelastic characteristics of epoxy resin in the curing and cooling processes.

2) The laminate was molded under the same molding condition and the residual stresses in the laminate were measured carefully using the layer removal method. The experimental results almost agree with the theoretical ones. It is clarified that the generation mechanism of residual stress due to the curing shrinkage as well as thermal shrinkage can be quantitatively obtained by assuming that the reciprocation law of time, temperature and degree of curing.

REFERENCES

[1] Miyano,Y., Shimbo,M., Kunio,T., 1982, Exp. Mech., 22:310-315.
[2] Miyano,Y., Shimbo,M., Kunio,T., 1984,Exp. Mech., 24:75-80.
[3] Shimbo,M., Miyano,Y., Kunio,T., 1986,Pro.SEM Conf., Exp. Mech., 459-464.
[4] Maneschy,C.E., Shimbo,M., Miyano,Y., Woo,T.C., 1986, Exp. Mech., 26: 306-311.

[5] Shimbo,M., Sugimori,S., Miyano,Y.,etc, 1990,Pro.KSME/JSME Conf., 37-42.
[6] Miyano,Y., Sugimori,S., Shimbo,M., etc. 1990,Pro. 9th SEM Int Conf.,3: 277-282.
[7] Miyano,Y.,Kuroda,H., Shimbo,M., 1993,Pro. 50th An. SEM Conf., 502-507.
[8] Shimbo,M ,Shimizu T.,Miyano,Y., 1994,Pro. SEM Spring Conf. 849-854.

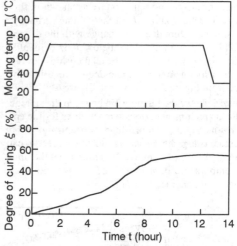

Fig.8 Process of molding temperature and degree of curing in inner layer of three layer laminates.

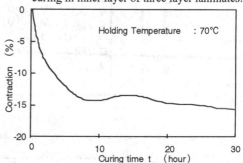

Fig.9 Contraction with curing time of EPIKOTE 604.

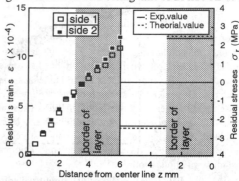

Fig.10 Residual strains and stresses distributions of EPIKOTE 604.

Experimental Mechanics, Allison (ed.) © *1998 Balkema, Rotterdam, ISBN 90 5809 014 0*

Comparative analysis of laser weldment properties by grating interferometry and digital speckle photography

L. Sałbut & M. Kujawińska
Institute of Micromechanics and Photonics, Warsaw University of Technology, Poland

D. Holstein & W. Jüptner
Bremer Institut für Angewandte Strahltechnik, Germany

ABSTRACT: The methodology and instrumentation for the determination of laser weldment properties based on automated grating interferometry and digital speckle photography are presented. The investigations are performed on a laser welded steel specimen under tensile loading. The local and global displacement/strain fields are analysed in elastic and plastic state of specimen. The results obtained have shown that the methods are complementary in the sense of the possible range and sensitivity of displacement/strain measurements.

1. INTRODUCTION

Introducing extensively laser beam welding into industry, it is necessary to provide extended knowledge about laser weldment properties. This task may be fullfield with experimental and numerical methods for the analysis of the behaviour of weld structure under various conditions. Here the narrow dimension and high hardness of laser weldments [Mayer and Bergman, 1994] must be taken into account, in order to determine the laser weldment properties with regard to the safety of laser welded structures, for example the resistance against brittle fracture. Applied measurement techniques have to provide displacements or strains with high spatial resolution due to the high gradients in the weld material and the heat affected zone. Furthermore, up to now the weldment inspection guidelines and standards of the classification societies hinder a better diffusion of laser beam welding into industry, e.g. into the shipbuilding industry. The high values of the hardness of laser welds are usually beyond those given as the maximum hardness in the guidelines and standards. However, the strength of the weldment is not seriously affected by this high hardness. Thus an improved and powerful knowledge about laser weldment properties obtained by non-destructive measurements allows to qualify laser welded joints for industrial applications.

In the paper we focus on complementary application of two optical full-field measurement methods: the grating (moiré) interferometry [Kujawińska and Sałbut, 1995] and digital speckle photography [Dainty, 1989] for laser weldment testing under tensile load. Both methods measure in-plane displacement fields, however the ranges of measurements, spatial resolutions, gradients allowed and the accuracies differ. This is the reason why special attention is paid for determination of the tasks for both methods in the process of laser weldment properties determination.

2. PRINCIPLES OF EXPERIMENTAL METHODS

The basic features of two optical full-field experimental methods for in-plane displacement / strain determination, namely: grating interferometry, GI and digital speckle photography DSP are given in Table 1.

The above features clearly indicate that grating interferometry is preferable for high accuracy, laboratory investigation of the displacement/strain distribution [Kujawińska and Sałbut, 1995, Kujawińska et al, 1996]. Digital speckle photography provides wider range of measurement and it is much more suitable to perform on-line quality control of the weldment, as it does not require any surface preparation and relies on a simple measurement arragement [Holstein et al, 1998)].

Table 1. Basic features of grating interferometry (GI) and digital speckle photography (DSP)

Method [1]	Lateral resolution	Accuracy of displacement determination	Range measurement for strain		Modification of sample	Complexity of setup
			min.	max.		
GI	10 μm	±20 nm	5×10^{-6}	-8000×10^{-6} [2]	grating replication	medium
DSP	100 μm	±250 nm	250×10^{-6}	to sample failure	white paint or not required [3]	low

[1] The properties of GI and DSP given under assumption that they have the field of view 5x5 mm^2
[2] iteratively the range may be extended up to the grating failure
[3] depending on the surface finish

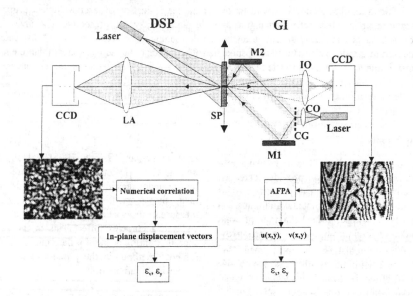

Fig.1. The scheme of the experiment setup simultaneously performing measurements by grating interferometry GI and digital speckle photography DSP. SP – specimen; LA – lens/aperture; M1,M2 – interferometer mirrors; CG – compensating grating; IO – imaging lens; AFPA – automatic fringe pattern analyser

As the grating interferometry setup, a portable, workshop interferometer (Fig.1) with the measuring head with compensating grating (CG) is applied [Czarnek, 1990]. This solution assures that GI is relatively insensitive to vibrations and can be used in unstable environment or directly on a loading machine.

A cross-pattern diffraction grating with frequency of 1200 l/mm is replicated at the sample (SP) and illuminated sequentially by two pairs of coherent beams. The deformations of x- and y- direction gratings introduce into the output interferograms

information about u or v in-plane displacements, respectively. The deformations may be caused by releasing of the residual stresses (no load required) or by the load applied to the sample. The intensities captured by CCD camera equal:

$$I(x, y) = a(x, y) + b(x, y)\cos\left\{\frac{4\pi}{d}u(x, y)\,or\,\frac{4\pi}{d}v(x, y)\right\} \quad (1)$$

where a(x, y) and b(x,y) are the background and contrast functions, respectively; d is the period of the specimen grating.

Interferograms obtained in GI system are analysed by automatic fringe pattern analyser AFPA based on spatial carrier frequency phase shifting method [Kujawińska and Sałbut, 1995]. It allows to determine displacements on the base of single interferogram with properly introduced carrier fringes. The displacement maps with the resolution 256 x 256 points over the measurement area are obtained. The displacement fields are then numerically differentiated in order to calculate ε_x, $\varepsilon_{\bar{y}}$ and γ_{xy} strain maps.

As an alternative to the grating interferometry, the digital speckle photography is investigated in order to determine the in-plane displacements and strains. In contrast to the GI this method does not require any preparation the surface of a specimen (in the case of metallic reflections, it may be covered with white paint). Furthermore, in comparison to other optical methods the digital speckle photography has the advantage of a simple set-up which is also insensitive to vibrations (Fig. 1). The specimen under investigation is coherently illuminated and the visible granular speckle pattern on the opaque surface of the specimen is stored electronically. This speckle pattern is a precise representation of the surface at the current load state. It appears due to interferences of the coherent light after reflection at the rough surface of the specimen.

If the speckle pattern of the reference state (I) and the speckle pattern at the considered loading state (I') are stored electronically by means of high resolution CCD camera, the evaluation of the displacements can be done numerically [Holstein et al, 1998]. The calculation is performed by cross correlations of subimages (usual size 64 x 64 pixels^2) from the whole available speckle images. Thereby the spectrum of one speckle subimage is multiplied with the conjugate spectrum of the other speckle subimage:

$$R_{II'} = F^{-1}\{F(I)^* F(I')\} \qquad (2)$$

where $R_{II'}$ is the cross correlation function, F the Fourier transformation, F^{-1} the inverse Fourier transformation and $*$ the complex conjugation. The mean displacement vector of the evaluated subimages is given by the location of the peak of the cross correlation function $R_{II'}$. Thus after the evaluation of all subimages the full in-plane displacement map is available.

In the experiment performed both methods were applied simultaneously from both sides of the same specimen with weld under tensile loading. The experimental setup and the scheme of further image processing is schematically demonstrated in Fig.1.

3. OBJECT OF MEASUREMENT

The specimens for the tensile test were cutted from laser welded steel plates (St 44-3N). The weldment was produced by a laser beam power of 10 kW using the steel wire S27572G3. The theoretical material properties of the sample are given in Table 2. The geometry of the specimen was prepared according to the standard DIN 50120 and its details are shown in Fig. 2a. The measurement areas were located at oposite sides of the specimen and equal: 5.4x3.3 mm^2 for GI and 10x10 mm^2 for DSP. Due to nonsymetrical shape of weldment (Fig. 2b), two orientation T and R were prepared by alternate attaching of the diffraction grating (1200 l/mm) and covering with white paint the opposite sides of specimens, as shown in Fig. 2c.

Table 2. Material properties of the base material

Material	St 44-3N
E [MPa]	210.000 MPa
ν	0.29

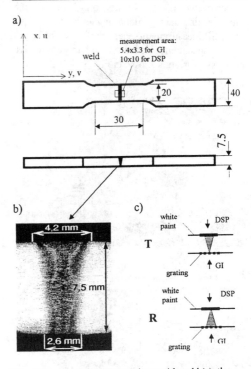

Fig.2. The geometry of the specimen with weld (a), the geometry of weld (b) and T and R – specimen orientation (c).

4. EXPERIMENTAL RESULTS

After adjusting the measurement setup (Fig.1), the sample was tensile loaded in the range up to 70 kN with the step of 5 kN. The examples of v(x,y) displacement maps obtained simultaneously for both DSP and GI methods for T specimen orientation are shown in Fig.3a and 3b. The area analysed by GI in respect with the DSP field of view is marked by dotted line. The displacement maps were measured for load 40 kN and calculated in respect to the reference load 1 kN (total stress is equal 260 MPa).

Next, the v(x,y) maps were differentiated for ε_y strain distribution determination. For GI the base of differentiation was equal to 0.064 mm. The ε_y strain distribution map obtained for stress 260 MPa is shown in Fig.3c.

For each load, the average values of ε_y strain were taken from the area of the base material and the weldment. Fig.4 shows strain/stress relations given by grating interferometry for T and R specimen orientations (base material: curves T-BM and R-BM and weldment T-W and R-W). The GI measurements were performed up to 45 kN only, due to the specimen grating damage under higher loads.

The DSP measurements were performed up to 70 kN for T specimen orientation and the respective strain/stress relation is shown in Fig. 5.

For elastic range (the first five steps of loads up to 25kN (160 MPa)) the differences between strains calculated for the base material and the weldment were not detected during these measurements. The material constants: Young's modulus, E, and Poisson ratio, v, determined for this range are given in Table 3. As it is shown, a good agreement between the results obtained by GI for T and R specimen orientation (difference in Young's modulus is app. 3%) was obtained. The average (R and T orientation) values of E is also in good agreement with theoretical E. The bigger difference (about 14.5%) is observed between GI and DSP results and between theoretical E. This difference may be caused by higher DSP measurement error obtained for lower loads. Also it can be caused by not perfect clamping of the sample which could cause its bending and introduce strain difference at the opposite side of the sample.

Tab.3. The comparison of Young's modulus E and Poisson ratio v for the base material.

Physical quality	GI R orientation	GI T orientation	DSP T orientation
E [MPa]	196340	202680	173300
v	0.30	0,33	0.33

a) DSP

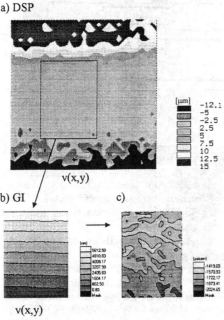

v(x,y)

b) GI

c)

v(x,y)

Fig.3. v(x,y) - displacement maps obtained by (a) DSP and (b) GI (tensile load P=40kN) and (c) ε_y (x,y) strain distribution calculated from v(x,y) shown in Fig. 3b.

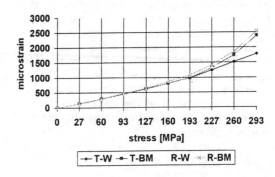

Fig. 4. The strain/stress relation obtained by GI for weldment (W) and base material (BM) with T or R specimen orientations

For the load higher than 30 kN (193 MPa) the difference between strains for base material and weldment are observed (Fig.4). Simultaneously, the relation between strain/stress is no more linear, what means that the material has reached the plastic range. The strains for weldment (curves T-W and R-W) are smaller than strains for base materials (curves T-BM and R-BM). It can be interpreted that the yield point for weldment is higher than that for base material.

Fig. 6 shows the details of the strain distribution for stress 293 MPa calculated by GI for T specimen orientation. Due to the presence of high plastic deformation in the base material (lower part of the specimen), the significant difference in the ε_y strain level in the weldment (-1780 μstrains) and base materials (-2390 μstrains) occurs, as shown in Fig. 7b. In order to monitor the phenomenon taking place during tensile loading specimen up to 60 kN (400 MPa, of the interferograms were recorded by video.

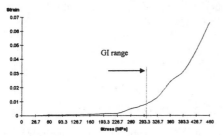

Fig. 5. The strain/stress relation obtained by DSP for T specimen orientaion

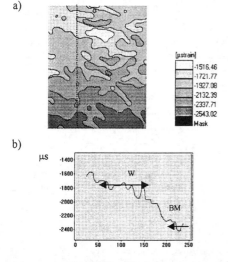

Fig.6. ε_y strain obtained by GI for load 445 kN (293 MPa): a) ε_y (x,y) grey level map and b) its vertical crossection indicated by dotted line in Fig. 6a.

Fig. 7 shows the sequence of the frames selected. Interferograms for the loads 45, 50, 55 and 60 kN clearly indicate the direction from which the plastic zone is coming and specificly the behaviour of the weldment and neighbourhood zones. At that range of loads the interferograms cannot be analysed by the spatial carrier phase shifting method due to low frequency of fringes (Fig. 7b), or even the fringe

pattern disappears due to damage of the grating or significant out-of-plane displacements of the specimen at the area where plastic deformation occurs (Fig. 7c, d). On the other hand, the range of loads for which the DSP quantitative results were obtained (Fig. 5) is much bigger (up to 70 kN) than that for GI. It proves that digital speckle photography is well suited for large range of displacement and strain values measurements.

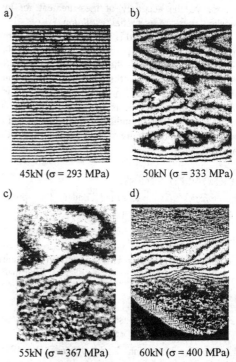

a)   b)

45kN (σ = 293 MPa)   50kN (σ = 333 MPa)

c)   d)

55kN (σ = 367 MPa)   60kN (σ = 400 MPa)

Fig.7. Interferograms obtained by GI for plastic range for T specimen orientation.

CONCLUSIONS

Grating interferometry and digital speckle photography are proposed as complement methods for determination of laser weldment properties. Grating interferometry is preferable for high sensitivity, local material properties determination within the elastic loading range. The further works should focus on proper methods of displacement data pre-processing (smoothing) in order to obtain more precise local information about the real material properties of the weld and the heat effected zone within elastic range of loading. Latter these values may be fed into FEM for proper modelling of a laser weld.

Digital speckle photography has currently a lower sensitivity to strain detection in comparison to GI, also it suffers for low accuracy while the specimen under low loads is measured. However DSP's measurement range is much higher than that of GI, so the single experimental setup enable to monitor the specimen behaviour under both, elastic and plastic loading range. Future works on DSP should focus on a subpixel sensitivity of its analysis in order to improve the accuracy of measurement for low loads and increase its spatial resolution.

In general, GI method is well suited for laboratory analysis of weldment properties. Its high quality results may be treated as the basis to determine DSP system requirements for the industrial measurement for the specific type of welded material and technological process. It is especially important as DSP is easy to apply in industrial environment and in principle does not require special preparation of the objects surface.

ACKNOWLEDGEMENTS

This work was performed under the project supported by Volkswagen-Stiftung, Hannover.

REFERENCES

Czarnek R. 1990. High sensitivity moiré interferometry with compact achromatic interferometer. Optics and Lasers in En. 13:93-101.

Dainty J.C. 1984. Laser speckle and related phenomena. Springer Verlag.

Holstein D., Hartmann H.-J., Jüptner W. 1998. *Investigation of laser welds by means of digital speckle photography*. Proc. SPIE. (in press)

Holstein D., Jüptner W., Kujawińska M, Sałbut L. 1997. *Investigations of residual stress induced by laser welding by means of optucal methods*. Proc. Conf. „Mechatronics", Warsaw. 789-796.

Kujawińska M., Sałbut L. 1995. New trends in instrumentation and application of automated grating interferometry. Proc. SPIE. 2545:2-6.

Kujawińska M., Sałbut L., Weise S., Jüptner W. 1996. Determination of laser beam weldment properties by grating interferometry method. Proc. SPIE. 2782:224-232.

Mayer S., Bergman H.W. 1994. Materials aspects of laser beam welding. Proc. of 5th ECLAT. DVS-bd. 163.

Experimental Mechanics, Allison (ed.)© 1998 Balkema, Rotterdam, ISBN 90 5809 014 0

Evaluation of residual stress inside tempered glass panel by PEM method

M. Saito & M. Takashi
Aoyama Gakuin University, Tokyo, Japan

J. Gotoh
Yokohama National University, Japan

ABSTRACT: When a polarized laser beam transmitted into a stressed glass block, a sort of optical birefringence occurs. However, since the optical sensitivity of a glass is not so high, it is somewhat difficult to measure photoelastic fringe pattern by the ordinary methods. It is well known that the Photo-Elastic Modulator method can measure not only minute retardation of birefringence less than a quarter wavelength but the orientation of principal birefringence simultaneously. Taking the distinctive feature of the method, a system of PEM will be used effectively for the simultaneous measurement of principal stress difference and principal stress direction in a transparent specimen which involves residual stresses. It is, however, difficult to determine the maximum or minimum values of stress or strain just near specimen surface accurately and precisely. An additional measurement of in-plane stresses on the surface is examined on the basis of the reflection technique.

1 INTRODUCTION

As well known, a glass is a typical example of brittle materials. When products of glass involving small crack or flaw are subjected to large external forces, they comes into dangerous situation easily. Recently, in many products made of glass, such as TV-tube glass panels and windows of buildings, high compressive residual stresses are often induced artificially at the outer surface by controlling temperature gradient in cooling process to keep its strength to stand with external forces. It is, therefore, very important to measure this type of residual stresses quantitatively to achieve the safe and fracture proof design.

A glass is highly clear and optically isotropic under unstressed state, but it shows optical anisotropy under stress. When a polarized laser beam is transmitted into a stressed glass block, a sort of optical birefringence is observed. Non-contact optical methods for measuring residual stresses, transmitted or scattered photoelasticity (Aben & Guillement 1993) and total reflection (Okamura et al. 1963), are currently used for this purpose. However, since the optical sensitivity of glasses is not so high, photoelastic fringe pattern can scarcely been recognized under ordinary conditions and circumstances. In this study, the authors attempts to evaluate the residual stresses inside a glass block using a PEM system which can give principal stress difference and principal stress direction simultaneously at a point.

This type of PEM system can measure primarily the small retardation of birefringence less than a quarter wavelength. The authors attempt to make sure that the PEM system can be applicable to the measurement of the retardation more than a quarter wavelength in high accuracy. The additional tests of the four-points bending on a rectangular bar specimen cut out from thick tempered glass panel, one is induced residual stresses and the other annealed, give us good results for the superposition of stress fields by the residual stress and the bending stress. Standing on the fairly good results of residual stress measurement in a rectangular bar specimen, the PEM method is also applied to a L-shaped corner specimen. However, any kinds of transmitted light technique are suffered from the difficulty in measuring stresses just on or near the surface due to unavoidable light scattering at edges and corners. Standing on the total reflection techniques, the authors discuss the possibility of the additional information for the determination of the maximum residual stress on the surface of a glass plate.

2 THEORY OF PHOTO-ELASTIC MODULATOR

2.1 *Characteristic features of PEM*

In the current techniques in photoelasticity, optical elements having the fixed characteristics for light, such as polarizers, quarter-waveplates and so on, are usually employed. On the other hand in the recent advances of polarimetry and optical devices, the element of so-called PEM has been developed (Bodoz et al. 1990), in which optical retardation can be electrically controlled by utilizing the nature of the crystal piezoelectricity.

An arbitrary birefringence of a crystal is controlled by voltage impression on crystal bar as an electrode. A set of PEM (PEM-90 HINDS INSTRUMENTS, INC.) which is applied AC voltage to the crystal frequency can be effectively utilized for accurate measurement of the retardation of birefringence.

The retardation of birefringence of the PEM δ_p is expressed as;

$$\delta_p = \delta_0 \sin \omega t \tag{1}$$

where ω is the resonant frequency (50kHz) of the PEM, t is time and δ_0 is the retardation of birefringence in proportion to amplitude of AC voltage impressed on the PEM. The PEM employed is operated by two high modulation frequencies, 50 kHz and 100 kHz.

2.2 Description of principal stress difference and direction

The application of the frequency modulation method (Niitsu et al. 1992 1993 1994 1997) gives high accuracy and high resolution for the measurement of the small retardation of birefringence (within a quarter-wavelength). Denoting unknown birefringence due to a model under stresses as δ^*, the superposed retardation of birefringence δ_c could be expressed as follows;

$$\delta_c = \delta^* + \delta_0 \sin \omega t \tag{2}$$

Now we consider a device which can obtain a transmitted light intensity proportional to the retardation of birefringence. The transmitted light intensity I is written as

$$I = I_0 \sin 2\delta_c \tag{3}$$

where I_0 is the incident light intensity and $\sin 2\delta_c$ is controlled by the PEM system. Thus, the retardation in the model can be evaluated by measuring the transmitted light intensity.

In the optical system adopted, a He-Ne laser light passes through a polarizer, a PEM, a quarter-wave plate, a birefringent TV-tube glass block, a quarter-wave plate and an analyzer. The direction of the polarizer to a reference axis, for example, is inclined +45 deg, the analyzer inclined −90 deg, the two quarter wave plates are inclined −45 deg to the axis of PEM, respectively. After passing through the analyzer, the intensity of light is detected by a photodetector. This detected light contains the DC voltage and two different AC voltages.

$$I = V_{DC} + V_{AC1} \sin \omega t + V_{AC2} \cos 2\omega t + \cdots \tag{4}$$

where, V_{DC} is the DC voltage, V_{AC1} the AC voltage for 50 kHz and V_{AC2} the AC voltage for 100 kHz oscillation. These entities are expressed as follows.

$$V_{DC} = \frac{\alpha I_0}{4} \{1 + \sin \gamma \sin 2\theta J_0(\delta_0)\} \tag{5}$$

$$V_{AC1} = A_1 \frac{-\alpha I_0}{2} J_1(\delta_0) \sin \gamma \cos 2\theta \tag{6}$$

$$V_{AC2} = A_2 \frac{\alpha I_0}{2} J_2(\delta_0) \sin \gamma \sin 2\theta \tag{7}$$

where, α is the transparency of the glass, θ the angle between the principal stress direction and the axis of the PEM, γ and δ_0 are the retardation of the glass specimen and the PEM, respectively. The angle θ of the principal axis of birefringence is taken as positive for counterclockwise. $J_0(\delta_0)$, $J_1(\delta_0)$ and $J_2(\delta_0)$ are Bessel functions, A_1 and A_2 are constants dependent on the equipment. The second term, $\sin \gamma \sin 2\theta J_0(\delta_0)$, in Eq. (5) can be eliminated by adjusting the PEM voltage as to be $J_0(\delta_0) \cong 0$. Thus, we can rewrite Eq. (5) as follows.

$$V_{DC} = \frac{\alpha I_0}{4} \tag{8}$$

Under the condition adjusted as above, we can derive the following equation from Eqs. (6) and (7).

$$\theta = \frac{1}{2} \tan^{-1} \left\{ -\frac{J_1(\delta_0) \cdot A_1 \cdot V_{AC2}}{J_2(\delta_0) \cdot A_2 \cdot V_{AC1}} \right\} \tag{9}$$

From Eqs. (6), (7) and (8), we find that V_{AC} is independent on I_0 and α. Then we have the following equation for the retardation γ,

$$\gamma = \sin^{-1} \sqrt{\left(\frac{V_{AC1}}{2 J_1(\delta_0) V_{DC} A_1} \right)^2 + \left(\frac{V_{AC2}}{2 J_2(\delta_0) V_{DC} A_2} \right)^2} \tag{10}$$

within the limitation of $0 < \gamma < \lambda/4$.
Standing on the Brewster's law in 2D-photoelasticity, γ is also related to the principal stress difference $\sigma_1 - \sigma_2$ as follows.

$$\gamma = \frac{2\pi d C (\sigma_1 - \sigma_2)}{\lambda} \tag{11}$$

where the λ is wavelength of He-Ne laser light and C is the stress-optical coefficient. Thus, the following equation is derived.

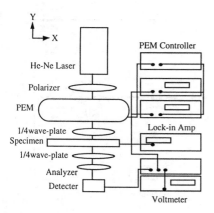

Fig. 1 A schematic diagram of the PEM apparatus

$$\sigma_1 - \sigma_2 = \frac{\lambda}{2\pi dc}\gamma$$

$$= \frac{\lambda}{2\pi dc}\sin^{-1}\sqrt{\left(\frac{V_{AC1}}{2J_1(\delta_0)V_{DC}A_1}\right)^2 + \left(\frac{V_{AC2}}{2J_2(\delta_0)V_{DC}A_2}\right)^2}$$

(12)

From Eqs. (9) and (12), we can easily calculate both the principal stress direction and the principal stress difference.

3 EXPERIMENT

3.1 *The PEM system adopted*

Fig. 1 shows a schematic diagram of the PEM apparatus. The light source is a He-Ne laser which has the intensity of 5 mW with the diameter of 0.05 mm. Although the intensity of the transmitted light involves DC and AC voltage components, they are separated each other by a signal conditioning module. The DC voltage is detected by a digital voltmeter and two AC voltages are detected by double-phase lock-in amplifiers, respectively. From Eq. (8), V_{DC} is always maintained constant. Therefore, the intensity of birefringence at a point depends only on V_{AC1} and V_{AC2}. Substituting V_{AC1} and V_{AC2} into Eq. (9), the principal stress direction is calculated.

3.2 *Specimens*

Fig. 2(a) and (b) show the geometries of specimen adopted cut out from a TV glass panel. (a) is a type of rectangular bar specimen cut out form the middle portion of the glass pane with 14 mm in width, 120 mm in length and 20 mm in thickness. One of the bar-type specimens is tempered and the other is completely annealed. (b) is an L-shaped specimen

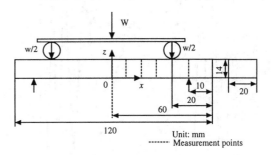

Fig. 2 (a) The geometry of the rectangular bar specimen of glass

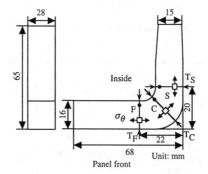

Fig. 2 (b) The geometry of the L-shaped specimen of glass

taken out from a corner portion of the same tempered panel. The location and direction of residual stress measurement is also shown in the figures. Since the PEM system is a point device, the minute laser point is scanned perpendicular to the surface with 0.05 mm interval, as shown in the figures. The stress-optic coefficient C is measured as 2.37×10^{-12} m²/N in both materials and independent on the heat treatment for tempering.

4 RESULTS AND DISCUSSIONS

4.1 *Distribution of retardation measured*

Fig. 3 shows an example the distribution of the retardation γ along z-axis inside a tempered glass. Referring to Eq. (12), the retardation γ measured repeats a hanging bell type increase and decrease within a range between 0 deg and 45 deg, when it comes more than a quarter-wavelength. Thus, a certain procedure of modification is necessary to obtain a master distribution over the whole region. Separating the hanging bell shaped γ curve at its top point, the oneside is rolled over from bottom to top. Then, we can construct a symmetrical master distribution of the retardation γ as shown.

1339

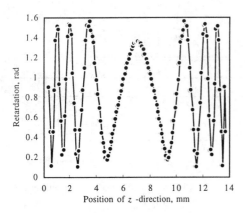

Fig. 3   The distribution of retardation γ ($x = 0$ mm)

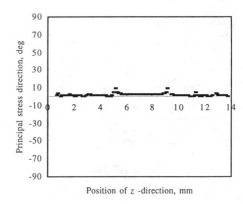

Fig. 5   The distribution of principal stress direction ($x = 0$)

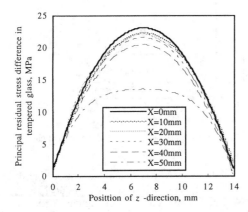

Fig. 4   The distribution of principal stress difference

4.2 Distribution of principal residual stress difference

4.2.1 Rectangular bar specimens

Fig. 4 shows an example of the master distributions of principal residual stress difference $\sigma_x - \sigma_z$ along z-axis at $x = 0$ (center), 10, 20, 30, 40, 50 mm, respectively, after the modification mentioned above. The principal stress difference shows the highest value at the middle layer and a symmetrical distribution within the specimen thickness, then decreases near both the top and bottom surfaces of the specimen. Near the surface layer, since it is usually difficult to get good results of residual stresses measurement because of light scattering by the effect of edge and corner, the master distributions are smoothly extrapolated up to the surfaces. The maximum value of $\sigma_x - \sigma_z$ is read as about 23 MPa at x = 0 mm (center).   As parting away from the center

of specimen, i.e. $x = 10$, 20...., the maximum value of principal residual stress difference decreases gradually by the end effect of stress relaxation due to edge cutting.

As shown in Fig. 5, the distribution of the direction angle of principal residual stress θ keeps almost constant of 0 deg over the whole thickness, except at several points at which $\gamma = \lambda/4$ and $V_{ACI} \cong$ 0 from Eq. (9).   The result suggest that the residual stress distribution in the glass panel consists only of normal stresses.   Thus, the residual normal stresses are easily separated each other, then the distribution of σ_x is obtained taking the balance of force in x-direction as in the usual technique.   Finally, the maximum residual stress are read as about 7 MPa tension at in the middle layer and about 18 MPa compression at both surfaces.

4.2.2 Four points bending superposition

In order to supplement the reliability of the residual stress measurement done in the previous section, a four point bending load is added to both tempered and annealed specimens.   In Fig. 6 (a) and (b), straight lines are the distributions of σ_x in the annealed specimen under four point bending loads of 260 and 520N.   We can observe the reasonable shifts of σ_x distribution due to the superposition of simple bending stress depending on the magnitude of loads. Then, good agreement of experimental results with theoretical estimation   on   the   shift   of   stress distribution is obtained for both cases of loading.   It would be emphasized that the residual stress induced in a transparent birefringent glass plate can be measured by the PEM method in high accuracy.

4.3 Corner specimen

Since the latest TV panel is required to have flat front

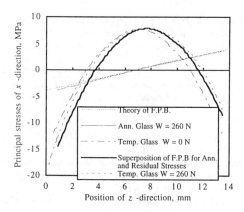

Fig. 6 (a) The distribution of principal stress under four points bending test (W = 260 N)

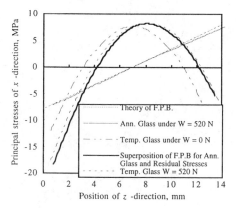

Fig. 6 (b) The distribution of principal stress under four points bending test (W = 520 N)

surface, very high residual stress has to be induced particularly around corner portion. In Fig. 7, the distributions of residual stress measured at three different locations around a corner are shown for comparison. The depth, r, in the direction from inside to outside, is normalized using the thickness corresponding to the location, denoting as T_i, where i = Front (16 mm), Center (22 mm) and Side (17 mm), in Fig. 2 (b). Although the maximum residual tensile stresses in the middle portion of thickness are almost same regardless to the location, not only the maxims compressive stresses at surface regions but the type of distribution are different each other. The largest compressive residual stress of -13 MPa is observed at the inside and outside of the front portion just adjacent to the corner. The flatter the front surface, the larger the bending stress is generated by vacuum inside TV tube. So, the safety of TV glass

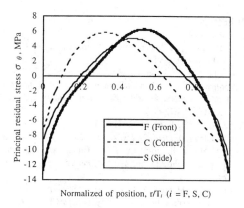

Fig. 7 The distribution of principal residual stress around a corner of tempered panel

tube has to be well studied taking the residual stress into account.

5 CONCLUDING REMARKS

To develop a convenient and accurate method for the measurement of residual stress induced in tempered glass plate, a photoelastic technique with a PEM system was examined. Although the system is usually good in the measurement of retardation within a quarter-wavelength of incident light, the applicability of the method is well expanded to the measurement of larger retardation in high accuracy. As the results, the distributions of residual stresses and principal stress direction in a tempered TV glass panel are successfully evaluated for two cases, namely a rectangle bar specimen cut out for the middle portion of the panel and an L-shaped corner specimen. The largest compressive residual stress is observed at the inside and out side of front surface just adjacent to a corner.

It should be pointed out that the PEM system could be applied and expanded to many other measurements of optical retardation in birefringent materials such as a technique of reflection photoelasticity with a very thin coat (Takeuchi et al. 1996).

ACKNOWLEDGMENTS

The authors appreciate the financial support of the Center for Science and Engineering Research Institute, Aoyama Gakuin University. The specimens used in this study were supplied by Asahi Glass Corporation. Our thanks are extended to Messrs. Y. Takeuchi and S. Yoneyama for their discussion.

REFERENCES

Aben, H. & C. Guillement 1993. *Photoelasticity of Glass*. Berlin: Springer-Verlag.

Badoz, J., M. P. Silverman & J. C. Canit 1990. Wave propagation through a medium with static and dynamic birefringence: theory of the photoelastic modulator. *J. Opt. Soc. Am.* 7: 672-682.

Canit, J. C. & J. Badoz 1990. New design for a photoelastic modulator. *Applied Optics*. 22: 592-594.

Korte, E.-H. & B. Jordanov 1989. Efficiency in circular polarization modulation. *Applied Spectroscopy*. 43: 289-293.

Niitsu, Y., K. Ichinose & K. Ikegami 1992. Stress measurement of transparent solid materials by polarized Laser. *Proc. ASME & JSME Joint conference on Electronics Packaging*. 2: 1005-1008.

Niitsu, Y., K. Ichinose & K. Ikegami 1993. A stress measuring method by polarized laser and photoelastic modulator. *Proc. ASME & JSME Joint conference on Electronics Packaging*. 1: 157-167.

Niitsu, Y., K. Ichinose & K. Ikegami 1994. Micro-stress measurement by laser photoelasticity. *Proc. ASME Mechanics Materials for Electronic Packaging*. 187: 29-35.

Niitsu, Y., K. Gomi, M. Sasaki & S. Tabei 1997. Development of scanning polarized laser microscope. *Prepr. of Jpn. Soc. Mech. Eng.*, 2: 75-76.

Okamura, H. 1963. 3-D stress measurement using the total reflection method. *The seminar of mechanical engineering in Jpn.*, 93-122.

Takeuchi, Y., S. Yoneyama & M. Takashi 1996. Elastic contact stress and strain analysis using the method photoelastic coatings. *Prepr. of Jpn. Soc. Mech. Eng.* A: 475-476.

Experimental Mechanics, Allison (ed.)© 1998 Balkema, Rotterdam, ISBN 90 5809 014 0

Complete residual stress measurement in axisymmetric glass articles

H. Aben, L. Ainola & J. Anton
Institute of Cybernetics, Tallinn Technical University, Estonia

ABSTRACT: Axisymmetric residual stress distribution is characterized by four stress components: radial stress σ_r, circumferential stress σ_θ, axial stress σ_z, and shear stress τ_{rz}. Integrated photoelasticity gives the possibility to determine directly only two of them, σ_z and τ_{rz}. However, often it is important to know also the values of other stress components. If stress gradient in the direction of the z axis is absent, i.e. $\tau_{rz} = 0$, all the stress components can be determined using besides measurement data the equilibrium equation and the classical sum rule $\sigma_z = \sigma_r + \sigma_\theta$. In the paper it is shown that the classical sum rule can be generalized for the case when axial stress gradient is present. Measurement data together with the equilibrium equation and the generalized sum rule allow for complete determination of the residual stresses.

1 INTEGRATED PHOTOELASTICITY FOR AXISYMMETRIC PROBLEMS

In integrated photoelasticity (Aben 1979, Aben & Guillemet 1993) the specimen is placed in an immersion bath and a beam of polarized light is passed through the specimen (Fig. 1). Transformation of the polarization of light in the specimen is measured on many rays. In certain cases this integrated optical information enables one to determine distribution of some components of the stress tensor.

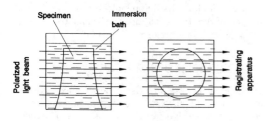

Figure 1. Experimental setup in integrated photoelasticity.

In the general case, due to the rotation of the principal stress directions on the light rays, optical phenomena are complicated and so are the algorithms for the interpretation of the measurement data (Aben 1979). However, if birefringence is weak, photoelastic measurements with a three-dimensional specimen can be carried out similarly to the two-dimensional case (Aben et al 1989). That is, on every ray it is possible to measure the parameter of the isoclinic φ and optical retardation Δ. The latter are related to components of the stress tensor on the ray by simple integral relationships

$$\Delta \cos 2\varphi = C \int (\sigma_z - \sigma_x) dy \qquad (1)$$

$$\Delta \sin 2\varphi = 2C \int \tau_{zx} dy \qquad (2)$$

Here C is the photoelastic constant and σ_x, σ_z and τ_{zx} are components of the stress tensor in the plane zx which is perpendicular to the wave normal y.

Using numerical experiments it has been shown (Aben et al. 1989) that in case of axial symmetry relationships (1) and (2) are valid if Δ is less than $3\lambda/4$ (λ denotes wavelength), and rotation of the principal directions is not strong.

2 CLASSICAL SUM RULE

Since residual stresses in glass have thermal origin, they can be considered as being caused by a fictitious temperature field (Bartenev 1970, Gardon 1980). Therefore, residual stresses in glass satisfy equations of thermoelasticity.

Thermal stresses in an axisymmetric body can be expressed as (Melan & Parkus 1953)

$$\sigma_r = 2G\left(\frac{\partial^2 F}{\partial r^2} - \Delta F\right)$$

$$+ \frac{2G}{1-2\mu}\frac{\partial}{\partial z}\left(\mu\Delta L - \frac{\partial^2 L}{\partial r^2}\right) \qquad (3)$$

$$\sigma_\theta = 2G\left(\frac{1}{r}\frac{\partial F}{\partial r} - \Delta F\right)$$

$$+ \frac{2G}{1-2\mu}\frac{\partial}{\partial z}\left(\mu\Delta L - \frac{1}{r}\frac{\partial L}{\partial r}\right) \qquad (4)$$

$$\sigma_z = 2G\left(\frac{\partial^2 F}{\partial z^2} - \Delta F\right)$$

$$+ \frac{2G}{1-2\mu}\frac{\partial}{\partial z}\left[(2-\mu)\Delta L - \frac{\partial^2 L}{\partial z^2}\right] \qquad (5)$$

$$\tau_{rz} = 2G\frac{\partial^2 F}{\partial r\partial z} + \frac{2G}{1-2\mu}\frac{\partial}{\partial r}\left[(1-\mu)\Delta L - \frac{\partial^2 L}{\partial z^2}\right]$$

$$(6)$$

where F is stress function and L Love's displacement function,

$$\Delta F = \frac{1+\mu}{1-\mu}\alpha T, \quad \Delta\Delta L = 0, \quad G = \frac{E}{2(1+\mu)},$$

$$\Delta = \frac{\partial^2}{\partial r^2} + \frac{1}{r}\frac{\partial}{\partial r} + \frac{\partial^2}{\partial z^2} \qquad (7)$$

Let us assume that a long cylinder or tube is manufactured by solidifying it in an axisymmetric temperature field without gradient in the axial direction. In this case the thermal (and residual) stresses are the same in all cross sections of the cylinder, except the parts near the ends of the latter. Now from Eqs. (3) to (5) follows the classical sum rule

$$\sigma_r + \sigma_\theta = \sigma_z \qquad (8)$$

The classical sum rule (8) was in a somewhat different way first derived by O'Rourke (O'Rourke 1950).

Let us express the stress components as power series:

$$\sigma_r = \sum_{k=0}^{m} a_{2k}\rho^{2k}, \quad \sigma_\theta = \sum_{k=0}^{m} b_{2k}\rho^{2k},$$

$$\sigma_z = \sum_{k=0}^{m} c_{2k}\rho^{2k}, \quad \tau_{rz} = \sum_{k=1}^{} d_{2k-1}\rho^{2k-1} \qquad (9)$$

where a_{2k}, b_{2k}, c_{2k} and d_{2k-1} are the coefficients to be determined, and ρ is dimensionless radius.

Inserting expansions (9) into the equilibrium equation

$$\frac{\partial\sigma_r}{\partial\rho} + \frac{\sigma_r - \sigma_\theta}{\rho} + \frac{\partial\tau_{rz}}{\partial z} = 0 \qquad (10)$$

and into the sum rule (8), we obtain (since $\partial\tau_{rz}/\partial z = 0$)

$$a_{2k} = \frac{1}{2(k+1)}c_{2k}, \quad b_{2k} = \frac{2k+1}{2(k+1)}c_{2k} \qquad (11)$$

Thus, in the absence of the stress gradient in axial direction, all the residual stress components can be determined since c_{2k} have been determined experimentally.

This method has been widely used for residual stress measurement in glass cylinders, axisymmetric fibers and fiber preforms (Aben & Guillemet 1993). As an example, in Figure 2 stress distribution in an axisymmetric graded index optical fiber preform is shown.

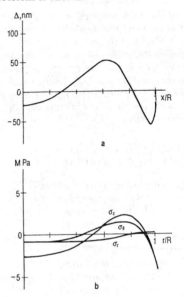

Figure 2. Radial distribution of optical retardation (a) and of the stresses (b) in an axisymmetric graded index optical fiber preform of quartz glass.

3 GENERALIZED SUM RULE

In the general case, in axisymmetric glass articles stress gradient in axial direction cannot be ignored. Let us derive from Eqs. (3) to (6) a relationship between stress components for that case.

If stress gradient in axial direction is present but smooth, we may write

$$\frac{\partial^2 F}{\partial z^2} = \frac{\partial^2 L}{\partial z^2} = 0 \qquad (12)$$

Now from Eqs. (3) to (6) follows

$$\sigma_r + \sigma_\theta = \sigma_z + \frac{2G}{1-2\mu}\frac{\partial}{\partial z}[3(\mu-1)\Delta L] \qquad (13)$$

Differentiating Eq. (6) relative to z and integrating along r reveals

$$\int_0^r \frac{\partial \tau_{rz}}{\partial z} dr = \frac{2G}{1-2\mu} \frac{\partial}{\partial z}[(1-\mu)\Delta L] + C(z) \quad (14)$$

where $C(z)$ is the integration constant.

From Eqs. (13) and (14) follows

$$\sigma_r + \sigma_\theta = \sigma_z - 3 \int_0^r \frac{\partial \tau_{rz}}{\partial z} dr + C(z) \quad (15)$$

The last relationship is named the generalized sum rule. It is valid when stress gradient in the axial direction is present with certain restrictions (12) upon the functions F and L. Actually, Eq. (15) is first approximation of the generalized sum rule. By handling Eqs. (3) to (6) asymptotically, higher approximations of the generalized sum rule can be obtained.

E.g., the next approximation of the generalized sum rule is

$$\sigma_r + \sigma_\theta = \sigma_z - 3 \int_0^r \frac{\partial \tau_{rz}}{\partial z} dr - \int_0^r \frac{1}{r} \left(\int_0^r r \frac{\partial^2 \sigma_z}{\partial z^2} dr \right) dr$$
$$+ D_1(z) \ln r + D_2(z) \quad (16)$$

4 NUMERICAL EXPERIMENTS

Let us analyze precision of the generalized sum rules in the case of a sphere in a symmetric temperature field

$$T = T_0 \left(\frac{r}{R} \right)^2 \quad (17)$$

From the wellknown solution (Timoshenko & Goodier 1951) we obtain formulas for stresses in dimensionless cylindrical coordinates ($\zeta = z/R$)

$$\sigma_\rho = 0.4 - 0.4\rho^2 - 0.8\zeta^2 \quad (18)$$
$$\sigma_\zeta = 0.4 - 0.8\rho^2 - 0.4\zeta^2 \quad (19)$$
$$\sigma_\theta = 0.4 - 0.8\rho^2 - 0.8\zeta^2 \quad (20)$$
$$\tau_{\rho\zeta} = 0.4\rho\zeta \quad (21)$$

Let us assume that, using integrated photoelasticity, stress components σ_ζ and $\tau_{\rho\zeta}$ are determined experimentally. Applying the generalized sum rule (15) and the equilibrium equation, we obtain after some transformations

$$\sigma_\rho = 0.45 - 0.45\rho^2 - 0.85\zeta^2 \quad (22)$$
$$\sigma_\theta = 0.45 - 0.95\rho^2 - 0.90\zeta^2 \quad (23)$$

Comparison of the stresses σ_ρ and σ_θ obtained with the generalized sum rule, with the exact values given by Eqs. (18) and (20), is given in Figure 3. It is interesting to mention that the second approximation of the generalized sum rule (16)

gives for σ_ρ and σ_θ exact values. The same result has been obtained when

$$T = T_0 \left(\frac{r}{R} \right)^4 \quad (24)$$

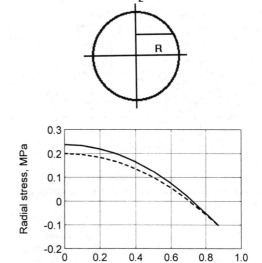

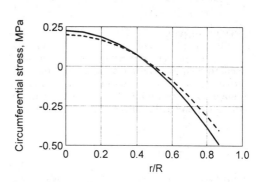

Figure 3. Comparison of stresses obtained with the generalized sum rule (———) with exact solution (– – –) at $z/R = 0.5$.

These results enable us to conclude that the way we have used to generalize the classical sum rule, is correct.

5 EXAMPLES

Using photoelastic measurement data, equilibrium equation (10), and the generalized sum rule (15), all components of the residual stress can be determined in an axisymmetric article.

As an example, residual stress distribution in a section of an axisymmetric article of optical glass is shown in Figure 4.

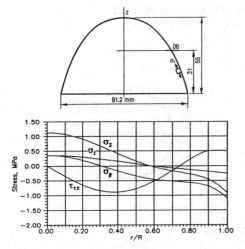

Figure 4. Geometry of the axisymmetric article of optical glass, and distribution of residual stresses in section 06.

As a second example, Figure 5 shows geometry of a part of a CRT glass bulb and stress distribution in two sections and on the external surface.

6 CONCLUSIONS

Classical sum rule has been generalized for the case when residual stress gradient in axial direction is present. The generalized sum rules (15) and (16) open up the possibility for complete residual stress measurement in axisymmetric glass articles.

7 ACKNOWLEDGEMENT

The support by Estonian Science Foundation under grant N° 2248 is greatly appreciated.

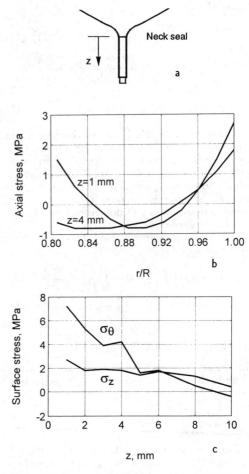

Figure 5. Geometry of the lower part of a CRT glass bulb (a), axial stress distribution in two sections (b), and on the external surface (c).

REFERENCES

Aben, H. 1979. *Integrated photoelasticity*. New York: McGraw-Hill.

Aben, H. & C.Guillemet 1993. *Photoelasticity of glass*. Berlin: Springer.

Aben, H., J.Josepson & K.-J.Kell 1989. The case of weak birefringence in integrated photoelasticity. *Optics and Lasers in Engineering*. 11:145–157.

Bartenew, G.M. 1970. *The structure and mechanical properties of glasses*. Groningen: Wolters-Nordhoff Publ.

Gardon, R. 1980. Thermal tempering of glass. In D.R.Uhlmann and N.J.Kreidl (eds) *Elasticity and strength of glass. Glass science and technology*, 5:146–217. New York: Academic Press.

Melan, E. & H.Parkus 1953. *Wärmespannungen infolge stationärer Temperaturfelder*. Wien: Springer.

O'Rourke, R.C. 1951. Three-dimensional photoelasticity. *J. Appl. Phys.* 22: 872–878.

Timoshenko, S & J.N.Goodier 1951. *Theory of elasticity*. New York et al: McGraw-Hill.

Creep

Experimental Mechanics, Allison (ed.)© 1998 Balkema, Rotterdam, ISBN 90 5809 014 0

Creep behaviour of pure aluminium at constant load and constant stress

K. Ishikawa, H. Okuda, M. Maehara & Y. Kobayashi
Toyo University, Kawagoe, Japan

ABSTRACT: We have carried out both a constant load and a constant stress creep test for pure aluminiums at a lower temperature, where neither diffusion nor recrystallization takes place. Whole creep curves are measured for 4N and 5N pure aluminiums at a room temperature. We discussed the difference between the constant load and the constant stress creep. The three regions, a primary, a steady state and a ternary creep, are observed under a constant load condition as well as under at a constant stress condition. Life to rupture is associated with the steady state creep. The difference of the creep damages between in a constant load creep and in a constant stress creep comes from the stress condition developed in the specimen.

1 INTRODUCTION

Creep behaviours of ductile materials are interesting subjects for understanding the mechanics of plastic deformation. Pure metals are so soft that the creep strain might be not neglected even at cryogenic temperatures (McDonald & Hartwing 1990). Creep behaviours of metals and alloys have been discussed from both the experimental and the theoretical viewpoint (Garofalo 1965). Nevertheless the experimental conditions were not specified in some publications, the theoretical analyses were often based on the constant stress conditions (Krajewski et al. 1997, Dunand & Jansen 1997). Therefore, there is considerable confusion as to the fundamental understanding of creep behaviours. Creep tests are classified largely into two loading systems (Garofalo et al. 1962). One is a constant applied load and another is a constant applied stress. A quasi-constant applied load is widely adopted for the simplicity (Argon 1996). Hence, we conducted both tests on 4N and 5N pure aluminiums at a room temperature. The discussion is focused on the difference between the creep behaviours, that is, the steady state creep rates, the creep mechanisms, the creep lives and the creep damages.

2 EXPERIMENTAL

2.1 *Materials*

Pure aluminiums have high electric-conductivity and good ductility at lower temperatures. The properties are advantageous to the application as a stabilizer for superconductors. The aluminiums used for this experiments have the chemical composition shown in Table 1. They were annealed at 773 K for 1 hour in air prior to the experiments. The tensile test was carried out at 293 K at the initial strain rate of 3×10^{-4} second^{-1}. The tensile strengths are 48 MPa and 37 MPa for 4N and 5N aluminium, respectively.

Table 1. The chemical composition of pure aluminiums tested (mass %).

Al	Si	Fe	Cu	Ti	Mg	Impurities
4N	0.0023	0.0025	0.0022	0.0052	------	0.0122
5N	0.0002	0.0001	0.0001	------	0.0001	0.0005

2.2 *Creep tests*

We conducted two kinds of uniaxial creep test in tension. One is a constant load test, where the initial applied load is kept constant but the applied stress is not constant during the whole test. The stress is just defined as the initial applied stress, the initial applied load divided by the initial cross section of the specimen. Another is a constant stress test, where the applied stress is kept constant during the experiment within the errors of 0.2%. The specimen dimension is 15 mm in the gauge length and 5 mm in the diameter of the round shape.

2.3 *Measurement of creep strain*

The elongation of specimens was measured on the gauge length with the laser extensometer. The mini-

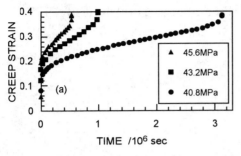

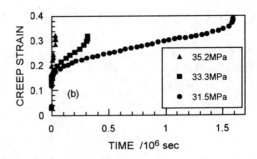

Figure 1. Variation of creep strain with time for 4N(a) and 5N(b) aluminiums under the condition of a constant applied load. The respective stresses are the initial applied values.

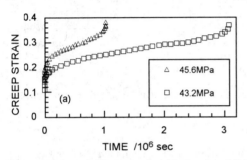

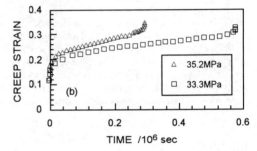

Figure 2. Variation of creep strain with time for 4N(a) and 5N(b) aluminiums under the condition of respective constant applied stresses.

mum resolution is less than 5 μ m. The test temperature is a room temperature, 293 K. The measured elongation was processed by the personal computer and transformed into the true strain against the creep time.

3 EXPERIMENTAL RESULTS

3.1 Creep curves at a constant load

Figure 1 (a) and (b) show the creep behaviours for 4N and 5N aluminiums at the respective initial applied stresses. The creep curves are divided into the three regions, that is, a primary, a steady state and a ternary creep. The curves at a lower stress have a longer steady state creep. The steady state creep disappears with further decreasing in the applied load (Ishikawa et al. 1997). In this experiment, we are concentrated on both a primary and a steady state creep. The reasons are that the primary creep strain is large for ductile metals and the steady state creep is associated with the creep life. The rupture strains would be similar, regardless of the applied load.

3.2 Creep curves at a constant stress

Figure 2 (a) and (b) show the creep curves of 4N and 5N alumimiums at a constant applied stress, respec-

tively. The basic features are similar to the curves at a constant load. The whole curves are different from those at a constant load. The total lives are quite longer but the rupture strain is similar each other. The reduction of the total life at a constant load would be resulted from the actual increment in the applied stress due to the reduced cross section. For a constant stress test, we could observe just the work hardening of the metals during the creep.

3.3 Creep strain rate at a constant load

Creep strain rate (creep velocity) is an important factor for the theoretical analysis of deformation mechanisms and the life prediction of structures. The strain rates are shown in Figure 3 (a) and (b) against the time for 4N and 5N aluminiums, respectively. The constant strain rates are clearly recognized under the condition of the lower applied initial stress. The duration of the constant strain rate is a greater part of the total life to failure. The strain rate depends upon the initial applied stress as usual. We have to say that the stress is not the real stress applied at the strain rate. The true applied stress is increasing with decreasing in the cross section of the specimen during the creep deformation.

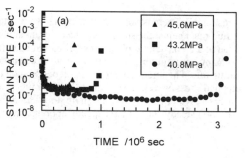
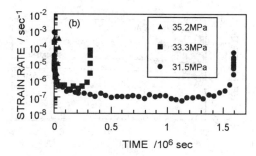

Figure 3. Strain rate-time curves for 4N(a) and 5N(b) aluminiums under the condition of a constant applied load. The respective stresses are the initial applied values.

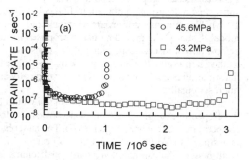
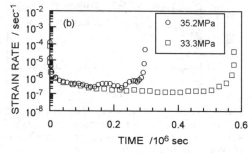

Figure 4. Strain rate-time curves for 4N(a) and 5N(b) aluminiums under the condition of respective constant applied stresses.

3.4 Creep strain rate at a constant stress

Creep strain rates are shown in Figure 4 (a) and (b) against the time for 4N and 5N aluminiums, respectively. We can also observe the extended constant strain rate region, which is a greater part of the life. The strain rate and the duration depend upon the applied stress, too. The higher the applied stress, the higher the former and the shorter the latter. With decreasing in the applied stress, the strain rate would decrease monotonically and the creep would stop finally (Ishikawa et al. 1997). The apparent steady state creep could be recognized at the higher applied stress close to the tensile strength.

When the stress is applied, the reduction of the strain rate begins owing to the work hardening of the material as well as at a constant load. After then, the constant strain rate is maintained for a long time. While the cross section of the specimen is reduced, the applied stress is kept constant. For a constant stress test, the apparent work hardening could not appear during the steady state creep. In other words, the microstructure developed in the metal during the steady state creep would be fairly stable and steady against the internal stress (Takeuchi & Argon 1967). The internal structure could have a highly mechanical stability .

4 DISCUSSION

4.1 Initial creep

The initial strain of the material consists of an instantaneous and a primary creep strain. The instantaneous strain is estimated as the elastic strain, which is constant against the time. The primary strain, however, is increasing with the creep time. Several time-laws are proposed for a primary creep (Cottrell 1963). There are two typical equations, that is, logarithmic ($\ln(\alpha t+1)$) and power ($t^{1/3}$) function (Cottrell 1996). We have scrutinized our experimental results with reference to the proposed equations. There is a better agreement with a logarithmic curve than a power law curve as shown in Figure 5. The constant α is unity for time in second. Face centered cubic metals have the similar expression at lower temperatures (Garofalo 1965).

4.2 Steady state creep

There are a lot of discussions on the existence of a constant creep strain rate, namely, steady state creep (Takeuchi & Argon 1967). It depends upon the expression of strain rate. When strain rate is drawn against the time as shown in Figure 3 and 4, the steady state creep is clearly observed as a greater part of the whole life. Another presentation of strain rate is given in Fig-

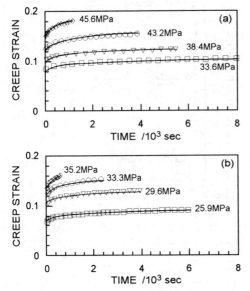

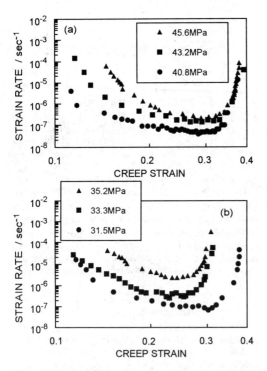

Figure 5. Comparision between the calculated and the experimental curves of the primary creep strain for 4N(a) and 5N(b) aluminiums under the condition of respective applied stresses.

ure 6 and 7. There is not clearly recognized a constant strain rate against the strain both at a constant applied load and a constant applied stress. It would be better to name a minimum strain rate rather than a constant strain rate. The minimum strain rate depends upon the applied stress as well. The strain to the minimum strain rate is a greater part of the total strain, too. The definition of a steady state creep depends upon the expression.

4.3 Macroscopic model and microscopic model

The stress-strain response of solid is modelled with a spring and a dash-pot combination (Felbeck & Atkins 1996). The famous systems are Maxwell and Voight model. Usually, the mixed model is proposed for the creep strain of visco-elastic material at a constant load as shown in Figure 8 (Felbeck & Atkins 1996). Plastic deformation is not homogeneous macroscopically and microscopically.

In consequence, a hard and a soft region would be developed in a single-phase metal deformed at a low temperature (Takeuchi & Argon 1967). The hard phase is assigned to Voigt model and the soft one to Maxwell model. The estimated viscosity coefficients are

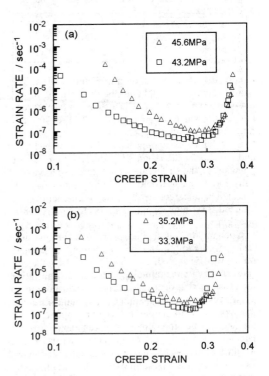

Figure 6. Strain rate v.s. creep strain curves of 4N(a) and 5N(b) aluminiums under the condition of a constant applied load. The respective stresses are the initial applied values.

Figure 7. Strain rate v.s. creep strain curves of 4N(a) and 5N(b) aluminiums under the condition of respective constant applied stresses.

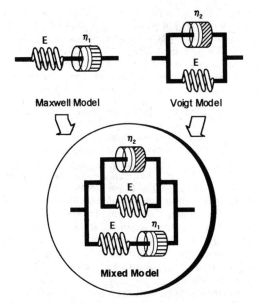

Maxwell Model   Voigt Model

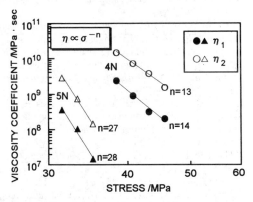

Mixed Model

Figure 8. Mixed model has a parallet combination of Maxwell model (soft phase) and Voigt model (hard phase).

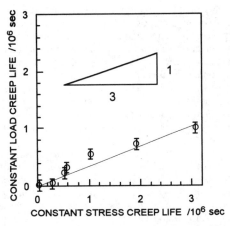

CONSTANT STRESS CREEP LIFE /10⁶ sec

Figure 10. Correlation between creep lives of pure aluminiums under the condition of constant applied load and stress.

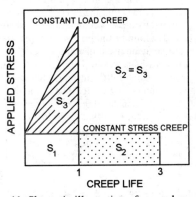

Figure 11. Shematic illustration of creep damage under the condition of a constant applied load and stress for pure aluminiums at lower temperatures.

Figure 9. Stress dependence of viscosity coefficients of 4N and 5N aluminiums under the condition of constant applied load. The stress refers to the initial applied value.

shown in Figure 9. They are functions of initial applied stress. The elastic constant, E is the value of aluminium. The stress dependence is similar for each material.

We could conclude that each elementary process is basically the same in both the phases for the respective metals. It would be a motion of dislocations, which are about ten times more mobile in the soft phase than in the hard phase.

4.4 Creep life

The prediction of creep life is essential information from the practical viewpoint. First of all, we made the comparison between the creep lives to rupture at a constant load and a constant stress. The true stress for the constant load test monotonically increased with the elongation of the specimen. The actual stress was not estimated precisely in this experiment. The ultimate stress for the creep would be evaluated with the knowledge of the strain at the initiation of the acceleration creep.

On the other hand, the applied stress is maintained constant even at the beginning of the acceleration creep for the constant stress test. The creep life for the constant stress test is not associated with the increasing in the applied stress. Therefore, another mechanism should be taken account of for the rupture at a constant stress. We investigated the fracture surfaces of the specimen crept at the constant stress. The scanning electron micrograph gave us the information of the fracture mechanism. A dimple pattern was observed on the surfaces. The void formation prior to the macroscopic reduction of the cross section takes place in the interior of the specimen. Since the applied stress was increased through the reduction of the real cross section, the acceleration creep initiates and the catastrophe comes. The both creep lives are shown in Figure 10.

There is a simple relationship between them. The lives for a constant stress creep are about three times longer than the lives for a constant load creep, regardless of the purity of the aluminiums. The similar result was reported for the high temperature creep of the steel (Garofalo 1965).

4.5 *Creep damage*

The rupture life is associated with the damage accumulated in the inside of the material. When a total creep life of material would be occupied by the steady state creep and the strain to failure is a material constant, the work done in the material is a product of the duration of the steady state creep and the actual applied stress. The schematic illustration is shown in Figure 11. The area refers to the tolerance to the creep rupture of metal at a constant load and a constant stress. The true stress at failure is six times higher for a constant load creep than for a constant stress creep. In addition, creep life could be estimated each other through the relationship.

5 CONCLUSIONS

The experimental study on the creep behaviours of high purity aluminiums leads to the following conclusion. The steady state creeps are observed for 4N and 5N pure aluminiums at a room temperature for both the constant load and the constant stress creep test. The steady state creep is associated with the creep life. The

total life of the constant stress creep is three times longer than that of the constant load creep. It would come from the difference in the damage mechanisms.

REFERENCES

Argon, A. S. 1960. Mechanical Properties of Single-Phase Crystalline Media: Deformation in the Presence of Diffusion, *Physical Metallurgy*. R. W. Cahn & P. Haasen (eds), Amstermdam: North-Holland.

Cottrell, A. H. 1963. *Dislocations and plastic flow in crystals*. Oxford: The Clarendon Press.

Cottrell, A. H. 1996. Strain hardening in Andrade creep. *Phil. Mag. Let.*. 74:375-379.

Dorn, J. E. 1957. *Creep and Recovery*. Ohio: ASM.

Dunand, D. C. & A. M. Jansen 1997. Creep of metals containing high volume fractions of unshearable dispersoids-Part I. *Acta mater.*. 45:4569-4581.

Felbeck, D. K. & A. G. Atkins 1996. *Strength and Fracture of Engineering Solids*. New Jersy:Prentice -Hall Inc.

Garofalo, F. , O. Richmond & W. F. Domis 1962. Design of Apparatus for Constant-Stress or Con stant-Load Creep Tests. *Journal of Basic Engineer ing*. 84:287-293.

Garofalo, F. 1996. *Fundamentals of Creep and Creep-Rupture in Metals*. N.Y.:Macmillan Co.

Ishikawa, K. , H. Okuda & Y. Kobayashi 1997. Creep behaviors of highly pure aluminium at lower temperatures. *Materials Science and Engineering*. A234-236: 154-156.

Krajewski, P. E. , J. E. Allison & J. W. Jones 1997. The effect of SiC particle reinforcement on the creep behavior of 2082 aluminum. *Metall. and Mat. Trans.*. 28A:611-620.

McDonald, L. C. & K. T. Hartwig 1990. Creep of pure aluminum at cryogenic temperatures. *Advances in Cryogenic Engineering (Materials)*. 36:1135-1141.

Takeuchi, S. & A. S. Argon 1967. Steady-state creep of single-phase crystalline matter at high temperature. *Journal of Materials Science*. 11:1542-1566.

Experimental Mechanics, Allison (ed.)© 1998 Balkema, Rotterdam, ISBN 90 5809 014 0

Creep behaviour of Glass-Fiber reinforced PES and its fiber volume fraction dependence

S. Somiya, T. Kawakami & K. K. Biswas
Keio University, Yokohama, Japan

ABSTRACT: The bending creep behavior of Polyether Sulphone (PES) and some kinds of glass fiber reinforced PES resin has been researched using three point bending test method. Here, the linearity of creep behavior has been discussed by drawing the master curve of creep compliance on each material and results comply with Arrhenius time-temperature reciprocation law. In this research, the effect of volume fraction of the reinforced fiber on creep behavior has also been studied.

1. INTRODUCTION

Although super engineering plastics with high heat-resistant mechanical properties are widely used in the fields including the aircraft and space industries, various types of FRP have been developed to improve the heat resistance. Even though those plastics and FRPs are superior in the high-temperature mechanical properties, they are viscoelastically deformed by using at high temperature for a long time. It is, therefore, expected to elucidate the characteristics of viscoelastic phenomena of those materials in detail and improve the estimation technique of long-term life based on this elucidation result for enhancing the reliability of the materials and aiming at the high degree of utilization. Although many researches concerning the creep behavior of FRPs have been made such as Donogue R.D., etc.1992 and Wang, J.Z., & Parvatareddy, H. etc.1995., a few researches have been done so far for investigating the linearity of the viscoelastic behavior for example, Horward, C.M. and Hollaway L. 1987 and Somiya, S. & Iwamoto, N. 1995. But there is few reports which discussed the relationship between the viscoelastic deformation and the amount of fiber contents in FRP such as Somiya, S. 1997.

In this study, the linearity of the viscoelastic behavior of FRPs was investigated by making an experiment of the bending creep deformation of polyether sulphone (PES) and its glass-fiber reinforced plastic in a temperature range from 160°C to 220°C. Moreover, the effects of the fiber volume fraction; V_f on the creep characteristics were investigated.

2. EXPERIMENTAL METHODS

2.1 Test specimens

Polyether sulphone resin (SUMIKAEXCEL®PES 4100G) and glass-fiber reinforced polyether sulphone (SUMIKAEXCEL® PES, GL20 and GL30 by Sumitomo Chem. Ind. Co.) are used as the test specimens. The glass transition temperature of the matrix is 225°C. The test specimens of rectangular strips 64 ×15 × 3 mm are cut out from an injection molded plate 64 × 64 × 3 mm with a diamond cutter. The test specimens with three different fiber weight fraction, W_f, (Volume fraction, V_f), of 0% (0%), 20% (12%) and 30% (19%) are prepared and called GF0, GF20 and GF30, respectively.

2.2 Bending test and bending creep test method

The bending test representing the static mechanical characteristics is made with an Instron type tester fixing at the span of 50 mm, the span ratio (span / specimen thickness) of approximately 16 and the cross head speed of 1 mm/min.

The creep characteristics are determined by the three point bending creep test. The load is fixed in such a way that the bending stress developed by the load is equalized to 1/10 of the static bending strength of each material. The bending strength and modulus were calculated by the maximum bending load and the slope from the original point of load - deflection curve. The loading method is a flatwise loading. The three point bending test instrument is installed in at thermostatic chamber, DH42, (room temperature to 300°C, temperature range: ±2.5°C). The test is made at the test temperatures, T, from 160°C to 220°C at intervals of 10°C. The temperature of the tester is adjusted to the test temperature for five minutes before beginning the creep test and then the load is applied. The testing time, t, is within 100 minutes. The variation of the deflection in creep, δ_c, with time is determined and the creep compliance, $D_c(t, T)$, is calculated from the result.

3. RESULTS AND DISCUSSIONS

3.1 *Fiber volume fraction dependence of static mechanical characteristics*

The relationship between the bending modulus obtained from the static three point bending test and the fiber volume fraction will be described hereafter. Table 1 lists the bending modulus and the bending strength of the test specimens.

The table reveals that the static mechanical characteristics including bending modulus and bending strength of the materials used in this study are improved with the increase of the fiber volume fraction. The relationship between the bending modulus, E_c, and V_f is approximated with a linear function and the gradient of a straight line is calculated by the method of least squares. The calculation result indicates that the coefficient of correlation of the bending modulus with V_f is 0.99 and the coefficient α is 0.49. Since the modulus of elasticity is equal to that of the resin itself when V_f is zero, the bending modulus is expressed by the following formula based on the mixture's law:

$$E_c = \alpha E_f V_f + E_m (1 - V_f) \qquad (1)$$

Where E_c modulus of elasticity of composites, E_f (modulus of glass fiber) = 72 GPa, E_m (modulus of matrix) = 121.5 MPa, and α = coefficient depending upon the shape, degree of orientation, etc. of reinforcing fiber. The relationship between the bending strength and the fiber volume fraction is not linear though the former is increased with the increase of the latter.

Table 1 Mechanical Properties of Used Materials

Materials	Vf(%)	Bending Modulus(GPa)	Bending Strength(MPa)
GF0	0	2.78	121.5
GF20	12	6.54	158.3
GF30	19	8.95	171.6

3.2 *Temperature dependency of creep deformation characteristics*

Fig.1 illustrates creep curve groups classified by the test temperatures of GF0 as an example. The figure reveals that the creep strain is increased with the passage of time. The result of the experiment made at intervals of 10°C indicates that the creep curves are

ranged according to temperature and the creep strain is steeply increased with the passage of time.

The relationship between the creep compliance, $D_c(t, T)$, calculated from the creep strain and the time, t were examined using the materials with various fiber volume fractions. Fig.2(a), Fig.3(a) and Fig.4(a) depict the variations of the creep compliance, $D_c(t,T)$, of GF0, GF20 and GF30 respectively. The creep compliance are increased by the increase of time and temperature. The figure reveals that the values of creep compliance are decreased and the variations of it according to the time and temperature are significantly mitigated with the increase of the fiber volume fraction, namely, the time and temperature dependence are reduced by the addition of fiber. This tendency is also observed in other fiber reinforced thermoplastics .

Examining the creep compliance curves of GF0, GF20 and GF30 in detail, the temperature conditions showing a little abnormal tendency in the order of size of the creep compliance curves at various temperatures are observed.   For example, the curves of GF0 at 170 and 180°C cross each other at a value on the short time side and slightly go into reverse. The similar tendency is observed in the curves at GF20 and GF30 though the temperature at the crossing is increased with the increase of fiber volume fraction. This may be attributed to the effect of a little content of water or an intermediate molecular weight polymer. Since this is an abnormal phenomenon in a specific narrow temperature range, the results obtained in the temperature range are omitted hereafter and only the comprehensive tendency will be investigated.

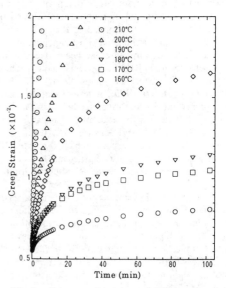

Fig.1   Creep strain curves of GF0

1356

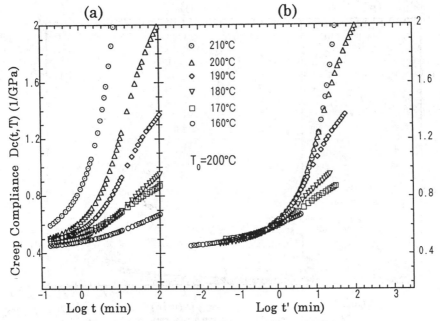

Fig.2　Creep compliance curves and master curve of GF0

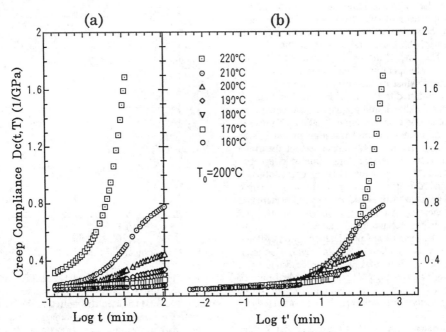

Fig.3　Creep compliance curves and master curve of GF20

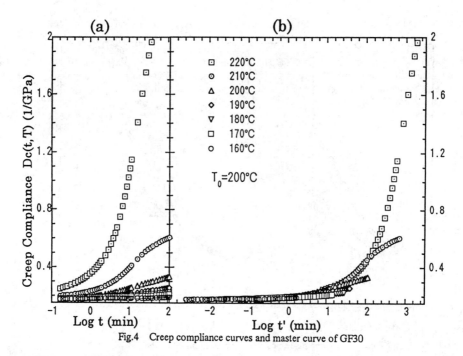

Fig.4 Creep compliance curves and master curve of GF30

3.3 Investigation of time and temperature reciprocation law of glass fiber reinforced PES

Fig.2 (b), Fig.3(b) and Fig.4(b) show the master curves at the reference temperature, 200°C of the creep compliance drawn from each creep curve group as shown in Fig.2(a), Fig.3(a) and Fig.4(a). In these figures, the horizontal axis show the physical time; t'. The time; t' means the converted time into the reference temperature from real time on real test temperatures. Most of the curves of GF0 plotted in Fig.2(a) are shifted to the short time side because the creep behavior shown in Fig.2(a) are observed on the shorter time side than the curve at 200°C. The creep compliance master curves of the matrix resin as shown in Fig.2(b) is smooth curve except the creep compliance curves on the long time side at each temperature. The creep compliance curve at each temperature deviates widely from the master curve with the passage of time and the deviation increases with the increase of time. This tendency is observed also in other materials. This maybe caused due to curing of the resin by the physical aging for example of the research on Wang, J.Z., etc. 1995. The creep behavior of PES is caused by the physical aging at a temperature as low as 160°C for a period of time as short as 100 minutes. It is, therefore, concluded that thermoplastic resins are largely affected by aging.

The master curves of GF20 and GF30 are illustrated in Fig.3(b) and Fig.4(b), respectively. The figure reveals that each master curve is a smooth one and shows the linear viscoelastic characteristics

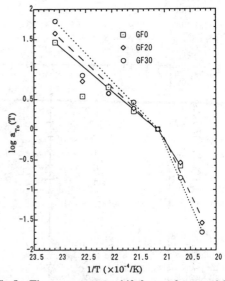

Fig.5 Time-temperature shift factors for materials

in the same way as the matrix shows. It is found from those two master curves that the values of creep compliance are decreased and shifted to the long time side with the increase of fiber volume fraction.

The relationship between the time-temperature shift factors, a_{T_0} (T), and the temperature is illustrated in Fig.5. The movements to the short time side and to the long time side are fixed to positive

1358

and negative, respectively. The relationship between the shift factor and the temperature except the abnormal result at 170°C is roughly expressed by two straight lines with different gradients at 200°C. This fact indicates that the linear viscoelastic behavior expressed by Arrhenius equation applies to PES and its GFRP in the same manner as to thermoplastic Polyimide resin and its CFRP as shown in Somiya. S., 1995.

3.4 Effect of fiber volume fraction on master curve

The relationship between the creep compliance mater curve and the fiber volume fraction will be mentioned hereafter. Fig.6 shows three master curves for GF0, GF20 and GF30 schematically. The figure reveals that the higher the fiber volume fraction, the more the master curve shifts downward and to the right side. That fact indicates that the creep behavior of fiber reinforced resins not only decreases the creep compliance at the same time as that of creep behavior but also delays the development of creep phenomenon.

Few papers discussing the relationship in the viscoelastic behavior between the materials with different fiber volume fractions have been reported by Somiya 1995. This paper investigates the effects of the fiber volume fraction on the creep deformation. The previous paper reported that the creep compliance master curves of FRPs with different fiber volume fractions can be drawn by shifting downward and to the right side. It reveals that the reinforcing fiber is effective for restraining the development of the deformation by viscosity as well as for transferring the load. The master curves of GF20 and GF30 are shifted respective proper distances to that of GF0 used as the basis. The result obtained by shifting them is illustrated in Fig.7. The relationships between the distances of the curves shifted in the directions of the axis of ordinates and abscissas in the graph and the fiber volume fraction are illustrated in Fig.8 and Fig.9, respectively. Fig.7 reveals that the master curve of each fiber volume fraction can be placed on that of resin. It is, therefore, elucidated that any master curve of glass-fiber reinforced polyether sulphone used in this study is controlled by the viscoelastic behavior of PES resin of the matrix regardless of the fiber volume fraction, and the viscoelastic characteristics are limited because the fiber apparently increases the viscosity coefficient.

The relationship between the shift factor of the master curve in the direction of the axis of coordinates and the fiber volume fraction illustrated in Fig.8 will be investigated using the mixture law described in the previous paper hereafter.

$$a_{T_{Dc}} = \frac{A}{\alpha E_f V_f + E_m (1 - V_f)} - \frac{A}{E_m} \quad (2)$$

Where A= experimental constant, and $\alpha = 0.49$ (obtained by the three point bending test).

Assuming that the constant of A is to be 1.18, theoretical curve is drawn in Fig.8. This curve complies with the experimental points. It is, therefore, concluded that the addition of the fiber affects the creep compliance on the axis of coordinates in the graph for the master curve according to the mixture law in the same manner as in the modulus.

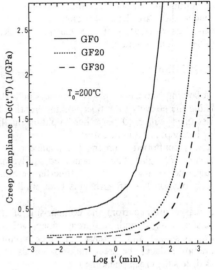

Fig.6 Master curves of creep compliance of materials

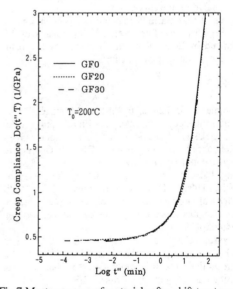

Fig.7 Master curves of materials after shift treatment

The relationship between the shift factor of the master curve in the direction of the axis of logarithmic time, $a_{Tt'}$, and the fiber volume fraction, V_f, is linear as shown in Fig.9. It reveals that the effect of the fiber volume fraction on the axis of time of the creep behavior is linear, namely, the variation of the viscosity term can be expressed by a linear function of the fiber volume fraction. Such a tendency agrees with the results obtained so far as shown in Somiya, S. 1997. Thus, it is elucidated that the creep compliance and shift factor in the direction of the axis of time of the master curve directly depend upon the fiber volume fraction, and accordingly they are basically controlled by the viscoelastic characteristics of PES matrix regardless of the fiber volume fraction.

4. CONCLUSIONS

The following conclusions were obtained by elucidating the effects of the fiber volume fraction in the glass-fiber reinforced polyether sulphone on the bending behaviors of it in this study:

(1) It was confirmed from the compliance master curves of glass-fiber reinforced polyether sulphone that the time-temperature reciprocation law of Arrhenius type applies to it.

(2) It was confirmed from the comparison of the master curves with different fiber volume fractions that the increase of fiber volume fraction reduces the creep deformation as well as delays the creep phenomenon.

(3) It was elucidated by using two shift factors, a_{TDc} and $a_{Tt'}$, that all of the master curves can be unified to one curve.

Thus, it was clarified that the creep behaviors of FRTPs with any fiber volume fractions can be estimated by determining the behavior of the resin itself in the test specimen of PES.

REFERENCE

Donogue, R.D., Peters, P.W.M. & Marci, G. 1992. The Influence of Mechanical Conditioning on the Viscoelastic Behaviour of Short-fiber Glass Reinforced Epoxy resin (GRP). *Composite Science and Technology* 44:43-55.

Wang, J.Z., Parvatareddy, H., Chang, T., Iyengar, N., Dilard, D.A. & Reifsnider, K.L. 1995. Physical Aging Behavior of High-performance Composites. *Composites Science and Technology* 54:405-415.

Horward, C.M. & Hollaway, L. 1987. The Characterization of the Non-linear Viscoelastic Properties of a Randomly Oriented Fiber/matrix Composite. *Composites* 18-4:317-323.

Somiya, S. & Iwamoto, N. 1995. Effect of Fiber Content on Creep Behavior of Composites

Engineering. *Proceeding of 2nd International Conference of Composites Engineering* :112-114.

Somiya, S. 1997. Fiber Volume Fraction Dependence of Creep Phenomena of Fiber Reinforced Plastics-Fundamental Research on Activation Energy. *Proceeding of International Conference on Advanced Technology in Experimental Mechanics* : 175-180

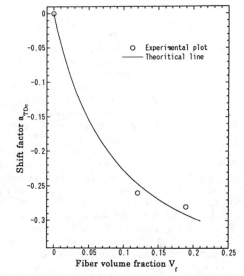

Fig.8 Shift factors a_{TDc} on creep compliance axis with fiber volume fraction

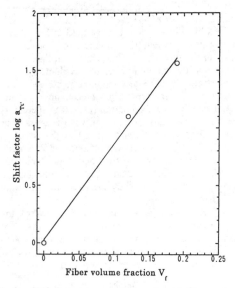

Fig.9 Shift factors $a_{Tt'}$ on time axis with fiber volume fraction

Composites

Experimental Mechanics, Allison (ed.)© 1998 Balkema, Rotterdam, ISBN 90 5809 014 0

Load direction effect on the damage in a quasi-isotropic CFRP laminate

Jianqiao Chen, Shigeo Takezono & Masamichi Nagata
Toyohashi University of Technology, Japan

ABSTRACT: Tensile tests are carried out for two kinds of specimens, made from a quasi-isotropic CFRP laminate, to investigate the influence of tensile direction on the damage pattern and final failure. A FEM code is used to analyze the stresses in the specimens, and an attempt is made to correlate the numerical analysis with the experimental observations.

1 INTRODUCTION

For a quasi-isotropic CFRP laminate, the elastic stress-strain behavior is expected to be the same in any directions. But Fracture and/or strength of it may strongly depend on the stacking sequence of laminae and/or the loading direction (Jones 1975, Xu 1995). This is because once matrix cracking or delamination takes place, the damaged laminate changes into an anisotropic one, and different fracture patterns may appear under different direction loading.

In this study, tensile tests are carried out for two kinds of specimens, cut out from a quasi-isotropic CFRP laminate in different directions, to investigate the influence of load direction on the damage pattern and tensile strength. A FEM code is used to analyze the stresses in the specimens, and the correlation between the numerical results and the actual observations is discussed.

2 TENSILE EXPERIMENT

2.1 *Experimental procedure*

The quasi-isotropic laminate [0/45/-45/90]s used in the experiment was made of carbon fiber reinforced epoxy lamina (Fiber: T800H, Epoxy: 3631, Toray Co., Japan). Figure 1 shows the definition of the principal material coordinates (123) and the specimen coordinates (xyz) for a lamina, in which x denotes the tensile direction. The fiber content of the material is about 60%. The 300mm *300mm 8 layers laminate has a thickness of 1.15mm. It was

cut into 250mm long and 25mm wide specimens. By cutting the laminate along its 0 layer and 45 layer respectively, two kinds of specimens were prepared as shown in Fig.2 (Type I: [0/45/-45/90]s, Type II: [-45/0/90/45]s).

Free edges of the specimens were polished up for observation of matrix cracking and delamination growth. GFRP tabs of 80mm long were glued to the gripped ends of the specimens. The gauge length was 80mm.

The specimens were tested in tension at a cross-head velocity of 0.5mm/min. A microscope was used to observe the damage development (matrix cracking and delamination growth) at the free edges of the specimens.

2.2 *Experimental results and discussions*

Typical results of damage development observation are listed in Tables 1 and 2. Photograph examples are shown in Fig.3 and Fig.4. Crack density is defined as number of cracks divided by gauge length. In Fig.3 (a), for example, we counted two cracks in –45 ply and one crack in 90 ply. The crack density in the Tables is the average value at the two free edges of a specimen.

It is found that type I specimen has larger tensile strength than type II (about 1.2 times), but matrix cracking and delamination appeared at lower stress in the former. There exist some differences in the damage growth pattern between the two kinds of specimens. In type I specimen, matrix cracks appeared first in the 90 layer, and then the cracking took place in the –45 layer. A few cracks also initiated in the 45 layer as applied stress increased.

Delamination has been observed at three interfaces: –45/90, 45/-45 and the midplane 90/90 (Fig.3). In type II specimen, however, matrix cracking took place mainly in the 90 layer. Only a small number of cracks were observed in the 45 layer. Delamination growth was found at 0/90 and 90/45 interfaces. At 45/45, the midplane of this type, no delamination has been observed (Fig.4).

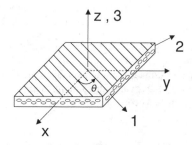

Figure 1. The principal material coordinates (123) and the specimen coordinates (xyz) for a lamina, in which x denotes the tensile direction.

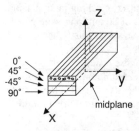

Type I : [0/45/-45/90]s

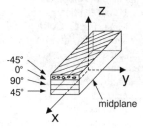

Type II : [-45/0/90/45]s

Figure 2. Two kinds of specimens and the specimen coordinates.

Table 1. Matrix cracking and delamination growth in [0/45/-45/90]s, type I specimen.

Stress	Crack density (1/cm)			Delamination			
(MPa)	45	-45	90	$\frac{0}{45}$	$\frac{45}{-45}$	$\frac{-45}{90}$	$\frac{90}{90}$
396	-	-	-	-	-	-	-
465	-	-	0.13	-	-	Y	-
490	-	0.81	*1.06	-	-	Y	Y
540	0.06	*3.13	*5.63	-	Y	Y	Y
#603	0.13	*5.31	*7.13	-	Y	Y	Y
756	Fracture						

- : No, Y : Yes, GL (gauge length) : 8cm
* Cracks distributed uniformly in x direction
# Delamination growth across the whole GL

- Average tensile strength: 796 MPa
- Standard deviation: 36 MPa
- Three specimens were used

Table 2. Matrix cracking and delamination growth in [-45/0/90/45]s, type II specimen.

Stress	Crack density (1/cm)			Delamination			
(MPa)	-45	90	45	$\frac{-45}{0}$	$\frac{0}{90}$	$\frac{90}{45}$	$\frac{45}{45}$
466	-	-	-	-	-	-	-
510	-	0.31	-	-	-	Y	-
546	-	3.44	-	-	Y	Y	-
570	-	*5.69	0.06	-	Y	Y	-
616	-	*11.83	1.03	-	Y	Y	-
681	Fracture						

- : No, Y : Yes, GL (gauge length) : 8cm
* Cracks distributed uniformly in x direction

- Average tensile strength: 663 MPa
- Standard deviation: 12 MPa
- Four specimens were used

Matrix crack and delamination seem to spread to a large range in type I specimen. Crack densities in 90 and –45 layers have close values (Table 1), and delamination took place at three interfaces as mentioned above. For type II, however, damage growth was localized in it: matrix cracking was concentrated in the 90 layer, and delamination was restricted at 0/90 and 90/45 interfaces (Table 2). Tables 1 and 2 also show that matrix cracks tended to distribute uniformly in x direction as stress increased. This uniform distribution in type I appeared at lower crack densities or lower stresses as compared with type II (Note the data with * in these Tables). We consider that the large-scale and relatively uniform damage distributions in type I specimen accounted for the higher tensile strength.

There exist some interactions between the matrix cracking and delamination. Crack density in some layer increased to be greater than a certain value ($\approx$ 1 1/cm in 90 layer, for example, in type I specimen) as delamination took place at both neighboring interfaces of the layer (Note the bold-faced data in Table 1 and Table 2).

3 STRESS ANALYSIS

3.1 *FEM analysis*

The classical lamination theory gives a good estimation of stresses in a laminate. But the edge effects, i.e., in-plane stresses making a change and interlaminar stresses developing at the free edges, should be considered (Pipes & Pagano 1970, Jones 1975). Many attempts have been made to investigate the edge effect, both theoretically and numerically (Puppo & Evensen 1970, Pagano 1974, Yin 1994). Most of these studies have dealt with very simple cases, such as [0/90]s or [45/-45]s, because of the complexity of analysis and/or the computation limitations. FEM analysis was made here in order to investigate the stress - damage growth relation.

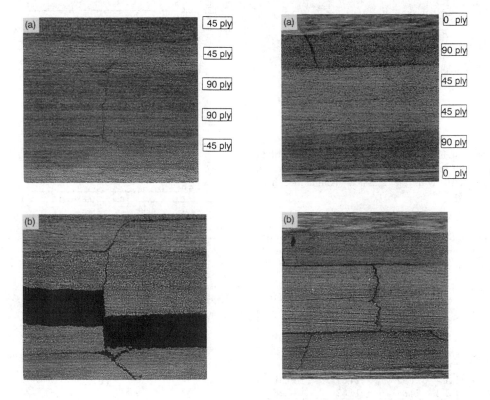

Figure 3. Photography of matrix cracking and delamination in [0/45/-45/90]s, type I specimen. The surface ply (0 ply) was not taken in the photo. (a): 490 MPa, (b): 540 MPa.

Figure 4. Photography of matrix cracking and delamination in [-45/0/90/45]s, type II specimen. The surface ply (-45 ply) was not taken in the photo. (a): 546 MPa, (b): 570 MPa.

In this study, a FEM code ANSYS was applied to investigate the stresses in the two kinds of specimens. The analytical FEM model is shown in Fig.5 where the sizes in y and z directions are enlarged. The orthotropic element SOLID was used in the calculation. The boundary conditions were given by

$$U (0,0,0)=0$$
$$V (0,0,0)=0$$
$$W (0,0,0)=0$$

and

$$W (x,y,0)=0$$

in which U, V and W are displacements in x, y and z directions, respectively. Tensile loading by given uniform longitudinal displacements at both ends of the specimen was considered.

The following data for the lamina were used in the calculation:

$$E_1 = 160.0\ GPa$$
$$E_2 = E_3 = 8.9\ GPa$$
$$G_{12} = G_{13} = G_{23} = 5.0\ GPa$$
$$v_{12} = v_{13} = v_{23} = 0.34$$

3.2 FEM Results and discussions

Stress analysis results are listed in Table 3 and shown in Figs.6 – 8. In the middle region of the specimen (except the free edges), the FEM results agree well with the classical lamination theory (Table 3).

Table 3. Stresses in the region except the free edges of specimen [0/45/-45/90]s at x=0. The strain of x direction is $\varepsilon_x = 6.25 \times 10^{-4}$.

		$\sigma_x (MPa)$	$\sigma_y (MPa)$	$\tau_{xy} (MPa)$
0°	A	100.03	0.15	0
	B	100.05	0.14	5.3×10^{-6}
45°	A	22.90	14.85	16.31
	B	22.88	14.82	16.28
-45°	A	22.90	14.85	-16.31
	B	22.88	14.82	-16.28
90°	A	5.00	-29.65	0
	B	5.00	-29.79	2.4×10^{-6}
A: Classical lamination theory				
B: FEM analysis				
The other stress components are zero				

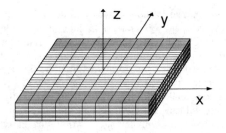

Figure 5. The analytical FEM model for upper half part of a specimen.

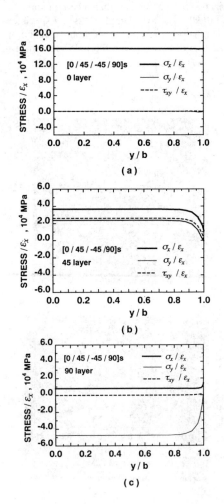

Figure 6. In-plane stresses in [0/45/-45/90]s laminate at x=0. These stresses are symmetrical about y=0. The specimen width is 2b. Stresses in the –45 layer are the same as (b) except the direction of τ_{xy} is opposite.

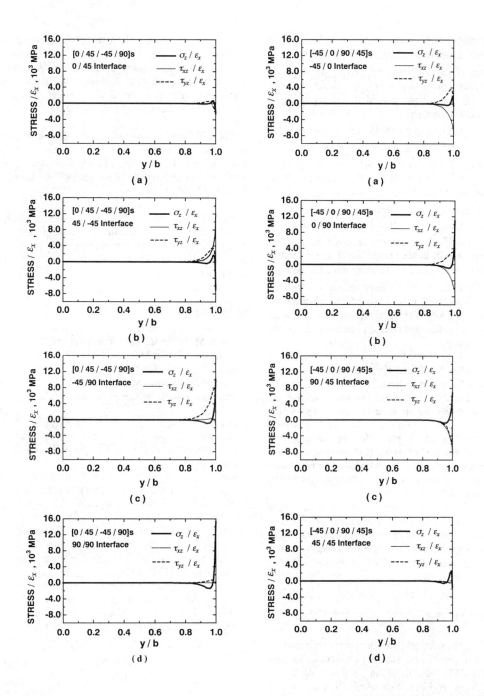

Figure 7. Interlaminar stresses in [0/45/-45/90]s laminate at x=0. The specimen width is 2b. The normal stress is symmetrical about y=0, while the shear stresses are antisymmetrical about y=0.

Figure 8. Interlaminar stresses in [-45/0/90/45]s laminate at x=0. The specimen width is 2b. The normal stress is symmetrical about y=0, while the shear stresses are antisymmetrical about y=0.

In the boundary region, however, this is not the case. For example, the stress σ_x in the 90 layer increases at the specimen edges (Fig.6), implying that matrix cracks are liable to initiate at those places. Figure 7 shows that in the edges of type I specimen, high normal stress σ_z develops at -45/90 and 90/90 interfaces. At 45/45, σ_z is compression, a harmless component, but a high shear stress τ_{xz} develops. At 0/45 interface, all stress components are very small. In type II (Fig.8), there exist large interlaminar stresses (both σ_z and τ_{xz}) at 0/90 and 90/45 interfaces. The interface –45/0 is the next severe one because of the high shear stress τ_{xz}, while at 45/45, the midplane of this type, stresses are very weak. It is seen from Figs.7 and 8 that the boundary condition ($\tau_{yz} = 0$, when $y = b$) is satisfied in all these cases. Since interlaminar stresses are considered to be the main cause of delamination, the FEM analytical results give a good explanation to the experimental observations mentioned in the previous Chapter.

In type I specimen, the 0 ply is the surface layer and all other layers are restricted by it. They all shear a definite part of load. In fractured specimens of this type, fiber break was observed in 45 and -45 layers, showing that these layers had borne with a large part of the applied stress. In type II, however, the restriction by 0 ply (the second layer) is weakened, stress concentration is more severe in it as compared with type I, resulting in lower tensile strength. Fiber break in 45 or –45 layer has hardly been found in this case.

It should be noted that delamination or matrix cracking results in stress redistribution and/or stress-relaxation in a laminate. Therefore, stress analysis of a cracked specimen is necessary in order to predict damage growth and correlate it to the laminate strength more precisely.

4 CONCLUSIONS

Tensile tests were carried out for two kinds of specimens made by cutting a quasi-isotropic laminate [0/45/-45/90]s along its 0 and 45 layers, respectively. A FEM code ANSYS was used to analyze the stresses in the specimens.

Experiments showed that in type I specimen, the surface layer (0 ply) restrict the other layers, and 45 and –45 layers, for example, bear a considerable part of tensile loading. Matrix cracks spread to a large range in it. In type II, however, the restriction by 0 ply is weakened and damage is localized as compared with type I, resulting in lower tensile strength. There exist some interactions between the matrix cracking and delamination. Crack density in a layer increases to a certain value as delamination takes place at both neighboring interfaces of the layer.

The FEM analysis showed that the interlaminar stress distributions are different between the two kinds of specimens. The numerical results agree with the experimental observations of damage growth at least qualitatively.

Acknowledgements - This work was partially supported by the Scientific Research Grant of Tatematsu Foundation. The authors wish to thank Mr. Ono and Mr. Kasai for their help in the experiment.

REFERENCES

Jones, R.M. 1975. *Mechanics of composite materials*. Tokyo: McGraw-Hill.

Pagano, N.J. 1974. On the calculation of interlaminar normal stress in composite laminate. *Journal of Composite Materials*. 8:65-81.

Pipes, R.B.& Pagano, N.J. 1970. Interlaminar stresses in composite laminates under uniform axial extension. *Journal of Composite Materials*. 4:538-548.

Puppo, A.H. & Evensen, H.A. 1970. Inmterlaminar shear in laminated composites under generalized plane stress. *Journal of Composite Materials*. 4:204-220.

Xu, L.Y. 1995. Influence of stacking sequence on the transverse matrix cracking in continuous fiber crossply laminates. *Journal of Composite Materials*. 29:1337-1358.

Yin, W.L. 1994. Simple solutions of the free-edge stresses in composite laminates under thermal and mechanical loads. *Journal of Composite Materials*. 28:573-586.

Experimental Mechanics, Allison (ed.) © 1998 Balkema, Rotterdam, ISBN 90 5809 014 0

Determination of stiffness characteristics of ±45° GFRP-laminates

H. Rapp & P. Wedemeyer
Institut für Aerospace-Technologie, Hochschule Bremen, Germany

ABSTRACT: Normal mode tests of composite structures, the properties of which are dominated by ±45° GFRP laminates, often show higher tension or bending stiffnesses than theoretical predictions. This especially comes true when the properties of the unidirectional layers are calculated by means of micromechanical relations. Young's modulus of ±45° GFRP laminates is determined experimentally at different load levels and loading frequencies. The stress-strain curves can be approximated very well by parabolic curves. What follows is a linear decrease of stiffness with strain and stress, respectively. Numerical results are presented for laminates made from different E-Glass-fibre prepregs. The results show that the real longitudinal stiffness of such laminates is up to 50 percent higher than predicted by micromechanical theory. It turns out that the reason for this is the inaccurate representation of the shear stiffness of the unidirectional layers by common micromechanical relations.

1 INTRODUCTION

Laminates are usually designed to be loaded in fibre direction. But there are some cases in which the individual unidirectional layers of a laminate are loaded in fibre direction as well as in shear direction. Examples of such structures are torsional tubes loaded by bending. As the primary loading in this case is torsion, the laminate fibre orientation is ±45°. Additional bending loads cause normal stresses in the laminate. According to the classical laminate theory this normal loading of a ±45° laminate results in a shear loading of the individual unidirectional layers.

In such cases bending stiffness and normal mode tests often show a remarkably higher stiffness than predicted by theory. This is especially the case when the properties of the individual unidirectional layers are calculated from fibre and matrix properties by means of micromechanical relations. The reason for this difference seems to be an inaccurate description of the shear stiffness of a unidirectional layer by micromechanical relations, especially at low strains.

To clarify this problem, static and dynamic stiffness tests are done. The properties of the used unidirectional layers as well as Young's modulus of ±45° GFRP-laminates are determined. Parameters for the tests are the fabric type, used for the specimen and the loading velocity. The test results are compared to results obtained by the classical laminate theory (CLT). Input data for the CLT are on the one hand the experimentally determined properties of the unidirectional layers and on the other hand the properties determined by the use of micromechanical relations.

2 STATIC AND DYNAMIC TENSION TESTS

Tension tests are done to determine the stiffness of ±45° GFRP-laminates. All specimens were made from prepregs consisting of E-glass fibres and 913 resin from Ciba Geigy. Different fabric types are considered: unidirectional prepreg EC9756, and fabric types 7781 and 120. Laminate plates measuring 300 mm x 300 mm are manufactured. The specimens are cut from these plates. All specimens have a length of 250 mm, a width of 25 mm and a thickness of about 2 mm. The exact thickness varies slightly due to different layer thickness of the prepregs. The fibre volume fraction for all specimen is between 55% and 63%.

All test results are converted to a reference fibre volume fraction of 60 %. This conversion is done by using the CLT and micromechanical relations:

$$E_{\text{Test }60\%} = E_{\text{Test }\varphi} \frac{E_{\text{CLT }60\%}}{E_{\text{CLT }\varphi}} \tag{1}$$

The difference between the real and the reference fibre volume fraction is only a few percent. It is assumed that due to this small variation of the fibre volume fraction all micromechanical theories (see chapter 3) yield very similar results, so that the special theory used for this conversion is insignificant.

2.1 Static tests

Static tension tests are performed with these specimens. The longitudinal and transverse strain measured by strain gauges on both sides of the specimen as well as the extension of the specimen, measured by two extension meters are recorded to determine the stress-strain curve. Figure 1 shows the test setup of the tension tests.

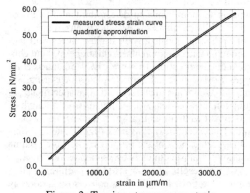

Figure 1: Test setup of the static and and dynamic tension tests

In figure 2 a typical stress-strain curve is given. The tension stress is plotted versus the mean value of the longitudinal strain measured by strain gauges on both sides of the specimen. The difference between the strains measured by strain gauges and by the extension meters is very small, so it can be neglected. Only the strain gauge data are taken into account for the further determination of the stiffness.

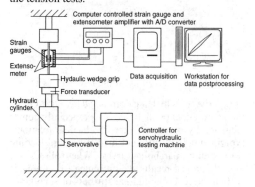

Figure 2: Tension stress versus strain

The thick line in figure 2 represents the measured data. The curvature of this curve shows a slight dependence of the longitudinal stiffness of a $\pm 45°$ GFRP-laminate upon the loading. In the considered load range, the stress-strain curve can be approximated very well by a parabolic curve. This means that Young's modulus depends linearly on the strain. In figure 2 the parabolic approximation is shown by the light gray curve inside the black curve.

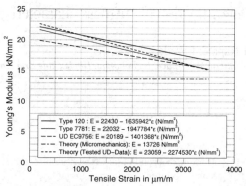

Figure 3: Measured Young's modulus of $\pm 45°$ GFRP laminates

From

$$\sigma = a_0 + a_1\varepsilon + a_2\varepsilon^2 \tag{2}$$

ensues for the tangential Young's modulus E:

$$E = \frac{\partial\sigma}{\partial\varepsilon}: \qquad E = a_1 + 2a_2\varepsilon. \tag{3}$$

Figure 3 gives the resulting tangential Young's modulus as a function of strain for the $\pm 45°$ laminates. The measured tangential Young's modulus shows a significant decrease with the loading for all types of $\pm 45°$ GFRP-laminates. At a strain value of about 0.0035 ($\approx 60\,\text{N/mm}^2$), the stiffness is only 75% of the stiffness at zero loading. The differences between the laminates made from different raw materials are small. The laminate made from the finest fabric (120) shows the highest, the laminate made from unidirectional prepregs (EC9756) shows the lowest stiffness.

In addition to the longitudinal strain, the transverse strain is measured by strain gauges. The Poisson's ratio of the $\pm 45°$ laminate as well as the shear modulus in the $\pm 45°$ direction (i.e. the shear modulus of the unidirectional layers) can be determined from these strains. The shear modulus follows from

$$G_\# = \frac{\sigma_{longitudinal}}{2(\varepsilon_{longitudinal} - \varepsilon_{transverse})}. \tag{4}$$

Further static tests are done to determine all stiffness data and Poisson's ratios of a unidirectional layer. The tests are performed with specimens cut from a unidirectional laminate in and perpendicular to the fibre direction. The experimentally obtained data are the basis for the calculation of Young's modulus of a laminate by using the classical laminate theory. Another way to determine the properties of a unidirectional layer are the use of micromechanical relations (see chapter 3).

Table 1 gives a comparison of tested and calculated properties of a unidirectional layer. For the calculations the equations given in N.N., LTH, 1974, are

Table 1: Material properties of a unidirectional laminate

Property		tested	micromech.
$E_{\parallel}$	N/mm^2	43500	45200
$E_{\perp}$	N/mm^2	13400	11595
$G_{\#\sigma=5\,N/mm^2}$	N/mm^2	7152	4363
$G_{\#\sigma=50\,N/mm^2}$	N/mm^2	5036	4363
$v_{\parallel\perp}$	-		0.29	0.27

$E_{\parallel}$:   Modulus in fibre direction,
$E_{\perp}$:   Modulus perpendicular to the fibre direction,
$G_{\#}$:   Shear modulus of a unidirectional layer,
$v_{\parallel\perp}$:   Poisson's ratio for loading in fibre direction.

used. In contrast to the shear modulus $G_{\#}$ Young's moduli $E_{\parallel}$ and $E_{\perp}$ scarcely depend on the loading.

In figure 3 Young's moduli of a $\pm 45°$ laminate determined by both procedures are given, too. The following can be stated after comparing the results:

1. If the stiffness of the unidirectional laminate is determined by test, the predicted Young's modulus of a $\pm 45°$ GFRP-laminate correspond well with the test results.

2. If the stiffness of the unidirectional laminate is determined by the use of micromechanical relations, there will be large differences between theoretical results and stiffness tests. The effective stiffness at zero loading is much higher than theoretical predictions.

2.2 Dynamic tests

Dynamic tests are done to obtain the influence of the loading velocity on the stiffness. A specimen made of 7781 type fabric is loaded in tension by a computer controlled servohydraulic testing machine. In a first test the specimen is loaded up to a maximum force of 3000 N (60 N/mm^2) at different loading velocities. Figure 4 shows the force plotted versus the time during these tests. The loading velocity ranges from 600 N/s up to 150000 N/s.

Figure 5 gives all stress-strain curves of these dynamic tests. Again, the stress is plotted versus the mean value of the strain, measured by strain gauges on both sides of the specimen. It can be seen from the figure that there is obviously no significant difference between the stiffnesses of the $\pm 45°$ GFRP-laminates when loaded with different velocities.

As done in the static tests, the ascending slopes of the stress-strain curves are approximated by parabolic curves. Young's moduli are derived from these curves. They are given in figure 6.

A second dynamic test is performed: First, the specimen is loaded up to a stress of 15 N/mm^2. It is then loaded by 5 cycles with an amplitude of 15

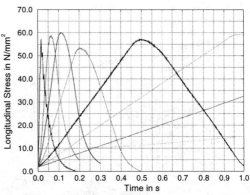

Figure 4: Dynamic loading of the tension specimen

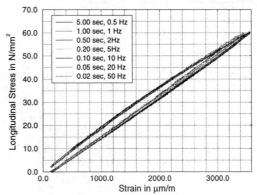

Figure 5: Stress versus strain for different loading velocities

N/mm^2. Then the load is increased to 25 and 35 N/mm^2. At both load steps the specimen is again dynamically loaded by 5 cycles with an amplitude of 25 and 35 N/mm^2, respectively. These loads result in a swelling loading with a minimum stress of about zero and a maximum stress of 30, 50, and 70 N/mm^2. Figure 7 shows this loading process.

Two ascendent load slopes are analysed from each load step. Young's modulus is determined by using the procedure described above. The resulting Young's moduli are shown figure 8. All curves correspond well to Young's moduli of the first dynamic test in figure 6. Young's modulus at the very low loading velocity (600 N/s or 0.2 Hz) shows a slightly higher decrease with strain than at higher velocities in both cases. This confirms the assumption that the stiffness increases with loading velocity.

As a result of the static and the dynamic tests, it can be concluded that the loading velocity has only a very small influence on the laminate stiffness. The results from static and dynamic tests correspond very well. But there is a large discrepancy between theoretical predictions and test results, if the basis for the predictions are common micromechanics. This is especially true for low loadings.

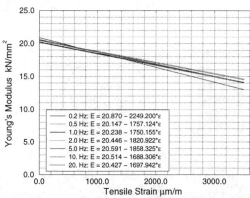

Figure 6: Young's modulus of a ±45° GFRP-laminate loaded at different loading velocities

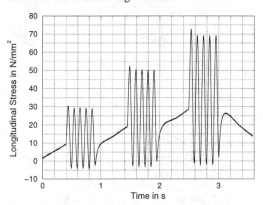

Figure 7: Stress-time function of second stiffness test

3 MICROMECHANICS AND CLASSICAL LAMINATE THEORY

The properties and the strength of laminates could be calculated by the use of the classical laminate theory, if the properties of the unidirectional layers were known (e.g. Tsai & Hahn, 1980). As already mentioned, these properties may be determined by means of tests or by micromechanics. Micromechanics are preferred since tests are expensive and take a long time, especially in the case of many different fibre volume fractions.

3.1 Classical laminate theory

Young modulus of a ±45° laminate can be derived from the classical laminate theory as follows:

$$E_{\pm 45°} = 4G_\# \left(1 - \frac{1}{1 + \frac{E_\| + E_\perp + 2E_\| v_{\perp\|}}{4G_\#(1 - v_{\|\perp}v_{\perp\|})}} \right) \quad (5)$$

According to this equation Young's modulus of a ±45° laminate mainly depends on the shear modulus $G_\#$ of the unidirectional layers the laminate is composed of. The factor in brackets depends on the fibre

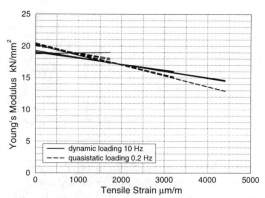

Figure 8: Young's modulus for the second dynamic test

material. For high modulus materials, such as high modulus carbon fibres, this factor is near unity, so that one can state: $E_{\pm 45°} \approx 4G_\#$. Whereas for glass fibre composites the factor has to be considered. Table 2 shows the factor for different fibre types.

Table 2: Young's modulus of ±45° laminates with different fibre materials

Fibre type	Young's modulus
E-Glass	$E_{\pm 45°} = 4G_\# \cdot 0.786$
S-Glass	$E_{\pm 45°} = 4G_\# \cdot 0.803$
T300	$E_{\pm 45°} = 4G_\# \cdot 0.888$
M40	$E_{\pm 45°} = 4G_\# \cdot 0.932$
GY70	$E_{\pm 45°} = 4G_\# \cdot 0.963$

3.2 Micromechanics

Many different micromechanical relations for the determination of the stiffness of a unidirectional layer perpendicular to the fibre direction and in shear direction are given in literature. The following equations show some relations for the shear modulus $G_\#$. In these equations, φ denotes the fibre volume fraction, G_M and G_F the shear modulus of the matrix and fibre, respectively.

(N.N. LTH, 1974):

$$G_\# = G_M \left[\left(1 - 2\sqrt{\frac{\varphi}{\pi}} \right) - \frac{\pi}{2\left(1 - \frac{G_M}{G_F}\right)} + \right.$$

$$\frac{2}{\left(1 - \frac{G_M}{G_F}\right)\sqrt{1 - 4\frac{\varphi}{\pi}\left(1 - \frac{G_M}{G_F}\right)^2}} \cdot$$

$$\left. \text{arctan} \sqrt{\frac{1 + 2\sqrt{\frac{\varphi}{\pi}\left(1 - \frac{G_M}{G_F}\right)}}{1 - 2\sqrt{\frac{\varphi}{\pi}\left(1 - \frac{G_M}{G_F}\right)}}} \right] \quad (6)$$

(Tsai & Hahn, 1980), (Moser 1992):

$$G_{\#} = \frac{\varphi + \frac{1}{2}\left(1 + \frac{G_M}{G_F}\right)(1 - \varphi)}{\left(\frac{\varphi}{G_F} + \frac{1}{2}\left(1 + \frac{G_M}{G_F}\right)\frac{1 - \varphi}{G_M}\right)} \qquad (7)$$

(Bergmann 1992):

$$G_{\#} = \frac{G_M G_F}{G_F(1 - \varphi) + G_M \varphi} \qquad (8)$$

(Puck 1967):

$$G_{\#} = G_M \frac{1 + 0.6\varphi^{0.5}}{(1 - \varphi)^{1.25} + \varphi \frac{G_M}{G_F}} \qquad (9)$$

(Ehrenstein 1981):

$$G_{\#} = G_M \frac{1 + 0.4\varphi^{0.5}}{(1 - \varphi)^{1.45} + \varphi \frac{G_M}{G_F}} \qquad (10)$$

The shear modulus can be determined with the material properties of fibre and matrix given in table 3. Both materials are treated as isotropic. The fibre volume fraction φ is 60%.

Figure 9 shows the theoretically predicted shear moduli as well as the shear moduli, determined by static tension tests at lower and higher shear stresses $\tau \, (= \sigma_{longitudinal}/2)$ of 5 N/mm^2 and 25 N/mm^2.

Table 3: Material properties of pure E-Glass fibres and epoxy resin

Property			fibre	matrix
Young's modulus	N/mm^2		73000	3500
Poisson's Ratio	-		0.28	0.35
Shear modulus	N/mm^2		28516	1296

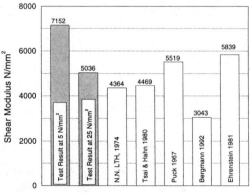

Figure 9: Shear modulus of a unidirectional layer, experimental and theoretical data

While the test result at the higher loading ($\tau = 25$ N/mm^2) and the theoretically predicted shear modulus correspond well to the formula by LTH, Tsai &

Hahn, Puck, and Ehrenstein, the test result at low loading is much higher than all theoretical values.

Accordingly, the stiffness behaviour of structures, dominated by $\pm 45°$ GFRP laminates can be predicted satisfactorily for higher loadings. However, at low load levels the real stiffness is much higher than theoretical predictions. Normal mode tests are usually done at low amplitudes and therefore at very low load levels. Therefore, the measured eigenfrequencies show a recognizably higher stiffness than static tests at higher loads. The tangent Young's modulus at nearly zero loading should be used for the theoretical analysis of normal mode tests.

As the reason for this is the incorrect prediction of the shear modulus of a unidirectional layer, the same applies to structures dominated by the shear behaviour e.g. bending beams composed of unidirectional GFRP-laminate loaded in torsion.

CONCLUSIONS

Young's modulus of $\pm 45°$ GFRP laminates depends on the loading. But there is nearly no dependence on the loading frequency in a range between zero and 20 Hz. The stress-strain curve can be approximated very well by a parabolic curve. Accordingly the stiffness decreases linearly with the loading.

Compared with theoretical predictions, got by a combination of micromechanics and the classical laminate theory, the real stiffness is remarkably higher. Especially near zero load conditions, which are often the case in experimental normal mode analysis, the stiffness is up to 50% higher than the theoretical predictions. The coincidence of test and theory is satisfactory at higher loads (> 60 N/mm^2 or > 3000 μm/m).

Therefore, the tangent modulus should be used for stiffness tests at low load levels (low stresses and strains, respectively). For higher loadings, the secant modulus should be taken into account.

Common micromechanical relations give a shear stiffness of the unidirectional laminates, which are mostly too low. Experimental results should be used instead.

To overcome these difficulties, simple micromechanical theories should be derived to describe the dependence of the matrix dominated properties (especially the shear modulus) on the loading correctly.

REFERENCES

Bergmann H.W.; *Konstruktionsgrundlagen für Faserverbundbauteile*, Springer-Verlag, Berlin, 1992,

Ehrenstein G.W.; *Glasfaserverstärkte Kunststoffe*, Expert Verlag, 1981

Moser K.; *Faserkunststoffverbund*; VDI Verlag, Düsseldorf, 1992

N.N.; *LTH - FVL; Faserverbund Leichtbau*; Industrie-Ausschuß Strukturberechnungsunterlagen (IASB), Germany, 1968 - 1997

Puck A.; *Zur Beanspruchung und Verformung von GFK-Mehrschichtverbunden-Bauelementen*, Kunstoffe, Vol. 57, Heft 4-7, 1967

Tsai S.W., Hahn H.T.; *Introduction to Composite Materials*, Technomic Publishing Co., Lancaster, Penn., 1980

Experimental Mechanics, Allison (ed.)© 1998 Balkema, Rotterdam, ISBN 90 5809 014 0

Mode II dynamic fracture toughness of unidirectional CFRP laminates

M. Arai, T. Adachi & H. Matsumoto
Tokyo Institute of Technology, Japan

T. Kobayashi
Graduate School of Tokyo Institute of Technology, Japan

ABSTRACT: In the present paper, Mode II dynamic interlaminer fracture toughnesses are evaluated for typical carbon fiber reinforced plastics (CFRP), those are CF/Epoxy, CF/Toughened Epoxy and CF/PEEK. In the experiments, split Hopkinson pressure bar method was applied to dynamic end notched flexure test. The mode II interlaminer fracture toughness of unidirectional CFRP was estimated by $\hat{J}$-integral from the measured impulsive load and reactions at the supported points by approximating ENF specimen with classical beam theory. The dependency of deflection rate on the dynamic interlaminar fracture toughness is discussed.

1 INTRODUCTION

Carbon fiber reinforced plastics (CFRP) have been known as advanced material for various engineering fields. CFRP composites should be carefully designed against transverse local loading such as low-velocity impact which would cause significant internal damage in the form of transverse cracks and delaminations (Morita *et al.* 1997). These examples are, for example, a tool dropping on a surface during maintenance,an airplane kicking up stones during take-off or landing, and a bird striking on an plane surface.

The interaction of matrix cracking and delamination plays an important role in damage initiation and development in composite structures subjected to transverse loading (Maikuma *et al.* 1989, 1990). In order to fundamentally understand the damage induced by transverse load in laminated composites, dynamic Mode II interlaminer fracture toughness should be determined precisely (Carlsson *et al.* 1986 ; Smiley *et al.* 1987 ; Mal *et al.* 1987 ; Kageyama *et al.* 1991).

The object of this investigation is to estimate them by split Hopkinson pressure bar (SHPB) method (Costin *et al.* 1977) with end notched flexure (ENF) test (Kurokawa *et al.* 1995). The dynamic behavior of the specimen is analyzed with classical beam theory based on Laplace transform.Fracture toughness G_c was estimated employing the scheme of J-integral with the measured impulsive load and reactions at the supported points. In the present study, three types of ENF spec-

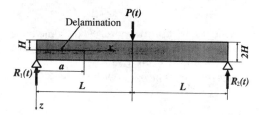

Figure 1. ENF Specimen

imens are used, that are made from CF-Epoxy, CF-Toughened Epoxy and CF-PEEK.

Through some experimental results, dynamic inter-laminar fracture toughness of unidirectional CFRP specimens is estimated and it is found that the fracture toughness indicates dependency on deflection rate.

2 THEORY

2.1 *Analysis of ENF specimen*

In the present study, dynamic Mode II fracture toughness of unidirectional CFRP laminates are estimated by ENF specimen with SHPB method. Classical beam theory is applied to analyze the dynamic deformation of ENF specimen as shown in figure 1. The ENF specimen is divided into three sections, where first section is a delamination part $(0 \leq x \leq a)$, and the others are no-delamination parts of $(a \leq x \leq L)$ and $(L \leq x \leq 2L)$ respectively.

Fundamental equations based on classical

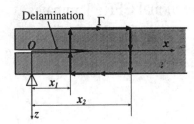

Figure 2. Contour of ENF specimen

beam theory for each sections are written as following forms.

$$D_i \frac{d^4 w_i}{dx^4} + 2\rho b H \frac{d^2 w_i}{dt^2} = 0 \quad (i = 1, 2, 3), \quad (1)$$

where w_i and D_i are deflection and bending stiffness for each sections. Bending stiffness D_i is given as

$$D_i = \begin{cases} bEH^3/6 & (i = 1, 0 \leq x \leq a) \\ 2bEH^3/3 & (i = 2, a \leq x \leq L) \\ 2bEH^3/3 & (i = 3, L \leq x \leq 2L). \end{cases} \quad (2)$$

ρ, b, E and t are density, width, Young's modulus and length of the beam. The definitions of length H, L are shown in figure 1.

Applying the Laplace transform to equation (1),

$$D_i \frac{d^4 \bar{w}_i}{dx^4} + 2\rho b H s^2 = 0, \quad (3)$$

where s is Laplace transform parameter, and notation '$-$' denotes quantities on the Laplace transformed domain. The general solution of the above expression is given as

$$\bar{w}_i = \sum_{j=1}^{4} C_{ij} \exp(\eta_j x), \quad (4)$$

where C_{ij} is unknown coefficient matrix, η_j are four independent roots of following equation.

$$(\eta^2 - \alpha)(\eta^2 + \alpha) = 0 \quad (5)$$

where α^4 is defined as $2\rho b H s^2 / D_i$.

Bending moments and reaction forces are prescribed on both supported points of the specimen as boundary conditions.

$$\left. \begin{array}{l} -D_1 \dfrac{d^2 \bar{w}_1}{dx^2} = 0 \\[2mm] -D_1 \dfrac{d^3 \bar{w}_1}{dx^3} = \bar{R}_1 \end{array} \right\} \quad \text{at} \quad x = 0, \quad (6)$$

$$\left. \begin{array}{l} -D_3 \dfrac{d^2 \bar{w}_3}{dx^2} = 0 \\[2mm] -D_3 \dfrac{d^3 \bar{w}_3}{dx^3} = -\bar{R}_2 \end{array} \right\} \quad \text{at} \quad x = 2L, \quad (7)$$

where $\bar{R}_1, \bar{R}_2$ denote the Laplace-transformed reaction forces.

The boundary condition at the crack tip ($x = a$) are written as

$$\left. \begin{array}{l} \bar{w}_1 = \bar{w}_2 \\[1mm] \dfrac{d\bar{w}_1}{dx} = \dfrac{d\bar{w}_2}{dx} \\[2mm] -D_1 \dfrac{d^2 \bar{w}_1}{dx^2} = -D_2 \dfrac{d^2 \bar{w}_2}{dx^2} \\[2mm] -D_1 \dfrac{d^3 \bar{w}_1}{dx^3} = -D_2 \dfrac{d^3 \bar{w}_2}{dx^3} \end{array} \right\} \quad \text{at} \quad x = a. \quad (8)$$

The boundary condition at the loading point are written as follows.

$$\left. \begin{array}{l} \bar{w}_2 = \bar{w}_3 \\[1mm] \dfrac{d\bar{w}_2}{dx} = \dfrac{d\bar{w}_3}{dx} \\[2mm] -D_2 \dfrac{d^2 \bar{w}_2}{dx^2} = -D_3 \dfrac{d^2 \bar{w}_3}{dx^2} \\[2mm] -D_2 \dfrac{d^3 \bar{w}_2}{dx^3} + D_3 \dfrac{d^3 \bar{w}_3}{dx^3} = \bar{P} \end{array} \right\} \quad \text{at} \quad x = L, \quad (9)$$

where $\bar{P}$ is Laplace-transformed impact load.

Substituting $\bar{R}_1, \bar{R}_2$ and $\bar{P}$ into equations (6), (8) and (9) respectively, unknown coefficients C_{ij} of equation (4), namely the deflections $\bar{w}_i$ can be determined on the Laplace transformed domain. Finally, employing inverse Laplace transform, deflection histories $w_i(t)$ are obtained.

The quantities obtained in the Laplace-transformed domain are numerically transformed into the time domain by the FFT algorithm according to Krings and Waller's method(1976).

$$f(t_k) = \frac{\exp(\gamma k \Delta t)}{T} \sum_{n=0}^{N_s - 1} \bar{f}(s_n) \exp \frac{j 2\pi n k}{N_s}, \quad (10)$$

where,

$$\Delta t = T/N_s, \quad \Delta \omega = 2\pi/T, \quad k = 0, 1, 2, \cdots, N_s - 1$$
$$\gamma = \alpha/T, \quad t_k = k\Delta t, \quad s_n = \gamma + jn\Delta\omega, \quad j = \sqrt{-1}$$

and T, N_s are the analyzed time range and the number of sample points respectively. α is specified as 6 in the present paper.

On the other hand, the impact load and reaction forces measured experimentally are numerically transformed with the following equation.

Table 1. Specification of strike, input and output bar

		Strike Bar	Input Bar	Output Bar
Length	[mm]	750	1500	1500
Diameter (Height × Width)	[mm]	10	10	3 × 1500
Material		Mild Steel(SS400)		
Young's Modulus	[GPa]	207		
Density	[kg/m³]	7.86 × 10³		

Table 2. Specification of CFRP specimens

			T300/#3631	T300/#2500	T800/#3900	APC2/AS4
Length	$2L_0$	[mm]	70			
Support Span	$2L_0$	[mm]	60			
Thickness	$2H$	[mm]	2.5	2.7	2.8	2.74
Delamination Length	a	[mm]	—	15		
Width	b	[mm]	10			
Young's Modulus	E	[GPa]	134	113	147	137
Density	ρ	[kg/m³]	1.6 × 10³	1.6 × 10³	1.56 × 10³	1.60 × 10³

$$\bar{f}(s_n) = \frac{T}{N} \sum_{k=0}^{N-1} f(t_k) \exp(-\gamma k \Delta t - \frac{j2\pi nk}{N}) \quad (11)$$

2.2 Evaluation of dynamic fracture toughness

In the present paper, $\hat{J}$-integral scheme is applied to evaluate the dynamic fracture toughness of ENF specimens. J-integral expression for the Mode II, along the path Γ shown in figure 2, is given as following form.

$$J = \int_\Gamma \left[\left\{ W - \left(\sigma_x \frac{du}{dx} + \tau_{xz} \frac{dw}{dx} \right) \right\} dz + \tau_{xz} \frac{dw}{dx} dx \right] (12)$$

where u and w denote the axial displacement and density of the strain energy. σ_x and τ_{xz} are axial stress and shear stress respectively. According to the classical beam theory, the above expression yields

$$J = \frac{EH^3}{12} \left\{ \left(\frac{d^2w}{dx^2} \right)^2 - \frac{d^3w}{dx^3} \frac{dw}{dx} \right\} \bigg|_{x_1}$$
$$- \frac{EH^3}{3} \left\{ \left(\frac{d^2w}{dx^2} \right)^2 - \frac{d^3w}{dx^3} \frac{dw}{dx} \right\} \bigg|_{x_2}, \quad (13)$$

namely the equation with only deflection w.

Taking account the inertia force of the area surrounded by Γ, $\hat{J}$-integral (Kishimoto et al. 1980) expression is given as

$$\hat{J} = J + \rho \int_{x_1}^{x_2} \frac{d^2w}{dt^2} \frac{dw}{dx} 2H dx, \quad (14)$$

where the integral terms of inertia force is integrated numerically in the present analysis.

In the static case, the deflection of the ENF specimen can be determined analytically. The deflections at x_1 and x_2 of the ENF specimen sup-

ported simply are written as follows.

$$w(x_1) = \frac{3P}{2bEH^3} \left(-\frac{x_1^3}{3} + \frac{L^3 + 3a^2L - a^3}{4L} x_1 \right) \quad (15)$$

$$w(x_2) = \frac{3P}{2bEH^3} \left(-\frac{x_2^3}{12} + \frac{L^3 - a^3}{4L} x_2 + \frac{a^3}{2} \right) \quad (16)$$

Substituting equations (15) and (16) into equation (13), the static J-integral representation is given as,

$$J = \frac{9P^2 a^2}{16Eb^2H^3}. \quad (17)$$

It is confirmed that equation (17) is equivalent to a equation derived from energy release rate presented by Russel(1985).

3 EXPERIMENTAL APPARATUS AND ENF SPECIMEN

Figure 3 shows experimental equipments of SHPB method. Strike bar, accelerated by pressure air, run into input bar contacting with the specimen. The impact load and reaction forces can be computed by one-dimensional wave propagation theory with strain histories on the input and output bars. The specification of the strike, input and output bars are shown in table 1.

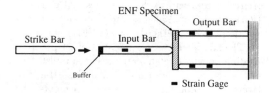

Figure 3. Schematic of dynamic ENF test equipment

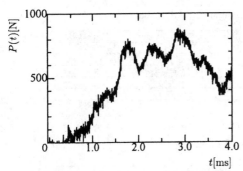

Figure 4. Impact load history with Pb buffer

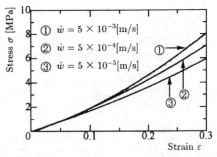

Figure 7. Stress-Strain curve of synthetic rubber

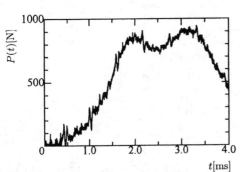

Figure 5. Impact load history with Teflon buffer

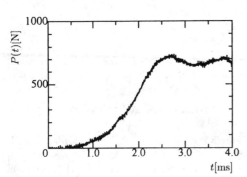

Figure 6. Impact load history with rubber buffer

In the present study, typical three types of carbon fiber reinforced plastics were investigated. T300/#2500 (Toray) has a epoxy matrix, the T800/#3900 (Toray) has a toughened epoxy matrix reinforced with Nylon particles. APC-2/AS4 (ICI) consists of AS4 graphite fiber and thermoplastic poly-ether-ether-ketone (PEEK) matrix.

The specimens were laminated of 24 prepregs unidirectionally. Delaminations at the midsurface of the specimen were embedded by inserting two layers of Kapton film. Prior to experiments, the delamination was slightly propagated with cutter knife to make its tip sharp.

4 EFFECT OF BUFFER FOR LOAD HISTORY

In the SHPB test, it is essential to remove the high-frequency undesirable wave which is included in the load history and obtain the smooth ramp-like load in order to reject the bounding of the specimen at the supported point.

In the present paper, three types of materials, lead, Teflon and synthetic rubber, are employed as the buffer. Impact load histories using them as the buffer are compared, where CF/Epoxy T300/#3631 beam with no-delamination are used for the study. The specimen size are shown in table 2. The thickness of buffers are equally specified to be 10mm approximately. In the case of lead only, small pieces of thickness from 1mm to 2mm were used stacking to total 10mm.

Under same launching pressure 0.28MPa, load histories obtained with each buffer are shown in figure 4, figure 5 and figure 6 respectively. Figure 4 shows that Pb-buffer does not give smooth load history. As shown in figure 5, although the high-frequency vibration is still found out, Teflon gives relatively smooth load history rather than Pb-buffer.

On the other hand, as shown in figure 6, the load history with synthetic rubber is very smooth with lamp-like shape, so synthetic rubber was employed as buffer for the following ENF test.

Stress-Strain curve of the present synthetic rubber is shown in figure 7. The compression test was executed with material testing machine (SHIMAZU:AG–100kNE), where the displacement rate are varied to three 5×10^{-5}, 5×10^{-4} and 5×10^{-3}m/s. Figure 7 shows that the synthetic rubber of the present study has a desirable characteristics since it has the strain rate dependency of the stress.

5 DYNAMIC ENF TESTING

Figure 8, 9 and 10 show the relation of Mode II dynamic fracture toughness and deflection rate $\dot{w}$

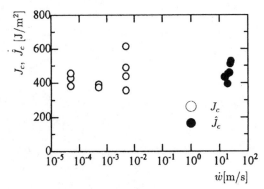

Figure 8. Relation between fracture toughness and deflection rate (T300/#2500)

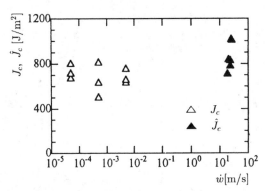

Figure 9. Relation between fracture toughness and deflection rate (T800/#3900)

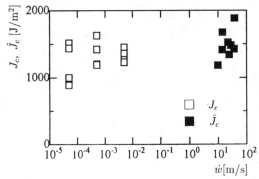

Figure 10. Relation between fracture toughness and deflection rate (APC-2/AS4)

of the loading point for T300/#2500, T800/#3900 and APC-2/AS4 respectively. In the present study, deflection rate $\dot{w}$ is employed for the index representing dynamic effect of the deformation.

Furthermore, experimental results of quasi-static ENF test are indicated also in these figure. The quasi-static test are executed with SHIMAZU : AG-100kNE employing the static

J-integral equation(17) since the effect of inertia is small sufficiently.

Generally, the fracture toughness has a tendency to increase with the deflection rate in the three specimens. It is observed that APC-2/AS4 gives the highest fracture toughness in the three, and T800/#3900 of toughened EPOXY has some advantage in the fracture toughness relative to T300/#2500 owing to mainly adding the Nylon particles in the interface.

6 CONCLUSION

In the present paper, dynamic Mode II interlaminar fracture toughnesses are evaluated for typical carbon fiber reinforced plastics (CFRP), those are CF/Epoxy, CF/Toughened Epoxy and CF/PEEK. Split Hopkinson pressure bar (SHPB) technique was applied to dynamic end notched flexure (ENF) test with $\hat{J}$-integral scheme approximating dynamic deformation of ENF specimen by classical beam theory. It was confirmed that Mode II dynamic fracture toughness of CFRP, of the three types of unidirectional beams, generally tend to increase with the deflection rate, and synthetic rubber is useful for the SHPB method because it gives smooth and lamp-like load history.

REFERENCES

Morita, H., Adachi, T., Tateishi, Y. and Matsumoto, H. 1997, *J. Reinforced Plastics and Composities*, Vol.16, No.2 : 376–385.

Maikuma, H., Gillespie, J. W. & Wilkins, D. J. 1989, *J. of Composite Materials*, Vol.23 : 124–149.

Maikuma, H., Gillespie, J. W. & Wilkins, D. J. 1990, *J. of Composite Materials*, Vol.24 : 124–149.

Carlsson, L. A., Gillepie, J. W. & Pipes, R. B. 1986, *Journal of Composite Materials*, Vol.20 : 594–604.

Smiley, A. J. & Pipes, R. B. 1987, *Composite Science and Technology*, Vol.29, : 1–15.

Mall, S., Law, G.E. & Katouzian, M. 1987, *J. of Composite Materials*, Vol.21, 1987 : 569–579.

Kageyama, K., Yanagisawa, N. & Kikuchi, M. 1991, *Composite Material: Fatigue and Fracture* (Third Volume), ASTM STP 1110 : 210.

Costin, L. S., Duffy, J. & Freund, L. B. 1977, ASTM STP 627 : 301–310

Kurokawa, T., Kusuda, T., Shimazaki, T. and Yamauchi, Y., 1995, *Constitutive Relation in High/Very High Strain Rates*, Proc. of IUTAM symposium Noda/Japan : 217–224.

Krings, W. & Waller, H. 1979, *Int. J. Numerical Methods in Engineering*, Vol.14 : 1183–1196.

Russel, A. J. & Street, K. N. 1985, ASTM STP 876 : 349–370.

Kishimoto, K., Aoki, K. & Sakata, M. 1980, *Eng. Fracture Mechanics*, Vol.13 : 387–394.

Experimental Mechanics, Allison (ed.) © 1998 Balkema, Rotterdam, ISBN 90 5809 014 0

A T-shaped specimen for the direct identification of the in-plane moduli of orthotropic composites

F. Pierron & Y. Surrel
Department of Mechanical and Materials Engineering, Ecole des Mines de Saint-Etienne, France

M. Grédiac
LERMES, Université Blaise Pascal Clermont-Ferrand II, France

ABSTRACT: This paper deals with the direct experimental identification of the four in-plane stiffness components of an orthotropic composite. First, the identification technique is briefly reviewed. Then, the experimental implementation is described and especially the whole-field optical grid method used for the measurement of the two components of the displacement field. Finally, the results are given for a woven glass epoxy composite and the identified values compare rather well with the expected ones determined by classical mechanical tests

1 INTRODUCTION

The identification of the elastic moduli of composites using classical mechanical testing requires the use of several specimens, which makes it expensive and time consuming. Moreover, these tests require the stress state within the tested section to be uniform for correct data interpretation. This is not always easy to achieve, particularly in shear and parasitic effects due to the material anisotropy can significantly bias the measurements.

Alternative methods have been proposed to identify the whole set of moduli from a single specimen that rely mostly on the measurement of the eigenfrequencies of free vibrating plates and on an iterative numerical identification process (Mota Soares et al., 1993, Fällström and Jonsson, 1991, Frederiksen, 1995). The main drawbacks of these techniques is that time-consuming iterative finite element computing is needed and that the convergence of the process is strongly dependent on an 'initial guess' of the moduli.

Recently, a novel direct identification approach has been proposed, first on static bending tests and more recently on free vibrating plates (Grédiac and Vautrin, 1990, Grédiac and Vautrin, 1993, Grédiac and Paris, 1996, Grédiac et al., 1998). The main feature of this approach is that the identification process is direct and does not require any iterations nor any finite element computing. However, extra experimental effort is needed since whole-field slope measurements are needed. These are performed by a laser deflectometry technique.

This paper presents the experimental implementation of an extension of this last method (Grédiac and Pierron, 1998) to the direct measurement of the four in-plane moduli of a woven glass fibre epoxy resin composite using a single T-shaped specimen and an optical whole-field in-plane displacement measurement technique based on the deformation of a bi-directional grid.

2 PRESENTATION OF THE IDENTIFICATION METHOD

The method relies on the global equilibrium of the structure. This equilibrium can be written using the Principle of Virtual Work (PVW):

$$\int_V \sigma_{ij}\epsilon_{ij}^* dV = \int_{\partial V} T_i u_i^* dS \tag{1}$$

where V is the volume of the system considered, σ the stress tensor, ϵ^* the virtual strain field, **T** the external loads, u^* the virtual displacement field associated to ϵ^* and ∂V is the boundary of the system. The convention of repeated indices for summation is adopted. This formula is valid for any set of virtual displacement/strain fields provided that they respect the boundary conditions. Therefore, by *chosing* specific virtual fields, equations connecting the differents stiffness components to the external loads and the actual strains will be derived. Since four stiffness components are to be identified, at least four independent virtual displacement fields are to be chosen.

The structure that is used for this work is a T-shaped specimen on which a bending-tension load is applied,

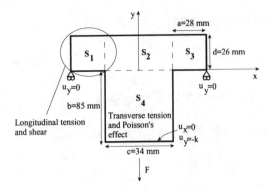

Figure 1. T-shaped specimen geometry

see Figure 1. The idea is that for areas S_1 and S_3, the in-plane shear and the longitudianl tension will be 'activated' and that in area S_4, transverse tension will be predominant. Therefore, by chosing the virtual displacement fields accordingly, the following set of four equations is derived:

$$
\mathbf{A} = \begin{bmatrix}
0 & \int_{S_4} \epsilon_y dS \\
0 & 0 \\
\int_{S_1} 2y\epsilon_x dS & 0 \\
\int_{S_4} y(y+b)\epsilon_x dS & 0
\end{bmatrix}
$$

$$
\begin{matrix}
\int_{S_4} \epsilon_x dS & 0 \\
0 & \int_{S_1} \epsilon_s dS \\
\int_{S_1} 2y\epsilon_y dS & 0 \\
\int_{S_4} y(y+b)\epsilon_y dS & \int_{S_4} x(2y+b)\epsilon_s dS
\end{matrix}
$$

$$
[\mathbf{A}] \begin{Bmatrix} Q_{xx} \\ Q_{yy} \\ Q_{xy} \\ Q_{ss} \end{Bmatrix} = \begin{Bmatrix} \dfrac{Fb}{e} \\ \dfrac{Fa}{2e} \\ \dfrac{Fa^2}{2e} \\ 0 \end{Bmatrix} \qquad (2)
$$

where a and b are geometrical parameters, e is the thickness, F the global applied load and Q_{ij} the stiffness tensor components in axes (x,y), see Figure 1. All the details about the method will be found in (Grédiac and Pierron, 1998).

The above system relates the stiffness components to be identified to a number of integrals involving the strain components and some geometrical parameters and the global applied load. Therefore, if the strain field can be measured over the whole specimen, then it will be possible to identify directly the four stiffness components. The method has been validated numerically (Grédiac and Pierron, 1998). The experimental implementation is described in the rest of the paper.

3 EXPERIMENTAL SET-UP

A T-shaped specimen has been cut using a diamond coated blade in a 2 mm thick woven glass epoxy panel. The fibres are aligned with the x and y axes of Figure 1. The panel was autoclaved from a 1037/XE85AI Hexcel prepreg. The material moduli have been measured by classical mechanical testing (Cerisier, 1998). These values will serve as a reference for the experimental identification. The dimensions of the 'T' are that of Figure 1. The dimensions have been chosen so that all the strain components to be measured in the relevant areas are of the same order of magnitude (Grédiac and Pierron, 1998) and that they are compatible with the experimental set-up.

The fixture is composed of a classical grip at the bottom of the machine and of a bending fixture at the top. The latter is such that the loading pins are fixed in rotation.

The measurement of the displacement field is made with a phase-stepped grid method. Two tranferable grids Mécanorma Normatex 1321 are bonded orthogonally onto the specimen with Vishay VPAC-1 glue. The grid lines act as a spatial carrier whose phase will be modulated by the x component of the displacement, following: $u_i = p \times \varphi/2\pi$, where p is the grid pitch, φ is the phase modulation and $i = x$ or y depending on whether the lines are vertical or horizontal. Phase stepping is realized by the lateral displacement of the CCD camera, which is placed on a computer-controlled Microcontrole UT100 translation stage. The two components of the displacement can be separated by an appropriate signal processing. To get the u_x (resp. u_y) displacement, the horizontal (resp. vertical) lines have to be removed from the grabbed images. This is done by averaging vertically (resp. horizontally) the grey levels over a distance corresponding to the grid pitch. This corresponds to a low-pass filtering of the vertical (resp. horizontal) spatial frequencies present in the images (Surrel and Zhao, 1994).

One of the major problem in the phase stepping detection is that the grid used has a binary black-and-white profile, instead of the sinusoidal profile of interference fringes for which the phase detection algorithms where first designed. So, special algorithms must be used which are insensitive to the extra harmonic content of the periodic intensity signal. In our experiments, we used the 'windowed discrete

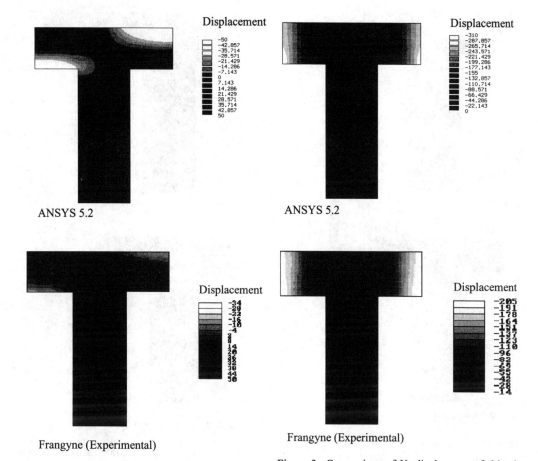

Displacement

| -50 |
| -42.857 |
| -35.714 |
| -28.571 |
| -21.429 |
| -14.286 |
| -7.143 |
| 0 |
| 7.143 |
| 14.286 |
| 21.429 |
| 28.571 |
| 35.714 |
| 42.857 |
| 50 |

ANSYS 5.2

Displacement

| -310 |
| -287.857 |
| -265.714 |
| -243.571 |
| -221.429 |
| -199.286 |
| -177.143 |
| -155 |
| -132.857 |
| -110.714 |
| -88.571 |
| -66.429 |
| -44.286 |
| -22.143 |
| 0 |

ANSYS 5.2

Displacement

Displacement

Frangyne (Experimental)

Frangyne (Experimental)

Figure 2. Comparison of U_x displacement fields obtained experimentally and by finite element

Figure 3. Comparison of U_y displacement fields obtained experimentally and by finite element

Fourier transform' (WDFT) algorithm (Surrel, 1996, Surrel, 1997) with 16 displacements of $p/8$ of the camera.

4 RESULTS

The specimen was placed in the fixture and a 1000 N load was applied to it. This value was chosen so that the shear stresses in the specimen remained in the linear behaviour zone. Images of the deformed grid were taken and the displacement field computed from it using the Frangyne software (Surrel and Zhao, 1994). These displacement fields are represented in Figures 2 and 3, together with the figures obtained from the finite element computations. The finite element model was developed with the ANSYS 5.2 finite element package using 2D plane42 plane stress elements (Grédiac and Pierron, 1998). As can be seen on these figures, the patterns of the experimental dis-

placements are very much like the finite element ones. However, the scales are differents because some experimental displacements are caused by the clearances in the fixture, some fixture deformation and also some non-linear material behaviour. Also, it can be seen on Figure 2 that horizontal u_x isodisplacement lines are apparent in the vertical bar of the 'T', which implies that some parasitic clockwise in-plane bending occurs due to non-simultaneous contact of the loading pins. This also explains why the positive displacements are of higher magnitude than the negative ones in the horizontal bar of the 'T'. However, considering the very small displacement values, the results are thought to be very satisfactory.

From the two measured in-plane displacement components, the three linearized strain components

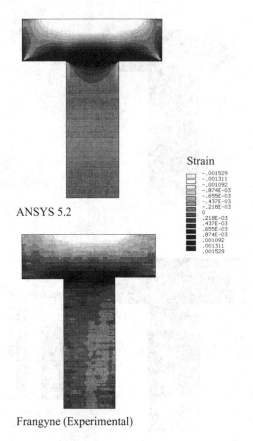

Figure 4. Comparison of ϵ_x strain fields obtained experimentally and by finite element

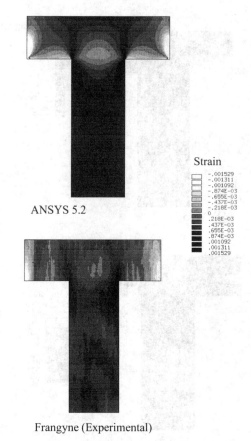

Figure 5. Comparison of ϵ_y strain fields obtained experimentally and by finite element

are derived by:

$$
\begin{cases}
\epsilon_x = \dfrac{\partial u_x}{\partial x} \\[2mm]
\epsilon_y = \dfrac{\partial u_y}{\partial y} \\[2mm]
\epsilon_s = \dfrac{\partial u_x}{\partial y} + \dfrac{\partial u_y}{\partial x}
\end{cases}
\tag{3}
$$

This is performed through numerical derivation of the experimental displacements. The procedure is that derivatives of the displacements are calculated over a certain number of pixels, which represents a spatial smoothing of the displacement fields. The number of pixels chosen is the minimum to obtain reasonable strain fields. Indeed, the smoothing results in a loss of spatial resolution. In our case, the derivation has been performed over 11 pixels. The results in terms of strains are represented in Figures 4, 5 and 6, together with the finite element results. As can be seen on these figures, the strain fields are much more prone

to experimental noise than the displacement values, which was expected. Also, as can be seen in Figure 4 for instance, the maximum strain values are lower experimentally than numerically. This can be caused by some material non-linearity but it is thought to be mainly the consequence of the loss of spatial resolution due to the derivation process. This effect is apparent on the three strain fields. It can be noted that the specimen geometry leads to equivalent strain component magnitudes (see the figures scales), as expected from (Grédiac and Pierron, 1998). Considering the small strain values (0.15%) and the fact that these measurements are the first attempt in the laboratory to measure the complete strain state over such a large specimen with the bi-directional grid method, the results were thought to be quite satisfactory.

From the results presented above, the strain fields are processed to calculate the integral values of equation 2. In order to do so, the integrals are approximated by a rectangle fit method (Grédiac and Pierron, 1998). However, it was found

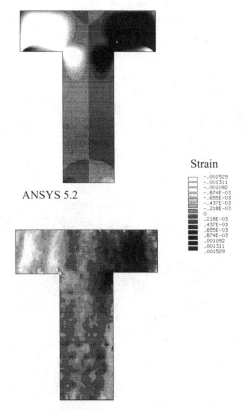

ANSYS 5.2

Strain

☐	-.001529
☐	-.001311
☐	-.001092
☐	-.874E-03
☐	-.655E-03
▨	-.437E-03
▨	-.218E-03
▨	0
▨	.218E-03
▨	.437E-03
▨	.655E-03
▨	.874E-03
▨	.001092
▨	.001311
■	.001529

Frangyne (Experimental)

Figure 6. Comparison of ϵ_s strain fields obtained experimentally and by finite element

Table 1. Identified stiffnesses and comparison with reference values

	Q_{xx}	Q_{yy}	Q_{xy}	Q_{ss}
Identified value (GPa)	35	35	24	5.2
Reference value (GPa)	27	27	8.2	4.8
Difference (%)	25	25	65	8

that the fourth line in the linear system of equation 2 was not well respected. Indeed, because of the quadratic weight function, this equation is very sensitive to measurement errors, as already mentioned in (Grédiac and Pierron, 1998). Therefore, it has been necessary to reduce the system of equation 2 by removing the fourth line and by assuming that $Q_{xx} = Q_{yy}$, since the reinforcement is a balanced glass fibre cloth. Using this assumption, the four stiffness components have been identified and the results are reported in Table 1. As expected, the identified values are all higher than the reference ones since the strain gradients are underestimated by the numerical deriva-

tion process. Also, though the strain fields seem to be better for ϵ_x and ϵ_y than for ϵ_s, the identified value of Q_{ss} is much closer to the reference than Q_{xx} and Q_{yy}. This is again due to the linear weighing functions in the third line of system 2. For the shear modulus, the equations are uncoupled and only involves a mean value of the shear strain over the S_1 and S_3 areas. Therefore, the critical parameters seem to be the errors on the location of the pixels more than the errors on the strains themselves. Finally, the large error on Poisson's component Q_{xy} is not surprising since this value is much smaller than the others and is well-known to be a difficult parameter to identify.

5 CONCLUSION

The present work is a first attempt to implement a direct identification method in order to measure the four independent moduli of an orthotropic material from a single T-shaped specimen, using a whole-field optical displacement measurement technique. The displacement fields obtained are quite satisfactory. However, the numerical derivation to get the in-plane strains involves a smoothing that tends to reduce the strain gradients. Also, the strain maps are much more perturbated than the displacement maps. Therefore, the identification process leads to an overestimation of the stiffnesses of 25% for the longitudinal stiffnesses but of only 8% on the shear modulus, Poisson's term being much more difficult to obtain accurately. One way to improve the above would be to have a direct measurement of the strains. Such a method is under development at Ecole des Mines de Saint-Etienne. Another problem that was found was that the quadratic weighing function equation could not provide reliable results because it was much more sensitive to measurement errors. The authors have devised a new virtual field that would lead to an additional linear weighing function equation that should give better results. For the future, an extension of the method to the identification of non-linear laws of behaviour is also considered.

REFERENCES

Cerisier, F. (1998). Conception d'une structure travaillante en composite et étude de ses liaisons. Phd thesis, Université Jean Monnet de Saint-Etienne, France. to be presented in March 1998.

Fällström, K. and Jonsson, M. (1991). A non-destructive method to determine material properties in anisotropic plates. Polymer Composites, 12(5):293–305.

Frederiksen, P. (1995). Identification of elastic constants including transverse shear moduli of thick orthotropic plates. Technical report, Danish Centre for Applied Mathematics and Mechanics. Report number 500.

Grédiac, M., Fournier, N., Paris, P.-A., and Surrel, Y. (1998). Direct identification of elastic constants of anisotropic plates from mode shape and natural frequency measurements. In ICEM XI, 11th International Conference on Experimental Mechanics, 24–28 August 1998, Oxford. Accepted.

Grédiac, M. and Paris, P.-A. (1996). Direct identification of elastic constants of anisotropic plates by modal analysis: theoretical and numerical aspects. Journal of Sound and Vibrations, 195(3):401–415.

Grédiac, M. and Pierron, F. (1998). A T-shaped specimen for the direct characterization of orthotropic materials. International Journal for Numerical Methods in Engineering. Accepted.

Grédiac, M. and Vautrin, A. (1990). A new method for determination of bending rigidities of thin anisotropic plates. Journal of Applied Mechanics, 57:964–968. Transactions of the American Society of Mechanical Engineers.

Grédiac, M. and Vautrin, A. (1993). Mechanical characterization of anisotropic plates : experiments and results. European Journal of Mechanics A/Solids, 12(6):819–838.

Mota Soares, C., Moreira de Freitas, M., Araujo, A., and Pedersen, P. (1993). Identification of material properties of composite plate specimen. Composite Structures, 25:277–285.

Surrel, Y. (1996). Design of algorithms for phase measurements by the use of phase-stepping. Applied Optics, 35(1):51–60.

Surrel, Y. (1997). Additive noise effect in digital phase detection. Applied Optics, 36(1):271–276.

Surrel, Y. and Zhao, B. (1994). Simultaneous u-v displacement field measurement with a phase-shifting grid method. In Pryputniewicz, R. J. and Stupnicki, J., editors, Interferometry '94: Photomechanics, volume SPIE 2342, pages 66–75.

Experimental Mechanics, Allison (ed.)© 1998 Balkema, Rotterdam, ISBN 90 5809 014 0

Anisotropic thermoviscoelastic constitutive equation of FRP

M. Shimizu
Graduate School of Tokyo Institute of Technology, Japan

T. Adachi, M. Arai & H. Matsumoto
Tokyo Institute of Technology, Japan

ABSTRACT In the present study, anisotropic thermoviscoelastic constitutive equation of fiber reinforced plastic (FRP) was identified from complex moduli measured experimentally. Complex moduli of laminae at $0°$ and $90°$ fiber angle were measured by dynamic viscoelastometer, where the angle was defined for carbon fiber direction, and the specimens were made of unidirectional carbon fiber/epoxy prepreg. The modulus were determined on the assumption that fiber was isotropic elastic and matrix resin was isotropic thermoviscoelastic. The elastic modulus of fiber and the thermoviscoelastic modulus of matrix resin were calculated from the experiments according to the law of mixture. Complex moduli of a CFRP lamina and natural frequency of a CFRP cantilever were predicted to confirm the constitutive equation. As a result, the predicted results agree with experimental ones. So it is shown that the constitutive equation is valid.

1. INTRODUCTION

Fiber reinforced Plastics (FRP) have been applied widely in industrial and engineering fields. FRP structures in aerospace are used under thermal environment from 93K to 573K. For these structures, foreign object damage (FOD) (Grszczuk 1975) is one of the most important problems. The authors (Shimizu et al. 1997) have obtained that impact damages in CFRP are influenced by thermal environments and the impact damages are related to thermoviscoelastic behavior of CFRP. Thus the thermoviscoelasticity of CFRP is important for mechanical design of aerospace structures and similar to predict stress and damage in CFRP under thermal environment. The relation between mechanical behavior of structure and thermo-viscoelastic characteristics of CFRP have been studied by some researchers. Sogabe (Sogabe et al. 1991) discussed temperature dependence of viscoelastic property subjected to impact load. Kasamori (Kasamori et al. 1992) researched the temperature and time dependence of CFRP Laminates from point of view of the mechanical behavior of the matrix resin, and Takada (Takada et al. 1992) studied time dependence of CFRP under isothermal condition.

The thermoviscoelastic constitutive equation for FRP is required to analyze stress field under thermal environment. Schapery (Schapery 1967) suggested the thermoviscoelastic constitutive equation of FRP, which were reinforced with short fiber. His constitutive equation was used several researches. And the equation for long fiber reinforced plastic was studied. Lou and Schapery (Lou & Schapery 1971) considered the constitutive equation to creep problem without temperature effect. Tuttle and Brinson (Tullte & Brinson 1986), Zhang and Xiang (Zhang & Xiang 1992) researched creep behavior, and Lin and Hwang (Lin & Hwang 1988) studied the numerical procedure based on Schapery's equation. Flaggs and Crossman (Flaggs & Crossman 1981) applied Schapery's equation to hygrothermal problems. In the identification of constitutive equation, complex moduli in constitutive equation is difficult and complicate generally, because the material is anisotropic and dependent on time and temperature. Then it is expected that the method of determination of complex moduli is simple.

In the present paper, the anisotropic thermo-viscoelastic constitutive equation for FRP laminated by the unidirectional prepreg is identified from complex moduli determined experimentally. The moduli in Schapery's type equation is determined on the assumption that the carbon fiber is isotropic elastic and the matrix resin is isotropic and linear thermoviscoelastic according to the law of mixture. Finally the validity of the equation is confirmed from the comparison with the experimental results about the lamina and the laminated for the CFRP.

2. MEASUREMENT OF THERMVISCOELASTIC BEHAVIOR

2.1 *Experimental Procedure*

Dynamic viscoelastic behavior of CFRP was measured under various thermal environments to obtain transversely isotropic thermoviscoelasticity. The CFRP specimens were laminated by unidirectional prepreg of carbon fiber/epoxy resin (Toray, T300/#2500). These specifications are denoted in Table 1. Dynamic viscoelastometer (Orientec, Rheovibron, DDV-III-EA) was used in the experiment, as shown in Figure.1. When the compulsory periodic tensile displacement with 3.5, 11, 35 and 110Hz was applied to the specimens at every 4K from -123K to 523K, the load was measured to determine the complex moduli.

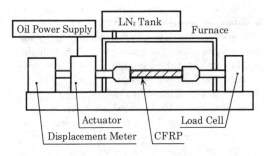

Figure 1. Equipment for measurement of dynamic viscoelastic behavior

Table 1. CFRP specimens for dynamic viscoelastic measurement

	Length [mm]	Thickness [mm]	Width [mm]
CF/ Epoxy $[0^{\circ}_{16}]$	76	2.5	1.0
CF/ Epoxy $[90^{\circ}_{16}]$	76	2.5	3.0

2.2 *Experimental Results*

The master curve of the complex moduli was made from experimental data under each temperature and frequency according to time-temperature superposition principle for linear thermoviscoelastic material. The master curves of the dynamic modulus E' and loss modulus E'' for the specimens are shown in Figure 2. The subscripts L and T of the moduli in Figure 2 denote longitudinal and transverse direction to fiber direction. The loss modulus E''_L is eliminated in Figure 2, since the modulus is extremely small compared with E'_L.

E'_T and E''_T have typically thermoviscoelastic properties with primary glass transition at 10^{-8}Hz and secondary glass transition at 10^5Hz under standard temperature 293K, though E'_L is

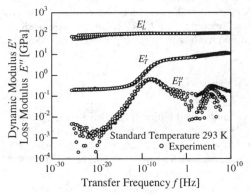

Figure 2. Master curves of dynamic and loss modulus

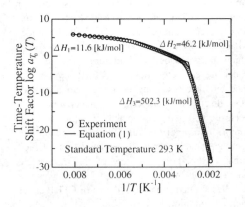

Figure 3. Time-temperature shift factors

approximately elastic independent of temperature. It is obtained the CFRP measured in the experiment is thermorheological simple because the complex moduli can be superposed by only temperature shift. The shift factor a_{T_0} for the master curves are shown in Figure 3. The shift factor a_{T_0} can be approximated by Arrhenius' equation

$$\log a_{T_0}(T) = \frac{\Delta H}{2.303R}\left(\frac{1}{T} - \frac{1}{T_0}\right) \qquad (1)$$

where ΔH, R, T and T_0 are activation energy, gas constant, temperature and standard temperature, respectively.

3. INDENTIFICATION OF CONSTITUTIVE EQUATION

The linear constitutive equation for anisotropic thermoviscoelastic material under isothermal condition is generally given as

$$\sigma_{ij}(t) = C_{ijkl}(0, a_{T_0})\varepsilon_{kl}(t) + \int_0^t C_{ijkl}(t - \tau, a_{T_0})\dot{\varepsilon}_{kl}(\tau)d\tau \qquad (2)$$

where $\sigma_{ij}(t)$, $\varepsilon_{kl}(t)$ and $C_{ijkl}(t, a_{T_0})$ denote components of stress, strain and complex moduli, respectively and t is time.

The Fourier Transform of Equation (2) is

$$\overline{\sigma}_{ij}(f) = C_{ijkl}(0, a_{T_0})\overline{\varepsilon}_{kl}(f) + \overline{C}_{ijkl}(f, a_{T_0})\overline{\varepsilon}_{kl}(f) \qquad (3)$$

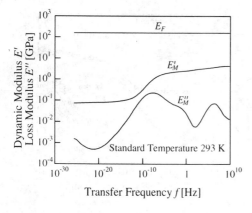

Figure 4. Determined Complex moduli of carbon fiber and matrix resin

where Fourier transformed quantity is shown as " ¯ ". And f is frequency.

A lamina of CFRP is transversely isotropic. Then according to correspondence principle (Schapery 1967) between elastic and viscoelastic materials in Fourier transformed domain, each transformed complex moduli $\overline{C}_{ijkl}(f)$ can be written as the following:

$$\overline{C}_{1111}(f) = \frac{1 - \overline{\nu}_{TT}(f)\nu_{TT}(f)}{\Delta}\overline{E}_L(f)$$

$$\overline{C}_{2222}(f) = \frac{1 - \overline{\nu}_{TL}(f)\overline{\nu}_{LT}(f)}{\Delta}\overline{E}_T(f)$$

$$\overline{C}_{1313}(f) = \overline{G}_{LT}(f)$$

$$\overline{C}_{1122}(f) = \frac{\overline{\nu}_{LT}(f) + \overline{\nu}_{TT}(f)\overline{\nu}_{LT}(f)}{\Delta}\overline{E}_T(f)$$

$$= \frac{\overline{\nu}_{TL}(f) + \overline{\nu}_{TL}(f)\overline{\nu}_{TT}(f)}{\Delta}\overline{E}_L(f) \qquad (4)$$

$$\overline{C}_{2233}(f) = \frac{\overline{\nu}_{TT}(f) + \overline{\nu}_{TL}(f)\overline{\nu}_{LT}(f)}{\Delta}\overline{E}_T(f)$$

and other $\overline{C}_{ijkl}(f) = 0$,

where

$$\Delta = 1 - 2\overline{\nu}_{LT}(f)\overline{\nu}_{TL}(f) - \overline{\nu}_{TT}(f)\overline{\nu}_{TT}(f)$$
$$- 2\overline{\nu}_{LT}(f)\overline{\nu}_{TL}(f)\overline{\nu}_{TT}(f)$$

And $\overline{E}_L(f)$, $\overline{E}_T(f)$, $\overline{G}_{LT}(f)$, $\overline{\nu}_{LT}(f)$, $\overline{\nu}_{TT}(f)$ are correspondent to engineering elastic coefficients.

From the experimental results in Figure 2, the thermoviscoelastic behavior of CFRP governs by the behavior of the matrix resin. Then it is assumed that the carbon fiber is isotropic elastic and matrix epoxy resin is isotropic thermoviscoelastic. The complex moduli of CFRP in Equation (4) can be computed according to the law of mixture. The engineering complex moduli in Equation (4) are yielded as follows:

$$\overline{E}_L(f) = E_F V_F + \overline{E}_M(f)(1 - V_F)$$

$$\frac{1}{\overline{E}_T(f)} = \frac{V_F}{E_F(f)} + \frac{(1 - V_F)}{\overline{E}_M(f)} \qquad (5)$$

where V_F is the volume fraction of carbon fiber

(63% for the specimen in the present paper). And the subscripts F and M denote the moduli for the fiber and the matrix resin.

$\overline{E}_M(f)$ is evaluated from Equation (5) with experimental results in Figure 2 and is approximated by Schapery's equation (Schapery 1967).

$$\overline{E}_M(f) = E'_{M0} - \sum_{n=1}^{N} E'_{Mn} \exp(-f / \rho_{Mn})$$
$$+ i \left\{ E''_{M0} - \sum_{n=1}^{N} E''_{Mn} \exp(-f / \rho_{Mn}) \right\} \qquad (6)$$

where E'_{Mn}, E''_{Mn} and ρ_{Mn} are constant values (denoted in appendix A). E_F is determined from elastic moduli E_L and E_T at room temperature. The results are denoted in Figure 4.

Other moduli can be evaluated from E_F, $\overline{E}_M(f)$ and the following equation

$$\frac{1}{\overline{G}_{LT}(f)} = \frac{V_F}{G_F(f)} + \frac{(1-V_F)}{\overline{G}_M(f)} \qquad (7)$$

where

$$G_F = \frac{E_F}{2(1+v_F)}$$

$$\overline{G}_M(f) = \frac{\overline{E}_M(f)}{2(1+v_M)} \qquad (8)$$

$$v_{LT} = v_F V_F + v_M (1 - V_F)$$

In Equation (8), Poisson's ratios are independent of temperature and frequency, these are

$$v_{LT} = 0.4, \quad v_{TT} = 0.5, \quad v_M = 0.3. \qquad (9)$$

The identified complex moduli are shown in Figure 5 as the results.

4. DISCUSSION

In order to confirmed the identified constitutive equation, the two kind of the experiments were carried out.

4.1 Complex Moduli of a Lamina at 45° fiber angle

The complex moduli of the lamina at 45° fiber angle were measured to compare with the moduli predicted by the identified constitutive equations. The specimen was a lamina stacked by the same prepreg as $[45°_{16}]$. Its thickness, width and length are 2.5mm, 1.0mm and 76mm, respectively. The

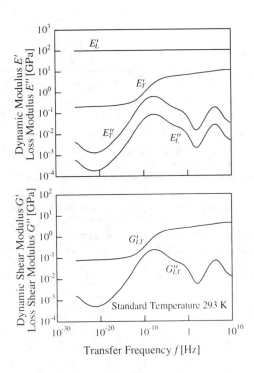

Figure 5. Identified complex moduli in equation (5) and (7)

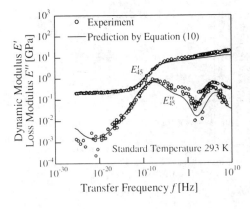

Figure 6. Complex moduli at 45° fiber angle of CFRP lamina

experimental procedure is the same as one in chapter 2 to determine thermoviscoelastic property.

The predicted complex moduli can be estimated by

$$\frac{4}{\overline{E}_{45}(f)} = \frac{1}{\overline{G}_{LT}(f)} + \frac{1}{\overline{E}_L(f)} + \frac{1}{\overline{E}_T(f)} - \frac{2\nu_{LT}}{\overline{E}_L(f)} \quad (10)$$

The comparison of the experimental and predicted results are shown in Figure 6. The predicted moduli agree with the experimental ones, especially well at low frequency.

4.2 Natural Frequency of a Laminated Beam

The natural frequencies for a laminated cantilever which were predicted from the identified constitutive equation, were compared with the experimental results. The stacking sequence of the specimen is $[90^{\circ}_2/0^{\circ}_2]_{SYM}$. The thickness, width and span length of it are 1.25mm, 20mm and 160mm, respectively. The specimen in a furnace was vibrated naturally with 10mm initial deflection. The strain history of the specimens was measured by a strain gage at 80mm away from the clamped end. The time range of the strain history was 8 sec. The temperature conditions of the furnace were selected as 293K, 343K, 393K and 423K. The first natural frequencies were estimated from Fourier transform of the measured strain history.

On the other hand, the natural frequencies were analyzed by the identified constitutive equation and the classical laminated beam theory(Jones 1975). The fundamental equation is given as follows:

$$\int_0^t \left\{ D(+0, a_{T_0}) + \frac{dD(t-\tau, a_{T_0})}{d(t-\tau)} \right\} \frac{\partial^4 w(x,t)}{\partial x^4} d\tau$$
$$+ \rho A \frac{\partial^2 w(x,t)}{\partial t^2} = 0 \quad (11)$$

where $D(t, a_{T_0})$, ρ, A and x are flexural rigidity which are evaluated from the Equation (2), the density, sectional area and axial coordinate, respectively.

From Equation (11) and the boundary condition at both ends, the natural frequencies of the laminated beam are computed. The experimental and analytical results are shown in Table 2. The experimental values are little smaller than the analytical ones, because in the experiments, the beam cannot be clamped exactly at the end and the rigidity of the specimen is reduced. The difference of both results is matter of no importance.

Therefore, it is obtained that the identified constitutive equation of CFRP are practically valid for the lamina and the laminate through the comparison with the experiments and the analysises. If more accurate constitutive equation is required, we consider that highly precise law of mixture or micromechanical approach should be applied to determine the constitutive equation.

5.CONCLUSION

In the present paper, the anisotropic thermo-viscoelastic constitutive equation for FRP was identified by the law of mixture and time-temperature superposition principle. The isotropic elastic moduli of fiber and the isotropic thermoviscoelastic matrix resin were determined experimentally. The complex moduli of the CFRP lamina with different fiber angle and the natural frequencies of the CFRP cantilever were measured to compare with the analytical results from the constitutive equation. Then it is confirmed that the identified constitutive equation is valid. The method suggested in the present paper is easy to identify the constitutive equation, because the law of mixture is simple. If the high accuracy of equation is required, the modified law of mixture or micromechanical approach should be considered in detail.

Table 2. First natural frequencies of CFRP laminated beam

	Analysis [Hz]	Experiment [Hz]
293K	30	28
343K	29	27
393K	25	25
423K	23	21

REFERENCES

Grszczuk, L.B. 1975. Foreign object impact damage to composite. ASTM STP 568.

Shimizu, M., Adachi, T., Arai, M. & Matsumoto, H. 1997. Influence of temperature on impact damage in CFRP laminates. Transactions of JSME, 63(607) : 603-609.

Sogabe, Y., Tsuduki, M., Sakai, K., Senda, T., Kishida, K. 1991. Temperature dependence of viscoelastic properties of CFRP subjected to impact loads. J. Soc.

Mat. Sci., Japan, 41(463) : 465-469.

Kasamori, M., Ohtsuka, T., Shimbo, M., Miyano, Y. 1992. Time-temperature dependences of mechanical properties of high temperature epoxy resin and CFRP laminates using the resin. J. Soc. Mat. Sci., Japan, 41(463) : 465-469.

Takada, S., Tsukui, K., Yoshioka, S., Hijikata, A. 1992. Evaluation of time-dependent deformation of high modulus CFRP under constant temperature condition. J. Japan Soc. Comp. Mat, 18(5) : 192-200.

Schapery, R.A. 1967. Stress analysis of viscoelastic composite materials. J. Comp. Mat., 1, 228-267.

Lou, Y.C. & Schapery, R.A. 1971. Viscoelastic characterization of nonlinear fiber-reinforced plastic. J. Comp. Mat., 5 : 208-234.

Tuttle, M.E. & Brinson, H.F. 1986. Prediction of the long-term creep compliance of general composite laminates. Experimental Mechanics, 26(1) : 89-102.

Zhang, S.Y. & Xiang, X.Y. 1992. Creep characterization of fiber reinforced plastic material. J. Reinforced Plastics and Composites, 11(10) : 1187-1194.

Lin, K.Y. & Hwang, I.H. 1988. Thermo-viscoelastic response of graphite/epoxy composites. J. Engineering Mat. and Tech., 110 : 113-116.

Flaggs, D.L. & Crossman, F.W. 1981. Analysis of the viscoelastic response of composite laminates during hygrothermal exposure. J. Comp. Mat., 15 : 21-40.

Jones, R.M. 1975. Mechanics of Composite Materials. Tokyo : McGraw-Hill Kogakusya. : 147-156

Appendix A. Constant Values E'_{Mn}, E''_{Mn} and ρ_{Mn}

n	E'_{Mn} [MPa]	E''_{Mn} [MPa]	ρ_{Mn} [Hz]
1	1.581	-0.6004	5.166×10^{24}
2	1.033	-0.3069	6.741×10^{23}
3	0.8633	-0.1587	8.797×10^{22}
4	1.110	-0.07431	1.148×10^{22}
5	1.104	-0.01639	1.498×10^{21}
6	0.5731	0.03601	1.955×10^{20}
7	0.2541	0.1007	2.551×10^{19}
8	1.025	0.2026	3.330×10^{18}
9	2.214	0.3907	4.345×10^{17}
10	2.463	0.7765	5.670×10^{16}
11	3.994	0.8049	7.400×10^{15}
12	5.372	2.258	9.657×10^{14}
13	7.074	3.758	1.260×10^{14}
14	11.18	5.908	1.645×10^{13}
15	17.17	16.66	2.146×10^{12}
16	44.90	27.63	2.801×10^{11}
17	71.47	44.48	3.655×10^{10}
18	144.4	63.10	4.770×10^{9}
19	244.0	53.86	6.224×10^{8}
20	316.9	14.46	8.123×10^{7}
21	334.7	-10.84	1.060×10^{7}
22	318.2	-49.07	1.383×10^{6}
23	251.5	-52.78	1.805×10^{5}
24	188.3	-31.92	2.356×10^{4}
25	108.4	-21.95	3.074×10^{3}
26	137.5	-8.834	4.012×10^{2}
27	101.7	-9.366	5.236×10^{1}
28	71.19	-14.28	6.833×10^{0}
29	97.20	-17.52	8.916×10^{-1}
30	135.3	-11.12	1.164×10^{-1}
31	228.8	-2.477	1.518×10^{-2}
32	143.8	1.636	1.982×10^{-3}
33	121.4	6.775	2.586×10^{-4}
34	171.4	17.99	3.375×10^{-5}
35	169.5	32.25	4.404×10^{-6}
36	214.4	19.68	5.747×10^{-7}
37	266.7	-30.61	7.500×10^{-8}
38	227.6	-29.32	9.788×10^{-9}
39	64.83	-7.354	1.277×10^{-9}
40	-43.49	-4.017	1.667×10^{-10}

Experimental Mechanics, Allison (ed.)© 1998 Balkema, Rotterdam, ISBN 90 5809 014 0

On the processing and testing of smart composites incorporating fiber optic sensors

A.L.Kalamkarov
Department of Mechanical Engineering, DalTech, Dalhousie University, Halifax, N.S., Canada

ABSTRACT: The processing, evaluation and experimental testing of the pultruded smart FRP composite materials are discussed. The specific application in view is the use of smart composite reinforcements for a monitoring of structures. The pultrusion technology for the fabrication of fiber reinforced polymer composites with embedded fiber optic sensors (Fabry-Perot and Bragg Grating) is developed. The optical sensor/composite material interaction is studied. The tensile and shear properties of the pultruded carbon/vinylester and glass/vinylester rods with and without optical fibers are determined. The microstructural analysis of the smart pultruded FRP is carried out. The interfaces between the resin matrix and the acrylate and polyimide coated optical fibers are examined and interpreted in terms of the coating's ability to resist high temperature and its compatibility with resin matrix. The strain monitoring inside the pultrusion die during the processing of smart FRP parts is performed using the embedded fiber optic sensors. The strain readings from the sensors and the extensometer are compared in mechanical tensile tests.

1 INTRODUCTION

Composite materials lend themselves as prime candidates for the rapidly expanding field of smart materials. A smart material is a structure which contains a built in sensing device to continuously monitor the current state and serviceability of the structure. This is referred to as a "passive" smart material which has many applications in civil engineering, see Kalamkarov and Kolpakov 1997. Composite are good candidates for making smart materials because their fabrication techniques inherently allow for the embeddment of sensors and communication lines. However, in the quest to advance composite materials applications into such new fields, it is desirable to reduce their cost through the use of automated production techniques. To date, hand layup in combination with vacuum bagging and autoclaving has been most often used to fabricate smart composites. This can be a labor intensive and time consuming process in which the quality of the final product is significantly affected by the skill and experience of the technician. Pultrusion which is the only continuous process, has received little attention in the area of smart composites, and there are currently only a few publications on this subject. However, by considering the costs and cycle times which are favorable criteria for selection of a manufacturing process, pultrusion is the fastest and most cost effective process. Pultrusion is also well suited to produce prestressing tendons and rebars, because it can provide the structures with a high degree of axial reinforcement. The pultrusion process however, inherently has the potential to generate residual stresses within a composite component for several reasons. The high output or production rate in feet per minute requires a fast cure rate as the raw materials travel through a pultrusion die which is typically just less than one meter in length. The resin matrix is thus subject to a dynamic cure profile created by strip heaters attached to the die surface. Accelerators and promoters are needed to cure the resin in addition to the normal catalysts. All considered, not much is known about the effect of these factors on the development of residual stresses. One must also consider that the infeeding of reinforcing fibers to the pultrusion die is also a dynamic process and problems associated with the balance and symmetry of the fiber distribution may occur. Once again, this effect may generate residual stress within the component. It is therefore useful to investigate the ability of embedded fiber optic sensors to monitor the strains during the processing and to measure the residual strains created by the pultrusion process.

In addition to the effects that the embeddment of the fiber optic sensors in the process of pultrusion might have on the mechanical properties of the composite itself, it is necessary to assess the performance of the sensors themselves under conditions of static and dynamic loading. In particular, it would be important to obtain data that would reflect on the repeatability and accuracy of the fiber optic strain measurements and also to compare the output of the pultruded sensors to that from conventional strain gauges such as external extensometers.

The objectives of the research reported herein are the following: to evaluate the residual strains induced during the pultrusion of FRP rods, to assess the behaviour of Fabry Perot and Bragg Grating sensors during pultrusion, to determine how the embeddment of optical fibers and their surface coatings affect the mechanical properties of the composite, to confirm the operation of the embedded sensors in the smart pultruded tendons, and to compare the output of the embedded sensors to that from an extensometer.

2 FIBER OPTIC SENSORS

In this study, two types of fiber optic sensors were embedded during the pultrusion of carbon and glass fiber reinforced rods: Fabry Perot and Bragg Grating sensors. The Fabry Perot sensor has been developed to use a broadband light source as opposed to laser light. It is highly sensitive and can make precise, linear, and absolute measurements, see Kalamkarov et al. (1997,a) for details.

The other type of fiber optic sensors used was of the Bragg Grating type. Bragg Grating sensors are based on creating a pattern of refractive index differentials directly onto the material of the fiber core. Fiber gratings selectively reflect certain wavelengths and transmit others. Which wavelengths are transmitted and which ones are reflected depend on both the refractive index of the core material as well as the spacing of the pattern. Changes in temperature or pressure will change the refractive index of the core material and hence cause a change in the wavelengths of peak reflection (or transmission). The presence of mechanical strain along the length of the fiber will have a similar effect since it will change the grating spacing. Measurements of these wavelength shifts provide the basis of operation of Bragg Grating sensors. One advantage of Bragg Grating sensors is that the shift in the wavelength of peak reflection and/or transmission is linear with temperature and axial strain. On the other hand, it is not possible to decouple the effects of temperature and strain with just one sensor. In addition, Bragg Grating sensors, unlike Fabry Perot sensors are quite sensitive to transverse strain because of the photoelastic effect. There are several techniques available to determine the wavelength shift, including optical spectrum analyzers and tunable filters. These sensors have a great potential in smart composites and structures.

The Bragg Grating sensors used in this study were of the single mode type operating at a Bragg wavelength λ_0 of about 1300 nm.

3 EXPERIMENTAL MATERIALS AND EQUIPMENT

Pultruded carbon and glass FRP rods were produced using a urethane modified bisphenol-A based vinyl ester resin system known for its good mechanical properties and excellent processability. Two types of organic peroxide catalysts were used to cure the resin, di-peroxydicarbonate and tert-butyl peroxybenzoate. Adequate release from the die was achieved using an internal lubricant. The mechanical properties of the resin system and of the carbon fiber rovings are given in Kalamkarov et al. (1997,b). The 9.5 mm diameter carbon rods were pultruded with 22 ends of rovings giving a volume fraction of 62%, while their glass counterparts were pultruded with 26 ends giving a volume fraction of 64%.

The Fabry Perot and Bragg Grating sensors were acquired as prepackaged assemblies. The sensing element was located at the front end of an optical fiber of approximately 2.0 - 5.0 meters total length. The fiber optic core and cladding were protected by a polyimide coating which acts as the contact surface between the optic fiber and the surrounding composite. Polyimide coating was used to ensure survival of the optical fiber when exposed to the high temperatures in the pultrusion die, see Kalamkarov et al. (1997,b). In these experiments the actual maximum die temperature was 150°C, whereas the polyimide provides protection to 350°C. The Fabry Perot sensors used in the experiments were rated for ±5000 or 0-10000 microstrain, while the Bragg grating ones were rated for ±5000 microstrain. The sensors were not temperature compensated.

Pultrusion was carried out on an experimental pultrusion line. To determine how the embeddment of optical fibers and their surface coatings affect the mechanical properties of composite materials a microstructural analysis was carried out on both the pultruded profile's cross section and on fracture surfaces obtained from mechanically fractured samples.

Finally, to assess the overall behaviour of the embedded fiber optic sensors, smart tendons were produced with the pultrusion process described above. Two lengths of samples were fabricated: short lengths of approximately 0.6-1.0 meters and long lengths of approximately 3.6-4.2 meters which would act as a test model for rebar or prestressing tendons. All of the short length samples were tested using an Instron servohydraulic load frame and an appropriate controller. The capacity of the installed load cell was 44 KN. Strain was measured externally on the samples using an extensometer. During the actual tests, four analog signals were read into a data acquisition system, one from the extensometer, one from the load cell, one from an LVDT, and a final strain signal from the Bragg Grating or Fabry Perot sensors.

4 EXPERIMENTAL AND DISCUSSION

The microstructural analysis showed that the polyimide coating on optical fiber results in a good interface between optical fiber and host material; whereas acrylate coating cannot withstand the harsh environment (high production temperature up to 150°C) and causes severe debonding of optical fiber and resin, see Figure 1.

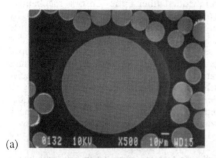

(a)

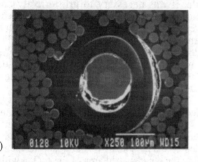

(b)

Figure 1. SEM micrographs show excellent interface between polyimide coated optical fiber and host material (a), and debonding between acrylate-coated optical fiber and host material (b).

Mechanical tests performed on glass and carbon FRP tendons, see Kalamkarov et al. (1997,b) indicated that the tensile strength and tensile modulus of the tendons were virtually unaffected by the embeddement of a single polyimide coated optical fiber. The suggested explanation is that the fiber reinforcement is the only significant factor directly affecting the tensile properties of a unidirectional composite, see Kalamkarov (1992). The embeddement of a single optical fiber slightly deteriorates the shear strength of glass rods, but no effect is evident in the case of carbon rods. However, as the number of embedded optical fibers in the composite is increased, the effect upon the shear strength is also increased.

As a first attempt at embedding a fiber optic sensor in a pultruded rod an unmodified Fabry Perot fiber optic sensor was added to the fiber feed side of the pultrusion process. The forward end of the sensor lead was bonded with a 5 minute epoxy to one of the carbon fiber rovings to ensure that it would feed into the die. From the location at which it was bonded, the sensor had to pass through two of the fiber feed cards before entering the die. The sensor was also located towards the outer surface of the carbon fiber rod. After the sensor had passed through the die and had been embedded in the composite rod, the pultrusion process was stopped to enable trimming away of several carbon fiber rovings in order to pass the pigtail and connector through the die.

The result of the first trial was a length of carbon fiber rod with an embedded Fabry Perot sensor. The end of the rod which was pulled dry during the shut down process contained the pigtail with connector. When the sensor was tested using the fiber optic readout unit, the readings tended to jump about from low (as expected) microstrains, to very high readings of strain far in excess of the 5000 microstrain limit. In tinkering with the sensor, one could bring the strain readings into a normal range by exerting some pressure on the actual connector. Two theories were postulated about the cause of fluctuations in the microstrain readings. One was that the harsh conditions of temperature, fiber compaction, or resin cure shrinkage in the pultrusion die damaged the sensor. The second theory was simply that the sensor was handled too roughly before or after processing, or that it may have been damaged by contact with the fiber feed cards or entrance into the die.

It was decided to conduct a series of experiments that would expose a Fabry Perot sensor separately to each of the variables in the pultrusion process. These are fiber compaction pressure, elevated temperature, liquid resin, and resin curing shrinkage stresses. Changes were also made to the fiber feed system to allow the sensor to be located more accurately in the center of the rod, and to protect the sensor from damage in the fiber feed system.

On account of the low survivability rates of Fabry Perot sensors in the pultrusion process, a novel method was developed to reinforce the sensors before being pultruded in order to offer more protection at the die entrance. The strains observed by this sensor were generally much lower than those of previous trials. There was a sudden negative strain as the sensor entered the die. This was most likely due to a slight bending of the micro-rod as it entered the die. The reading remained constant at approximately -100 microstrain until the sensor was 0.5 meters into the die, then it dropped suddenly to approximately 20 microstrain. From this point to the die exit the strain readings slowly went to zero. After the exit they were slightly negative.

Unlike Fabry Perot sensors, Bragg Grating sensors showed enhanced survivability during pultrusion and hence it was not considered necessary in this case to perform the various experiments described above whereby the sensors were exposed separately to each of the variables in the pultrusion process. It was not also necessary to reinforce Bragg Grating sensors. Nevertheless, we did subject Bragg Grating sensors to a "dry" pultrusion run (passing of the sensors and glass fibers through a heated die but with the fiber rovings not soaked in resin), primarily in order to compare the strain output with that from Fabry Perot sensors. In all experiments performed and described in this paper, the sensor was located at the center of the rod using a single glass fiber roving that traveled a straight line through the feed system. The observed strain readings followed the temperature variation in the die quite closely. These strains which reach a peak of around 1450 microstrain are most likely due to thermal expansion of the sensor. Note that the peak strain is well below the 5000 microstrain capability of the sensor, see Figure 2. Comparing the strain outputs for dry pultrusion runs from both Bragg Grating and Fabry Perot sensors, it is observed that the plots have a very similar shape as expected since they both closely follow the temperature profile in the die. The strains recorded by the Bragg Grating sensor were about 2.5 times larger than the corresponding strains from the Fabry Perot sensor, and this may be attributed to the difference in the coefficients of thermal expansion of the two sensor types.

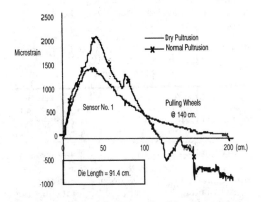

Figure 2. Comparison of output from Bragg Grating sensor during normal and "dry" pultrusion.

Subsequently to the dry runs, a number of normal pultrusion experiments with different Bragg Grating sensors were performed, just as it was previously done with Fabry Perot sensors. The overall profile of the plot is very similar to that for the dry run. Figure 2 shows the strain plots from the "dry" and normal pultrusion runs superimposed. The difference between the two curves is due to the curing of the resin. For example, the peak strain during normal pultrusion is much higher than the equivalent strain during "dry" pultrusion. This is likely to have been caused by an increased thermal expansion of the sensor

due to the exothermic reaction accompanying pultrusion. Also to be noted is the fact that as the product exits the die, the difference between the two curves in Figure 2 represents the residual strains induced by the pultrusion process. As well, Figure 2 also shows that during normal pultrusion the strains do not go back to zero as they do in "dry' pultrusion, because of the residual strains which are "locked in".

To assess and characterize the overall behavior of the embedded fiber optic sensors, mechanical testing of the pultruded tendons was carried out by applying various loads to the tendons while continuously monitoring strain via the embedded optical sensors and a standard extensometer clipped to the pultruded rod. The loading and test programs were carefully designed so that the strain capacity of the fiber optic sensors was not exceeded.

After pultrusion of the smart tendons, a fiber optic lead exited from their cross sections. Each lead was fitted with a connector which was required to plug into the associated electronics that process the fiber optic signals.

In order to test the samples in the load frame available, it was necessary to grip directly onto each end of the test sample. As mentioned, care was taken not to damage the fiber optic lead which exited through one end of the smart tendon. This precluded using the standard type of wedge grips that were purchased as an accessory for the load frame. Gripping was accomplished using a specialized type of fixture known as an open spelter socket. The socket is open at both ends so that the pultruded tendon is easily slipped inside. The inner dimensions of the socket are such that it forms a cone shaped cavity around the inserted section of pultruded rod. The socket was attached to the rod by potting the cone shaped area of the socket with a strong epoxy adhesive. The forked end of the spelter socket is affixed to the load frame via an attachment pin and one other adapter. The spelter socket is ideal because it allows the sample to be gripped without disturbing the fiber optic lead, it does not place unwanted strain near the sensing element, and it is strong enough to allow loading of the samples to the desired proof loads.

The smart FRP tendons were subjected to two basic waveforms in order to evaluate their performance and meet the objectives of the research. The first waveform was a trapezoidal waveform whereby the load was ramped from a low value (typically 100N) to a peak value of about 3000 to 4500 N, at a slow rate of 90 Newtons per second. The tendons were held at this load for 20 seconds and then ramped back down to the initial load at the same rate. The second waveform to which the smart tendons were subjected was a sinusoidal one. The frequency was one cycle per minute (0.0167 Hertz), and a typical range through which the load was cycled was from 400 to 5000 N.

The glass FRP tendon containing an embedded Bragg Grating was first subjected to a trapezoidal waveform as described above. A graph which illustrates the result of this test is shown in Figure 3. The data is plotted as load versus microstrain, for the strain signals received from both the extensometer and the Bragg Grating sensor. The output from the fiber optic strain gage was higher than the extensometer value over the entire range of applied loads. The maximum strain values read by the extensometer and the fiber optic sensor were 1400 microstrain and 1700 microstrain respectively. Inspection of the test sample however, revealed that one of the spelter sockets was not perfectly aligned with the composite tendon. Hence, in addition to the tensile stress, the loading applied by the load frame imparted a small bending stress to the tendon.

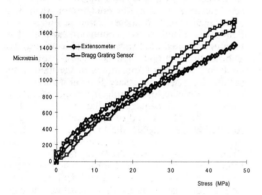

Figure 3. Strain from extensometer and embedded Bragg Grating sensor in a glass FRP tendon subjected to a trapezoidal load.

On repeating the above test it was observed that the the strain data from both the Bragg Grating sensor and the extensometer was repeatable although the output from sensor was consistently higher than that from the extensometer in all cases. In evaluating the performance of the Bragg Grating sensor, it is interesting to note that it returned to a value of zero strain at the end of each test. The resolution of the device if we compare strain values during the ramping stages of the tests, appears to be in the range of 50-100 microstrain.

The glass FRP tendon with the embedded Bragg Grating was next subjected to a sinusoidal waveform, also described above. The strain at the maximum applied load is 1700 microstrain for the extensometer and 2500 for the Bragg Grating sensor. Once again, the repeatability of the Bragg Grating sensors was high. Each of the five cycles produced consistent maximum and minimum strain values. The Bragg Grating exhibited more fluctuation in strain values over the five cycles as compared to the extensometer which produced a fairly linear data line.

In addition to the glass FRP tendon, a carbon FRP tendon containing an embedded Bragg Grating sensor was also subjected to the sinusoidal load.

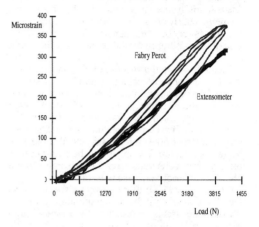

Figure 4. Strain from extensometer and embedded Fabry Perot sensor in a carbon FRP tendon subjected to a sinusoidal load.

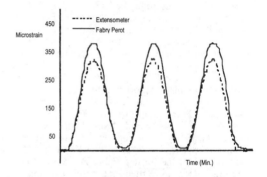

Figure 5. Strain vs. time plot from extensometer and embedded Bragg Grating sensor in carbon FRP tendon subjected to sinusoidal load.

The pertinent data indicates that the Bragg Grating strain output was fairly consistent over the five load cycles, consistently reaching approximately 800 microstrain at each peak loading. However, as with the trapezoidal waveform, the Bragg Grating output was significantly higher than that from the extensometer at the high loading points, being in fact about 400 microstrain greater at the peak load. At the minimum load points in the cycle, there was good agreement between the two measuring devices.

The smart FRP tendon containing an embedded Fabry Perot sensor was tested in a similar manner. The sensor was embedded in a carbon fiber tendon and subjected to a sinusoidal tensile waveform. In the present case, the waveform was cycled three times between loads of zero and 950 lbs. The data is presented in Figures 4 and 5 which respectively plot stress versus microstrain and microstrain versus time The data illustrates that the Fabry Perot sensor was fairly consistent over the three cycles of loading. It reached the same strain values at the peak and minimum load points over the three cycles.

Figure 5 shows that both the Fabry Perot sensor and the extensometer exhibited a consistent sinusoidal waveform. The values read by the Fabry Perot sensor were only about 14% higher than those from the extensometer at the peak loads.

The Fabry Perot sensor tested above performed better than the Bragg Grating sensors. The fact that the sensor was reinforced prior to being pultruded into a carbon FRP tendon may have played some role in its superior performance.

CONCLUSIONS

Bragg Grating and Fabry Perot fiber optic sensors have been successfully embedded into FRP composite parts during pultrusion. The Bragg Grating sensors show a greater survivability in the pultrusion process than the Fabry Perot sensors and hence there was no need to reinforce them. Dry pultrusion runs were performed with Bragg Grating sensors and the thermal strain output obtained conformed quite well with the temperature profile within the die. Similar experiments performed with Fabry Perot sensors showed that these sensors also experience thermal strain that followed the temperature variations in the die. However, since the coefficient of thermal expansion of Bragg Grating sensors is greater than that of the Fabry Perot counterparts, the thermal strain experienced by the latter was lower than that experienced by the former. Fabry Perot and Bragg Grating sensors were then incorporated in normal pultrusion runs. The different sensors showed strain outputs combining thermal and residual strains, which had similar basic profile even though the absolute values of the strains varied from sensor to sensor. The information provided by these experiments yields valuable insight to the specifics of the pultrusion process.

Pertinent microscopic analysis indicated that polyimide coating on optical fibers results in a good interface between the optical fiber and the host material. On the other hand, acrylate coating cannot withstand the harsh environment (high production temperature) characterizing the pultrusion process, and causes severe debonding of optical fiber and resin. Therefore the polyimide-coated optical fibers should be a preferential selection for pultrusion of smart FRP reinforcements.

It was also determined that embedded optical fibers have no significant effect on the tensile properties of the pultruded FRP, but they slightly deteriorate the shear strength of the composites. This slight decrease in shear strength was evident in the case of glass rods but not in the case of carbon rods. As the number of embedded optical fibers in the composite is increased, the effect upon the shear strength is also increased.

Mechanical testing was carried out in order to assess the overall behaviour of the smart tendons with the embedded Fabry Perot and Bragg Grating sensors. The Bragg Grating sensor embedded in glass FRP tendon was subjected to trapezoidal and sinusoidal load inputs. Although in both cases, the strain from the sensor was higher than that from the extensometer over the entire range of the applied loads, the discrepancy was larger for the higher loads. Subsequently, the Bragg Grating sensor embedded in carbon FRP tendon was also subjected to the same load regime. Once again, the sensor strain readings were higher than the corresponding extensometer readings. Finally, the Fabry Perot sensor embedded in carbon FRP tendon was subjected to a sinusoidal waveform. The strain readings from the sensor conformed quite well with those from the extensometer except at the peak loads where the Fabry Perot output was higher than the extensometer output by about 14%.

ACKNOWLEDGMENT

This work was supported by ISIS - CANADA, the Intelligent Sensing for Innovative Structures Canadian Network of Centres of Excellence through the Project T3.4 on Smart Reinforcements and Connectors.

REFERENCES

A.L.Kalamkarov 1992. *Composite and reinforced elements of construction*. Chichester, New-York: John Wiley & Sons.

A.L.Kalamkarov, S.Fitzgerald, and D.MacDonald 1997,a. On the Processing and Evaluation of Smart Composite Reinforcement, *Proceedings of the SPIE* . 2361, in press.

A.L.Kalamkarov and A.G.Kolpakov 1997. *Analysis, design and optimization of composite structures*. Chichester, New-York: John Wiley & Sons.

A.L.Kalamkarov, D.MacDonald and P.Westhaver 1997,b. "On pultrusion of Smart FRP composites", *Proceedings of the SPIE*. 3042: 400-409.

Experimental Mechanics, Allison (ed.)© 1998 Balkema, Rotterdam, ISBN 90 5809 014 0

Determination of tensile fatigue life of unidirectional CFRP

Yasushi Miyano, Masayuki Nakada & Hiroshi Kudoh
Materials System Research Laboratory, Kanazawa Institute of Technology, Japan

Rokuro Muki
Civil and Environmental Engineering Department, University of California, Los Angeles, Calif., USA

ABSTRACT: This paper concerns the determination of time and temperature dependent fatigue life in the longitudinal direction of unidirectional carbon fiber reinforced plastics (CFRP). Applying the strand-testing method with the newly-devised temperature chamber and grip ends to carbon fibers/epoxy composite strands (CF/Ep strand), the master curve for constant strain-rate tensile strength was constructed and tensile fatigue strength were measured at various frequencies and temperatures. Same strengths were measured also by the conventional split-disk method, in which carbon fibers/epoxy composite rings (CF/Ep ring) were used as the specimens. Similarities and differences between the two results, one by the proposed method and another by the split-disk method, are discussed. We believe that the proposed strand-testing method is excellent for the determination of tensile fatigue life in the longitudinal direction of unidirectional CFRP because of its small size of the apparatus and unidirectional loading for the entire specimen. It is clear from the results of strand test that the tensile fatigue strength of CFRP shows a characteristic time and temperature dependence.

1 INTRODUCTION

In recent years materials that possess high specific strength and specific modulus were developed to fulfill the need for advanced structures. Carbon fiber reinforced plastics (CFRP) are materials that have these properties, and are being used in structures as primary as well as secondary load carrying members. Prediction of life for such CFRP structures is of great importance and concern.

It is well known that the mechanical behavior of polymer resins exhibits time and temperature dependence, called viscoelastic behavior, not only above the glass transition temperature T_g but also below T_g. Thus, it can be presumed that the mechanical behavior of CFRP using polymer resins as matrices also significantly depends on time and temperature. It has been confirmed that the viscoelastic behavior of polymer-matrix resin controls the time and temperature dependence of the mechanical behavior of CFRP [1-3].

The purpose of this paper is to accurately determine the time and temperature dependence of tensile fatigue life in the longitudinal direction of unidirectional CFRP. It is difficult to obtain reliable values for the tensile fatigue strength at temperature above room temperature by conventional testing method using a laminated specimen because the failure occurs frequently at grips rather than at the central portion of specimen. Therefore, we propose the strand testing method with the newly-devised temperature chamber and grip ends to hold the strand. Small diameter of strands permits the design of a very compact testing machine. This strand-testing method supplies the reliable strength in both constant strain-rate (CSR) and fatigue tests since the fracture occurs at the central portion of all specimens tested.

The CSR and fatigue tests for carbon fiber/epoxy composite strands (CF/Ep strand) were carried on the testing machine just described. The results were compared with those measured by the conventional split-disk method [4] in which carbon fibers/epoxy composite rings (CF/Ep ring) were used as the specimens. The split-disk method is conveniently adopted because of its capability of measuring fatigue strength at high temperature and its absence of grips. However, the uniform tensile-strain distribution is impossible to maintain at the test section of the specimen.

2 EXPERIMENT

The time and temperature dependence of tensile fatigue strength in the longitudinal direction of unidirectional CFRP were determined by two methods: the strand-testing method with the newly-devised temperature chamber and grip ends and the conventional split-disk method. The specimens used in these methods are, respectively, CF/Ep strand and

CF/Ep ring.

2.1 *Preparation of specimens*

CF/Ep strand consists of high strength carbon fibers, TORAYCA T400-3K (Toray) and a general purpose epoxy resin, EPIKOTE 828 (Yuka Shell Epoxy). The CF/Ep strand was produced by filament winding method, was cured at 70°C for 12 hours, and postcured at 150°C for 4 hours and at 190°C for 2 hours. The glass transition temperature T_g of the epoxy resin is 112°C. The fiber volume fraction of CF/Ep strand is approximately 60%. The diameter and length of CF/Ep strands are 1mm and 310mm, respectively.

CF/Ep ring produced by filament winding method consists of same carbon fibers and epoxy resin mentioned above. The curing conditions are the same to those for CF/Ep strand specimens. The fiber volume fraction of CF/Ep ring is approximately 60%. The diameter, width, and thickness of CF/Ep ring are 200mm, 6.4mm, and 1mm, respectively.

2.2 *Test procedures*

The tensile test specimens of CF/Ep strand were prepared according to Japanese Industrial Standard (JIS R 7601) as shown in Fig.1. The CF/Ep strand specimens were attached to the stainless-steel end tabs by an epoxy adhesive. The CSR and fatigue tests for the CF/Ep strand were conducted by using a small temperature chamber with the specimen grips outside the chamber as shown in this figure. Temperature in the central 70mm-region of specimen was held within 2°C of the indicated temperature.

The tensile CSR tests were conducted on an Instron type testing machine and the tensile fatigue tests on an electro-hydraulic servo testing machine. The tensile CSR tests were carried out for six temperatures between 50 and 150°C. Loading rates (cross-head speeds) were 0.01, 1, and 100mm/min. The tensile fatigue tests were carried out for three temperatures T= 50, 110, and 150°C at a frequency f= 2Hz, and for two temperatures T= 80 and 130°C at f= 0.02Hz while stress ratio (minimum stress/maximum stress) was kept at 0.1.

The tensile CSR and fatigue strengths σ_s (v, T) and σ_f (N_f, f, T) are defined by

$$\sigma_s(v,T) = P_s(v,T)\frac{\rho}{t_e} \qquad (1)$$

$$\sigma_f(N_f,f,T) = P_f(N_f,f,T)\frac{\rho}{t_e} \qquad (2)$$

where P_s and P_f are maximum load [N], while ρ, t_e, v, f, t, and N_f stand for, in this order, density of strand [kg/m³], tex of strand [kg/m], loading rate [mm/min], frequency [Hz], testing temperature [°C], and number of cycles to failure. These tensile strengths of CF/Ep strand obtained by Eqn.(1) and Eqn.(2) are those of

carbon fiber bundle.

The tensile test specimens of CF/Ep ring were prepared according to ASTM D 2290 as shown in Fig.2. The tensile CSR and fatigue tests were carried out for exactly same combinations of temperature, loading rate, frequency, and stress ratio as done for the tests of strands.

2.3 *Improvements on strand testing method*

Pulling-out of strand from the grip ends during loading occurred initially in both CSR and fatigue testing. This pulling-out was suppressed by two improvements: use of the small temperature chamber as shown in Fig.1 and use of specimen with taper-shaped configuration at both ends fixed by adhesive resin in the grips. After the improvements, fracture of all specimens tested for both CSR and fatigue loadings occurred within the central 70mm region of the specimen.

3 RESULTS AND DISCUSSION

3.1 *Tensile CSR strength*

The temperature dependence of tensile CSR strength σ_s for CF/Ep strand is shown at 3 loading rates in the upper side of Fig.3, while that for CF/Ep ring is depicted at 3 loading rates in the lower side. The dotted lines in the lower side indicate the σ_s obtained from those for CF/Ep strand considering the fiber volume fraction of CF/Ep ring. The dependence of the CSR strength upon temperature and loading rate has similar trend for both strand and ring specimens, but the values of CSR strength for low to moderate temperature are much less for the ring specimens. The low strength for ring specimens may be due to nonuniform distribution of strain in the test section.

In the upper left side of Fig.4, the dependence of the σ_s for CF/Ep strand upon t_s, the time period from initial loading to the maximum load P_s, is presented for various temperatures. The associated master curve at a reference temperature T_0=50°C was constructed by shifting σ_s at various constant temperatures along the log scale of t_s, so that they overlapped smoothly each other as shown in the upper right side of Fig.4. For materials that obey time-temperature superposition principle as for CF/Ep strand, there exists the time-temperature shift factor

$$a_{T_0}(T) = \frac{t_s}{t_s'} \qquad (3)$$

where t_s and t_s' are the time to failure, respectively, at temperature T and at reference temperature T_0=50°C in which the time, t_s', is called the reduced time. The lower side of Fig.4 shows the similar results for CF/Ep ring which also obeys the time-temperature superposition principle.

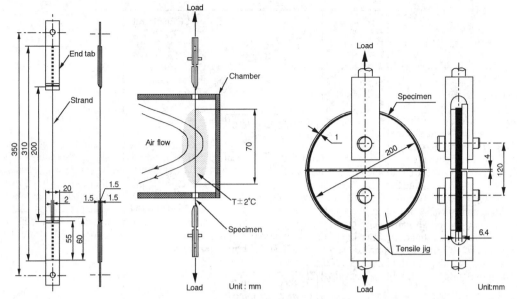

Fig.1 Configuration of strand specimen and testing method

Fig.2 Configuration of ring specimen and testing method

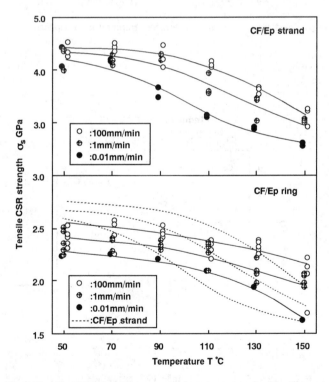

Fig.3 Tensile CSR strength versus temperature at various loading rates for CF/Ep strand and ring

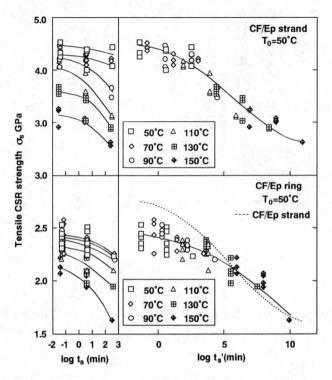

Fig.4 Master curves of the tensile CSR strength for CF/Ep strand and ring

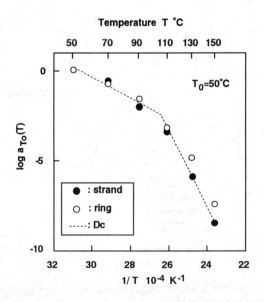

Fig.5 Time-temperature shift factors of the tensile CSR strength for CF/Ep strand and ring

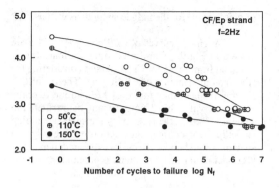

Fig.6 Tensile fatigue strength versus number of cycles to failure for CF/Ep strand

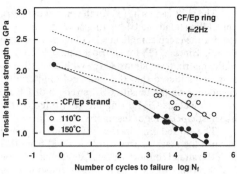

Fig.9 Tensile fatigue strength versus number of cycles to failure for CF/Ep ring

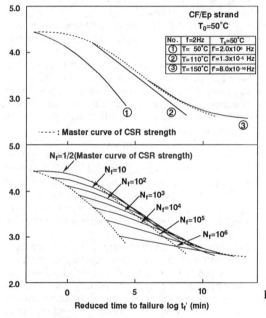

Fig.7 Master curves of the tensile fatigue strength for CF/Ep strand

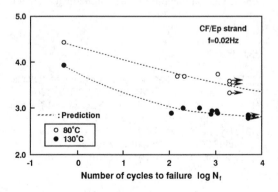

Fig.8 Prediction of the tensile fatigue strength for CF/Ep strand by using the master curve of the tensile fatigue strength

The shift factors $a_{T_0}(T)$ for the master curves of σ_s of CF/Ep strand and CF/Ep ring are compared with those for the creep compliance D_c of the matrix resin in Fig.5. Since $a_{T_0}(T)$ for CF/Ep strand, CF/Ep ring, and the matrix resin almost agree with each other, it can be considered that the time and temperature dependence on the tensile CSR strength of CF/Ep strand as well as CF/Ep ring are controlled by the viscoelastic behavior of the matrix resin.

The dotted line in the lower side of Fig.4 indicates the master curve of σ_s obtained from that for CF/Ep strand considering the fiber volume fraction of CF/Ep ring. In the region of low temperature or short reduced time to failure, σ_s for CF/Ep ring is lower than that for CF/Ep strand, although σ_s for both testing methods agree well with each other in the region of high temperature or long reduced time to failure, where the ultimate strain is smaller than that at the region of low temperature and short reduced time to failure. The difference of σ_s by both testing methods in the region of low temperature and short reduced time to failure may be due to nonuniform distribution of strain in the test section of the CF/Ep ring.

3.2 Tensile fatigue strength

Figure 6 exhibits the tensile fatigue strength σ_f versus number of cycles to failure N_f (σ_f - N_f curve) for CF/Ep strand at a frequency f= 2Hz and a stress ratio R=0.1. It has been assumed in this figure that the σ_f at N_f = 1/2 corresponds to the tensile CSR strength σ_s at t_s=1/(2f).

As preliminary to construct the master curves for σ_f, we define the reduced frequency and the reduced time of fatigue failure by

$$f' = f \cdot a_{T_0}(T) \quad , \quad t_f' = \frac{t_f}{a_{T_0}(T)} = \frac{N_f}{f'} \qquad (4)$$

We consider two sets of the master curves of σ_f at the reference temperature T_0=50°C. They are functions of the reduced time: one is for the fixed reduced frequency which follows immediately from Fig.6 and Eqn.(4) as shown in the upper half of Fig.7 and another is for fixed N_f which are constructed by connecting the points of the same N_f on the curves of same reduced frequency as shown in the lower half of Fig.7. From the lower half of Fig.7, we see that the σ_f-curves tends to become flat as the reduced time to failure t_f' or the N_f increases. The asymptotic value of σ_f for large t_f' or N_f is the tensile fatigue limit in the longitudinal direction of unidirectional CFRP.

Figure 8 displays test results for σ_f-N_f at f=0.02Hz and two temperature levels for CF/Ep strand and prediction curves constructed from the master curves in the lower side of Fig.7. Since the prediction curves for fatigue strength based on the time-temperature superposition principle for CSR strength capture test data satisfactorily, we consider that the

principle holds for the fatigue strength as well as CSR strength.

The σ_f - N_f curves of CF/Ep ring at f= 2Hz are shown in Fig.9. The dotted lines indicate the σ_f obtained from those for CF/Ep strand considering the fiber volume fraction of CF/Ep ring. The σ_f for CF/Ep ring is lower than that for CF/Ep strand over the entire interval. This difference may be attributed to several effects in the ring specimen such as bending stress in the specimen, the friction between jig and specimen, and bundle effect since the specimen cross-sectional ratio of ring to strand is about eight.

4 CONCLUSION

This paper concerns the determination of time and temperature dependent fatigue life in the longitudinal direction of unidirectional CFRP. The results may be summarized as follows:
(1) The strand-testing method with the newly-devised temperature chamber and grip ends is proposed which is suitable for the determination of temperature-dependent CSR strength and fatigue life in the longitudinal direction of unidirectional CFRP.
(2) The results obtained by the strand method were compared with those measured by the conventional split-disk method. Causes of discrepancy of the two results are discussed.
(3) The shift factor for the creep compliance of the matrix resin holds also for the tensile CSR strength as well as the tensile fatigue strength of CF/Ep strand and ring. Thus, the viscoelastic behavior of matrix resin controls that of CFRP and the same time-temperature superposition principle apply to mechanical properties of the matrix resin and to the strength of CFRP.
(4) The master curves of fatigue strength for fixed reduced frequencies or fixed numbers of cycles to failure can be constructed from the fatigue strength measured at a single frequency and various temperatures based on the time-temperature superposition principle. The asymptotic value of fatigue strength for large reduced time to failure is the tensile fatigue limit in the longitudinal direction of unidirectional CFRP.

REFERENCES

[1]Miyano,Y., M.K.McMurray, J.Enyama, M.Nakada, 1994. J. of Composite Materials, 28, 1250~1260.
[2]Miyano,Y., M.K.McMurray , N.Kitade, M.Nakada and M.Mohri, 1995. Composites, 26, 713~717.
[3]Miyano,Y., M.Nakada, M.K.McMurray and R.Muki,1997. J. of Composite Materials, 31, 619~638.
[4]The American Society for Testing and Materials, ASTM D 2290.

Experimental Mechanics, Allison (ed.) © 1998 Balkema, Rotterdam, ISBN 90 5809 014 0

Experimental study of a composite plate subjected to concentrated overturning moments

L. B. Kelly
General Motors Truck Group, Pontiac, Mich., USA

I. Miskioglu & J. B. Ligon
Department of Mechanical Engineering-Engineering Mechanics, Michigan Technological University, Houghton, Mich., USA

ABSTRACT: This investigation focused on the design, construction and evaluation of a test fixture to study a bolted composite panel subjected to a concentrated overturning couple moment. Verification of the fixture consisted of an investigation into the strain fields developed in an aluminum plate that was loaded in the fixture. A combination of photoelastic and strain gage methods were used for this analysis.

A brief study was conducted to examine the effectiveness of the method. Composite panels of continuous random oriented glass fibers in a polyurethane SRIM material were tested to failure with three different bolt washer diameters. The strength and stiffness of the joint increased as the washer diameter increased for the cases considered. Additionally, the failure mechanism was not influenced by the washer diameter.

1. INTRODUCTION

Glass fiber reinforced polymer composite panels have a wide range of current and potential applications in structural design. The effective and efficient use of such panels in structural applications requires a complete understanding of both the composite material and the interactions of the material with the mechanisms used to attach the material to other structural components. The most common joining methods used with composites include chemical, such as adhesives, and mechanical, such as bolts or rivets.

Mechanical fastening systems are often required because of the long term reliability of a bolted or riveted joint such as in the case of a seat belt attached to a composite panel in an automobile. Typical loads that these types of joints may receive include transverse out-of-plane load (P), in-plane torque (T), in-plane bearing load (F_1), and an overturning couple moment (F_2*d) as shown in Figure 1. Because of the inherent complexity of a bolted joint, design evaluation has to rely heavily on experimental methods to augment numerical techniques. One of the most complex loading situations involves overturning moments.

This study focuses on the design and verification of a test fixture for applying overturning couple moments (F_2*d) to composite plates as shown in Figure 1. This investigation is part of a continuing study of the special class of mechanical fastening problems whereby concentrated loads are transferred through a bolted joint to a composite panel.

In the area of joining composite materials, typical references include Adams and Wake [1] for adhesive joints, and Eisenmann and Learhardt [2] and Crews [3] for bolted joints. The majority of the work presented in the references deal with in-plane loads.

Hughes [4], Dykstra [5] and Polk [6] investigated the problem of thermoset polymer composite plates subjected to an out-of-plane transverse load P (see Figure 1) where they developed numerical and theoretical models which were compared with experiments.

Upon reviewing the literature of bolted joint design of composite materials, it is apparent that the special case of a panel subjected to an overturning couple moment has received little attention. The problem of fastening composites differs from that of joining metals in that the strengths of the members and the bolts can differ significantly. With metal to metal joints, the failure of both the material and the bolt is of interest. With composite panels, the failure generally occurs only in the composite material with the bolt remaining intact. This failure typically occurs along the washer boundary of the bolted connection due to the stress concentrations that occur in that region. Therefore, the failure is influenced by the geometry of the connection and its ability to reduce the stress concentrations near the joint boundary.

This paper deals with a test apparatus and method that was developed to help understand those factors that effect the magnitude of the stress concentrations in a composite plate subjected to the overturning couple moment.

2. OVERTURNING MOMENT FIXTURE DESIGN

The test fixture developed to apply the overturning moment to a plate is composed of five primary elements: the load frame, the base and the clamping ring, the hydraulic actuators, the load cell to measure the clamping force, and the moment application apparatus.

2.1 Load frame

The load frame consisted of two 203.2 mm x 203.2 mm I-section columns which were bolted to a test bed on the bottom and had steel plates welded to the top, see Figure 2. A 203.2 mm x 127 mm I-section steel beam was bolted across the span between the two columns which provided a horizontal platform for attaching components. In the center of the span, two 50.8 mm square steel tubes were bolted vertically to which the base and the clamping ring were attached. The horizontal beam and the floor supported the hydraulic actuators mounted vertically as shown in Figure 2.

2.2 Base and clamping ring

The composite plate was secured between a clamping ring and a base. The clamping ring and base were mounted vertically to the two 50.8 mm square steel tubes in the center of the column span. Both the base and the ring were made of 25.4 mm thick steel plates. The base was a 304.8 mm x 304.8 mm square with a 203.2 mm hole in the center. The clamping ring had a 304.8 mm outer diameter and 203.2 mm inner diameter and was secured to the base by 24 equally spaced 9.5 mm bolts. The bolts were tightened to a torque of 54.2 N-m. This provided the clamped boundary conditions for the composite plate.

2.3 Actuators

Two hydraulic actuators were used to apply the forces that produced the couple moment on the composite plate. One actuator was attached to the horizontal platform beam while the second one was bolted to the test bed. Both actuators were equipped with 4.45 kN load cells, each having its own controller. The controllers were operated in a load control mode. The synchronization between the channels was achieved with a function generator. The function generator sent a linear ramp input wave form to the controllers to provide the couple moment loading function. The frequency of the loading function was 0.001 Hz.

2.4 Clamping force load cell

Consistent test results require that the clamping force exerted on the plate by the bolted connection be consistently applied and measured. A schematic of the connection is shown in Figure 3. A custom load cell with a maximum capacity of 66.75 kN was built to measure the force exerted on the plate by the bolted connection.

2.5 Moment application apparatus

The moment application apparatus consisted of three components - the yoke, the connection rod and a threaded rod as shown in Figures 2 and 3. Steel yokes were attached to each load cell by the threaded connection on each cell. Each yoke consisted of a U-shaped fixture with a hardened steel pin used to apply the couple load to the connecting rod. The ends of the pin were press fit into the inner race of radial roller bearings whose outer races were press fit through the sides of the yoke. The bearings were needed to allow free rotational movement of the pin along the connection rod as the actuators moved during the loading process.

The high-carbon steel connection rods were placed through the yokes and secured to the threaded rod passing through and bolted to the composite plate, Figure 3. The purpose of the connection rods were to provide a rigid structural mechanism of transferring each load from the actuators to the threaded rod bolted connection to produce the overturning moment.

2.6 Deflection and rotation measurement

Each actuator was equipped with an LVDT. The

output of the LVDT was used to measure the displacement of the point that the yoke pin contacted the connection rod. Using the yoke displacement and the connection rod moment arm length, the angle of rotation of the center of the panel was known for each load increment.

The applied moment and corresponding center panel rotation were recorded with a data acquisition system.

2.7 *Fixture verification*

The alignment of the load fixture to apply a concentrated couple moment to the test plate was calibrated by both photoelastic coatings and strain gages. When the loads were properly applied, the strain field would be symmetric with respect to both the horizontal and the vertical axes passing through the center of the plate.

Two 1.6 mm thick aluminum plates were used for the verification tests. For the photoelastic tests, a disk of Measurement Group PS-1 photoelastic coating was bonded to the surface of one of the aluminum plates with PC-1 adhesive. The outside and inside diameters of the coating were adjusted to allow sufficient clearance between the coating and the clamping ring on the outside and the washer on the inside.

After the load fixture was aligned to obtain symmetric isochromatic fringe patterns, four 3-element rectangular rosettes (Measurement Group-EA-062RB-120) were bonded to the second aluminum plate along a reference diameter as shown in Figure 4. Strain gage rosettes were located at a radius of 28.6 mm from the center of the plate. Strain measurements were taken with the reference diameter aligned with 0° (horizontal), 45°, 90° (vertical) and 135° orientations in the test fixture. This was accomplished by rotating, clamping, and testing the plate at each of the four angle orientations of the reference diameter. The strains at location 1 and 4, and locations 2 and 3 were compared for each plate orientation to evaluate the alignment and operation of the test fixture. The comparison of measurements at the four locations for the 90° orientation is presented in Figure 5.

3. OVERTURNING MOMENT TEST OF A COMPOSITE PANEL

A brief study of the effects that the washer diameters

had on the failure of a composite plate was conducted to assess the performance of the apparatus for its intended use. A random oriented, continuous stran glass fiber reinforced thermoset polymer composite panel was chosen for the test.

Three different washer diameters were tested to determine what, if any, influence the washer diameter had upon the failure of the composite plate. Each plate was tested with the same rate of loading to enable comparisons to be made between the different tests. Two plates were tested for each washer diameter.

4. RESULTS

Reflective photoelasticity was used to align the actuators and connections rods until a symmetric isochromatic fringe pattern was observed.

A comparison of the strain measurements at four strain gage rosette locations revealed that the strains at locations 1 and 4 and those at locations 2 and 3 appeared to be the same for each of the orientations of the plate, see Figure 5.

The overturning moment test results are shown in Figure 6.

Examination of the plots of the tests reveals that the diameter of the washer did have an effect on the behavior of the plate. An increase in washer diameter resulted in an increase in the stiffness and strength of the assembly. For all plates tested, failure of the joint was observed to initiate at the edge of the washer.

5. CONCLUSIONS

A test fixture to apply a concentrated couple moment to a composite panel was designed, manufactured and its operation verified by reflection photoelasticity and strain gages.

A cursory investigation was conducted to determine the effect that three different washer sizes had on the failure strength of a bolted joint in a composite plate subjected to the overturning moment. The results demonstrated the effectiveness of the test fixture. In addition, the tests indicated that an increase in the washer diameters tested resulted in an increase in the failure strength of the composite joint. The failure initiation location, at the edge of the washer, was not influenced by washer diameter.

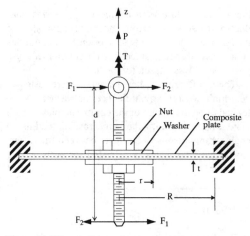

Figure 1. Plate subjected to general loading condition.

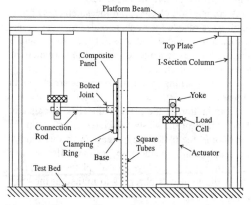

Figure 2. Overturning couple moment test frame.

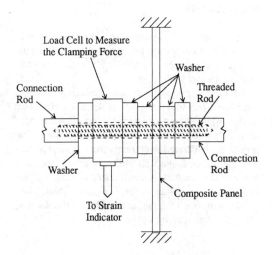

Figure 3. Bolted joint on composite plate.

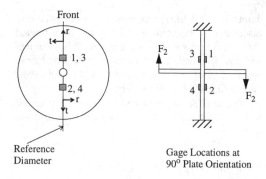

Figure 4. Relative locations of rosettes on aluminum plate.

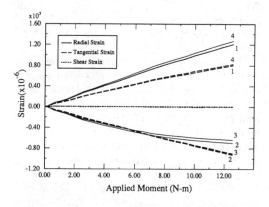

Figure 5. Strains at locations 1-4 and 2-3 for 90° plate orientation.

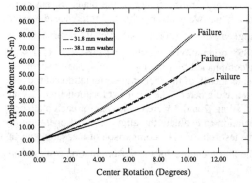

Figure 6. Overturning moment results for composite plates.

6. ACKNOWLEDGMENT

The authors wish to think the Michigan Materials Processing Institute, General Motors Corporation, and Chrysler Motor Company for providing financial support for the ongoing project under which this apparatus was developed.

REFERENCES

Adams, R.D. & W.C. Wake. 1984. *Structural Adhesive Joints in Engineering*. Elsevier Science Publishing Co., Inc., New York.

Eisenmann, J.R. & J.L. Leonhardt. 1981. Improving Composite Bolted Joint Efficiency by Laminate Tailoring. *Joining of Composite Materials, ASTM SPT 749*, K.T. Kedward, Ed., Am. Soc. for Testing and Materials: 117-130.

Crews, J.H., Jr., 1981. Bolt-Bearing Fatigue of a Graphite/Epoxy Laminate. *Joining of Composite Materials, ASTM SPT 749*, K.T. Kedward (ed), Am. Soc. for Testing and Materials: 131-144.

Hughes, C.F. & D.L. Sikarskie. 1993. Concentrated Load Transfer on Composite Plates. *Proc. Am. Soc. for Composites, 8th. Tech. Conf., Cleveland, OH*: 943-952.

Dykstra, D.R., Ligon, J.B. & I. Miskioglu. 1997. Transverse Load Effects on Repaired Composites. *Journal of Reinforced Plastics and Composites*, 16(15): 1425-1434.

Polk, D.A., Ligon, J.B. & I. Miskioglu. 1997. Fastener Design for Transversely Loaded Composite Plates. *Proc. Am. Soc. for Composites: 12th Tech. Conf., Dearborn, MI:* 667-676.

Experimental Mechanics, Allison (ed.) © 1998 Balkema, Rotterdam, ISBN 90 5809 014 0

Strain measurements in FRP composite materials by means of neural networks

D. Grimaldi
Department of Electronics, Computer and System Science, University of Calabria, Italy

R. S. Olivito & L. Surace
Department of Structural Engineering, University of Calabria, Italy

ABSTRACT: This paper shows the experimental results of specimens of glass fibres unidirectionally arranged in an epoxy resin matrix and subjected to tensile tests. In particular, the strain response of such specimens, subjected to mechanical and thermal loads is analyzed by using Artificial Neural Networks (ANNs). The results of these tests are compared with those obtained analytically and the advantages of using ANNs, as compared to the use of traditional temperature compensation, are shown.

1 INTRODUCTION

The sector concerning the application of composite materials in structural engineering is particularly interested in the accurate measuring of strain states by separating the influence of thermal variations from the response of the sensor used. In fact, thermal variations can cause mistakes in the evaluation of the stress state existing in the structure and the influence of the thermal variations on the sensor can lead to mistakes that could seriously involve both the experiments and the planning of the component itself.

Traditionally, strain gauge temperature compensation is realized by means of an external compensator applied to a specimen of the same material under test. In the case of composite materials the utilization of the previous external compensator can produce additional errors resulting from the different orientation of the strain gauge temperature compensation, which may give a residual thermal response.

Moreover, it should be noted that the usual strain gauges utilized are calibrated for omogeneous and isotropic materials, and that their utilization for composite materials can cause additional errors.

To overcome all these difficulties, and to compensate for both the thermal influence on the strain gauge and calibration problems, a technique based on ANNs (Artificial Neural Networks) was used in this paper.

ANN have generated great interest in different disciplines, and there are many specialised applications throughout the fields of technology and engineering Their success is directly dependent upon their ability to simplify and accomplish with speed many tasks not feasible using conventional techniques.

Artificial Neural Networks are employed to solve both problems of system identification and signal processing. They show high immunity to disturbance or noise superimposed on the signal.

The ANN is not programmed, but it is trained to carry out assignments submitted to it. The training consists in iterative operations which modify the coefficients (weight), that characterize the neural network itself. The difference between a classical algorithm and a neural network, therefore, is that in the first case it is necessary to know the functional bonds by analizing a series of examples, constituted by pairs of input-output values (traning set) whereas in the second it is not.

This interesting behaviour on the part of the neural network is particularly useful in cases where it is not possible to know the functional relationships between the input and output data. Once trained the neural network is also able to operate with accurate precision on data not belonging to the whole training set.

This method allows the construction of a scheme for the neural pre-elaboration of the signal by a strain sensor with the purpose of reducing the influence of variations in environmental temperature at a wide interval of values.

This paper shows the results of experiments which compare the response of specimens of glass fibres unidirectionally arranged in an epoxy resin matrix subjected to tensile test by using artificial neural networks and temperature compensation strain gauges. In particular, the response under the action of thermal loads ranging from -20°C to 80°C is analyzed.

2 THEORETICAL BACKGROUND TO ANNs

ANN architectures imitate the behaviour of the human brain and are based on the properties and features of the neurones and synapses of the neurobiological systems. Neurones are the base units in which all processing activities are performed and are characterised by a summing property and a non-linear transfer characteristic, similar to the sigmoidal one. Synapses are the connections which transmit the information. These can be transmitted with modification by means of the synapse weights.

The key feature of the ANN structure consists of a parallel distributed network, comprising many artificial neurones interconnected by weighted connections. Each artificial neurone is characterised by an input vector, a single output value and a non-linear transfer characteristic. It processes information in a predetermined manner and furnishes the results either as the ANN's output or to the input of another neurone. The weighted connections store the information and the value of the weights is pre-defined or determined by a learning algorithm. The cross-connections between the neurones form: (i) a set of input layer neurones, (ii) a set of output layer neurones, and (iii) a set of hidden layer neurones. The choice of the number of neurones in the input and output layers is closely connected to the level of accuracy desired. The number and dimensions of the hidden layers depend only on the ANN's performance achieved in terms of model fidelity and operating speed.

Amongst the numerous ANN architectures available in the literature, the feed-forward networks with one or more hidden layers and the back-propagation learning algorithm are the most common in measurement applications. Indeed, when compared with other architectures, they offer several successful applications in pattern classification, pattern matching and function approximation, as well as, in the learning of any non-linear mapping to any desired degree of accuracy.

With the aim of providing detailed knowledge of particulars, the feed-forward networks with one hidden layer and the back-propagation learning algorithm are considered in this paper.

As is well known, the use of this ANN involves three separate phases which have to be followed, and which are depicted in Figure 1:

a) the learning phase (Fig. 1a), in which the ANN is forced to furnish the desired outputs (out) in correspondence to a determined input (in) (the learning set). In this phase, the ANN learning level is verified by means of suitable performance indices (output error ε), and the objective function, defined coherently by the data of the learning set;

b) the validation phase (Fig. 1b), in which the ANN's generalisation capability is verified by means of data (the validation set) which are completely different from the data used in the previous phase;

c) the production phase (Fig. 1c), in which the ANN is capable of providing the required outputs (y) that correspond to any input.

In order to train the ANN to model physical phenomena it is necessary: (i) to specify the learning set constituted by an adequate number of input and target output vector pairs, and (ii) to minimise the objective function, defined as the relative difference (normally in the Euclidean sense), between the actual ANN output and the target outputs. It is possible to minimise the objective function by using a learning algorithm. The strategy of the learning algorithm is devoted: (i) to determining the input/output transfer characteristic of each neurone by a supervised learning section, (ii) to modifying the connection weights and the neurone bias by means of an adaptive process (error back propagation) which minimises the output neurone errors. This phase ends when the ANN furnishes the outputs necessary for the learning set. Progress in the ANN learning phase is monitored through a decrease in the maximum relative error. Some effort must be devoted to determining the learning set in order to obtain high ANN accuracy.

In the validation phase, the ANN accuracy is tested using the validation set. These vector pairs are different from those of the learning phase, but have similar characteristics. If the ANN's performance does not reach the desired level of accuracy a new learning phase and a modified learning set are necessary. The modified learning set must take into account the additional new information obtained by the previous set.

After the learning and validation phases, the production phase begins and the ANN is used to provide the corresponding outputs required for any input.

Trained in this manner, the ANN represents a valid model of the non-linear phenomena for which it has been trained. The main cost will be the time taken to set up the learning phase.

Some of the drawbacks connected with the training are: difficulties in establishing opportune constituent data values for the training set; the possibility of reverting to a situation of local minimum that does not allow the network to reach the final training with the necessary precision; the elevated time required by the training phase. To overcome these drawbacks a method has recently been developed which allows the pre-elaboration of the measurement signals by opportune structures of calculation constituted from neural artificial networks.

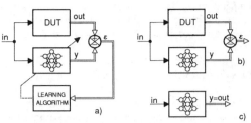

Figure 1. Neural modeling phases of the Device Under Test (DUT): a) learning phase; b) validation phase; c) production phase.

3 THERMOELASTIC BEHAVIOUR

The temperature effects in an orthotropic material, which is a fibrous composite, are generally different along the principal directions of orthotropy; all this can be taken into account by introducing different linear coefficients of thermal expansion α_i corresponding to the strain components.

By considering the case of a plate subjected to a stress condition acting in the xy-plane, the classical elastic constitutive equations referred to the x-y axes, can be written in the following form:

$$\{\varepsilon\} = [C]\{\sigma\} + \{\alpha\}\Delta T \qquad (1)$$

where:

$$\{\varepsilon\} = \begin{Bmatrix} \varepsilon x \\ \varepsilon y \\ \varepsilon xy \end{Bmatrix}, \quad \{\alpha\} = \begin{Bmatrix} \alpha x \\ \alpha y \\ \alpha xy \end{Bmatrix}$$

are the vectors of the elastic strains and of the thermal expansion coefficients respectively.

The thermal expansion coefficients referred to the x-y axes can be given as a function of the principal 1-2 axes by:

$$\alpha x = \alpha_1 \cos^2 \theta + \alpha_2 \sin^2 \theta \qquad (2)$$

$$\alpha y = \alpha_1 \sin^2 \theta + \alpha_2 \cos^2 \theta \qquad (3)$$

$$\alpha xy = 2(\alpha_1 - \alpha_2)\sin\theta\cos\theta \qquad (4)$$

where θ is the orientation angle of the fibres with respect to the reference x axis. Since the composite material examined has unidirectional fibres parallel among them with $\theta = 0°$, from equations (2), (3) and (4)

$$\alpha x = \alpha_1, \; \alpha y = \alpha_2, \; \alpha xy = 0 \qquad (5)$$

is obtained.

The thermal expansion coefficient along the two principal directions can be determined by the expressions:

$$\alpha_1 = (\alpha_f E_f + \alpha_m E_m (1 - \eta_f))/(E_f \eta_f \\ + E_m (1 - \eta_f)) \qquad (6)$$

$$\alpha_2 = (1 + \nu_m)\alpha_m (1 - \eta_f) + (1 + \nu_f)\alpha_f \eta_f \\ - \alpha_1 (\nu_f \eta_f + \nu_m (1 - \eta_f)) \qquad (7)$$

where E and ν are Young's modulus and Poisson's ratio of the fibres (f) and the matrix (m) and η_f is the fibre volume percentage.

Equations (6) and (7) point out that the thermal expansion behaviour of the composite examined depends on the thermoelastic properties of the material components (E_f, E_m, ν_f, ν_m, α_f, α_m) and on the parameters of production technology.

The elastic parameters are generally certified by the company and have a negligible degree of uncertanity as far as the determination of the strain state is concerned. On the contrary, the thermoelastic parameters show a high degree of uncertainty.

Thermal expansion behaviour of the matrix is greatly influenced by several factors, the most important of which is the synthesis process and the successive fibre impregnation. It would seem therefore that strain measurements in a composite material subjected to mechanical and thermal loads can be carried out more accurately by experiments using strain gauges.

In a unidirectional composite material subjected to such loads the total strain given by the strain gauge is:

$$\varepsilon_t = \varepsilon_m + \varepsilon_{\Delta t} + \varepsilon_r \qquad (8)$$

where ε_t is the total strain of specimen tested, ε_m is the mechanical component, $\varepsilon_{\Delta t}$ is the thermal component and ε_r is the thermal component of the strain gauge used.

In order to determine only the strain ($\varepsilon_m + \varepsilon_{\Delta t}$) it is necessary to eliminate the thermal component ε_r of the strain gauge used from the total deformation ε_t.

4 MATERIALS AND METHODS

The experimental investigation was carried out on commercial FRP composite materials obtained by an automatic poltrusion process, with glass fibres unidirectionally arranged in an epoxy resin matrix.

$260 \times 40 \times 7$ mm specimens, respecting ASTM D3039-74 standard, were obtained by cutting the 1m flat bars, to achieve a sufficiently wide isothermic zone.

The physical and mechanical properties of the tested material are show in Table 1.

Table 1. Mechanical and physical properties of the component materials.

Material	E [MPa]	ν	Volume [%]	Ultimate Strength [MPa]
Fibre Glass	70,000	0.20	75	1700
Epoxy Resin	3500	0.34	25	85
Composite	53,500	0.26	100	950

As a function of these properties and the thermal coefficient α_f and α_m from equation (1) the characteristic stress-strain curve of the composite was obtained for temperatures of -20°C, 20°C and 80°C.

The experimental tests were carried out using the equipment illustrated in Figure 2. This is composed of: an electromechanical test machine, a data acquiring device, a climatic cell and a personal computer.

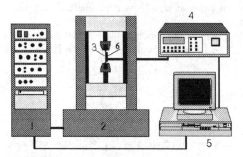

1: Control device
2: Universal testing machine
3: Sample
4: Data acquiring device
5: Personal computer
6: Climatic cell

Figure 2. Mechanical tests setup

With this setup monotonic tensile tests were conducted up to 50% of the ultimate load, and stress-strain diagrams were obtained for temperatures of -20°C, 20°C and 80°C.

Electric strain gauges were placed at the centre of one side of the specimen to measure strain, while temperature was kept constant for the duration of the test by a climatic cell. The strain gauge reading was effected without a temperature compensation gauge.

RESULTS AND CONCLUSIONS

The paper presents a new method to compensate the temperature influence on the strain gauge output based on the use of an ANN. The training set of the ANN has been organized as follows:

(i) the input data are both the deformations obtained from.the output of the strain gauge and the material temperature;

(ii) the output data are the load obtained from the analytical model of the material.

Figure 3 shows both the experimental and analytical data utilised for the training set. In this figure are plotted the load versus deformation curves in the range from -20°C to 80°C for both the experimental and analytical evaluation.

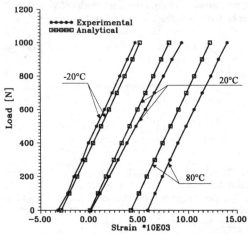

Figure 3. Comparison between analytical and experimental σ–ε curves

Figure 4 shows the load versus deformation plots in the range from -20°C to 80°C after the compensation by means of the ANN compared with the experimental and analytical ones. The high level of compensation is a consequence of the fact that the ANN is able to compensate:

(i) the alignment error between the strain gauge and the material fibres,

(ii) the calibration error, as consequence of the

fact that the strain gauge is calibrated by using iso-tropic materials.

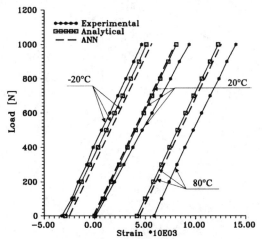

Figure 4. Comparison among experimental, analytical and compensated by ANN σ–ε curves.

The progress of the research is devoted to investigate the possibility (i) to compensate the environmental influence on the strain gauge and (ii) to compare the accuracy of the compensation by means of ANN with the classical ones.

REFERENCES

Ajovalasit, A. 1995. Tecniche estensimetriche per l'analisi delle tensioni nei compositi. *XXIV Convegno Nazionale AIAS*: 17-35. Parma: Italy.
Ciaccia, P. & Mario, D. 1989. Reti neurali: proprietà e problematiche aperte. *Alta Frequenza*: I: 185-210.
Daponte, P. & Grimaldi, D. Artificial Neural Networks in Measurement, submitted to Measurement.
Daponte, P., Grimaldi, D. & Michaeli, L. 1996. Signal processing in measurement instrumentation using analog neural networks. *Proc. IEEE IMTC'96*: 1004-1009. Brussels: Belgium.
Daponte, P., Grimaldi, D. & Michaeli, L. 1997. Analog neural networks for pre-processing in measurement sensor. *XIV IMEKO World Congress*. Tampere: Finland.
Daponte, P., Grimaldi, D., Piccolo, A. & Villacci, D. 1995. A neural diagnostic system for the monitoring of transformer heating. *Measurement*: 18: 35-46.
Egawa, K. 1992. Strain measurement on CFRP.FW cylinder as a structural member of space structure. *Proc. of the Int. Symp. on Nondestructive Testing & Stress-Strain Measurement*: II: 674- 680. Tokio.
Grimaldi, D., Olivito, R. S. & Surace, L. 1997. Applicazioni di tecniche neurali in misure di deformazione in materiali compositi FRP. *XXVI Convegno Nazionale AIAS*: 701-707. Catania: Italy.
Grimaldi, D., Olivito, R. S. & Surace, L. 1998. Damage eva-luation in FRP composites by means of neural networks. *To appear on Proc. of the Fifth Int. Conf. On Composites Engineering, ICCE/5*. Las Vegas: USA.
Hecht, R. & Nielsen, A. 1990. Neurocomputing. *Addison Welsey Publisshing Company*.
Hudson, W.B. 1996. Introduction and Overview of Artificial Neural Networks. *Instrumentation and Measurement Application, IEEE Technology UPDATE SERIES Neural Networks Theory, Technology, and Applications*: 746-749. New York.
Maren, A., Harston, C. & Pap, R. 1990. Handbook of neural computing application. *Accademic Press Inc*.
Olivito, R. S., Stumpo, P. & Surace, L. 1996. Nondestructive evaluation of damage in composite materials. *Proc. Advanced Composite Materials in Bridges and Structures. International Conference*: 109-116. Montreal: Quebec: Canada.
Perry, C.C. 1984. The resistance strain gage revisited. *Experimental Mechanics*: 24: 286-299.
Tuttle, H.E. & Brinson, H. F. 1986. Resistance strain gage technology as applied to composite materials. *Experimental Mechanics*: 26:153-154.
Tuttle, M. E. 1985. Error in strain measurement using strain gages on composites. *Proc. 1985 SEM Fall Conf. on Exp. Mech. Traducer Technology for Physical Measurement*: 170-179. SEM Bethel (CT): USA.

1415

Experimental Mechanics, Allison (ed.)© 1998 Balkema, Rotterdam, ISBN 90 5809 014 0

Improvement of mechanical properties of aramid short fiber reinforced polyester

A.K.M.Masud, E.Nakanishi, K.Isogimi & J.Suzuki
Department of Mechanical Engineering, Mie University, Tsu, Japan

ABSTRACT: Present research is conducted to fabricate FRPs by reinforcing Unsaturated Polyester (UP) resin with Aramid short fiber. The interfacial shear strength is determined for this fiber/matrix combination by performing single fiber pull out test and it is found that this combination possesses an average shear stress of 9.107MPa. Moreover, the critical length of the fiber is determined for which the embedded fiber doesn't pull out from the matrix material. Some chemical and physical surface treatments of the fibers are investigated to improve the mechanical properties of the fabricated FRPs. Chemical treatments hardly effect the composite strength but temperature treatment improves the fiber strength as well as the properties of the composite material.

1. INTRODUCTION

Considerable work has been conducted to fabricate high performance composites by long fiber reinforcement and their theoretical interpretation and micromechanics. The stress-strain curves of the epoxy matrix and their typical unidirectional composites in fiber direction and transverse direction are clearly reported by Isaac M.Daniel & Ori Ishai (1994). The behavior of unidirectional composites in the transverse to the fiber direction, specially the strength, is dominated by the matrix properties. The stress-strain curves of typical unidirectional composites in the transverse to the fiber direction, shows that these materials exhibit quasilinear behavior with relatively low ultimate strains and strengths compared to those in the fiber direction and even to that of the matrix material. Therefore, their anisotropic behavior (Daniel & Ishai 1994) as well as their poor machinability and inability to make complicated shaped parts, reduces their sophisticated applications. On the other hand, the short fiber reinforced composites are usually fabricated by injection molding, where the short fibers are oriented toward the mold fill direction; so, these injection molded FRPs are also anisotropic (Voss & Friedrich 1986) and further limits their applications. Therefore, present research is conducted as a first step to produce a high performance isotropic composite material that can be molded to any complicated shape.

In this paper, aramid short fibers, which are drawing attention as new strengthening and reinforcing composites, were used for the FRP fabrication. There are a few researchers who are using the aramid short fiber as reinforcing element for the epoxy matrix and

it is reported by Tanaka, K. & Yamaguchi, M. (1994) that the uniform dispersion of the aramid fiber in to epoxy resin is possible up to the amount of 2.44wt%. So, attention was given to raise the fiber content rather than the fiber dispersion in the foregoing research(Masud 1997). A special molding process was employed in the research. Tension test and three point bending test were carried out to investigate the mechanical properties of the produced FRPs.

Further, mechanical behavior of composites are influenced by the properties of fiber/matrix interface and considerable works (Hsueh 1990, Jiang & Penn 1992, Banbaji 1988) have been conducted to determine the interfacial properties. The interfacial shear strength and the critical length for which the fiber fractures out leaving the intact fiber/matrix interface, were also determined by the single fiber pull-out test for the materials employed in this research. It is realized that some surface treatment of the aramid fibers could improve the interface properties, so as to improve the mechanical properties of the fabricated FRPs. It is reported by A.R. Bunsell (1988) that the H_2SO_4 of moderate concentration do not affect the fiber strength too much and it is also clearly stated (Minoshima 1997) that drying of Kevlar fiber makes the surface wavy in nano scale. So the effects of these fiber treatments on the material properties are investigated in this research.

2. MATERIALS AND EXPERIMENTAL

Aramid short fiber of Kevlar 49, by DuPont Toray Co. Ltd., is used as reinforcing material and Unsaturated polyester(UP), 8285 AP, by Takeda Yakuhin Co. Ltd, as matrix material. Hardener is Permeck (N), by Nihon Yushi Co. Ltd., Japan. Two different lengths of aramid short fibers were employed for FRP fabrication, i.e., 1mm and 3mm as specified by the manufacturer. Table 1 shows some measured properties of the fiber compared to that of the standard values. To determine the strength of the aramid fiber, long fiber is selected and mounted on construction paper tabs, following a modified version of ASTM D3379 like H.D.Wagner & S.L.Phoenix (1984). The specimens were tested in tension under a cross head speed of 3.5mm/min. As it is stated by H.D.Wagner & S.L.Phoenix (1984) that diameter of the aramid fiber varies from fiber to fiber, so the diameter was measured by SEM photography and all the test results of this research are arranged by taking the diameters into consideration. The lengths of the short fibers are measured by optical microscopy.

Table 1 Some properties of aramid fiber

	Measured value	Standard value
Diameter measured by SEM		
Number of measurements	202	
Mean diameter (μm)	13.31	12*
Standard deviation (μm)	0.95	
Tensile strength of the fiber		
Number of tests	25	
Mean breaking load (gf)	34.76	
Standard deviation (gf)	4.307	
Mean tensile strength (MPa)	2450	2758**
Length of the short fibers		
Number of measurements(each)	200	
Mean length of 1mm fiber (mm)	1.101	1.0*
Mean length of 3mm fiber (mm)	2.912	3.0*

* Specified by the manufacturer(DuPont Toray Co. LTD.) and ** A.R. Bunsell (1988).

3. FIBER TREATMENT AND FRP FABRICATION

3.1 Fiber treatment process and their fracture properties

As we know that the fiber/matrix bonding of this combination is very weak. So, it is thought that some fiber surface treatment could rather increase the interface strength and would make the micro crack initiation delay at the interface. But the stress required for crack propagation would remain the same and the delayed crack initiation would rather increase the material strength.

Therefore, some chemical and physical surface treatments were carried out. The Kevlar fibers were immersed into the 10% H_2SO_4 (for 99hrs, 120hrs and 144hrs), 20% H_2SO_4 (for 120hrs) and 30% H_2SO_4 (for 96 hrs and 120hrs). After the acid treatment long water wash of the fibers was provided to remove the remaining acid on the fiber and was verified by using litmus sheet. The adhesive Aron Alfa (Konishi Co. Ltd., Japan) has the property to emit some gases when

it exposes in atmosphere. And it was found that these emitted gases affect the fiber surface. So the fibers are treated for 24 hours into the emitted gas condition. The fibers and the Aron Alfa are kept separately at the up side and bottom side of an air tight chamber separating by a permeable net. The surface conditions of the treated fibers are observed by SEM. Figures 1a, b show the surface appearance of the Kevlar fibers on the H_2SO_4 and the Aron Alfa treatment. During the acid treatment, somewhat local surface defects occur as shown in Figure 1a. For 30% (96 hrs) H_2SO_4 treatment, about 180 defects are found in a single 18cm long fiber by careful SEM observation. No defects however are observed for the 10% H_2SO_4 treatment. On the other hand, a large number of surface defects are observed on the Aron Alfa treated fiber as shown in Figure 1b. There may have some chemical residue on the surface of the fiber that might affect the interface.

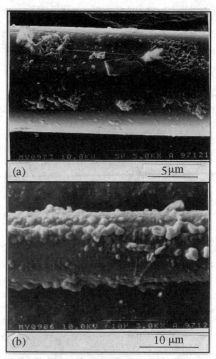

Figure 1 SEM photos of the surface of (a) 30% H_2SO_4 (96 hrs) and (b) Aron Alfa treated aramid fiber.

Further, K.Minoshima et al. (1997) show drying of Kevlar 49 fiber makes the fiber surface curl with a maximum amplitude of 100nm. He also showed some AFM images of fiber surface of Kevlar 49, which was conditioned in hot air for one month. Therefore, another surface treatment was done by drying the Kevlar fiber into a drying oven at 353K for 60 days. The single fiber tension test was carried out in order to know the effect of the surface treatment on the fiber property. The test condition is followed as described in section 2. The fiber fracture load is plotted in Figure 2, for different fiber treatments.

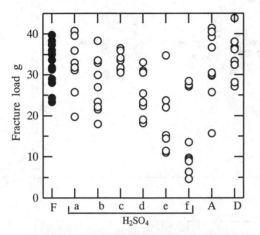

Figure 2 Fracture load of the fresh fiber (F), the H_2SO_4 treated fibers (a: 10% H_2SO_4 for 99 hrs, b: 10% H_2SO_4 for 120 hrs, c: 10% H_2SO_4 for 144 hrs, d: 20% H_2SO_4 for 120 hrs, e: 30% H_2SO_4 for 96 hrs, f: 30% H_2SO_4 for 120 hrs), Aron Alfa treated fiber (A) and dried fiber (D).

It is reported that the fiber fracture load increases about 10% (Minoshima 1997) when the fiber is conditioned at 353K for a month. Although the data obtained has large scatter, it can be think from Figure 2 that the fracture load of the temperature treated fiber increases a little compared to that of fresh fiber. The fracture load of the 10% H_2SO_4 treated fibers remain almost the same as the fresh fiber but increase of acid concentration and immersion time reduces the fiber strength as shown in Figure 2.

3.2 *Fabrication process of FRPs*

The foregoing paper (Masud 1997) clearly describes the molding process as well as the molding device for the FRP fabrication. Aramid short fiber is mixed into degassed UP with the hardener (1wt% of UP) and sufficient pressure is applied to the mixture to get the desired shape and possible homogeneity for 24 hrs (for complete hardening). To remove the residual strain, the produced FRPs are annealed at 353 K for 3 hrs in a drying oven.

The fibers treated by 10% (144 hrs) and 30% (96 hrs) H_2SO_4, emitted gases of Aron Alfa and the fiber drying are selected to fabricate the FRPs. The tension and bending tests are carried out to investigate the effect of the fiber treatments on the material properties.

4. EFFECT OF SURFACE TREATMENT ON MECHANICAL PROPERTIES

4.1 *Tension test*

The tension test was carried out under the cross head speed of 0.027 mm/sec and the load responses were recorded by using a xy recorder.

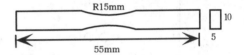

Figure 3 Dimension of the tension test specimen

Figure 3 shows the dimension of the test specimen . Figure 4a shows the tension test results for the fabricated FRP of 1mm and 3mm fiber content for the fresh fiber and the treated fibers. The results for both the matrix material and the nontreated fresh fibers were obtained

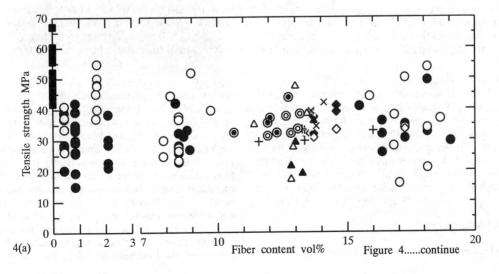

4(a) Fiber content vol% Figure 4......continue

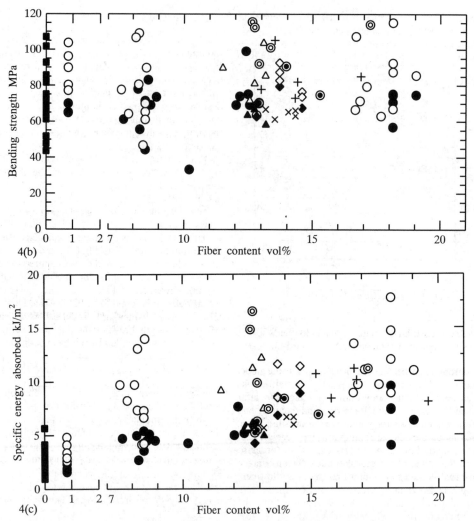

Figure 4a, b, c Tension and bending test results of FRPs of fresh fiber and the treated fibers [■ Matrix material, ● 1mm fresh fiber, ○ 3mm fresh fiber, ◆ 1mm treated fiber(10% H_2SO_4), ◇ 3mm treated fiber(10% H_2SO_4), ✕ 1mm treated fiber(30% H_2SO_4), ✛ 3mm treated fiber(30% H_2SO_4), ▲ 1mm treated fiber(Aron Alfa), △ 3mm treated fiber(Aron Alfa), ◉ 1mm treated fiber(353K for 60days) and ◎ 3mm treated fiber(353K for 60days).]

in the foregoing paper.

Although, the data obtained on experiment has large scattering, Figure 4a shows sudden fall of the tensile strength at very low fiber content for both the fiber lengths. And, further fiber enrichment increases the tensile strength gradually. It seems that the acid and the Aron Alfa fiber treatments have almost no effect on the tensile strength. From Figure 4(a) however, we find that the tensile strength of the FRPs of the temperature treated fiber content has low scatter.

4.2 Bending Test

Three point bending test was carried out under the same operating conditions those were applied for the tension test. The dimension of the bending test specimen is 10mm X 5mm X 55mm (as fabricated). Figure 4b, c show the bending test results for the fabricated FRP of 1mm and 3mm fiber content for the fresh fiber and treated fiber. From Figure 4b, we can find that the bending strength is hardly raised with fiber content for the FRPs of the fresh fiber and chemically treated fiber content. The bending strength of the temperature treated fiber content FRPs increase a little compared to that of the other fiber content.

Figure 4c shows the specific energy absorption capacity before complete fracture, calculated by computing the load response diagrams during the bending test. Here, also specific energy absorbed increases for the temperature fiber treatment.

5. FURTHER TEST OF INTERNAL PROPERTIES

5.1 Model test 1st (Single fiber pull-out test)

Pull-out test was carried out by making a simple model specimen as shown in Figure 5a, b.

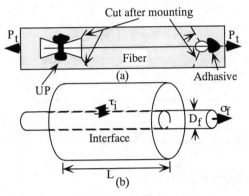

Figure 5 (a) Schematic diagram of single fiber pull-out test specimen and (b) Geometry of the pull-out test.

Construction paper tabs are selected to make the model specimen. The embedded length L of the fiber was measured by using the optical microscopy after embeddedment into the UP matrix. And the other end is firmly fixed by a strong adhesive (Aron Alfa). Then, after mounting the specimen on the tension test machine, the dashed portion is cut off, in order to make the fiber free. Here, the extension of the construction paper is assumed to have very low and equal values for each specimen, so it is neglected. Tensile load P_t required for the fiber pull-out and the fiber fracture are shown in Figure 6.

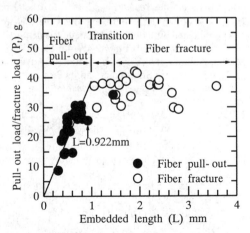

Figure 6 Load required for fiber fracture and pull-out vs. embedded length of the fiber into the UP matrix.

The Figure 6 shows a rise in pull out load with the embedded length. As the bonding between the fiber and matrix is very low, during the pull out process interfacial slippage occurs rather to yield or fracture the matrix material. So the pull out load increases with the embedded length. Therefore, the data of the fiber pull-out region are fitted by the method of least squares and obtained the P_t/L as 38.86 g/mm. And this value is used for calculating the interfacial shear strength using the following Equation 1, with the assumption that the shear stress is constant in the parts of the fiber/matrix interface.

$$\tau_i = \frac{\sigma_f D_f}{4L} = \frac{P_t}{\pi D_f L} \qquad (1)$$

Here, D_f is the fiber diameter and the mean measured value of the fiber diameter (Table 1) is used for calculation. The σ_f is the stress developed in the fiber for different embedded length(L) and is defined as P_t/A_f, where P_t and A_f are the load required to pull out (Fig. 6) the fiber for different embedded lengths and the fiber cross sectional area. Therefore, the interfacial shear strength is found to be 9.107 MPa. Further, the critical embedded length L_c is defined as the shortest embedded length for which the fiber does not fracture out and it is found around 1mm by experiment.

5.2 Model test 2nd(Crack initiated FRPs)

Two small sharp crack initiations were employed at each side of the neck of the tension test specimen. This is in part, simulates the condition that a small short crack is initiated at some interfacial place under the critical tensile load value. And the length of the crack initiation is measured by using optical microscopy.

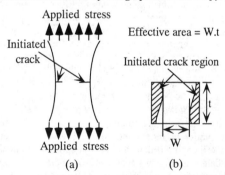

Figure 7(a) Tensile test specimen with two crack initiation in each side (b) Realization of effective area.

Then, the specimen was tested in tension with a cross head speed of 0.071 mm/min and the fracture load is recorded. The substantial section areas were estimated as an effective sectional area by careful observation of the fractured surface and used for calculating the fracture strength of the specimen. Repeated experiment shows the fairly same value of the fracture strength, i.e. 13.39

MPa. The schematic representation of the specimen and the effective area on the fractured surface are shown in Figure 7.

Based on the experimental results obtained through these two cases (model test 1st and 2nd), we can derive the following investigations;

From the test 1st, it is clear that the interfacial shear strength is much lower than the tensile strength (Fig.4a) of the matrix material, so, it can be expected that peeling of the fiber from the matrix occurs and it causes crack initiation at the interface when the applied stress exceeds the interfacial strength. Further, from test 2nd it is found that the stress required to propagate the initiated crack has considerably lower value. At very low fiber content of 0.45vol%, it is expected that the number of 1mm fibers should be 40 per cubic mm. Therefore, when the applied stress exceeds the interfacial strength, the number of micro crack initiations should be as many as 1000 at the cross section of the specimen for perfect dispersion. And when the applied stress exceeds the stress required to propagate the cracks (test 2nd) and ideally should be fractured out. In reality, as the fibers are not perfectly dispersed and the propagated crack stops for a while at the interfaces, the stress required for propagation will be a bit higher than that is obtained in the test 2nd. Further, fiber enrichment acts as jamming to the crack propagation and will cause gradually increase of tensile strength. Therefore, the strength of the FRPs can be expected to improve more for the physically treated fiber enrichment. The same mechanisms is also true for 3mm fiber content FRP.

On the other hand, it seems that the chemical treatments hardly affect the material properties. During the acid treatment, some kind of surface defect occurs as shown in Figure 1a. We think the bonding becomes stronger at the defect regions but due to the presence of unchanged fiber surface portion, the occurrence of micro cracks would be the same as described for the fresh fiber. Once the crack is initiated, it would propagate according to model test 2nd. The phenomena are also true for aron alfa treatment.

From Figure 4, it can be thought that the physical fiber surface treatment improves the material properties as mentioned above. K.Minoshima et al. (1997) reported that the drying of fibers makes their whole surface wavy (nano scale). Therefore it can be thought that the interface might be stronger than that for the untreated fiber (Model test 1st) and might delay the micro crack initiations at the interface, so as to increase the material strength. Although the stress required for crack propagation remains same (as describe in the model test 2nd), the specific absorbed energy increases a bit due to the delay of crack initiations from the interfaces.

CONCLUSIONS

For the fresh fiber, the tensile strength of the fabricated FRP has sudden decrease at very low fiber content and further increase of the fiber content leads to increase the tensile strength. The bending strength is hardly improved with fiber content.

The interfacial shear strength is considerably low compared to the tensile strength of the matrix material. So, it is expected that some fiber treatment might increase the interface strength and could rather delay the micro crack initiation at the interface. It is found that the temperature treatment makes fiber surface wavy (Minoshima 1997) and increases the fiber strength as well as the composite property.

The H_2SO_4 fiber surface treatment reduces the fiber strength. The tensile strength of the fiber increases about 10% of the fresh fiber, for the temperature treatment. Further the chemical treatments hardly affect the material property where the temperature treatment do improve a little.

REFERENCE

Banbaji, J. 1988. On a More Generalized Theory of the Pull-Out Test from an Elastic Matrix. *Composites Sci. and Tech*. 32: 183-193.

Bunsell, A.R. (edited) 1988. *Fibre Reinforcements For Composite Materials. Composite Materials Series*. Elsevier 2: 275,279-282.

Chun-Hway Hsueh 1990. Interfacial Debonding and Fiber pull-out Stresses of Fiber-reinforced Composites. *Mater. Sci. Engg*. A123: 1-11.

Daniel Wagner, H. & Leigh Phoenix, S. 1984. A Study of Statistical Variability in the Strength of Single Aramid Filaments. *J. Comp. Mater*. 18: 312-338.

Isaac M. Daniel & Ori Ishai.1994 *Engineering Mechanics of Composite Materials*. Oxford University Press. 26-36.

Jiang, K.R. & Penn, L.S. 1992. Improved analysis and experimental evaluation of the single filament pull-out test. *Composite Sci. and Tech*. 45: 89-103.

Masud, A.K.M. et al.1997. Fabrication of Advanced FRP with Aramid Short Fiber and Its Properties. *Int.Conf. on Adv. Tech. in Experimental Mechanics*.181-186.

Minoshima, K. et al. 1997. Influence of Vacuum and Water on Tensile Fracture Behavior of Aramid Fibers. *Int. Conf. on Adv. Tech. in Expe.l Mechanics*. 373-378.

Tanaka, K. & Yamaguch, M. 1994. Dynamic mechanical properties of multi-functional epoxy resin filled with aramid short fibers. *Adv. Composite Mater*. 3 (No.3): 209-222.

Voss, H. & Friedrich, K. 1986. Influence of short fiber reinforcement on the fracture behaviour of a bulk liquid crystal polymer. *Journal of material science*. 21: 2889-2900.

Experimental Mechanics, Allison (ed.) © 1998 Balkema, Rotterdam, ISBN 90 5809 014 0

Author index